FORCE (WEIGHT)

1 newton = 0.2247 pounds
1 kilogram = 2.2046 pounds
1 kilogram = 9.81 newtons
1 metric ton = 1000 kilograms
1 metric ton = 9.81 kilonewtons
1 metric ton = 2204.6 pounds

1 pound = 4.45 newtons
1 pound = 0.4536 kilograms

1 ton = 2000 pounds
1 ton = 0.9072 metric tons

LENGTH (DISTANCE)

1 millimeter = 0.03937 inches
1 centimeter = 0.3937 inches
1 meter = 3.281 feet
1 kilometer = 0.6215 miles
1 meter = 1.0936 yards

1 inch = 25.4 millimeters
1 inch = 2.54 centimeters
1 foot = 0.3048 meters
1 mile = 1.609 kilometers
1 yard = 0.9144 meters

MASS

1 kilogram = 2.2046 pounds
1 kilogram = 0.0685 slugs

1 pound = 0.4536 kilograms
1 slug = 14.6 kilograms

POWER

1 watt = 1 joule per second
1 watt = 1 newton-meter per second
1 kilowatt = 1.341 horsepower

1 horsepower = 550 foot-pound per second
1 horsepower = 746 watts
1 horsepower = 0.746 kilowatts

PRESSURE

1 pascal = 1 newton per square meter
1 kilopascal = 1000 pascal
1 kilopascal = 0.145 lb per square inch

1 lb per square inch = 6.8972 kilopascals
1 lb per square inch = 6897.2 pascals

ENGINEERING HYDROLOGY

PRINCIPLES AND PRACTICES

VICTOR MIGUEL PONCE

San Diego State University

PRENTICE HALL, Upper Saddle River, New Jersey 07458

Library of Congress Cataloging-in-Publication Data

Ponce, Victor Miguel.
    Engineering hydrology : principles and practices / Victor Miguel
  Ponce.
      p.   cm.
  Bibliography: p.
  Includes index.
  ISBN 0-13-277831-9
    1. Hydraulic engineering. 2. Water resources development.
  I. Title.
  TC145.P66 1989
  627--dc19   88-29265
                                                                  CIP

Editorial/production supervision: *Nancy Menges*
Cover design: *Wanda Lubelska Design*
Manufacturing buyer: *Mary Noonan*
Cover figure: Adapted from Eagleson, *Dynamic Hydrology*, McGraw-Hill, 1970, with
  permission.

©1989 by Prentice-Hall, Inc.
A Pearson Education Company
Upper Saddle River, NJ 07458

Printed in the United States of America

10 9 8 7 6 5 4 3 2 1

ISBN   0-13-315466-1

Prentice-Hall International (UK) Limited,London
Prentice-Hall of Australia Pty. Limited, Sydney
Prentice-Hall Canada Inc., Toronto
Prentice-Hall Hispanoamericana, S.A., Mexico
Prentice-Hall of India Private Limited, New Delhi
Prentice-Hall of Japan, Inc., Tokyo
Pearson Education Asia Pte. Ltd., Singapore
Editora Prentice-Hall do Brasil, Ltda., Rio de Janeiro

CONTENTS

Contents

Contents

Contents

PREFACE

This book aims to provide, under one cover, a balanced treatment of basic principles and current practices in engineering hydrology. It is primarily intended for undergraduate students in civil engineering and related disciplines who are beginning a course of study in engineering hydrology. First-year graduate students and practicing engineers will find the comprehensive and up-to-date coverage to be intellectually stimulating as well as of practical value.

In planning this book, I have consciously departed from the classical order of coverage, which organizes the topics according to the various phases of the hydrologic cycle, i.e., precipitation, hydrologic abstractions, and so on. The hydrologic sense of this approach may be plausible, but its engineering significance is lacking. Instead, I have relied on a catchment-scale framework, by which procedures applicable to small catchments are presented first, followed by those of midsize catchments, and finally, by those of large catchments. The advantage of this catchment-scale approach to engineering hydrology is that coverage proceeds from the simple to the complex, i.e., from the rational method to the unit hydrograph to the routing methodologies. Of course, overland flow is the exception that confirms the rule.

In engineering hydrology, the concept of catchment scale has been recognized for some time, but only recently has it acquired an importance of its own. This has closely paralleled a change in our perception of what constitutes a small, a midsize, or a large catchment. For instance, the upper limit for the applicability of the rational method has been steadily decreasing in the last three decades, from 13 km^2 to as little as 0.65 km^2 today. Perhaps more importantly, the upper limit for the applicability of the unit hydrograph has also been steadily decreasing, from up to 5000 km^2 in the 1930s to as little as 50 to 250 km^2 today. It is surmised that the computer has played a major role in this gradual change in our perception of catchment scale. As our computational capabilities continue to increase, so does our use of more refined methods of analysis. It is therefore both timely and appropriate to expound not only on how, but also on *why* we do things the way we do.

Chapter 1 sets the stage for the remainder of the book, describing the hydrologic cycle and the uses of and approaches to engineering hydrology. Chapter 2, *Basic Hydrologic Principles,* contains an introductory treatment of hydrologic principles, necessary to provide a common ground. Chapter 3, *Hydrologic Measurements,* is included at this early stage to expose the reader to the reality of hydrologic data collection. The quantitative coverage of engineering hydrology begins in Chapter 4, *Hydrology of Small Catchments,* and continues in Chapter 5, *Hydrology of Midsize Catchments.* These two chapters dwell on both hydrologic fundamentals and practical applications. Chapter 6, *Frequency Analysis,* and Chapter 7, *Regional Analysis,* describe the use of statistics and probability in the solution of problems of engineering hydrology. Chapters 8, 9, and 10 deal with routing methodologies applicable to reservoirs, stream channels, and catchments. Chapter 11, *Subsurface Water,* and Chapter 12, *Snow Hydrology,* complement the subjects of Chapters 4 to 10. The concepts, methods, and techniques described in Chapters 4 to 12 constitute the building blocks of the catchment models, described in Chapter 13. Chapter 14 provides an overview of hydrologic practices in current use by federal agencies. Chapter 15 rounds out the study of engineering hydrology by focusing on the important subject of sediment.

This book is designed to be used either (a) as a textbook for upper-division undergraduate study, (b) as a reference source for first-year graduate students with no previous exposure to engineering hydrology, or (c) as a self-teaching refresher for practicing engineers. Typically, a one-semester syllabus may cover 50 to 70 percent of the book, depending on the pace of coverage. There is enough material to provide ample flexibility in the choice of depth and breadth of subject coverage.

In writing this book, I wish to acknowledge the lasting influence of my teachers, colleagues, and students. *Zbig Osmolski,* Pima County Department of Transportation and Flood Control District, Tucson, Arizona, reviewed the entire manuscript and provided numerous suggestions for its improvement. The publisher's reviewers, *Paul Chan,* New Jersey Institute of Technology; *Hanif Chaudhry,* Washington State University; and *Richard Weisman,* Lehigh University, made significant contributions to both content and style. The following persons reviewed individual chapters of the book: *Carlos Alonso,* USDA Agricultural Research Service, Fort Collins, Colorado; *Cat Cecilio,* Pacific Gas and Electric Company, San Francisco; *Howard Chang,* San Diego State University; *Don Newton,* Tennessee Valley Authority, Knoxville; *Emilio Rios,* Sato and Associates, Denver; *David Rockwood,* Portland, Oregon; *Fred Theurer* and *Don Woodward,* USDA Soil Conservation Service, Washington, D.C. The following persons provided me with information that eventually was included in the book: *Hugo Benito,* Organization of American States, Washington, D.C.; *Newton Carvalho,* ELETROBRAS, Rio de Janeiro, Brazil; *Walter Crampton,* Group Delta Consultants, San Diego; *Roger Cronshey, Bill Merkel* and *Dave Ralston,* USDA Soil Conservation Service, Washington, D.C.; *Miguel Coutinho* and *Emidio Santos,* Technical University of Lisbon, Portugal; *Edward Davis,* U.S. Army Corps of Engineers North Pacific Division, Portland; *Mark Dawson,* Kern County Department of Public Works, Bakersfield; *Phil Demery,* Santa Barbara County Flood Control and Water Conservation District, Santa Barbara; *Douglas Fenn,* NOAA National Weather Service, Silver Spring, Maryland; *Richard Harness,* International Technology Corporation, Martinez, California; *Leon Hyatt,*

U.S. Bureau of Reclamation, Denver; *Marshall Jennings*, U.S. Geological Survey, Stennis Space Center, Mississippi; *Sam McCown*, NOAA National Climatic Data Center, Asheville, North Carolina; *Wilbert Thomas, Jr.*, U.S. Geological Survey, Reston, Virginia; *Joseph Tram*, Maricopa County Flood Control District, Phoenix, Arizona; and *David Williams*, WEST Consultants, Inc., San Diego. I am deeply grateful to all.

Victor Miguel Ponce
San Diego, California

INTRODUCTION

This book deals with principles of hydrologic science and their application to the solution of hydraulic, hydrologic, environmental, and water resources engineering problems.

This introductory chapter is divided into six sections. Section 1.1 defines hydrology and engineering hydrology. Section 1.2 describes the hydrologic cycle, a fundamental tenet of hydrologic science. Section 1.3 describes the closely related concepts of catchment and hydrologic budget. Section 1.4 explains the use of hydrologic knowledge in the solution of typical problems of hydraulic and hydrologic engineering. Section 1.5 elaborates on the various approaches used to solve problems of engineering hydrology. Section 1.6 discusses surface runoff, flood hydrology, and catchment scale. The concept of catchment scale is used in this book to develop a framework for the study of hydrologic models, methods, and techniques.

1.1 DEFINITION OF HYDROLOGY AND ENGINEERING HYDROLOGY

Hydrology is one of the earth sciences. It studies the waters of the earth, their occurrence, circulation and distribution, their chemical and physical properties, and their relation to living things. Hydrology encompasses surface water hydrology and groundwater hydrology; the latter, however, is considered to be a subject in itself. Other related earth sciences include climatology, meteorology, geology, geomorphology, sedimentology, geography, and oceanography.

Engineering hydrology is an applied earth science. It uses hydrologic principles in the solution of engineering problems arising from human exploitation of the water resources of the earth. In its broadest sense, engineering hydrology seeks to establish relations defining the spatial, temporal, seasonal, annual, regional, or geographical variability of water, with the aim of ascertaining societal risks involved in sizing hydraulic structures and systems.

1.2 THE HYDROLOGIC CYCLE

The hydrologic cycle describes the continuous recirculatory transport of the waters of the earth, linking atmosphere, land, and oceans. The process is quite complex, containing many subcycles. To explain it briefly, water evaporates from the ocean surface, driven by energy from the sun, and joins the atmosphere, moving inland. Once inland, atmospheric conditions act to condense and precipitate water onto the land surface, where, driven by gravitational forces, it returns to the ocean through streams and rivers.

Figure 1-1 shows a pictorial representation of the hydrologic cycle. A schematic view, including the interaction between the various phases and water-holding elements, is shown in Fig. 1-2 [3]. This schematic includes all physical processes relevant to engineering hydrology. Precipitation and other liquid-transport phases are represented by straight arrows, while evaporation and other vapor-transport phases are represented by wavy arrows.

The water-holding elements of the hydrologic cycle are

1. Atmosphere
2. Vegetation
3. Snowpack and icecaps
4. Land surface
5. Soil
6. Streams, lakes, and rivers
7. Aquifers
8. Oceans

Liquid-transport phases of the hydrologic cycle are

1. Precipitation from the atmosphere onto land surface
2. Throughfall from vegetation onto land surface
3. Melt from snow and ice onto land surface
4. Surface runoff from land surface to streams, lakes, and rivers, and from streams, lakes, and rivers to oceans
5. Infiltration from land surface to soil
6. Exfiltration from soil to land surface
7. Interflow from soil to streams, lakes, and rivers and vice versa
8. Percolation from soil to aquifers

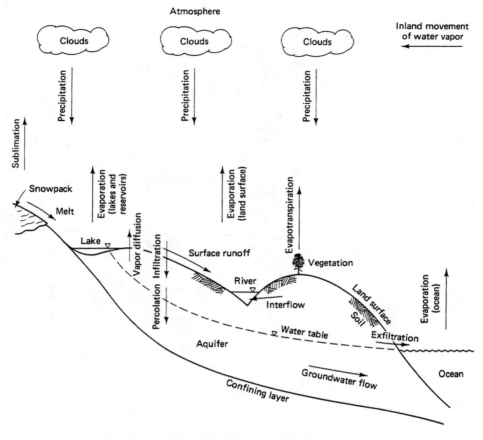

Figure 1-1. The hydrologic cycle.

9. Capillary rise from aquifers to soil
10. Groundwater flow from streams, lakes, and rivers to aquifers and vice versa and from aquifers to oceans and vice versa

Vapor-transport phases of the hydrologic cycle are

1. Evaporation from land surface, streams, lakes, rivers, and oceans to the atmosphere
2. Evapotranspiration from vegetation to the atmosphere
3. Sublimation from snowpack and icecaps to the atmosphere
4. Vapor diffusion from soil to land surface

1.3 THE CATCHMENT AND ITS HYDROLOGIC BUDGET

A catchment is a portion of the earth's surface that collects runoff and concentrates it at its furthest downstream point, referred to as the *catchment outlet*. The runoff con-

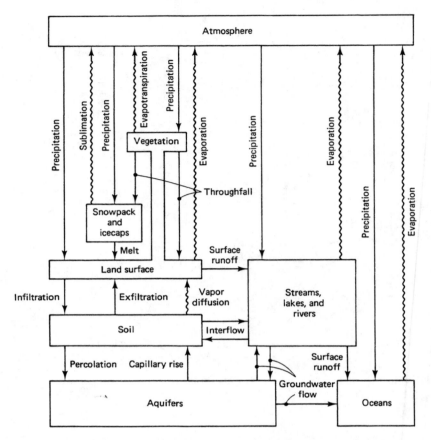

Figure 1-2. Schematic view of the hydrologic cycle (By permission from "Dynamic Hydrology," P. S. Eagleason, 1970, McGraw-Hill [3]).

centrated by a catchment flows either into a larger catchment or into the ocean. The place where a stream enters a larger stream or body of water is referred to as the *mouth*.

In United States hydrologic practice, the terms *watershed* and *basin* are commonly used to refer to catchments. Generally, watershed is used to describe a small catchment (stream watershed), whereas basin is reserved for large catchments (river basins). In this book, the word *catchment* is used without a specific connotation of scale, whereas the use of the words watershed and basin follow established practice.

The interpretation of the hydrologic cycle within the confines of a catchment leads to the concept of hydrologic budget. The hydrologic budget refers to an accounting of the various transport phases of the hydrologic cycle within a catchment, with the aim of ascertaining their relative magnitudes. The following is a hydrologic budget equation that considers both surface water and groundwater:

$$\Delta S = P - (E + T + G + Q) \tag{1-1}$$

in which ΔS = change in storage, P = precipitation, E = evaporation, T = evapotranspiration, G = groundwater flow out of the catchment, and Q = surface runoff.

Within an appropriate time span, the change in water volume remaining in storage in a catchment is the difference between precipitation and the sum of evaporation, evapotranspiration, groundwater outflow, and surface runoff. In hydrologic practice, the terms of Eq. 1-1 are expressed in units of water depth, i.e., a water volume uniformly distributed over the catchment area.

A hydrologic budget equation that accounts only for surface water is

$$\Delta S = P - (E + T + I + Q) \tag{1-2}$$

in which I = infiltration and all other terms are as defined previously. Assuming ΔS = 0 (no change in storage within a given time span), Eq. 1-2 reduces to

$$Q = P - L \tag{1-3}$$

in which L = *losses*, or hydrologic abstractions, equal to the sum of evaporation, evapotranspiration, and infiltration. Equation 1-3 states that runoff is equal to precipitation minus the aggregate of all losses. This concept is the basis of many practical methods for runoff computations (Chapters 4 and 5).

1.4 USES OF ENGINEERING HYDROLOGY

Engineering hydrology seeks to answer questions of the following type:

1. What is the maximum probable flood at a proposed dam site?
2. How does a catchment's water yield vary from season to season and from year to year?
3. What is the relationship between a catchment's surface water and groundwater resources?
4. When evaluating low flow characteristics, what flow level can be expected to be exceeded 90 percent of the time?
5. Given the natural variability of streamflows, what is the appropriate size of an instream storage reservoir?
6. What hydrologic hardware (e.g., rainfall sensors) and software (computer models) are needed for real-time flood forecasting?

In seeking answers to these questions, engineering hydrology uses analysis and measurement. Hydrologic analysis aims to develop a methodology to quantify a certain phase or phases of the hydrologic cycle—for instance, precipitation, infiltration, or surface runoff. The unit hydrograph technique (Chapter 5) is a good example of a time-tested method of hydrologic analysis. Field measurements such as stream gaging (Chapter 3) complement and verify the analysis. Statistical methods, e.g., linear regression (Chapter 7), supplement hydrologic analysis and/or measurement.

Generally, the hydrologic engineer is interested in describing either flow rates or volumes, including their spatial, temporal, seasonal, annual, or regional variability. Flow rates (discharges) are commonly expressed in cubic meters per second or cubic feet per second; volumes are expressed in cubic meters, cubic hectometers, or acre-feet. In engineering hydrology, volumes are often expressed in depth units (millimeters, centimeters, or inches), intended to represent a uniform water depth over the catchment area.

1.5 APPROACHES TO ENGINEERING HYDROLOGY

There are many approaches to engineering hydrology. These can be thought of as models seeking to represent the behavior of the prototype (i.e., the real world). Generally, models can be classified as either (a) material, or (b) formal. A material model is a physical representation of a prototype, simpler in structure and with properties similar to those of the prototype. A formal model is a mathematical abstraction of an idealized situation that preserves the important structural properties of the prototype [4].

Material models can be either iconic or analog. Iconic models are simplified representations of real-world hydrologic systems, such as lysimeters, rainfall simulators, and experimental watersheds. Analog models are those that base their measurements on substances different from those of the prototype, such as the flow of electrical current to represent the flow of water.

In engineering hydrology, all formal models are mathematical in nature—hence the use of the term *mathematical model* to refer to all formal models. Unless specifically stated otherwise, the term *model* will be used here to refer to a mathematical model. The latter is by far the most widely used model type in engineering hydrology.

Mathematical models can be either (1) theoretical, (2) conceptual, or (3) empirical. A theoretical model is founded on a set of general laws; conversely, an empirical model is largely based on inferences derived from the analysis of data. A conceptual model lies somewhere in between theoretical and empirical models.

In engineering hydrology, four types of mathematical models are in current use: (1) deterministic, (2) probabilistic, (3) conceptual, and (4) parametric. A deterministic model is formulated by using laws of physical or chemical processes, as described by differential equations. A probabilistic model, whether statistical or stochastic, is governed by laws of chance or probability. Statistical models deal with observed samples, whereas stochastic models focus on the random properties of certain hydrologic time series—for instance, daily streamflows [5]. A conceptual model is a simplified representation of the physical processes, obtained by lumping spatial and/or temporal variations, and described in terms of either ordinary differential equations or algebraic equations. A parametric model represents hydrologic processes by means of algebraic equations that contain key parameters to be determined by empirical means.

Methods of analysis in engineering hydrology can generally be classified under one of the four model types just mentioned. For instance, the kinematic wave-routing technique (Chapters 4 and 9) is deterministic, being governed by a partial differential equation describing the mass and momentum balance of fluid mechanics. The Gumbel method of flood frequency analysis (Chapter 6) is probabilistic, being based on an extreme value probability law. The cascade of linear reservoirs (Chapter 10) is conceptual, seeking to simulate the complexities of catchment response by means of a series of hypothetical linear reservoirs. The rational method (Chapter 4) is parametric, with peak flow (for a given frequency) estimated on the basis of an empirically determined runoff coefficient.

In principle, deterministic models mimic physical processes and should, therefore, be closest to reality. In practice, however, the inherent complexity of physical phenomena generally limits the deterministic approach to well defined cases for which a clear cause-effect relationship can be demonstrated. Probabilistic methods are used to fit measured data (i.e., statistical hydrology) and to model random components (stochastic hydrology) in cases where their presence is readily apparent [2]. Where simplicity is desired, conceptual and parametric methods and models continue to play an important role in hydrologic engineering practice.

Hydrologic models can be either *lumped* or *distributed*. Lumped models can describe temporal variations but cannot describe spatial variations. A typical example of a lumped hydrologic model is the unit hydrograph (Chapter 5), which describes a catchment's unit response without regard to the response of individual subcatchments.

Unlike lumped models, distributed models have the capability to describe both temporal and spatial variations. Distributed models are much more computationally intensive than lumped models and are therefore ideally suited for use with a computer. A typical example of a distributed hydrologic model is an overland flow computation using routing techniques (Chapter 10). In this case, equations of mass and momentum (or surrogates thereof) are used to compute temporal variations of discharge and flow depth at several locations *within* a catchment.

Solutions to hydrologic models can be either analytical or numerical. Analytical solutions are obtained by using classical tools of applied mathematics, such as Laplace transforms, perturbation theory, and the like. Numerical solutions are obtained by discretizing differential equations into algebraic equations and solving them, usually with the aid of a computer. Examples of analytical solutions are the linear models used in hydrologic systems analysis [1]. Examples of numerical solutions abound, such as those used in hydrologic routing techniques (Chapters 8 and 9) and in the computer models in current use (Chapter 13).

1.6 SURFACE RUNOFF, FLOOD HYDROLOGY, AND CATCHMENT SCALE

Surface runoff occurs when rainfall intensity exceeds the abstractive capability of the catchment. Eventually, large amounts of surface runoff concentrate to produce large flow rates referred to as floods. The study of floods, their occurrences, causes, transport, and effects is the subject of flood hydrology.

In nature, rainfall varies in space and time. In engineering hydrology, rainfall can be assumed to be either (1) constant in both space and time, (2) constant in space but varying in time, or (3) varying in both space and time. The catchment scale helps determine which one of these assumptions is justified on practical grounds. Generally, *small catchments* are those in which runoff can be modeled by assuming constant rainfall in both space and time. *Midsize catchments* are those in which runoff can be modeled by assuming rainfall to be constant in space but to vary in time. *Large catchments* are those in which runoff can be modeled by assuming rainfall to vary in both space and time.

In flood hydrology, small catchments are usually modeled with a simple empirical approach, such as the rational method (Chapter 4). For midsize catchments, a lumped conceptual model such as the unit hydrograph is preferred by most engineers in practice (Chapter 5). For large catchments, temporal and spatial variations of rainfall and runoff may dictate the use of a distributed modeling approach, including reservoir and stream channel routing (Chapters 8 and 9). Figure 1-3 shows a matrix depicting the relationship between catchment scale and three commonly used approaches to flood hydrology.

The larger the catchment, the more likely it is to be gaged, i.e., to possess a streamflow record. Conversely, the smaller the catchment, the more unlikely it is to be gaged. This fact dictates that the probabilistic approach (Frequency Analysis, Chapter 6) is primarily applicable to large catchments, particularly to those possessing a fairly long record period. For ungaged catchments or for gaged catchments with relatively short record periods, statistical techniques can be used to develop parametric models having regional applicability (Regional Analysis, Chapter 7). The subjects of catchment routing (Chapter 10) and catchment modeling (Chapter 13) span the gamut of hydrologic applications, from small to large catchments (Fig. 1-3).

Catchment scale

	Small	Midsize	Large
Rational method	Usually	Not applicable	Not applicable
Unit hydrograph	Not applicable	Usually	Sometimes
Routing methodologies	Sometimes	Sometimes	Usually

(Method or approach)

Figure 1-3. Relationship between catchment scale and three commonly used approaches to flood hydrology.

QUESTIONS

1. What is the hydrologic cycle?
2. Name the liquid-transport phases of the hydrologic cycle.
3. Name the vapor-transport phases of the hydrologic cycle.
4. What is a catchment?
5. Give two examples of engineering problems (different from those mentioned in the text) where hydrologic knowledge is necessary to obtain a solution.
6. What is material model? A formal model?
7. What is an iconic model? An analog model?
8. What is a deterministic model? A lumped model?
9. Contrast conceptual and parametric models.
10. Contrast analytical and numerical solutions.
11. What is a small catchment from the flood hydrology standpoint? A midsize catchment? A large catchment?

PROBLEMS

1-1. During a given year, the following hydrologic data were collected for a 2500-km^2 basin: total precipitation, 620 mm; total combined loss due to evaporation and evapotranspiration, 320 mm; estimated groundwater outflow (including groundwater depletion), 100 mm; and mean surface runoff, 150 mm. What is the change in volume of water (in hm^3) remaining in storage in the basin during the elapsed year?

1-2. During 1987, the following hydrologic data were collected for a 85-mi^2 watershed: total precipitation, 27 in.; total combined loss due to evaporation and evapotranspiration, 10 in.; estimated groundwater outflow (including groundwater depletion), 7 in.; and mean surface runoff, 9 in. What is the change in volume of water (in ac-ft) remaining in storage in the watershed during 1987?

1-3. During a given year, the following hydrologic data were collected for a certain 350-km^2 catchment: total precipitation, 850 mm; combined evaporation and evapotranspiration, 420 mm; and surface runoff, 225 mm. Calculate the volume of infiltration (in hm^3), neglecting changes in surface water storage and groundwater effects.

1-4. During a given year, the following hydrologic data were measured for a certain 60-mi^2 watershed: total precipitation, 35 in.; and estimated losses due to evaporation, evapotranspiration, and infiltration, 28 in. Calculate the mean annual runoff (in ft^3/s). Neglect changes in surface water storage and groundwater effects.

REFERENCES

1. Agricultural Research Service, U.S. Department of Agriculture. (1973). "Linear Theory of Hydrologic Systems," *Technical Bulletin No.* 1468. (J.C.I. Dooge, author). Washington, D.C., October.
2. Bras, R., and I. Rodriguez-Iturbe. (1985). *Random Functions and Hydrology*. Reading, Mass.: Addison-Wesley.
3. Eagleson, P. S. (1970). *Dynamic Hydrology*. New York: McGraw-Hill.
4. Woolhiser, D. A., and D. L. Brakensiek. (1982). "Hydrologic Modeling of Small Watersheds," Chapter 1 in *Hydrologic Modeling of Small Watersheds,* edited by C. T. Haan et al. ASAE Monograph No. 5, St. Joseph, Michigan.
5. Yevjevich, V. (1972). *Stochastic Processes in Hydrology*. Fort Collins, Colo.: Water Resources Publications.

BASIC HYDROLOGIC PRINCIPLES

Engineering hydrology takes a quantitative view of the hydrologic cycle. Generally, equations are used to describe the interaction between the various phases of the hydrologic cycle. As shown in Chapter 1, the following basic equation relates precipitation and surface runoff:

$$Q = P - L \tag{2-1}$$

in which Q = surface runoff, P = precipitation; and L = losses, or hydrologic abstractions. The latter term is interpreted as the summation of the various precipitation-abstracting phases of the hydrologic cycle.

Rainfall is the liquid form of precipitation; snowfall and hail are the solid forms. In common usage, the word *rainfall* is often used to refer to precipitation. Exceptions are the cases where a distinction between liquid and solid precipitation is warranted.

Generally, the catchment has an abstractive capability that acts to reduce total rainfall into effective rainfall. The difference between total rainfall and effective rainfall is the losses or hydrologic abstractions. The abstractive capability is a characteristic of the catchment, varying with its level of stored moisture. Hydrologic abstractions include interception, infiltration, surface storage, evaporation, and evapotranspiration. The difference between total rainfall and hydrologic abstractions is called runoff. Therefore, the concepts of effective rainfall and runoff are equivalent.

The terms in Eq. 2-1 can be expressed as rates (millimeters per hour, centimeters per hour, or inches per hour), or when integrated over time, as depths (millimeters, centimeters or inches). In this sense, a given depth of rainfall or runoff is a volume of water uniformly distributed over the catchment area.

This chapter is divided into four sections. Section 2.1 deals with precipitation, its meteorological aspects, quantitative description, spatial and temporal variations, and data sources. Section 2.2 discusses hydrologic abstractions that are important in engineering hydrology: interception, infiltration, surface storage, evaporation, and evapotranspiration. Section 2.3 defines geometric and other catchment properties relevant to hydrologic analysis. Section 2.4 deals with runoff analysis, both in a qualitative and quantitative way. The concepts presented in this chapter are of an introductory nature, intended to provide the necessary background for the more specialized study that will follow.

2.1 PRECIPITATION

Meteorological Aspects

The earth's atmosphere contains water vapor. The amount of water vapor can be conveniently expressed in terms of a depth of precipitable water. This is the depth of water that would be realized if all the water vapor in the air column above a given area were to condense and precipitate on that area.

There is an upper limit to the amount of water vapor in an air column. This upper limit is a function of the air temperature. The air column is considered to be saturated when it contains the maximum amount of water vapor for its temperature. Lowering the air temperature results in a reduction of the air column's capacity for water vapor. Consequently, an unsaturated air column, i.e., one that has less than the maximum amount of water vapor for its temperature, can become saturated without the actual addition of moisture if its temperature is lowered to a level at which the actual amount of water vapor will produce saturation. The temperature to which air must be cooled, at constant pressure and water vapor content, to reach saturation is called the *dewpoint*. Condensation usually occurs at or near saturation of the air column.

Cooling of Air Masses. Air can be cooled by many processes. However, adiabatic cooling by reduction of pressure through lifting is the only natural process by which large air masses can be cooled rapidly enough to produce appreciable precipitation. The rate and amount of precipitation are a function of the rate and amount of cooling and of the rate of moisture inflow into the air mass to replace the water vapor that is being converted into precipitation.

The lifting required for the rapid cooling of large air masses can be produced by either (1) horizontal convergence, (2) frontal lifting, or (3) orographic lifting. More than one of these processes is usually active in the lifting associated with the heavier precipitation rates and amounts.

Horizontal convergence, or simply convergence, occurs when the pressure and wind fields act to concentrate inflow of air into a particular area, such as a low-pressure area. If this convergence takes place in the lower layers of the atmosphere, the tendency to pile up forces the air upward, resulting in its cooling.

Frontal lifting takes place when relatively warm air flowing towards a colder (hence denser) air mass is forced upward, with the cold air acting as a wedge. Cold air overtaking warmer air will produce the same result by *wedging* the latter aloft. The surface of separation between the two different air masses is called a *frontal surface*. A frontal surface always slopes upward toward the colder air mass; the intersection of the frontal surface with the ground is called a *front* (Fig. 2-1).

Orographic lifting occurs when air flowing toward an orographic barrier (i.e., mountain) is forced to rise in order to pass over it. The slopes of orographic barriers are usually steeper than the steepest slopes of frontal surfaces. Consequently, air is cooled much more rapidly by orographic lifting than by frontal lifting.

Condensation of Water Vapor into Liquid or Solid Form. Condensation is the process by which water vapor in the atmosphere is converted into liquid droplets or, at low temperatures, into ice crystals. The results of the process are often, but not always, visible in the form of clouds, which are airborne liquid water droplets or ice crystals or a mixture of these two.

Saturation does not necessarily result in condensation. Condensation nuclei are required for the conversion of water vapor into droplets. Among the more effective condensation nuclei are certain products of combustion and salt particles from the sea. There are usually enough condensation nuclei in the air to produce condensation when the water vapor reaches saturation point.

Growth of Cloud Droplets and Ice Crystals to Precipitation Size. When air is cooled below its initial saturation temperature and condensation continues to take place, liquid droplets or ice crystals tend to accumulate in the resulting cloud. The rate at which this excess liquid and solid moisture is precipitated from the cloud depends upon (1) the speed of the upward current producing the cooling, (2) the rate of growth of the cloud droplets into raindrops heavy enough to fall through the upward current, and (3) a sufficient inflow of water vapor into the area to replace the precipitated moisture.

Water droplets in a typical cloud usually average about 0.01 mm in radius and weigh so little that an upward current of only 0.0025 m/s is sufficient to keep them from falling [73]. Although no definite drop size can be said to mark the boundary between cloud and raindrops, a radius of 0.1 mm has been generally accepted. The radius of most raindrops reaching the ground is usually much greater than 0.1 mm and may reach 3 mm. Drops larger than this tend to break into smaller drops because the surface tension is insufficient to withstand the distortions the drop undergoes in falling through the air. Drops of 3 mm radius have a terminal velocity of about 10 m/s; therefore, an unusually strong upward current would be required to keep a drop of this size from falling.

Various theories have been advanced to explain the growth of a cloud element into a size that can precipitate. The two principal processes in the formation of precip-

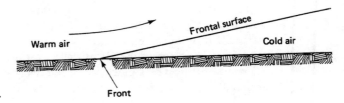

Figure 2-1. Frontal lifting of air masses.

Basic Hydrologic Principles   Chap. 2

itation are (1) the ice crystal process, and (2) the coalescence process [26]. These two processes may operate together or separately. The ice crystal process involves the presence of ice crystals in a supercooled (cooled to below freezing) water cloud. Due to the fact that saturation vapor pressure over water is greater than that over ice, there is a vapor pressure gradient from water drops to ice crystals. This causes the ice crystals to grow at the expense of the water drops and, under favorable conditions, to attain precipitation size. The ice crystal process is operative only in supercooled water clouds, and it is most effective at about $-15\ °C$.

The coalescence process is based on the difference in fall velocities and consequent collisions to be expected between cloud elements of different sizes. The rate of growth of cloud elements by coalescence depends upon the initial range of particle sizes, the size of the largest drops, the drop concentration, and the sizes of the aggregated drops. The electric field and drop charge may affect collision efficiencies and may therefore be important factors in the release of precipitation from clouds [63]. Unlike the ice crystal process, the coalescence process occurs at any temperature, its effectiveness varying from solid to liquid particles.

Forms of Precipitation. Precipitation occurs primarily in the form of drizzle, rain, snow, or hail. Drizzle consists of tiny liquid water droplets, usually between 0.1 and 0.5 mm in diameter, falling at intensities rarely exceeding 1 mm/h. Rain consists of liquid water drops, mostly larger than 0.5 mm in diameter. Rainfall refers to amounts of liquid precipitation. Rainfall intensities can be classified as: light, up to 3 mm/h; moderate, from 3 to 10 mm/h; and heavy, over 10 mm/h. A rainstorm is a rainfall event lasting a clearly defined duration.

Snow is composed of ice crystals, primarily in complex hexagonal form and often aggregated into snowflakes that may reach several millimeters in diameter. Snowfall is precipitation in the form of snow. A snowstorm is a snowfall event with a clearly defined duration. Snowpack is the volume of snow accumulated on the ground after one or more snowstorms. Snowmelt, or melt, is the volume of snow that has changed from solid to liquid state and is available for runoff.

Hail is composed of solid ice stones or hailstones. Hailstones may be spheroidal, conical, or irregular in shape and may range from about 5 to over 125 mm in diameter. A hailstorm is a precipitation event in the form of hail.

Quantitative Description of Rainfall

A rainfall event, or storm, describes a period of time having measurable and significant rainfall, preceded and followed by periods with no measurable rainfall. The time elapsed from start to end of the rainfall event is the rainfall duration. Typically, rainfall duration is measured in hours. However, for very small catchments it may be measured in minutes, while for very large catchments it may be measured in days.

Rainfall durations of 1, 2, 3, 6, 12, and 24 h are common in hydrologic analysis and design. For small catchments, rainfall durations can be as short as 5 min. Conversely, for large river basins, durations of 2 d and longer may be applicable [78]. Rainfall depth is measured in mm, cm, or in., considered to be uniformly distributed over the catchment area. For instance, a 60-mm, 6-h rainfall event produces 60 mm of depth over a 6-h period.

Rainfall depth and duration tend to vary widely, depending on geographic location, climate, microclimate, and time of the year. Other things being equal, larger rainfall depths tend to occur more infrequently than smaller rainfall depths. For design purposes, rainfall depth at a given location is related to the frequency of its occurrence. For instance, 60 mm of rainfall lasting 6 h may occur on the average once every 10 y at a certain location. However, 80 mm of rainfall lasting 6 h may occur on the average once every 25 y at the same location.

Average rainfall intensity is the ratio of rainfall depth to rainfall duration. For example, a rainfall event producing 60 mm in 6 h represents an average rainfall intensity of 10 mm/h. Rainfall intensity, however, varies widely in space and time, and local or instantaneous values are likely to be very different from the spatial and temporal average. Typically, rainfall intensities are in the range 0.1-30.0 mm/h, but can be as large as 150 to 350 mm/h in extreme cases.

Rainfall frequency refers to the average time elapsed between occurrences of two rainfall events of the same depth and duration. The actual elapsed time varies widely and can therefore be interpreted only in a statistical sense. For instance, if at a certain location a 100-mm rainfall event lasting 6 h occurs on the average once every 50 y, the 100-mm, 6-h rainfall frequency for this location would be 1 in 50 years, 1/50, or 0.02.

The reciprocal of rainfall frequency is referred to as *return period* or recurrence interval. In the case of the previous example, the return period corresponding to a frequency of 0.02 is 50 y.

Generally, larger rainfall depths tend to be associated with longer return periods. The longer the return period, the longer the historical record needed to ascertain the statistical properties of the distribution of annual maximum rainfall. Due to the paucity of long rainfall records, extrapolations are usually necessary to estimate rainfall depths associated with long return periods.

These extrapolations entail a certain measure of risk. When the risk involves human life, the concepts of rainfall frequency and return period are no longer considered adequate for design purposes. Instead, a reasonable maximization of the meteorological factors associated with extreme precipitation is used, leading to the concept of probable maximum precipitation (PMP). For a given geographic location, catchment area, event duration, and time of the year, the PMP is the theoretically greatest depth of precipitation. In flood hydrology studies, the PMP is used as a basis for the calculation of the probable maximum flood (PMF).

For certain projects, a precipitation depth less than the PMP may be justified for economic reasons. This leads to the concept of standard project storm (SPS). The SPS is taken as an appropriate percentage of the applicable PMP and is used to calculate the standard project flood (SPF) (Chapter 14).

Temporal and Spatial Variation of Precipitation

Temporal Rainfall Distribution. Rainfall intensities for events of short duration (1 h or less) can usually be expressed as an average value, obtained by dividing rainfall depth by rainfall duration. For longer events, instantaneous values of rainfall intensity are likely to become more important, particularly for flood-peak determinations.

The temporal rainfall distribution depicts the variation of rainfall depth within a storm duration. It can be expressed in either discrete or continuous form. The discrete form is referred to as a *hyetograph*, a histogram of rainfall depth (or rainfall intensity) with time increments as abscissas and rainfall depth (or rainfall intensity) as ordinates (Fig. 2-2(a)).

The continuous form is the temporal rainfall distribution, a function describing the rate of rainfall accumulation with time. Rainfall duration (abscissas) and rainfall depth (ordinates) can be expressed in percentage of total value (Fig. 2-2(b)). The dimensionless temporal rainfall distribution is used to convert a storm depth into a hyetograph, as shown in the following example.

Example 2-1.

Using the dimensionless temporal rainfall distribution shown in Fig. 2-3, calculate a hyetograph for a 15-cm, 6-h storm.

For convenience, a time increment of 1 h, or 1/6 of the storm duration is chosen. The cumulative rainfall percentages (at increments of 1/6 of storm duration) obtained from Fig. 2-3 are the following: 10, 20, 40, 70, 90, and 100%. Therefore, the incremental percentages are: 10, 10, 20, 30, 20, and 10%. For a 15-cm total storm depth, the incremental (hourly) rainfall depths are the following: 1.5, 1.5, 3.0, 4.5, 3.0, and 1.5 cm.

Spatial Rainfall Distribution. Rainfall varies not only temporally but also spatially, i.e., the same amount of rain does not fall uniformly over the entire catchment. *Isohyets* are used to depict the spatial variation of rainfall. An isohyet is a contour line showing the loci of equal rainfall depth (Fig. 2-4(a)).

Individual storms may have a spatial distribution or pattern in the form of concentric isohyets of approximately elliptic shape (Fig. 2-4(b)). This gives rise to the term *storm eye* to depict the center of the storm. In general, storm patterns are not static, moving gradually in a direction approximately parallel to that of the prevailing winds.

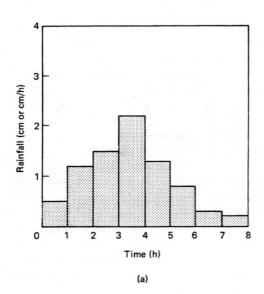

(a)

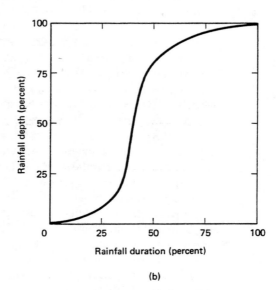

(b)

Figure 2-2. (a) Hyetograph; (b) dimensionless temporal rainfall distribution.

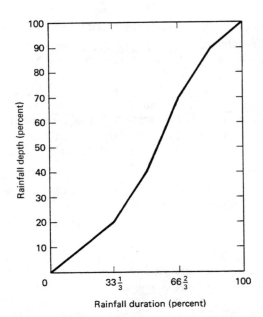

Figure 2-3. Dimensionless temporal rainfall distribution: Example 2-1.

Rainfall duration (percent)

For regional rainfall mapping, isohyets are commonly referred to as *isopluvials*. Isopluvial maps for the United States are published by the National Weather Service [50, 51, 74–78]. These maps show contours of equal rainfall depth, applicable for a range of durations, frequencies, and catchment sizes; see, for example, Fig. 2-5.

For large catchments, highly intensive storms (thunderstorms) may cover only a fraction of the whole basin, yet they may lead to severe flooding in certain localized areas. The role of thunderstorms in determining the flood potential of large basins is usually assessed on an individual basis.

Average Precipitation Over an Area. A precipitation (or rainfall) amount is measured with rain gages. During a given storm, it is likely that the depth measured by two or more rain gages of the same type will not be the same. In hydrologic analysis, it is often necessary to determine a spatial average of the rainfall depth over the catchment. This is accomplished by either of the following methods: (1) average rainfall, (2) Thiessen polygons, and (3) isohyetal method.

In the average rainfall method, the rainfall depths measured by the rain gages located within the catchment are tabulated. These rainfall depths are then averaged to find the average precipitation over the catchment (see Example 2-2(a)).

In the *Thiessen polygons* method, the locations of the rain gages are plotted on a scale map of the catchment and surrounding area. The locations (stations) are joined with straight lines in order to form a pattern of triangles, preferably with sides of approximately equal length. Perpendicular bisectors to the sides of these triangles are drawn to enclose each station within a polygon called a Thiessen polygon, circumscribing an area of influence. The average precipitation over the catchment is calculated by weighing each station's rainfall depth in proportion to its area of influence (see Example 2-2(b)).

In the isohyetal method, the locations of the rain gages are plotted on a scale

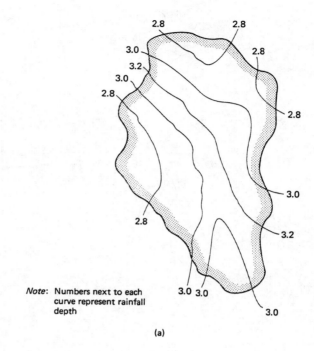

Note: Numbers next to each curve represent rainfall depth

(a)

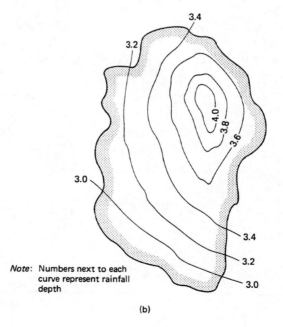

Note: Numbers next to each curve represent rainfall depth

(b)

Figure 2-4. (a) Isohyets (contours of equal rainfall depth); (b) a storm eye.

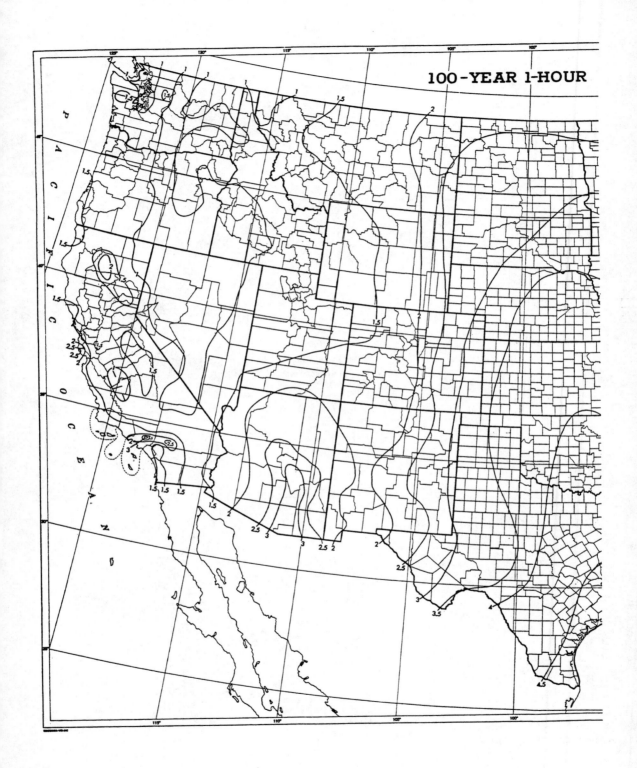

100-YEAR 1-HOUR

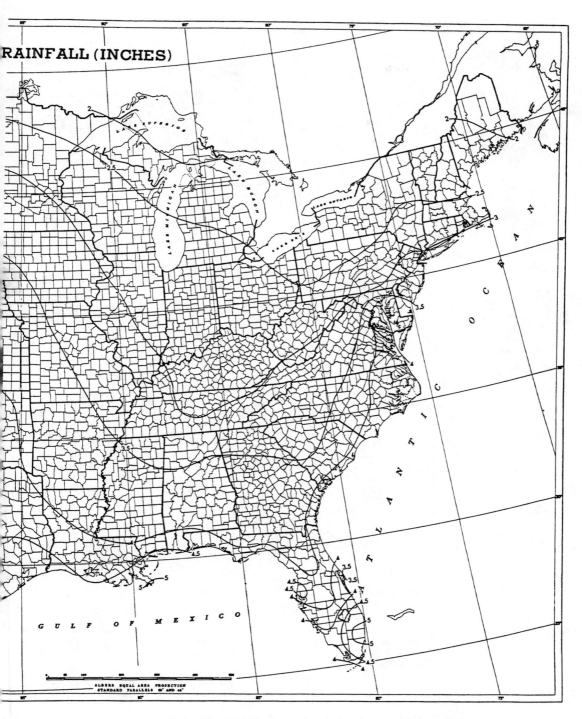

RAINFALL (INCHES)

Figure 2-5. (a) 100-y frequency 1-h duration isopluvial map [76].

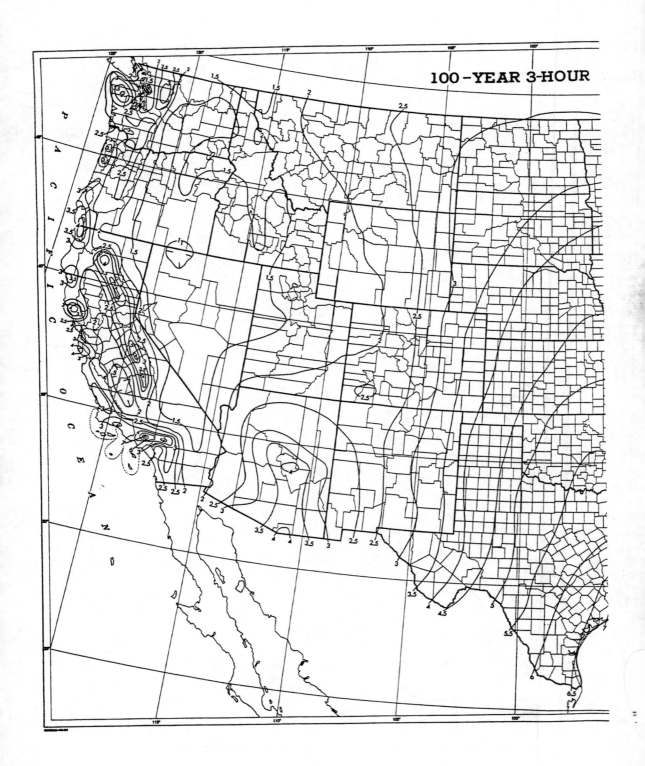

100-YEAR 3-HOUR

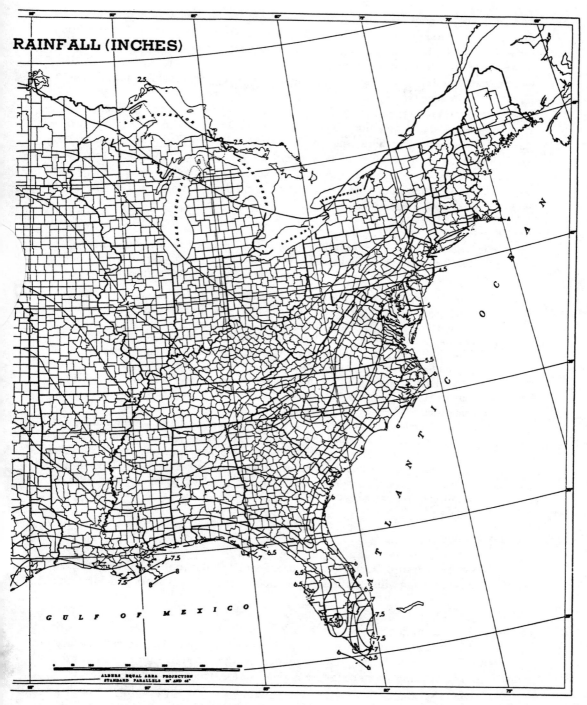

RAINFALL (INCHES)

Figure 2-5. (b) 100-y frequency 3-h duration isopluvial map [76].

map of the catchment and surrounding area. Each station's rainfall depth is used to draw isohyets throughout the catchment in a manner similar to that used in the preparation of topographic contour maps. The mid-distance between two adjacent isohyets is used to delineate the area of influence of each isohyet. The average precipitation over the catchment is calculated by weighing each isohyetal increment in proportion to its area of influence (see Example 2-2(c)).

The isohyetal method is regarded as more accurate than either the Thiessen polygons or average rainfall methods. This is particularly the case when averaging precipitation over catchments where orographic effects have a significant influence on the local storm pattern. The Thiessen polygons method is generally more accurate than the average rainfall method. The increase in accuracy is likely to be more marked when averaging precipitation over catchments with widely varying rainfall depths or large differences in areas of influence.

Example 2-2.

With reference to the catchment (28.6 km^2) and rain gage locations depicted in Fig. 2-6, calculate the average precipitation over the catchment following: (a) the average rainfall method, (b) the Thiessen polygons method, and (c) the isohyetal method.

(a) *Average rainfall method.* The measured precipitation depths for the seven stations (A to G) are summed, and the sum divided by 7. The average precipitation over the catchment is 17.5/7 = 2.50 cm.

(b) *Thiessen polygons method.* Each stations's area of influence is found by drawing perpendicular bisectors to the sides of triangles formed by joining the stations with straight lines, as shown in Fig. 2-6(b). The sum of the products $(P_i A_i)$ is calculated (P is the precipitation depth, A is the area of influence, and i denotes each individual station). The average precipitation over the catchment is calculated by dividing the sum of the products by the catchment area: 71.27/28.6= 2.49 cm.

(c) *Isohyetal method.* Isohyets (contours of equal depth) are drawn as shown in Fig. 2-6 (c). The mid-distance between two adjacent isohyets serves to delineate the area of influence of each isohyet. The sum of the products $(P_i A_i)$ is calculated (P is the value of each isohyet, A is its area of influence, and i denotes each isohyetal increment). The average precipitation over the catchment is calculated by dividing the sum of the products by the catchment area: 70.325/28.6 = 2.46 cm.

Storm Analysis

Storm Depth and Duration.   Storm depth and duration are directly related, storm depth increasing with duration. An equation relating storm depth and duration is

$$h = ct^n \qquad (2\text{-}2)$$

in which h = storm depth in centimeters; t = storm duration in hours; c = a coefficient; and n = an exponent (a positive real number less than 1). Typically n varies between 0.2 and 0.5, indicating that storm depth increases at a lesser rate than storm duration. By analyzing storm data on a regional or local basis, Eq. 2-2 can be used to predict storm depth as a function of storm duration. The applicability of such an equation, however, is limited to the regional or local conditions for which it was derived.

Equation 2-2 can also be used to study the characteristics of extreme rainfall

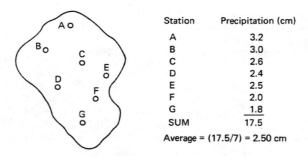

Station	Precipitation (cm)
A	3.2
B	3.0
C	2.6
D	2.4
E	2.5
F	2.0
G	1.8
SUM	17.5

Average = (17.5/7) = 2.50 cm

(a) Average rainfall method

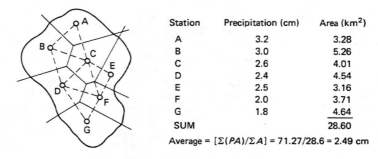

Station	Precipitation (cm)	Area (km^2)
A	3.2	3.28
B	3.0	5.26
C	2.6	4.01
D	2.4	4.54
E	2.5	3.16
F	2.0	3.71
G	1.8	4.64
SUM		28.60

Average = $[\Sigma(PA)/\Sigma A]$ = 71.27/28.6 = 2.49 cm

(b) Thiessen polygons method

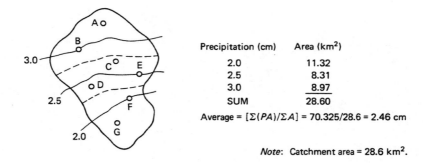

Precipitation (cm)	Area (km^2)
2.0	11.32
2.5	8.31
3.0	8.97
SUM	28.60

Average = $[\Sigma(PA)/\Sigma A]$ = 70.325/28.6 = 2.46 cm

Note: Catchment area = 28.6 km^2.

(c) Isohyetal method

Figure 2-6. Average precipitation over an area: Example 2-2.

events. A logarithmic plot of depth-duration data for the world's greatest observed rainfall events (Table 2-1) results in the following enveloping line:

$$h = 39t^{0.5} \tag{2-3}$$

in which h = rainfall depth in centimeters and t = rainfall duration in hours. The data of Table 2-1 are plotted in Fig. 2-7, including the enveloping line, Eq. 2-3.

Storm Intensity and Duration.   Storm intensity and duration are inversely related. From Eq. 2-2, an equation linking storm intensity and duration, can be ob-

TABLE 2-1 WORLD'S GREATEST OBSERVED RAINFALL EVENTS [38]

Duration	Depth (cm)	Location	Date
1 min	3.8	Barot, Guadeloupe	Nov. 26, 1970
8 min	12.6	Fussen, Bavaria	May 25, 1920
15 min	19.8	Plumb Point, Jamaica	May 12, 1916
42 min	30.5	Holt, MO	June 22, 1947
2 h 10 min	48.3	Rockport, WV	July 18, 1889
2 h 45 min	55.9	D'Hanis, TX (17 mi NNW)	May 31, 1935
4 h 30 min	78.2	Smethport, PA	July 18, 1942
9 h	108.7	Belouve, Reunion	Feb. 28, 1964
12 h	134.0	Belouve, Reunion	Feb. 28–29, 1964
18 h 30 min	168.9	Belouve, Reunion	Feb. 28–29, 1964
24 h	187.0	Cilaos, Reunion	Mar. 15–16, 1952
2 d	250.0	Cilaos, Reunion	Mar. 15–17, 1952
3 d	324.0	Cilaos, Reunion	Mar. 15–18, 1952
4 d	372.1	Cherrapunji, India	Sept. 12–15, 1974
5 d	385.4	Cilaos, Reunion	Mar. 13–18, 1952
6 d	405.5	Cilaos, Reunion	Mar. 13–19, 1952
7 d	411.0	Cilaos, Reunion	Mar. 12–19, 1952
15 d	479.8	Cherrapunji, India	June 24–30, 1931
31 d	930.0	Cherrapunji, India	July 1861
3 mo	1637.0	Cherrapunji, India	May–July 1861
6 mo	2245.0	Cherrapunji, India	Apr.–Sept. 1861
1 y	2646.0	Cherrapunji, India	Aug. 1860–July 1861
2 y	4077.0	Cherrapunji, India	1860–1861

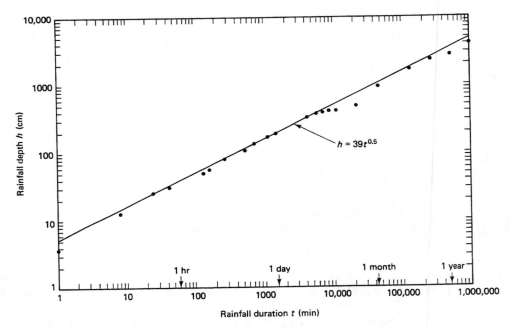

Figure 2-7. Depth-duration data for the world's greatest observed rainfall events.

Basic Hydrologic Principles   Chap. 2

tained by differentiating rainfall depth with respect to duration, to yield:

$$\frac{dh}{dt} = i = cnt^{n-1} \tag{2-4}$$

in which i = storm intensity. Simplifying,

$$i = \frac{a}{t^m} \tag{2-5}$$

in which $a = cn$ and $m = 1 - n$. Since n is less than 1, it follows that m is also less than 1.

Another intensity-duration model is the following:

$$i = \frac{a}{t + b} \tag{2-6}$$

in which a and b are constants to be determined by regression analysis (Chapter 7). A general intensity-duration model combining the features of Eqs. 2-5 and 2-6 is

$$i = \frac{a}{(t + b)^m} \tag{2-7}$$

For $b = 0$, Eq. 2-7 reduces to Eq. 2-5; for $m = 1$, Eq. 2-7 reduces to Eq. 2-6.

Intensity-Duration-Frequency. For small catchments, it is often necessary to determine several intensity-duration curves, each for a different frequency or return period. A set of *intensity-duration-frequency curves* is referred to as IDF curves, with duration plotted in the abscissas, intensity in the ordinates, and frequency (or alternatively, return period) as curve parameter. Either arithmetic (Fig. 2-8 (a)) or logarithmic (Fig. 2-8 (b)) scales are used in the construction of IDF curves. Such curves are developed by government agencies for use in urban storm-drainage design and other applications (Chapter 4).

A formula for IDF can be obtained by assuming that the constant a in Eqs. 2-5 through 2-7 is related to return period as follows:

$$a = kT^n \tag{2-8}$$

in which k = a coefficient; T = return period; and n = an exponent (not related to that of Eq. 2-2). This leads to

$$i = \frac{kT^n}{(t + b)^m} \tag{2-9}$$

The values of k, b, m, and n are evaluated from measured data or local experience.

Storm Depth and Catchment Area. Generally, the greater the catchment area, the smaller the spatially averaged storm depth. This variation of storm depth with catchment area has led to the concept of *point depth*, defined as the storm depth associated with a given *point area*. A point area is the smallest area below which the

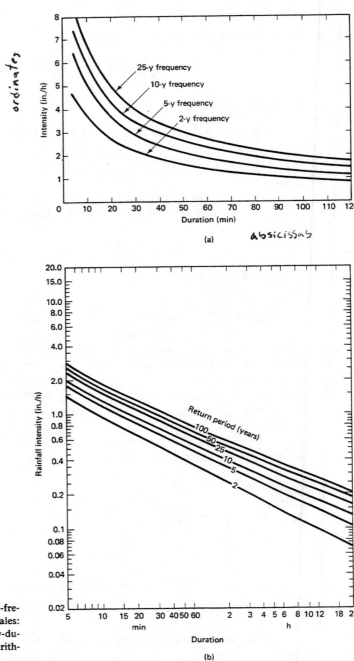

ordinates

absicissas

Figure 2-8. (a) An intensity-duration-frequency function using arithmetic scales: Davenport, Iowa [2]; (b) An intensity-duration-frequency function using logarithmic scales: San Jose, California [72].

Basic Hydrologic Principles   Chap. 2

variation of storm depth with catchment area can be assumed to be negligible. In the United States, the point area is usually taken as 25 km² (10 mi²).

The point depth applies for all areas less than the point area. For areas greater than the point area, a reduction in point depth is necessary to account for the decrease of storm depth with catchment area. This depth reduction is accomplished with a depth-area reduction chart, a function relating catchment area (abscissas) to point depth percentage (ordinates). Storm duration is usually a curve parameter in a depth-area reduction chart.

Generalized depth-area reduction charts applicable to the contiguous United States, for areas up to 1000 km² (400 mi²) and durations from 30 min to 10 d have been published by the National Weather Service (Figs. 2-9(a) and (b)). Regional and locally derived depth-area reduction charts may differ from these generalized charts (see Section 14.1).

Depth-Duration-Frequency. For midsize catchments, hydrologic analysis shifts its focus to rainfall depth. Isopluvial maps depicting storm depths, applicable for a range of durations, frequencies and catchment areas, are available for the entire United States [50, 51, 74–78]. These maps show point depth values and are therefore subject to depth-area reduction by the use of an appropriate chart.

Depth-Area-Duration. Another way of describing the relation between storm depth, duration and catchment area is the technique known as *depth-area-duration* (DAD) analysis. This technique is basically an alternate way of portraying the reduction of storm depth with area, with duration as a third variable.

To construct a DAD chart, a storm having a single major center (storm eye) is identified. Isohyetal maps showing maximum storm depths for each of several typical durations (6-h, 12-h, 24-h, etc.) are prepared. For each map, the isohyets are taken as boundaries circumscribing individual areas. For each map and each individual area, a spatially averaged rainfall depth is calculated by dividing the total rainfall volume by the individual area. This procedure provides DAD data sets used to construct a chart showing depth versus area, with duration as a curve parameter (Fig. 2-10).

DAD analysis can also be used to study regional rainfall characteristics. Table 2-2 shows maximum DAD data for the United States, based on four extreme events. The data confirm that storm depth increases with duration and decreases with catchment area.

Probable Maximum Precipitation. For large projects, storm analysis using depth-duration-frequency data is not sufficient to eliminate the likelihood of failure. In such cases, the concept of PMP is used instead. In the United States, PMP estimates are developed following guidelines included in the HMR (NOAA Hydrometeorological Reports) series [42–48] and related publications [74–76]. These reports contain methodologies and maps for the estimation of PMP for a given geographic location, range of durations and catchment sizes, and time of the year (Chapter 14).

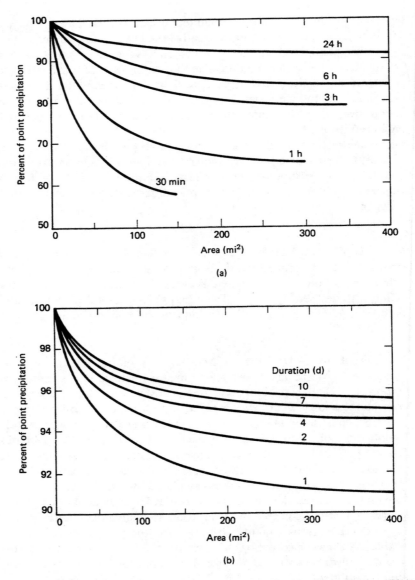

Figure 2-9. NWS generalized depth-area reduction charts: (a) 30-min to 24-h duration; (b) 1- to 10-d duration.

Geographic and Seasonal Variations of Precipitation

Precipitation varies not only temporally and spatially but also with geographic location and climate. Average annual precipitation—the total depth of precipitation that accumulates in one year, on the average, at a given location—is used to classify climates as (1) arid, (2) semiarid, or (3) humid.

An arid climate prevails in regions having less than 400 mm of average annual precipitation. A semiarid climate is found in regions having between 400 and 750 mm

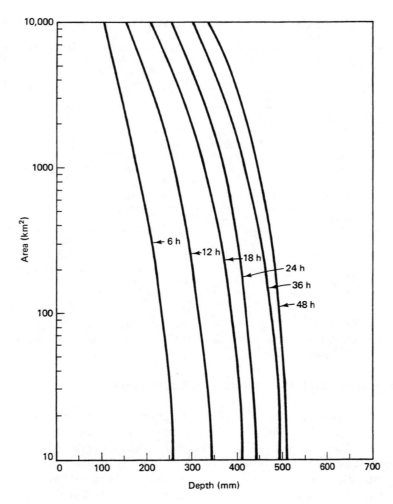

Figure 2-10. A depth-area-duration curve.

of average annual precipitation. A humid climate typifies regions with more than 750 mm of average annual precipitation [60]. While overall precipitation amounts are greater for humid regions than for semiarid or arid regions, the latter are subject to greater precipitation variability.

In arid regions, a drought prevails for many months within an ordinary year. A drought is an extended period with no measurable or significant precipitation. Rainfall occurs mostly during the summer, with storms being typically isolated events, showing a wide variability in time and space. Temperature is high and annual precipitation amounts are low.

A semiarid region is subject to seasonal precipitation, with little or no precipitation in other parts of the year. Rainfall patterns vary widely from region to region and, within a certain region, rainfall is also apt to vary widely. Temperature is high and annual precipitation amounts are moderate.

A humid region is characterized by large amounts of precipitation, with little

TABLE 2-2 MAXIMUM DEPTH-AREA-DURATION DATA FOR THE UNITED STATES [76]
(Average depth in centimeters, storm indicated by superscripted letter)

Area (km²)	Duration (h)						
	6	12	18	24	36	48	72
25	62.7[a]	75.7[b]	92.2[c]	98.3[c]	106.2[c]	109.5[c]	114.8[c]
250	49.8[b]	66.8[c]	82.5[c]	89.4[c]	96.3[c]	98.8[c]	103.1[c]
500	45.5[b]	65.0[c]	79.8[c]	86.9[c]	93.2[c]	95.8[c]	99.6[c]
1250	39.1[b]	62.5[c]	75.4[c]	83.0[c]	88.9[c]	91.4[c]	94.7[c]
2500	34.0[b]	57.4[c]	69.6[c]	76.7[c]	83.6[c]	85.6[c]	88.6[c]
5000	28.4[b]	45.0[c]	57.1[c]	63.0[c]	69.3[c]	72.1[c]	75.4[c]
12500	20.6[b]	28.2[b]	35.8[b]	39.4[c]	47.5[d]	52.6[d]	62.0[d]

Storm	Date	Storm Center
a	July 17–18, 1942	Smethport, PA
b	September 8–10, 1921	Thrall, TX
c	September 3–7, 1950	Yankeetown, FL
d	June 27–July 1, 1899	Hearne, TX

variability in time or space. Droughts are infrequent, with precipitation occurring throughout the year in amounts largely predictable by analyzing relatively short records. Temperatures are low, and annual precipitation amounts are high.

Precipitation Data Sources and Interpretation

Precipitation data are obtained by measurement using rain gages (Chapter 3). The National Climatic Data Center publishes precipitation data for about 11,000 stations in the United States. A large number of additional gages are operated by other federal, state, and local agencies and by individuals. Federal agencies collecting precipitation data on a regular basis include the U.S. National Weather Service, the U.S. Army Corps of Engineers, the USDA Soil Conservation Service, the USDA Forest Service, the Bureau of Reclamation, and the Tennessee Valley Authority.

Precipitation data can be assembled using either hourly, daily, monthly, or yearly intervals. In the United States, the National Climatic Data Center publishes hourly precipitation data and maximum 15-min duration amounts in the publication *Hourly Precipitation Data.* Daily and monthly precipitation data are found in the publication *Climatological Data.* Precipitation frequency atlases (U.S. Weather Bureau Technical Paper No. 40 [76], NOAA Technical Memorandum NWS Hydro-35 [51], and Precipitation Frequency Atlas of the Western United States [50]) are available from the National Climatic Data Center, Federal Building, Asheville, North Carolina. Monthly and seasonal precipitation maps are found in *Weekly Weather and Crop Bulletin,* available through NOAA/USDA Joint Agricultural Weather Facility, USDA South Building, Washington, D.C. Additional sources of precipitation data are given in *Annotated Bibliography of NOAA Publications of Hydrometeorological Interest,* which is updated at irregular intervals by the National Weather Service, and in *Selective Guide to Climatic Data Sources,* updated at irregular intervals by the National Climatic Data Center.

Filling In Missing Records. Incomplete records of rainfall are sometimes possible due to operator error or equipment malfunction. In this case, it is often necessary to estimate the missing record. Assume that a certain station X has a missing record. A procedure to fill in the missing record is to identify three index stations (A, B, and C) having complete records, located as close to and as evenly spaced around station X as possible. The mean annual rainfall for each of the stations X, A, B, and C is evaluated. If the mean annual rainfall at each of the index stations A, B, or C is within 10 percent of that of station X, a simple arithmetic average of the rainfall values at the index stations provides the missing value at station X.

If the mean annual rainfall at any of the index stations differs by more than 10% from that of station X, the normal ratio method is used [55]. In this method, the missing precipitation value at station X is the following:

$$P_X = (1/3)\left[\left(\frac{N_X}{N_A}\right)P_A + \left(\frac{N_X}{N_B}\right)P_B + \left(\frac{N_X}{N_C}\right)P_C\right] \qquad (2\text{-}10)$$

in which P = precipitation, N = mean annual rainfall, and the subscripts X, A, B, and C refer to the respective stations.

An alternate method for filling in missing precipitation data has been developed by the National Weather Service [49]. The method requires data for four index stations A, B, C, and D, each located closest to the station X of interest, and in each of four quadrants delimited by north-south and east-west lines drawn through station X (Fig. 2-11). The estimated precipitation value at station X is the weighted average of the values at the four index stations. For each index station, the applicable weight is the reciprocal of the square of its distance L to station X.

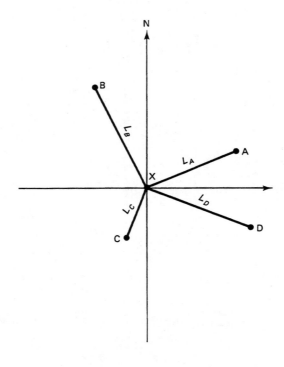

Figure 2-11. Position of station X and index stations A, B, C, and D.

The procedure is described by the following formula:

$$P_X = \frac{\displaystyle\sum_{i=1}^{4} (P_i/L_i^2)}{\displaystyle\sum_{i=1}^{4} (1/L_i^2)} \qquad (2\text{-}11)$$

in which P = precipitation; L = distance between index stations and station X; and i refers to each one of the index stations A, B, C, and D.

Double-mass Analysis. Changes in the location or exposure of a rain gage may have a significant effect on the amount of precipitation it measures, leading to inconsistent data (data of different nature within the same record).

The consistency of a rainfall record is tested with double-mass analysis. This method compares the cumulative annual (or, alternatively, seasonal) values of station Y with those of a *reference* station X. The reference station is usually the mean of several neighboring stations. The cumulative pairs (double-mass values) are plotted in an *xy* arithmetic coordinate system, and the plot is examined for trend changes. If the plot is essentially linear, the record at station Y is consistent. If the plot shows a break in slope, the record at station Y is inconsistent and should be corrected. The correction is performed by adjusting the records prior to the break to reflect the new state (after the break). To accomplish this, the rainfall records prior to the break are multiplied by the ratio of slopes after and before the break (see Fig. 2-12).

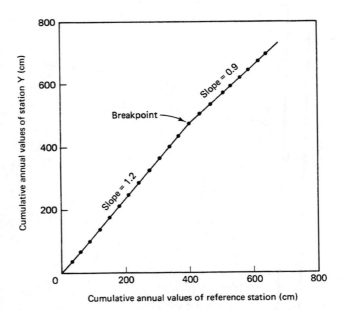

Figure 2-12.   Double-mass analysis.

Basic Hydrologic Principles   Chap. 2

2.2 HYDROLOGIC ABSTRACTIONS (Losses)

Hydrologic abstractions are the processes acting to reduce total precipitation into effective precipitation. Effective precipitation eventually produces surface runoff. The difference between total and effective precipitation is the depth abstracted by the catchment.

The processes by which precipitation is abstracted by the catchment are many. Those important in engineering hydrology are the following: (1) interception, (2) infiltration, (3) surface or depression storage, (4) evaporation, and (5) evapotranspiration.

Interception

Interception is the process by which precipitation is abstracted by vegetation or other forms of surface cover. Interception loss is the fraction of precipitation that is retained by the vegetative cover or other surface and either absorbed by it or eventually returned to the atmosphere through evaporation. *Throughfall* is that part of precipitation that reaches the ground by first passing through the vegetative cover. Interception losses are a function of (1) storm character, including intensity, depth, and duration, (2) type, extent, and density of vegetative cover, and (3) time of the year (season).

Interception is usually the first abstractive process to act during a storm. Light storms are substantially abstracted by the interception process. Light storms occur frequently and therefore constitute the majority of the storms. The interception loss accumulated in one year, primarily from light storms, amounts to about 25 percent of the average annual precipitation.

For moderate storms, interception losses are apt to vary widely, being greater during the growing season and smaller at other times of the year. Studies have shown that interception values are likely to vary from 7 to 36 percent of total precipitation during the growing season, and from 3 to 22 percent during the remainder of the year [10].

For heavy storms, interception losses usually amount to a small fraction of the total rainfall. For long-duration or infrequent storms, the effect of interception on the overall process of abstraction is likely to be small. In certain cases, particularly for flood hydrology studies, the neglect of interception is generally justified on practical grounds.

The interception loss comprises two distinct elements [22]. The first is the interception storage, i.e., the depth (or volume) retained in the foliage against the forces of wind and gravity. The second is the evaporation loss from the foliage surface, which takes place throughout the duration of the storm. The combination of these two processes leads to the following formula for estimating interception loss [10]:

$$L = S + KEt \tag{2-12}$$

in which L = interception loss, in millimeters; S = interception storage depth in millimeters, usually varying from 0.25 to 1.25 mm; K = ratio of evaporating foliage surface to its horizontal projection; E = evaporation rate in millimeters per hour; and t = storm duration in hours.

Infiltration

Infiltration is the process by which precipitation is abstracted by seeping into the soil below the land surface. Once below the surface, the abstracted water moves either laterally, as *interflow,* into streams, lakes, and rivers or vertically, by *percolation,* into aquifers. The water in streams, lakes, and rivers is subject to gravitational forces, moving toward the oceans as surface flow. The water held in aquifers moves as groundwater flow, driven primarily by gravitational forces, eventually flowing into a stream or reaching the ocean.

Infiltration is a complex process. It is described by either an instantaneous infiltration rate or an average infiltration rate, both measured in millimeters per hour. The total infiltration depth, in millimeters, is obtained by integrating the instantaneous infiltration rate over the storm duration. The average infiltration rate is obtained by dividing the total infiltration depth by the storm duration.

Infiltration rates vary widely, depending on (1) the condition of the land surface (crust), (2) the type, extent, and density of vegetative cover; (3) the physical properties of the soil, including grain size and gradation; (4) the storm character, i.e., intensity, depth, and duration; (5) the water temperature; and (6) the water quality, including chemical constituents and other impurities.

Infiltration Formulas. For a given storm, infiltration rates tend to vary in time. The initial infiltration rate is the rate prevailing at the beginning of the storm. This rate is likely to be the maximum rate for the given storm, gradually decreasing as the storm progresses in time. For storms of long duration, the infiltration rate eventually reaches a constant value, referred to as final (or equilibrium) infiltration rate. This process led Horton [24] to the following formula to describe the variation of infiltration rate with time:

$$ f = f_c + (f_o - f_c)e^{-kt} \qquad (2\text{-}13) $$

in which f = instantaneous infiltration rate; f_o = initial infiltration rate; f_c = final infiltration rate; k = a constant; and t = time in hours. The units of k are h^{-1}. For $t = 0$, $f = f_o$; and for $t = \infty$, $f = f_c$ (see Fig. 2-13).

Equation 2-13 has three parameters: (1) initial infiltration rate; (2) final infiltration rate; and (3) the k value describing the rate of decay of the difference between initial and final infiltration rates. Field measurements are necessary in order to determine appropriate values of these parameters. A plot of infiltration rate versus time enables the estimation of the final rate. With a knowledge of the final rate, two sets of f and t are obtained from the plot and used, together with Eq. 2-13, to solve simultaneously for f_c and k.

Integrating Eq. 2-13 between $t = 0$ and $t = \infty$, leads to

$$ F = \frac{(f_o - f_c)}{k} \qquad (2\text{-}14) $$

in which F = the total infiltration depth above the $f = f_c$ line. Equation 2-14 enables the calculation of the total infiltration depth, assuming that the storm lasts long enough for the equilibrium rate to be attained.

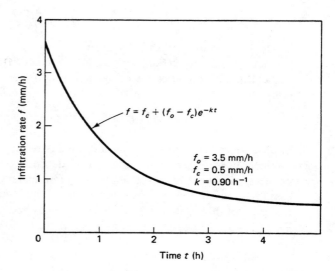

Figure 2-13. Horton's infiltration formula.

$$f = f_c + (f_o - f_c)e^{-kt}$$

$f_o = 3.5 \text{ mm/h}$
$f_c = 0.5 \text{ mm/h}$
$k = 0.90 \text{ h}^{-1}$

Example 2-3.

Assuming $f_o = 10$ mm/h, $f_c = 5$ mm/h, and $k = 0.95$ h^{-1}, calculate the total infiltration depth for a storm lasting 6 h.

After 6 h, the difference between instantaneous and final rates is negligible. Therefore, the total infiltration depth is: (10 mm/h − 5 mm/h)/0.95 h^{-1} + (5 mm/h × 6 h) = 35.26 mm.

Typical infiltration rates at the end of 1 h (f_1) are shown in Table 2-3. Generally, these values are reasonable approximations of final (i.e., equilibrium) infiltration rates.

Recent developments in infiltration theory have sought to improve on the Horton model. Philip [58] has proposed the following model:

$$f = (1/2)\, st^{-1/2} + A \tag{2-15}$$

in which f = instantaneous infiltration rate; s = an empirical parameter related to the rate of penetration of the wetting front (the wetting surface characterized by a very high potential gradient); A = an infiltration value that is close to the value of saturated hydraulic conductivity at the surface; and t = time.

In Eq. 2-15, for $t = 0$, $f = \infty$; and for $t = \infty$, $f = A$. In practice, the initial infiltration rate has a finite value. In spite of this limitation, the Philip formula seems to be a good fit to experimental data. Integration of the Philip equation leads to

TABLE 2-3 TYPICAL f_1 VALUES [1]

Soil Group	f_1 (mm/h)
Low (clays, clay loam)	0.25–2.50
Intermediate (loams, clay, silt)	2.50–12.50
High (sandy soils)	12.50–25.00

$$F = st^{1/2} + At \qquad (2\text{-}16)$$

in which F = total depth of infiltration.

An infiltration model with a sound theoretical basis is the Green and Ampt formula [18]. This equation describes infiltration rate under ponded water conditions as follows:

$$f = K \left(1 + \frac{H + P_f}{Z_f} \right) \qquad (2\text{-}17)$$

in which f = infiltration rate in millimeters per hour; K = saturated hydraulic conductivity in millimeters per hour; H = depth of ponded water in millimeters; P_f = capillary pressure at the wetting front in millimeters; and Z_f = vertical depth of saturated zone in millimeters. In practice, however, it may be difficult to measure some of the terms of this equation. Recent progress has been achieved by grouping the terms in Eq. 2-17 into predictable parameters linked to the physical processes [33].

Infiltration Indexes.   Practical evaluations of infiltration have been hampered by its spatial and temporal variability. This has led to the use of infiltration indexes, which model the infiltration process in an approximate yet practical way.

Infiltration indexes assume that infiltration rate is constant throughout the storm duration. This assumption tends to underestimate the higher initial rate of infiltration while overestimating the lower final rate. For this reason, infiltration indexes are best suited for applications involving either long-duration storms or catchments with high initial soil moisture content. Under such conditions, the neglect of the variation of infiltration rate with time is generally justified on practical grounds.

For moderate storms, the use of infiltration indexes is largely an empirical procedure, with attention being focused on matching the prevailing soil moisture condition and storm duration in order to effect a proper balance of rainfall and runoff amounts.

In practice, the most commonly used infiltration index is the ϕ-index, defined as the (constant) infiltration rate to be subtracted from the prevailing rainfall rate in order to obtain the runoff volume that actually occurred [11]. The computation of the ϕ-index requires a storm pattern, i.e., a plot of rainfall intensity versus time, and a measured runoff volume (or depth). The computation involves a trial-and-error procedure, which can be readily computerized.

Example 2-4.

The following rainfall distribution was measured during a 6-h storm.

Time (h)	0	1	2	3	4	5	6
Rainfall Intensity (cm/h)		0.5	1.5	1.2	0.3	1.0	0.5

The runoff depth has been estimated at 2 cm. Calculate the ϕ-index.

From the rainfall distribution, the total rainfall is 5 cm. Therefore, the depth abstracted by infiltration is $(5 - 2) = 3$ cm. With reference to Fig. 2-14, the ϕ-index is calculated by trial and error. By inspection, a value of ϕ between 0.5 and 1.0 cm/h is assumed. A mass balance leads to

$$(1.5 - \phi) \times 1 + (1.2 - \phi) \times 1 + (1.0 - \phi) \times 1 = 2 \text{ cm} \qquad (2\text{-}18)$$

From Eq. 2-18, solving for ϕ gives: $\phi = 0.567$ cm/h, verifying that the assumed range for ϕ was correct. Had the assumed range been wrong, the calculated ϕ-value would have been out of that range. In Fig. 2-14, the 2 cm of runoff are above the ϕ-index line; the 3 cm of abstracted rainfall are below the ϕ-index line.

Another widely used infiltration index is the W-index [11], which, unlike the ϕ-index, takes explicit account of interception loss and depth of surface storage. The formula for the W-index is the following:

$$W = \frac{P - Q - S}{t_f} \qquad (2\text{-}19)$$

in which W = W-index in millimeters per hour; P = rainfall depth in millimeters; Q = runoff depth in millimeters; S = the sum of interception loss and depth of surface storage in millimeters; and t_f = the total time (hours) during which rainfall intensity is greater than W.

The W_{min} index is the W-index calculated for extremely wet conditions. It is derived using data from the last of a series of storms and is used in estimating maximum flood potential. In this sense, the W_{min} index approaches a spatially averaged value of the final infiltration rate. For such extreme conditions, the values of W_{min} and ϕ are almost identical.

Infiltration Rates Derived from Rainfall-Runoff Data. Infiltration formulas depict the variation of infiltration rates with time. Infiltration rates, however, vary not only temporally but also spatially. Unless the field measurements and related parameter estimation are fairly good representations of the spatial variability, the rates calculated by infiltration formulas are likely to be different from reality.

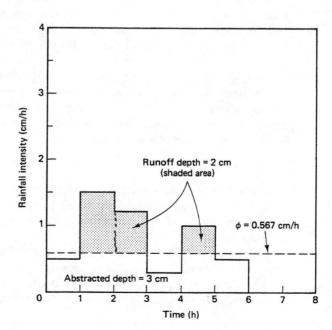

Figure 2-14. Calculation of ϕ-index: Example 2-4.

This difficulty is circumvented by calculating infiltration rates indirectly, from concurrent rainfall-runoff measurements. Such a calculation provides a temporal and spatial average of infiltration rate, amounting to a ϕ-index, with its associated advantages and disadvantages.

Infiltration and Catchment Size.   For midsize and large catchments, the natural variability of infiltration rates makes it necessary to resort to the evaluation of total infiltration depth. In practice, total infiltration depths are derived from rainfall-runoff analysis. However, for each data set, the calculation is highly dependent on the level of soil moisture antecedent to the storm. The catchment moisture level is referred to as the *antecedent moisture condition,* or AMC (Chapter 5). Initial infiltration rates and, consequently, total infiltration depths are a function of prevailing antecedent moisture condition.

Surface or Depression Storage

Surface (or depression) *storage* is the process by which precipitation is abstracted by being retained in puddles, ditches, and other natural or artificial depressions on the land surface. Water held in depressions either evaporates or eventually contributes to soil moisture by infiltration. The spatial variability of storage in surface depressions precludes its precise calculation.

Intuitively, the milder the catchment's relief, the greater the effect of depression storage. Field data reported by Viessman [71] showed conclusively that depression storage is inversely related to catchment slope. Usually, an equivalent depth of depression storage can be estimated based on experience. For instance, Hicks [20] has used depression storage depths of 5, 3.75, and 2.5 mm for sand, loam, and clay, respectively. Tholin and Keifer [66] have used values of 6.25 mm in pervious urban areas and 1.5 mm for paved areas. Where accurate estimations are difficult, depression storage amounts can be lumped together with other more tractable hydrologic abstractions such as interception or infiltration.

An alternate way of accounting for depression storage is the use of a peak-flow correction factor, as in the SCS TR-55 graphical method (Section 5.3).

Typically, the effect of depression storage varies in time and, consequently, with storm duration. At the beginning of a storm, depression storage usually plays an active role in abstracting precipitation amounts. As time progresses, depression storage volumes are eventually filled, with any additional water going on to constitute runoff. This has led to the following conceptual model of depression storage:

$$V_s = S_d(1 - e^{-kP_e}) \tag{2-20}$$

in which V_s = equivalent depth of depression storage in millimeters; P_e = precipitation excess, defined as total precipitation depth minus interception loss minus total infiltration depth; S_d = depression storage capacity in millimeters; and k = a constant.

Linsley et al. [38] have suggested that values of S_d for most catchments are in the range of 10-50 mm. The value of the constant k is estimated by assuming that for very small values of precipitation excess (P_e close to 0), essentially all the precipitation goes into depression storage ($dV_s/dP_e = 1$). This leads to $k = 1/S_d$.

Evaporation

Evaporation is the process by which water accumulated on the land surface (including that held in surface depressions and water bodies such as lakes and reservoirs) is converted into vapor state and returned to the atmosphere. Evaporation occurs at the evaporating surface, the contact between water body and overlying air. At the evaporating surface, there is a continuous exchange of liquid water molecules into water vapor and vice versa. In engineering hydrology, evaporation refers to the net rate of water transfer (loss) into vapor state.

Evaporation is expressed as an evaporation rate in millimeters per day (mm/d), centimeters per day (cm/d), or inches per day (in./d). Evaporation rate is a function of several meteorological and environmental factors. Those important from an engineering standpoint are (1) net solar radiation, (2) the saturation vapor pressure, (3) the vapor pressure of the air, (4) air and water surface temperatures, (5) wind velocity, and (6) atmospheric pressure.

Evaporation rates are significantly affected by climate. Studies have shown that evaporation rates are high in arid and semiarid regions and low in humid regions. For instance, mean annual lake evaporation in the United States varies from 20 in. (508 mm) in the Northeast (Maine) and Northwest (Washington) to 86 in. (2184 mm) in the Southwest (California and Arizona) (Fig. 2-15 [16]).

The effect of climate on evaporation has a substantial impact on water resources development. The planning and design of storage reservoirs in arid and semiarid regions requires a detailed evaluation of the potential for reservoir evaporation. These calculations determine to a large extent the feasibility of building surface water storage projects on regions subject to high evaporation rates.

Unlike other phases of the hydrologic cycle, lake evaporation cannot be measured directly. Therefore, several approaches have been developed to calculate evaporation. These vary in nature and are based on either (1) a water budget, (2) an energy budget, or (3) mass-transfer techniques.

Water Budget Method for Determining Reservoir Evaporation. The water budget method assumes that all relevant water-transport phases can be evaluated for a time period Δt, and expressed in terms of volumes. Reservoir or lake evaporation is calculated as follows:

$$E = P + Q - O - I - \Delta S \qquad (2\text{-}21)$$

in which E = volume evaporated from the reservoir, P = precipitation falling directly onto the reservoir, Q = surface runoff inflow into the reservoir, O = outflow from the reservoir, I = net volume infiltrated from the reservoir into the ground, and ΔS = change in stored volume. All terms in Eq. 2-21 refer to a time period Δt, usually taken as 1 wk or greater.

Most terms in Eq. 2-21 can be evaluated directly. Precipitation is readily measured, and inflow and outflow can be obtained by integrating the flow records. The change in stored volume is determined by means of water stage recorders. Net infiltration, however, can be evaluated only indirectly, either by measuring soil permeability or monitoring changes in groundwater level in nearby wells. The difficulty in measur-

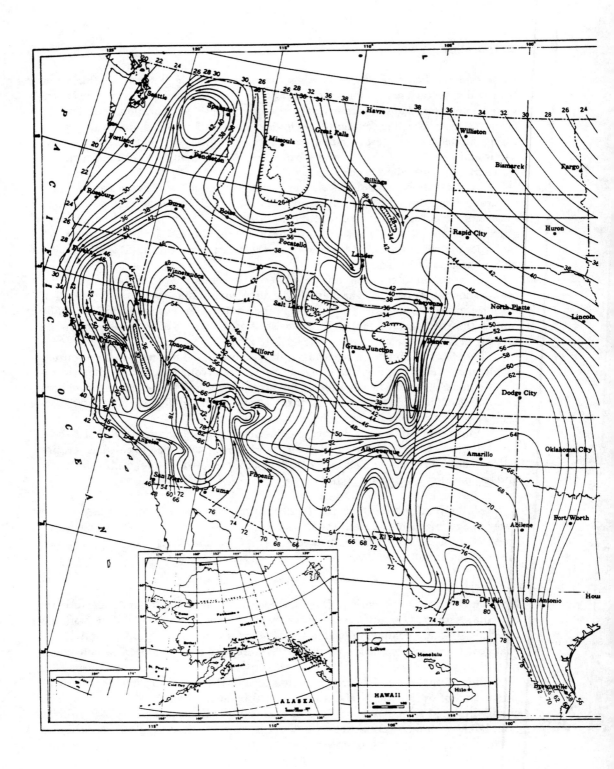

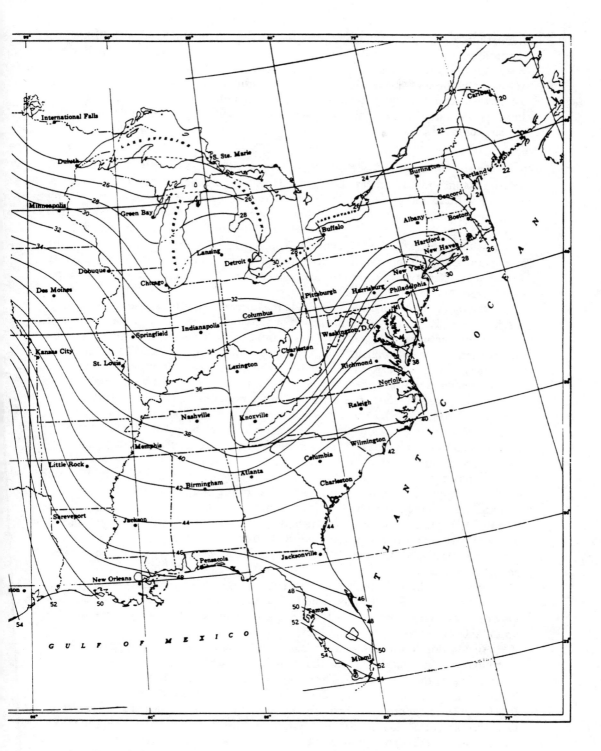

Figure 2-15. Mean annual lake evaporation, in inches [16].

ing net infiltration generally limits the water budget method to areas with little or no net infiltration. In spite of this limitation, the water budget method has been found to work reliably under certain idealized conditions. Water budget studies from Lake Hefner, Oklahoma, show that the method can provide evaporation volumes within a 10 percent accuracy about two-thirds of the time [70]. Conditions at Lake Hefner, however, were highly selective, and lesser accuracy is to be expected under more typical circumstances.

Energy Budget Method for Determining Reservoir Evaporation.

During evaporation, significant energy exchanges occur at the evaporating surface. A balance of these energy exchanges leads to the *energy budget method* of calculating reservoir evaporation. The amount of heat required to convert one gram of water into vapor, i.e., the heat of vaporization, varies with temperature. For instance, at 20°C the heat of vaporization is 586 calories (Table A-1, Appendix A). To maintain the temperature of the evaporating surface, large quantities of heat must be supplied by radiation, by heat transfer from the atmosphere, and from energy stored in the water body.

Radiation is the emission of energy in the form of electromagnetic waves from all bodies above 0 K. Solar radiation received on the earth's surface is a major component of the energy balance. Solar radiation reaches the outer surface of the atmosphere at a nearly constant flux of about 1.94–2.0 cal/cm²/min, or langleys/min (1 langley (ly) = 1 cal/cm²), measured perpendicular to the incident radiation. Nearly all this radiation is of wavelengths in the range 0.3–3.0 μm, with about half of it of wavelengths in the visible range (0.4–0.7 μm). The earth also emits radiation, but since its surface temperature is about 300 K, this terrestrial radiation is of much lower intensity and greater wavelength (3.0–50.0 μm) than solar radiation. Since there is little overlap between these two radiation spectra, it is customary to refer to solar radiation as *short-wave* radiation and to terrestrial radiation as *long-wave* radiation [21].

In passage through the atmosphere, solar radiation changes both its flux and spectral composition. Some of it is reflected back to space, and some of it is absorbed and scattered by the atmosphere. The fraction (about half) of the original solar radiation flux that reaches the earth's surface is called *direct solar radiation*. The fraction of the radiation reflected and scattered by the atmosphere that reaches the ground is called *sky radiation*. The sum of direct solar radiation and sky radiation is called *global radiation*.

Albedo is the reflectivity coefficient of a surface toward shortwave radiation. This coefficient varies with color, roughness, and inclination of the surface and is of the order of 0.10 for water, 0.10–0.30 for vegetated areas, 0.15–0.40 for bare soil, and up to 0.90 for snow-covered areas [21].

In addition to the shortwave radiation balance, there is also a long-wave radiation balance. The earth's surface emits radiation, part of which is absorbed and reflected back by the atmosphere. The difference between outgoing and incoming fluxes is called *long-wave radiation loss*. During the day, long-wave radiation may be a small fraction of the total radiation balance, but at night, in the absence of solar radiation, long-wave radiation dominates the radiation balance. *Net radiation* is equal to the net shortwave (solar) radiation minus the long-wave (terrestrial) radiation *loss*.

In the energy budget method, the incoming energy can be expressed as

$$Q_i = Q_s (1 - A) - Q_b + Q_a \tag{2-22}$$

in which Q_i = incoming energy; Q_s = global radiation (shortwave radiation from sun and sky); A = albedo; Q_b = long-wave radiation loss by water body; and Q_a = net energy advected into the water body by streams, rain, snow, and the like.

The energy expenditure, which must balance the incoming energy, is expressed as

$$Q_o = Q_h + Q_e + Q_t \tag{2-23}$$

in which Q_o = energy expenditure; Q_h = sensible heat transfer from water body to the atmosphere by convection and conduction; Q_e = energy expended in the evaporation process; and Q_t = increase in energy stored in the water body. The value of Q_e is negative when condensation is taking place. All terms in Eqs. 2-22 and 2-23 are given in calories per square centimeter per day, or langleys per day. The energy used in evaporation is converted into equivalent evaporation rate by the following formula:

$$Q_e = \rho \ell E \tag{2-24}$$

in which ρ = density of water in grams per cubic centimeter; ℓ = heat of vaporization, a function of temperature (see Table A-1, Appendix A), in calories per gram (cal/g); and E = evaporation rate, in centimeters per day.

The terms Q_h and Q_e in Eq. 2-23 are difficult to evaluate directly. Bowen has suggested that their ratio is more tractable and can be evaluated by means of the following equation:

$$B = \frac{Q_h}{Q_e} = \gamma \, \frac{T_s - T_a}{e_s - e_a} \, \frac{p}{1000} \tag{2-25}$$

in which B = Bowen's ratio; γ = a psychrometric constant equal to 0.66 mb/°C; T_s = water surface temperature in degrees Celsius; T_a = overlying air temperature in degrees Celsius; e_s = saturation vapor pressure at the water surface temperature in millibars; e_a = vapor pressure of the overlying air in millibars; and p = atmospheric pressure in millibars.

A balance of incoming energy (Eq. 2-22) and energy expenditure (Eq. 2-23), taking into account Eqs. 2-24 and 2-25 leads to:

$$E = \frac{Q_s(1 - A) - Q_b + Q_a - Q_t}{\rho \ell (1 + B)} \tag{2-26}$$

The quantities $Q_s(1 - A)$ and Q_b can be measured with radiometers, which are instruments designed to measure radiation. The quantity Q_a can be determined by measuring volumes and temperatures of the water flowing into and out of the body, and Q_t is evaluated by periodic measurements of water temperatures. An example of the application of the energy budget method to a large lake is the study of evaporation in Lake Ontario by Bruce and Rodgers [8].

Mass-transfer Approach. Evaporation rates are dependent on the temperature of the water surface and the prevailing atmospheric pressure. Higher water temperatures induce more vigorous molecular action and result in higher evaporation rates. However, a higher atmospheric pressure limits the movement of water molecules and results in lower evaporation rates. The overall effect of atmospheric pressure in reducing evaporation is small and can usually be neglected on practical grounds.

Studies have shown that evaporation rates are a function of the difference between the saturation vapor pressure at the water surface temperature and the vapor pressure of the overlying air (partial vapor pressure). The saturation vapor pressure is a function of temperature (see Tables A-1 and A-2, Appendix A). The partial vapor pressure can be calculated by multiplying the saturation vapor pressure at the air temperature by the relative humidity of the air (in percentage) and dividing by 100. As the process of mass transfer continues, the lowest layer of the atmosphere eventually becomes saturated, and the net evaporation rate tends to diminish and even reverse (condensation). Thus an agent such as the wind, which carries away the water molecules as they leave the water surface, is necessary for continuous evaporation. The recognition of these processes led Dalton [13] to formulate the classical law bearing his name:

$$E = f(u)(e_s - e_a) \qquad (2\text{-}27)$$

in which E = evaporation rate; $f(u)$ = a function of the horizontal wind speed; e_s = the saturation vapor pressure at the water surface temperature; and e_a = the vapor pressure of the overlying air.

Several empirical equations of the type of Eq. 2-27 have been developed over the years. Collectively, they are referred to as *mass-transfer equations*. A commonly used mass-transfer equation is that of Meyer [39]:

$$E = C(e_o - e_a)\left[1 + \left(\frac{W}{10}\right)\right] \qquad (2\text{-}28a)$$

in which E = evaporation rate in inches per month; C = a coefficient varying from 15 for small ponds to 11 for large lakes and reservoirs; e_o = saturation vapor pressure at the mean monthly air temperature in inches of mercury; e_a = vapor pressure of the air at the mean monthly air temperature in inches of mercury; and W = mean monthly wind speed at 25 ft height in miles per hour.

Another version of the Meyer equation is the following [40, 71]:

$$E = C(e_s - e_a)\left[1 + \left(\frac{W}{10}\right)\right] \qquad (2\text{-}28b)$$

in which E = evaporation rate, in inches per day; C = a coefficient varying from 0.50 for small ponds to 0.36 for large lakes and reservoirs; e_s = saturation vapor pressure at the water surface temperature in inches of mercury; e_a = vapor pressure of the air in inches of mercury; and W = daily mean wind speed at 25 ft height, in miles per hour.

A set of mass-transfer equations developed in connection with the Lake Hefner evaporation studies [70] is the following:

$$E = 0.00304(e_s - e_2)v_4 \qquad (2\text{-}29a)$$

$$E = 0.00241(e_s - e_8)v_8 \qquad (2\text{-}29b)$$

in which E = evaporation rate in inches per day; e_s = saturation vapor pressure at the water surface temperature in inches of mercury; e_2 and e_8 are partial (air) vapor pressures over the lake at 2- and 8-m heights, respectively, in inches of mercury; and v_4 and v_8 are wind speeds over the lake at 4- and 8-m heights, respectively, in miles per day. If e_2 and v_4 are taken upwind from the lake, the constant in Eq. 2-29a reduces to 0.0027. These formulas were carefully developed using water budget data from Lake Hefner, with a surface area of 2500 ac (1012 ha). They have since been tested in other reservoirs, including Lake Mead and others in the western United States [9].

Combination Methods for Determining Reservoir Evaporation. The use of both energy budget and mass-transfer approaches leads to an alternate way of determining reservoir evaporation. Penman [56] combined these two concepts to develop a formula for practical use. An approximate energy balance (neglecting variations of energy by the water body, $Q_a = 0$, and $Q_t = 0$, in Eqs. 2-22 and 2-23) led Penman to

$$Q_s(1 - A) - Q_b = Q_h + Q_e \qquad (2\text{-}30)$$

The left side of this equation is the net radiation, or Q_n. The right side can be expressed as $Q_e(1 + B)$. Therefore,

$$Q_n = Q_e(1 + B) \qquad (2\text{-}31)$$

By using Eq. 2-24, Eq. 2-31 is converted to evaporation rate units (centimeters per day):

$$E_n = E(1 + B) \qquad (2\text{-}32)$$

in which E_n is the net radiation (in evaporation rate units) and E is the evaporation rate.

For $p = 1000$ mb (close to atmospheric pressure at sea level, equal to 1013.2 mb), Bowen's ratio (Eq. 2-25) reduces to

$$B = \gamma \frac{T_s - T_a}{e_s - e_a} \qquad (2\text{-}33)$$

A saturation vapor-pressure gradient between water and air temperatures is defined as

$$\Delta = \frac{e_s - e_o}{T_s - T_a} \qquad (2\text{-}34)$$

in which e_s = saturation vapor pressure at the water surface temperature T_s and e_o = saturation vapor pressure at the overlying air temperature T_a.

The Dalton formula (Eq. 2-27) enables the calculation of E_a/E, i.e., the ratio of mass-transfer evaporation E_a (assuming that the temperatures of water surface and overlying air are equal) to the evaporation rate E:

$$\frac{E_a}{E} = \frac{e_o - e_a}{e_s - e_a} \qquad (2\text{-}35)$$

Combining Eqs. 2-32 through 2-35 and using some algebraic manipulation, the Penman equation is obtained:

$$E = \frac{\Delta E_n + \gamma E_a}{\Delta + \gamma} \qquad (2\text{-}36)$$

in which E (evaporation rate), E_n (net radiation in evaporation rate units) and E_a (mass-transfer evaporation rate) are given in centimeters per day; and Δ and γ are given in millibars per degree Celsius.

From Eq. 2-36, it is apparent that Δ and γ are weighting factors, affecting the net radiation and mass-transfer evaporation rates, respectively. The weighting factor γ is equal to 0.66 mb/°C. The weighting factor Δ is a function of the air temperature and can be estimated by the following formula [38]:

$$\Delta = (0.00815 T_a + 0.8912)^7 \qquad (2\text{-}37)$$

in which Δ is given in millibars per degree Celsius and T_a = air temperature in degrees Celsius. This formula is applicable for air temperatures greater than $-25°C$.

Equation 2-36 can also be expressed as follows:

$$E = \frac{\alpha E_n + E_a}{\alpha + 1} \qquad (2\text{-}38)$$

in which $\alpha = \Delta/\gamma$, a function of air temperature. Values of α (with Δ based on Eq. 2-37) are shown in Table 2-4.

The mass-transfer evaporation rate E_a is evaluated with an appropriate mass-transfer equation. For instance, the following formula has been proposed by Dunne [15]:

$$E_a = (0.013 + 0.00016\, v_2)e_o\, \frac{100 - RH}{100} \qquad (2\text{-}39)$$

in which E_a = mass-transfer evaporation rate, in centimeters per day; v_2 = wind velocity, measured at a 2-m depth, in kilometers per day; e_o = saturation vapor pressure at the overlying air temperature in millibars; and RH = relative humidity in percent.

Other Penman-type equations have been developed over the years. The National Weather Service uses a Penman-type equation to develop correlations for estimating evaporation, based on mean daily air temperature and dew point, wind movement per day, and solar radiation [34]. The method has been computerized by Lamoreaux [36].

TABLE 2-4 VALUES OF PENMAN'S RATIO α AS A FUNCTION OF AIR TEMPERATURE

Air Temperature T_a (°C)	$\alpha = \Delta/\gamma$
0	0.68
5	0.93
10	1.25
15	1.66
20	2.19
25	2.86
30	3.69
35	4.73
40	6.00

Example 2-5.

Calculate evaporation rate by the Penman method, for the following atmospheric conditions: air temperature $T_a = 20°C$; net radiation $Q_n = 550$ cal/cm^2/d; wind speed (at 2 m above surface) $v_2 = 200$ km/d; RH $= 70\%$.

From Table A-1 (Appendix A), the saturation vapor pressure at the air temperature of 20°C is $e_o = 23.37$ millibars. The mass-transfer evaporation rate is calculated by Eq. 2-39: $E_a = 0.316$ cm/d. From Table A-1, the heat of vaporization at 20°C is $\ell = 586$ cal/g. Equation 2-24 is used to convert the net radiation to evaporation rate units: $E_n = 550$ cal/cm^2/d /(0.998 g/cm^3 $\times$ 586 cal/g) = 0.94 cm/d. Penman's ratio for $T_a = 20°C$ is obtained from Table 2-4: $\alpha = 2.19$. The evaporation rate is calculated by Eq. 2-38: $E = 0.74$ cm/d.

Evaporation Determinations Using Pans. Uncertainty in the applicability of the various evaporation formulas has led to the indirect measurement of evaporation using evaporation pans. An *evaporation pan* is a device designed to measure evaporation by monitoring the loss of water in the pan during a given time period, usually 1 d. It provides a measurement of the integrated effect of net radiation, wind, temperature and humidity on the evaporation from an open surface.

Evaporation pans vary widely in size, shape, materials, and exposure. The pan measurement is likely to be somewhat different from the actual amount of lake evaporation. The ratio of pan-to-lake evaporation is an empirical constant referred to as the *pan coefficient*. Evaporation measurements using pans are discussed in Chapter 3.

Evapotranspiration

Evapotranspiration is the process by which water in the land surface, soil, and vegetation is converted into vapor state and returned to the atmosphere. It consists of evaporation from water, soil, vegetative, and other surfaces and includes transpiration by vegetation. In this sense, evapotranspiration encompasses all the water converted into vapor and returned to the atmosphere, and therefore it is an important component in the long term water balance of a catchment.

Transpiration is the process by which plants transfer water from the root zone to the leaf surface, where it eventually evaporates into the atmosphere. The process by which transpiration takes place can be described as follows: Osmotic pressures at the root zone act to move water into the roots. Once inside the root, water is transported through the plant stem to the intercellular spaces located within the leaves. Air enters the leaves through small surface openings called *stomata*. *Chloroplasts* within the leaves use carbon dioxide from the air and a small portion of the available water to manufacture the carbohydrates necessary for plant growth. As air enters the leaf, water escapes through the open stomata and reaches the leaf surface, where it becomes available for evaporation. The ratio of water transpired and eventually evaporated to that actually used in plant growth is very large, up to 800:1 or more [38].

Transpiration is a part of plant life, and therefore it is a continuous process, occurring with or without the presence of precipitation. During a storm, however, interception amounts may use some of the energy available for evaporation, thereby reducing the amount of transpiration. The extent of this effect varies with vegetation type.

Transpiration is also limited by the rate at which moisture becomes available to the plants. Some authorities believe that transpiration is independent of the available soil moisture as long as the latter is above the *permanent wilting point,* i.e., the soil moisture at which permanent wilting would occur. Others assume that transpiration is roughly proportional to the prevailing soil moisture.

Transpiration rates and amounts vary widely, depending on vegetation type, depth of root zone, and extent and density of vegetative cover. Measurements of transpiration are difficult and are usually possible only under highly controlled circumstances. Since transpiration results in evaporation, transpiration amounts are a function of the same meteorological and climatic factors that control evaporation rates. In practice, transpiration is combined with evaporation and expressed as evapotranspiration, which includes all the water converted into vapor and returned to the atmosphere.

In evapotranspiration studies, the concept of *potential evapotranspiration* attributed to Thornthwaite [67] is widely used. Potential evapotranspiration (PET) is the amount of evapotranspiration that would take place under the assumption of an ample supply of moisture at all times. Therefore, PET is an indication of optimum crop water requirements.

Doorenbos and Pruitt [14] introduced the concept of *reference crop evapotranspiration* (ET₀), which is similar to that of potential evapotranspiration. Reference crop evapotranspiration is the rate of evapotranspiration from an extended surface of 8- to 15-cm tall green grass cover of uniform height, actively growing, completely shading the ground, and not short of water. Therefore, the reference crop evapotranspiration can be taken as the potential evapotranspiration of the reference crop (short green grass).

Potential evapotranspiration is equivalent to the evaporation that would occur on a free water surface of extended proportions but of negligible heat storage capacity [35]. Therefore, methods used to calculate potential evapotranspiration resemble the methods used to calculate evaporation. Like evaporation, there are many methods of calculating potential evapotranspiration, each having its own range of applicability. Data requirements vary widely, reflecting the assumptions used in their development.

Most potential evapotranspiration formulas are empirical, dependent upon the known correlation between potential evapotranspiration and one or more meteorological or climatic variables such as radiation, temperature, wind velocity, and vapor pressure difference. Other formulas relate evapotranspiration to direct measurements of water losses using evaporation pans. Models of evapotranspiration and potential evapotranspiration can be grouped into (1) temperature models, (2) radiation models, (3) combination models, and (4) pan-evaporation models.

When applied to a given set of conditions, the various potential evapotranspiration formulas usually give different estimates. These, however, do not vary widely, with the ratio of maximum and minimum estimates fluctuating throughout the year and rarely exceeding 2:1. As with any such calculation, regional or local experience should be taken into account when choosing an appropriate method to calculate PET.

Temperature Models to Estimate Evapotranspiration. The Blaney-Criddle formula [4, 5] is typical of the temperature models for the estimation of evapotranspiration. The formula has been widely used to estimate crop water requirements.

Its original version, applicable on a monthly basis, has the following form:

$$F = PT \qquad (2\text{-}40)$$

in which F = evapotranspiration for a given month in inches; P = a day-length variable, the ratio of the total daytime hours for a given month to the total daytime hours in the year, a function of latitude; and T = mean monthly temperature, in degrees Fahrenheit.

In SI units, applicable on a daily basis, the Blaney-Criddle formula is the following:

$$f = p(0.46t + 8.13) \qquad (2\text{-}41)$$

in which f = daily consumptive use factor in millimeters; p = the ratio of mean daily daytime hours for a given month to the total daytime hours in the year as a percent, a function of latitude (Table A-3, Appendix A); and t = mean daily temperature for a given month in degrees Celsius.

For a given crop, the consumptive water requirement is the amount of water required to meet its evapotranspiration needs without being limited by lack of water. The consumptive water requirement is equal to the product of consumptive use factor f times an empirical consumptive use crop coefficient k_c.

Consumptive water requirements vary widely between climates having similar air temperatures and day lengths. Therefore, the effect of climate on crop water requirements is not fully described by the consumptive use factor f. The effect of climate can be incorporated into the crop coefficient k_c. Generally, the value of k_c is time- and place-dependent, with local field experiments normally required to determine its proper value.

Doorenbos and Pruitt [14] have proposed a modification of the original Blaney and Criddle formula to account for the effect of actual insolation time (ratio n/N between actual and maximum possible bright sunshine hours), minimum relative humidity, and daytime wind speed. Their equation is

$$ET_o = a + bf \qquad (2\text{-}42)$$

in which ET_o = reference crop evapotranspiration and a and b are constants. The relationship between ET_o and f is shown in Fig. 2-16 for three levels of actual insolation time, low (less than 0.6), medium (0.6–0.8), and high (more than 0.8); three levels of minimum relative humidity, low (less than 20%), medium (20–50%), and high (more than 50%); and three levels of daytime wind speed (at a 2-m depth), light (0–2 m/s), moderate (2–5 m/s), and strong (greater than 5 m/s).

The consumptive water requirement for a given crop, ET_c, can be calculated as follows:

$$ET_c = k_c ET_o \qquad (2\text{-}43)$$

Approximate range of seasonal crop coefficients are shown in Table 2-5.

Example 2-6.

Calculate the reference crop evapotranspiration by the Blaney-Criddle method during the month of March for a location at 35°N, with mean daily temperature of 18°C. Assume medium actual insolation time, medium minimum relative humidity, and moderate daytime wind speed.

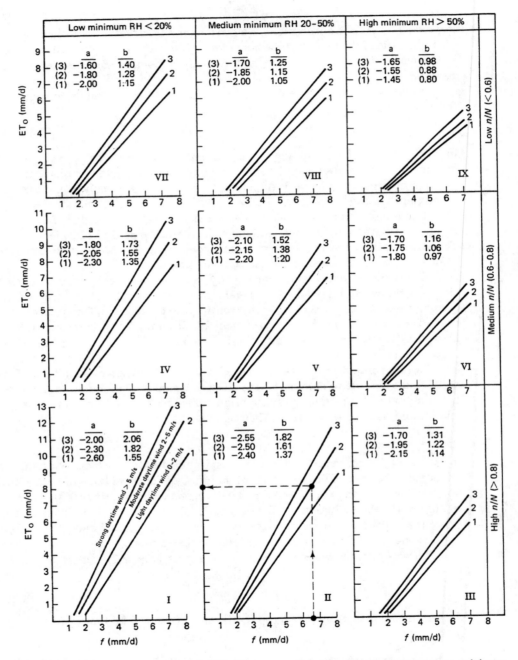

Figure 2-16. Correction to Blaney-Criddle formula to account for actual insolation time, minimum relative humidity, and daytime wind speed [14].

Basic Hydrologic Principles   Chap. 2

TABLE 2-5 APPROXIMATE RANGE OF
SEASONAL CROP COEFFICIENTS [45]

Crop	Range in crop coefficient k_c
Alfalfa	0.90–1.05
Avocado	0.65–0.75
Bananas	0.90–1.05
Beans	0.20–0.25
Cocoa	0.95–1.10
Coffee	0.95–1.10
Cotton	0.50–0.65
Dates	0.85–1.10
Deciduous trees	0.60–0.70
Flax	0.55–0.70
Grains (small)	0.25–0.30
Grapefruit	0.70–0.85
Maize	0.30–0.45
Oil seeds	0.25–0.40
Onions	0.25–0.40
Orange	0.60–0.75
Potatoes	0.25–0.40
Rice	0.45–0.65
Sisal	0.65–0.75
Sorghum	0.30–0.45
Soybeans	0.30–0.45
Sugar beets	0.50–0.65
Sugarcane	1.05–1.20
Sweet potatoes	0.30–0.45
Tobacco	0.30–0.35
Tomatoes	0.30–0.45
Vegetables	0.15–0.30
Vineyards	0.30–0.55
Walnuts	0.65–0.75

From Table A-3 (Appendix A), $p = 0.27$. Using Eq. 2-41, $f = 4.43$ mm/d. From Fig. 2-16, for $f = 4.43$ and medium actual insolation time, medium minimum relative humidity, and moderate daytime wind speed (graph V, curve 2), $ET_o = 4.0$ mm/d.

The Thornthwaite method is another widely used temperature model to estimate potential evapotranspiration [67]. The method is based on an annual temperature efficiency index J, defined as the sum of 12 monthly values of heat index I. Each index I is a function of the mean monthly temperature T, in degrees Celsius, as follows:

$$I = \left(\frac{T}{5}\right)^{1.514} \tag{2-44}$$

Evapotranspiration is calculated by the following formula:

$$\text{PET}(0) = 1.6 \left(\frac{10T}{J}\right)^c \tag{2-45}$$

in which PET(0) = potential evapotranspiration at 0° latitude in centimeters per month; and c is an exponent to be evaluated as follows:

$$c = 0.000000675J^3 - 0.0000771J^2 + 0.01792J + 0.49239 \qquad (2\text{-}46)$$

At latitudes other than 0°, potential evapotranspiration is calculated by

$$PET = K \ PET(0) \qquad (2\text{-}47)$$

in which K is a constant for each month of the year, varying as a function of latitude (see Table A-4, Appendix A).

Radiation Models. Priestley and Taylor [59] proposed that potential evapotranspiration be taken as the radiation part of the Penman equation (Eq. 2-36 with $E_a = 0$) affected with an empirical constant. Priestley and Taylor's formula is the following:

$$PET = \frac{1.26\Delta(Q_n/\rho\ell)}{\Delta + \gamma} \qquad (2\text{-}48)$$

in which PET = potential evapotranspiration in centimeters per day; Q_n = net radiation, in calories per square centimeter per day; Δ is defined by Eqs. 2-34 and 2-37; and γ is the psychrometric constant ($\gamma = 0.66$ mb/°C). Equation 2-48 can also be expressed as follows:

$$PET = \frac{1.26\alpha(Q_n/\rho\ell)}{\alpha + 1} \qquad (2\text{-}49)$$

in which the constant α can be obtained from Table 2-4.

Example 2-7.

Calculate the potential evapotranspiration rate by the Priestley and Taylor formula, assuming an air temperature of 20°C and net radiation of 600 cal/cm^2/d.

For the given temperature, the heat of vaporization is $\ell = 586$ cal/g (Table A-1, Appendix A). Therefore, net radiation in evaporation units is $600/(0.998 \times 586) = 1.026$ cm/d. From Table 2-4, the value of α is 2.19. Using Eq. 2-49, the potential evapotranspiration is PET = 0.89 cm/d.

Combination Models. The Penman model is typical of the combination models (combining energy budget and mass-transfer approaches) for calculating potential evapotranspiration. The original Penman model provided an estimate of evaporation from a free water surface. Experimental values of crop coefficients were initially suggested by Penman to relate free water surface evaporation to evapotranspiration [57]. These coefficients (0.6 in the winter and 0.8 in the summer) were intended to be multiplied by the evaporation rate determined by Eq. 2-36 in order to obtain the equivalent evapotranspiration rate. Other studies have suggested that free water surface evaporation and potential evapotranspiration are nearly equal and that Eq. 2-36 slightly overestimates lake evaporation [9].

The question of whether free water surface evaporation and potential evapotranspiration can be calculated by the same formula is a matter of great practical interest. Differences in nature and behavior of the surfaces may be considered to affect either the radiation or mass transfer terms of Eq. 2-36. For instance, the differences in albedo can be substantial. For most farm crops, the value of albedo is close to 0.25 [41]. The albedo of a free water surface, however, is only about 0.05 to 0.07 for sun altitudes above 55° [69]. This alone would justify greater values for the radiation term of free water surface evaporation as compared to that of evapotranspiration. On the other hand, judging from turbulence theory, the mass transfer term of Eq. 2-36 is greater for rougher surfaces such as crops and vegetation and smaller for smoother surfaces such as those of free water. The two terms of Eq. 2-36 may well compensate each other at least partially. The radiation term would be greater for free water surface evaporation, whereas the mass-transfer term appears to be greater for potential evapotranspiration [9].

The enhanced physical basis and considerable flexibility of the combination models have resulted in their wide use in the calculation of both free surface water evaporation and potential evapotranspiration. Studies have shown that potential evapotranspiration correlates well with combination data. For instance, an equation relating potential evapotranspiration to solar radiation and wind velocity, and applicable to arid and semiarid climates, is the following [52]:

$$PET = a + Q_r + bW \tag{2-50}$$

in which PET = potential evapotranspiration, Q_r = solar radiation, W = wind velocity, and a and b are empirical constants to be determined on a local or regional basis.

Pan-evaporation Models. Evaporation pans provide a measurement of the integrated effect of radiation, wind, temperature, and humidity on evaporation from an open surface. Plants and vegetation respond to the same climatic conditions, but several factors produce significant differences in the loss of water. The albedo from water surfaces is in the range 0.05-0.07, whereas that of most vegetative surfaces is in the range 0.20-0.25. Daytime storage of heat within the pan can be appreciable and may cause almost equal distribution of evaporation between night and day, whereas most crops experience 95 percent of their evaporation during daytime hours. In addition, the siting of the pan and the pan environment influence the measured evaporation, especially when the pan is placed in cropped rather than fallow fields. Notwithstanding these deficiencies, with proper siting and maintenance and the use of standard equipment, evaporation pans are still warranted for the prediction of crop water requirements.

The basic pan-evaporation formula is the following:

$$PET = K_p E_p \tag{2-51}$$

in which PET = potential evapotranspiration; K_p = pan coefficient; and E_p = pan evaporation.

The most common evaporation pans are the NWS Class A pan and the Colorado sunken pan. The NWS Class A pan is circular, 122 cm in diameter and 25.4 cm deep, made of galvanized iron (22 gage) or Monel® Metal (0.8 mm). The pan is mounted on a wooden open-frame platform with its bottom 15 cm above ground level. The soil is built up to within 5 cm of the bottom of the pan, and the pan must be level. It is filled with water 5 cm below the rim, and water level is maintained within 7.5 cm below the rim.

Colorado sunken pans are sometimes preferred in crop water-requirement studies because these pans have a water level 5 cm below the rim at soil level height and yield a better estimate of reference crop evapotranspiration than the NWS Class A pan. The Colorado sunken pan is 92 cm square and 46 cm deep, made of galvanized iron and set in the ground with the rim 5 cm above ground level. The water inside the pan is maintained at or slightly below ground level.

The pan evaporation approach has been widely used in the determination of potential evapotranspiration. Stanhill [65], for example, concluded that the NWS Class A evaporation pan was the most promising method for estimating potential evapotranspiration. Doorenbos and Pruitt [14] have given guidelines for choosing an appropriate value of pan coefficient for several climatic and site conditions (see Table 2-6).

2.3 CATCHMENT PROPERTIES

Surface runoff in catchments occurs as a progression of the following forms: (1) overland flow, (2) rill flow, (3) gully flow, (4) streamflow, and (5) river flow. Overland flow is runoff that occurs during or immediately after a storm, in the form of sheet flow over

TABLE 2-6 PAN COEFFICIENT K_p FOR NWS CLASS A PAN FOR DIFFERENT GROUND COVER AND LEVELS OF MEAN RELATIVE HUMIDITY AND 24-HOUR WIND SPEED [14]

Relative Humidity (%)		Pan Surrounded by Short Green Crop				Pan Surrounded by Dry Fallow Land[1]		
		Low 40	Medium 40–70	High 70		Low 40	Medium 40–70	High 70
Wind Speed (km/d)	Upwind Distance of Green Crop (m)				Upwind Distance of Dry Fallow (m)			
Light (less than 175)	0	0.55	0.65	0.75	0	0.70	0.80	0.85
	10	0.65	0.75	0.85	10	0.60	0.70	0.80
	100	0.70	0.80	0.85	100	0.55	0.65	0.75
	1000	0.75	0.85	0.85	1000	0.50	0.60	0.70
Moderate (175–425)	0	0.50	0.60	0.65	0	0.65	0.75	0.80
	10	0.60	0.70	0.75	10	0.55	0.65	0.70
	100	0.65	0.75	0.80	100	0.50	0.60	0.65
	1000	0.70	0.80	0.80	1000	0.45	0.55	0.60
Strong (425–700)	0	0.45	0.50	0.60	0	0.60	0.65	0.70
	10	0.55	0.60	0.65	10	0.50	0.55	0.65
	100	0.60	0.65	0.70	100	0.45	0.50	0.60
	1000	0.65	0.70	0.75	1000	0.40	0.45	0.55
Very Strong (greater than 700)	0	0.40	0.45	0.50	0	0.50	0.60	0.65
	10	0.45	0.55	0.60	10	0.45	0.50	0.55
	100	0.50	0.60	0.65	1000	0.40	0.45	0.50
	1000	0.55	0.60	0.65	1000	0.35	0.40	0.45

[1]For extensive areas of bare fallow soil and no agricultural development, pan coefficients are reduced by 20% under hot windy conditions and by 5–10% for moderate wind, temperature, and humidity conditions.

the land surface. Rill flow is runoff that occurs in the form of small rivulets, primarily by concentration of overland flow. Gully flow is runoff that has concentrated into depths large enough so that it has the erosive power to carve its own deep and narrow channel (gully). Streamflow is concentrated runoff originating in overland flow, rill flow, or gully flow and is characterized by well defined channels or streams of sizable depth. Streams carry their flow into larger streams, which flow into rivers to constitute river flow.

A catchment can range from as little as 1 ha (or acre) to hundreds of thousands of square kilometers (or square miles). Small catchments (small watersheds) are those where runoff is controlled by overland flow processes. Large catchments (river basins) are those where runoff is controlled by storage processes in the river channels. Between small and large catchments, there is a wide range of catchment sizes with runoff characteristics falling somewhere between those of small and large catchments. Depending on their relative size, midsize catchments are referred to as either *watersheds* or *basins*.

The hydrologic characteristics of a catchment are described in terms of the following properties: (1) area, (2) shape, (3) relief, (4) linear measures, and (5) drainage patterns.

Catchment Area

Area, or *drainage area,* is perhaps the most important catchment property. It determines the potential runoff volume, provided the storm covers the whole area. The catchment divide is the loci of points delimiting two adjacent catchments, i.e., the collection of high points separating catchments draining into different outlets. Due to the effect of subsurface flow (interflow and groundwater flow), the hydrologic catchment divide may not strictly coincide with the topographic catchment divide. The hydrologic divide, however, is less tractable than the topographic divide; therefore, the latter is preferred for practical use.

The topographic divide is delineated on a quadrangle sheet or other suitable topographic map. High points are identified at the outset, and contour lines are examined to determine the direction of surface runoff. Runoff originates at high points and moves toward lower points in a direction perpendicular to the contour lines. The area enclosed within the topographic divide is the catchment area.

In general, the larger the catchment area, the greater the amount of surface runoff and, consequently, the greater the surface flows. Several formulas have been proposed to relate peak flow to catchment area (Chapter 7). A basic formula is

$$Q_p = cA^n \tag{2-52}$$

in which Q_p = peak flow, A = catchment area, and c and n are parameters to be determined by regression analysis. Other peak flow methods base their calculations on peak flow per unit area, for instance, the TR-55 method (Chapter 5).

Catchment Shape

Catchment shape is the outline described by the horizontal projection of a catchment. Horton [25] described the outline of a normal catchment as a pear-shaped ovoid.

Large catchments, however, vary widely in shape. A quantitative description is provided by the following formula [23]:

$$K_f = \frac{A}{L^2} \qquad (2\text{-}53)$$

in which K_f = form ratio, A = catchment area, and L = catchment length, measured along the longest watercourse. Area and length are given in consistent units such as square kilometers and kilometers, respectively.

An alternate description is based on catchment perimeter rather than area. For this purpose, an equivalent circle is defined as a circle of area equal to that of the catchment. The compactness ratio is the ratio of the catchment perimeter to that of the equivalent circle. This leads to

$$K_c = \frac{0.282P}{A^{1/2}} \qquad (2\text{-}54)$$

in which K_c = compactness ratio, P = catchment perimeter, and A = catchment area, with P and A given in any consistent set of units.

Catchment response refers to the relative concentration and timing of runoff. The role of catchment shape in catchment response has not been clearly established. Other things being equal, a high form ratio (Eq. 2-53) or a compactness ratio close to 1 (Eq. 2-54) describes a catchment having a fast and peaked catchment response. Conversely, a low form ratio or a compactness ratio much larger than 1 describes a catchment with a delayed runoff response. However, many other factors, including catchment relief, vegetative cover, drainage density, and so on, are usually more important than catchment shape, with their combined effect not being readily discernible.

Catchment Relief

Relief is the elevation difference between two reference points. Maximum catchment relief is the elevation difference between the highest point in the catchment divide and the catchment outlet. The principal watercourse (or main stream) is the central and largest watercourse of the catchment and the one conveying the runoff to the outlet. Relief ratio is the ratio of maximum catchment relief to the catchment's longest horizontal straight distance measured in a direction parallel to that of the principal watercourse . The relief ratio is a measure of the intensity of the erosional processes active in the catchment.

The overall relief of a catchment is described by *hypsometric analysis* [37]. This refers to a dimensionless curve showing the variation with elevation of the catchment subarea above that elevation. To develop this curve, the elevation of the highest or maximum point in the catchment divide, corresponding to 0 percent area, is identified. Also, the elevation of the lowest or minimum point of the catchment, corresponding to 100 percent area, is identified. Subsequently, several elevations located between maximum and minimum are selected, and the subareas above each one of these elevations determined by measuring along the respective topographic contour lines. The elevations are converted to height above minimum elevation and expressed in percentage of the maximum height. Likewise, the subareas above each one of the elevations are expressed as percentages of total catchment area. The hypsometric curve (Fig. 2-

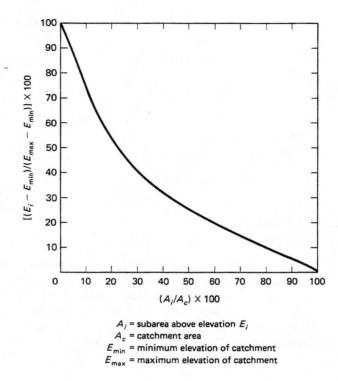

A_i = subarea above elevation E_i
A_c = catchment area
E_{min} = minimum elevation of catchment
E_{max} = maximum elevation of catchment

Figure 2-17. Hypsometric curve.

17) shows percent area in the abscissas and percent height in the ordinates. The median elevation of the catchment is obtained from the percent height corresponding to 50 percent area.

The hypsometric curve is used when a hydrologic variable such as precipitation, vegetative cover, or snowfall shows a marked tendency to vary with altitude. In such cases, the hypsometric curve provides the quantitative means to evaluate the effect of altitude.

Other measures of catchment relief are based on stream and channel characteristics. The longitudinal profile of a channel is a plot of elevation versus horizontal distance. At a given point in the profile, the elevation is usually a mean value of the channel bed. Between any two points, the channel gradient (or channel slope) is the ratio of elevation difference to horizontal distance separating them.

In the absence of geologic controls (rock outcroppings), longitudinal profiles of streams and rivers are usually concave upward, i.e., they show a persistent decrease in channel gradient in the downstream direction (Fig. 2-18) as the flow moves from mountain streams to river valleys and into the ocean. The reason for this downstream decrease in channel gradient is not readily apparent; however, it is known that channel gradients are directly related to bottom friction and inversely related to flow depth. Typically, small mountain streams have high values of bottom friction (due to the presence of cobbles and boulders in the stream bed) and small depths. Conversely, large rivers have comparatively lower values of bottom friction and larger depths. This interaction of channel gradient and bottom friction helps explain the typical decrease in channel gradient in the downstream direction.

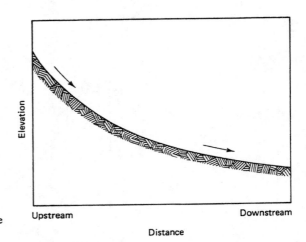

Figure 2-18. Typical longitudinal profile of streams and rivers.

Channel gradients vary widely, from higher than 0.10 for very steep mountain streams to as low as 0.000006 for certain tidal rivers [17]. The channel gradient of a principal watercourse is a convenient measure of catchment relief. A longitudinal profile defines the maximum and minimum elevations and the horizontal distance between them. The channel gradient obtained directly from the maximum and minimum elevations is referred to as the S_1 slope.

A somewhat more representative measure of channel gradient is the S_2 slope, defined as the constant slope that makes the shaded area above it (Fig. 2-19) equal to the shaded area below it. An expedient way to calculate the S_2 slope is to equate the total area below it to the total area below the longitudinal profile (Fig. 2-19).

A measure of channel gradient which takes into account the basin response time is the *equivalent slope*, or S_3. To calculate this slope the channel is divided into n subreaches, and a slope is calculated for each subreach. Based on Manning's equation (Section 2.4), the time of flow travel through each subreach is assumed to be inversely proportional to the square root of its slope. Likewise, the time of travel through the whole channel is assumed to be inversely proportional to the square root of the equivalent slope. This leads to the following equation:

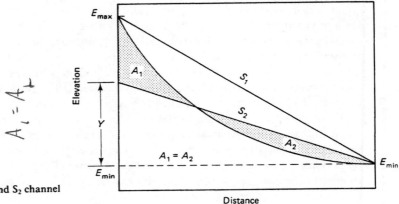

Figure 2-19. Sketch of S_1 and S_2 channel gradients.

Basic Hydrologic Principles   Chap. 2

$$S_3 = \left[\frac{\sum\limits_{i=1}^{n} L_i}{\sum\limits_{i=1}^{n} (L_i/S_i^{1/2})} \right]^2 \qquad (2\text{-}55)$$

in which S_3 = equivalent slope, L_i = each i of n subreach lengths, and S_i = each i of n subreach slopes.

Grid methods are often used to obtain measures of land surface slope for runoff evaluations in small and midsize catchments. For instance, the Soil Conservation Service [68] determines average surface slope by overlaying a square grid pattern over the topographic map of the watershed. The maximum surface slope at each grid intersection is evaluated, and the average of all values calculated. This average is taken as the representative value of surface slope (Fig. 2-20).

Example 2-8.

Given a longitudinal profile with the following elevations and distances, calculate the slopes S_1, S_2 and S_3.

Distance (m)	0	5,000	10,000	15,000	20,000
Elevation (m)	900	910	930	960	1000

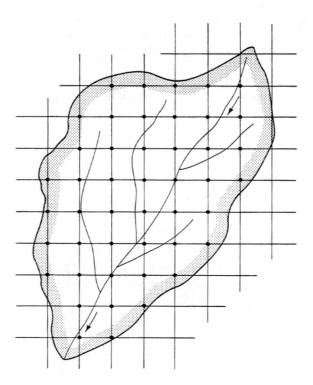

Figure 2-20. Grid overlay to determine land surface slope.

Average land surface slope = average of maximum surface slopes at grid intersections

The maximum and minimum elevations are 1000 and 900 m, respectively. The horizontal distance between them is 20,000 m. Therefore, $S_1 = 100/20,000 = 0.005$. With reference to Fig. 2-21, $S_2 = Y/20,000$. The area under the longitudinal profile is 750,000 m². The area under S_2 is 10,000 Y. Therefore, $Y = 75$ m, and $S_2 = 0.00375$. The individual reaches are all 5000 m long, and the individual slopes are 0.002, 0.004, 0.006, and 0.008 respectively. The application of Eq. 2-55 leads to $S_3 = 0.0041$.

Linear Measures

Linear measures are used to describe the one-dimensional features of a catchment. For instance, for small catchments, the overland flow length is the distance of surface runoff that is not confined to any clearly defined channel.

The catchment length (or hydraulic length) is the length measured along the principal watercourse (Fig. 2-22). The principal watercourse (or main stream) is the central and largest watercourse of the catchment and the one conveying runoff to the outlet.

The length to catchment centroid is the length measured along the principal watercourse, from the catchment outlet to a point located closest to the catchment centroid (Fig. 2-22). In practice, the catchment centroid is estimated as the common point to two or more straight lines that bisect the catchment area in approximately equal subareas.

The concept of stream order is essential to the hierarchical description of streams within a catchment. Overland flow can be thought of as a hypothetical stream of zero order. A first-order stream is that receiving flow from zero-order streams, i.e. overland flow. Two first-order streams combine to form a second-order stream. In general, two m-order streams combine to form a stream of order $m + 1$ (Fig. 2-23). The catchment's stream order is the order of the main stream.

A catchment's stream order is directly related to its size. Large catchments have

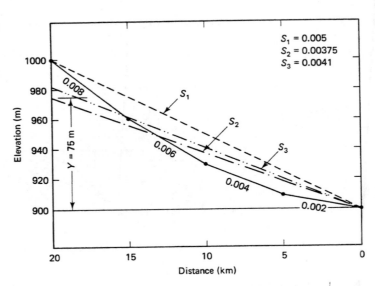

Figure 2-21. Calculation of channel gradients: Example 2-8.

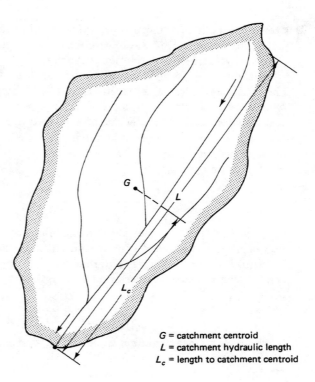

Figure 2-22. Linear measures of a catchment.

G = catchment centroid
L = catchment hydraulic length
L_c = length to catchment centroid

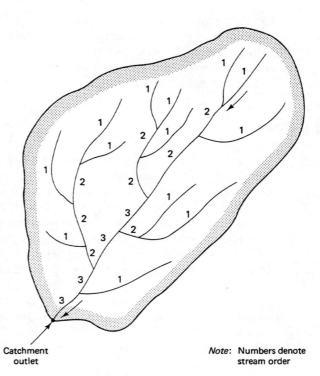

Figure 2-23. Concept of stream order.

Catchment
outlet

Note: Numbers denote
stream order

stream orders of 10 or more. The evaluation of stream order is highly sensitive to map scale. Therefore, considerable care is required when using stream order analysis in comparative studies of catchment behavior.

The lengths of all streams can be added to determine the total stream length. The catchment's drainage density is the ratio of total stream length to catchment area. A high drainage density reflects a fast and peaked runoff response, whereas a low drainage density is characteristic of a delayed runoff response.

The mean overland flow length is approximately equal to half the mean distance between stream channels. Therefore, it can be approximated as one-half of the reciprocal of drainage density:

$$L_o = \frac{1}{2D} \tag{2-56}$$

in which L_o = mean overland flow length and D = drainage density. This approximation neglects the effect of ground and channel slope, which makes the actual mean overland flow length longer than that estimated by Eq. 2-56. The following equation can be used to estimate overland flow length more precisely:

$$L_o = \frac{1}{2D[1 - (S_c/S_s)]^{1/2}} \tag{2-57}$$

in which S_c = mean channel slope and S_s = mean surface slope.

Drainage Patterns

Drainage patterns in catchments vary widely. The more intricate patterns are an indication of high drainage density. Types of drainage patterns that are recognizable on aerial photographs are shown in Fig. 2-24 [27]. These patterns reflect geologic, soil, and vegetation effects and are often related to hydrologic properties such as runoff response or annual water yield.

2.4 RUNOFF

Surface runoff, or simply runoff, refers to all the waters flowing on the surface of the earth, either by overland sheet flow or by channel flow in rills, gullies, streams, or rivers. Surface runoff is a continuous process by which water is constantly flowing from higher to lower elevations by the action of gravitational forces. Small streams combine to form larger streams which eventually grow into rivers. In time, rivers carry their flow into the ocean, completing the hydrologic cycle.

Runoff is expressed in terms of volume or flow rate. The units of runoff volume are cubic meters or cubic feet. Flow rate (or discharge) is the volume per unit of time passing through a given area. It is expressed in cubic meters per second or cubic feet per second. Flow rate usually varies in time; therefore, its value at any time is the instantaneous or local flow rate. The local flow rate can be averaged over a period of time to give the average value for that period. The local flow rate can be integrated over a period of time to give the accumulated runoff volume.

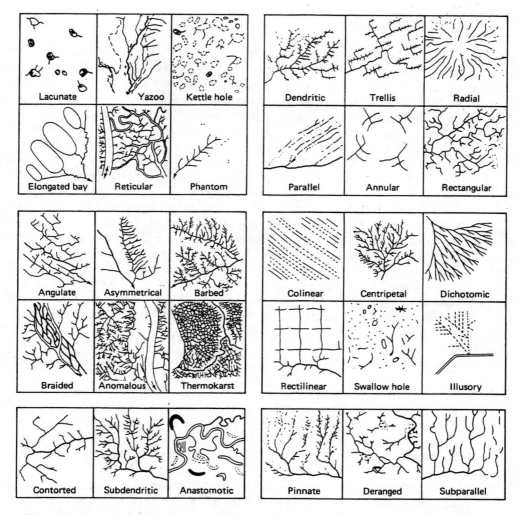

Figure 2-24. Drainage patterns recognizable on aerial photographs (By permission from "Photogrametric Engineering and Remote Sensing," copyright 1960, American Society of Photogrametry and Remote Sensing [27]).

In engineering hydrology, runoff is commonly expressed in depth units. This is accomplished by dividing the runoff volume by the catchment area to obtain an equivalent runoff depth distributed over the entire catchment.

For certain applications, runoff is alternatively expressed in terms of either (1) peak flow per unit drainage area, (2) peak flow per unit runoff depth, or (3) peak flow per unit drainage area per unit runoff depth. In the first case, the units are cubic meters per second per square kilometer; in the second case, cubic meters per second per centimeter; in the third case, cubic meters per second per square kilometer per centimeter.

Runoff Components

Runoff consists of water from three sources: (1) surface flow, (2) interflow, and (3) groundwater flow. Surface flow is the product of effective rainfall, i.e., total rainfall minus hydrologic abstractions. Surface flow is also called *direct runoff*. Direct runoff has the capability to produce large flow concentrations in a relatively short period of time. Therefore, direct runoff is largely responsible for flood flows.

Interflow is subsurface flow, i.e., flow that takes place in the unsaturated soil layers located beneath the ground surface. Interflow consists of the lateral movement of water and moisture toward lower elevations, and it includes some of the precipitation abstracted by infiltration. It is characteristically a slow process, but eventually interflow volumes flow into streams and rivers.

Groundwater flow takes place below the groundwater table (Chapter 11) in the form of saturated flow through alluvial deposits and other water-bearing formations located beneath the soil mantle. Groundwater flow includes the portion of infiltrated volume that has reached the water table by percolation from the overlying soils. Like interflow, groundwater flow is characteristically a slow process. Like surface runoff, groundwater flow is a continuous process, with water constantly moving to lower elevations (or to zones of lower potential) and eventually making its way to the sea. Where the groundwater table is close to the surface, groundwater flow may be intercepted by streams and rivers and discharge into them.

Stream Types and Baseflow

Streams can be grouped into three types: (1) perennial, (2) ephemeral, and (3) intermittent. Perennial streams are those that always have flow. During dry weather (i.e., absence of rain), the flow of perennial streams is *baseflow,* consisting of interflow and groundwater flow intercepted by the stream. Streams that feed from groundwater reservoirs are called *effluent streams*. Perennial and effluent streams are typical of humid regions.

Ephemeral streams are those that have flow only in direct response to effective precipitation, i.e., during and immediately following a major storm. Ephemeral streams do not intercept groundwater flow and therefore have no baseflow. Instead, ephemeral streams usually contribute to groundwater by seepage through their porous channel beds. Streams that feed water into groundwater reservoirs are called *influent streams*. Channel abstractions from influent streams are referred to as *channel transmission losses*. Ephemeral and influent streams are typical of arid and semiarid regions.

Intermittent streams are those of mixed characteristics, behaving as perennial at certain times of the year and ephemeral at other times. Depending on seasonal conditions, these streams may feed to or from the groundwater.

Baseflow estimates are important in dry weather hydrology—for instance, in the calculation of the total runoff volume produced by a catchment in a year, referred to as the annual water yield. In flood hydrology, baseflow is used to separate surface runoff into direct and indirect runoff. *Indirect runoff* is surface runoff originating in interflow and groundwater flow. Baseflow is a measure of indirect runoff.

Antecedent Moisture

Surface runoff is directly related to effective precipitation, and effective precipitation is inversely related to the hydrologic abstractions. During rainy periods, infiltration plays a major role in abstracting total precipitation. Actual infiltration rates and amounts vary widely, being highly dependent on the initial level of soil moisture. For a given storm, the latter is referred to as *antecedent moisture*, or antecedent moisture condition. Catchments with low initial soil moisture, i.e., dry catchments, are not conducive to high runoff response. Conversely, catchments with high initial moisture, i.e., wet catchments, are likely to produce large quantities of runoff.

The recognition that direct runoff is a function of antecedent moisture has led to the concept of antecedent precipitation index (API). The average moisture level in a catchment varies daily, being replenished by precipitation and depleted by evaporation and evapotranspiration. The assumption of a logarithmic depletion rate leads to a catchment's API for a day with no rain:

$$I_i = KI_{i-1} \tag{2-58}$$

in which I_i = index for day i, I_{i-1} = index for day $i - 1$, and K = a recession factor normally in the range 0.85-0.98 [38]. If rain occurs in any day, the rainfall depth is added to the index. The index at day zero (initial value) would have to be estimated. Likewise, the applicable value of K is determined from either data or experience.

The API is directly related to runoff depth. The greater the value of the index, the greater the amount of runoff. In practice, regression and other statistical tools are used to relate runoff to API. These relations are invariably empirical and therefore strictly applicable only to the situation for which they were derived.

Other measures of catchment moisture have been developed over the years. For instance, the Soil Conservation Service (SCS) uses the concept of antecedent moisture condition (AMC) (Chapter 5), grouping catchment moisture into three levels: AMC I, a dry condition; AMC II, an average condition; and AMC III, a wet condition. Moisture conditions ranging from AMC II to AMC III are normally used in hydrologic design.

Another example of the use of the concept of antecedent moisture is that of the SSARR model (Chapter 13). The SSARR model computes runoff volume based on a relationship linking runoff percent to a *soil-moisture index* (SMI), with precipitation intensity as a third variable. Runoff percent is the ratio of runoff to rainfall, multiplied by 100. Such runoff-moisture-rainfall relation is empirical and, therefore, is limited to the basin for which it was derived.

Rainfall-Runoff Relations

Rainfall can be measured in a relatively simple way. However, runoff measurements usually require an elaborate stream-gaging procedure (Chapter 3). This difference has led to rainfall data being more widely available than runoff data. The typical catchment has many more rain gages than stream-gaging stations, with the rainfall records likely to be longer than the streamflow records.

The fact that rainfall data is more voluminous than runoff data has led to the calculation of runoff by relying on rainfall data. Although this is an indirect procedure, it has proven its practicality in a variety of applications.

A basic linear model of rainfall-runoff is the following:

$$Q = b(P - P_a) \tag{2-59}$$

in which Q = runoff depth, P = rainfall depth, P_a = rainfall depth below which runoff is zero, and b = slope of the line (Fig. 2-25). Rainfall depths smaller than P_a are completely abstracted by the catchment, with runoff starting as soon as P exceeds P_a. To use Eq. 2-59, it is necessary to collect several sets of rainfall-runoff data and to perform a linear regression (Chapter 7) to determine the values of b and P_a. The simplicity of Eq. 2-59 precludes it from taking into account other important runoff-producing mechanisms such as rainfall intensity, infiltration rates, or antecedent moisture. In practice, the correlation usually shows a wide range of variation, limiting its predictive ability.

The effect of infiltration rate and antecedent moisture on runoff is widely recognized. Several models have been developed in an attempt to simulate these and other related processes. Typical of such models is the SCS runoff curve number model, which has had wide acceptance in engineering practice. The SCS model is based on a nonlinear rainfall-runoff relation that includes a third variable (curve parameter) called the *runoff curve number*, or *CN*. In a particular application, the *CN* value is determined by a detailed evaluation of soil type, vegetative and land use patterns, antecedent moisture, and hydrologic condition of the catchment surface. The SCS runoff curve number method is described in Chapter 5.

Runoff Concentration

An important characteristic of surface runoff is its concentration property. To describe it, assume that a storm falling on a given catchment produces a uniform effective rainfall intensity distributed over the entire catchment area. In such a case, sur-

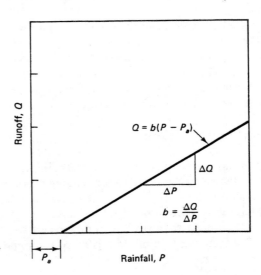

Figure 2-25. Basic linear model of rainfall-runoff.

face runoff eventually concentrates at the catchment outlet, provided the effective rainfall duration is sufficiently long. _Runoff concentration_ implies that the flow rate at the outlet will gradually increase until rainfall from the entire catchment has had time to travel to the outlet and is contributing to the flow at that point. At that time, the maximum, or equilibrium, flow rate is reached, implying that the surface runoff has _concentrated_ at the outlet. The time that it takes a parcel of water to travel from the farthest point in the divide to the catchment outlet is referred to as the _time of concentration,_ or concentration time.

The equilibrium flow rate is calculated by multiplying the effective rainfall intensity by the catchment area:

$$Q_e = 2.78 I_e A \qquad (2\text{-}60)$$

in which Q_e = equilibrium flow rate in liters per second; I_e = effective rainfall intensity in millimeters per hour; and A = catchment area in hectares; and 2.78 is the conversion factor for the indicated units. In U.S. customary units, with Q_e in cubic feet per second, I_e in inches per hour, and A in acres, the conversion factor in Eq. 2-60 is 1.008.

The process of runoff concentration can lead to three distinct types of catchment response. The first type occurs when the effective rainfall duration is equal to the concentration time. In this case, the runoff concentrates at the outlet, reaching its maximum (equilibrium) rate after an elapsed time equal to the concentration time. Rainfall stops at this time, and subsequent flows at the outlet are no longer concentrated because not all the catchment is contributing. Therefore, the flow gradually starts to recede back to zero. Since it takes the concentration time for the farthest runoff parcels to travel to the outlet, the recession time is approximately equal to the concentration time, as sketched in Fig. 2-26(a). (In practice, due to nonlinearities, actual recession flows are usually asymptotic to zero). This type of response is referred to as _concentrated_ catchment flow.

The second type of catchment response occurs when the effective rainfall duration exceeds the concentration time. In this case, the runoff concentrates at the outlet, reaching its maximum (equilibrium) rate after an elapsed time equal to the concentration time. Since rainfall continues to occur, the whole catchment continues to contribute to flow at the outlet, and subsequent flows remain concentrated and equal to the equilibrium value. After rainfall stops, the flow gradually recedes back to zero. Since it takes the concentration time for the farthest runoff parcels to travel to the outlet, the recession time is approximately equal to the concentration time (see Fig. 2-26(b)). This type of response is referred to as _superconcentrated_ catchment flow.

The third type of response occurs when the effective rainfall duration is shorter than the concentration time. In this case the flow at the outlet does not reach the equilibrium value. After rainfall stops, the flow recedes back to zero. The requirements that volume be conserved and recession time be equal to the concentration time lead to the idealized flattop response shown in Fig. 2-26(c). This type of response is referred to as _subconcentrated_ catchment flow.

In practice, concentrated and superconcentrated flows are typical of small catchments, i.e., those likely to have short concentration times. On the other hand, subconcentrated flows are typical of midsize and large catchments, i.e., those with longer concentration times.

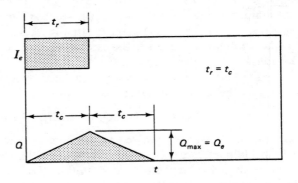

(a) Concentrated

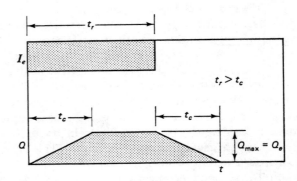

(b) Superconcentrated

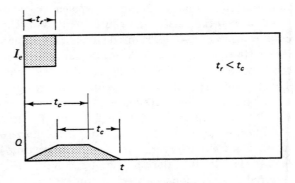

(c) Subconcentrated

Figure 2-26. Runoff concentration—
types of catchment response: (a) concentrated; (b) superconcentrated; (c) subconcentrated.

Basic Hydrologic Principles   Chap. 2

Concentration Time.   Hydrologic procedures for small catchments usually require an estimate of concentration time (Chapter 4). However, accurate estimates are generally difficult to make. For one thing, concentration time is a function of runoff rate; therefore, an estimate can only represent a certain flow level, whether it be low flow, average flow, or high flow.

Several formulas for calculating concentration time as a function of selected catchment parameters are available. Most formulas are empirical in nature and therefore of somewhat limited value. An alternate approach to calculate concentration time is to divide the principal watercourse into several subreaches and to assume an appropriate flow level for each subreach. Subsequently, a steady flow formula such as the Manning equation is used to calculate the mean flow velocity and associated travel time through each subreach. The concentration time through the reach is the sum of the subreach travel times. This procedure, while practical, is based on several assumptions, including a flow-rate level and Manning n values.

A limitation of the steady flow approach to the calculation of concentration time is the fact that the flow is generally unsteady. In practice, this means that the speed of travel of the wavelike features of the flow may be greater than the mean flow velocity calculated using steady flow principles. For instance, for turbulent flow, kinematic wave theory (Chapters 4 and 9) can justify a wave speed as much as 5/3 the mean flow velocity, with the associated reduction in travel time. Yet in most cases, the ratio between wave speed and mean flow velocity is likely to be smaller than 5/3:1. In practice, uncertainties involved in the computation of concentration time have contributed to a blurring of the distinction between the two speeds.

Concentration Time Formulas.   Notwithstanding the inherent complexities, calculations of concentration time continue to be part of the routine practice of engineering hydrology. Concentration time is a key ingredient in the rational method (Chapter 4) and other methods used to calculate the runoff response of small watersheds. Formulas vary, but most relate concentration time to suitable slope, length, rainfall, and roughness parameters [54]. A well-known formula relating concentration time to slope and length parameters is the Kirpich formula [31], applicable to small agricultural watersheds with drainage areas of less than 80 ha (200 ac). In SI units, the Kirpich formula is

$$t_c = \frac{0.06628 L^{0.77}}{S^{0.385}} \qquad (2\text{-}61)$$

in which t_c = concentration time in hours; L = length of the principal watercourse, from outlet to divide, in kilometers; and S = slope between maximum and minimum elevation (S_1 slope), in meters per meter.

In U.S. customary units, with t_c in minutes, L in feet, and slope in feet per foot, the coefficient of Eq. 2-61 is 0.0078.

Another formula for estimating concentration time is that of Hathaway [19]:

$$t_c = \frac{0.606(Ln)^{0.467}}{S^{0.234}} \qquad (2\text{-}62)$$

in which n is a roughness factor and all other terms are the same as in Eq. 2-61, expressed in SI units. Applicable values of n are given in Table 2-7.

Sec. 2.4   Runoff   **69**

TABLE 2-7 VALUES OF ROUGHNESS FACTOR n FOR USE IN EQUATION 2-62 [9]

Type of Surface	Value of n
Smooth impervious	0.02
Smooth bare-picked soil	0.10
Poor grass, row crops or moderately rough bare soil	0.20
Pasture	0.40
Deciduous timber land	0.60
Coniferous timber land, or deciduous timber land with deep litter or grass	0.80

Actual estimates of concentration time using empirical formulas are bound to vary. Therefore, reasonable care should be exercised when choosing an appropriate formula.

Example 2-9.

Use the Kirpich and Hathaway formulas to estimate concentration time for a catchment with the following characteristics: $L = 0.75$ km, $S = 0.008$, and moderately rough bare soil.

The application of Eq. 2-61 leads to $t_c = 0.34$ h. With $n = 0.2$, the application of Eq. 2-62 leads to $t_c = 0.77$ h. For comparison purposes, with $n = 0.1$, Eq. 2-62 gives $t_c = 0.56$ h.

A physically based approach to the calculation of concentration time is possible by means of overland flow techniques (Chapter 4). As a first approximation, concentration time can be taken as the time to equilibrium of overland flow. Formulas for the calculation of time to equilibrium are given in Section 4.2. Detailed calculations of concentration time based on overland flow techniques can be found in the literature of kinematic flow modeling [53].

Runoff Diffusion and Streamflow Hydrographs

In nature, catchment response shows a more complex behavior than that which may be attributed solely to runoff concentration. For one thing, it is unlikely that effective rainfall intensity would remain constant in time and space throughout the storm duration. Theory and experimental evidence have shown that runoff rates are governed by natural processes of convection and diffusion. Convection refers to runoff concentration as described in the previous paragraphs. *Diffusion* is the mechanism acting to spread the flow rates in time and space.

The net effect of runoff diffusion is to reduce the flow rates to levels below those that could be attained by convection only. In practice, diffusion acts to smooth out catchment response. The resulting response function is usually continuous, and it is referred to as the *streamflow hydrograph*, runoff hydrograph, or simply the hydrograph. Typical single-storm hydrographs have a shape similar to that shown in Fig. 2-27(a). They are usually produced by storms with effective rainfall duration less than

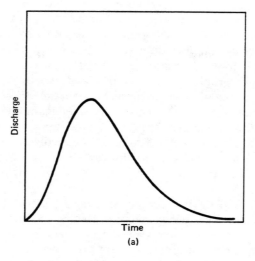

(a)

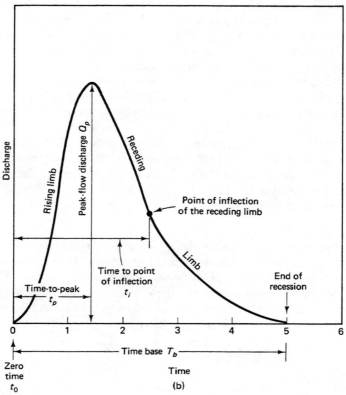

(b)

Figure 2-27. (a) Typical single-storm hydrograph; (b) elements of single-storm hydrograph.

concentration time. Therefore, they resemble subconcentrated catchment flow, albeit with the addition of a small but perceptible amount of diffusion.

The various elements in a typical single-storm hydrograph are shown in Fig. 2-27(b). The zero time (or starting time) depicts the beginning of the hydrograph. The hydrograph peak describes the maximum flow rate. The time-to-peak is measured from zero time to the time at which the peak flow is attained. The rising limb is the part of the hydrograph between zero time and time-to-peak. The recession (or receding limb) is the part of the hydrograph between time-to-peak and time base. The time base is measured from zero time to a time defining the end of the recession. The recession is logarithmic in nature, approaching zero flow in an asymptotic way. For practical applications, the end of the recession is usually defined in an arbitrary manner. The point of inflection of the receding limb is the point corresponding to zero curvature. The hydrograph volume is obtained by integrating the flow rates from zero time to time base.

The shape of the hydrograph, showing a positive skew (Chapter 6), with recession time greater than rising time, is caused by the essentially different responses of surface runoff, interflow, and groundwater flow. Indeed, the runoff hydrograph can be thought of as consisting of the sum of three hydrographs, as shown in Fig. 2-28(a). The fast and peaked hydrograph is produced by surface runoff, while the other two are the result of interflow and groundwater flow. The superposition of these hydrographs (Fig. 2-28(b)) results in a runoff hydrograph exhibiting a long tail (positive skew).

The feature of positive skew allows the definition of few additional geometric hydrograph properties. The time-to-centroid is measured from zero time to the time separating the hydrograph into two equal volumes (Fig. 2-29). The volume-to-peak is obtained by integrating the flow rates from zero time to time-to-peak. In synthetic unit hydrograph analysis (Chapter 5), the ratio of volume-to-peak to hydrograph volume is used as a measure of hydrograph shape.

Hydrographs of perennial streams may include substantial amounts of baseflow. The separation of direct runoff from baseflow can be accomplished by resorting to one of several hydrograph separation techniques (Chapter 5). These techniques can also be used in the analysis of multiple-storm hydrographs, which typically exhibit two or more peaks and valleys.

Analytical Hydrographs. Analytical expressions for streamflow hydrographs are sometimes used in hydrologic studies. The simplest formula is based on either a sine or cosine function. These, however, have zero skew (Chapter 6) and therefore do not properly describe the shape of natural hydrographs.

An analytical hydrograph that is often used to simulate natural hydrographs is the gamma function, expressed as follows:

$$Q = Q_b + (Q_p - Q_b) \left(\frac{t}{t_p}\right)^m e^{[(t_p - t)/(t_g - t_p)]} \tag{2-63}$$

in which Q = flow rate; Q_b = baseflow; Q_p = peak flow; t_p = time-to-peak; t_g = time-to-centroid; t = time; and $m = t_p/(t_g - t_p)$. For values of t_g greater than t_p, Eq. 2-63 exhibits positive skew.

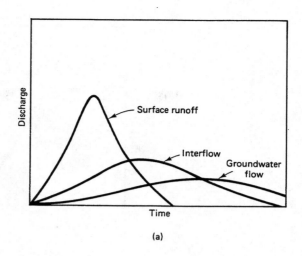

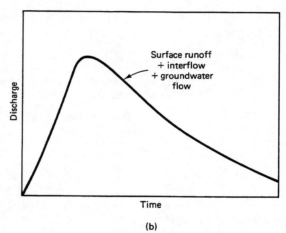

Figure 2-28. Components of runoff hydrograph.

(a)

(b)

Example 2-10.

Use Eq. 2-63 to calculate streamflow hydrograph ordinates at hourly intervals, with the following data: $Q_b = 100$ m³/s; $Q_p = 500$ m³/s; $t_p = 3$ h; $t_g = 4.5$ h.

The application of Eq. 2-63 leads to

$$Q = 100 + 400 \left(\frac{t}{3}\right)^2 e^{[(3-t)/(4.5-3)]} \tag{2-64}$$

The hydrograph ordinates at hourly intervals are shown in Table 2-8. It is seen that the flow rate at the start is 100 m³/s, it reaches a peak of 500 m³/s at 3 h, and it takes 15 h to recede back to 103 m³/s.

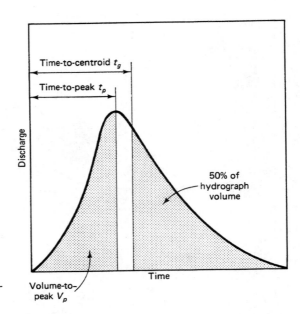

Figure 2-29. Additional single-storm hydrograph properties.

TABLE 2-8   CALCULATED GAMMA HYDROGRAPH
ORDINATES: EXAMPLE 2-10

Time (h)	Flow (m^3/s)	Time (h)	Flow (m^3/s)	Time (h)	Flow (m^3/s)
					. . .
0	100				
1	269	6	317	11	126
2	446	7	251	12	116
3	500	8	201	13	110
4	465	9	166	14	106
5	393	10	142	15	103

Flow in Stream Channels

Streamflow hydrographs flow in stream channels that are formed on the land surface. The following properties are used to describe stream channels: (1) cross-sectional dimensions, (2) cross-sectional shape, (3) longitudinal slope, and (4) boundary friction.

The channel cross section has the following geometric and hydraulic elements: flow area, top width, wetted perimeter, hydraulic radius, hydraulic depth, and aspect ratio. The flow area is the area of the cross section occupied by the flow. The top width is the channel width at the elevation of the water surface. The wetted perimeter is the perimeter of the flow area excluding the top width. The hydraulic radius is the ratio of flow area to wetted perimeter. The hydraulic depth is the ratio of flow area to top width. The aspect ratio, a measure of cross-sectional shape, is the ratio of top width to hydraulic depth.

Channel top widths vary widely, ranging from a few meters for small mountain streams to several kilometers for very large rivers. Mean flow depths range from as low as a fraction of a meter for small mountain streams to as high as 50-80 m for very large

Basic Hydrologic Principles   Chap. 2

rivers. Aspect ratios vary widely in nature; however, most streams and rivers have aspect ratios in excess of 10. Very wide streams (e.g., braided streams) may have aspect ratios exceeding 100.

The longitudinal channel slope is the change in elevation with distance. The mean bed elevation is generally used to calculate channel slope. For short reaches or mild slopes, slope calculations may be hampered by the difficulty of accurately establishing the mean bed elevation. A practical alternative is to use the water surface slope as a measure of channel slope. The water surface slope, however, varies in space and time as a function of the flow nonuniformity and unsteadiness. The steady equilibrium (i.e., uniform) water surface slope is usually taken as a measure of channel slope. Therefore, mean bed slope and steady equilibrium water surface slope are often treated as synonymous. Generally, the longer the channel reach, the more accurate the calculation of channel slope.

Boundary friction refers to the type and dimensions of the particles lining the channel cross section below the waterline. In alluvial channels, geomorphic bed features such as ripples and dunes may represent a substantial contribution to the overall friction (Chapter 15). Particles lying on the channel bed may range from large boulders for typical mountain streams to silt particles in the case of large tidal rivers.

For small streams, particles on the channel banks may be as large as the particles on the bottom. River banks, however, are more likely to consist of particles of different sizes than those on the channel bottom. The high aspect ratio of rivers generally results in the banks contributing only a small fraction of the total boundary friction. Therefore, the boundary friction is often taken as synonymous with bed or bottom friction.

Uniform-flow Formulas. Flow in streams and rivers is evaluated by using empirical formulas such as the Manning or Chezy equations. The Manning formula is

$$V = \left(\frac{1}{n}\right) R^{2/3} S^{1/2} \tag{2-65}$$

in which V = mean flow velocity in meters per second; R = hydraulic radius in meters; S = channel slope; and n = Manning friction coefficient (with V in feet per second and R in feet, the right side of Eq. 2-65 is multiplied by 1.486). In natural channels, values of n usually range from as low as 0.024 for large rivers with cross sections devoid of vegetation (e.g., Columbia River at Vernita, Washington [3]), to as large as 0.079 for small streams with bed composed of large, angular boulders and banks consisting of exposed rock, boulders, and trees (e.g., Cache Creek near Lower Lake, California [3]). Typical values of n for natural streams and rivers are in the range 0.03–0.05.

The Chezy equation is

$$V = C(RS)^{1/2} \tag{2-66}$$

in which C = Chezy coefficient in square root of meters per second; other terms are the same as for Eq. 2-65. Chezy coefficients calculated for the same conditions as above range from 79 $m^{1/2}/s$ for large rivers to 11 $m^{1/2}/s$ for small streams. Typical values of C are in the range 40–70 $m^{1/2}/s$.

Equation 2-66 can be expressed in dimensionless form, as follows:

$$V = (C/g^{1/2})(gRS)^{1/2} \tag{2-67}$$

in which $C/g^{1/2}$ = dimensionless Chezy coefficient and g is the gravitational acceleration. For the same conditions given earlier, the dimensionless Chezy coefficient varies in the range of 3.5 for large rivers to 25 for small streams.

For certain applications, Eq. 2-67 can be readily transformed into a formula with an enhanced physical meaning. For wide channels, the top width and wetted perimeter can be assumed to be approximately the same. This implies that the hydraulic depth *(D)* can be substituted for the hydraulic radius *(R)*, leading to

$$S = fF^2 \tag{2-68}$$

in which f = a dimensionless friction factor equal to g/C^2 and F = Froude number, equal to $V/(gD)^{1/2}$. It can be shown that the friction factor f in Eq. 2-68 is equal to 1/8 of the Darcy-Weisbach friction factor used to calculate head losses in closed conduits. For the preceding conditions, values of f vary in the range of 0.0016 for large streams to 0.081 for small streams. Typical values of f are in the range 0.002–0.006.

Equation 2-68 states that for wide channels, the channel slope is proportional to the square of the Froude number, with the friction factor as the proportionality coefficient. In practice, Eq. 2-68 can be used as a convenient predictor of any of these three dimensionless parameters, once the other two are known. Furthermore, it implies that if one of the three parameters is kept constant, a change in one of the other two causes a corresponding change in the third.

Notwithstanding the theoretical appeal of Eqs. 2-66 and 2-68, the Manning equation has had wider acceptance in practice. This is attributed to the fact that in natural channels, the Chezy coefficient is not constant, tending to increase with the hydraulic radius. The comparison of Eqs. 2-65 and 2-66 leads to

$$C = (1/n)R^{1/6} \tag{2-69}$$

which implies that the Manning n value is a constant. Experience has shown, however, that the value of n may vary with discharge, especially in the case of alluvial rivers.

River Stages. At any location along a river, the river stage is the elevation of the water surface above a given datum. This datum can be either an arbitrary one or the NGVD, the *national geodetic vertical datum*, a standard measure of mean sea level.

River stages are a function of flow rate. Flow rates can be grouped into (1) low flow, (2) average flow, and (3) high flow. Low flow is typical of the dry season, when streamflow is largely composed of baseflow originating in contributions from interflow and groundwater flow. High flow occurs during the wet season, when streamflow is primarily due to contributions from surface runoff. Average flow usually occurs in midseason and may have mixed contributions from surface runoff, interflow, and groundwater flow.

The study of low flows is necessary when determining minimum flow rates, below which a certain use would be impaired. Examples of such uses are irrigation water requirements, hydropower generation, and minimum flows needed for compliance with water pollution regulations. Average flows play an important role in the calculation of monthly and annual volumes available for storage and use. Applications are usually found in connection with the sizing of storage reservoirs.

The study of high flows is related to the subject of floods and flood hydrology. Typically, during high flows, natural streams and rivers have the tendency to overflow their banks, with stages reaching above bank-full stage, or flood stage. In such cases, the flow area includes a portion of the land located on both sides of the river. In alluvial valleys, the land that is subject to inundation during periods of high flow is referred to as the *flood plain*. The evaluation of high flows is necessary for flood forecasting and flood control.

Rating Curves.   It is known that river stage varies as a function of discharge, but the exact nature of the relationship is not readily apparent. Given a long and essentially prismatic channel reach, a single-valued relationship between stage and discharge at a cross section defines the *equilibrium rating* curve. For steady uniform flow, the rating curve is unique, i.e., there is a single value of stage for each value of discharge and vice versa. In this case, the equilibrium rating curve can be calculated with either the Chezy or Manning equation. In open channel hydraulics, this property of uniqueness of the rating qualifies the channel reach as a *channel control*.

However, other flow conditions, specifically nonuniformity (gradually varied steady flow) and unsteadiness (e.g. gradually varied unsteady flow), can cause deviations from the steady equilibrium rating. These deviations are less tractable. In particular, flood wave theory justifies the presence of a loop in the rating, as shown in Fig. 2-30. Intuitively, the rising limb of the flood-wave hydrograph has a steeper water surface slope than that of equilibrium flow, leading to greater flows and lower stages. Conversely, the receding limb has a milder water surface slope, resulting in smaller flows and higher stages—thus the rationale for the loop's presence. The loop effect, however, is likely to be small and is usually neglected on practical grounds. Where

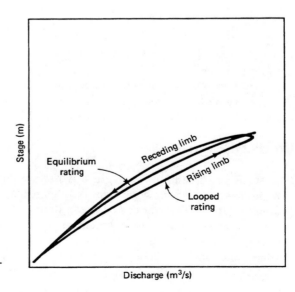

Figure 2-30. Equilibrium and looped rating curves.

increased accuracy is required, unsteady flow modeling (Chapter 9) can be used to determine the actual size of the loop.

Two other mechanisms have a bearing in the evaluation of stage-discharge relations: the short-term and long-term sedimentation effects. The short-term effects are due to the fact that the amount of boundary friction varies with flow rate. Rivers flowing on loose boundaries composed of gravel, sand, and silt constantly try to minimize their changes in stage. This is accomplished through the following mechanism: During low flow, the bed friction consists not only of grain friction but also of form friction, caused by bed features such as ripples and dunes (Chapter 15). During high flows, the swiftness of the current acts to obliterate the bed features, reducing the form friction to a minimum, with only the grain friction remaining. The reduced friction during high flows gives rivers the capability to carry a greater discharge for a given stage. This explains the demonstrated shift from low-flow rating to high-flow rating in natural river channels.

The long-term sedimentation effect is due to the fact that rivers continuously subject their boundaries to endless cycles of erosion and deposition, depending on the sediment load they carry (Chapter 15). Some very active rivers may be eroding; others may be aggrading. Yet some geomorphologically active rivers may substantially change their cross sections during major floods. Invariably, shifts in rating are the net result of these natural geomorphic processes.

Rating-curve Formulas.   In spite of the apparent complexities, rating curves are a useful and practical tool in hydrologic analysis, allowing the direct conversion of stage to discharge and vice versa. Discharge can be obtained from the rating by the simple procedure of measuring the stage. Conversely, if discharge is known, for instance, at a catchment outlet, stage at the outlet can be readily determined from a suitable rating.

There are several ways to determine an equation for the rating. Invariably, they are based on curve-fitting stage-discharge data. A widely used equation is the following [30]:

$$Q = a(h - h_o)^b \qquad (2\text{-}70)$$

in which Q = discharge; h = gage height; h_o = reference height; and a and b are constants. Several values of reference height are tried. The proper value of reference height is that which makes the stage-discharge data plot as close as possible to a straight line on logarithmic paper. Subsequently, the values of the constants a and b are determined by regression analysis (Chapter 7).

Streamflow Variability

The study of streamflow variability is the cornerstone of engineering hydrology. Streamflow and river flow vary not only seasonally but also annually and with geographic location. The amount of streamflow and river flow is directly related to the amount of moisture in the catchment or basin. Precipitation supplies the moisture, but whether this moisture travels to the catchment outlet remains to be determined by further analysis.

On a global annual basis, streamflow data indicate that total runoff volume is approximately equal to 30 percent of the total precipitation volume. The difference is due to the catchment's abstractive processes, i.e., interception, infiltration, surface storage, evaporation, and evapotranspiration. The figure of 30 percent is a global annual average; therefore, it does not account for seasonal or geographical variability. Streamflow data for individual basins show a substantial variability in the runoff volume for a given period, as compared to the rainfall volume for the same period.

Seasonal Variability. The typical catchment in a temperate region shows runoff volumes varying during the year, being low during the dry season and high during the wet season. Catchments in more extreme climates, however, may show different behaviors. For instance, in the ephemeral streams typical of arid regions, runoff volumes are nonexistent during periods of no precipitation. For these streams, runoff occurs only in direct response to precipitation. On the other hand, in humid and tropical climates, rivers show substantial amounts of runoff throughout the year, with little variability between the seasons.

The reason for the seasonal variability of streamflow lies in the relative contributions of direct and indirect runoff. During the wet season in temperate regions, indirect runoff is a small, but nevertheless measurable, fraction of direct runoff. In arid regions, particularly for ephemeral streams, indirect runoff is either negligible or does not exist. For humid regions, indirect runoff is substantial throughout the year and may be of the same order of magnitude as direct runoff.

The phenomena can be further explained in the following way: Groundwater reservoirs serve as the mechanism for the storage of large amounts of moisture, which are slowly transported to lower elevations. Some of the moisture is eventually released back to the surface waters. The process is slow and, is therefore, subject to a substantial amount of diffusion. The net effect is that of a permanent contribution from groundwater to surface water in the form of dry-weather flow. To evaluate the seasonal variability of streamflow, it is therefore necessary to examine the relation between surface water and groundwater.

Annual Variability. Year-to-year streamflow variability shows some of the same features as those of seasonal streamflow variability. For instance, large catchments show runoff variability from one year to the next as a function of the state of moisture at the end of the first year and of the precipitation amounts added during the second year. As in the case of seasonal variability, annual streamflow variability is linked to the relative contributions of direct and indirect runoff. During dry years, rainfall goes on to replenish the catchment's stored moisture, with little of it showing as direct runoff. This results in the low levels of runoff that characterize dry years. Conversely, during wet years, the catchment's storage capacity fills up quickly, and any additional precipitation is almost entirely converted into surface runoff. This produces the high streamflow levels that are characteristic of wet years.

Streamflow variability is therefore intrinsically linked to the relative contributions of direct and indirect runoff. A possible line of inquiry is to focus on the mechanics of surface and groundwater flow, while accounting for the temporal and spatial variability of the various processes involved. However, the absence of reliable data for all the relevant phases of the hydrologic cycle makes the evaluation of streamflow variability using purely mechanistic principles a rather complex process.

A practical alternative that has enjoyed wide acceptance is the reliance on statistical tools to compensate for the incomplete knowledge of the physical processes. This has led to the concept of flow frequency or flood frequency, expressed as the average period of time (the return period) that it will take a certain flow value to recur at a given location. An annual flood series is abstracted from daily discharge measurements at the given station. This is accomplished either by selecting the maximum daily flow for each of n years of record (the annual maxima series), or by selecting the n greatest flow values in the entire n-year record, regardless of when they occurred (the annual exceedence series). The statistical analysis of the flood series (Chapter 6) allows the calculation of the flow rate associated with one or more selected frequencies.

The procedure is relatively straightforward, but it is limited by the record length. Its predictive capability decreases sharply when used to evaluate floods with return periods substantially in excess of the record length. An advantage of the method is its reproducibility, which means that two persons are likely to arrive at the same result when using the same methodology. This is a significant asset when comparing the relative merits of competing water resources projects. Methods for flood frequency analysis are discussed in Chapter 6.

Daily-flow Analysis. The variability of streamflow can also be expressed in terms of the day-to-day fluctuation of flow rates at a given station. Some streams show great variability from day to day, with high peaks and low valleys succeeding one another endlessly. Other streams show very little day-to-day variability, with high flows being not very different from low flows.

The reason for this difference in behavior can be attributed to differences in the nature of catchment response. Small and midsize catchments are likely to have steep gradients and therefore to concentrate flows with negligible runoff diffusion, producing hydrographs that show a large number of high peaks and corresponding low valleys. Conversely, large catchments are likely to have milder gradients and therefore to concentrate flows with substantial runoff diffusion. The diffusion mechanism acts to spread the flows in time and space, resulting in a succession of smooth hydrographs showing low peaks and comparatively high valleys.

Daily flow data may not be sufficient to allow calculation of the runoff volumes produced by small watersheds. In cases where accuracy is required, hourly flows (or perhaps flows measured at 3-h intervals) may be necessary to describe adequately the temporal variability of the flow.

In the past two decades, the development of stochastic models of streamflow variability has resulted in a substantial body of knowledge referred to as *stochastic hydrology*. For a detailed treatment of this subject, see [7, 62, 79].

Flow-duration Curve. A practical way to evaluate day-to-day streamflow variability is the flow-duration curve. To determine this curve for a particular location, it is necessary to obtain daily flow data for a certain period of time, either 1 y or a number of years. The length of the record indicates the total number of days in the series. The daily flow series is sequenced in decreasing order, from the highest to the lowest flow value, with each flow value being assigned an order number. For instance, the highest flow value would have order number one; the lowest flow value would have the last order number, equal to the total number of days. For each flow value, the percent time is defined as the ratio of its order number to the total number of days, expressed in percentage. The flow-duration curve is obtained by plotting flow versus percent time, with percent time in the abscissas and flow in the ordinates.

A flow-duration curve allows the evaluation of the permanence of characteristic low-flow levels. For instance, the flow expected to be exceeded 90 percent of the time can be readily determined from a flow-duration curve (Fig. 2-31). The permanence of low flows is increased with streamflow regulation. The usual aim is to be able to assure the permanence of a certain low-flow level 100 percent of the time. Regulation causes a shift in the flow-duration curve by increasing the permanence of low flows while decreasing that of high flows (Fig. 2-31). Streamflow regulation is accomplished with storage reservoirs.

The flow-duration curve is helpful in the planning and design of water resources projects. In particular, for hydropower studies, the flow-duration curve serves to determine the potential for firm power generation. In the case of a run-of-the-river plant, with no storage facilities, the firm power is usually assumed on the basis of flow available 90 to 97 percent of the time.

Flow-mass Curve. Another way to evaluate day-to-day (and seasonal) streamflow variability is the flow-mass curve. A mass curve of daily values of a variable is a plot of time in the abscissas versus cumulative values of the variable in the ordinates. When using flow values, such a plot is referred to as the *flow-mass curve*.

For daily flow records in cubic meters per second, the ordinates of the flow mass curve are in cubic meters or cubic hectometers. For any given day, the ordinate of the flow-mass curve is the accumulated runoff volume up to that day. According to Chow [10], the flow-mass curve is believed to have been first suggested by Rippl [61]; hence

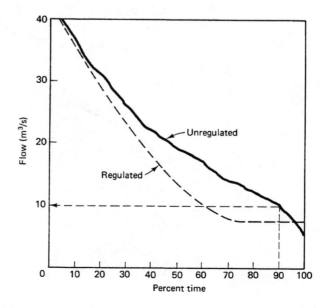

Figure 2-31. Flow-duration curve.

the name Rippl curve. The shape of the flow-mass curve resembles that of the letter S (Fig. 2-32); therefore, it is also referred to as the S curve.

Applications of flow-mass curves are to reservoir design and operation, including the determination of reservoir capacity and the establishment of operating rules for storage reservoirs. Figure 2-32 shows a typical flow-mass curve. At any given time, the slope of the mass curve is a measure of the instantaneous flow rate. The slope of the line PQ, drawn between the points P and Q, represents the average flow between the two points. The slope of the line AB, drawn between the starting point A and the ending point B, is the average flow for the entire period.

To use the flow-mass curve for reservoir design, two lines parallel to line AB and tangent to the flow-mass curve are drawn. The first one, $A'B'$, is tangent to the mass curve at the highest tangent point C. The second one, $A''B''$ is tangent to the mass curve at the lowest tangent point D. The vertical difference between these two tangent lines, in cubic meters, is the storage volume required to release a constant flow rate. This constant release rate is equal to the slope of the line AB. A reservoir with a volume equal to AA'' at the start would be full at C and empty at D, with no spill (excess volume) or shortage (deficit). A reservoir that is empty at the start has water while the S curve remains above the AB line and is empty (show a deficit) when the S curve moves below that line. A reservoir that is full at the start will spill water (excess volume) as long as the inflow remains greater than the outflow (from A to C).

The draft rate (or demand rate) is the release rate required to fulfill downstream needs, such as irrigation or power generation. A line having a slope equal to the draft

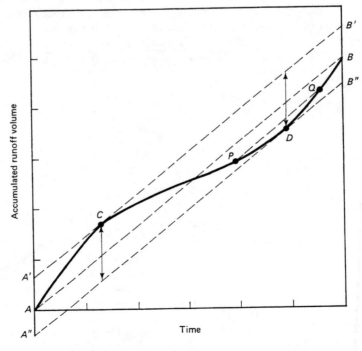

Figure 2-32. Flow-mass curve.

rate is the draft line. The draft rate need not be necessarily constant. In practice, reservoir withdrawals are variable, leading to a variable draft rate and variable draft line, which amounts to an outflow mass curve. The superposition of inflow and out-flow mass curves enables the detailed analysis of reservoir storage.

The *residual mass curve* is a plot of the differences between the S curve ordinates and the corresponding ordinates from line AB. The ordinates of the residual mass curve can be either positive or negative. The residual mass curve accentuates the peaks and valleys of the cumulative flow record.

Range is the difference between the maximum and minimum ordinates of the residual mass curve for a given period. Range analysis was pioneered by Hurst [28, 29], who proposed the following formula for the calculation of maximum range:

$$R = s \left(\frac{N}{2} \right)^{0.73} \tag{2-71}$$

in which R = reservoir storage volume required to guarantee a constant release rate equal to the mean of the data (annual runoff volume) over a period of N years and s = the standard deviation of the data (annual runoff volume) (Chapter 6).

Equation 2-71 was derived by curve-fitting data for a wide variety of natural phenomena. The exponent 0.73 was the mean of values varying between 0.46 and 0.96. A theoretical analysis based on the normal probability distribution (Chapter 6) showed that the exponent of Eq. 2-71 should be 0.5 instead of 0.73. This apparent discrepancy between theory and data, known as the *Hurst phenomenon,* has been the subject of numerous studies [32].

Geographical Variability of Streamflow.

Streamflow varies from one catchment to another and from one geographic region of a certain climate to another of a different climate. Two variables help describe the geographical variability of streamflow: (1) catchment area and (2) mean annual precipitation. Intuitively, the volume available for runoff is directly proportional to the catchment area. This, how-ever, is limited by the available precipitation. The mean annual precipitation deter-mines the type of climate and therefore has a direct bearing on the variability of streamflow.

The catchment area is important, not only because of the potential runoff vol-ume but also because larger catchments have milder overall gradients. This causes increased runoff diffusion, increasing the chances for infiltration and loss of surface water to groundwater. The net effect is a decrease in peak discharge per unit area.

The above reasoning is supported by data showing peak flows to be directly related to catchment area as in Eq. 2-52. Consequently, the peak discharge per unit area is

$$q_p = \frac{c}{A^m} \tag{2-72}$$

in which q_p = peak discharge per unit area, in cubic meters per second per square kilometer (or alternatively, in cubic feet per second per square mile); A = catchment area in square kilometers (or square miles); and c and m are empirical constants, with $m = 1 - n$. Since n (Eq. 2-52) is generally less than 1, it follows that m is also generally less than 1. Equation 2-72 confirms that peak discharge per unit area is

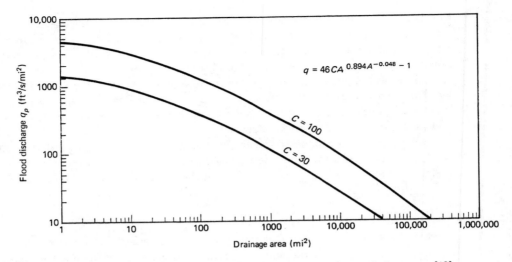

Figure 2-33. Creager curves: flood discharge per unit area versus drainage area [12].

inversely related to drainage area. An example of such a trend is given by the Creager curves, shown in Fig. 2-33 [12]:

$$q_p = 46CA^{0.894A^{-0.048}-1} \tag{2-73}$$

Values of C in the range 30–100 encompass most of the flood data compiled by Creager et al. [12]. This range can be taken as a measure of the regional variability of flood discharges. Equation 2-73, however, limits itself to providing a peak discharge per unit area, with no connotation of frequency attached to the calculated values.

QUESTIONS

1. Describe the frontal lifting of air masses. What is orographic lifting?
2. Describe the concept of rainfall frequency. What is the PMP? What is the PMF?
3. In what case is the isohyetal method preferred over the Thiessen polygons method?
4. When is an IDF curve used? When is a Depth-Duration-Frequency value used?
5. How does average annual precipitation affect climate?
6. When is the normal ratio method used to fill in missing precipitation records? What is a double-mass analysis?
7. What type of storm is likely to be subtantially abstracted by interception?
8. What factors affect the process of infiltration?
9. Compare the Horton and Philip infiltration formulas.
10. What type of application justifies the use of a ϕ-index?
11. In what case is depression storage likely to be important in runoff evaluation?
12. What is the basis of the energy budget method for determining reservoir evaporation?
13. What is albedo?
14. What assumptions did Penman use in deriving his evaporation formula?
15. What is transpiration? Why is it considered a hydrologic abstraction?

Basic Hydrologic Principles   Chap. 2

16. What is potential evapotranspiration? What is reference crop evapotranspiration?

17. What is the rationale for using evaporation formulas in the evaluation of evapotranspiration?

18. What are the various types of surface flow that can occur in nature?

19. What is a hypsometric curve? When is it used?

20. Derive the formula for equivalent slope (Eq. 2-55).

21. What is interflow? What is groundwater flow?

22. What is direct runoff? What is indirect runoff?

23. How does an ephemeral stream differ from an intermittent stream?

24. Why is the catchment's antecedent moisture important in flood hydrology?

25. What is catchment response? What is runoff concentration? What is runoff diffusion?

26. Why do single-storm streamflow hydrographs generally exhibit a long tail?

27. Why is the Manning equation preferred over the Chezy equation in practice? What is the advantage of the Chezy equation?

28. Discuss low flows and high flows in connection with arid and humid climates.

29. What is a rating curve? What are the various processes likely to affect a rating?

30. How can seasonal and annual streamflow variability be explained? What is the reason for the high peaks and low valleys of typical daily streamflows of small upland catchments?

31. What is a flow-duration curve? For what is it used?

32. What is a flow-mass curve? For what is it used?

33. What is the Hurst phenomenon?

34. How does peak discharge per unit area vary with catchment size? Why?

PROBLEMS

2-1. A 465-km^2 catchment has mean annual precipitation of 775 mm and mean annual flow of 3.8 m^3/s. What percentage of total precipitation is abstracted by the catchment?

2-2. A 9250-km^2 catchment has mean annual precipitation of 645 mm and mean annual flow of 37.3 m^3/s. What is the precipitation depth abstracted by the catchment?

2-3. Using the dimensionless temporal rainfall distribution shown in Fig. 2-3, calculate a hyetograph for an 18-cm, 12-h storm, defined at 1-h intervals.

2-4. A 100-km^2 catchment is instrumented with 13 rain gages located as shown in Fig. P-2-4. Immediately after a certain precipitation event, the rainfall amounts accumulated in each gage are as shown. Calculate the average precipitation over the catchment by the following methods: (a) average rainfall, (b) Thiessen polygons, and (c) isohyetal method.

2-5. A certain catchment experienced a rainfall event with the following incremental depths:

Time (h)	0–3	3–6	6–9	9–12
Rainfall (cm)	0.4	0.8	1.6	0.2

Determine: (a) the average rainfall intensity in the first 6 h, (b) the average rainfall intensity for the entire duration of the storm.

2-6. The following dimensionless temporal rainfall distribution has been determined for a local storm:

Time (%)	0	10	20	30	40	50	60	70	80	90	100
Depth (%)	0	5	10	25	50	75	90	95	97	99	100

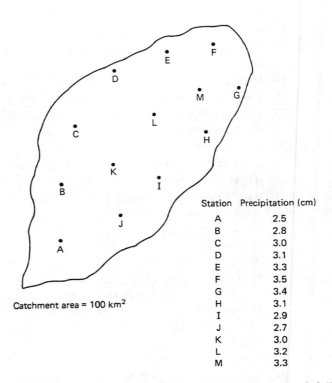

Station	Precipitation (cm)
A	2.5
B	2.8
C	3.0
D	3.1
E	3.3
F	3.5
G	3.4
H	3.1
I	2.9
J	2.7
K	3.0
L	3.2
M	3.3

Catchment area = 100 km²

Figure P-2-4. Catchment and rain-gage location: Problem 2-4.

Calculate a design hyetograph for a 12-cm, 6-h storm. Express in terms of hourly rainfall depths.

2-7. Given the following intensity-duration data, find the a and m constants of Eq. 2-5.

Intensity (mm/h)	50	30
Duration (h)	0.5	1.0

2-8. Given the following intensity-duration data, find the constants a and b of Eq. 2-6.

Intensity (mm/h)	60	40
Duration (h)	1	2

2-9. Construct a depth-area curve for the 6-h duration isohyetal map shown in Fig. P-2-9.

2-10. The precipitation gage for station X was inoperative during part of the month of January. During that same period, the precipitation depths measured at three index stations A, B, and C were 25, 28, and 27 mm, respectively. Estimate the missing precipitation data at X, given the following average annual precipitation at X, A, B, and C: 285, 250, 225, and 275 mm, respectively.

2-11. The precipitation gage for station Y was inoperative during a few days in February. During that same period, the precipitation at four index stations, each located in one of four quadrants (Fig. 2-11), is the following:

Quadrant	Precipitation (mm)	Distance (km)
I	25	8.5
II	28	6.2
III	27	3.7
IV	30	15.0

Estimate the missing precipitation data at station Y.

Basic Hydrologic Principles   Chap. 2

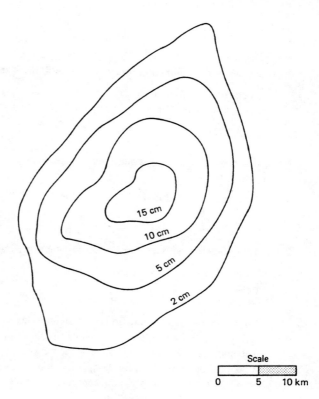

Figure P-2-9. Six-hour isohyetal map:
Problem 2-9.

Scale

0   5   10 km

2-12. The annual precipitation at station Z and the average annual precipitation at 10 neighboring stations are as follows:

Year	Precipitation at Z (mm)	10-station Average (mm)
1972	35	28
1973	37	29
1974	39	31
1975	35	27
1976	30	25
1977	25	21
1978	20	17
1979	24	21
1980	30	26
1981	31	31
1982	35	36
1983	38	39
1984	40	44
1985	28	32
1986	25	30
1987	21	23

Use double-mass analysis to correct for any data inconsistencies at station Z.

2-13. Calculate the interception loss for a storm lasting 30 min, with interception storage 0.3 mm, ratio of evaporating foliage surface to its horizontal projection $K = 1.3$, and evaporation rate $E = 0.4$ mm/h.

Problems

2-14. Show that $F = (f_o - f_c)/k$ in which F is the total infiltration depth above the $f = f_c$ line, Eq. 2-13.

2-15. Fit a Horton infiltration formula to the following measurements:

Time (h)	f (mm/h)
1	2.35
3	1.27
∞	1.00

2-16. Given the following measurements, determine the parameters of the Philip infiltration equation.

Time (h)	f (mm/h)
2	1.7
4	1.5

2-17. The following rainfall distribution was measured during a 12-h storm:

Time (h)	0–2	2–4	4–6	6–8	8–10	10–12
Rainfall Intensity (cm/h)	1.0	2.0	4.0	3.0	0.5	1.5

Runoff depth was 16 cm. Calculate the ϕ-index for this storm.

2-18. Using the data of Problem 2-17, calculate the W-index, assuming the sum of interception loss and depth of surface storage is $S = 1$ cm.

2-19. A certain catchment has a depression storage capacity of $S_d = 2$ mm. Calculate the equivalent depth of depression storage for the following values of precipitation excess: (a) 1 mm, (b) 5 mm, and (c) 20 mm.

2-20. Use the Meyer equation to calculate monthly evaporation for a large lake, given the following data: month of July, mean monthly air temperature 70°F, mean monthly relative humidity 60%, monthly mean wind speed at 25-ft height, 20 mi/h.

2-21. Derive the Penman equation (Eq. 2-36).

2-22. Use the Penman method to calculate the evaporation rate for the following atmospheric conditions: air temperature, 25°C; net radiation, 578 cal/cm^2/d, wind speed at 2-m above the surface, $v_2 = 150$ km/d; relative humidity, 50%.

2-23. Use the Penman method (together with the Meyer equation) to calculate the evaporation rate (in inches per day) for the following atmospheric conditions: air temperature, 70°F, water surface temperature, 50°F, daily mean wind speed at 25 ft height, $W = 15$ mi/h, relative humidity 30%, net radiation, $Q_n = 15$ Btu/in.^{2}/d. Assume a large lake to use Eq. 2-28b.

2-24. Use the Blaney-Criddle method (with corrections due to Doorenbos and Pruitt) to calculate reference crop evapotranspiration during the month of July for a geographic location at 40°N, with mean daily temperature of 25°C. Assume high actual insolation time, 70% minimum relative humidity, and 1 m/s daytime wind speed.

2-25. Use the Thornthwaite method to calculate the potential evapotranspiration during the month of May for a geographic location at 35°N, with the following mean monthly temperatures, in degrees Celsius.

Jan	Feb	Mar	Apr	May	Jun	Jul	Aug	Sep	Oct	Nov	Dec
6	8	10	12	15	20	25	20	16	12	10	8

2-26. Use the Priestley and Taylor formula to calculate the potential evapotranspiration for a site with air temperature of 15°C and net radiation of 560 cal/cm^2/d.

2-27. The following data have been obtained by planimetering a 135-km^2 catchment.

Elevation (m)	Subarea above indicated elevation (km^2)
1010	135
1020	85
1030	65
1040	30
1050	12
1060	4
1070	0

Calculate a hypsometric curve for this catchment.

2-28. Given the following longitudinal profile of a river channel, calculate the following slopes: (a) S_1, (b) S_2, and (c) S_3.

Distance (km)	0	50	100	150	200	250	300
Elevation (m)	10	30	60	100	150	220	350

2-29. The bottom of a certain 100-km reach of a river can be described by the following longitudinal profile:

$$y = 100e^{-0.00001x}$$

in which y = elevation with reference to an arbitrary datum in meters; and x = horizontal distance measured from upstream end of reach in meters. Calculate the S_2 slope.

2-30. Given the following record of daily precipitation, calculate the antecedent precipitation index. Assume the starting value of the index to be equal to 0 and recession constant $K = 0.85$.

Day	Precipitation (cm)
1	0.0
2	0.1
3	0.3
4	0.4
5	0.2
6	0.0
7	0.0
8	0.7
9	0.8
10	0.9
11	1.2
12	0.5
13	0.0
14	0.0

2-31. A 35-ha catchment experiences 5 cm of precipitation, uniformly distributed in 2 h. If the time of concentration is 1 h, what is the maximum possible flow rate at the catchment outlet?

2-32. Calculate hourly ordinates of a gamma hydrograph with the following characteristics: peak flow, 1000 m^3/s; baseflow, 0 m^3/s; time-to-peak, 3 h; and time-to-centroid, 6 h.

2-33. The following data have been measured in a river: mean velocity V = 1.8 m/s, hydraulic radius R = 3.2 m, channel slope S = 0.0005. Calculate the Manning and Chezy coefficients.

2-34. The Chezy coefficient for a wide channel is 49 m$^{1/2}$/s and the bottom slope is 0.00037. What is the Froude number of the uniform (i.e., steady equilibrium) flow?

2-35. The flow duration characteristics of a certain stream can be expressed as follows: $Q = (950/T) + 10$, in which $Q =$ discharge in cubic meters per second, and $T =$ percent time, restricted to the range 1–100%. What flow can be expected to be exceeded (a) 90% of the time, (b) 95% of the time, and (c) 100% of the time?

2-36. A reservoir has the following average monthly inflows, in cubic hectometers:

Jan	Feb	Mar	Apr	May	Jun	Jul	Aug	Sep	Oct	Nov	Dec
30	34	35	48	72	85	72	55	51	40	34	32

Determine the reservoir storage volume required to release a constant draft rate throughout the year.

2-37. The analysis of 43 y of runoff data at a reservoir site in a large river has led to the following: mean annual runoff volume, 24 km³; standard deviation, 7 km³. What is the reservoir storage volume required to guarantee a constant release rate equal to the mean of the data?

2-38. Calculate the peak discharge for a 1000-mi² drainage area using the Creager formula (Eq. 2-73) with (a) $C = 30$ and (b) $C = 100$.

REFERENCES

1. American Society of Civil Engineers. (1949). "Hydrology Handbook," *Manual of Engineering Practice No.* 28.
2. American Society of Civil Engineers. (1960). "Design and Construction of Sanitary and Storm Sewers," *Manual of Engineering Practice No.* 37.
3. Barnes, H. H., Jr. (1967). "Roughness Characteristics of Natural Channels," *U.S. Geological Survey Water Supply Paper No.* 1849.
4. Blaney, H. F., and W. D. Criddle. (1950). "Determining Water Requirements in Irrigated Areas from Climatological and Irrigation Data," *USDA Irrigation and Water Conservation, SCS TP-96,* August.
5. Blaney, H. F., and W. D. Criddle. (1962). "Determining Consumptive Use of Irrigation Water Requirements," *USDA Technical Bulletin No.* 1275, Washington, D.C.
6. Bowen, I. S. (1926). "The Ratio of Heat Losses by Conduction and by Evaporation from Any Water Surface," *Physics Review,* Vol. 27, pp. 779–787.
7. Bras, R., and I. Rodriguez-Iturbe. (1985). *Random Functions and Hydrology.* Reading, Mass.: Addison-Wesley.
8. Bruce, J. P., and G. K. Rodgers. (1962). "Water Balance in the Great Lakes System, Great Lakes Basin," *American Association for the Advancement of Science, Publication No.* 71, Washington, D.C.
9. Bruce, J. P., and R. H. Clark. (1966). *Introduction to Hydrometeorology.* Elmsford, NY: Pergamon Press.
10. Chow, V. T. (1964). *Handbook of Applied Hydrology.* New York: McGraw-Hill.
11. Cook, H. L. (1946). "The Infiltration Approach to the Calculation of Surface Runoff," *Transactions, American Geophysical Union,* Vol. 27, No. V, October, pp. 726–747.
12. Creager, W. P., J. D. Justin, and J. Hinds. (1945). *Engineering for Dams.* Vol. 1, *General Design.* New York: John Wiley.
13. Dalton, J. (1802). "Experimental Essays on the Constitution of Mixed Gases; on the Force of Steam or Vapor from Water and Other Liquids, Both in a Torricellian Vacuum and in Air; on Evaporation; and on the Expansion of Gases by Heat," *Manchester Literary and Philosophical Society Proceedings,* Vol. 5, pp. 536–602.
14. Doorenbos J., and W. O. Pruitt. (1977). "Guidelines for Predicting Crop Water Requirements," *Irrigation and Drainage Paper No.* 24, FAO, Rome.

15. Dunne, T., and L. B. Leopold. (1978). *Water in Environmental Planning*. San Francisco: Freeman and Co.

16. Environmental Data Service, Environmental Science Services Administration, U.S. Department of Commerce, "Climatic Atlas of the United States," 1968.

17. Fread, D. L. (1985). "Channel Routing," in *Hydrological Forecasting*, M. G. Anderson and T. P. Burt, eds. New York: John Wiley.

18. Green, W. H. and G. A. Ampt. (1911). "Studies on Soil Physics. 1. The Flow of Air and Water Through Soils," *Journal of Agricultural Soils*, Vol. 4, pp. 1-24.

19. Hathaway, G. A. (1945). "Design of Drainage Facilities," *Transactions*, ASCE, Vol. 110, pp. 697-730.

20. Hicks, W. I. (1944). "A Method of Computing Urban Runoff," *Transactions*, ASCE, Vol. 109, pp. 1217-1233.

21. Hillel, D. (1971). *Soil and Water, Physical Principles and Processes*. New York: Academic Press.

22. Horton, R. E. (1919). "Rainfall Interception," *Monthly Weather Review*, Vol. 47, September, pp. 603-623.

23. Horton, R. E. (1932). "Drainage Basin Characteristics," *Transactions, American Geophysical Union*, Vol. 13, pp. 350-361.

24. Horton, R. E. (1933). "The Role of Infiltration in the Hydrologic Cycle," *Transactions, American Geophysical Union*, Vol. 14, pp. 446-460.

25. Horton, R. E. (1941). "Sheet Erosion, Present and Past," *Transactions, American Geophysical Union*, Vol. 22, pp. 299-305.

26. Houghton, H. G. (1959). "Cloud Physics," *Science*, Vol. 129, No. 3345, February, pp. 307-313.

27. Howe, R. H. L. (1960). "The Application of Aerial Photographic Interpretation to the Investigation of Hydrologic Problems," *Photogrametric Engineering*, Vol. 26, pp. 85-95.

28. Hurst, H. E. (1951). "Long-term Storage Capacity of Reservoirs," *Transactions, ASCE*, Vol. 116, pp. 770-799.

29. Hurst, H. E. (1956). "Methods of Using Long-term Storage in Reservoirs," *Proceedings, Institution of Civil Engineers*, London, England, Vol. 5, pt. 1, No. 5, September, pp. 519-543.

30. Kennedy, E. J. (1984). "Discharge Ratings at Gaging Stations," U.S. Geological Survey, *Techniques of Water Resources Investigations*, Book 3, Chapter A10.

31. Kirpich, Z. P. (1940). "Time of Concentration of Small Agricultural Watersheds," *Civil Engineering*, Vol. 10, June, p. 362.

32. Klemes, V. (1974). "The Hurst Phenomenon: A Puzzle?" *Water Resources Research*, Vol. 10, No. 4, pp. 675-688.

33. Knapp, B. J. (1978). "Infiltration and Storage of Soil Water," in *Hillslope Hydrology*, M. J. Kirkby, ed. New York: John Wiley.

34. Kohler, M. A., T. J. Nordenson, and W. E. Fox. (1955). "Evaporation from Pans and Lakes," Weather Bureau, U.S. Department of Commerce, *Research Paper No. 38*.

35. Kohler, M. A., and M. M. Richards. (1962). "Multicapacity Basin Accounting for Predicting Runoff from Storm Precipitation," *Journal of Geophysical Research*, Vol. 67, pp. 5187-5197.

36. Lamoreaux, W. W. (1962). "Modern Evaporation Formula Adapted to Computer Use," *Monthly Weather Review*, Vol. 90, No. 1, pp. 26-28.

37. Langbein, W. B., et al. (1947). "Topographic Characteristics of Drainage Basins," *U.S. Geological Survey Water Supply Paper No.* 968-C.

38. Linsley, R. K., M. A. Kohler, and J. L. H. Paulhus. (1982). *Hydrology for Engineers*, 3d. Ed. New York: McGraw-Hill.

39. Meyer, A. F. (1915). "Computing Runoff from Rainfall and Other Physical Data," *Transactions*, ASCE, Vol. 79, pp. 1056-1224.

40. Meyer, A. F. (1944). "Evaporation from Lakes and Reservoirs," *Minnesota Resources Commission,* St. Paul, Minnesota, June. 41. Monteith, J. L. (1959). "The Reflection of Short Wave Radiation by Vegetation," *Quarterly Journal of the Royal Meteorological Society,* Vol. 85, pp. 586–592.

42. NOAA *Hydrometeorological Report No. 39.* (1963). "Probable Maximum Precipitation in the Hawaiian Islands."

43. NOAA *Hydrometeorological Report No. 43.* (1966). "Probable Maximum Precipitation, Northwest States."

44. NOAA *Hydrometeorological Report No. 36.* (1969). "Interim Report, Probable Maximum Precipitation in California," Oct. 1961, reprinted with revisions of Oct. 1969.

45. NOAA *Hydrometeorological Report No. 49.* (1977). "Probable Maximum Precipitation Estimates, Colorado River and Great Basin Drainages."

46. NOAA *Hydrometeorological Report No. 51.* (1978). "Probable Maximum Precipitation Estimates, United States East of the 105th Meridian."

47. NOAA *Hydrometeorological Report No. 52.* (1982). "Application of Probable Maximum Precipitation Estimates, United States East of the 105th Meridian."

48. NOAA *Hydrometeorological Report No. 55.* (1984). "Probable Maximum Precipitation Estimates-United States Between the Continental Divide and 103rd Meridian."

49. NOAA National Weather Service. (1972). "National Weather Service River Forecast System. Forecast Procedures," *Technical Memorandum NWS-HYDRO 14,* Dec., pp. 3.1–3.14.

50. NOAA National Weather Service. (1973). "Atlas 2: Precipitation Atlas of the Western United States."

51. NOAA National Weather Service. (1977). "Five- to 60-Minute Precipitation Frequency for the Eastern and Central United States," *Technical Memorandum NWS HYDRO-35.*

52. Osmolski, Z. (1985). "Estimating Potential Evapotranspiration from Climatological Data in an Arid Environment," Ph.D. Diss., School of Renewable and Natural Resources, University of Arizona, Tucson.

53. Overton, D. E., and M. E. Meadows. (1976). *Stormwater Modeling.* New York: Academic Press.

54. Papadakis, C. N., and M. N. Kazan. (1987). "Time of Concentration in Small Rural Watersheds," *Proceedings of the Engineering Hydrology Symposium,* ASCE, Williamsburg, Virginia, August 3–7, pp. 633–638.

55. Paulhus, J. L. H., and M. A. Kohler. (1952). "Interpolation of Missing Precipitation Records," *Monthly Weather Review,* Vol. 80, No. 8, August, pp. 129–133.

56. Penman, H. L. (1948). "Natural Evaporation from Open Water, Bare Soil and Grass," *Proceedings of the Royal Society,* London, Vol. 193, pp. 120–145.

57. Penman, H. L. (1952). "The Physical Basis of Irrigation Control," *Proceedings, 13th International Horticulture Congress,* London.

58. Philip, I. R. (1957), (1958). "The Theory of Infiltration," *Soil Science,* Vol. 83, 1957, pp. 345–357; and 1958, pp. 435–458.

59. Priestley, C. H. B., and R. J. Taylor. (1972). "On the Assessment of Surface Heat Flux and Evaporation Using Large Scale Parameters," *Monthly Weather Review,* Vol. 100, pp. 81–92.

60. Ragunath, H. M. (1985). *Hydrology.* New Delhi: Halsted Press.

61. Rippl, W. (1883). "The Capacity of Storage Reservoirs for Water Supply," *Proceedings, Institution of Civil Engineers,* London, England, Vol. 71, pp. 270–278.

62. Salas, J. D., J. W. Delleur, V. Yevjevich, and W. L. Lane. (1980). *Applied Modeling of Hydrologic Time Series.* Littleton, Colo.: Water Resources Publications.

63. Sartor, D. (1954). "A Laboratory Investigation of Collision Efficiencies, Coalescence and Electrical Charging of Simulated Cloud Particles," *Journal of Meteorology,* Vol. 11, No. 2, April, pp. 91–103.

64. Schumm, S. A. (1956). "Evolution of Drainage Systems and Slopes in Perth Amboy, New Jersey," *Geological Society of America Bulletin*, Vol. 67, pp. 597–646.

65. Stanhill, G. (1965). "The Concept of Potential Evapotranspiration in Arid Zone Agriculture," *Proceedings, Montpelier Symposium in Arid Zone Research*, UNESCO, Paris, Vol. 25, pp. 109–171.

66. Tholin, A. L., and C. J. Keifer. (1960). "The Hydrology of Urban Runoff," *Transactions, ASCE*, Vol. 125, pp. 1308–1379.

67. Thornthwaite, C. W., H. G. Wilm, et al. (1944). "Report of The Committee on Transpiration and Evaporation, 1943–1944," *Transactions, American Geophysical Union*, Vol. 25, pt. V, pp. 683–693.

68. USDA Soil Conservation Service. (1985). *National Engineering Handbook, No. 4: Hydrology*, Washington, D.C.

69. U.S. Geological Survey. (1952). "Water Loss Investigations: Vol. 1-Lake Hefner Studies," *Circular No.* 229.

70. U.S. Geological Survey. (1954). "The Water Budget Control," in Water Loss Investigations, Lake Hefner Studies, *Professional Paper No.* 269.

71. Viessman, W., J. W. Knapp, G. L. Lewis, and T. E. Harbaugh. (1977). *Introduction to Hydrology*, 2d. ed. New York: Harper & Row.

72. Weather Bureau, U.S. Department of Commerce. (1955). "Rainfall Intensity-Duration-Frequency Curves," *Technical Paper No.* 25.

73. Weather Bureau, U.S. Department of Commerce. (1960). "Generalized Estimates of Probable Maximum Precipitation for the United States West of the 105th Meridian," *Technical Paper No.* 38.

74. Weather Bureau, U.S. Department of Commerce. (1962). "Generalized Estimates of Probable Maximum Precipitation and Rainfall Frequency Data for Puerto Rico and Virgin Islands for Areas to 400 Square Miles, Durations to 24 Hours, and Return Periods from 1 to 100 Years," *Technical Paper No.* 42.

75. Weather Bureau, U.S. Department of Commerce. (1962). "Rainfall Frequency Atlas of Hawaiian Islands for Areas to 200 Square Miles, Durations to 24 Hours, and Return Periods from 1 to 100 Years," *Technical Paper No.* 43.

76. Weather Bureau, U.S. Department of Commerce. (1963). "Rainfall Frequency Atlas of the United States for Durations from 30 Minutes to 24 Hours and Return Periods from 1 to 100 Years," *Technical Paper No.* 40.

77. Weather Bureau, U.S. Department of Commerce. (1963). "Probable Maximum Precipitation and Rainfall Frequency Data for Alaska for Areas to 400 Square Miles, Durations to 24 Hours, and Return Periods from 1 to 100 Years," *Technical Paper No.* 47.

78. Weather Bureau, U.S. Department of Commerce. (1964). "Two-to-Ten Day Precipitation for Return Periods of 2 to 100 Years in the Contiguous United States," *Technical Paper No.* 49.

79. Yevjevich, V. (1972). *Stochastic Processes in Hydrology*. Fort Collins, Colo.: Water Resources Publications.

HYDROLOGIC MEASUREMENTS

Engineering hydrology is based on analysis and measurements. Measurements are necessary in order to complement and verify the analysis. Hydrologic measurements are usually performed in the field, using equipment and techniques specifically designed to measure a variable characterizing a certain phase of the hydrologic cycle. For instance, rainfall is measured with rain gages, evaporation is measured with evaporation pans, and streamflow is measured using stream-gaging techniques.

Measurements are closely related to hydrologic analysis. In some cases they are an integral part of it; in others, they serve to support it. For instance, statistical hydrology is not possible without measurements. In flood frequency analysis, a historical flow record is necessary in order to define the properties of the predictive equations. With parametric models, measurements aid in parameter estimation, increasing model reliability. Deterministic and conceptual models also benefit from hydrologic measurements.

This chapter is divided into five sections. Section 3.1 describes precipitation measurements and Section 3.2 deals with snowpack measurements. Section 3.3 describes evaporation and evapotranspiration measurements and Section 3.4 discusses infiltration and soil moisture measurements. Streamflow measurements are discussed in Section 3.5.

3.1 PRECIPITATION MEASUREMENTS

Precipitation is measured with rain gages. A rain gage is an instrument that captures precipitation and measures its accumulated volume during a certain time period. The precipitation depth for the given period is equal to the accumulated volume divided by the collection area of the gage. The average precipitation intensity is equal to the precipitation depth divided by the length of the period.

Any receptacle that has vertical sides and is open to the air is a defacto rain gage and can provide valuable information on accumulated rainfall during a storm. Two such measurements, however, are not directly comparable unless the receptacles are of the same size and shape and similarly exposed. To increase the utility of the measurements, it is necessary to use standard equipment and procedures.

Rain gages can be of two types: (1) nonrecording or (2) recording. A *nonrecording gage* measures the total rainfall depth accumulated during one time period, usually 1 d. In the United States, the standard nonrecording rain gage used by the National Weather Service has a funnel-shaped collector element or receiver of 8-in. top diameter located inside an overflow can. Rain is caught by the collector and funneled into a measuring tube. The cross-sectional area of the measuring tube is one-tenth that of the collector. Therefore, rainfall depths are amplified ten times as they pass from the collector into the measuring tube, increasing the accuracy of the measurement.

A recording rain gage records the time it takes for rainfall depth accumulation. Therefore, it provides not only a measure of rainfall depth but also of rainfall intensity. The slope of the curve showing accumulated rainfall depth versus time is a measure of the instantaneous rainfall intensity. Recording rain gages rely on one of the following devices: (1) a tipping bucket, (2) a weighing mechanism, or (3) a float chamber.

The *tipping-bucket gage* features a two-compartment receptacle (i.e., the bucket) pivoted on a knife edge. The device is calibrated so that when one of the compartments is full (with a fixed amount of rain) and the other is empty, the bucket overbalances and tips. At the start, rain is funneled into one of the compartments, which is positioned for filling. As rainfall continues to fill this first compartment, the second remains empty. When the first compartment is full, the bucket tips, emptying its contents into a reservoir and at the same time placing the second compartment in filling position. The tipping closes an electric circuit, which drives a pen that records on a strip chart affixed to a clock-driven revolving drum. Thus, each electrical contact representing a specific amount of rain is recorded. The alternate filling and emptying of the two compartments continues until rainfall ceases.

The tipping-bucket gage has a few disadvantages. During periods of intense rainfall, some of the rain may not be measured while the bucket is tipping. In addition, the record consists of a series of steps rather than being a smooth curve, and the gage is not suitable for measuring snow. Nevertheless, tipping-bucket gages are durable, simple to operate, and of good overall reliability.

A *weighing gage* has a device that weighs the rain or snow collected in a bucket (Fig. 3-1). As it fills with precipitation, the bucket moves downward and its movement is transmitted to a pen on a strip-chart recorder. This type of gage is useful in cold climates where it is necessary to record both rainfall and snowfall. However, weighing gages have some disadvantages. Among them are wind action on the bucket, which produces erratic traces on the recording chart, and the overall lack of sensitivity of the measurement.

Float gages are essentially water-level gages. A float located inside a chamber is

Figure 3-1 Weighing rain gage (photo courtesy of NOAA).

connected to a pen on a strip-chart recorder. The float rises as the collected rainwater enters the chamber, and the rise of the float is recorded on the chart. Some float gages are limited to the capacity of the chamber. Others are equipped with a self-starting siphoning device that empties the chamber when it becomes full and returns the pen to the zero position on the strip chart. The use of float gages is limited to nonfreezing ambient temperatures, although heaters and other similar devices have been used in an attempt to overcome the problem of freezing. Oil and mercury, which have freezing temperatures below that of water, have also been used inside the chamber. The siphoning action of the float gage can cause serious losses of rain during severe storms.

Errors in Measuring Precipitation Data

The water collected by a rain gage is only a small sample of the precipitation that has fallen in a certain area. Whether this sample is representative of the average precipitation over the area remains to be determined by further analysis.

A series of rain gages located within a drainage area constitutes a rain-gage network. The density of the network is the number of rain gages per square kilometer (or square mile). The error of rainfall measurements can be investigated by studying the spatial averages computed from networks of different densities [1, 13, 23]. In general, sampling errors increase with an increase in rainfall depth. Conversely, sampling errors decrease with an increase in network density, storm duration, and catchment area.

An important question in engineering hydrology is whether errors in precipitation measurement can serve to compound the errors inherent in the use of rainfall-runoff simulation models. The answer to this question is elusive. Limited data by Johanson [11] indicates that the error variability in precipitation measurements is likely to be less than the error variability in model calibration. Calibration is the process by which model parameters are adjusted to match measured and simulated flows.

Precipitation Measurements Using Telemetry

Self-reporting rain gages (or rainfall sensors) have automatic data transmittal capabilities. These rain gages use automatic radio transmitters (telemeters) to broadcast rainfall measurements from a remote station to a central station in real time, i.e., during the storm event. The advantage of a telemetric station is that it shortens the time that would otherwise be required to gather rainfall data. In certain cases, especially when speed of processing is of utmost importance, a network of rainfall sensors linked by telemetry may be the only practical means of collecting rainfall data. Applications of telemetric rainfall sensors are usually found in connection with operational hydrology and real-time flood forecasting.

A typical design of a remote self-reporting rain-gage station consists of a cylindrical pipe 12 ft high and 1 ft in diameter (Fig. 3-2). A tipping-bucket mechanism contained within the pipe generates a digital input signal whenever 1 mm of rainfall depth drains through the funnel assembly. The signal is transmitted to a central station, and time of reception is recorded.

The link between the remote station and the central station is usually established by radio, telephone, or a combination of both. Where radio frequencies are scarce, telephone lines can be used to transmit the data. The remote rainfall station is interfaced to a telephone line either through a modem (MOdulator-DEModulator) or an acoustic coupler. The latter enables the remote station to be called from any telephone, whether it is equipped with a modem or not.

Radio transmission may take the form of a very high frequency (VHF) or ultrahigh frequency (UHF) link for short distances, or high frequency (HF) for very long distances. VHF and UHF frequencies behave in a manner similar to light and, therefore, cannot travel far beyond the horizon. The best reception is obtained when the transmitting and receiving antennas are within line of sight of each other. With the use of high masts, a span of 40 km or more can be achieved. Transmitter power ranges from 5 W for short distances to 25 W for longer distances.

Very long distance transmissions require repeating stations at 30- to 60-km intervals, but these are expensive and difficult to maintain. An alternative is to use HF radio, by which great distances can be spanned through a series of reflections between ionosphere and ground. Normal transmissions have a span of a few hundred kilometers. These HF links, however, are subject to variations in signal strength and are susceptible to interference with other transmitters, making their use more difficult than VHF and UHF. Special techniques can be used to improve the quality of the transmission, for instance those used in connection with real-time flood forecasting in the Pantanal region of Mato Grosso, Brazil [25].

Another way to transmit data through radio waves is by the use of satellites. A remote station can transmit data to a satellite for retransmittal to a ground receiving station. The radio link operates in the UHF range and requires only a few watts of power. To transmit data via a satellite, the station is linked to a commercially made data collection platform. This device stores the day's data in its solid state memory for transmission every 24 h, although hourly transmissions are also possible.

Precipitation Measurements Using Radar. Weather radar systems are a potentially powerful tool for measuring the temporal and spatial variability of rain-

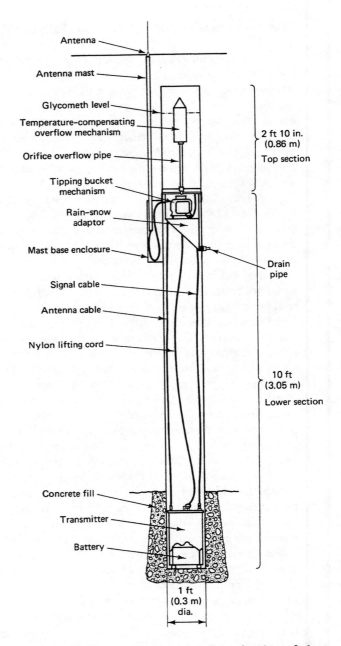

Figure 3-2 Self-reporting rain gage station.

storms. A radar system operates by emitting a regular succession of pulses of electromagnetic radiation from its antenna. The pulses are on the order of 1 μs, and the system emits approximately 1000 of these pulses every second. Between pulses, the system's antenna becomes a receiver of the energy of the emitted pulses scattered by various targets. These returned signals are transformed into a visual display on the radar scope.

For spherical objects (e.g. raindrops), the power received can be expressed as follows:

$$P = \frac{K \Sigma n D^6}{\lambda^4 R^2}$$ (3-1)

in which P = power received, n = number of drops, D = diameter of drops, λ = wavelength of the radiation, R = distance (range) from the radar, and K = a factor that depends on the power of the transmitted signal, antenna size and shape, and properties of the scattering particles.

Weather radars have wavelengths in the range of 3 to 10 cm. Following Eq. 3-1, a 3-cm radar returns about 120 times as much power as that returned by a 10-cm radar. Hence a 3-cm radar can detect weak targets such as the small droplets associated with very light rain, whereas a 10-cm radar can be used to sense much heavier rains.

Attenuation, caused by absorption and scattering by clouds and precipitation, can also affect the performance of the radar. Attenuation is a function of radar wavelength, being greater for shorter wavelengths.

The reduction in power received P with distance R (range) is a constant for a given system and can be adjusted to make distant targets show the same brightness as closer targets of similar character. Since power P is proportional to the sixth power of the drop diameter D, radars in the 3- to 10-cm wavelength size can easily sense rain-size droplets and not sense other size particles at all. A 10-cm radar is used for detecting highly intensive storms, which are likely to produce extreme floods. For light rains or snow sensing, a shorter wavelength radar is preferable.

The radar reflectivity ($\Sigma n D^6$) can be empirically related to precipitation intensity as follows:

$$Z = AI^B$$ (3-2)

in which Z = radar reflectivity; I = precipitation intensity; and A and B are empirical constants. The values of A and B depend upon the type of precipitation being observed. Many values have been specified; those most often used are $A = 200$ and $B = 1.6$ [4].

Several types of error are possible when using radar to sense precipitation. For instance, the radar beam can overshoot shallow precipitation at long ranges, missing the target. Another source of error is the presence of low level evaporation beneath the radar beam, as well as several other meteorological factors [7]. Uncertainties in radar sensing of precipitation can be resolved by calibrating the system with a rain gage. This is usually accomplished by fixing exponent B (in Eq. 3-2) at a certain value (for instance $B = 1.6$), and using rain-gage data to derive a value of coefficient A.

3.2 SNOWPACK MEASUREMENTS

Measurements of snow include both newly fallen snow and snow accumulation, i.e., snowpack. Snowpack measurements are expressed in terms of *water equivalent*, i.e., the depth of water that is obtained after melting a certain depth of snowpack. Water equivalent is a measure of the amount of water remaining in storage in the snowpack. Water-equivalent data are useful in water-yield forecasts, since they integrate in one measurement both snowfall and snowmelt.

As with rainfall measurements, snowfall or its water equivalent must be measured by sampling at several points and averaging point values to obtain a representative value of the snowpack. A simple method for measuring snowfall is to use a *snowboard*. The snowboard is placed on the ground or on old snow surface to permit the accumulation of new snow on top of it. An inverted rain-gage cylinder is used to isolate a core of the new snow, which is then melted and measured in the same way as rainfall. By measuring each fall of snow in this manner and replacing the clean board ready to receive fresh snowfalls, accumulated total snowfall throughout the season may be known at any time. Such measurements are fairly reliable, provided that they are taken soon after each snowfall and that the snow on the board has not been subjected to drifting, melting, or evaporation.

The density of a snow sample is the ratio of the volume of melt water to the initial volume of the sample, expressed as a percentage. Snow density in the typical snowpack varies widely, both within the vertical structure of the snowpack and with time.

Snow stakes are often used to measure snow accumulation. Water equivalent of the snowpack can be determined from depth measurements by using known densities of snow, obtained under similar environmental conditions. Snow stakes are also used where it is impractical to obtain water equivalents by direct sampling.

Direct sampling of water equivalent is accomplished by the use of snow samplers. The *Mount Rose sampler* is commonly used in the United States. It consists of a tube fitted with a cutter on one edge. The tube has an inside diameter of 1.485 in., so that a core weighing 1 oz is equivalent to 1 in. of water. Sampling consists of pushing the tube vertically into the snow to full snow depth, withdrawing the tube with the snow content, and weighing the contents.

Snow Courses

When performing snow measurements, common practice is to sample water equivalent at a number of points along an established line called a *snow course*. Snow courses are selected with the objective of obtaining representative data from a given area. The number of snow courses varies, depending on terrain features and meteorological characteristics. Site selection considers the following aspects: (1) meteorological conditions related to storm experience, (2) position with respect to large-scale topographic features, (3) position with regard to local environmental features, such as wind, exposure, orientation, and ground slope, and (4) site conditions, including local drainage and presence of brush or rocks. In addition, snow courses are positioned so that they are representative not only of snowfall but also of snowmelt.

The number of sample points varies depending upon the consistency in the spatial distribution of snow. Sample points should avoid the effect of trees, boulders, and other obstructions. If there is little protection from the wind, sample points are spread over a wide area to average out variations due to drifting. In general, five snow-course sample points are adequate for well-positioned snow courses that have a minimum of irregularities caused by drifting or wind erosion and a smooth ground surface clear of obstructions. When conditions are less than ideal, additional snow-course points are required for adequate sampling of the water equivalent.

Radioisotope Snow Measurements

Specially designed radioisotope snow gages with telemetering capabilities are used to measure the water equivalent of snowpack at remote, unattended sites. The equipment measures water equivalent by correlating it with the attenuation of a gamma ray emission (cobalt 60) as it travels through the snow from source to detector. The original equipment had the source positioned at ground level, with the detector 15 ft above it [28]. Later models were reversed to minimize temperature effects on the detector, with the detector positioned at ground level and the source 15 ft above it [16]. Recalibration of radioisotope snow equipment at regular intervals is necessary to guarantee the accuracy of the measurement.

Profiling radioactive snow gages are used to determine the variation of water equivalent within the snowpack depth [21]. These gages consist of a gamma photon source and detector, which are moved synchronously through vertical tubes located about 60 cm apart in the snowpack. The gages are used for measuring temporal and spatial variations in snowpack properties.

Determination of Catchment Water Equivalent

Point values from all snow courses representative of an area are used to determine *catchment water equivalent*. The relationship between catchment water equivalent and point values is dependent upon the location of the snow courses. When snow courses are distributed equally throughout the range of elevations, an arithmetic average of point values usually provides a satisfactory value of catchment water equivalent. Refinements can be obtained by weighing data from each snow course in proportion to the percentage of catchment area covered by it.

Elevation is an important factor in converting point measurements into catchment water equivalent. Generally, snow courses tend to be concentrated at higher elevations, and therefore an arithmetic average is not appropriate. An alternative is to develop a *snow chart*, a plot showing the variation of water equivalent with elevation. This chart is used together with the catchment's area-elevation curve (i.e., the hypsometric curve, Section 2.3). The catchment's elevation difference is divided into several equal increments. For each elevation increment, a subarea is obtained from the area-elevation curve, and a corresponding water equivalent is obtained from the snow chart. A water equivalent value representative of the entire catchment can be obtained by weighing the individual water equivalents in proportion to their respective subareas.

Sources of Snow Survey Data

The Federal-State-Private Cooperative Snow Survey System publishes snow survey data on a regular basis. The system is coordinated by the Soil Conservation Service, with jurisdiction in the Western United States. The State of California, however, maintains its own snow survey system, administered by its Department of Water Resources. In the Eastern United States, several federal, state, and private agencies perform snow surveys. These are used for various purposes, including fish and wildlife management, recreational uses, and highway maintenance. Federal agencies collecting snow survey data are the National Weather Service and the U.S. Geological Survey.

3.3 EVAPORATION AND EVAPOTRANSPIRATION MEASUREMENTS

A practical way to measure evaporation directly is by the use of an evaporation pan. The pan exposes a free water surface to the air, and the evaporation rate is determined by measuring the water loss during one time period, usually 1 d. The evaporation rate measured by the pan, however, is generally not the same as that of a lake or reservoir exposed to similar meteorological conditions.

The difference is attributed to the pan installation and exposure. For example, a pan installed on supports above the ground surface is subject to extra radiation on the sides. On the other hand, a buried pan is subject to appreciable heat exchange between it and the surrounding soil. These and other factors contribute to make the overall heat balance for the pan a complex phenomenon.

The factors responsible for the discrepancy usually combine to produce a pan measurement that is greater than the actual lake or reservoir evaporation. Therefore, a correction factor is applied to the pan evaporation measurement in order to arrive at the actual value of lake or reservoir evaporation. This correction factor is referred to as the *pan coefficient*.

NWS Class A Pan

The National Weather Service Class A pan is the most widely used evaporation pan in the United States. This pan has been recommended as a standard for evaporation measurements by the World Meteorological Organization.

The Class A pan is made of unpainted galvanized iron, has a diameter of 122 cm (4 ft) and a height of 25.4 cm (10 in.) and is mounted about 15 cm (6 in.) above the ground on supports which permit a free flow of air around and under the pan (Fig. 3-3). Water loss is determined by daily measurements of water level using a micrometer hook gage installed in a stilling well set inside the pan. The pan is initially filled to a height of 20 cm (8 in.) and is refilled when the water level has fallen below 17.5 cm (7 in.). Daily evaporation is computed as the difference between two successive observations, corrected to account for any intervening precipitation (measured in a nearby gage). An alternate procedure is to add a measured amount of water daily to bring the water level in the pan up to a fixed point in the stilling well. This procedure permits a more accurate measurement of water loss and assures that the pan has the proper water level at all times.

Because of interception of solar radiation by the sides, the Class A pan, like other similarly exposed pans, usually exaggerates the actual lake or reservoir evapora-

Figure 3-3 Class A evaporation pan (photo courtesy of NOAA).

tion. Therefore, its pan coefficient is less than 1, with an annual average of approximately 0.7 (see Table 2-6).

Little is known about the spatial variability of evaporation. However, it seems likely that it is not as large as that of precipitation. If this is the case, a network of much lesser density would be required for a correct assessment of evaporation. For general-purpose and preliminary evaporation estimates, a density of one station per 5000 km² appears to be sufficient [14].

Evapotranspirometers

Evapotranspirometers are instruments designed to measure potential evapotranspiration. An evapotranspirometer consists of a central tank and at least two other watertight soil tanks. The soil tanks are open to the air above them and are connected through underground pipes to collecting cans located in the central tank. The soil tanks support a continuous vegetative cover, such as grass. Water can enter the soil tanks only from above, either as natural or artificial precipitation, and can leave the tanks only through the bottom pipes directly into the collecting cans in the central tank. During one time period, the difference between the amount of water entering each soil tank and the amount of water accumulated in the respective collecting can is the water lost to evapotranspiration, provided the proper allowance is made for changes in moisture storage in the soil tank. If the soil moisture in the tank is maintained at field capacity, the measured difference represents PET.

Evapotranspirometer measurements are made daily. The soil tanks are sprinkled with a known quantity of water, which varies depending on the time of the year (season) and on the amount of precipitation that fell on the previous day. To minimize soil leaching, the water that percolated the day before is mixed with the irrigation water. Evapotranspiration depth is equal to precipitation depth plus irrigation depth minus percolation depth. Soil moisture, however, rarely remains constant from day to day. For instance, it increases greatly during and immediately after precipitation. Therefore, on rainy days, the measurement usually indicates a high PET value, whereas the following day it indicates a low, and perhaps even a negative PET value. Only during a prolonged period with no precipitation would the measurements give an accurate indication of the day-to-day variations of PET. However, long-term, i.e., monthly or seasonal, values are likely to be fairly good representations of PET.

Lysimeters

Lysimeters are instruments designed to measure actual evapotranspiration. Actual evapotranspiration is much more difficult to measure than potential evapotranspiration. During the summer months when soil moisture may be considerably depleted, actual rates of evapotranspiration fall well below the potential rate. The actual rate is determined not only by climatic factors but also by the ability of the plant to extract water from the soil and by the speed of movement of soil moisture to the plant roots.

A properly constructed lysimeter must be representative of the surrounding area. Vegetation cover, surface conditions, soil structure, porosity, stratification, and water-flow relationships (infiltration, permeability, and capillarity) must remain as true to the prototype as possible. Ideal conditions are rarely obtained, particularly when actual evapotranspiration is markedly less than PET.

Exact duplication within the soil tanks of the natural conditions usually requires that the size of a lysimeter tank be greater than that of an evapotranspirometer. The larger the tanks, the less the influence of edge effects and the greater the likelihood that the root systems in the tank will simulate natural conditions. Inevitably, the greater the tanks, the heavier they become, and the more cumbersome and difficult they are to handle. For instance, the set of weighing monolith lysimeters at Coshocton, Ohio [9], are 2.4 m (8 ft) deep and 3.1 m (10 ft) in diameter. Other lysimeter experiments have been reported in the literature (see, for example, [15]).

3.4 INFILTRATION AND SOIL-MOISTURE MEASUREMENTS

Infiltration rates vary greatly, both in time and space. Therefore, care should be taken to ensure that a measurement or a series of measurements are representative of the area under study. In practice, infiltration rates are determined either by the use of infiltrometers or by the analysis of rainfall-runoff data from natural catchments.

Infiltrometers

Infiltrometers are instruments designed to measure the rate at which water is absorbed by the soil surface enclosed within a small, clearly defined area. There are two types of infiltrometers: (1) flooding and (2) sprinkler.

A flooding infiltrometer consists of two concentric metal rings that are inserted a distance of 2 to 5 cm into the ground. Usually, the diameter of the inner ring is about 25 cm, whereas that of the outer ring is about 35 cm. The rate at which water must be applied to the inner ring to maintain a constant head of 0.5 cm is taken as a measure of infiltration rate. In order to prevent the water from spreading laterally below the ground surface, the same head of water is maintained in the annular space between the rings.

Many factors contribute to make the infiltration rate measured with a flooding infiltrometer different from actual infiltration rate. For one thing, the insertion of the rings disturbs the soil immediately around them, leading to an increase in infiltration rate. Differences in head between inner and annular spaces are also likely to cause divergence. Moreover, the flooding condition is not representative of actual conditions. Usually, all these factors combine to produce a high estimate of infiltration rate from a flooding infiltrometer. In addition, a large number of these tests are required for assessment of the spatial variability of infiltration.

A sprinkler infiltrometer is designed to avoid some of the pitfalls of the flooding infiltrometer. In the sprinkler infiltrometer, a simulated rainfall condition is applied over a small plot by using sprinklers. A common plot in use is the F plot, which is 1.8 m wide and 3.6 m long. In the F plot, large drops are applied to the plot and surrounding areas from two rows of special nozzles mounted along each long side of the plot. These nozzles direct their spray upward and slightly inward to cover the plot with relatively uniform rainfall intensities of about 4.5, 9.0, and 13.5 cm/h, depending on how many sets of nozzles are used. The drops reach a height of 2 m above the plot surface and are therefore able to produce erosion and surface conditions resembling those of natural rain [17]. The simulated rainfall is continued for as long as necessary to attain an equilibrium runoff condition at the plot outlet. Average infiltration rate is calculated as the difference between the constant rainfall rate and the constant (i.e., equilibrium) runoff rate.

Due to the spatial and temporal variability of infiltration, field measurements can provide only qualitative information, best suited for comparative studies. Quantitative estimates that are representative of actual conditions are more likely to be obtained from methods based on rainfall-runoff analysis.

Infiltration Rates from Rainfall-Runoff Data

The use of rainfall-runoff data to determine infiltration rates represents an extension of the sprinkler infiltrometer technique. For a storm with a single runoff peak, the procedure resembles that of the calculation of a ϕ-index (Section 2.2). The rainfall hyetograph is integrated to calculate the total rainfall volume. Likewise, the runoff hydrograph is integrated to calculate the runoff volume. The infiltration volume is obtained by subtracting runoff volume from rainfall volume. The average infiltration rate for the given storm is obtained by dividing infiltration volume by rainfall duration.

The procedure can be extended to complex storms consisting of several sub-storms and related runoff peaks. First, it is necessary to use hydrograph separation or similar techniques to isolate the various substorms and their corresponding hydrographs (Sections 5.2 and 11.5). By calculating sets of rainfall and runoff volumes for each substorm, a measure of how the infiltration rate varies from one substorm to the next can be obtained. The procedure works best in cases where the hydrographs can be readily separated, particularly for upland catchments. The procedure is also recommended for cases where interception and surface storage are negligible as compared to infiltration. Interception can be assumed to be insignificant for highly intensive storms, whereas surface storage is likely to be small in catchments with high relief. When used judiciously, this type of analysis can provide a reliable means of determining infiltration rates.

A factor to take into account when using rainfall-runoff analysis to determine infiltration rates is the effect of long-term storage. For large basins, the time elapsed between rainfall and runoff may be so great that it may be practically impossible to determine the amount of runoff produced by a storm event within a reasonable length of time. In practice, this limits infiltration analysis based on rainfall-runoff data to basins with negligible long-term storage.

Measurements of Soil Moisture

Two levels of soil moisture are used in engineering hydrology: (1) field capacity and (2) permanent wilting point. The field capacity is the maximum amount of moisture that the soil structure can hold against the force of gravity. It characterizes the upper level of moisture above which additional infiltration will tend to pass rather rapidly through the soil. The permanent wilting point is the soil moisture content at which permanent wilting of plants starts to occur.

Soil moisture can be measured directly or indirectly. The direct measurement involves the determination of the weight loss from several oven-dried field samples. Each sample is weighed before and after being dried at a temperature of 105°C. The moisture content is the ratio of the weight of water loss to the weight of the dry soil, as a percentage. These measurements, however, are time consuming and do not provide a record of the continuous change of soil moisture with time.

Indirect measurements of soil moisture involve the use of tensiometers to measure the suction force with which water is held in moist soil. The instrument consists of a tube full of water, with a porous cup at the bottom and a stopper on top. The tube is connected to a mercury manometer or vacuum gage. When the tube is inserted into the soil, water moves through the porous cup to the surrounding soil, causing a pressure drop to register in the manometer. The drier the soil, the greater the amount of water leaving the tube and, consequently, the greater the pressure decrease. Tensiometer tests can be performed *in situ*, but their application is restricted within a limited range of soil moisture [18, 19].

The *neutron probe* is a device used for indirect measurement of soil moisture content in the field [12, 24]. The method is based on the fact that fast neutrons are scattered and slowed down when they collide with the protons of hydrogen atoms. The probe consists of a fast neutron source and a slow neutron counter, which registers a high count when the soil moisture is high and a low count when the soil moisture is low. A calibration curve relates the neutron count to the moisture content of the soil. A clear advantage of the neutron probe is the speed of measurement; however, a disadvantage is that with the neutron probe it may be difficult to ascertain changes of soil moisture with depth.

The *water balance method* is another indirect way of determining soil moisture.

The method is based on the assumption that soil moisture can be represented as the difference between precipitation (input) and evapotranspiration (output). Thus, moisture content can be evaluated directly from readily available precipitation and evapotranspiration data. When rainfall exceeds evapotranspiration, the soil moisture increases; conversely, when evapotranspiration exceeds rainfall, the soil moisture decreases. The water balance, however, must include the change in evapotranspiration with moisture availability. When soil moisture is high, evapotranspiration can take place at the potential rate; when soil moisture is low, the actual evapotranspiration rate is much lower than the potential rate. Thus, to use this method it is necessary to relate actual evapotranspiration to soil moisture. When sufficient data are available, this method can provide useful and accurate results [26, 27].

3.5 STREAMFLOW MEASUREMENTS

The discharge at a given location along a stream can be evaluated in two ways: either by measuring the stage and using a known rating to obtain the discharge from it or by directly measuring the cross-sectional flow area and mean velocity of the stream. The point along the stream where the measurements are made is called the *gaging site,* or gaging station. The measurement of discharge and stage is referred to as *stream gaging.*

The development of a good rating curve is crucial to the accurate determination of discharge from stage. The quality of the rating is evaluated in terms of its stability and permanence. A stable rating remains constant in time, i.e., the effects of flow nonuniformity, unsteadiness, or erosion and sedimentation are negligible. A permanent rating is one that is not likely to be disturbed by human activities.

A gaging site should be located at a point along the stream where there is a high correlation between stage and discharge. Stated in other terms, the stage-discharge rating should be close to being single-valued, i.e., featuring a one-to-one correspondence between stage and discharge. Either *section* or *channel control* is necessary for the rating to be single-valued.

A rapid or fall located immediately downstream of a gaging site forces critical flow through it, providing a section control. In the absence of a natural section control, an artificial control—for instance, a concrete weir—can be built to force the rating to become single-valued. This type of control is very stable under low and average flow conditions.

A long downstream channel of relatively uniform cross-sectional shape, constant slope, and bottom friction provides a channel control. However, a gaging site relying on channel control requires periodic recalibration to check its stability. To improve channel control, the gaging site should be located far from downstream backwater effects caused by reservoirs, large river confluences, or tides.

Stage Measurements

Manual Gages. The simplest type of manual gage is the *vertical staff gage.* This is a scale reading to centimeters (or tenths of a foot), which is vertically attached to a fixed feature such as a bridge pier or a pile. The scale must be positioned so that all possible water levels can be read promptly and accurately. Where this is not feasible, several sectional staff gages are placed in such a way that one of them is always accesible for measurement.

Another type of manual gage is the *wire gage*. Wire gages consist of a reel holding a length of light cable with a weight affixed to the end of the cable. The reel is mounted in a fixed position—for instance, on a bridge span—and the water level is measured by unreeling the cable until the weight touches the water surface. Each revolution of the reel unwinds a specific length of cable, permitting the calculation of the distance to the water surface.

Manual gages are used where stages do not vary greatly from one measurement to the next. They are impractical in small or flashy streams, where substantial changes in stage may occur between readings.

Recording Gages. A *recording gage* measures stages continuously and records them on a strip chart. The mechanism of a recording gage is usually either float-actuated or pressure-actuated. In the float-actuated recorder, a pen recording the water level on a strip chart is actuated by a float on the surface of the water. The recorder and float are housed in a suitable enclosure on top of a stilling well connected to the stream by two intake pipes (two pipes are used in case one of them becomes clogged). The stilling well protects the float from debris and ice and dampens the effect of wave action. This type of gage is commonly used for continuous measurements of water levels in rivers and lakes.

The *pressure-actuated recorder* eliminates the need for a stilling well. The sensing element of the recorder is a diaphragm, which is submerged in the stream. The changing water level produces a change in pressure on the diaphragm, which is transmitted to the recorder. Another type of pressure-actuated recorder is the *bubble gage*, developed by the U.S. Geological Survey [3]. The bubble gage consists of a specially designed servomanometer, gas-purge system, and recorder. Nitrogen fed through a tube bubbles freely into the stream through an orifice positioned at a fixed location below the water surface. The pressure in the tube, equal to that of the piezometric head above the orifice, is transmitted to the servomanometer, which converts changes in pressure in the gas-purge system into pen movements on a strip-chart recorder. In this way, a continuous record of stage is obtained.

Telemetric Gages. Gages with automatic data transmittal capabilities are called self-reporting gages, or stage sensors. These instruments use telemeters to broadcast stage measurements in real time, from a stream-gaging location to a central site. This type of gage is ideally suited for applications where speed of processing is of utmost importance, e.g., for operational hydrology or real-time flood forecasting.

Self-reporting gages are of the float-actuated or pressure-actuated type. The float-actuated type is used in stream and lake installations where a concrete abutment (e.g., a bridge pier) is already in place and where sediment loading is minimal. A typical station consists of a top section, which houses a transmitter and float-type water-level sensor, a stilling well with antenna mounting bracket, antenna cable, and connectors.

Bubble-type water-level sensors are used in applications where a stilling well is either impractical or too expensive and where the stream carries a heavy sediment load. In a typical installation, the bubbler orifice is anchored in the stream bed, and a plastic tube connects this orifice to a dry air or dry nitrogen supply and to a fluid manometer measuring assembly. Changes in river stage cause changes in piezometric head at the bubbler orifice. These changes in pressure are recorded by the manometer assembly. Data are transmitted automatically to a central station for further processing.

Other self-reporting stage sensors use a solid-state pressure transducer to sense pressure changes. They are used for measuring river stages where installation of a standpipe and stilling well is not feasible and where a highly sensitive device is not required. This type of sensor continues to provide readings even if a small amount of sediment builds up around the orifice.

Discharge Measurements

A discharge measurement at a stream cross section requires the determination of flow area and mean velocity for a given stage. The cross section should be perpendicular to the flow, and mean velocity should be based on a sufficient number of velocity measurements across the section.

In a typical stream-gaging procedure, each of several depth soundings, usually 20 to 30, defines the position of a vertical. Each depth sounding is associated with a partial section of the stream. A partial section is a rectangle of depth equal to the sounding and of width equal to half the difference of the distances to adjacent verticals (Fig. 3-4). At each vertical, the following observations are made: (1) the distance to a reference point on the stream or river bank, (2) the flow depth, and (3) the velocity as measured by a *current meter* at one or two points along the vertical. In the two-point method, the current meter is positioned at 0.2 and 0.8 of the flow depth. In the one-point method, the current meter is positioned at 0.6 of the flow depth, measured from the water surface. The average of the velocities at 0.2 and 0.8 depth or the single velocity at 0.6 depth is taken as the mean velocity in the vertical. Where a two-point measurement is impractical (e.g., in very shallow streams), the one-point method is recommended.

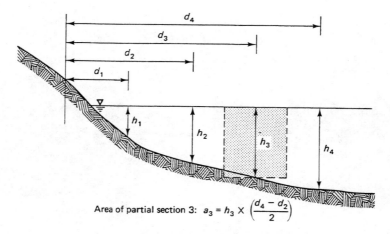

Area of partial section 3: $a_3 = h_3 \times \left(\dfrac{d_4 - d_2}{2}\right)$

Figure 3-4   Partial section in stream gaging.

For each partial section, the discharge is calculated as:

$$q = va \qquad (3\text{-}3)$$

in which q = discharge, v = mean velocity, and a = flow area. The total stream discharge Q is the sum of the discharges of each partial section.

Current Meters. Current meters measure flow velocity by counting the number of revolutions per second of the meter assembly. The rotation can be around a vertical axis, leading to the cup meter, or around a horizontal axis, leading to the propeller meter.

Cup meters are widely used in the United States. The most common type of cup meter is the Price current meter, which has six cups mounted on a vertical axis (Fig. 3-5). The flow velocity is proportional to the angular velocity of the meter rotor. The flow velocity is determined by counting the number of revolutions per second of the rotor and consulting the meter calibration table.

Calibration of a current meter is performed in a rating station, which has a reinforced concrete basin 20 m long, 1.8 m deep, and 1.8 m wide. On top of the vertical side walls of the basin and extending along its entire length are steel rails designed to carry an electrically driven rating car. Generally, the basin is full with still water. To calibrate the current meter, it is affixed to the rating car and positioned below the still water surface. The rating car is displaced at constant speed across the length of the basin. Paired observations of the velocity of the car versus the number of revolutions per second of the meter rotor lead to the rating expressed as:

$$V = KN + C \qquad (3\text{-}4)$$

in which V = flow velocity as measured by the current meter; N = number of revolutions per second of the rotor assembly; and K and C are calibration constants [22].

Older current meters required periodic recalibration in order to minimize errors in velocity measurement due to wear or accidental damage. More recent models, however, have cups made of plastic and are so much alike that one rating can be used for several meters.

Figure 3-5 Price AA current meter (photo courtesy of U.S. Geological Survey).

Price current meters types AA and A are used for two-point velocity measurements in streams with flow depths above 0.75 m and for one-point measurements in streams with depths ranging from 0.45 to 0.75 m. The dwarf (or pigmy) current meter is used for one-point measurements in shallow streams or laboratory flumes with depths in the range 0.10 to 0.45 m.

Current-meter Measurements.
Techniques for measuring stream velocity with a current meter vary with stream size. If the stream is wadable, the meter is affixed to a graduated depth rod. If the stream is too deep to wade, the meter is suspended on a cable and is held in the water with a sounding weight. The weights are made of various sizes, from 6.8 kg to 135 kg. Measurements using cable suspension are made from bridges, cableways, or boats. For the heavier sounding weights or when using a boat, a sounding reel may be required.

The current meter is affixed to a rod on top of the sounding weight. Soundings are made by setting the meter at the water surface and lowering it until the sounding weight rests on the stream bed. Velocity measurements are made either by the two-point or the one-point method. The size of the sounding weight is a function of the depth and swiftness of the stream. If the velocity of the stream is too high for the weight, the latter can drift downstream and greatly exaggerate the flow depth. In this case, it is necessary to reduce the measured depth to correct for downstream drift [5].

Chemical Methods for Measuring Velocity.
Several chemical methods have been developed to measure stream velocity. They are normally used in cases where it is impractical to use current meters. Such is the case for shallow streams, very large rivers, or tidal flow. These methods can be grouped into (1) tracer and (2) dilution methods.

A *tracer* is a substance that is not normally present in the stream and that is not likely to be lost by chemical reaction with other substances. Salt, fluorescein dye, and radioactive materials are commonly used as tracers. Small quantities of the tracer are injected into the stream at a source, and the time of travel to one or more downstream points is monitored.

When salt is used as a tracer, it is introduced at several points across a section. Electrodes are mounted at the two ends of a uniform reach beginning a short distance downstream from the salt source. These electrodes are connected to a recording galvanometer and are used to monitor the passage of the salt solution. The velocity of the stream is the length of the reach divided by the time that it takes the bulk of the salt solution to travel through the reach.

An expedient measurement of velocity can be obtained by timing the travel of floats. A surface float travels with a speed that is about 1.2 times the mean velocity. Floats extending well below the surface usually travel with a speed close to the mean velocity.

In the dilution method, a concentrated solution of a substance is introduced at a constant rate at a source point. Further downstream, after complete mixing has taken place, the flow is sampled to determine the equilibrium concentration of the mixture. A mass balance of flow and substance leads to the following equation:

$$C_s Q_s = C_e (Q + Q_s) \qquad (3\text{-}5)$$

in which C_s = concentration of the substance solution at the source; C_e = equilibrium concentration of mixture at the sampling point; Q_s = rate of inflow of substance solution at the source; and Q = stream discharge. Solving for Q from Eq. 3-5 leads to:

$$Q = \left(\frac{C_s}{C_e} - 1\right) Q_s \tag{3-6}$$

The dilution method is particularly useful for very turbulent flows, which can provide complete mixing within a relatively short distance. It is also applicable when the cross section is so rough that alternative methods are unfeasible. The method requires the assurance of complete mixing and an accurate determination of equilibrium concentration of the mixture.

Physical Methods for Measuring Velocity.   The ultrasonic and electromagnetic flowmeters are examples of physical devices to measure flow velocity. In the ultrasonic method, two sonic pulses are emitted and received, each at opposite banks of the river. The instruments are not located directly across from each other on the banks but rather on a diagonal line, making a 45° angle with the flow direction. Therefore, one of the pulses travels with the current and the other against it. The difference in travel time between the two pulses is related to the longitudinal flow velocity. The method is applicable to large rivers where current metering or other direct techniques are not feasible. Its accuracy is claimed to be within 2 percent [10, 20].

The electromagnetic flowmeter is based on the fact that a conductor moving in an electromagnetic field generates a current within it. The river flow is a conductor cutting through the vertical component of the earth's magnetic field. The current can be measured by two electrodes, which are set at right angles to flow and magnetic field directions, respectively. The devices are applicable to measurement of tidal velocities, preferably in large rivers [2].

Indirect Determination of Peak Discharge: The Slope-Area Method

The high stages and swift currents that prevail during floods combine to increase the risk of accident and bodily harm. Therefore, it is generally not possible to measure discharge during the passage of a flood. An estimate of peak discharge can be obtained indirectly by the use of open channel flow formulas. This is the basis of the *slope-area* method.

To apply the slope-area method for a given river reach, the following data are required: (1) the reach length, (2) the *fall*, i.e., the mean change in water surface elevation through the reach, (3) the flow area, wetted perimeter, and velocity head coefficients at upstream and downstream cross sections, and (4) the average value of Manning n for the reach [6].

The following guidelines are used in selecting a suitable reach: (1) high-water marks should be readily recognizable, (2) the reach should be sufficiently long so that fall can be measured accurately, (3) the cross-sectional shape and channel dimensions should be relatively constant, (4) the reach should be relatively straight, although a contracting reach is preferred over an expanding reach, and (5) bridges, channel bends, waterfalls, and other features causing flow nonuniformity should be avoided.

The accuracy of the slope-area method improves as the reach length increases. A suitable reach should satisfy one or more of the following criteria: (1) the ratio of reach length to hydraulic depth should be greater than 75, (2) the fall should be greater than or equal to 0.15 m, and (3) the fall should be greater than either of the velocity heads computed at the upstream and downstream cross sections [8].

The procedure consists of the following steps:

1. Calculate conveyance K at upstream and downstream sections:

$$K_u = \left(\frac{1}{n}\right)A_u R_u^{2/3} \tag{3-7a}$$

$$K_d = \left(\frac{1}{n}\right)A_d R_d^{2/3} \tag{3-7b}$$

in which K = conveyance; A = flow area; R = hydraulic radius; n = reach Manning coefficient; and u and d denote upstream and downstream, respectively (Eq. 3-7 is given in SI units).

2. Calculate the reach conveyance, equal to the geometric mean of upstream and downstream conveyances:

$$K = (K_u K_d)^{1/2} \tag{3-8}$$

in which K = reach conveyance.

3. Calculate the first approximation to the energy slope:

$$S = \frac{F}{L} \tag{3-9}$$

in which S = first approximation to the energy slope; F = fall; and L = reach length.

4. Calculate the first approximation to the peak discharge:

$$Q_i = KS^{1/2} \tag{3-10}$$

in which Q_i = first approximation to the peak discharge.

5. Calculate the velocity heads [6]:

$$h_{vu} = \frac{\alpha_u(Q_i/A_u)^2}{2g} \tag{3-11a}$$

$$h_{vd} = \frac{\alpha_d(Q_i/A_d)^2}{2g} \tag{3-11b}$$

in which h_{vu} and h_{vd} are the velocity heads at upstream and downtream sections, respectively; α_u and α_d are the velocity head coefficients at upstream and downstream cross sections, respectively; and g = gravitational acceleration.

6. Calculate an updated value of energy slope:

$$S_i = \frac{F + k(h_{vu} - h_{vd})}{L} \tag{3-12}$$

in which S_i = updated value of energy slope, and k = loss coefficient. For expanding flow, i.e., $A_d > A_u$, $k = 0.5$; for contracting flow, i.e., $A_u > A_d$, $k = 1$.

7. Calculate an updated value of peak discharge:

$$Q_i = KS_i^{1/2} \tag{3-13}$$

8. Go back to step 5 and repeat steps 5 to 7. In step 5, use the updated value of peak discharge obtained in the previous step 7. In step 6, use the updated values of velocity heads obtained in step 5. In step 7, use the updated value of energy slope obtained in step 6. The procedure is terminated when the difference between two successive values of peak discharge obtained in step 7 is negligible. In practice, this is usually accomplished within three to five iterations.

Example 3-1.

Use the slope-area method to calculate the peak discharge for the following data: reach length = 500 m; fall = 0.5 m; Manning n = 0.04; upstream flow area = 1050 m²; upstream wetted perimeter = 400 m; upstream velocity head coefficient = 1.10; downstream flow area = 1000 m²; downstream wetted perimeter = 375 m; downstream velocity head coefficient = 1.12.

The hydraulic radius and conveyance at the upstream section are R_u = 2.625 m and K_u = 49,952 m³/s, respectively. The hydraulic radius and conveyance at the downstream section are R_d = 2.667 m and K_d = 48,075 m³/s, respectively. The reach conveyance (Eq. 3-8) is K = 49,005 m³/s. The first approximation to the energy slope (Eq. 3-9) is S = 0.5/500 = 0.001. The first approximation to peak discharge (Eq. 3-10) is Q_i = 1550 m³/s. Since A_u is greater than A_d, $k = 1$. The remaining computations (steps 5 through 7) are summarized in Table 3-1. After three iterations, the final value of the peak discharge is obtained: Q_p = 1526 m³/s.

TABLE 3-1 PEAK DISCHARGE COMPUTATION BY THE SLOPE-AREA METHOD: EXAMPLE 3-1

Iteration No.	h_{vu} (m)	h_{vd} (m)	Energy Slope (m/m)	Peak Discharge (m³/s)
1			0.00100	1550
2	0.122	0.137	0.00097	1526
3	0.118	0.133	0.00097	1526

QUESTIONS

1. Describe the various types of recording rain gages. When is the use of telemetry necessary?
2. Explain the physical basis for radar measurements of precipitation. What radar wavelength should be used to sense heavy rains?
3. What is the water equivalent of the snowpack? What is a snow course?
4. Explain the procedure to determine a catchment's water equivalent using a snow chart.
5. Why is the pan-evaporation measurement likely to be greater than the actual lake evaporation?

6. What is an evapotranspirometer? On what principle is it based? What is a lysimeter?

7. Describe the two types of equipment to measure infiltration rates in the field.

8. How does the catchment size affect the analysis of infiltration rates from rainfall-runoff data?

9. How is soil moisture determined by the water balance method?

10. What are the two properties of a good rating curve? Describe the two types of control in open channel flow. How does control affect the rating? Explain.

11. Describe two types of recording gages to measure stage. When is a telemetric gage needed?

12. Describe the various means to carry out stream velocity measurements using current meters.

13. When are chemical and physical methods to measure stream velocity applicable? What is the crucial assumption in the dilution method to measure stream velocity?

14. What is the slope-area method? When is it used? What is the recommended minimum fall to preserve accuracy?

PROBLEMS

3-1. A snow sample 20 cm high melted into 3 cm of water. What was the density of the snow sample?

3-2. What is the water equivalent of a snow accumulation measuring 9 in. with a density of 8%?

3-3. The following elevation–area–snow-water-equivalents have been measured in a certain catchment:

Elevation (m)	2000	2500	3000	3500	4000
Cumulative area (km²)	0	255	432	519	605
Snow-water equivalent (mm)	0	0	8	22	30

Determine the catchment's overall snow-water equivalent.

3-4. The following snow-chart and hypsometric data have been measured in a certain catchment:

Elevation (%)	0	10	20	30	40	50	60	70	80	90	100
Cumulative Area (%)	0	22	39	54	64	76	81	88	94	97	100
Snow-water equivalent (mm)	3	3	4	4	5	5	5	7	7	8	8

Determine the catchment's overall snow-water equivalent.

3-5. Given the following stream gaging data, calculate the discharge.

Vertical no.	1	2	3	4	5	6	7	8	9	10	11
Distance to reference point (m)	15	20	25	30	35	40	45	50	55	60	65
Sounding depth (m)	0.0	0.5	0.8	1.2	1.5	2.5	3.0	2.0	1.2	0.8	0.0
Velocity at 0.2 depth (m/s)	0.0	0.5	0.7	0.9	1.2	1.4	1.7	1.3	0.9	0.7	0.0
Velocity at 0.8 depth (m/s)	0.0	0.4	0.6	0.7	0.8	1.1	1.3	1.0	0.7	0.6	0.0

3-6. A certain substance is introduced at point A of a stream at the rate of 50 L/s with a concentration of 12,000 ppm. At a downstream point B, after complete mixing, the concentration of the substance is measured to be 15 ppm. Calculate the stream discharge.

3-7. Calculate the flood discharge of a certain stream by the slope-area method, given the following data: upstream flow area $A_u = 402$ m^2, upstream wetted perimeter $P_u = 98$ m, upstream $\alpha_u = 1.11$, downstream flow area $A_d = 453$ m^2, downstream wetted perimeter $P_d = 105$ m, downstream $\alpha_d = 1.13$, fall $F = 0.5$ m, reach length $L = 870$ m, and reach Manning $n = 0.04$.

3-8. Calculate the flood discharge of a certain stream by the slope-area method, given the following data: upstream flow area $A_u = 3522$ m^2, upstream wetted perimeter $P_u = 650$ m, upstream $\alpha_u = 1.17$, downstream flow area $A_d = 3259$ m^2, downstream wetted perimeter $P_d = 621$ m, downstream $\alpha_d = 1.21$, fall $F = 0.35$ m, reach length $L = 1,250$ m, and reach Manning $n = 0.028$.

REFERENCES

1. Alvarez, F., and W. K. Henry. (1970). "Rain Gage Spacing and Reported Rainfall," *Bulletin, International Association for Scientific Hydrology,* Vol. 15, No. 3, March, pp. 97–107.

2. Barron, E. G. (1959). "New Instruments of the U.S. Geological Survey for the Measurement of Tidal Flow," *Proceedings, ASCE Hydraulics Division Eighth National Conference,* Colorado State University, Fort Collins, Colorado.

3. Barron, E. G. (1963). "New Instruments for Surface Water Investigations," in *Selected Techniques for Water Resources Investigations,* U.S. Geological Survey Water Supply Paper No. 1692-Z, pp. Z-4 to Z-8.

4. Battan, L. J. (1974). *Radar Observation of the Atmosphere.* University of Chicago Press.

5. Buchanan, T. J., and W. P. Somers. (1965). "Discharge Measurements at Gaging Stations," U.S. Geological Survey, *Surface Water Techniques,* Book 1, Chapter 11.

6. Chow, V. T. (1959). *Open Channel Hydraulics.* New York: McGraw-Hill.

7. Collier, C. G. (1985). "Remote Sensing for Hydrologic Forecasting," in *Facets of Hydrology II,* John C. Rodda, ed. New York: John Wiley.

8. Dalrymple, T., and M. A. Benson. (1976). "Measurements of Peak Discharge by the Slope-Area Method," *Techniques of Water Resources Investigations of the United States Geological Survey,* Book 3, Chapter A2.

9. Harrold, L. L., and F. R. Dreibelbis. (1951). "Agricultural Hydrology as Evaluated by Monolith Lysimeters," *USDA Technical Bulletin No.* 1050, pp. 174–175.

10. Herschy, R. W. (1976). "New Methods of River Gaging," in *Facets of Hydrology,* J. C. Rodda, ed. New York: John Wiley.

11. Johanson, R. C. (1971). "Precipitation Network Requirements for Streamflow Estimation," Stanford University, Department of Civil Engineering, Technical Report No. 147, August.

12. Holmes, J. W., and K. G. Turner. (1958). "The Measurement of Water Content of Soils by Neutron Scattering: A Portable Apparatus for Field Use," *Journal of Agricultural Engineering Research,* Vol. 3, pp. 199–204.

13. Huff, F. A. (1970). "Sampling Errors in Measurement of Mean Precipitation," *Journal of Applied Meteorology,* Vol. 9, No. 1, February, pp. 35–44.

14. Linsley, R. K. (1958). "Techniques for Surveying Water Resources," World Meteorological Organization, *Technical Note No.* 26.

15. McIlroy, I. C., and C. J. Sumner. (1961). "A Sensitive High Capacity Balance for Continuous Automatic Weighing in the Field," *Journal of Agricultural Engineering Research,* Vol. 6, pp. 252–258.

16. McKean, G. A. (1967). "A Nuclear Radiation Snow Gage," University of Idaho Engineering Experiment Station Bulletin, No. 13, August.

17. Musgrave, G. W., and H. N. Holtan. (1964). "Infiltration," in *Handbook of Applied Hydrology*. V. T. Chow, ed. New York: McGraw-Hill.

18. Richards, L. A. (1928). "The Usefulness of Capillary Potential to Soil Moisture and Plant Investigators," *Journal of Agricultural Research*, Vol. 37, pp. 719–742.

19. Richards, L. A. (1949). "Methods for Measuring Soil-Moisture Tension," *Soil Science*. Vol. 68, pp. 95–112.

20. Schuster, J. C. (1975). "Measuring Water Velocity by Ultrasonic Flowmeter," *Journal of the Hydraulics Division*, ASCE, Vol. 101, pp. 1503–1517.

21. Smith, J. L., H. G. Halverson, and R. A. Jones. (1970). "The Profiling Radioactive Snow Gage," *Transactions, Isotopic Snow Gage Information Meeting*. Idaho Nuclear Energy Commission and Soil Conservation Service, Sun Valley, Idaho, October.

22. Smoot, G. F., and C. E. Novak. "Calibration and Maintenance of Vertical-Axis Type Current Meters," *Techniques of Water Resources Investigations of the United States Geological Survey*. Book 8, Chapter B2.

23. Stephenson, P. M. (1968). "Objective Assessment of Adequate Numbers of Raingages for Estimating Areal Rainfall Depths," *International Association for Scientific Hydrology*. Publication No. 78, pp. 252–264.

24. Stone, J. F., D. Kirkham, and A. A. Read. (1958). "Soil Moisture Determination by a Portable Neutron Scattering Moisture Meter," *Proceedings, Soil Science Society of America*. Vol. 19, pp. 419–423.

25. Strangeways, I. C. (1985). "Automatic Weather Stations," in *Facets of Hydrology II*. John C. Rodda, ed. New York: John Wiley.

26. Thornthwaite, C. W., and J. R. Mather. (1954). "The Computation of Soil Moisture," in *Estimating Soil Tractionability from Climatic Data*. Publications in Climatology, Vol. 7, pp. 397–402.

27. Thornthwaite, C. W., and J. R. Mather. (1955). "The Water Balance," *Publications in Climatology*. Vol. 8.

28. U.S. Army Corps Of Engineers, North Pacific Division. (1956). "Snow Hydrology, Summary Report of the Snow Investigations," Portland, Oregon, June.

HYDROLOGY OF SMALL CATCHMENTS

The following characteristics describe a small catchment: (1) rainfall can be assumed to be uniformly distributed in time, (2) rainfall can be assumed to be uniformly distributed in space, (3) storm duration usually exceeds concentration time, (4) runoff is primarily by overland flow, and (5) channel storage processes are negligible.

Catchments possessing some or all of the above properties are *small* in a hydrologic sense. Their runoff response may be described using relatively simple parametric methods, which lump all the relevant hydrologic processes into a few key descriptors such as rainfall intensity and catchment area. When increased accuracy is required, small catchments may be analyzed using complex deterministic methods such as overland flow analysis. For routine applications, however, all that is usually required is the simple parametric approach.

It is difficult to define the upper limit of a small catchment without being arbitrary to some degree. Given the natural variability in catchment slopes, vegetation cover, and so on, no single value is universally applicable. In practice, both concentration time and catchment area have been used to define the upper limit of a small catchment. Some authorities regard a catchment with concentration time of 1 h or less as a small catchment. For others, a catchment of less than 2.5 km² is considered small. Invariably, any such limit is likely to be somewhat arbitrary.

This chapter deals with the hydrology of small catchments. It is divided into two sections. Section 4.1 describes the rational method and its application to urban storm drainage. Section 4.2 discusses overland flow theory and applications.

4.1 RATIONAL METHOD

The rational method is the most widely used method for the analysis of runoff response from small catchments. It has particular application in urban storm drainage, where it is used to calculate peak runoff rates for the design of storm sewers and small drainage structures. The popularity of the rational method is attributed to its simplicity, although reasonable care is necessary in order to use the method correctly.

The rational method takes into account the following hydrologic characteristics or processes: (1) rainfall intensity, (2) rainfall duration, (3) rainfall frequency, (4) catchment area, (5) hydrologic abstractions, (6) runoff concentration, and (7) runoff diffusion (Section 2.4).

In general, the rational method provides only a peak discharge, although in the absence of runoff diffusion it is possible to obtain an isosceles-triangle-shaped runoff hydrograph. The peak discharge is the product of (1) runoff coefficient, (2) rainfall intensity, and (3) catchment area, with all processes being lumped into these three parameters. Rainfall intensity contains information on rainfall duration and frequency. In turn, rainfall duration is related to concentration time, i.e., to the runoff concentration properties of the catchment. The runoff coefficient accounts for both hydrologic abstractions and runoff diffusion and may also be used to account for frequency. In this way, all the major hydrologic processes responsible for runoff response are embodied in the rational formula.

The rational method does not take into account the following characteristics or processes: (1) spatial or temporal variations in either total or effective rainfall, (2) concentration time much greater than rainfall duration, and (3) a significant portion of runoff occurring in the form of streamflow. In addition, the rational method does not explicitly account for the catchment's antecedent moisture condition, although the latter may be implicitly accounted for by varying the runoff coefficient.

The above conditions dictate that the rational method be restricted to small catchments. For one thing, the assumption of constant rainfall in space and time is strictly valid only for small catchments. Furthermore, for small catchments storm duration usually exceeds concentration time. Finally, in small catchments, runoff occurs primarily as overland flow rather than as streamflow.

There is no consensus regarding the upper limit of a small catchment. Values ranging from 0.65 to 12.5 km^2 have been quoted in the literature [2, 23]. The current trend is to use 1.3 to 2.5 km^2 as the upper limit for the applicability of the rational method. There is no theoretical lower limit, however, and catchments as small as 1 ha may be analyzed by the rational method.

The rational method is based on the following formula:

$$Q_p = CIA \tag{4-1}$$

in which Q_p = peak discharge corresponding to a given rainfall intensity, duration, and frequency; C = runoff coefficient, a dimensionless empirical coefficient related to the abstractive and diffusive properties of the catchment; I = rainfall intensity, averaged in time and space; and A = catchment area.

For rainfall intensity in millimeters per hour, catchment area in square kilometers, and peak discharge in cubic meters per second, the formula for the rational method is the following:

$$Q_p = 0.278CIA \qquad (4\text{-}2)$$

For rainfall intensity in millimeters per hour, catchment area in hectares, and peak discharge in liters per second, the formula is:

$$Q_p = 2.78CIA \qquad (4\text{-}3)$$

For rainfall intensity in inches per hour, catchment area in acres, and peak discharge in cubic feet per second, the formula is:

$$Q_p = 1.008CIA \qquad (4\text{-}4)$$

The unit conversion coefficient 1.008 is usually neglected on practical grounds.

Methodology

The first requirement of the rational method is that the catchment be small. Once the size requirement has been met, the three components of the formula are evaluated separately. The catchment area is determined by planimetering or other suitable means. Boundaries may be established from topographic maps or aerial photographs. The drainage area survey should also include (1) land use and land use changes; (2) percentage of imperviousness; (3) characteristics of soil and vegetative cover that may affect the runoff coefficient; and (4) general magnitude of ground slopes and catchment gradient necessary to determine time of concentration.

The evaluation of rainfall intensity is a function of several factors. First, it is necessary to determine the time of concentration of the catchment. Normally, this is accomplished either (1) by using an empirical formula, (2) by assuming a flow velocity based on hydraulic properties and calculating the travel time through the catchment's hydraulic length, or (3) by calculating the steady equilibrium flow velocity (using the Manning equation) and associated travel time through the hydraulic length. Procedures to calculate time of concentration are not very well defined, often involving crucial assumptions such as flow level, channel shape, friction coefficients, and so on. Nevertheless, a value of time of concentration can usually be developed for practical use.

For urban storm-sewer design, time of concentration at a point is the sum of two parts: (1) inlet time and (2) time of flow in the storm sewer up to that point. Inlet time is the longest time required for runoff to flow over the catchment surface to the nearest sewer inlet. Time of flow in the sewer, from inlet to point of interest, is calculated using hydraulic flow formulas.

Once time of concentration has been determined, storm duration is made equal to time of concentration. This amounts to an assumption of concentrated catchment flow (Section 2.4). Then, a rainfall frequency applicable to the given design condition is chosen. Design frequencies (and return periods) vary with the type of project and degree of protection desired. Commonly used return periods are (1) 5 to 10 y for storm sewers in residential areas, (2) 10 to 50 y for storm sewers in commercial areas, and (3) 50 to 100 y for flood protection works. The size and importance of the project, as well as design criteria established by federal, state and local agencies, have a bearing in the selection of design frequency. The longer the return period (i.e., the smaller the frequency), the greater the peak discharge calculated by the rational formula.

The question of whether rainfall frequency and peak flow frequency are equivalent is an elusive one. The rational method bases the calculation of peak flow on a chosen rainfall frequency. In nature, however, the frequencies of storms and floods are not necessarily the same, largely due to the effect of antecedent moisture condition, variability in channel transmission losses, overbank storage, and the like. In practice, runoff coefficients are usually adjusted upward to reflect postulated decreases in runoff frequency. This procedure, while empirical, has seemed to work well.

Once rainfall duration and frequency have been determined, the corresponding rainfall intensity is obtained from the appropriate IDF curve (Fig. 2-8). The applicable curve can usually be obtained from cognizant government agencies. Where IDF curves are nonexistent, they can be developed from regional isopluvial maps containing depth-duration-frequency data. These maps are published by the National Weather Service [19, 20, 25].

Due to the hyperbolic nature of the intensity-duration curve, an error in rainfall duration causes an error of opposite sign in rainfall intensity. For instance, if the rainfall duration is too long (i.e., time of concentration too long), the calculated rainfall intensity will be too low and vice versa.

Once rainfall intensity and catchment area have been obtained, a runoff coefficient applicable to the given design condition is selected. Runoff coefficients are theoretically restricted to the range 0.0 to 1.0. In practice, values of runoff coefficient in the range 0.05 to 0.95 are usually adopted. The runoff coefficient accounts for the processes of (1) hydrologic abstractions and (2) runoff diffusion. Hydrologic abstractions include interception, infiltration, surface storage, evaporation, and evapotranspiration (Section 2.2). Runoff diffusion is a measure of the catchment's ability to attenuate the flood peaks (Section 2.4).

In essence, the runoff coefficient is the ratio of the actual (i.e., calculated) peak runoff rate to the maximum possible runoff rate. For instance, for $C = 1$, the calculated peak discharge is equal to the maximum possible discharge. Typical values of runoff coefficients for a wide variety of conditions are given in design manuals and other reference books; see for instance Tables 4-1(a) and (b). These values reflect the reduction in peak runoff that is likely to be produced by several combinations of rainfall abstraction and runoff diffusion. For instance, in Table 4-1(a), a lawn with a steep gradient might have $C = 0.2$; but a lawn with a mild gradient might have $C = 0.05$. On the other hand, an asphaltic street (no abstraction) might have $C = 0.95$. Likewise, in Table 4-1(b), a decrease in terrain slope in rural areas shows a corresponding decrease in runoff coefficient.

The runoff coefficients shown in Table 4-1(a) are applicable to storms of 5- to 10-y return period. Less frequent storms (e.g., 50 y) require the use of higher coefficients because infiltration and other abstractions have a reduced role in runoff generation for the larger storms. The coefficients shown in Table 4-1(a) represent average antecedent moisture conditions and are not designed to account for multiple storms or storms of very long duration. Special design cases usually warrant the use of higher runoff coefficients to simulate the existence of wet antecedent moisture conditions in the catchment. Experimental evidence has shown that runoff coefficients tend to increase from one storm to another occurring shortly thereafter, with runoff coefficients tending to increase with storm duration.

Design values of runoff coefficients are usually a function of rainfall intensity

**TABLE 4-1(a) AVERAGE RUNOFF COEFFICIENTS
FOR URBAN AREAS:
5-Y AND 10-Y DESIGN FREQUENCY**

Description of Area	Runoff Coefficients
Business	
Downtown areas	0.70 to 0.95
Neighborhood areas	0.50 to 0.70
Residential	
Single-family areas	0.30 to 0.50
Multiple units, detached	0.40 to 0.60
Multiple units, attached	0.60 to 0.75
Residential (suburban)	0.25 to 0.40
Apartment-dwelling areas	0.50 to 0.70
Industrial	
Light areas	0.50 to 0.80
Heavy areas	0.60 to 0.90
Parks, cemeteries	0.10 to 0.25
Playgrounds	0.10 to 0.25
Railroad yard areas	0.20 to 0.40
Unimproved areas	0.10 to 0.30

Character of Surface	Runoff Coeficients
Streets	
Asphaltic	0.70 to 0.95
Concrete	0.80 to 0.95
Brick	0.70 to 0.85
Drives and walks	0.70 to 0.85
Roofs	0.75 to 0.95
Lawns, sandy soil	
Flat (2 percent)	0.05 to 0.10
Average (2 to 7 percent)	0.10 to 0.15
Steep (7 percent)	0.15 to 0.20
Lawns, heavy soil	
Flat (2 percent)	0.13 to 0.17
Average (2 to 7 percent)	0.18 to 0.22
Steep (7 percent)	0.25 to 0.35

Source: Design and Construction of Sanitary and Storm Sewers.
ASCE Manual of Engineering Practice, no. 37, 1960.

and, therefore, of rainfall frequency. Higher values of runoff coefficient are applicable for higher values of rainfall intensity and return period. A typical C versus I curve is shown in Fig. 4-1 [5]. Alternate ways of expressing the variation of runoff coefficient with rainfall frequency are shown in Figs. 4-2 and 4-3 [6, 22].

With runoff coefficient, rainfall intensity, and catchment area determined, the peak discharge is calculated by Eq. 4-1. The apparent simplicity of the procedure, however, is misleading. For one thing, there is a range of possible runoff coefficients for each surface condition. Therefore, the chosen C value is usually based on additional field information or designer's experience. The effect of frequency and/or antecedent moisture condition needs to be evaluated carefully. Furthermore, there is no absolute certainty that the calculated time of concentration (and therefore, the rainfall duration) is correct or even that it remains constant throughout the range of possi-

TABLE 4-1(b) AVERAGE RUNOFF COEFFICIENTS FOR RURAL
AREAS

Topography and Vegetation	Soil Texture		
	Open Sandy Loam	Clay and Silt Loam	Tight Clay
Woodland[1]			
Flat	0.10	0.30	0.40
Rolling	0.25	0.35	0.50
Hilly	0.30	0.50	0.60
Pasture			
Flat	0.10	0.30	0.40
Rolling	0.16	0.36	0.55
Hilly	0.22	0.42	0.60
Cultivated Land			
Flat	0.30	0.50	0.60
Rolling	0.40	0.60	0.70
Hilly	0.52	0.72	0.82

[1]*Note:* Flat (0–5% slope); rolling (5–10%); hilly (10–30%).

Source: Schwab, R. J. et al. (1971). *Elementary Soil and Water Engineering.* 2d. ed. New York: John Wiley.

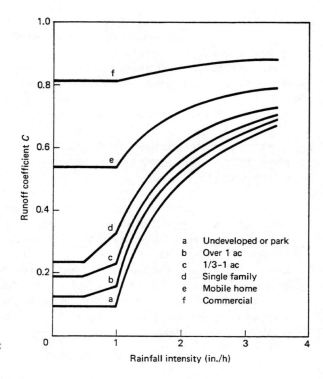

Figure 4-1 Variation of runoff coefficient with rainfall intensity [5].

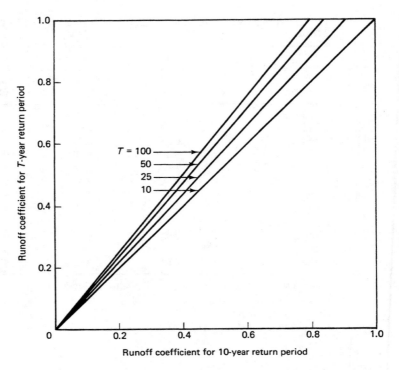

Figure 4-2 Variation of runoff coefficient with rainfall frequency [6].

ble frequencies. In fact, since larger flows generally travel with greater velocities, time of concentration tends to decrease with an increase in return period. Notwithstanding these complexities, the rational method remains a practical way to calculate peak discharge for small catchments based on a few relevant hydrologic parameters.

Theory of the Rational Method

The rational method is based on the principles of runoff concentration and diffusion. For simplicity, the process can be explained in two parts: (1) concentration without diffusion and (2) concentration with diffusion.

Runoff Concentration Without Diffusion. In the absence of diffusion, a catchment concentrates the flow at the outlet, attaining the maximum possible flow (i.e., the equilibrium flow rate) at a time equal to the concentration time. By forcing the design rainfall duration to be equal to the concentration time, concentrated catchment flow is obtained at the outlet. Since there is no diffusion, the method gives not only a peak flow but also a hydrograph corresponding to that of concentrated catchment flow (Fig. 4-4(a)), with recession time equal to rising time. The runoff coefficient is then simply the ratio of effective rainfall to total rainfall. A mass balance of effective rainfall and runoff leads to:

$$V_r = I_e t_r A = CIA \, t_r \tag{4-5}$$

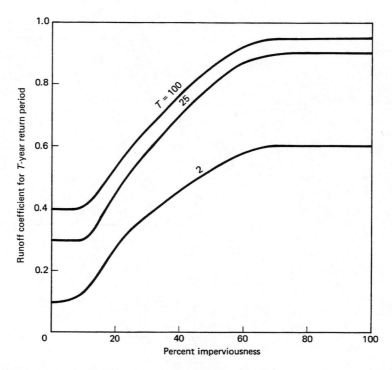

Figure 4-3 Variation of runoff coefficient with percent imperviousness and rainfall frequency [22].

in which V_r = runoff volume; I_e = effective rainfall; and t_r = rainfall duration (either effective or total). Equation 4-5 leads to:

$$C = C_a = \frac{I_e}{I} \qquad (4\text{-}6)$$

in which C_a = runoff coefficient due only to abstraction ($C_a \leq 1$).

Runoff concentration without diffusion is typical of steep catchments where runoff diffusion is negligible compared to runoff concentration. For catchments of mild slope, diffusion plays a larger role. For a hypothetical catchment of zero ground slope, the diffusion effect is theoretically the only one present.

Runoff Concentration With Diffusion. When diffusion is present, the rational method accounts for it in the runoff coefficient. Thus, the runoff coefficient is used to model not only abstraction but also diffusion. Diffusion modifies the catchment response in such a way as to increase the recession time and decrease the peak flow. Therefore, a hydrograph shape can no longer be obtained directly from a mass balance as in the case of runoff concentration without diffusion. The lack of a hydrograph shape does not impede the use of the rational method, because the diffusion can be represented directly in the peak flow formula, by lowering the runoff coefficient below that due only to abstraction (Fig. 4-4(b)).

The reduction in the runoff coefficient amounts to:

$$C = C_d C_a \qquad (4\text{-}7)$$

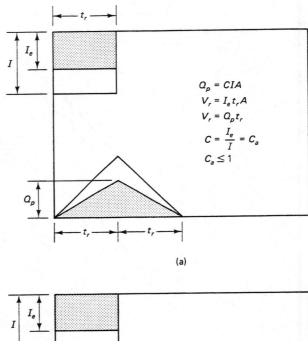

$$Q_p = CIA$$
$$V_r = I_e t_r A$$
$$V_r = Q_p t_r$$
$$C = \frac{I_e}{I} = C_a$$
$$C_a \leq 1$$

(a)

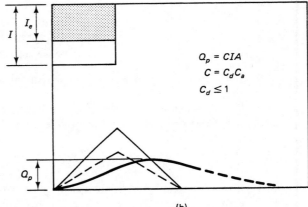

$$Q_p = CIA$$
$$C = C_d C_a$$
$$C_d \leq 1$$

(b)

Figure 4-4 Rational method—abstraction and diffusion processes: (a) Flow concentration without diffusion; (b) Flow concentration with diffusion.

in which C = runoff coefficient and C_d = component of runoff coefficient accounting only for diffusion ($C_d \leq 1$).

The question of whether the peak is reached before, at, or after the time of concentration, as Fig. 4-4(b) shows, is immaterial, since the method does not provide the shape of the hydrograph, limiting itself to providing a peak discharge.

In the absence of diffusion, $C_d = 1$ and $C = C_a$. Likewise, in the absence of abstraction, $C_a = 1$ and $C = C_d$. In the absence of abstraction and diffusion (e.g., a very steep catchment with an impermeable surface), $C_a = 1$, $C_d = 1$, and, therefore, $C = 1$.

In practice, no quantitative distinction is made between the two components of the runoff coefficient. Usage, however, reflects the fact that runoff diffusion is being implicitly considered; see, for instance, the marked change in runoff coefficient with surface slope shown in Table 4-1. The indicated range in runoff coefficients may be attributed to the diffusion effect. Within the range shown in the table, the higher C

values correspond to steeper catchments, whereas the lower C values are associated with catchments of milder slope.

Further Developments

Attempts to analyze the behavior of the rational method have led to the concept of peak flow per unit area [24]:

$$q_p = \frac{Q_p}{A} = CI \tag{4-8}$$

in which q_p = peak flow per unit area. Rainfall intensity varies with rainfall duration and frequency. Likewise, runoff coefficient also varies with rainfall duration and frequency. Therefore, a relation linking peak flow per unit area to rainfall duration and frequency can be obtained:

$$q_p = f(t_r, T) \tag{4-9}$$

in which T = return period.

Another approach is based on expressing the rational formula in the following form:

$$C = \frac{Q_p}{IA} \tag{4-10}$$

where now the runoff coefficient can be interpreted as dimensionless peak flow, or peak flow per unit area per unit rainfall intensity. It follows that dimensionless peak flow is related to the abstractive and diffusive properties of the catchment.

A similar concept is used in the Soil Conservation Service TR-55 method (Section 5.3). In this method, a unit peak flow is defined as the peak flow per unit area per unit rainfall depth. In the graphical method included in TR-55, the unit peak flow is a function of time of concentration, abstraction parameter, and temporal storm pattern. The fact that unit peak flow is a function of temporal storm pattern qualifies the TR-55 graphical method as an extension of the rational method to midsize catchments. While no upper limit to catchment size is indicated, the method is restricted to concentration times less than or equal to 10 h.

Applications of the Rational Method

Relation Between Runoff Coefficient and ϕ-index. The runoff coefficient can be related to total rainfall intensity and ϕ-index, provided the following assumptions are satisfied: (1) catchment response occurs under negligible diffusion and (2) total and effective rainfall intensities are constant in time. The first assumption is valid for steep catchments, whereas the second assumption is implicit in the application of the rational method. For catchment response without diffusion:

$$C = C_a = \frac{I_e}{I} \tag{4-11}$$

For constant rainfall intensities:

$$I_e = I - \phi \tag{4-12}$$

Combining Eqs. 4-11 and 4-12:

$$C = \frac{I - \phi}{I} \tag{4-13}$$

Areal Weighing of Runoff Coefficients.   Values of runoff coefficients may vary within a given catchment. When a clear pattern of variation is apparent, a weighted value of runoff coefficient should be used. For this purpose, the individual subcatchments are delineated and their respective runoff coefficients are identified. The weighted value is obtained by weighing the runoff coefficients in proportion to the areas of their respective subcatchments. This leads to

$$Q_p = 0.278I(\sum_i C_i A_i) \tag{4-14}$$

in which C_i = runoff coefficient of ith subcatchment and A_i = drainage area of ith subcatchment. Applicable units are those of Eq. 4-2.

Composite Catchments.   A *composite catchment* is one that drains two or more adjacent subareas of widely differing characteristics. For instance, assume that a catchment has two subareas A and B with times of concentration t_A and t_B, respectively, with t_A being much less than t_B (Fig. 4-5).

To apply the rational method to this composite catchment, several rainfall durations are chosen, ranging from t_A to t_B, in suitable increments. The calculation proceeds by trial and error, with each trial associated with each rainfall duration. To calculate the partial contribution from subarea B, an assumption must be made regarding the rate at which the flow is concentrated at the catchment outlet. The rainfall duration that gives the highest combined peak flow *(A plus B)* is taken as the design rainfall duration. The procedure is illustrated by the following example.

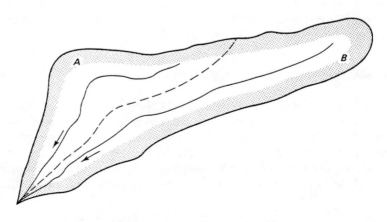

Figure 4-5   Rational method: composite catchments.

Example 4-1.

Calculate the peak discharge by the rational method for a 1-km² composite catchment with the following characteristics:

	Subarea A	Subarea B
Area (km²)	0.4	0.6
Runoff Coefficient	0.6	0.3
Time of Concentration (min)	20	60

Assume a return period $T = 10$ y and the following IDF function:

$$I = \frac{1000\, T^{0.2}}{(t_r + 20)^{0.7}} \qquad (4\text{-}15)$$

in which I = rainfall intensity in millimeters per hour; T = return period in years; and t_r = rainfall duration in minutes. To compute the contribution of subarea B, assume that the flow concentrates linearly at the outlet, i.e., each equal increment of time causes an equal increment of area contributing to the flow at the outlet.

First, choose rainfall durations between 20 min and 60 min at 10-min intervals. For each rainfall duration, rainfall intensity is calculated by Eq. 4-15. The assumption of linear concentration for subarea B leads to the following:

Rainfall Duration (min)	Rainfall Intensity (mm/h)	Contributing Area of B (km²)
20	119.83	0.2
30	102.50	0.3
40	90.22	0.4
50	80.99	0.5
60	73.76	0.6

For $t_r = 20$ min, the peak flow is (Eq. 4-14):

$$Q_p = 0.278 \times 119.83[(0.6 \times 0.4) + (0.3 \times 0.2)] = 10.0 \text{ m}^3/\text{s}$$

Successive trials for rainfall durations of 30, 40, 50, and 60 min result in lower peak flows. Therefore, the peak flow is 10.0 m³/s and the design rainfall duration is 20 min.

Effect of Catchment Shape. The rational method is suited to catchments where drainage area increases more or less linearly with catchment length. If this is not the case, the peak flow may not increase with an increase in catchment area. To illustrate, take the catchment shown in Fig. 4-6. The time of concentration to point A is t_A; the time of concentration to point B is t_B; and t_B is greater than t_A. Therefore, I_A is greater than I_B. The drainage area to point A is A_A, and the drainage area to point B is A_B, and A_B is greater than A_A. Assuming the same runoff coefficient for the partial area (to point A) and the total area (to point B), the peak flow at A is $Q_{pA} = CI_A A_A$. Likewise, the peak flow at B is $Q_{pB} = CI_B A_B$. For Q_{pB} to be greater than Q_{pA}, it is necessary that A_B/A_A be greater than I_A/I_B. In other words, the drainage area must grow in the downstream direction at least as fast as the decrease in corresponding rainfall intensity. Otherwise, the peak discharge at A would be greater than that at B. The situation is illustrated by the following example.

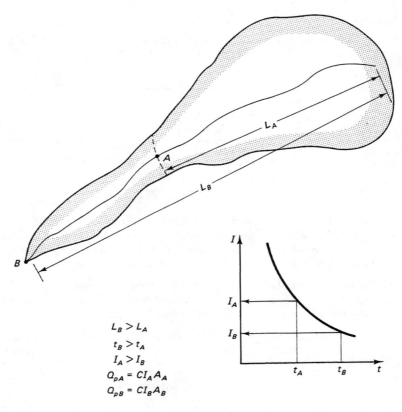

$$L_B > L_A$$
$$t_B > t_A$$
$$I_A > I_B$$
$$Q_{pA} = CI_A A_A$$
$$Q_{pB} = CI_B A_B$$

Figure 4-6 Rational method: effect of catchment shape.

Example 4-2.

Assume that the drainage area at A (Fig. 4-6) has a time of concentration such that the applicable rainfall intensity is 50 mm/h, and that from point A to point B the time of concentration increases, thereby decreasing the applicable rainfall intensity for the drainage area at B to 40 mm/h. Assume that the drainage area at A is 0.8 km² and at B is 0.9 km². Compute the peak flow at points A and B. Assume $C = 0.5$.

The peak flow at A is (Eq. 4-2): $Q_{pA} = 0.278 \times 0.5 \times 50 \times 0.8 = 5.56$ m³/s. The peak flow at B is: $Q_{pB} = 0.278 \times 0.5 \times 40 \times 0.9 = 5.00$ m³/s. It is seen that for this case the peak flow decreases from A to B. This is because the ratio of drainage areas 0.9/0.8 = 1.125 is less than the inverse ratio of rainfall intensities 50/40 = 1.25.

Modified Rational Method. The application of the rational method to large urban catchments, i.e., those featuring well-defined conveyance channels and drainage areas greater than 1.3 km² but less than 2.5 km², requires special techniques. For one thing, the flow is likely to vary widely along the main channel, ranging from small at the upstream reaches to larger at the downstream reaches. In this case, it may be difficult to determine an average value of concentration time.

An alternative is to apply the rational method incrementally, using a technique known as the *modified rational method*. The method requires the subdivision of the catchment into several subcatchments, as shown in Fig. 4-7. First, the time of concen-

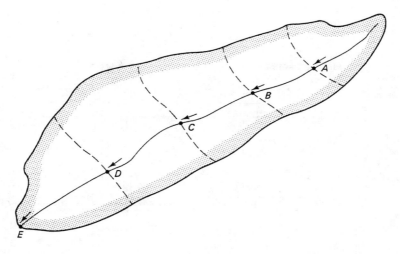

Figure 4-7 Catchment subdivision in modified rational method.

tration t_A is estimated and used to calculate the peak flow Q_{pA} at A, using Eq. 4-2. With the aid of open channel flow formulas, Q_{pA} is conveyed through the main channel from A to B, and the travel time t_{AB} calculated. The time of concentration, $t_B = t_A + t_{AB}$, is used to calculate the peak flow Q_{pB} at B, again using Eq. 4-2. The procedure continues in the downstream direction until the peak flow Q_{pE} is calculated. If the runoff coefficients are different for each subcatchment, Eq. 4-14 can be used in lieu of Eq. 4-2. While the procedure is relatively straightforward, it may result in peak flows decreasing in the downstream direction (due to the effect of catchment shape).

Application to Storm-sewer Design. A typical plan for design of a small storm-sewer project is shown in Fig. 4-8. Table 4-2 shows a summary of the computations illustrating the application of the rational method to determine design flows. The example is based on the following conditions:

1. Runoff Coefficients
   (a) Residential area: $C = 0.3$
   (b) Business area: $C = 0.6$
   (c) Areal weighing of runoff coefficients where required.
2. Intensity-Duration-Frequency curve shown in Fig. 2-8(a). Selected design frequency: 5 y.
3. Inlet time: 20 min.
4. Manning n in sewer: 0.013.
5. Free outfall to river at elevation 80.
6. A drop of 0.1 ft across each manhole where no change in pipe size occurs (to account for head losses). When a change in pipe size occurs, set the elevation of 0.8 of pipe depths equal, and provide corresponding fall in manhole invert. (*Note:* In larger systems, a more rigorous analysis of hydraulic losses through manholes, transitions, and changes in direction is required for adequate hydraulic design).

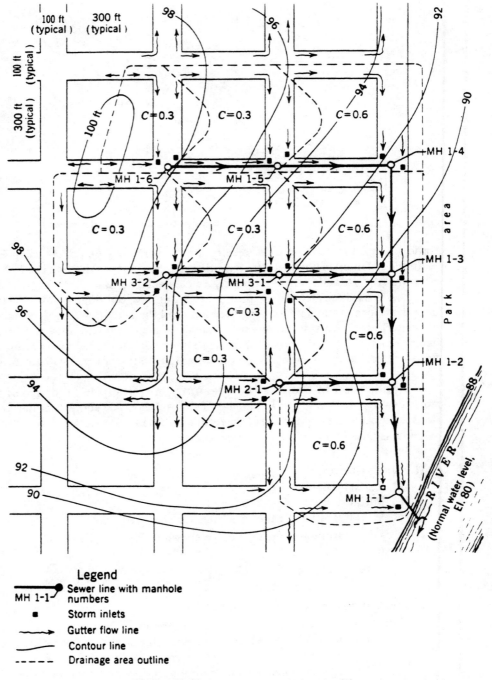

Figure 4-8 Typical storm-sewer design plan [2].

TABLE 4-2 SUMMARY OF COMPUTATIONS ILLUSTRATING APPLICATION OF THE RATIONAL METHOD TO STORM-SEWER DESIGN

(1)	(2)	(3)	(4)	(5)	(6)	(7)	(8)	(9)	(10)	(11)	(12)	(13)	(14)	(15)	(16)	(17)	(18)	(19)	(20)	(21)	(22)	(23)	(24)
1	1-6	1-5	400	2.64	2.64	20.0	1.4	0.3	3.7	1.11	2.93	0.85	12	3.3	4.0	4.6	9	...	3.40	93.0	89.6	98.4	94.9
1	1-5	1-4	400	3.61	6.25	21.4	1.2	0.3	3.6	1.08	6.75	0.75	18	9.2	5.1	5.6	11	0.40	3.00	89.2	86.2	94.9	91.8
1	1-4	1-3	400	3.88	10.13	22.6	1.2	0.42	3.4	1.43	14.50	0.45	24	15.2	4.8	5.6	18	0.40	1.80	85.8	84.0	91.8	89.7
3	3-2	3-1	400	5.55	5.55	20.0	1.1	0.3	3.7	1.11	6.16	1.00	15	6.4	5.1	5.9	12	...	4.00	91.0	87.0	96.2	92.3
3	3-1	1-3	400	6.43	11.98	21.1	1.1	0.3	3.6	1.08	12.92	0.60	24	17.5	5.5	6.1	15	0.60	2.40	86.4	84.0	92.3	89.7
1	1-3	1-2	400	3.92	26.03	23.8	1.10	0.39	3.3	1.29	33.60	0.30	36	37.0	5.1	5.9	26	0.80	1.20	83.2	82.0	89.7	89.5
2	2-1	1-2	400	2.52	2.52	20.0	1.4	0.3	3.7	1.11	2.80	0.90	12	3.2	4.1	4.7	9	...	3.6	87.5	83.9	92.7	89.5
1	1-2	1-1 Out-	400	3.86	32.41	24.9	1.1	0.41	3.2	1.31	42.50	0.24	42	50.0	5.2	5.9	29	0.4	0.96	81.6	80.64	89.5	88.5
1	1-1	fall	125	5.44	37.85	26.0	...	0.44	3.2	1.41	53.20	0.30	42	56.0	5.7	6.6	33	0.1	0.38	80.54	80.16	88.5	...

KEY

(1) Line; (2) Manhole, From; (3) Manhole, To; (4) Length, ft; (5) Area, increment, acres; (6) Area, total, acres; (7) Flow time, to upper end, minutes; (8) Flow time, in section, minutes; (9) Average runoff coefficient; (10) Rainfall, in./h; (11) Runoff, ft^3/s/ac; (12) Total runoff, ft^3/s; (13) Slope of sewer, percentage; (14) Diameter, in.; (15) Capacity, full, ft^3/s; (16) Velocity, full, ft/s; (17) Design flow velocity, ft/s; (18) Depth of flow, in.; (19) Manhole invert drop, ft; (20) Fall in sewer, ft; (21) Sewer invert, lower end; (22) Sewer invert, upper end; (23) Ground elevation, upper end; (24) Ground elevation, lower end.

This example is extracted from *Design and Construction of Sanitary and Storm Sewers,* ASCE Manual of Engineering Practice no. 37, 1960 [2].

4.2 OVERLAND FLOW

Overland flow is surface runoff that occurs in the form of sheet flow on the land surface without concentrating in clearly defined channels. This type of flow is the first manifestation of surface runoff, since the latter occurs first as overland flow before it has a chance to flow into channels and become streamflow.

Overland flow theory uses deterministic methods to describe surface runoff in overland flow planes. The theory is based on established principles of fluid mechanics such as laminar and turbulent flow, mass and momentum conservation, and unsteady free surface flow. The spatial and temporal description leads to differential equations and to their solution by analytical or numerical means. In certain cases, simplified conceptual models can be developed for practical use.

Overland flow theory seeks to find an answer to the problem of catchment response; i.e., what is the hydrograph that will be produced at a catchment's outlet, subject to a given effective rainfall? In overland flow applications, effective rainfall is also referred to as *rainfall excess.* Unlike the rational method, which generally does not produce a hydrograph, overland flow models have the capability to account not only for runoff concentration but also for runoff diffusion. Another advantage of overland flow models is their distributed nature, i.e., the fact that rainfall excess can be allowed to vary in space and time if necessary. Overland flow models, then, are a more powerful tool than parametric models such as the rational method. However, the complexity increases in direct relation to their increased level of detail.

As with the rational method, a question that must be addressed at the outset is the following: What size catchment can be analyzed with overland flow techniques? Here again, the answer is not very clear. Intuitively, overland flow computations should be applicable to small catchments, primarily because overland flow is the main surface-flow feature of small catchments. The method, however, is not necessarily restricted to small catchments. Midsize catchments may also benefit from the increased detail of overland flow models. The actual limit is a practical one. Computations need to be performed in modules of relatively small size; otherwise, it is likely that the terrain's topographic, frictional, and vegetative features will not be properly represented in the overland flow model. In practice, overland flow techniques are restricted to catchments for which the surface features can be adequately represented within the model's topological structure. Otherwise, the amount of lumping introduced (i.e., temporal and spatial averaging) would interfere with the method's ability to predict the occurrence of flows in a distributed context.

Unlike the rational method, overland flow applications may require substantial computer facilities. Overland flow techniques often form part of computer models that simulate all relevant phases of the hydrologic cycle. These models use overland flow techniques in their catchment routing component. With the widespread use of computers, it is likely that overland flow techniques will play an increasing role in routine hydrologic engineering practice.

The fundamentals of overland flow theory are presented here. Catchment routing methods are described in Chapter 10.

Overland Flow Theory

The mathematical description of overland flow begins with the equation of mass conservation of fluid mechanics, also referred to as the *continuity equation*. In one-dimensional flow, this equation states that the change in flow per unit length (in the flow direction) in a control volume is balanced by the change in flow area per unit time:

$$\frac{\partial Q}{\partial x} + \frac{\partial A}{\partial t} = 0 \tag{4-16}$$

This equation does not include sources or sinks. Inclusion of the latter leads to:

$$\frac{\partial Q}{\partial x} + \frac{\partial A}{\partial t} = q_L \tag{4-17}$$

in which q_L = lateral inflow or outflow (inflow positive, outflow negative), or net lateral flow per unit length, in L^2T^{-1} units.

In small-catchment hydrology, overland flow is assumed to take place on the overland flow plane. This is a plane of length L (in the flow direction), slope S_o (Fig. 4-9), and of theoretically infinite width. Therefore, a unit-width analysis is appropriate. For a unit width, Eq. 4-17 is converted to:

$$\frac{\partial q}{\partial x} + \frac{\partial h}{\partial t} = i \tag{4-18}$$

in which q = flow rate per unit width; h = flow depth; and i = lateral inflow (rainfall excess), or inflow per unit area, in LT^{-1} units. While lateral inflow can vary in time and space, a first approximation is to consider it constant.

Flow over the plane can be described as follows: As excess rainfall begins, water accumulates on the plane surface and begins to flow out of the plane at its lower end.

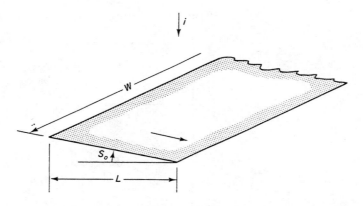

Figure 4-9 Overland flow plane.

Flow at the outlet (i.e., the outflow) increases gradually from zero, while the total volume of water stored over the plane also increases gradually. Eventually, if rainfall excess continues, both outflow and total volume of water stored over the plane reach a constant value. These constants are referred to as *equilibrium outflow* and *equilibrium storage volume*. For continuing rainfall excess, outflow and storage volume remain constant and equal to the equilibrium value. Immediately after excess rainfall ceases, outflow begins to draw water from storage, gradually decreasing while depleting the storage volume. Eventually, outflow returns to zero as the storage volume is completely drained.

The process is depicted in Fig. 4-10. The flow from start to equilibrium is called the *rising limb* of the overland flow hydrograph. The flow from equilibrium back to zero is called the *receding limb* of the hydrograph. The equilibrium outflow can be calculated by recognizing that at equilibrium state, the outflow must equal the inflow (i.e., rainfall excess). Therefore,

$$q_e = \left(\frac{i}{3600}\right)L \tag{4-19}$$

in which q_e = equilibrium outflow in liters per second per meter; i = rainfall excess in millimeters per hour; and L = plane length in meters.

Equation 4-19 is essentially a statement of runoff concentration, similar to Eq. 2-60 or to Eq. 4-1 with $C = 1$. Whether the flow actually does concentrate and reach its equilibrium value will depend upon the duration of the rainfall excess t_r and the time t_e required to reach equilibrium. If t_r is greater than t_e, equilibrium is reached.

The volume of storage is the area below the line $q = q_e$ and above the rising limb of the overland flow hydrograph, as shown in Fig. 4-10. As a first approximation, the (shaded) area above the rising limb can be assumed to be equal to the area below the rising limb. In this case, the equilibrium storage volume is:

$$S_e = \frac{q_e t_e}{2} \tag{4-20}$$

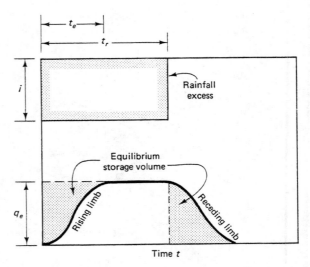

Figure 4-10 Sketch of overland flow hydrograph.

Hydrology of Small Catchments Chap. 4

in which S_e = equilibrium storage volume in liters per meter; q_e = equilibrium outflow in liters per second per meter; and t_e = time to equilibrium in seconds.

In practice, surface and other irregularities cause the equilibrium state to be approached asymptotically, and therefore the actual time to equilibrium is not clearly defined. A value of t corresponding to $q = 0.98 q_e$ may be taken as a practical measure of t_e. Then, Eq. 4-20 is only an approximation of the actual storage volume.

The equation of continuity, Eq. 4-18, can also be expressed in the following form:

$$\left(\frac{1}{i}\right) \frac{\partial h}{\partial t} + \left(\frac{u}{i}\right) \frac{\partial h}{\partial x} + \left(\frac{h}{i}\right) \frac{\partial u}{\partial x} = 1 \qquad (4\text{-}21)$$

in which $u = q/h$ = mean velocity.

The value of equilibrium outflow was obtained from Eq. 4-19 based on continuity considerations. However, the shape of the rising and receding limbs and the time to equilibrium remain to be determined. This can be obtained through the equation of momentum conservation (or equation of motion), following established principles of unsteady open channel flow [3, 9, 17]. The equation of motion, however, is a nonlinear partial differential equation. A form of this equation with u and h as dependent variables is [17]:

$$\left(\frac{1}{g}\right) \frac{\partial u}{\partial t} + \left(\frac{u}{g}\right) \frac{\partial u}{\partial x} + \frac{\partial h}{\partial x} + S_f - S_o + \frac{iu}{gh} = 0 \qquad (4\text{-}22)$$

in which S_f = friction slope, S_o = plane slope, g = gravitational acceleration, and all other terms have been previously defined. All terms in Eqs. 4-21 and 4-22 are dimensionless.

The solution of Eqs. 4-21 and 4-22 can be attempted in a variety of ways. Analytical solutions are usually based on an assumption of linearity [1, 21]. Numerical solutions have been extensively applied to stream and river flow problems [15, 17]. To date, overland flow problems have been solved with the following approaches: (1) storage concept, (2) kinematic wave technique, (3) diffusion wave technique, and (4) dynamic wave technique.

The storage concept is similar to that used in reservoir routing (Chapter 8). The kinematic wave technique simulates runoff concentration in the absence of runoff diffusion. The diffusion wave technique simulates runoff concentration in the presence of small amounts of runoff diffusion. The dynamic wave technique solves the complete set of governing equations, Eqs. 4-21 and 4-22, including runoff concentration, diffusion, and dispersion processes. For practical applications, the storage concept and kinematic and diffusion wave techniques can be shown to be useful approximations to the complete equations (Chapter 9).

In principle, the kinematic wave is an improvement over the storage concept. In turn, the diffusion wave is an improvement over the kinematic wave, whereas the dynamic wave is an improvement over the diffusion wave. Invariably, the effort involved in obtaining a solution increases in direct relation to the complexity of the governing equations, including initial and boundary conditions. The storage and kinematic wave techniques are described in the following sections. A brief introduction to the diffusion wave technique is also given, but the dynamic wave solution for overland flow [4] is outside the scope of this book.

Overland Flow Solution Based on Storage Concept

Early approaches to solve the overland flow problem are attributed to Horton [10] and Izzard [13,14]. In particular, Horton noticed that experimental data justified a relationship between equilibrium outflow and equilibrium storage volume of the following form:

$$q_e = aS_e^m \qquad (4\text{-}23)$$

in which a and m are empirical constants. A mean flow depth h_e is defined in the following way:

$$h_e = \frac{S_e}{L} \qquad (4\text{-}24)$$

Combining Eqs. 4-23 and 4-24:

$$q_e = bh_e^m \qquad (4\text{-}25)$$

in which $b = aL^m$, another constant. The value of the exponent m is a function of flow regime, depending on whether the latter is laminar, turbulent (either Manning or Chezy), or mixed laminar-turbulent. Typical values of m are shown in Table 4-3.

An estimate of time to equilibrium can be obtained by combining Eqs. 4-20 and 4-23 and solving for t_e:

$$t_e = \frac{2}{q_e^{(m-1)/m}a^{1/m}} \qquad (4\text{-}26)$$

For laminar flow conditions, $b = aL^m = C_L$, where C_L is defined as [3]:

$$C_L = \frac{gS_o}{3\nu} \qquad (4\text{-}27)$$

TABLE 4-3 TYPICAL VALUES OF EXPONENT m IN EQS. 4-23 or 4-25

Flow regime	m
Laminar	3.0
Turbulent	
based on Manning formula	1.667
based on Chezy formula	1.5
Mixed laminar-turbulent	
(based on Manning formula)	1.667 to 3.0
75% turbulent	2.0
50% turbulent	2.333
25% turbulent	2.667
Mixed laminar-turbulent	
(based on Chezy formula)	1.5 to 3.0
75% turbulent	1.875
50% turbulent	2.25
25% turbulent	2.625

and ν = kinematic viscosity, a function of water temperature (see Tables A-1 and A-2, Appendix A). The units of C_L are $L^{-1}T^{-1}$. Furthermore, with $q_e = iL$, Eq. 4-26 reduces to the following for the case of $m = 3$ (laminar flow):

$$t_e = \frac{2L^{1/3}}{i^{2/3}C_L^{1/3}} \qquad (4\text{-}28)$$

in which t_e = time to equilibrium, in seconds; L = length of overland flow plane, in meters; and i = rainfall intensity, in meter per second.

For turbulent Manning flow conditions, $b = aL^m = (1/n)S_o^{1/2}$, in which n is the Manning constant. With $q_e = iL$, Eq. 4-26 reduces to the following for the case of mixed laminar-turbulent flow (m in the range 5/3 to 3):

$$t_e = \frac{2(nL)^{1/m}}{i^{(m-1)/m}S_o^{1/(2m)}} \qquad (4\text{-}29)$$

with the same units as Eq. 4-28 (t_e in seconds, L in meters, i in meters per second). Notice that (as expected) time to equilibrium increases with friction and plane length and decreases with rainfall intensity and plane slope.

Equation 4-29 was developed by assuming the Manning formula in the rating. Therefore, it is strictly applicable only to $m = 5/3$. In practice, however, this equation is also used for mixed laminar-turbulent flows with m in the range 5/3 to 3. In addition, the very shallow depths that usually prevail in overland flow calculations result in a substantial increase in friction. These differences are accounted for by using an effective roughness parameter N in lieu of the Manning constant [11]. Values of N are given in Table 4-4.

The Horton-Izzard solution to the overland flow problem is based on the assumption that Eq. 4-23 is valid not only at equilibrium but also at any other time:

$$q = aS^m \qquad (4\text{-}30)$$

in which q = outflow at time t and S = storage volume at time t. This assumption is convenient because it allows an analytical solution for the shape of the overland flow hydrograph. Equation 4-30 is the formula for a nonlinear reservoir, i.e., a function relating outflow and storage volume in a nonlinear way ($m \neq 1$). Therefore, a nonlin-

TABLE 4-4 EFFECTIVE ROUGHNESS PARAMETER FOR OVERLAND FLOW [11]

Type of Surface	N
Dense growth	0.40–0.50
Pasture	0.30–0.40
Lawns	0.20–0.30
Bluegrass sod	0.20–0.50
Short-grass prairie	0.10–0.20
Sparse vegetation	0.05–0.13
Bare clay-loam soil (eroded)	0.01–0.03
Concrete-asphalt	
Very shallow depths, less than 6 mm	0.10–0.15
Shallow depths, 6 mm or more	0.05–0.10

ear reservoir is being used to model the equation of motion, Eq. 4-22. In essence, a deterministic model (Eq. 4-22) has been replaced by a conceptual model (Eq. 4-30).

Equation 4-16 can be expressed in one space increment Δx to yield:

$$I - O = \frac{dS}{dt} \tag{4-31}$$

in which I = inflow to the control volume, O = outflow from the control volume, and dS/dt = rate of change of storage in control volume (Fig. 4-11). For the overland flow case, $I = iL$ and $O = q$. Therefore:

$$iL - q = \frac{dS}{dt} \tag{4-32}$$

which through Eqs. 4-19, 4-23, and 4-30 leads to:

$$aS_e^m - aS^m = \frac{dS}{dt} \tag{4-33}$$

Integrating Eq. 4-33:

$$t = \frac{1}{a} \int \frac{1}{S_e^m - S^m} \, dS \tag{4-34}$$

and through additional algebraic manipulation:

$$t = \frac{1}{a^{1/m} q_e^{(m-1)/m}} \int \frac{d(q/q_e)^{1/m}}{1 - (q/q_e)} \tag{4-35}$$

Equation 4-35 was solved by Horton for $m = 2$, which describes a flow regime that is 75% turbulent (between laminar, for which $m = 3$, and turbulent Manning, for which $m = 5/3$). Horton's solution can be expressed as follows [1]:

$$\frac{q}{q_e} = \tanh^2 \left[2\left(\frac{t}{t_e}\right) \right] \tag{4-36}$$

in which t_e = time to equilibrium.

The dimensionless rising hydrograph for the case of $m = 3$ is shown in Fig. 4-12 [14]. With the aid of this figure, Eqs. 4-19 for q_e, and Eq. 4-28 for t_e, the rising limb of the overland flow hydrograph for $m = 3$ can be calculated.

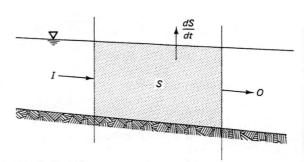

Figure 4-11 Inflow, outflow, and rate of change of storage in a control volume.

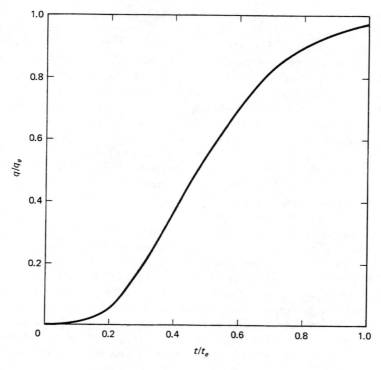

Figure 4-12 Dimensionless rising hydrograph based on storage concept, $m = 3$ [13].

Several assumptions limit the applicability of the Horton-Izzard solution. The most important one is the nonlinear form of the storage equation, Eqs. 4-23 and 4-30. The overall validity of this assumption remains to be established. Izzard has suggested that the method should be restricted to cases where the product of rainfall intensity (in millimeters per hour) and plane length (in meters) *(iL)* does not exceed 3000. Notwithstanding the apparent limitations, the Horton-Izzard solution of overland flow has been extensively used, particularly in the design of airport drainage [3, 7].

Overland Flow Solution Based on Kinematic Wave Theory

According to this theory, Eq. 4-22 can be approximated by a single-valued flow-depth rating at any point, leading to

$$q = bh^m \tag{4-37}$$

in which b and m are constants analogous to those of Eq. 4-25. Unlike Horton's approach, which bases the rating on the storage volume on the entire plane (Eqs. 4-23 and 4-30), the kinematic wave approach bases the rating on flow depths at individual cross sections. This difference has substantial implications for computer modeling because whereas the Horton approach is lumped in space, the kinematic approach is not, and therefore it is better suited to distributed computation.

The difference between a rating based on storage volume and one based on flow depth merits further consideration. There are two distinct features in free surface flow in natural catchments: (1) reservoirs and (2) channels. In an ideal reservoir, the water surface slope is zero (Fig. 4-13(a)), and, therefore, outflow and storage volume are uniquely related. If an outflow rating is desired, storage volume can be uniquely related to stage; consequently, outflow can be uniquely related to stage and flow depth. On the other hand, in an ideal channel, the water surface slope is nonzero (Fig. 4-13(b)), and, in general, storage is a nonunique function of inflow and outflow. If a flow rating is desired, the only practical way of obtaining it is to relate flow to its depth.

In the Horton approach, outflow is related to storage volume and, by extension, to the mean flow depth on the overland plane. In the kinematic wave approach, outflow is related to the outflow depth. Since the typical overland flow problem has a nonzero water surface slope, it is more likely to behave as a channel rather than a reservoir. Therefore, it would appear that the kinematic wave approach is a better model of the physical process than the storage concept. Further analysis has shown, however, that while the kinematic wave lacks diffusion, the storage concept does not. In this sense, the storage concept may well be a better model than the kinematic wave approach for cases featuring significant amounts of runoff diffusion.

The kinematic wave assumption, Eq. 4-37, amounts to substituting a uniform flow formula (such as Manning's) for the equation of motion, Eq. 4-22. In essence, it says that as far as momentum is concerned, the flow is steady. The unsteadiness of the phenomena, however, is preserved through the continuity equation, Eq. 4-18 (or Eq. 4-21). The implication of the kinematic wave assumption is that unsteady flow is visualized as a succession of steady uniform flows, with the water surface slope remaining constant at all times. This, of course, can be reconciled with reality only if the flow

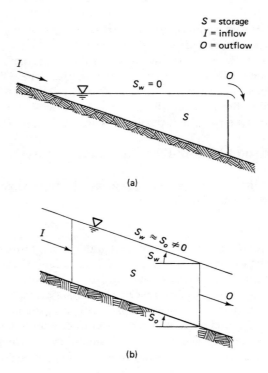

Figure 4-13 (a) Ideal reservoir; (b) ideal channel.

unsteadiness is very mild; that is, if the changes in stage occur very gradually. In practice, a necessary condition for the applicability of Eq. 4-37 to unsteady flows is that the changes in momentum be negligible compared to the force driving the steady flow, i.e., gravity (the plane or channel slope). The *kinematic flow number* used in overland flow applications serves as a quantitative measure of the *kinematicity* of a given unsteady flow condition, that is, of the extent to which Eq. 4-37 is a good surrogate of Eq. 4-22 and, therefore, a valid description of the unsteady flow phenomena.

The application of kinematic wave theory to the overland flow problem begins with Eq. 4-18, repeated here:

$$\frac{\partial q}{\partial x} + \frac{\partial h}{\partial t} = i \qquad (4\text{-}18)$$

Differentiating Eq. 4-37 with respect to flow depth, assuming that b and m are constants (a wide channel of constant friction) gives:

$$\frac{\partial q}{\partial h} = mbh^{m-1} = m\left(\frac{q}{h}\right) = mu = c \qquad (4\text{-}38)$$

in which c = celerity of a kinematic wave. Since m is generally greater than 1 (Table 4-3), the celerity of a kinematic wave is generally greater than the mean flow velocity.

Multiplying Eqs. 4-18 and 4-38 and using the chain rule,

$$\frac{\partial q}{\partial t} + c\frac{\partial q}{\partial x} = ci \qquad (4\text{-}39)$$

which is a form of the kinematic wave equation with q as the dependent variable, applicable to overland flow.

Using the same approach, an expression for kinematic flow in terms of flow depth can be derived:

$$\frac{\partial h}{\partial t} + c\frac{\partial h}{\partial x} = i \qquad (4\text{-}40)$$

With $c = dx/dt$, i.e. the slope of the *characteristic* lines on an xt plane (Fig. 4-14), the left side of Eqs. 4-39 and 4-40 denotes total differentials. Therefore,

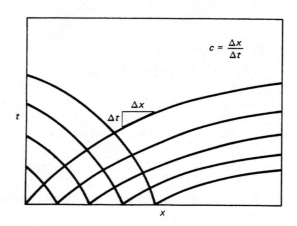

Figure 4-14 Characteristic lines on xt plane.

$$\frac{dq}{dt} = ci \qquad (4\text{-}41)$$

$$\frac{dq}{dx} = i \qquad (4\text{-}42)$$

$$\frac{dh}{dt} = i \qquad (4\text{-}43)$$

$$\frac{dh}{dx} = \frac{i}{c} \qquad (4\text{-}44)$$

In particular, Eq. 4-43 can be integrated to yield:

$$h = it \qquad (4\text{-}45)$$

which implies that the flow depth at any point along the plane increases linearly with time—provided, of course, that rainfall excess i remains constant.

The overland flow solution under the kinematic wave assumption resembles that of the storage concept, with a rising limb, an equilibrium state, and a receding limb. Although the equilibrium flow is the same $(q_e = iL)$, the time to equilibrium is different, as is the shape of rising and receding limbs.

To derive the kinematic wave solution, Eq. 4-37 is expressed in terms of equilibrium outflow:

$$q_e = bh_e^m \qquad (4\text{-}46)$$

in which, unlike in Eqs. 4-24 and 4-25, h_e is now interpreted as the equilibrium flow depth at the outlet. Dividing Eq. 4-37 by Eq. 4-46 leads to:

$$\frac{q}{q_e} = \left(\frac{h}{h_e}\right)^m \qquad (4\text{-}47)$$

Since $h = it$ (Eq. 4-45), this leads to:

$$\frac{q}{q_e} = \left(\frac{t}{t_k}\right)^m \qquad (4\text{-}48)$$

in which t_k = kinematic time parameter, defined as:

$$t_k = \frac{h_e}{i} \qquad (4\text{-}49)$$

Equation 4-48 is applicable for $t \leq t_k$. Otherwise, the flow would exceed the equilibrium value, which is clearly a physical impossibility. In this sense, the kinematic time parameter may be interpreted as a kinematic time-to-equilibrium.

With Eq. 4-46, and since $q_e = iL$ and $b = (1/n)S_o^{1/2}$ (for turbulent Manning friction in wide channels), the kinematic time parameter can be expressed as follows:

$$t_k = \frac{(nL)^{1/m}}{i^{(m-1)/m} S_o^{1/(2m)}} \qquad (4\text{-}50)$$

which is the same as Eq. 4-29 but without the factor 2. In other words, the analytical solution to the kinematic wave equation is a parabola (Eq. 4-48), whereas the analytical solution of the storage concept is a hyperbolic trigonometric function (Eq. 4-36).

The kinematic approach to the solution of the overland flow problem was studied in detail by Iwagaki [12], Henderson and Wooding [8], Wooding [26, 27, 28], and Woolhiser and Liggett [29], among others. Wooding developed the concept of the *open book* shown in Fig. 4-15, which has been extensively used in catchment modeling (see Chapter 10). The open book is formed by two overland flow planes. The outflow from the planes is lateral inflow to the channel, which conveys the flow to the catchment outlet.

Woolhiser and Liggett calculated the shape of the rising hydrograph under kinematic flow. Furthermore, they established the limit for the kinematicity of the flow in terms of the kinematic flow number, defined as follows:

$$K = \frac{S_o L}{F^2 h_o} \tag{4-51}$$

in which K = kinematic flow number, a dimensionless number; F = Froude number corresponding to the equilibrium flow at the outlet; and h_o = equilibrium flow depth at the outlet (i.e., h_e). Values of K greater than 20 describe kinematic flow, while lower values do not [18]. In other words, for low K values, Eq. 4-37 is no longer a good approximation to Eq. 4-22.

In particular, since S_o is likely to vary within a wider range than either L, F, or h_o, Eq. 4-51 could be interpreted to mean that the property of kinematicity is directly related to plane slope: the steeper the slope, the greater the K value and the more kinematic the resulting flow is. Conversely, the milder the slope, the lower the K value and the less kinematic the flow is. The reason for this inability of the kinematic wave solution to account for a wide range of slopes is apparent from the nature of Eq. 4-22. The kinematic wave solution accounts only for friction slope and plane slope. All other terms are excluded from the formulation and are, therefore, absent from the solution. For very mild plane slopes, the importance of these terms may be promoted to the point where neglecting them is no longer justified.

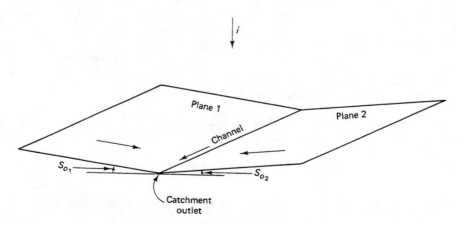

Figure 4-15 Wooding's open-book catchment schematization [26].

While this appears to impose stringent limitations, the situation in practice is quite different. Most overland flow problems have steep slopes, on the order of $S_o = 0.01$ or more, resulting in the flow being essentially kinematic, as confirmed by their kinematic flow number (Eq. 4-51). However, much milder slopes, around $S_o = 0.0001$, result in very low kinematic flow numbers; for these cases, the kinematic wave solution is not sufficient.

A source of complexity in kinematic wave solutions arises from the fact that the wave celerity in Eq. 4-39 varies with the flow, making it a nonlinear (i.e., quasilinear) equation. At first, this appears to be an advantage. Further examination, however, reveals that this property may lead to a steepening tendency of the wave. Analytical solutions, if carried long enough (i.e., in very long channels), invariably lead to the phenomenon called *kinematic shock*, the steepening of the kinematic wave to the point where it attains an almost vertical face. In nature, small amounts of diffusion and other irregularities usually act in such a way as to control and arrest shock development. An analytical solution, however, has no such imperfections. In essence, the total absence of diffusion in the analytical solution permits the uncontrolled development of the kinematic shock. Numerical solutions, however, usually have small amounts of diffusion and are, therefore, not perfect, with the shock being a rare occurrence in this case. This fact gives rise to practical implications for catchment and stream channel routing. A detailed discussion on this subject is given in Chapters 9 and 10.

Overland Flow Solution Based on Diffusion Wave Theory

According to diffusion wave theory, the flow depth gradient $\partial h/\partial x$ in Eq. 4-22 is largely responsible for the diffusion mechanism, which is naturally present in unsteady free surface flows. Therefore, its inclusion in the analysis should provide runoff concentration with diffusion.

The inclusion of the depth gradient term substantially increases the difficulty of obtaining a solution. An established procedure is to linearize the governing equations, Eqs. 4-21 and 4-22, around reference flow values. This entails using small-perturbation theory to develop linear analogs of these equations. This procedure, while heuristic, has worked well in a number of applications.

Following Lighthill and Whitham [16], the linear analogs of Eqs. 4-21 and 4-22, neglecting the inertia terms (the first two terms of Eq. 4-22) and the momentum source term for simplicity, are, respectively,

$$\frac{\partial h}{\partial t} + u_o \frac{\partial h}{\partial x} + h_o \frac{\partial u}{\partial x} = i \tag{4-52}$$

$$\frac{\partial h}{\partial x} + S_o \left[2\left(\frac{u}{u_o}\right) - (4/3)\left(\frac{h}{h_o}\right) \right] = 0 \tag{4-53}$$

The coefficients of Eq. 4-52 are constant, being the reference flow velocity u_o and depth h_o, respectively. The second term of Eq. 4-53 represents a combined linear version of the friction slope S_f and channel slope S_o. Furthermore, in Eq. 4-53, friction obeys the Manning friction formula. If the Chezy formula is used, the factor $4/3$ is replaced by 1.

Differentiating Eq. 4-53 with respect to space and eliminating the term $\partial u / \partial x$, from the resulting equation (with the aid of Eq. 4-52), the following is obtained:

$$\frac{\partial h}{\partial t} + \left(\frac{5}{3}\right) u_o \frac{\partial h}{\partial x} = i + \left(\frac{u_o h_o}{2 S_o}\right) \frac{\partial^2 h}{\partial x^2} \tag{4-54}$$

A similar expression with discharge as dependent variable can also be derived, albeit at the cost of increased algebraic manipulation (Chapter 9).

Unlike the kinematic wave equation (Eq. 4-40), which has no second-order term, the diffusion wave equation (Eq. 4-54) has a second-order term. Therefore, the latter can describe not only runoff concentration but also runoff diffusion. Equation 4-54 can be expressed in the following form:

$$\frac{\partial h}{\partial t} + c \frac{\partial h}{\partial x} = i + \nu_h \frac{\partial^2 h}{\partial x^2} \tag{4-55}$$

in which

$$c = (5/3) u_o \tag{4-56}$$

is the reference celerity of a diffusion wave (for turbulent Manning friction in wide channels) and

$$\nu_h = \frac{u_o h_o}{2 S_o} \tag{4-57}$$

is the *hydraulic diffusivity*, or channel diffusivity, to differentiate it from kinematic viscosity (ν).

Equation 4-54 implies that the diffusion component (second-order term) is small compared with the concentration component (first-order terms) and that the channel diffusivity controls the diffusion contribution to the flow. In effect, for very low values of S_o, channel diffusivity is very large. In the limit, as channel slope approaches zero, channel diffusivity grows unbounded and Eq. 4-54 is no longer applicable. However, for realistic mild channel slopes (approximately in the range 0.001–0.0001), the contribution of the diffusion term can be quite significant. Diffusion wave theory, then, applies for the milder slopes for which kinematic wave theory is not sufficient. Its application to catchment routing and hydrologic models of overland flow is discussed in Chapter 10.

In summary, overland flow techniques can provide more detail than the rational method in describing the peak and timing of outflow hydrographs from small catchments. The increase in detail, however, is invariably associated with an increase in complexity. The overland flow approach is particularly suited to computer applications, where numerous computations can be readily made in an effective way. In practice, an abstraction module must be coupled with the overland flow module in order to arrive at a meaningful model. Ideally, the abstraction module should be as detailed as the overland flow module. In practice, however, due to the complex spatial and temporal variability of the abstractive processes, this consistency between model components is seldom met.

QUESTIONS

1. Name three properties that characterize a small catchment. Explain each one of them.
2. What hydrologic processes does the rational method account for? Explain how they affect runoff.
3. What processes are not considered in the rational method? Explain.
4. How are the frequencies of storms and floods related? How does the rational method account for this difference?
5. What processes are included in the runoff coefficient?
6. Under what assumption does the rational method provide the shape of the outflow hydrograph?
7. Describe the phenomenon of overland flow. Contrast overland flow analysis with the rational method approach.
8. What crucial assumption makes the storage solution of overland flow different from the kinematic wave solution?
9. Why is overland flow likely to be of mixed laminar-turbulent nature? How is this modeled in practice?
10. What is an ideal reservoir? An ideal channel? Contrast these two concepts.
11. What is the kinematic wave celerity? Why is it generally greater than the flow velocity?
12. What is the kinematic flow number? What does it describe? For what kind of channel slope is it likely that the kinematic wave solution would not be applicable?
13. What is kinematic shock? Why does it often occur in analytic kinematic wave solutions while it is seldom present in numerical solutions?
14. Contrast kinematic and diffusion wave approaches to overland flow.
15. Why does Eq. 4-54 describe runoff diffusion, but Eq. 4-40 does not? What is channel diffusivity?

PROBLEMS

4-1. Rain falls on a 150-ha catchment with intensity 2 cm/h and duration 2 h. Use the rational method to calculate the peak runoff, assuming runoff coefficient $C = 0.6$ and time of concentration $t_c = 1.5$ h.

4-2. Rain falls on a 300-ac watershed with intensity 0.5 in./h and duration 2 h. Use the rational method to calculate the peak runoff, assuming runoff coefficient $C = 0.4$ and time of concentration $t_c = 2$ h.

4-3. Rain falls on a 545-ha catchment with intensity 45 mm/h and duration 1 h. Use the rational method to calculate the peak runoff for the following conditions: (a) natural, with time of concentration 2 h and $C = 0.4$; (b) improved, partially paved area, with time of concentration $t_c = 1$ h and $C = 0.7$. State any assumptions used.

4-4. Rain falls on a 1.5-km² watershed with intensity 20 mm/h and duration 2 h. Use the rational method to calculate the peak runoff for the following two conditions: (a) vegetated (natural) watershed with time of concentration $t_c = 3$ h and $C = 0.3$; and (b) improved, partially paved area with time of concentration $t_c = 2$ h and $C = 0.6$. State any assumptions used.

4-5. Rain falls on a catchment with intensity 35 mm/h and duration 2 h. The catchment area is 250 ha, with time of concentration $t_c = 2$ h and $\phi = 15$ mm/h. Calculate the peak runoff. State any assumptions used.

4-6. Rain falls on a watershed with intensity 1 in/h and duration 3 h. The watershed area is 500 ac, with time of concentration $t_c = 2$ h and $\phi = 0.3$ in./h. Calculate the peak runoff. State any assumptions used.

4-7. Rain falls on a watershed with intensity 30 mm/h and duration 1 h. The watershed area is 0.8 km² with time of concentration $t_c = 2$ h and $\phi = 15$ mm/h. Use the rational method to calculate the peak runoff. State any assumptions used.

4-8. Rain falls on a 125-ha catchment with the following characteristics: (1) 20%, $C = 0.3$; (2) 30%, $C = 0.4$; (3) 50%, $C = 0.6$. Calculate the peak runoff due to a storm of 45 mm/h intensity lasting 1 h. Assume time of concentration $t_c = 30$ min.

4-9. Rain falls on a 90-ha catchment with the following characteristics: (1) 12 ha, $C = 0.3$; (2) 48 ha, $C = 0.7$; (3) 30 ha, $C = 0.9$. Calculate the peak runoff from a 50 mm/h storm lasting 2 h. Assume time of concentration $t_c = 2$ h.

4-10. Rain falls on a 300-ha composite catchment which drains two subareas, as follows: (1) subarea A, steep, draining 20%, with time of concentration 10 min and $C = 0.8$; and (2) subarea B, milder steep, draining 80%, with time of concentration 60 min and $C = 0.4$. Calculate the peak runoff corresponding to the 25-y-frequency. Use the following IDF function:

$$I = \frac{800\, T^{0.2}}{(t_r + 15)^{0.7}}$$

in which I = rainfall intensity in millimeters per hour, T = return period in years, and t_r = rainfall duration in minutes. Assume linear flow concentration at the catchment outlet.

4-11. Rain falls on a 150-ha composite catchment, which drains two subareas, as follows: (1) subarea A, steep, draining 30%, with time of concentration 20 min; and (2) subarea B, milder steep, draining 70%, with time of concentration 60 min. The hydrologic abstraction is given in terms of $\phi = 25$ mm/h. Calculate the 10-y-frequency peak flow. Use the following IDF function:

$$I = \frac{650\, T^{0.22}}{(t_r + 18)^{0.75}}$$

in which I = rainfall intensity in millimeters per hour, T = return period in years, and t_r = rainfall duration in minutes. Assume linear flow concentration at the catchment outlet. State any other assumptions used.

4-12. Rain falls on a composite catchment, which drains two subareas, as follows: (1) subarea A, draining 84 ha, $C = 0.4$, time of concentration 30 min; (2) subarea B, draining 180 ha, $C = 0.6$, time of concentration 60 min. The runoff concentration for subarea B is a nonlinear function expressed as follows:

% of time of concentration	0	10	20	30	40	50	60	70	80	90	100
% of maximum discharge	0	5	10	20	30	50	70	80	90	95	100

Calculate the 50-yr-frequency peak flow. Use the following IDF function:

$$I = \frac{52\, T^{0.24}}{(t_r + 22)^{0.8}}$$

in which I = rainfall intensity in centimeters per hour, T = return period in years, and t_r = rainfall duration in minutes.

4-13. A developed catchment is divided into five subareas, as sketched in Fig. 4-7, with the following data:

Collection Point	Subarea Increment (ha)	Travel Time (min)	C
A	25	10	0.6
B	40	15	0.6
C	60	20	0.5
D	50	25	0.5
E	25	20	0.4

Calculate the 10-y-frequency peak flow. Use the following IDF function:

$$I = \frac{500\, T^{0.18}}{(t_r + 20)^{0.78}}$$

in which I = rainfall intensity in millimeters per hour, T = return period in years, and t_r = rainfall duration in minutes.

4-14. A developed catchment is divided into five subareas, as sketched in Fig. 4-7, with the following data:

Collection Point	Subarea Increment (ha)	Travel Time (min)	C
A	15	5	0.7
B	30	10	0.6
C	20	15	0.4
D	10	15	0.7
E	15	15	0.9

Calculate the 5-y-frequency peak flow. Use the following IDF function:

$$I = \frac{750\, T^{0.2}}{(t_r + 25)^{0.7}}$$

in which I = rainfall intensity in millimeters per hour, T = return period in years, and t_r = rainfall duration in minutes.

4-15. The length of an overland flow plane is $L = 90$ m. Determine the equilibrium outflow corresponding to a rainfall excess $i = 35$ mm/h.

4-16. An overland flow plane is 100 m long and 200 m wide, with time to equilibrium equal to 1 h. Estimate the equilibrium storage volume (in cubic meters) for a rainfall excess $i = 54$ mm/h.

4-17. Use the storage concept to calculate the time to equilibrium for an overland flow plane with the following characteristics: 100% laminar flow, plane length $L = 75$ m, plane slope $S_o = 0.01$, rainfall excess $i = 72$ mm/h, water temperature 20°C.

4-18. Calculate the mean overland flow depth (at equilibrium) under a laminar flow regime for a plane length $L = 80$ m, rainfall excess $i = 30$ mm/h, and plane slope $S_o = 0.012$. Use water temperature $T = 15°C$. What would be the mean overland flow depth if the water temperature increased to 25°C?

4-19. Use the storage concept to calculate the time to equilibrium for an overland flow plane with the following characteristics: turbulent Manning friction with $n = 0.06$, plane length $L = 50$ m, plane slope $S_o = 0.02$, rainfall excess $i = 72$ mm/h.

4-20. Calculate the rising limb of an overland flow hydrograph using Horton's equation (Eq. 4-36), assuming 75% turbulent flow. Use: Manning $n = 0.06$, plane length $L = 60$ m, plane slope $S_o = 0.015$, rainfall excess $i = 30$ mm/h.

4-21. Derive the formula for the kinematic time parameter for 100% laminar flow ($m = 3$).

4-22. Using the formula derived in Problem 4-21, calculate the rising limb of an overland flow hydrograph using the kinematic wave approach, assuming plane length $L = 100$ m, plane slope $S_o = 0.01$, rainfall excess $i = 25$ mm/h, water temperature $T = 20°C$.

4-23. An overland flow plane has the following characteristics: plane length $L = 35$ m, plane slope $S_o = 0.008$, Manning $n = 0.08$, rainfall excess $i = 55$ mm/h. Determine if the kinematic wave approximation is applicable to this set of overland flow conditions.

4-24. Calculate the hydraulic (channel) diffusivity for each of the following two flow conditions: (1) bed slope $S_o = 0.005$, mean flow depth $h_o = 0.01$ m, and mean velocity $u_o = 0.05$ m/s; and (2) bed slope $S_o = 0.00005$, mean flow depth $h_o = 0.02$ m, and mean velocity $u_o = 0.1$ m/s.

REFERENCES

1. Agricultural Research Service, U.S. Department of Agriculture. (1973). "Linear Theory of Hydrologic Systems," *Technical Bulletin No.* 1468, (J. C. I. Dooge, author). Washington, D.C.

2. American Society of Civil Engineers. (1960). "Design and Construction of Sanitary and Storm Sewers," *Manual of Engineering Practice No.* 37, also *Water Pollution Control Federation Manual of Practice No.* 9.

3. Chow, V. T. (1959). *Open Channel Hydraulics.* New York: McGraw-Hill.

4. Chow, V. T., and A. Ben-Zvi. (1973). "Hydrodynamic Modeling of Two-dimensional Watershed Flow," *Journal of the Hydraulics Division,* ASCE, Vol. 99, No. HY11, Nov. pp. 2023–2040.

5. County of Kern, California. (1985). "Revision of Coefficient of Runoff Charts," Office Memo, dated January 18, 1985.

6. County of Solano, California. (1977). "Hydrology and Drainage Design Procedure," prepared by Water Resources Engineers, Inc., Walnut Creek, California, October.

7. Hathaway, G. A. (1945). "Design of Drainage Facilities," in Military Airfields, A Symposium, *Transactions, American Society of Civil Engineers,* Vol. 110, pp. 697–733.

8. Henderson, F. M., and R. A. Wooding. (1964). "Overland Flow and Groundwater Flow from a Steady Rainfall of Finite Duration," *Journal of Geophysical Research,* Vol. 69, No. 8, April, pp. 1531–1540.

9. Henderson, F. M. (1965). *Open Channel Flow,* New York: MacMillan.

10. Horton, R. E. (1938). "The Interpretation and Application of Runoff Plot Experiments with Reference to Soil Erosion Problems," *Proceedings, Soil Science Society of America,* Vol. 3, pp. 340–349.

11. Hydrologic Engineering Center, U.S. Army Corps of Engineers. (1985). "HEC-1, Flood Hydrograph Package, Users Manual," September 1981, revised January 1985.

12. Iwagaki, Y. (1955). "Fundamental Studies on the Runoff Analysis by Characteristics," *Bulletin, Disaster Prevention Research Institute,* Vol. 10, Kyoto University, December.

13. Izzard, C. F. (1944). "The Surface Profile of Overland Flow," *Transactions, American Geophysical Union,* Vol. 25, No. 6, pp. 959–968.

14. Izzard, C. F. (1946). "Hydraulics of Runoff from Developed Surfaces," *Proceedings, Highway Research Board,* Washington, D.C., Vol. 26, pp. 129–146.

15. Lai, C. (1986). "Numerical Modeling of Unsteady Open Channel Flow," in *Advances in Hydroscience,* Vol. 14. Orlando: Academic Press.

16. Lighthill, M. H., and G. B. Whitham. (1955). "On Kinematic Waves. I. Flood Movement in Long Rivers," *Proceedings, Royal Society,* London, Vol. A229, May, pp. 281–316.

17. Mahmood, K., and Yevjevich, V. (1975). *Unsteady Flow in Open Channels,* in 3 volumes. Fort Collins, Colo.: Water Resources Publications.

18. Miller, W. A., and Cunge, J. A. (1975). "Simplified Equations of Unsteady Flow," in *Unsteady Flow in Open Channels,* K. Mahmood and V. Yevjevich, eds., Fort Collins, Colo.: Water Resources Publications.

19. NOAA National Weather Service. (1973). *Atlas 2: Precipitation Atlas of the Western United States.*

20. NOAA National Weather Service. (1977). "Five- to 60-Minute Precipitation Frequency for the Eastern and Central United States," *Technical Memorandum NWS HYDRO-35.*

21. Ponce, V. M., and D. B. Simons. (1977). "Shallow Wave Propagation in Open Channel Flow," *Journal of the Hydraulics Division,* ASCE, Vol. 103, No. HY12, December, pp. 1461–1476.

22. Rantz, S. E. (1971). "Suggested Criteria for Hydrologic Design of Storm-drainage Facilities in the San Francisco Bay Region, California," *U.S. Geological Survey Open-file Report,* Menlo Park, California.

23. San Diego County. (1985). *Hydrology Manual,* San Diego, California, revised January.

24. Schaake, Jr., J. C., Geyer, and J. W. Knapp. (1967). "Experimental Examination of the Rational Method," *Journal of the Hydraulics Division,* ASCE, Vol. 93, No. HY6, November, pp. 353–370.

25. Weather Bureau, U.S. Dept. of Commerce. (1963). "Rainfall Frequency Atlas of the United States for Durations from 30 Minutes to 24 Hours and Return Periods from 1 to 100 Years," *Technical Paper No. 40.*

26. Wooding, R. A. (1965). "A Hydraulic Model for the Catchment-stream Problem, I. Kinematic Wave Theory," *Journal of Hydrology,* Vol. 3, Nos. 3/4, pp. 254–267.

27. Wooding, R. A. (1965). "A Hydraulic Model for the Catchment-stream Problem, II. Numerical Solutions," *Journal of Hydrology,* Vol. 3, Nos. 3/4, pp. 268–282.

28. Wooding, R. A. (1965). "A Hydraulic Model for the Catchment-stream Problem, III. Comparison with Runoff Observations," *Journal of Hydrology,* Vol. 4, pp. 21–37.

29. Woolhiser, D. A., and Liggett, J. A. (1967). "Unsteady One-dimensional Flow Over a Plane: The Rising Hydrograph," *Water Resources Research,* Vol. 3, No. 3, pp. 753–771.

HYDROLOGY OF MIDSIZE CATCHMENTS

The following characteristics describe a midsize catchment: (1) rainfall intensity varies *within* the storm duration; (2) rainfall can be assumed to be uniformly distributed in space; (3) runoff is by overland flow and stream channel flow; and (4) channel storage processes are negligible.

Catchments possessing some or all the above properties are *midsize* in a hydrologic sense. Since rainfall intensity varies within the storm duration, catchment response is described by methods that take explicit account of the temporal variation of rainfall intensity. The most widely used method to accomplish this is the unit hydrograph technique. In a nutshell, it consists of deriving a hydrograph for a unit storm and using it as a building block to develop the hydrograph corresponding to the actual storm hyetograph.

In unit hydrograph analysis, the unit hydrograph duration is usually a fraction of the concentration time. The increase in concentration time is due to the larger drainage area and the associated reduction in overall catchment gradient. The latter has the effect of increasing runoff diffusion.

The assumption of uniform spatial distribution of rainfall is a characteristic of midsize catchment analysis. This assumption allows the use of a lumped method such as the unit hydrograph.

Unlike midsize catchments, for large catchments rainfall is likely to vary spatially, either as a general storm of concentric isohyetal distribution covering the entire catchment with moderate rainfall or as a highly intensive local storm (thunderstorm) covering only a portion of the catchment.

An important feature of large catchments that sets them apart from midsize catchments is their substantial capability for channel storage. Channel storage processes act to attenuate the flows while in transit in the river channels. Attenuation can be due either to longitudinal storage (for inbank flows) or to lateral storage (for overbank flows). In the first case, the storage amount is largely controlled by the slope of the main channel. For catchments with mild channel slopes, channel storage is substantial; conversely, for catchments with steep channel slopes, channel storage is negligible. Since large catchments are likely to have mild channel slopes, it follows that they have a substantial capability for channel storage.

In practice, this means that large catchments cannot be analyzed with spatially lumped methods such as the unit hydrograph, since these methods do not take explicit account of channel storage processes. Therefore, unlike for midsize catchments, for large catchments it is necessary to use channel routing (Chapter 9) to account for the expanded role of river flow in the overall runoff response.

As with the limit between small and midsize catchments, the limit between midsize and large catchments is not immediately apparent. For midsize catchments, runoff response is primarily a function of the characteristics of the storm hyetograph, with concentration time playing a secondary role. Therefore, the latter is not well suited as a descriptor of catchment scale. Values ranging from 100 to 5000 km^2 have been variously used to define the limit between midsize and large catchments. While there is no consensus to date, the current trend is toward the lower limit. In practice, it is likely that there would be a range of sizes within which both midsize and large catchment techniques are applicable. However, the larger the catchment area, the less likely it is that the lumped approach is able to provide the necessary spatial detail.

It should be noted that the techniques for midsize and large catchments are indeed complementary. A large catchment can be viewed as a collection of midsize subcatchments. Unit hydrograph techniques can be used for subcatchment runoff generation, with channel routing used to connect streamflows in a typical dendritic network fashion (Fig. 5-1). Such a computationally intensive procedure is ideally suited to solution with the aid of a computer. Examples of hydrologic computer models using the network concept are HEC-1 of the U.S. Army Corps of Engineers and TR-20 of the USDA Soil Conservation Service. These and other computer models are described in Chapter 13.

In practice, channel-routing techniques are not necessarily restricted to large catchments. They can also be used for midsize catchments and even for small catchments. However, the routing approach is considerably more complicated than the unit hydrograph technique. The routing approach is applicable to cases where an increased level of detail is sought, above that which the unit hydrograph technique is able to provide—for instance, when the objective is to describe the temporal variation of streamflow at several points *inside* the catchment. In this case, the routing approach may well be the only way to accomplish the modeling objective.

The hydrologic description of midsize catchments consists of two processes: (1) rainfall abstraction and (2) hydrograph generation. This chapter focuses on a method of rainfall abstraction that is widely used for hydrologic design in the United States: the Soil Conservation Service (SCS) runoff curve number method. Other rainfall abstraction procedures used by existing computer models are discussed in Chapter 13.

With regard to hydrograph generation, this chapter centers on the unit hydrograph technique, which is a defacto standard for midsize catchments, having been

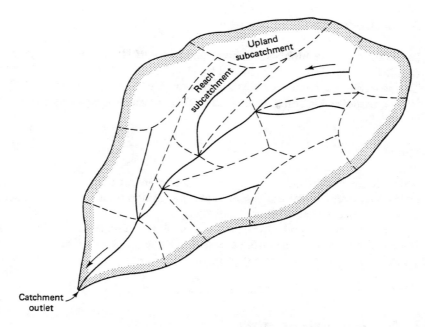

Figure 5-1 Subdivision of large catchment into midsize upland and reach subcatchments.

used extensively throughout the world. The SCS TR-55 method, also included in this chapter, has peak flow and hydrograph generation capabilities and is applicable to small and midsize urban catchments with concentration time in the range 0.1-10.0 h. The TR-55 method is based on the runoff curve number method, unit hydrograph techniques, and simplified stream channel routing procedures.

This chapter is divided into three sections. Section 5.1 describes the runoff curve number method. Section 5.2 discusses unit hydrograph techniques, including unit hydrographs derived from measured data and synthetic unit hydrographs. Section 5.3 deals with the TR-55 graphical method for peak-discharge determinations.

5.1 RUNOFF CURVE NUMBER METHOD

The runoff curve number method is a procedure for hydrologic abstraction developed by the USDA Soil Conservation Service [21]. In this method, runoff depth (i.e., effective rainfall depth) is a function of total rainfall depth and an abstraction parameter referred to as runoff curve number, curve number, or *CN*. The curve number varies in the range 1 to 100, being a function of the following runoff-producing catchment properties: (1) hydrologic soil type, (2) land use and treatment, (3) ground surface condition, and (4) antecedent moisture condition.

The runoff curve number method was developed based on 24-h rainfall-runoff data. It limits itself to the calculation of runoff depth and does not explicitly take into account temporal variations of rainfall intensity. The temporal rainfall distribution is introduced at a later stage, during the generation of the runoff hydrograph, by means of the convolution of the unit hydrograph (Section 5.2).

Runoff Curve Number Equation

In the runoff curve number method, actual runoff is referred to as Q, and potential runoff (total rainfall) is represented by P, with $P \geq Q$. The actual retention is $P - Q$. The potential retention (or potential maximum retention) is S, with $S \geq P - Q$.

The method is based on an assumption of proportionality between retention and runoff:

$$\frac{P - Q}{S} = \frac{Q}{P} \tag{5-1}$$

which states that the ratio of actual retention to potential retention is equal to the ratio of actual runoff to potential runoff. This assumption underscores the conceptual basis of the runoff curve number method.

For practical applications, Eq. 5-1 is improved by reducing the potential runoff by an amount equal to the initial abstraction. The initial abstraction consists mainly of interception, infiltration, and surface storage, all of which occur before runoff begins.

$$\frac{P - I_a - Q}{S} = \frac{Q}{P - I_a} \tag{5-2}$$

in which I_a = initial abstraction.

Solving for Q from Eq. 5-2:

$$Q = \frac{(P - I_a)^2}{P - I_a + S} \tag{5-3}$$

which is physically subject to the restriction that $P \geq I_a$ (i.e., the potential runoff minus the initial abstraction cannot be negative).

To simplify Eq. 5-3, initial abstraction is related to potential maximum retention as follows:

$$I_a = 0.2S \tag{5-4}$$

This relation was obtained based on rainfall-runoff data from small experimental watersheds. The coefficient 0.2 has been subjected to wide scrutiny. For instance, Springer et al. [18] evaluated small humid and semiarid catchments and found that the coefficient in Eq. 5-4 varied in the range 0.0 to 0.26. Nevertheless, 0.2 is the standard initial abstraction coefficient recommended by SCS [21]. For research applications and particularly when warranted by field data, it is possible to consider the initial abstraction coefficient as an additional parameter in the runoff curve number method. In general:

$$I_a = KS \tag{5-5}$$

in which K = initial abstraction parameter.

With Eq. 5-4, Eq. 5-3 reduces to:

$$Q = \frac{(P - 0.2S)^2}{P + 0.8S} \tag{5-6}$$

which is subject to the restriction that $P \geq 0.2S$.

Since potential maximum retention varies widely, it is more appropriate to express it in terms of a runoff curve number, an integer varying in the range 1 to 100, in the following form:

$$S = \frac{1000}{CN} - 10 \qquad (5\text{-}7)$$

in which CN is the runoff curve number (dimensionless) and S, 1000 and 10 are given in inches. To illustrate, for $CN = 100$, $S = 0$; and for $CN = 1$, $S = 990$ in. Therefore, the catchment's capability for rainfall abstraction is inversely proportional to the runoff curve number. For $CN = 100$ no abstraction is possible, with runoff being equal to total rainfall. On the other hand, for $CN = 1$ practically all rainfall would be abstracted, with runoff being essentially equal to zero.

With Eq. 5-7, Eq. 5-6 can be expressed in terms of CN:

$$Q = \frac{[CN(P + 2) - 200]^2}{CN[CN(P - 8) + 800]} \qquad (5\text{-}8)$$

which is subject to the restriction that $P \geq (200/CN) - 2$. In Eq. 5-8, P and Q are given in inches. In SI units, the equation is:

$$Q = \frac{R[CN(P/R + 2) - 200]^2}{CN[CN(P/R - 8) + 800]} \qquad (5\text{-}9)$$

which is subject to the restriction that $P \geq R[(200/CN) - 2]$. With $R = 2.54$ in Eq. 5-9, P and Q are given in centimeters.

For a variable initial abstraction, Eq. 5-8 is expressed as follows:

$$Q = \frac{[CN(P + 10K) - 1000K]^2}{CN\{CN[P - 10(1 - K)] + 1000(1 - K)\}} \qquad (5\text{-}10)$$

which is subject to the restriction that $P \geq (1000K/CN) - 10K$. An equivalent equation in SI units is:

$$Q = \frac{R[CN(P/R + 10K) - 1000K]^2}{CN\{CN[P/R - 10(1 - K)] + 1000(1 - K)\}} \qquad (5\text{-}11)$$

which is subject to the restriction that $P \geq R[(1000K/CN) - 10K]$.

A graph of Eqs. 5-8 and 5-9 is shown in Fig. 5-2. This figure is applicable only for the standard initial abstraction value, $I_a = 0.2S$. If this condition is relaxed, as in Eqs. 5-10 and 5-11, Fig. 5-2 has to be modified appropriately.

Estimation of Runoff Curve Number From Tables

With rainfall P and runoff curve number CN, the runoff Q can be determined by either Eq. 5-8 or Eq. 5-9 or from Fig. 5-2.

For ungaged watersheds, estimates of runoff curve numbers are given in tables supplied by federal agencies (SCS, Forest Service) and local city and county departments. Tables of runoff curve numbers for various hydrologic soil-cover complexes are widely available. The hydrologic soil-cover complex describes a specific combination of hydrologic soil group, land use and treatment, hydrologic surface condition, and

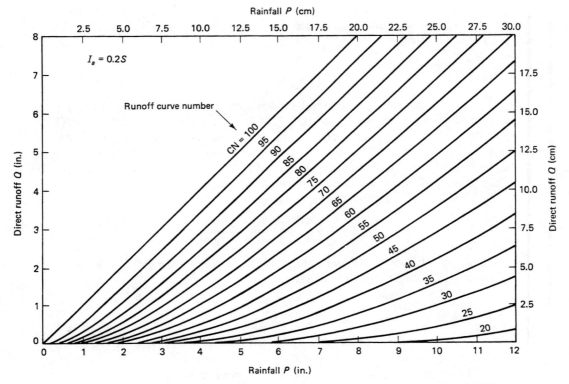

Figure 5-2 Direct runoff as a function of rainfall and runoff curve number [21].

antecedent moisture condition. All these have a direct bearing on the amount of runoff produced by a watershed. The hydrologic soil group describes the type of soil. The land use and treatment describes the type and condition of vegetative cover. The hydrologic condition refers to the ability of the watershed surface to enhance or impede direct runoff. The antecedent moisture condition accounts for the recent history of rainfall, and consequently it is a measure of the amount of moisture stored by the catchment.

Hydrologic Soil Groups. All soils are classified into four hydrologic soil groups of distinct runoff-producing properties. These groups are labeled A, B, C, and D.

Group A consists of soils of low runoff potential, having high infiltration rates even when wetted thoroughly. They are primarily deep, very well drained sands and gravels, with a characteristically high rate of water transmission.

Group B consists of soils with moderate infiltration rates when wetted thoroughly, primarily moderately deep to deep, moderately drained to well drained, with moderately fine to moderately coarse textures. These soils have a moderate rate of water transmission.

Group C consists of soils with slow infiltration rate when wetted thoroughly, primarily soils having a layer that impedes downward movement of water or soils of moderately fine to fine texture. These soils have a slow rate of water transmission.

Group D consists of soils of high runoff potential, having very slow infiltration rates when wetted thoroughly. They are primarily clay soils with a high swelling potential, soils with a permanent high water table, soils with a clay layer near the surface, and shallow soils overlying impervious material. These soils have a very slow rate of water transmission.

Maps showing the geographical distribution of hydrologic soil types for most areas in the United States are available either directly from SCS or from pertinent local agencies. Additional detail on U.S. soils and their hydrologic soil groups can be found in NEH-4 [21].

Land Use and Treatment. The effect of the surface condition of a watershed is evaluated by means of land use and treatment classes. Land use pertains to the watershed cover, including every kind of vegetation, litter and mulch, fallow (bare soil), as well as nonagricultural uses such as water surfaces (lakes, swamps, and so on), impervious surfaces (roads, roofs, and the like), and urban areas. Land treatment applies mainly to agricultural land uses, and it includes mechanical practices such as contouring or terracing and management practices such as grazing control and crop rotation. A class of land use/treatment is a combination often found in a catchment.

The runoff curve number method distinguishes between cultivated land, grasslands, and woods and forests. For cultivated lands, it recognizes the following land uses and treatments: fallow, row crop, small grain, close-seed legumes, rotations (from poor to good), straight-row fields, contoured fields, and terraced fields. Additional detail on these land use and treatment classes can be found in NEH-4 [21].

Hydrologic Condition. Grasslands are evaluated by the hydrologic condition of native pasture. The percent of areal coverage by native pasture and the intensity of grazing are visually estimated. A poor hydrologic condition describes less than 50 percent areal coverage and heavy grazing. A fair hydrologic condition describes 50 to 75 percent areal coverage and medium grazing. A good hydrologic condition describes more than 75 percent areal coverage and light grazing.

Woods are small isolated groves or trees being raised for farm or ranch use. The hydrologic condition of woods is visually estimated as follows: (1) poor—heavily grazed or regularly burned woods, with very little litter and few shrubs, (2) fair—grazed but not burned, with moderate litter and some shrubs, and (3) good—protected from grazing, with heavy litter and many shrubs covering the surface.

Runoff curve numbers for forest conditions are based on guidelines developed by the U.S. Forest Service. The publication *Forest and Range Hydrology Handbook* [23] describes the determination of runoff curve numbers for national and commercial forests in the eastern United States. The publication *Handbook of Methods for Hydrologic Analysis* [24] is used for curve number determinations in the forest-range regions in the western United States.

Antecedent Moisture Condition. The runoff curve number method has three levels of antecedent moisture, depending on the total rainfall in the 5-d period preceding a storm (see Table 5-1). The dry antecedent moisture condition (AMC I) has the lowest runoff potential, with the soils being dry enough for satisfactory plowing or cultivation to take place. The average antecedent moisture condition (AMC II) has an

TABLE 5-1 SEASONAL RAINFALL LIMITS FOR
THREE LEVELS OF ANTECEDENT MOISTURE
CONDITION (AMC) [21]

	Total 5-d Antecedent Rainfall (cm)	
AMC	Dormant Season	Growing Season
I	Less than 1.3	Less than 3.6
II	1.3 to 2.8	3.6 to 5.3
III	More than 2.8	More than 5.3

Note: This table was developed using data from the midwestern United States. Therefore, caution is recommended when using the values supplied in this table for AMC determinations in other geographic or climatic regions.

average runoff potential. The wet antecedent moisture condition (AMC III) has the highest runoff potential, with the watershed practically saturated from antecedent rainfalls. The AMC can be estimated from information such as that of Table 5-1 or other similar regionally derived tables.

Tables of runoff curve numbers for various hydrologic soil-cover complexes are in current use. Table 5-2(a) shows runoff curve numbers for urban areas, Table 5-2(b) shows them for cultivated agricultural areas, Table 5-2(c) shows them for other agricultural lands, and Table 5-2(d) shows them for arid and semiarid rangelands. Runoff curve numbers shown in these tables are for the average AMC II condition. Corresponding runoff curve numbers for AMC I and AMC III conditions are shown in Table 5-3.

Using Eq. 5-7, Hawkins et al [8] have expressed the values in Table 5-3 in terms of potential maximum retention. They correlated the values of potential maximum retention for AMC I and III with those of AMC II and found the following ratios to be a good approximation:

$$\frac{S_I}{S_{II}} \cong \frac{S_{II}}{S_{III}} \cong 2.3 \tag{5-12}$$

This led to the following relationships:

$$CN_I = \frac{CN_{II}}{2.3 - 0.013\ CN_{II}} \tag{5-13}$$

$$CN_{III} = \frac{CN_{II}}{0.43 + 0.0057 CN_{II}} \tag{5-14}$$

which can be used in lieu of Table 5-3 to calculate runoff curve numbers for AMC I and AMC III in terms of the AMC II value.

Estimation of Runoff Curve Numbers from Measured Data

The runoff curve number method was developed primarily for design applications in ungaged catchments and was not intended for simulation of actual recorded hydrographs. However, where rainfall-runoff data are available, estimations of runoff curve

TABLE 5-2(a)   RUNOFF CURVE NUMBERS FOR URBAN AREAS[1] [22]

Cover Description		Curve Numbers for Hydrologic Soil Group:			
Cover Type and Hydrologic Condition	Average Percent Impervious Area[2]	A	B	C	D
Fully developed urban areas (vegetation established)					
Open space (lawns, parks, golf courses, cemeteries, etc.)[3]:					
Poor condition (grass cover less than 50%)		68	79	86	89
Fair condition (grass cover 50 to 75%)		49	69	79	84
Good condition (grass cover greater than 75%)		39	61	74	80
Impervious areas:					
Paved parking lots, roofs, driveways, etc. (excluding right-of-way)		98	98	98	98
Streets and roads:					
Paved; curves and storm sewers (excluding right-of-way)		98	98	98	98
Paved; open ditches (including right-of-way)		83	89	92	93
Gravel (including right-of-way)		76	85	89	91
Dirt (including right-of-way)		72	82	87	89
Western desert urban areas:					
Natural desert landscaping (pervious areas only)[4]		63	77	85	88
Artificial desert landscaping (impervious weed barrier, desert shrub with 1- to 2-in. sand or gravel mulch and basin borders)		96	96	96	96
Urban districts:					
Commercial and business	85	89	92	94	95
Industrial	72	81	88	91	93
Residential districts by average lot size:					
$\frac{1}{8}$ ac. or less (town houses)	65	77	85	90	92
$\frac{1}{4}$ ac.	38	61	75	83	87
$\frac{1}{3}$ ac.	30	57	72	81	86
$\frac{1}{2}$ ac.	25	54	70	80	85
1 ac.	20	51	68	79	84
2 ac.	12	46	65	77	82
Developing urban areas					
Newly graded areas (pervious areas only, no vegetation)[5]		77	86	91	94
Idle lands (curve numbers (CNs) are determined using cover types similar to those in Table 5-2(c)).					

Notes:

[1]Average antecedent moisture condition and $I_a = 0.2S$.

[2]The average percent impervious area shown was used to develop the composite *CN*s. Other assumptions are as follows: Impervious areas are directly connected to the drainage system; impervious areas have a $CN = 98$; and pervious areas are considered equivalent to open space in good hydrologic condition. *CN*s for other combinations of conditions may be computed using Fig. 5-16 or 5-17.

[3]*CN*s shown are equivalent to those of pasture. Composite *CN*s may be computed for other combinations of open space cover type.

[4]Composite CN's for natural desert landscaping should be computed using Figs. 5-16 or 5-17 based on the impervious area percentage ($CN = 98$) and the pervious area *CN*. The pervious area *CN*s are assumed equivalent to desert shrub in poor hydrologic condition.

[5]Composite *CN*s to use for the design of temporary measures during grading and construction should be computed using Figs. 5-16 or 5-17, based on the degree of development (impervious area percentage) and the CNs for the newly graded pervious areas.

TABLE 5-2(b) RUNOFF CURVE NUMBERS FOR CULTIVATED AGRICULTURAL LANDS[1] [22]

Cover Description			Curve Numbers for Hydrologic Soil Group:			
Cover Type	Treatment[2]	Hydrologic Condition[3]	A	B	C	D
Fallow	Bare soil	—	77	86	91	94
	Crop residue cover (CR)	Poor	76	85	90	93
		Good	74	83	88	90
Row crops	Straight row (SR)	Poor	72	81	88	91
		Good	67	78	85	89
	SR + CR	Poor	71	80	87	90
		Good	64	75	82	85
	Contoured (C)	Poor	70	79	84	88
		Good	65	75	82	86
	C + CR	Poor	69	78	83	87
		Good	64	74	81	85
	Contoured and terraced (C&T)	Poor	66	74	80	82
		Good	62	71	78	81
	C&T + CR	Poor	65	73	79	81
		Good	61	70	77	80
Small grain	SR	Poor	65	76	84	88
		Good	63	75	83	87
	SR + CR	Poor	64	75	83	86
		Good	60	72	80	84
	C	Poor	63	74	82	85
		Good	61	73	81	84
	C + CR	Poor	62	73	81	84
		Good	60	72	80	83
	C&T	Poor	61	72	79	82
		Good	59	70	78	81
	C&T + CR	Poor	60	71	78	81
		Good	58	69	77	80
Close-seeded or broadcast legumes or rotation meadow	SR	Poor	66	77	85	89
		Good	58	72	81	85
	C	Poor	64	75	83	85
		Good	55	69	78	83
	C&T	Poor	63	73	80	83
		Good	51	67	76	80

Notes:

[1]Average antecedent moisture condition and $I_u = 0.2S$.

[2]*Crop residue cover* applies only if residue is on at least 5% of the surface throughout the year.

[3]Hydrologic condition is based on combination of factors that affect infiltration and runoff, including: (1) density and canopy of vegetated areas; (2) amount of year-round cover; (3) amount of grass or close-seeded legumes in rotation; (4) percent of residue cover on the land surface (good hydrologic condition is greater than or equal to 20%); and (5) degree of surface roughness. *Poor:* Factors impair infiltration and tend to increase runoff. *Good:* Factors encourage average and better than average infiltration and tend to decrease runoff.

TABLE 5-2(c) RUNOFF CURVE NUMBERS FOR OTHER AGRICULTURAL LANDS[1] [22]

| Cover Description | | Curve Numbers for Hydrologic Soil Group: | | | |
Cover Type	Hydrologic Condition	A	B	C	D
Pasture, grassland, or range-continuous forage for grazing[2]	Poor	68	79	86	89
	Fair	49	69	79	84
	Good	39	61	74	80
Meadow-continuous grass, protected from grazing and generally mowed for hay	—	30	58	71	78
Brush—brush-weed grass mixture with brush being the major element[3]	Poor	48	67	77	83
	Fair	35	56	70	77
	Good	30[4]	48	65	73
Woods—grass combination (orchard or tree farm)[5]	Poor	57	73	82	86
	Fair	43	65	76	82
	Good	32	58	72	79
Woods.[6]	Poor	45	66	77	83
	Fair	36	60	73	79
	Good	30[4]	55	70	77
Farmsteads—buildings, lanes, driveways, and surrounding lots.	—	59	74	82	86

Notes:

[1]Average antecedent moisture condition and $I_a = 0.2S$.

[2]*Poor:* less than 50% ground cover on heavily grazed with no mulch.
Fair: 50 to 75% ground cover and not heavily grazed.
Good: more than 75% ground cover and lightly or only occasionally grazed.

[3]*Poor:* less than 50% ground cover.
Fair: 50 to 75% ground cover.
Good: more than 75% ground cover.

[4]Actual curve number is less than 30; use $CN = 30$ for runoff computations.

[5]*CN*s shown were computed for areas with 50% woods and 50% grass (pasture) cover. Other combinations of conditions may be computed from the *CN*s for woods and pasture.

[6]*Poor:* Forest litter, small trees, and brush are destroyed by heavy grazing or regular burning.
Fair: Woods are grazed but not burned, and some forest litter covers the soil.
Good: Woods are protected from grazing, and litter and brush adequately cover the soil.

numbers can be obtained directly from data. These values complement and in certain cases may even replace the information obtained from tables.

To estimate runoff curve numbers from data, it is necessary to assemble corresponding sets of rainfall-runoff data for several events occurring individually. As far as possible, the selected events should be of constant intensity and should uniformly cover the catchment. A recommended procedure is to select events that correspond to annual floods [13]. Inclusion of events of greater frequency may lead to more conservative (higher) values of runoff curve number [18]. The selected sets should encompass a wide range of antecedent moisture conditions, from dry to wet.

For each event, a value of P, total rainfall depth, is identified. The associated direct runoff hydrograph is integrated to obtain the direct runoff volume. This runoff volume is divided by the catchment area to obtain Q, the direct runoff depth (in centi-

TABLE 5-2(d) RUNOFF CURVE NUMBERS FOR ARID AND SEMIARID RANGELANDS[1] [22]

Cover Description		Curve Numbers for Hydrologic Soil Group:			
Cover Type	Hydrologic Condition[2]	A[3]	B	C	D
Herbaceous—mixture of grass, weeds, and low-growing brush, with brush the minor element.	Poor		80	87	93
	Fair		71	81	89
	Good		62	74	85
Oak-aspen—mountain brush mixture of oak brush, aspen, mountain mahogany, bitter brush, maple, and other brush.	Poor		66	74	79
	Fair		48	57	63
	Good		30	41	48
Pinyon-juniper—pinyon, juniper, or both; grass understory.	Poor		75	85	89
	Fair		58	73	80
	Good		41	61	71
Sagebrush with grass understory.	Poor		67	80	85
	Fair		51	63	70
	Good		35	47	55
Desert shrub—major plants include saltbrush, greasewood, creosotebush, blackbrush, bursage, palo verde, mesquite, and cactus.	Poor	63	77	85	88
	Fair	55	72	81	86
	Good	49	68	79	84

Notes:
[1]Average antecedent moisture condition and $I_a = 0.2S$. For range in humid regions, use Table 5-2(c).
[2]*Poor:* less than 30% ground cover (litter, grass, and brush overstory).
Fair: 30 to 70% ground cover.
Good: more than 70% ground cover.
[3]Curve numbers for group A have been developed only for desert shrub.

meters or inches). The values of P and Q are plotted on Fig. 5-2 and a corresponding value of CN is identified. The procedure is repeated for all events, and a CN value is obtained for each event, as shown in Fig. 5-3. In theory, the AMC II runoff curve number is that which separates the data into two equal groups, with half of the data plotting above the line and half below it. The AMC I runoff curve number is the curve number that envelopes the data from below. The AMC III runoff curve number is the curve number that envelopes the data from above (see Fig. 5-3).

Assessment of Runoff Curve Number Method

The positive features of the runoff curve number method are its simplicity and the fact that runoff curve numbers are related to the major runoff producing properties of the watershed, such as soil type, vegetation type and treatment, surface condition, and antecedent moisture. The method is used in practice to determine runoff depths based on rainfall depths and curve numbers, with no explicit account of rainfall intensity and duration.

A considerable body of experience has been accumulated on the runoff curve number method. Publications continue to appear in the literature either to augment the already extensive experience or to examine critically the applicability of the method to individual situations. For best results, however, the method should be used judiciously, with particular attention paid to its capabilities and limitations.

TABLE 5-3 CORRESPONDING RUNOFF CURVE NUMBERS FOR THREE AMC CONDITIONS [21]

AMC II	AMC I	AMC III	AMC II	AMC I	AMC III
100	100	100	60	40	78
99	97	100	59	39	77
98	94	99	58	38	76
97	91	99	57	37	75
96	89	99	56	36	75
95	87	98	55	35	74
94	85	98	54	34	73
93	83	98	53	33	72
92	81	97	52	32	71
91	80	97	51	31	70
90	78	96	50	31	70
89	76	96	49	30	69
88	75	95	48	29	68
87	73	95	47	28	67
86	72	94	46	27	66
85	70	94	45	26	65
84	68	93	44	25	64
83	67	93	43	25	63
82	66	92	42	24	62
81	64	92	41	23	61
80	63	91	40	22	60
79	62	91	39	21	59
78	60	90	38	21	58
77	59	89	37	20	57
76	58	89	36	19	56
75	57	88	35	18	55
74	55	88	34	18	54
73	54	87	33	17	53
72	53	86	32	16	52
71	52	86	31	16	51
70	51	85	30	15	50
69	50	84			
68	48	84	25	12	43
67	47	83	20	9	37
66	46	82	15	6	30
65	45	82	10	4	22
64	44	81	5	2	13
63	43	80	0	0	0
62	42	79			
61	41	78			

Experience with the method has shown that results are sensitive to curve number. This stresses the importance of an accurate estimation of curve number to minimize the variance in runoff determinations. The standard tables provide helpful guidelines, but local experience is recommended for increased accuracy. Typical runoff curve numbers used in design are in the range 50 to 95.

Closely associated with the method's sensitivity to runoff curve number is its sensitivity to antecedent moisture. Since runoff curve number varies with antecedent moisture, markedly different results can be obtained for each of the three levels of

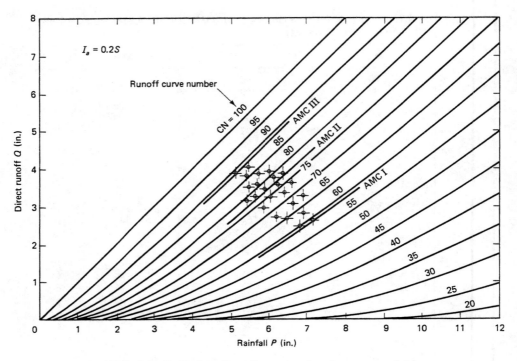

Figure 5-3 Estimation of runoff curve numbers from measured data.

antecedent moisture. At first, this appears to be a limitation; however, closer examination reveals that runoff is indeed a function of antecedent moisture, with the method's sensitivity to AMC reflecting the conditions likely to prevail in nature. Hjelmfelt et al. [9] attached a probability meaning to AMC, with AMC I corresponding to 10 percent probability of exceedence, AMC II to 50 percent, and AMC III to 90 percent. This may help explain why practical enveloping curves to determine AMC I and AMC III usually do not encompass all the data.

The popularity of the runoff curve number method is largely due to its simplicity, although proper care is necessary to use the method correctly. The method is essentially a conceptual model to estimate runoff volumes based on established hydrologic abstraction mechanisms, with the effect of antecedent moisture taken in a probability context. In practice, (average) AMC II describes a typical design condition. When warranted, other antecedent moisture conditions, including those intermediate between I, II, and III, can be considered. An example of regional practice is given in Table 5-4.

Experience with the runoff curve number method has shown that the curve numbers obtained from Table 5-2 tend to be conservative (i.e., too high) for large catchments, especially those located in semiarid and arid regions. Often this is due to the fact that these large catchments have additional sources of hydrologic abstraction, in particular, channel transmission losses, not accounted for by the tables. In this case it is necessary to perform a separate evaluation of the effect of channel abstractions on the quantity of surface runoff.

While the applicability of the runoff curve number procedure appears to be in-

	SCS Antecedent Moisture Condition			
Design Frequency	Coast	Foothills	Mountains	Desert
5–35 y	1.5	2.5	2.0	1.5
35–150 y	2.0	3.0	3.0	2.0

Source: San Diego County Hydrology Manual.

dependent of catchment scale, its indiscriminate use for catchments in excess of 250 km^2 (100 mi^2) without catchment subdivision is generally not recommended. The runoff curve number was originally developed by SCS for use in midsize rural watersheds. Subsequently, the method was applied to small and midsize urban catchments (the TR-55 method). Therefore, its extension to large basins requires considerable judgment.

Example 5-1.

A certain catchment experiences 12.7 cm of total rainfall. The catchment is covered by pasture with medium grazing, and 32 percent of B soils and 68 percent of C soils. This event has been preceded by 6.35 cm of rainfall in the last 5 d. Following the SCS methodology, determine the direct runoff for the 12.7 cm rainfall event.

A fair hydrologic condition is chosen for pasture with medium grazing. From Table 5-2(c), the runoff curve numbers for pasture with fair hydrologic condition are $CN = 69$ for B soils, and $CN = 79$ for C soils. The applicable CN is a weighted value:

$$CN = (69 \times 0.32) + (79 \times 0.68) = 76 \qquad (5\text{-}15)$$

Since this event has been preceded by a substantial amount of moisture in the last few days, AMC III is chosen. From Table 5-3, for AMC II $CN = 76$, AMC III $CN = 89$. From Eq. 5-9 or Fig. 5-2, with $CN = 89$ and $P = 12.7$ cm (5 in.), a value of $Q = 9.58$ cm (3.77 in.) is obtained as the direct runoff for this event.

5.2 UNIT HYDROGRAPH TECHNIQUES

The concept of unit hydrograph, originated by Sherman [14], is used in midsize catchment analysis as a means to develop a hydrograph for any given storm. The word *unit* is normally taken to refer to a unit depth of effective rainfall or runoff. However, it should be noted that Sherman first used the word to describe a unit depth of runoff (1 cm or 1 in.) lasting a unit increment of time (i.e., an indivisible increment). The unit increment of time can be either 1-h, 3-h, 6-h, 12-h, 24-h, or any other suitable duration. For midsize catchments, unit hydrograph durations from 1 to 6 h are common.

The unit hydrograph is defined as the hydrograph produced by a unit depth of runoff uniformly distributed over the entire catchment and lasting a specified duration. To illustrate the concept of unit hydrograph, assume that a certain storm produces 1 cm of runoff and covers a 50-km^2 catchment over a period of 2 h. The hydrograph measured at the catchment outlet would be the 2-h unit hydrograph for this 50-km^2 catchment (Fig. 5-4).

A unit hydrograph for a given catchment can be calculated either (1) directly, by

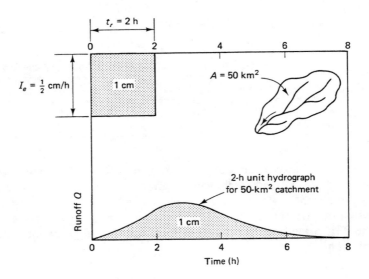

Figure 5-4 Concept of unit hydrograph.

using rainfall-runoff data for selected events, or (2) indirectly, by using a synthetic unit hydrograph formula. While both methods can be used for gaged catchments, only the latter method is appropriate for ungaged catchments.

Since a unit hydrograph has meaning only in connection with a given storm duration, it follows that a catchment can have several unit hydrographs, each for a different rainfall duration. Once a unit hydrograph for a given duration has been determined, other unit hydrographs can be derived from it by using one of the following methods: (1) superposition method and (2) S-hydrograph method.

Two assumptions are crucial to the development of the unit hydrograph technique. These are the principles of *linearity* and *superposition*. Given a unit hydrograph, a hydrograph for a runoff depth other than unity can be obtained by simply multiplying the unit hydrograph ordinates by the indicated runoff depth (linearity). This, of course, is possible only under the assumption that the time base remains constant regardless of runoff depth (Fig. 5-5(a)).

The time base of all hydrographs obtained in this way is equal to that of the unit hydrograph. Therefore, the procedure can be used to calculate hydrographs produced by a storm consisting of a series of runoff depths, each lagged in time one increment of unit hydrograph duration (Fig. 5-5(b)). The summation of the corresponding ordinates of these hydrographs (superposition) allows the calculation of the composite hydrograph (Fig. 5-5(c)). The procedure depicted in Fig. 5-5 is referred to as the *convolution* of a unit hydrograph with an effective storm hyetograph.

In essence, the procedure amounts to stating that the composite hydrograph ordinates are a linear combination of the unit hydrograph ordinates, while the composite hydrograph time base is the sum of the unit hydrograph time base minus the unit hydrograph duration plus the storm duration.

The assumption of linearity has long been considered one of the limitations of unit hydrograph theory. In nature, it is unlikely that catchment response will always follow a linear function. For one thing, discharge and mean velocity are nonlinear

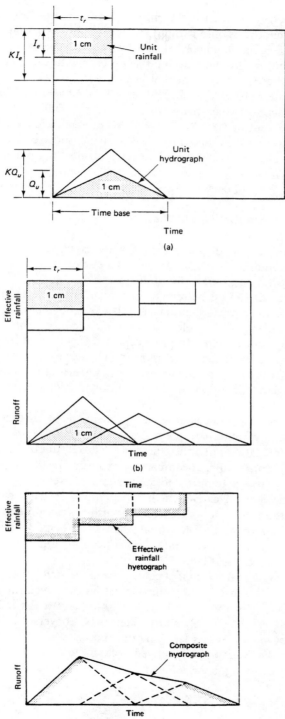

Figure 5-5 Unit hydrograph properties: (a) linearity; (b) lagging; (c) superposition.

functions of flow depth and stage. In practice, the linear assumption provides a convenient means of calculating runoff response without the complexities associated with nonlinear analysis. More recent developments of unit hydrograph theory [1, 4, 15] have relaxed the linear assumption. Methods accounting for the nonlinearity of runoff response constitute what is known as nonlinear unit hydrograph theory.

The upper limit of applicability of the unit hydrograph is not very well defined. Sherman [14] used it in connection with basins varying from 1300 to 8000 km². Linsley et al. [10] mention an upper limit of 5000 km² in order to preserve accuracy. More recently, the unit hydrograph has been linked to the concept of midsize catchment, i.e., greater than 2.5 km² and less than 250 km². This certainly does not preclude the unit hydrograph technique from being applied to catchments larger than 250 km², although overall accuracy is likely to decrease with an increase in catchment area.

Development of Unit Hydrographs: Direct Method

To develop a unit hydrograph by the direct method it is necessary to have a gaged catchment, i.e., a catchment equipped with rain gages and a stream gage at the outlet, and adequate sets of corresponding rainfall-runoff data.

The rainfall-runoff records should be screened to identify storms suitable for unit hydrograph analysis. Ideally, a storm should have a clearly defined duration, with no rainfall preceding it or following it. The selected storms should be of uniform rainfall intensity both temporally and spatially. In practice, the difficulty in meeting this latter requirement increases with catchment size. As catchment scale grows from midsize to large, the requirement of spatial rainfall uniformity in particular is seldom met. This limits unit hydrograph development by the direct method to midsize catchments.

Catchment Lag. The concept of *catchment lag,* basin lag, or lag time is central to the development of unit hydrograph theory. It is a measure of the time elapsed between the occurrence of unit rainfall and the occurrence of unit runoff. Catchment lag is a global measure of response time, encompassing hydraulic length, catchment gradient, drainage density, drainage patterns, and other related factors.

There are several definitions of catchment lag, depending on what particular instant is taken to describe the occurrence of either unit rainfall or runoff. Hall [7] has identified seven definitions, shown in Fig. 5-6. Among them the T2 lag, defined as the time elapsed from the centroid of effective rainfall to the peak of runoff, is the most commonly used definition of catchment lag.

In unit hydrograph analysis, the concept of catchment lag is used to characterize the catchment response time. Because runoff volume must be conserved (i.e., runoff volume should equal 1 unit of effective rainfall depth), short lags result in unit responses featuring high peaks and relatively short time bases; conversely, long lags result in unit responses showing low peaks and long time bases.

In practice, catchment lag is empirically related to catchment characteristics. A general expression for catchment lag is the following:

$$t_l = C\left(\frac{LL_c}{S^{1/2}}\right)^N \tag{5-16}$$

in which t_l = catchment lag; L = catchment length (length measured along the main

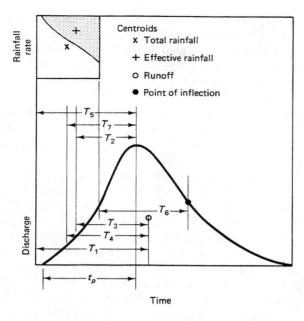

Figure 5-6 Alternate definitions of catchment lag [7].

stream from outlet to divide); L_c = length to catchment centroid (length measured along the main stream from outlet to a point located closest to the catchment centroid); S = a weighted measure of catchment slope, usually taken as the S_2 channel slope (Chapter 2); and C and N are empirical parameters. The parameter L describes length, L_c is a measure of shape, and S relates to relief.

Methodology. In addition to the requirements of uniform rainfall intensity in time and space, storms suitable for unit hydrograph analysis should be of about the same duration. The duration should lie between 10 percent to 30 percent of catchment lag. The latter requirement implies that runoff response is of the subconcentrated type, with rainfall duration less than time of concentration. Indeed, subconcentrated flow is a characteristic of midsize catchments.

For increased accuracy, direct runoff should be in the range 0.5 to 2.0 units (usually centimeters or inches). Several individual storms (at least five events) should be analyzed to assure consistency. The following steps are applied to each individual storm:

1. Separation of the measured hydrograph into direct runoff hydrograph (DRH) and baseflow (BF), following the procedures explained below.
2. Calculation of direct runoff volume (DRV) by integrating the direct runoff hydrograph (DRH).
3. Calculation of direct runoff depth (DRD) by dividing the direct runoff volume (DRV) by the catchment area.
4. Calculation of unit hydrograph (UH) ordinates by dividing the ordinates of the direct runoff hydrograph (DRH) by the direct runoff depth (DRD).
5. Estimation of the unit hydrograph duration.

The catchment unit hydrograph is obtained by averaging the unit hydrograph ordinates obtained from each of the individual storms and averaging the respective unit hydrograph durations. Minor adjustments in hydrograph ordinates may be necessary to ensure that the volume under the unit hydrograph is equal to one unit of runoff depth.

Hydrograph Separation. Only the direct runoff component of the measured hydrograph is used in the computation of the unit hydrograph. Therefore, it is necessary to separate the measured hydrograph into its direct runoff and baseflow components. Interflow is usually included as part of baseflow.

Procedures for baseflow separation are usually arbitrary in nature. First, it is necessary to identify the point in the receding limb of the measured hydrograph where direct runoff ends. Generally, this ending point is located in such a way that the receding time up to that point is about 2 to 4 times the time-to-peak (Fig. 5-7). For large basins, this multiplier may be greater than 4. As far as possible, the location of the ending point should be such that the time base is an even multiple of the unit hydrograph duration. A common assumption is that baseflow recedes at the same rate as prior to the storm until the peak discharge has passed and then gradually increases to the ending point P in the receding limb, as illustrated by line a in Fig. 5-7. If a stream and groundwater table are hydraulically connected (Fig. 5-8), water infiltrates during the rising limb, reducing baseflow, and exfiltrates during the receding limb, increas-

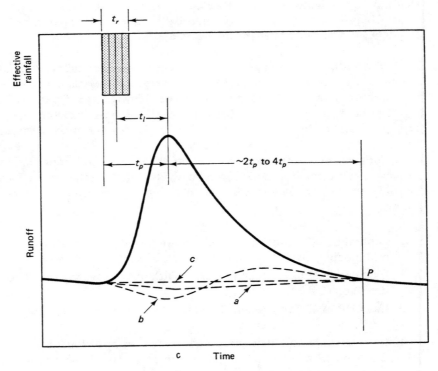

Figure 5-7 Procedures for baseflow separation.

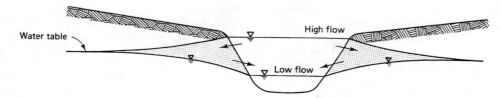

Figure 5-8 Hydraulically connected stream and water table.

ing baseflow, as shown by line *b* in Fig. 5-7 [5]. The most expedient assumption for baseflow separation is a straight line from the start of the rising limb to the ending point, as shown by line *c*. Differences in baseflow due to the various separation techniques are likely to be small when compared to the direct runoff hydrograph volume. Additional methods for hydrograph separation and baseflow recession are described in Chapter 11.

Example 5-2.

A unit hydrograph is to be developed for a 37.8-km² catchment with a lag time of 12 h. A 2-h rainfall produced the following streamflow data:

Time (h)	0	2	4	6	8	10	12	14	16	18	20	22	24
Streamflow (m³/s)	2	1	3	5	9	8	7	6	5	4	3	1	1

Develop a unit hydrograph for this catchment.

A summary of the calculations is shown in Table 5-5. Columns 1 and 2 show time and measured streamflow, respectively. Baseflow is established by examining the measured streamflow. Since the hydrograph rise starts at 2 h and ends at 22 h, a value of baseflow equal to 1 m³/s appears reasonable. (In practice, a more detailed analysis as described in Section 11.5 may be necessary.) Column 3 shows the ordinates of the DRH obtained by substracting baseflow from the measured streamflow. To calculate direct runoff depth,

TABLE 5-5 DEVELOPMENT OF UNIT HYDROGRAPH: DIRECT METHOD, EXAMPLE 5-2

(1)	(2)	(3)	(4)	(5)	(6)	(7)
Time (h)	Streamflow (m³/s)	DRH (m³/s)	Simpson's coefficients	Volume	UH (m³/s)	Verification
0	2	—	—	—	—	—
2	1	0	1	0	0.00	0.00
4	3	2	4	8	2.50	10.00
6	5	4	2	8	5.00	10.00
8	9	8	4	32	10.00	40.00
10	8	7	2	14	8.75	17.50
12	7	6	4	24	7.50	30.00
14	6	5	2	10	6.25	12.50
16	5	4	4	16	5.00	20.00
18	4	3	2	6	3.75	7.50
20	3	2	4	8	2.50	10.00
22	1	0	1	0	0.00	0.00
24	1	—	—	—	—	—
Sum				126		157.50

DRH is integrated numerically following Simpson's rule. The Simpson's rule coefficients are shown in Col. 4. Column 5 shows the weighted ordinates obtained by multiplying Col. 3 by Col. 4. Summing up the weighted ordinates (Col. 5), a value of 126 m³/s is obtained. Since the integration interval is 2 h, the DRV (according to Simpson's rule) is DRV = (126 m³/s × 7200 seconds)/3 = 302,400 m³. The DRD is obtained by dividing DRV by the catchment area (37.8 km²) to yield: DRD = 0.8 cm. The unit hydrographs ordinates (Col. 6) are calculated by dividing the DRH ordinates (Col. 3) by DRD. To verify the calculations, the unit hydrograph shown in Col. 6 is integrated by multiplying Col. 4 times Col. 6 to obtain Col. 7. The sum of Col. 7 is 157.5 m³/s. It is verified that the ratio of DRV to unit hydrograph volume is indeed 0.8 (126/157.5 = 0.8). Finally, it is confirmed that the unit hydrograph duration (2 h) is an appropriate percentage (17 percent) of the lag time (12 h).

Development of Unit Hydrographs: Indirect Method

In the absence of rainfall-runoff data, unit hydrographs can be derived by synthetic means. A synthetic unit hydrograph is a unit hydrograph derived following an established formula, without the need for rainfall-runoff data analysis.

The development of synthetic unit hydrographs is based on the following principle: Since the volume under the hydrograph is known (volume is equal to catchment area multiplied by 1 unit of runoff depth), the peak discharge can be calculated by assuming a certain unit hydrograph shape. For instance, if a triangular shape is assumed (Fig. 5-9), the volume is equal to:

$$V = \frac{Q_p T_{bt}}{2} = A \times (1) \tag{5-17}$$

in which V = volume under the triangular unit hydrograph; Q_p = peak flow; T_{bt} = time base of the triangular unit hydrograph; A = catchment area; and (1) = one unit of runoff depth. From Eq. 5-17:

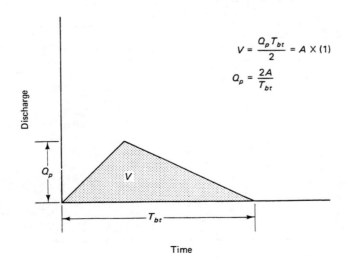

$$V = \frac{Q_p T_{bt}}{2} = A \times (1)$$

$$Q_p = \frac{2A}{T_{bt}}$$

Figure 5-9 Triangular unit hydrograph.

$$Q_p = \frac{2A}{T_{bt}} \qquad (5\text{-}18)$$

Synthetic unit hydrograph methods usually relate time base to catchment lag. In turn, catchment lag is related to the timing response characteristics of the catchment, including catchment shape, length, and slope. Therefore, catchment lag is a fundamental parameter in synthetic unit hydrograph determinations.

Several methods are available for the calculation of synthetic unit hydrographs. Two widely used methods, the Snyder and the Soil Conservation Service methods, are described here. The Clark method, also widely used, is based on catchment routing techniques; therefore, it is described in Chapter 10.

Snyder's Synthetic Unit Hydrograph

In 1938, Snyder [17] introduced the concept of synthetic unit hydrograph. The analysis of a large number of hydrographs from catchments in the Appalachian region led to the following formula for lag:

$$t_l = C_t(LL_c)^{0.3} \qquad (5\text{-}19)$$

in which t_l = catchment or basin lag in hours, L = length along the mainstream from outlet to divide, L_c = length along the mainstream from outlet to point closest to catchment centroid, and C_t = a coefficient accounting for catchment gradient and associated catchment storage. With distances L and L_c in kilometers, Snyder gave values of C_t varying in the range 1.35 to 1.65, with a mean of 1.5. With distances L and L_c in miles, the corresponding range of C_t is 1.8 to 2.2, with a mean of 2.

Snyder's formula for peak flow is:

$$Q_p = \frac{C_p A}{t_l} \qquad (5\text{-}20)$$

which when compared with Eq. 5-18 reveals that

$$C_p = \frac{2}{\dfrac{T_{bt}}{t_l}} \qquad (5\text{-}21)$$

is an empirical coefficient relating triangular time base to lag. Snyder gave values of C_p in the range 0.56 to 0.69, which are associated with T_{bt}/t_l ratios in the range 3.57 to 2.90. The lower the value of C_p (i.e., the lower the peak flow), the greater the value of T_{bt}/t_l and the greater the capability for catchment storage.

In SI units, Snyder's peak flow formula is:

$$Q_p = \frac{2.78 C_p A}{t_l} \qquad (5\text{-}22)$$

in which Q_p = unit hydrograph peak flow corresponding to 1 cm of effective rainfall, in cubic meters per second; A = catchment area, in square kilometers; and t_l = lag, in hours. In U.S. customary units, Snyder's peak flow formula is

$$Q_p = \frac{645 C_p A}{t_l} \qquad (5\text{-}23)$$

in which Q_p = unit hydrograph peak flow corresponding to 1 in. of effective rainfall in cubic feet per second; A = catchment area in square miles; and t_l = lag in hours.

In Snyder's method, the unit hydrograph duration is a linear function of lag:

$$t_r = \tfrac{2}{11} t_l \qquad (5\text{-}24)$$

in which t_r = unit hydrograph duration.

In applying the procedure to flood forecasting, Snyder recognized that the actual duration of the storm is usually greater than the duration calculated by Eq. 5-24. Therefore, he devised a formula to increase the lag in order to account for the increased storm duration. This led to:

$$t_{lR} = t_l + \frac{t_R - t_r}{4} \qquad (5\text{-}25)$$

in which t_{lR} is the adjusted lag corresponding to a duration t_R.

Assuming uniform effective rainfall for simplicity, the unit hydrograph time-to-peak is equal to one-half of the storm duration plus the lag (Fig. 5-7). Therefore, the time-to-peak in terms of the lag is:

$$t_p = \tfrac{12}{11} t_l \qquad (5\text{-}26)$$

When calculating the actual time base of the unit hydrograph, Snyder included interflow as part of direct runoff. This results in a longer time base than that corresponding only to direct runoff. Snyder's formula for actual time base is the following:

$$T_b = 72 + 3t_l \qquad (5\text{-}27)$$

in which T_b = actual unit hydrograph time base (including interflow) in hours and t_l = lag in hours. For a 24-h lag, this formula gives $T_b/t_l = 6$, which is a reasonable value considering that interflow is being included in the calculation. For smaller lags, however, Eq. 5-27 gives unrealistically high values of T_b/t_l. For instance, for a 6-h lag, $T_b/t_l = 15$. For midsize catchments, and excluding interflow, experience has shown that values of T_b/t_p around 5 (corresponding to values of T_b/t_l around 5.45) may be more realistic.

The Snyder method gives peak flow (Eq. 5-22), time-to-peak (Eq. 5-26), and time base (Eq. 5-27) of the unit hydrograph. These values can be used to sketch the unit hydrograph, adhering to the requirement that unit hydrograph volume should equal 1 unit of runoff depth. Snyder gave a *distribution chart* (Fig. 5-10) to aid in plotting the unit hydrograph ordinates, but cautioned against the exclusive reliance on this graph to develop the shape of the unit hydrograph.

The Snyder method has been extensively used by the U.S. Army Corps of Engineers. Their experience has led to two empirical formulas that aid in determining the shape of the Snyder unit hydrograph [20]:

$$W_{50} = \frac{5.87}{(Q_p/A)^{1.08}} \qquad (5\text{-}28)$$

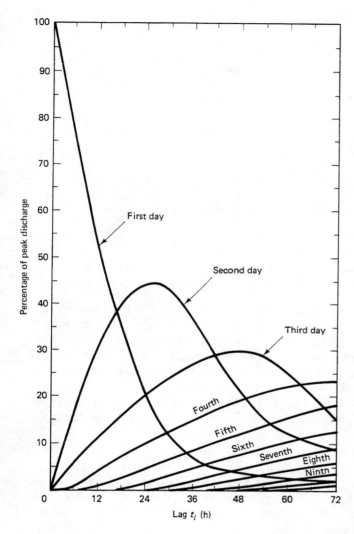

The chart shows curves labeled: First day, Second day, Third day, Fourth, Fifth, Sixth, Seventh, Eighth, Ninth. The y-axis is "Percentage of peak discharge" (0 to 100), and the x-axis is "Lag t_l (h)" (0 to 72).

Figure 5-10 Snyder's distribution chart for plotting unit hydrograph ordinates [17].

$$W_{75} = \frac{3.35}{(Q_p/A)^{1.08}} \qquad (5\text{-}29)$$

in which W_{50} = width of unit hydrograph at 50 percent of peak discharge in hours; W_{75} = width of unit hydrograph at 75 percent of peak discharge in hours; Q_p = peak discharge in cubic meters per second; and A = catchment area in square kilometers (Fig. 5-11). These time widths should be proportioned in such a way that one-third is located before the peak and two-thirds after the peak.

Snyder cautioned that lag may tend to vary slightly with flood magnitude and that synthetic unit hydrograph calculations are likely to be more accurate for fan-shaped catchments than for those of highly irregular shape. He recommended that the coefficients C_t and C_p be determined on a regional basis.

Examination of Eq. 5-19 reveals that C_t is largely a function of catchment slope,

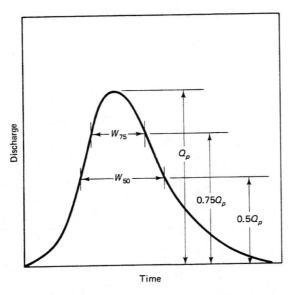

Figure 5-11 Snyder's synthetic unit hydrograph widths: W_{50} and W_{75} [20].

since both length and shape have already been accounted for in L and L_c, respectively. Since Eq. 5-19 was derived empirically, the actual value of C_t depends on the units of L and L_c. Furthermore, Eq. 5-19 implies that when the product of (LL_c) is equal to 1, the lag is equal to C_t. Since for two catchments of the same size, lag is a function of slope, it is unlikely that C_t is a constant. To give an example, an analysis of 20 catchments in the north and middle Atlantic United States [19] led to: $C_t = 0.6/S^{1/2}$. A similar conclusion is drawn from Eq. 5-16. Therefore, values of C_t have regional meaning, in general being a function of catchment slope. Values of C_t quoted in the literature reflect the natural variability of catchment slopes.

The parameter C_p is dimensionless and varies within a narrow range. In fact, it is readily shown that the maximum possible value of C_p is $\frac{11}{12}$. Since triangular time base cannot be less than twice the time-to-peak (otherwise, runoff diffusion would be negative, clearly a physical impossibility), it follows that in the limit (i.e., absence of runoff diffusion), $T_{bt} = 2t_p$, and, therefore, $C_p = t_l/t_p = \frac{11}{12}$. In practice, triangular time base is usually about 3 times the time-to-peak. For $T_{bt} = 3t_p$, a similar calculation leads to: $C_p = 0.61$, which lies approximately in the middle of Snyder's data (0.56–0.69).

Since C_t increases with catchment storage and C_p decreases with catchment storage, the ratio C_t/C_p can be directly related to catchment storage. Furthermore, the reciprocal ratio (C_p/C_t) can be directly related to extent of urban development, since the latter usually results in a substantial reduction in the catchment's storage capability [25].

Example 5-3.

Calculate the properties of a Snyder unit hydrograph using the following data: $L = 25$ km, $L_c = 10$ km, $A = 400$ km^2, $C_t = 1.5$, and $C_p = 0.61$.

Using Eq. 5-19, $t_l = 7.86$ h. From Eq. 5-21, solving for T_{bt}, $T_{bt} = 25.77$ h. Using Eq. 5-22, $Q_p = 86.3$ m^3/s. Using Eq. 5-24, $t_r = 1.43$ h. Using Eq. 5-26, $t_p = 8.57$ h. The time base calculated by Eq. 5-27 is $T_b = 95.58$ h. This is too high a value. Instead, assume time base $T_b = 5t_p$; then: $T_b = 42.85$ h. Using Eq. 5-28, $W_{50} = 30.8$ h; using Eq. 5-29, $W_{75} = 17.6$ h. The actual unit hydrograph is drawn primarily on the basis of Q_p, t_p and T_b, with the remaining values used as guidelines.

SCS Synthetic Unit Hydrograph

The SCS synthetic unit hydrograph is the dimensionless unit hydrograph developed by Victor Mockus in the 1950s and described in NEH-4 [21]. This hydrograph was developed based on the analysis of a large number of natural unit hydrographs from a wide range of catchment sizes and geographic locations. The method has come to be recognized as the SCS synthetic unit hydrograph and has been applied to midsize catchments throughout the world.

The method differs from Snyder's in that it uses a constant ratio of triangular time base to time-to-peak, $T_{bt}/t_p = \frac{8}{3}$, which implies that C_p is constant and equal to 0.6875. Unlike Snyder's method, the SCS method uses a constant ratio of actual time base to time-to-peak, $T_b/t_p = 5$. In addition, it uses a dimensionless hydrograph function to provide a standard unit hydrograph shape.

To calculate catchment lag (the T2 lag), the SCS method uses the following two methods: (1) the curve number method and (2) the velocity method. The curve number method is limited to catchments of areas less than 8 km^2 (2000 ac), although recent evidence suggests that it may be extended to catchments up to 16 km^2 (4000 ac) [11].

In the curve number method, the lag is expressed by the following formula:

$$t_l = \frac{L^{0.8}(2540 - 22.86\,CN)^{0.7}}{14{,}104\,CN^{0.7}\,Y^{0.5}} \tag{5-30}$$

in which t_l = catchment lag in hours; L = hydraulic length (length measured along principal watercourse) in meters; CN = runoff curve number; and Y = average catchment land slope in meters per meter. In U.S. customary units, the formula is:

$$t_l = \frac{L^{0.8}(1000 - 9\,CN)^{0.7}}{1900\,CN^{0.7}\,Y^{0.5}} \tag{5-31}$$

in which t_l is in hours, L is in feet, and Y is in percent. In the curve number method, the average catchment land slope is obtained by superimposing a square grid pattern over the catchment topographic map, evaluating the maximum land slope at each grid intersection within the catchment, and averaging these values to obtain a representative value of catchment land slope. Equations 5-30 and 5-31 are restricted to curve numbers in the range 50 to 95.

The velocity method is used for catchments larger than 8 km^2, or for curve numbers outside of the range 50 to 95. The main stream is divided into reaches, and the 2-y flood (or alternatively the bank-full discharge) is estimated. In certain cases it may be desirable to use discharges corresponding to 10-y frequencies or more. The mean velocity is computed, and the reach concentration time is calculated by using the reach valley length (straight distance). The sum of the concentration time for all reaches is the concentration time for the catchment. The lag is estimated as follows:

$$\frac{t_l}{t_c} = \frac{6}{10} \tag{5-32}$$

in which t_l = lag and t_c = concentration time. SCS experience has shown that this ratio is typical of midsize catchments [21].

In the SCS method the ratio of time-to-peak to unit hydrograph duration is fixed at

$$\frac{t_p}{t_r} = 5 \tag{5-33}$$

which is close to Snyder's ratio of 6. Assuming uniform effective rainfall for simplicity, the time-to-peak is by definition equal to

$$t_p = \frac{t_r}{2} + t_l \tag{5-34}$$

Eliminating t_r from Eqs. 5-33 and 5-34, leads to

$$\frac{t_p}{t_l} = \frac{10}{9} \tag{5-35}$$

Therefore:

$$\frac{t_r}{t_l} = \frac{2}{9} \tag{5-36}$$

and

$$\frac{t_r}{t_c} = \frac{2}{15} \tag{5-37}$$

To derive the SCS unit hydrograph peak flow formula, the ratio $T_{bt}/t_p = \frac{8}{3}$ is used in Eq. 5-18, leading to

$$Q_p = \frac{\frac{3}{4}A}{t_p} \tag{5-38}$$

In SI units, the peak flow formula is:

$$Q_p = \frac{2.08A}{t_p} \tag{5-39}$$

in which Q_p = unit hydrograph peak flow for 1 cm of effective rainfall in cubic meters per second; A = catchment area in square kilometers; and t_p = time-to-peak in hours. In U.S. customary units, the SCS peak flow formula is:

$$Q_p = \frac{484A}{t_p} \tag{5-40}$$

in which Q_p = unit hydrograph peak flow for 1 in. of effective rainfall; A = catchment area in square miles; and t_p = time-to-peak in hours.

Given Eqs. 5-32 and 5-34, the time-to-peak can be readily calculated as follows: $t_p = 0.5t_r + 0.6t_c$. Once t_p and Q_p have been determined, the SCS dimensionless unit hydrograph (Fig. 5-12) is used to calculate the unit hydrograph ordinates. The shape of the dimensionless unit hydrograph is more in agreement with unit hydrographs that are likely to occur in nature than the triangular shape ($T_{bt}/t_p = \frac{8}{3}$) used to develop the

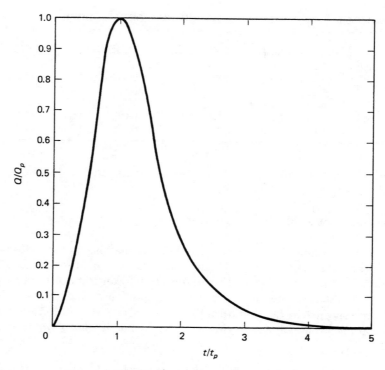

Figure 5-12 SCS dimensionless unit hydrograph [21].

peak flow value. The dimensionless unit hydrograph has a value of $T_b/t_p = 5$. Values of SCS dimensionless unit hydrograph ordinates at intervals of 0.2 (t/t_p) are given in Table 5-6.

The SCS method provides a unit hydrograph shape and therefore leads to more reproducible results than the Snyder method. However, the ratio T_{bt}/t_p is kept constant and equal to $\frac{8}{3}$. Also, when lag is calculated by the velocity method, the ratio t_l/t_c is kept constant and equal to $\frac{6}{10}$. Although these assumptions are based on a wide range of data, they render the method inflexible in certain cases.

In particular, values of T_{bt}/t_p other than $\frac{8}{3}$ may lead to other shapes of unit hydrographs. Larger values of T_{bt}/t_p (equivalent to lower values of C_p in the Snyder method) imply greater catchment storage. Therefore, since the SCS method fixes the value of T_{bt}/t_p, it should be limited to midsize catchments in the lower end of the spectrum (2.5–250 km²). The Snyder method, however, by providing a variable T_{bt}/t_p, may be used for larger catchments (from 250 up to 5000 km²) [10].

Efforts to extend the range of applicability of the SCS method have led to the relaxation of the T_{bt}/t_p ratio. It can be shown that the ratio p of volume-to-peak (volume under the rising limb of the triangular unit hydrograph) to the triangular unit hydrograph volume is the reciprocal of the ratio T_{bt}/t_p. For instance, in the case of the standard SCS synthetic unit hydrograph, $T_{bt}/t_p = \frac{8}{3}$, and $p = \frac{3}{8}$. In terms of p, Eq. 5-38 can be expressed as follows:

$$Q_p = \frac{2pA}{t_p} \tag{5-41}$$

TABLE 5-6   SCS DIMENSIONLESS UNIT HYDROGRAPH ORDINATES

t/t_p	Q/Q_p	t/t_p	Q/Q_p	t/t_p	Q/Q_p	t/t_p	Q/Q_p	t/t_p	Q/Q_p
0.0	0.00								
0.2	0.10	1.2	0.93	2.2	0.207	3.2	0.040	4.2	0.0100
0.4	0.31	1.4	0.78	2.4	0.147	3.4	0.029	4.4	0.0070
0.6	0.66	1.6	0.56	2.6	0.107	3.6	0.021	4.6	0.0030
0.8	0.93	1.8	0.39	2.8	0.077	3.8	0.015	4.8	0.0015
1.0	1.00	2.0	0.28	3.0	0.055	4.0	0.011	5.0	0.0000

which converts the SCS method into a two-parameter model like the Snyder method, thereby increasing its flexibility.

Other Synthetic Unit Hydrographs

The Snyder and SCS methods base their calculations on the following properties: (1) catchment lag, (2) ratio of triangular time base to time-to-peak, and (3) ratio of actual time base to time-to-peak. In addition, the SCS method specifies a gamma function for the shape of the unit hydrograph. Many other synthetic unit hydrographs have been reported in the literature [16]. In general, any procedure defining geometric properties and hydrograph shape can be used to develop a synthetic unit hydrograph.

Example 5-4.

Calculate the SCS synthetic unit hydrograph for a 6.42 km² catchment with the following data: Hydraulic length $L = 2204$ m; runoff curve number $CN = 62$; average land slope $Y = 0.02$.

Using Eq. 5-30, $t_l = 1.8$ h. Therefore: $t_r = 0.4$ h; $t_p = 2$ h; $T_b = 10$ h. Using Eq. 5-39, $Q_p = 6.68$ m³/s. Using Table 5-6, the ordinates of the unit hydrograph are calculated as shown in Table 5-7.

Change in Unit Hydrograph Duration

A unit hydrograph, whether derived by direct or indirect means, is valid only for a given (effective) storm duration. In certain cases, it may be necessary to change the duration of a unit hydrograph. For instance, if an X-hour unit hydrograph is going to be used with a storm hyetograph defined at Y-hour intervals, it is necessary to convert the X-hour unit hydrograph into a Y-hour unit hydrograph.

In general, once a unit hydrograph of a given duration has been derived for a catchment, a unit hydrograph of another duration can be calculated. There are two methods to change the duration of unit hydrographs: (1) the superposition method and (2) the S-hydrograph method. The superposition method converts an X-hour unit hydrograph into a nX-hour unit hydrograph, in which n is an integer. The S-hydrograph method converts an X-hour unit hydrograph into a Y-hour unit hydrograph, regardless of the ratio between X and Y.

Superposition Method.   This method allows the conversion of an X-hour unit hydrograph into a nX-hour unit hydrograph, in which n is an integer. The proce-

TABLE 5-7 UNIT HYDROGRAPH ORDINATES:
EXAMPLE 5-4
($Q_p = 6.68$ m³/s; $t_p = 2$ h)

t/t_p	Q/Q_p	t (h)	Q (m³/s)
0.0	0.00	0.0	0.000
0.2	0.10	0.4	0.668
0.4	0.31	0.8	2.071
0.6	0.66	1.2	4.410
0.8	0.93	1.6	6.212
1.0	1.00	2.0	6.680
1.2	0.93	2.4	6.212
1.4	0.78	2.8	5.210
1.6	0.56	3.2	3.740
1.8	0.39	3.6	2.605
2.0	0.28	4.0	1.870
2.2	0.207	4.4	1.382
2.4	0.147	4.8	0.982
2.6	0.107	5.2	0.714
2.8	0.077	5.6	0.514
3.0	0.055	6.0	0.367
3.2	0.040	6.4	0.267
3.4	0.029	6.8	0.194
3.6	0.021	7.2	0.140
3.8	0.015	7.6	0.100
4.0	0.011	8.0	0.073
4.2	0.010	8.4	0.067
4.4	0.007	8.8	0.047
4.6	0.003	9.2	0.020
4.8	0.0015	9.6	0.010
5.0	0.0000	10.0	0.000

dure consists of lagging n X-hour unit hydrographs in time, each for an interval equal to X hours, summing the ordinates of all n hydrographs, and dividing the summed ordinates by n to obtain the nX-hour unit hydrograph. The volume under X-hour and nX-hour unit hydrographs is the same. If T_b is the time base of the X-hour hydrograph, the time base of the nX-hour hydrograph is equal to $T_b + (n - 1)X$. The procedure is illustrated by the following example.

Example 5-5.

Use the superposition method to calculate the 2-h and 3-h unit hydrographs of a catchment, based on the following 1-h unit hydrograph:

Time (h)	0	1	2	3	4	5	6	7	8	9	10	11	12
Flow (m³/s)	0	100	200	400	800	700	600	500	400	300	200	100	0

The calculations are shown in Table 5-8. Column 1 shows the time in hours. Column 2 shows the ordinates of the 1-h unit hydrograph. Column 3 shows the ordinates of the 1-h unit hydrograph, lagged 1 h. Column 4 shows the ordinates of the 1-h unit hydrograph, lagged 2 h. Column 5 shows the ordinates of the 2-h unit hydrograph, obtained by summing the ordinates of Cols. 2 and 3 and dividing by 2. Column 6 shows the ordinates of the 3-h unit hydrograph, obtained by summing the ordinates of Cols. 2, 3, and 4, and dividing by 3. The sum of ordinates for 1-h, 2-h, and 3-h unit hydrographs is the same:

TABLE 5-8 CHANGE IN UNIT HYDROGRAPH DURATION,
SUPERPOSITION METHOD: EXAMPLE 5-5

(1)	(2)	(3)	(4)	(5)	(6)
Time (h)	1-h UH	Lagged 1 h	Lagged 2 h	2-h UH	3-h UH
0	0	0	0	0	0
1	100	0	0	50	33
2	200	100	0	150	100
3	400	200	100	300	233
4	800	400	200	600	467
5	700	800	400	750	633
6	600	700	800	650	700
7	500	600	700	550	600
8	400	500	600	450	500
9	300	400	500	350	400
10	200	300	400	250	300
11	100	200	300	150	200
12	0	100	200	50	100
13	0	0	100	0	33
14	0	0	0	0	0
Sum	4300			4300	4299

4300 m^3/s. The time base of the 1-h unit hydrograph is 12 h, whereas the time base of the 2-h unit hydrograph is 13 h and the time base of the 3-h unit hydrograph is 14 h.

S-Hydrograph Method. The S-hydrograph method allows the conversion of an X-hour unit hydrograph into a Y-hour unit hydrograph, regardless of the ratio between X and Y. The procedure consists of the following steps:

1. Determine the X-hour S-hydrograph (Fig. 5-13). The X-hour S-hydrograph is derived by accumulating the unit hydrograph ordinates *at intervals equal to X*.
2. Lag the X-hour S-hydrograph by a time interval equal to Y hours.
3. Subtract ordinates of the two previous S-hydrographs.
4. Multiply the resulting hydrograph ordinates by X/Y to obtain the Y-hour unit hydrograph.

The volume under X-hour and Y-hour unit hydrographs is the same. If T_b is the time base of the X-hour unit hydrograph, the time base of the Y-hour unit hydrograph is $T_b - X + Y$.

Example 5-6.

For the 2-h unit hydrograph calculated in the previous example, derive the 3-h unit hydrograph by the S-hydrograph method. Use this 3-h unit hydrograph to derive the 2-h unit hydrograph, confirming the applicability of the S-hydrograph method, regardless of the ratio between X and Y.

The calculations are shown in Table 5-9. Column 1 shows the time in hours. Column 2 shows the 2-h unit hydrograph ordinates calculated in the previous example. Column 3 is

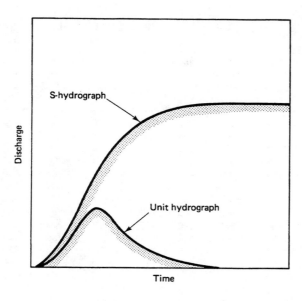

S-hydrograph

Unit hydrograph

Discharge

Time

Figure 5-13 Sketch of unit hydrograph and corresponding S-hydrograph.

the 2-h S-hydrograph, obtained by accumulating the ordinates of Col. 2 at intervals of X = 2 h. Column 4 is the S-hydrograph of Col. 3 lagged $Y = 3$ h. Column 5 is equal to Col. 3 minus Col. 4. Column 6 is the product of Col. 5 times $X/Y = \frac{2}{3}$. Column 6 is the 3-h unit hydrograph. Its sum is 4299 m³/s, the same as the sum of Col. 2, confirming that is contains a unit volume. The time base of the 2-h unit hydrograph is 13 h, and the time base of the 3-h unit hydrograph is 14 h. Column 7 is the 3-h S-hydrograph, obtained by accumulating the ordinates of Col. 6 at intervals of $X = 3$ h. Column 8 is the S-hydrograph of Col. 7 lagged $Y = 2$ h. Column 9 is equal to Col. 7 minus Col. 8. Column 10 is the product of Col. 9 times $X/Y = \frac{3}{2}$. Column 10 is the 2-h unit hydrograph, and it is confirmed to be the same as that of Col. 2.

Minor errors in unit hydrograph ordinates may often lead to errors (i.e., undesirable oscillations) in the resulting S-hydrograph. In this case, a certain amount of smoothing may be required to achieve the typical S-shape (Fig. 5-13).

Convolution and Composite Hydrographs

The procedure to derive a composite or flood hydrograph based on a unit hydrograph and an effective storm hyetograph is referred to as hydrograph convolution. This technique is based on the principles of linearity and superposition. The volume under the composite hydrograph is equal to the total volume of the effective rainfall. If T_b is the time base of the X-hour unit hydrograph and the storm consists of n X-hour intervals, the time base of the composite hydrograph is equal to $T_b - X + nX = T_b + (n - 1) X$. The convolution procedure is illustrated by the following example.

Example 5-7.

Assume that the following 1-h unit hydrograph has been derived for a certain watershed:

Time (h)	0	1	2	3	4	5	6	7	8	9
Flow (m³/s)	0	100	200	400	800	600	400	200	100	0

TABLE 5-9 CHANGE IN UNIT HYDROGRAPH DURATION, S-HYDROGRAPH METHOD: EXAMPLE 5-6

(1)	(2)	(3)	(4)	(5)	(6)	(7)	(8)	(9)	(10)
Time (h)	2-h UH	2-h SH	Lagged 3 h	Col. 3 −Col. 4	3-h UH	3-h SH	Lagged 2 h	Col. 7 −Col. 8	2-h UH
0	0	0	0	0	0	0	0	0	0
1	50	50	0	50	33	33	0	33	50
2	150	150	0	150	100	100	0	100	150
3	300	350	0	350	233	233	33	200	300
4	600	750	50	700	467	500	100	400	600
5	750	1100	150	950	633	733	233	500	750
6	650	1400	350	1050	700	933	500	433	650
7	550	1650	750	900	600	1100	733	367	550
8	450	1850	1100	750	500	1233	933	300	450
9	350	2000	1400	600	400	1333	1100	233	350
10	250	2100	1650	450	300	1400	1233	167	250
11	150	2150	1850	300	200	1433	1333	100	150
12	50	2150	2000	150	100	1433	1400	33	50
13	0	2150	2100	50	33	1433	1433	0	0
14	0	2150	2150	0	0	1433	1433	0	0
Sum	4300				4299				4300

A 6-h storm with a total of 5 cm of effective rainfall covers the entire watershed and is distributed in time as follows:

Time (h)	0		1		2		3		4		5		6
Effective rainfall (cm)		0.1		0.8		1.6		1.2		0.9		0.4	

Calculate the composite hydrograph using the convolution technique.

The calculations are shown in Table 5-10. Column 1 shows the time in hours, and Col. 2 shows the unit hydrograph ordinates in cubic meters per second. Column 3 shows the product of the first-hour rainfall depth times the unit hydrograph ordinates. Column 4 shows the product of the second-hour rainfall depth times the unit hydrograph ordinates, lagged 1 h with respect to Col. 3. The computational pattern established by Cols. 3 and 4 is the same for Cols. 5-8. Column 9, the sum of Cols. 3 through 8, is the composite hydrograph for the given storm pattern. The sum of Col. 2 is 2800 m^3/s and is equivalent to 1 cm of net rainfall. The sum of Col. 9 is verified to be 14,000 m^3/s, and, therefore, the equivalent of 5 cm of effective rainfall. The time base of the composite hydrograph is $T_b = 9 + (6 - 1) \times 1 = 14$ h.

Unit Hydrographs from Complex Storms

The convolution procedure enables the calculation of a storm hydrograph based on a unit hydrograph and a storm hyetograph. In theory, the procedure can be reversed to allow the calculation of a unit hydrograph for a given storm hydrograph and storm hyetograph.

Method of Forward Substitution. The unit hydrograph can be calculated directly due to the banded property of the convolution matrix (see Table 5-10). With

TABLE 5-10 COMPOSITE HYDROGRAPH BY CONVOLUTION: EXAMPLE 5-7

(1)	(2)	(3)	(4)	(5)	(6)	(7)	(8)	(9)
Time (h)	UH (m^3/s)	$0.1\times$ UH	$0.8\times$ UH	$1.6\times$ UH	$1.2\times$ UH	$0.9\times$ UH	$0.4\times$ UH	Composite Hydrograph
0	0	0	—	—	—	—	—	0
1	100	10	0	—	—	—	—	10
2	200	20	80	0	—	—	—	100
3	400	40	160	160	0	—	—	360
4	800	80	320	320	120	0	—	840
5	600	60	640	640	240	90	0	1670
6	400	40	480	1280	480	180	40	2500
7	200	20	320	960	960	360	80	2700
8	100	10	160	640	720	720	160	2410
9	0	0	80	320	480	540	320	1740
10	—	—	0	160	240	360	240	1000
11	—	—	—	0	120	180	160	460
12	—	—	—	—	0	90	80	170
13	—	—	—	—	—	0	40	40
14	—	—	—	—	—	—	0	0
Sum	2800							14,000

m = number of nonzero unit hydrograph ordinates, n = number of intervals of effective rainfall, and N = number of nonzero storm hydrograph ordinates, the following relation holds:

$$N = m + n - 1 \tag{5-42}$$

Therefore:

$$m = N - n + 1 \tag{5-43}$$

By elimination and back substitution, the following formula can be developed for the unit hydrograph ordinates u_i as a function of storm hydrograph ordinates q_i and effective rainfall depths r_k:

$$u_i = \frac{q_i - \sum_{j=i-1,1}^{k=2,n} u_j r_k}{r_1} \tag{5-44}$$

for i varying from 1 to m. In the summation term, j *decreases* from $i - 1$ to 1, and k *increases* from 2 up to a maximum of n.

This recursive equation allows the direct calculation of a unit hydrograph based on hydrographs from complex storms. In practice, however, it is not always feasible to arrive at a solution because it may be difficult to get a perfect match of storm hydrograph and effective rainfall hyetograph (due to errors in the data). For one thing, the measured storm hydrograph would have to be separated into direct runoff and baseflow before attempting to use Eq. 5-44.

The uncertainties involved have led to the use of the least square technique. In this technique, rainfall-runoff data *(r, h)* for a number of events are used to develop a set of average values of u using statistical tools [12]. Other methods to derive unit hydrographs for complex storms are discussed by Singh [16].

Example 5-8.

Use Eq. 5-44 and the storm hydrograph obtained in the previous example to calculate the unit hydrograph.

Since $N = 13$ and $n = 6$: $m = 8$. The first ordinate is $u_1 = q_1/r_1 = 10/0.1 = 100$. The second ordinate is $u_2 = (q_2 - u_1r_2)/r_1 = (100 - 100 \times 0.8)/0.1 = 200$. The third ordinate is $u_3 = [q_3 - (u_2r_2 + u_1r_3)]/r_1 = [360 - (200 \times 0.8 + 100 \times 1.6)]/0.1 = 400$. The fourth ordinate is $u_4 = [q_4 - (u_3r_2 + u_2r_3 + u_1r_4)]/r_1 = [840 - (400 \times 0.8 + 200 \times 1.6 + 100 \times 1.2)]/0.1 = 800$. The remaining ordinates are obtained in a similar way.

5.3 SCS TR-55 METHOD

The TR-55 method is a collection of simplified procedures developed by the USDA Soil Conservation Service to calculate peak discharges, flood hydrographs, and stormwater storage volumes in small and midsize urban watersheds [22]. It consists of two main procedures: (1) a graphical method, and (2) a tabular method. The graphical method is used to calculate peak discharges, whereas the tabular method calculates flood hydrographs by using simplified routing procedures. These methods were developed based on information obtained with the SCS TR-20 hydrologic computer model (Chapter 13). They are designed to be used in cases where their applicability can be clearly demonstrated, in lieu of more elaborate techniques. Whereas TR-55 does not specify watershed size, the graphical method is limited to catchments with concentration times in the range 0.1 to 10.0 h. Likewise, the tabular method is limited to catchments with concentration times in the range 0.1 to 2.0 h.

The first version of TR-55 dates back to 1975. The second version (1986) incorporates the experience gained with the use of the first version. The graphical method (1986 version) is described in this section. For details on the tabular method, reference is made to the original source [22].

TR-55 Storm, Catchment and Runoff Parameters

Rainfall in TR-55 is described in terms of total rainfall depth and one of four standard 24-h temporal rainfall distributions: type I, type IA, type II, and type III (Fig. 5-14). Type I applies to California (south of the San Francisco Bay area) and Alaska; type IA applies to the Pacific Northwest and Northern California; type III applies to the Gulf Coast states; and type II applies everywhere else within the contiguous United States (Fig. 5-15 and Chapter 13).

The duration of these rainfall distributions is 24 h. This constant duration was selected because most rainfall data is reported on a 24-h basis. Rainfall intensities corresponding to durations shorter than 24 h are contained within the SCS distributions. For instance, if a 10-y 24-h rainfall distribution is used, the 1-h period with the most intense rainfall corresponds to the 10-y 1-h rainfall depth.

TR-55 uses the runoff curve number method (Section 5.1) to abstract total rainfall depth and calculate runoff depth. The abstraction procedure follows the guide-

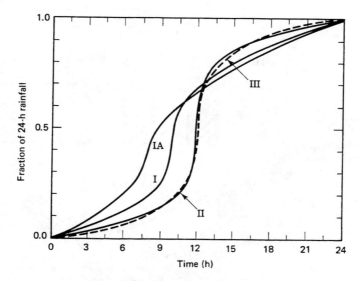

Figure 5-14 SCS 24-h rainfall distributions [22].

lines established in NEH-4 [21], with extensions to account for runoff curve numbers applicable to urban areas. In addition, TR-55 includes procedures to determine concentration time for the following types of surface flow: (1) overland flow, (2) shallow concentrated flow, and (3) streamflow. Shallow concentrated flow is a type of surface flow of characteristics in between those of overland flow and streamflow.

TR-55 Procedures

When using TR-55, there is a choice between graphical or tabular method. The graphical method gives only a peak discharge, whereas the tabular method provides a flood hydrograph. The tabular method is also recommended for catchments consisting of subcatchments of dissimilar characteristics, where hydrograph combination and channel routing may be necessary for increased accuracy.

The primary objective of TR-55 is to provide simplified techniques, thereby reducing the effort involved in routine hydrologic calculations. The potential accuracy of the method is less than that which could be obtained with more elaborate procedures. The method is strictly applicable to surface flow and should not be used to describe flow properties in underground conduits.

Selection of Runoff Curve Number

To estimate runoff curve numbers for urban catchments, TR-55 defines two types of areas: (1) pervious and (2) impervious. Once pervious and impervious areas are delineated, composite curve numbers are calculated by areal weighing (see Section 4.1 for an example of areal weighing).

Impervious areas are of two types: (1) connected and (2) unconnected. *Connected impervious areas* are those in which runoff flows directly into the drainage system, or where runoff (from the impervious area) flows over a pervious area as shallow

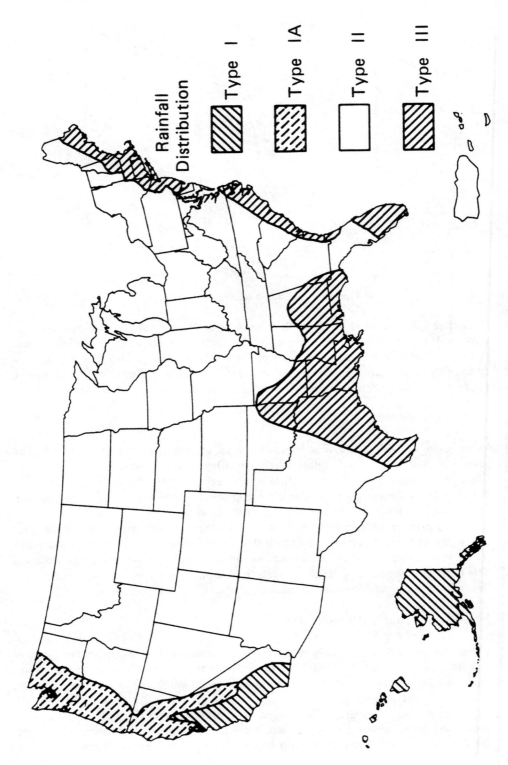

Figure 5-15 Approximate geographical boundaries for SCS rainfall distributions [22].

concentrated flow such as in a grass-lined swale. *Unconnected impervious areas* are those in which runoff (from the impervious area) flows over a pervious area (as overland flow) before it enters the drainage system.

Table 5-2(a) shows urban runoff curve numbers for different classes of pervious areas and connected impervious areas. Tables 5-2(b), (c), and (d) show runoff curve numbers for agricultural lands, and arid and semiarid rangelands.

Figure 5-16 is used in lieu of Table 5-2 if the impervious area percentages or pervious area classes are other than those shown in the tables. When the impervious areas are unconnected, Fig. 5-16 is used in cases where the total impervious area exceeds 30 percent of the catchment. Figure 5-16 gives a composite CN as a function of percent of imperviousness and pervious area CN.

Figure 5-17 is used to determine the composite runoff curve number when all or portions of the impervious areas are unconnected and the total impervious area is 30 percent or less. Figure 5-17 gives a composite curve number as a function of percent of imperviousness, ratio of unconnected impervious area to total impervious area, and pervious area curve number.

Travel Time and Concentration Time

For any reach or subreach, *travel time* is defined as the ratio of flow length to average flow velocity. At any given point, the concentration time is the sum of travel times through the individual subreaches.

For overland flow, TR-55 uses the following formula for travel time:

$$t_t = \frac{0.007(nL)^{0.8}}{P_2^{0.5} S^{0.4}} \tag{5-45}$$

in which t_t = travel time in hours; n = Manning n; L = flow length, in feet; P_2

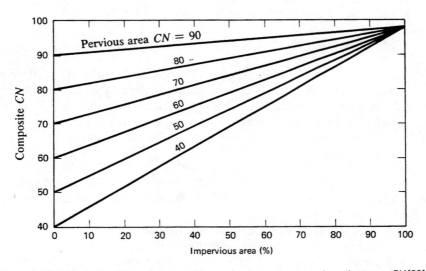

Figure 5-16 Composite CN as a function of impervious area percent and pervious area CN [22].

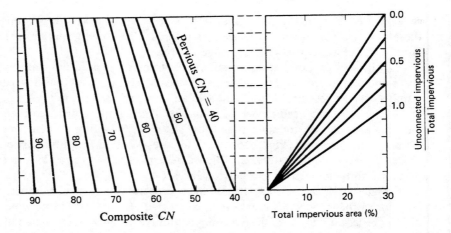

Figure 5-17 Composite CN as a function of total impervious area percent, ratio of unconnected impervious area to total impervious area, and pervious area CN [22].

$= 2\text{-}y$ 24-h rainfall depth in inches; and $S =$ average land slope, in feet per foot. In SI units, this equation is:

$$t_t = \frac{0.0288(nL)^{0.8}}{P_2^{0.5} S^{0.4}} \tag{5-46}$$

in which L is given in meters; P_2, in centimeters; S, in meters per meter; and the remaining terms are the same as in Eq. 5-45. TR-55 values of Manning n applicable to overland flow are given in Table 5-11.

Overland flow lengths over 300 ft (90 m) lead to a form of surface flow referred to as *shallow concentrated flow*. In this case, the average flow velocity is determined from Fig. 5-18. For streamflow, the Manning equation (Eq. 2-65) can be used to calculate average flow velocities. Values of Manning n applicable to open channel flow are obtained from standard references [2, 3, 6].

TR-55 Graphical Method

The TR-55 graphical method calculates peak discharge based on the concept of unit peak flow. The unit peak flow is the peak flow per unit area, per unit runoff depth. In TR-55, unit peak flow is a function of (1) concentration time, (2) ratio of initial abstraction to total rainfall, and (3) storm type.

Peak discharge is calculated by the following formula:

$$Q_p = q_u A Q F \tag{5-47}$$

in which $Q_p =$ peak discharge in $L^3 T^{-1}$ units; $q_u =$ unit peak flow in T^{-1} units; $A =$ catchment area in L^2 units; $Q =$ runoff depth in L units; and $F =$ surface storage correction factor (dimensionless).

To use the graphical method, it is first necessary to evaluate the catchment flow type and to calculate the concentration time assuming either overland flow, shallow concentrated flow, or streamflow. The runoff curve number is determined from either Table 5-2, Fig. 5-16, or Fig. 5-17. A flood frequency is selected, and an appropriate

TABLE 5-11 TR-55 MANNING n VALUES FOR OVERLAND
FLOW [22]

Surface Description	Manning n
Smooth surfaces	
(concrete, asphalt, gravel, or bare soil)	0.011
Fallow (no residue)	0.05
Cultivated ground	
(residue cover less than or equal to 20%)	0.06
(residue cover greater than 20%)	0.17
Grass	
Range, short prairie	0.15
Dense	0.24
Bermuda	0.43
Range	0.13
Woods	
Light underbrush	0.40
Dense underbrush	0.80

Note: Dense grass includes weeping lovegrass, bluegrass, buffalo grass,
blue gamma grass, native grass mixture, alfalfa, and the like.

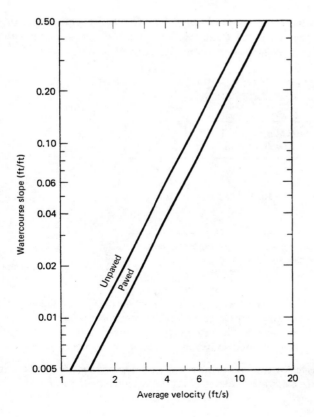

Figure 5-18 Average velocities for estimating travel time for shallow concentrated flow [22].

rainfall map (depth-duration-frequency) is used to determine the rainfall depth for the 24-h duration and the chosen frequency. With the rainfall depth P and the CN, the runoff depth Q is determined using either Fig. 5-2, Eqs. 5-8, or 5-9.

The initial abstraction is calculated by combining Eqs. 5-4 and 5-7 to yield:

$$I_a = \frac{200}{CN} - 2 \qquad (5\text{-}48)$$

in which I_a = initial abstraction, in inches. The equivalent SI formula is:

$$I_a = \frac{508}{CN} - 5.08 \qquad (5\text{-}49)$$

in which I_a is given in centimeters.

The surface storage correction factor F is obtained from Table 5-12 as a function of the percentage of pond and swamp areas. With concentration time t_c, ratio I_a/P, and storm type (either I, IA, II, or III), Fig. 5-19 is used to determine the unit peak flow in cubic feet per second per square mile per inch. Interpolation can be used for values of I_a/P different than those shown in Fig. 5-19. For values of I_a/P outside of the range shown in Fig. 5-19, the maximum (or minimum) value should be used. To obtain unit peak flow in cubic meters per second per square kilometer per centimeter, the unit peak flow values obtained from Fig. 5-19 are multiplied by 0.0043. Peak discharge is calculated by Eq. 5-47 as a function of unit peak flow, catchment area, runoff depth, and surface storage correction factor.

The TR-55 graphical method is limited to runoff curve numbers greater than 40, with concentration time in the range 0.1 to 10.0 h, and surface storage areas spread throughout the catchment and covering less than 5 percent of it.

TABLE 5-12 TR-55 SURFACE STORAGE CORRECTION FACTOR F [22]

Percentage of Pond and Swamp Areas	F
0.0	1.00
0.2	0.97
1.0	0.87
3.0	0.75
5.0	0.72

Note: Pond and swamp areas should be spread throughout the catchment.

Example 5-9.

Calculate the 10-y peak flow by the TR-55 graphical method using the following data: catchment area 4 km²; total impervious area 0.8 km²; unconnected impervious area 0.6 km²; pervious area curve number $CN = 70$; storm type II; concentration time 1.5 h; 10-y rainfall $P = 9$ cm; percentage of pond and swamp areas, $F = 1$ percent.

Since there are unconnected impervious areas and the total impervious area amounts to less than 30 percent of the catchment, Fig. 5-17 is used to calculate the composite curve number. With total impervious area (20 percent), ratio of unconnected impervious to total impervious (0.75), and pervious CN (70), the composite curve number from Fig. 5-17 is $CN = 74$. The runoff depth (Eq. 5-9) is $Q = 3.23$ cm. The initial abstraction (Eq. 5-49) is $I_a = 1.78$ cm, and the ratio $I_a/P = 0.2$. From Fig. 5-19(c) (storm type II), concentration time 1.5 h, and $I_a/P = 0.2$, the unit peak flow is 250 ft^3/(s-mi^2-in.) or $250 \times 0.0043 = 1.075$ m^3/(s-km^2-cm). From Table 5-12, $F = 0.87$. From Eq. 5-47, with $q_u = 1.075$ m^3/(s-km^2-cm); $A = 4$ km^2; $Q = 3.23$ cm; and $F = 0.87$, the peak discharge is $Q_p = 12.08$ m^3/s.

Example 5-10.

Calculate the 25-y peak flow by the TR-55 graphical method using the following data: (1) urban watershed, area $A = 15$ mi^2; (2) surface flow is shallow concentrated, paved, hydraulic length $L = 34,560$ ft; slope $S = 0.014$; (3) 26 percent of the watershed is $\frac{1}{3}$-ac lots, 30 percent impervious, soil group B; (4) 42 percent of the watershed is $\frac{1}{2}$-ac lots, with lawns in fair hydrologic condition, 36 percent impervious, soil group C; (5) 32 percent of the watershed is $\frac{1}{2}$-ac lots, with lawns in good hydrologic condition, 24 percent total impervious, 50 percent (of it) unconnected, soil group C; (6) storm type I, 25-y rainfall $P = 5$ in.; 0.2% ponding.

From Fig. 5-18, the average velocity along the hydraulic length is 2.4 ft/s. Therefore, the time of concentration is $t_c = 4$ h. For the 26 percent subarea, $\frac{1}{3}$-ac lots, 30 percent impervious, B soil, the curve number is obtained directly from Table 5-2(a): $CN = 72$. For the 42 percent subarea, $\frac{1}{2}$-ac lots, 36 percent impervious, C soil, first the pervious CN is obtained from Table 5-2(a) (open space in fair hydrologic condition): $CN = 79$. Then, the composite CN is obtained from Fig. 5-16: $CN = 86$. For the 32 percent subarea, $\frac{1}{2}$-ac lots, 24 percent total impervious, 50 percent (of it) unconnected, soil group C, first the pervious CN is obtained from Table 5-2(a) (open space in good hydrologic condition): $CN = 74$. Then, the composite CN is obtained from Fig. 5-17: $CN = 78$. The composite CN for the three subareas is: $CN = (0.26 \times 72) + (0.42 \times 86) + (0.32 \times 78) = 80$. The runoff depth (Eq. 5-8) is $Q = 2.9$ in. The initial abstraction (Eq. 5-48) is $I_a = 0.5$ in.; then, the ratio $I_a/P = 0.1$. The unit peak flow (Fig. 5-19(a)) is $q_u = 90$ ft^3/(s-mi^2-in.). The surface storage correction factor (Table 5-12) is $F = 0.97$. Finally, the peak flow (Eq. 5-47) is $Q_p = 3800$ ft^3/s.

Assessment of TR-55 Graphical Method

The TR-55 graphical method provides peak discharge as a function of unit peak flow, catchment area, runoff depth, and surface storage correction factor. The unit peak flow is a function of concentration time, abstraction parameter I_a/P, and storm type. The runoff depth is a function of total rainfall depth and runoff curve number.

In the TR-55 graphical method, concentration time accounts for both runoff concentration and runoff diffusion. From Fig. 5-19, it is seen that unit peak flow decreases with concentration time, implying that the longer the concentration time, the greater the catchment storage and peak flow attenuation.

The parameter I_a/P is related to the catchment's abstractive properties. The greater the curve number, the lesser the value of I_a/P and the greater the unit peak flow. The surface storage correction factor F reduces the peak discharge to account for additional runoff diffusion caused by surface storage features typical of low relief catchments (i.e., ponds and swamps). The geographical location and associated storm type is accounted for by the four standard SCS temporal storm distributions. Therefore, the TR-55 graphical method accounts for hydrologic abstraction, runoff concen-

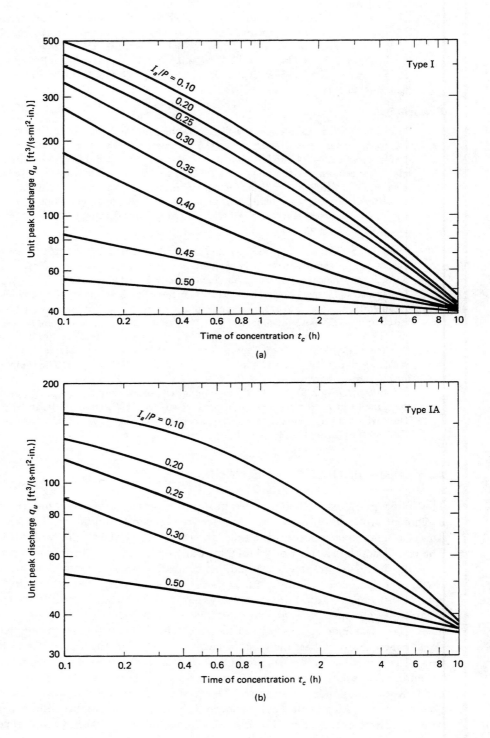

(a)

(b)

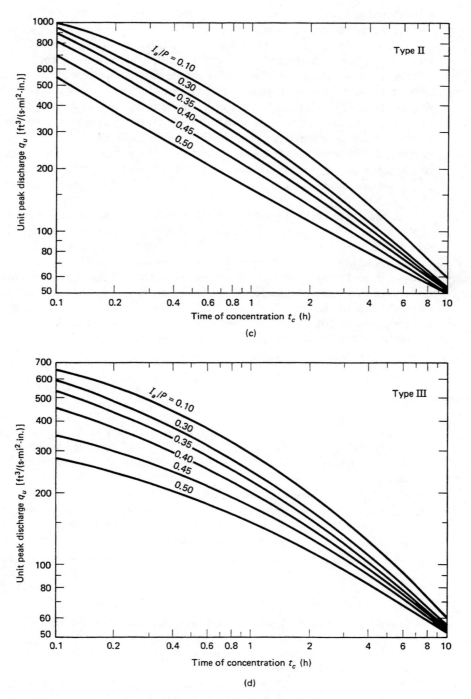

Figure 5-19 Unit peak discharge in TR-55 graphical method: (a) SCS type I rainfall distribution; (b) SCS type IA rainfall distribution; (c) SCS type II rainfall distribution; (d) SCS type III rainfall distribution [22].

tration and diffusion, geographical location and type of storm, and the additional surface storage of low-relief catchments.

The TR-55 graphical method can be considered an extension of the rational method to midsize catchments. The unit peak flow used in the graphical method is similar in concept to the runoff coefficient of the rational method. However, unlike the latter, the TR-55 graphical method includes runoff curve number and storm type and is applicable to midsize catchments with concentration times of up to 10 h.

The unit values of catchment area, runoff depth, and concentration time can be used to provide a comparison between the TR-55 graphical method and the rational method. To illustrate, assume a catchment area of 1 mi^2 (640 ac), time of concentration 1 h, and corresponding rainfall intensity 1 in./h, and runoff coefficient $C = 0.95$ (the maximum practicable value). A calculation by Eq. 4-4 gives a peak discharge of $Q_p = 613$ ft^3/s.

A calculation with the TR-55 graphical method, using the lowest possible value of abstraction for comparison purposes ($I_a/P = 0.10$), gives the following: For storm type I, 200 ft^3/s; type IA, 108 ft^3/s; type II, 360 ft^3/s; and type III, 295 ft^3/s. This example shows the effect of regional storm hyetograph on the calculated peak discharge. It also shows that the TR-55 graphical method generally gives lower peak flows than the rational method. This may be attributed to the fact that the TR-55 method accounts for runoff diffusion in a somewhat better way than the rational method. However, it should be noted that the peak discharges calculated by the two methods are not strictly comparable, since the value of $I_a/P = 0.1$ does not correspond exactly to $C = 0.95$.

QUESTIONS

1. What catchment properties are used in estimating a runoff curve number? What significant rainfall characteristic is absent from the SCS runoff curve number method?
2. What is the antecedent moisture condition in the runoff curve number method? How is it estimated?
3. What is hydrologic condition in the runoff curve number method? How is it estimated?
4. Describe the procedure to estimate runoff curve numbers from measured data. What level of antecedent moisture condition will cause the greatest runoff? Why?
5. What is a unit hydrograph? What does the word *unit* refer to?
6. Discuss the concepts of linearity and superposition in connection with unit hydrograph theory.
7. What is catchment lag? Why is it important in connection with the calculation of synthetic unit hydrographs?
8. In the Snyder method of synthetic unit hydrographs, what do the parameters C_t and C_p describe?
9. Compare lag, time-to-peak, time base, and unit hydrograph duration in the Snyder and SCS synthetic unit hydrograph methods.
10. What is the shape of the triangle used to develop the peak flow formula in the SCS synthetic unit hydrograph method? What value of Snyder's C_p matches the SCS unit hydrograph?
11. What elements are needed to properly define a synthetic unit hydrograph?

12. What is the difference between superposition and S-hydrograph methods to change unit hydrograph duration? In developing S-hydrographs, why are the ordinates summed up only at intervals equal to the unit hydrograph duration?

13. What is hydrograph convolution? What assumptions are crucial to the convolution procedure?

14. What is an unconnected impervious area in the TR-55 methodology? What is unit peak flow?

15. Given the similarities between the TR-55 graphical method and the rational method, why is the former based on runoff depth while the latter is based on rainfall intensity?

PROBLEMS

5-1. An agricultural watershed has the following hydrologic characteristics: (1) a subarea in fallow, with bare soil, soil group B, covering 32%; and (2) a subarea planted with row crops, contoured and terraced, in good hydrologic condition, soil group C, covering 68%. Determine the runoff Q, in centimeters, for a 10.5-cm rainfall. Assume an AMC II antecedent moisture condition.

5-2. A rural watershed has the following hydrologic characteristics: (1) a pasture area, in fair hydrologic condition, soil group B, covering 22%; (2) a meadow, soil group B, covering 55%; and (3) woods, poor hydrologic condition, soil group B–C, covering 23%. Determine the runoff Q, in centimeters, for a 12-cm rainfall. Assume an AMC III antecedent moisture condition.

5-3. Rain falls on a 9.5-ha urban catchment with an average intensity of 2.1 cm/h and duration of 3 h. The catchment is divided into (1) business district (with 85% impervious area), soil group C, covering 20%; and (2) residential district, with $\frac{1}{3}$-ac average lot size (with 30% impervious area), soil group C. Determine the total runoff volume, in cubic meters, assuming an AMC II antecedent moisture condition.

5-4. Rain falls on a 950-ha catchment in a semiarid region. The vegetation is desert shrub in fair hydrologic condition. The soils are: 15% soil group A; 55% soil group B, and 30% soil group C. Calculate the runoff Q, in centimeters, caused by a 15-cm storm on a wet antecedent moisture condition. Assume that field data support the use of an initial abstraction parameter $K = 0.3$.

5-5. The hydrologic response of a certain 10-mi² agricultural watershed can be modeled as a triangular-shaped hydrograph, with peak flow and time base defining the triangle. Five events encompassing a wide range of antecedent moisture conditions are selected for analysis. Rainfall-runoff data for these five events are as follows:

Rainfall P (in.)	Peak flow Q_p (ft³/s)	Time base (h)
7.05	3100	12.
6.41	3700	14.
5.13	4100	13.
5.82	4500	12.
6.77	3500	14.

Determine a value of AMC II runoff curve number based on the above data.

5-6. The following rainfall-runoff data were measured in a certain watershed:

Rainfall P (cm)	Runoff Q (cm)
15.2	12.3
10.5	10.1
7.2	4.3
8.4	5.2
11.9	9.1

Assuming that the data encompass a wide range of antecedent moisture conditions, estimate the AMC II runoff curve number.

5-7. The following rainfall distribution was observed during a 6-h storm:

Time (h)	0		2		4		6
Intensity (mm/h)		10		15		12	

The runoff curve number is $CN = 76$. Calculate the ϕ-index.

5-8. The following rainfall distribution was observed during a 12-h storm:

Time (h)	0		2		4		6		8		10		12
Intensity (mm/h)		5		10		13		18		3		10	

The runoff curve number is $CN = 86$. Calculate the ϕ-index.

5-9. The following rainfall distribution was observed during a 6-h storm:

Time (h)	0		2		4		6
Intensity (mm/h)		18		24		12	

The ϕ-index is 10 mm/h. Calculate the runoff curve number.

5-10. The following rainfall distribution was observed during a 24-h storm:

Time (h)	0	3	6	9	12	15	18	21	24
Intensity (mm/h)		5	8	10	12	15	5	3	6

The ϕ-index is 4 mm/h. Calculate the runoff curve number.

5-11. A unit hydrograph is to be developed for a 29.6-km^2 catchment with a 4-h T2 lag. A 1-h rainfall has produced the following runoff data:

Time (h)	0	1	2	3	4	5	6	7	8	9	10	11	12
Flow (m^3/s)	1	2	4	8	12	8	7	6	5	4	3	2	1

Based on this data, develop a 1-h unit hydrograph for this catchment. Assume baseflow is 1 m^3/s.

5-12. A unit hydrograph is to be developed for a 190.8-km^2 catchment with a 12-h T2 lag. A 3-h rainfall has produced the following runoff data:

Time (h)	0	3	6	9	12	15	18	21	24
Flow (m^3/s)	15	20	55	80	60	48	32	20	15

Based on this data, develop a 3-h unit hydrograph for this catchment. Assume baseflow is 15 m^3/s.

5-13. Calculate a set of Snyder synthetic unit hydrograph parameters for the following data: catchment area $A = 480$ km^2; $L = 28$ km; $L_c = 16$ km; $C_t = 1.45$; and $C_p = 0.61$.

5-14. Calculate a set of Snyder synthetic unit hydrograph parameters for the following data: catchment area $A = 950$ km^2; $L = 48$ km; $L_c = 21$ km; $C_t = 1.65$; and $C_p = 0.57$.

5-15. Calculate an SCS synthetic unit hydrograph for the following data: catchment area $A = 7.2$ km²; runoff curve number $CN = 76$; hydraulic length $L = 3.8$ km; and average land slope $Y = 0.012$.

5-16. Calculate an SCS synthetic unit hydrograph for the following data: catchment area (natural catchment) $A = 48$ km²; runoff curve number $CN = 80$; hydraulic length $L = 9$ km; and mean velocity along hydraulic length $V = 0.25$ m/s.

5-17. Calculate the peak flow of a triangular SI unit hydrograph (1 cm of runoff) having a volume-to-peak to unit-volume ratio $p = \frac{3}{10}$. Assume basin area $A = 100$ km², and time-to-peak $t_p = 6$ h.

5-18. Given the following 1-h unit hydrograph for a certain catchment, find the 2-h unit hydrograph using (a) the superposition method and (b) the S-hydrograph method.

Time (h)	0	1	2	3	4	5	6
Flow (ft³/s)	0	500	1000	750	500	250	0

5-19. Given the following 3-h unit hydrograph for a certain catchment, find the 6-h unit hydrograph using (a) the superposition method and (b) the S-hydrograph method.

Time (h)	0	3	6	9	12	15	18	21	24
Flow (m³/s)	0	5	15	30	25	20	10	5	0

5-20. Given the following 2-h unit hydrograph for a certain catchment, find the 3-h unit hydrograph. Using this 3-h unit hydrograph, calculate the 1-h unit hydrograph.

Time (h)	0	1	2	3	4	5	6	7
Flow (m³/s)	0	25	75	87.5	62.5	37.5	12.5	0

5-21. Given the following 4-h unit hydrograph for a certain catchment, find the 6-h unit hydrograph. Using this 6-h unit hydrograph, calculate the 4-h unit hydrograph, verifying the computations.

Time (h)	0	2	4	6	8	10	12	14	16	18	20	22	24
Flow (m³/s)	0	10	30	60	100	90	80	70	50	40	20	10	0

5-22. Given the following 4-h unit hydrograph for a certain catchment: (a) Find the 6-h unit hydrograph; (b) using the 6-h unit hydrograph, calculate the 8-h unit hydrograph; (c) using the 8-h unit hydrograph, calculate the 4-h unit hydrograph, verifying the computations.

Time (h)	0	2	4	6	8	10	12	14	16	18	20
Flow (m³/s)	0	10	25	40	50	40	30	20	10	5	0

5-23. The following 2-h unit hydrograph has been developed for a certain catchment:

Time (h)	0	2	4	6	8	10	12
Flow (ft³/s)	0	100	200	150	100	50	0

A 6-h storm covers the entire catchment and is distributed in time as follows:

Time (h)	0	2	4	6
Total rainfall (in./h)		1.0	1.5	0.5

Calculate the composite hydrograph for the effective storm pattern, assuming a runoff curve number $CN = 80$.

5-24. The following 3-h unit hydrograph has been developed for a certain catchment:

Time (h)	0	3	6	9	12	15	18	21	24
Flow (m³/s)	0	10	20	30	25	20	15	10	0

A 12-h storm covers the entire catchment and is distributed in time as follows:

Time (h)	0	3	6	9	12
Total rainfall (mm/h)		6	10	18	2

Calculate the composite hydrograph for the effective storm pattern, assuming a runoff curve number $CN = 80$.

5-25. A certain basin has the following 2-h unit hydrograph:

Time (h)	0	1	2	3	4	5	6	7	8	9	10	11	12	13
Flow (m³/s)	0	5	15	30	60	75	65	55	45	35	25	15	5	0

Calculate the flood hydrograph for the following effective rainfall hyetograph:

Time (h)	0	3	6
Effective rainfall (cm/h)		1.0	2.0

5-26. Given the following flood hydrograph and effective storm pattern, calculate the unit hydrograph ordinates by the method of forward substitution.

Time (h)	0	1	2	3	4	5	6	7	8	9	10	11	12
Flow (m³/s)	0	5	18	46	74	93	91	73	47	23	9	2	0

Time (h)	0	1	2	3	4	5	6
Effective rainfall (cm/h)		0.5	0.8	1.0	0.7	0.5	0.2

5-27. Using TR-55 procedures, calculate the time of concentration for a watershed having the following characteristics: (1) overland flow, dense grass, length $L = 100$ ft, slope $S = 0.01$, 2-y 24-h rainfall $P_2 = 3.6$ in.; (2) shallow concentrated flow, unpaved, length $L = 1400$ ft, slope $S = 0.01$; and (3) streamflow, Manning $n = 0.05$, flow area $A = 27$ ft², wetted perimeter $P = 28.2$ ft, slope $S = 0.005$, length $L = 7300$ ft.

5-28. Using TR-55 procedures, calculate the time of concentration for a watershed having the following characteristics: (1) overland flow, bermuda grass, length $L = 50$ m, slope $S = 0.02$, 2-y 24-h rainfall $P_2 = 9$ cm; and (2) streamflow, Manning $n = 0.05$, flow area $A = 4.05$ m², wetted perimeter $P = 8.1$ m, slope $S = 0.01$, length $L = 465$ m.

5-29. A 250-ac watershed has the following hydrologic soil-cover complexes: (1) soil group B, 75 ac, urban, $\frac{1}{2}$-ac lots with lawns in good hydrologic condition, 25% connected impervious; (2) soil group C, 100 ac, urban, $\frac{1}{2}$-ac lots with lawns in good hydrologic condition, 25% connected impervious; and (3) soil group C, 75 ac, open space in good condition. Determine the composite runoff curve number.

5-30. A 120-ha watershed has the following hydrologic soil-cover complexes: (1) soil group B, 40 ha, urban, $\frac{1}{2}$-ac lots with lawns in good hydrologic condition, 35% connected impervious; (2) soil group C, 55 ha, urban, $\frac{1}{2}$-ac lots with lawns in good hydrologic condition, 35% connected impervious; and (3) soil group C, 25 ha, open space in fair condition. Determine the composite runoff curve number.

5-31. A 90-ha watershed has the following hydrologic soil-cover complexes: (1) soil group C, 18 ha, urban, 1/3-ac lots with lawns in good hydrologic condition, 30% connected impervious; (2) soil group D, 42 ha, urban, $\frac{1}{3}$-ac lots with lawns in good hydrologic condition, 40% connected impervious; and (3) soil group D, 30 ha, urban, $\frac{1}{3}$-ac lots with lawns in fair hydrologic condition, 30% total impervious, 25% of it unconnected impervious area. Determine the composite runoff curve number.

5-32. Use the TR-55 graphical method to compute the peak discharge for a 250-ac watershed, with 25-y 24-h rainfall $P = 6$ in., time of concentration $t_c = 1.53$ h, runoff curve number $CN = 75$, and Type II rainfall.

5-33. Use the TR-55 graphical method to calculate the peak discharge for a 960-ha catchment, with 50-y 24-h rainfall $P = 10.5$ cm, time of concentration $t_c = 3.5$ h, runoff curve number $CN = 79$, type I rainfall, and 1% pond and swamp areas.

5-34. Calculate the 25-y peak flow by the TR-55 graphical method for the following watershed data: (1) urban watershed, area $A = 9.5$ km^2; (2) surface flow is shallow concentrated, paved; hydraulic length $L = 3850$ m; slope $S = 0.01$; (3) 42% of watershed is $\frac{1}{3}$-ac lots, lawns with 85% grass cover, 34% total impervious, soil group C; (4) 58% of the watershed is $\frac{1}{3}$-ac lots, lawns with 95% grass cover, 24% total impervious, 25% of it unconnected, soil group C; (5) Pacific Northwest region, 25-y 24-h rainfall $P = 10$ cm; 1% ponding.

5-35. Solve problem 5-32 using the computer program EH500 included in Appendix D.

5-36. Solve problem 5-33 using the computer program EH500 included in Appendix D.

REFERENCES

1. Amorocho, J., and G. T. Orlob. (1961). "Nonlinear Analysis of Hydrologic Systems," University of California Water Resources Center, *Contribution No. 40*, November.

2. Barnes, H. H. Jr. (1967). "Roughness Characteristics of Natural Channels," *U.S. Geological Survey Water Supply Paper No.* 1849.

3. Chow, V. T. (1959). *Open-Channel Hydraulics.* New York: McGraw-Hill.

4. Diskin, M. H. (1964). "A Basic Study of the Nonlinearity of Rainfall-Runoff Processes in Watersheds," Ph.D. Diss., University of Illinois, Urbana.

5. Freeze, R. A., and J. A. Cherry. (1979). *Groundwater,* Englewood Cliffs, N.J.: Prentice-Hall.

6. French, R. H. (1986). *Open-Channel Hydraulics.* New York: McGraw-Hill.

7. Hall, M. J. (1984). *Urban Hydrology.* London: Elsevier Applied Science Publishers.

8. Hawkins, R. H., A. T. Hjelmfelt, and A. W. Zevenbergen. (1985). "Runoff Probability, Storm Depth, and Curve Numbers," *Journal of the Irrigation and Drainage Division,* ASCE, Vol. 111, No. 4, December, pp. 330–340.

9. Hjelmfelt, A. T., K. A. Kramer, and R. E. Burwell. (1981). "Curve Numbers as Random Variables," *Proceedings, International Symposium on Rainfall-Runoff Modeling,* Mississippi State University, (also Water Resources Publications, Littleton, Colorado).

10. Linsley, R. K., M. A. Kohler, and J. L. H. Paulhus. (1962). *Hydrology for Engineers,* 3d. ed. New York: McGraw-Hill.

11. McCuen, R. H., W. J. Rawls, and S. L. Wong. (1984). "SCS Urban Peak Flow Methods," *Journal of Hydraulic Engineering,* ASCE, Vol. 110, No. 3, March, pp. 290–299.

12. Newton, D. J., and J. W. Vineyard. (1967). "Computer-Determined Unit Hydrographs from Floods," *Journal of the Hydraulics Division,* ASCE, Vol. 93, No. HY5, pp. 219–236.

13. Rallison, R.E., and R. G. Cronshey. (1979). Discussion of "Runoff Curve Numbers with Varying Soil Moisture," *Journal of the Irrigation and Drainage Division,* ASCE, Vol. 105, No. IR4, pp. 439–441.

14. Sherman, L. K. (1932). "Streamflow from Rainfall by Unit-Graph Method," *Engineering News-Record,* Vol. 108, April 7, pp. 501–505.

15. Singh, K. P. (1962). "A Nonlinear Approach to the Instantaneous Unit Hydrograph," Ph.D. Diss., University of Illinois, Urbana.

16. Singh, V. P. (1988). *Hydrologic Systems. Vol.* 1: Rainfall-Runoff Modeling. Englewood Cliffs, N.J.: Prentice-Hall.

17. Snyder, F. F. (1938). "Synthetic Unit-Graphs," *Transactions, American Geophysical Union,* Vol. 19, pp. 447–454.

18. Springer, E. P., B. J. McGurk, R. H. Hawkins, and G. B. Coltharp. (1980). "Curve Numbers from Watershed Data," *Proceedings, Symposium on Watershed Management,* ASCE, Boise, Idaho, July, pp. 938–950.

19. Taylor, A. B., and H. E. Schwarz. (1952). "Unit Hydrograph Lag and Peak Flow Related to Basin Characteristics," *Transactions, American Geophysical Union,* Vol. 33, pp. 235–246.

20. U.S. Army Corps of Engineers. (1959). "Flood Hydrograph Analysis and Computations," *Engineering and Design Manual EM* 1110-2-1405, Washington, D.C.

21. USDA Soil Conservation Service. (1985). *SCS National Engineering Handbook, Section 4: Hydrology,* Washington, D.C.

22. USDA Soil Conservation Service. (1986). "Urban Hydrology for Small Watersheds," *Technical Release No.* 55 (TR-55), Washington, D.C.

23. U.S. Forest Service. (1959). *Forest and Range Hydrology Handbook,* Washington, D.C.

24. U.S. Forest Service. (1959). *Handbook on Methods of Hydrologic Analysis,* Washington, D.C.

25. Van Sickle, D. (1969). "Experience with the Evaluation of Urban Effects for Drainage Design," in Effects of Watershed Changes on Streamflow, *Proceedings, Water Resources Symposium No.* 2, University of Texas, Austin, pp. 229–254.

SUGGESTED READINGS

Sherman, L. K. (1932). "Streamflow from Rainfall by Unit-Graph Method," *Engineering News-Record,* Vol. 108, April 7, pp. 501–505.

Snyder, F. F. (1938). "Synthetic Unit-Graphs," *Transactions, American Geophysical Union,* Vol. 19, pp. 447–454.

USDA Soil Conservation Service. (1985). *SCS National Engineering Handbook, Section 4: Hydrology,* Washington, D.C.

USDA Soil Conservation Service. (1986). "Urban Hydrology for Small Watersheds," *Technical Release No.* 55 (TR-55), Washington, D.C.

FREQUENCY ANALYSIS

The term *frequency analysis* refers to the techniques whose objective is to analyze the occurrence of hydrologic variables within a statistical framework, i.e., by using measured data and basing predictions on statistical laws. These techniques are applicable to the study of statistical properties of either rainfall or runoff (flow) series. In engineering hydrology, however, frequency analysis is commonly used to calculate flood discharges.

In principle, techniques of frequency analysis are applicable to gaged catchments with long periods of streamflow record. In practice, these techniques are primarily used for large catchments, because these are more likely to be gaged and have longer record periods. Frequency analysis is also applicable to midsize catchments, provided the record length is adequate. For ungaged catchments (either midsize or large), frequency analysis can be used in a regional context to develop flow characteristics applicable to *hydrologically homogeneous* regions. These techniques comprise what is referred to as regional analysis (Chapter 7).

The question to be answered by flow frequency analysis can be stated as follows: Given *n* years of daily streamflow records for stream S, what is the maximum (or minimum) flow *Q* that is likely to recur with a frequency of once in *T* years on the average? Or, what is the maximum flow *Q* associated with a *T*-year return period? Alternatively, frequency analysis seeks to answer the inverse question: What is the return period *T* associated with a maximum (or minimum) flow *Q*?

In more general terms, the preceding questions can be stated as follows: Given *n* years of streamflow data for stream *S* and *L* years of design life of a certain structure, what is the probability *P* of a discharge *Q* being exceeded at least once during the design life L? Alternatively, what is the discharge Q which has the probability P of being exceeded during the design life L?

This chapter is divided into three sections. Section 6.1 contains a review of statistics and probability concepts useful in engineering hydrology. Section 6.2 describes techniques of flood frequency analysis. Section 6.3 discusses low-flow frequency and droughts.

6.1 CONCEPTS OF STATISTICS AND PROBABILITY

Frequency analysis uses random variables and probability distributions. A *random variable* follows a certain probability distribution. A *probability distribution* is a function that expresses in mathematical terms the relative chance of occurrence of each of all possible outcomes of the random variable. In statistical notation, $P(X = x_1)$ is the probability P that the random variable X takes on the outcome x_1. A shorter notation is $P(x_1)$.

An example of random variable and probability distribution is shown in Fig. 6-1. This is a discrete probability distribution because the possible outcomes have been arranged into groups (or classes). The random variable is discharge Q; the possible outcomes are seven discharge classes, from 0–100 m³/s to 600–700 m³/s. In Fig. 6-1, the probability that Q is in the class 100–200 m³/s is 0.25. The sum of probabilities of all possible outcomes is equal to 1.

A cumulative discrete distribution, corresponding to the discrete probability distribution of Fig. 6-1, is shown in Fig. 6-2. In this figure, the probability that Q is in a class less than or equal to the 100–200 class is 0.40. The maximum value of probability of the cumulative distribution is 1.

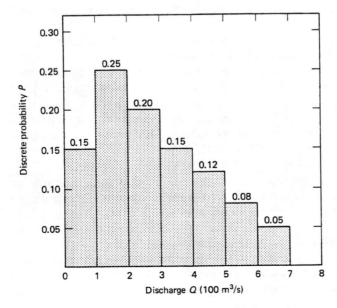

Figure 6-1 Discrete probability distribution.

Frequency Analysis   Chap. 6

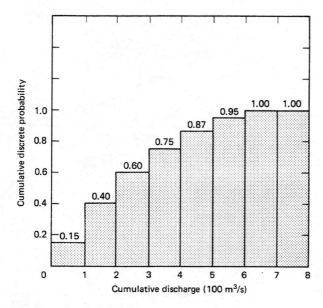

Figure 6-2 Cumulative discrete probability distribution.

Properties of Statistical Distributions

The properties of statistical distributions are described by the following measures: (1) central tendency, (2) variability, and (3) skewness. Statistical distributions are described in terms of *moments*. The first moment describes central tendency, the second moment describes variability, and the third moment describes skewness. Higher-order moments are possible but are seldom used in practical applications.

The first moment about the origin is the *arithmetic mean,* or mean. It expresses the distance from the origin to the centroid of the distribution (Fig. 6-3(a)):

$$\bar{x} = \frac{1}{n} \sum_{i=1}^{n} x_i \tag{6-1}$$

in which $\bar{x}$ is the mean, x_i is the random variable, and n is the number of values.

The *geometric mean* is the nth root of the product of n terms:

$$\bar{x}_g = (x_1 \, x_2 \, x_3 \cdots x_n)^{1/n} \tag{6-2}$$

The logarithm of the geometric mean is the mean of the logarithms of the individual values. The geometric mean is to the lognormal probability distribution what the arithmetic mean is to the normal probability distribution.

The *median* is the value of the variable that divides the probability distribution into two equal portions (or areas) (Fig. 6-3(b)). For certain skewed distributions (i.e. one with third moment other than zero), the median is a better indication of central tendency than the mean. Another measure of central tendency is the *mode,* defined as the value of the variable that occurs most frequently (Fig. 6-3(c)).

Statistical moments can be defined about axes other than the origin. The second moment about the mean is the *variance,* defined as

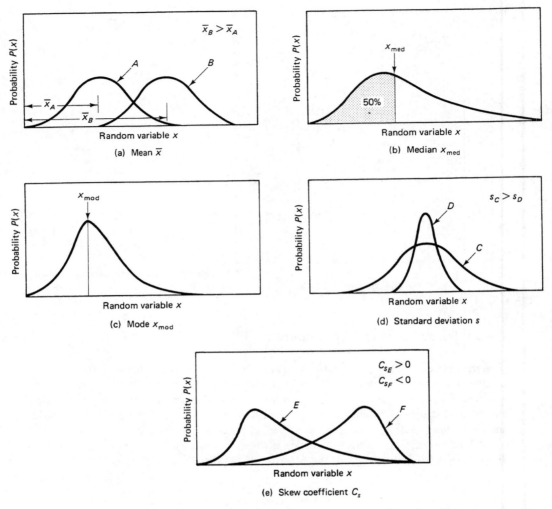

Figure 6-3 Properties of statistical distributions: (a) mean $\bar{x}$; (b) median x_{med}; (c) mode x_{mod}; (d) standard deviation s; (e) skew coefficient C_s.

$$s^2 = \frac{1}{n-1} \sum_{i=1}^{n} (x_i - \bar{x})^2 \qquad (6\text{-}3)$$

in which s^2 is the variance. The square root of the variance, s, is the *standard deviation.* The *variance coefficient* (or coefficient of variation) is defined as

$$C_v = \frac{s}{\bar{x}} \qquad (6\text{-}4)$$

The standard deviation and variance coefficient are useful in comparing relative variability among distributions. The larger the standard deviation and variance coefficient, the larger the spread of the distribution (Fig. 6-3(d)).

Frequency Analysis   Chap. 6

The third moment about the mean is the *skewness*, defined as follows:

$$a = \frac{n}{(n-1)(n-2)} \sum_{i=1}^{n} (x_i - \bar{x})^3 \tag{6-5}$$

in which a is the skewness. The *skew coefficient* is defined as

$$C_s = \frac{a}{s^3} \tag{6-6}$$

For symmetrical distributions, the skewness is 0 and $C_s = 0$. For right skewness (distributions with the long tail to the right), $C_s > 0$; for left skewness (long tail to the left), $C_s < 0$ (Fig. 6-3(e)).

Another measure of skewness is *Pearson's skewness*, defined as the ratio of the difference between mean and mode to the standard deviation.

Example 6-1.

Calculate the mean, standard deviation, and skew coefficient for the following flood series: 4580, 3490, 7260, 9350, 2510, 3720, 4070, 5400, 6220, 4350, and 5930 m³/s.

The calculations are shown in Table 6-1. Column 1 shows the year and Col. 2 shows the annual maximum flows. The mean (Eq. 6-1) is calculated by summing up Col. 2 and dividing the sum by $n = 11$. This results in $\bar{x} = 5171$ m³/s. Column 3 shows the flow deviations from the mean, $x_i - \bar{x}$. Column 4 shows the square of the flow deviations, $(x_i - \bar{x})^2$. The variance (Eq. 6-3) is calculated by summing up Col. 4 and dividing the sum by $(n-1) = 10$. This results in: $s^2 = 3,780,449$ m⁶/s². The square root of the variance is the standard deviation: $s = 1944$ m³/s. The variance coefficient (Eq. 6-4) is $C_v = 0.376$. Column 5 shows the cube of the flow deviations, $(x_i - \bar{x})^3$. The skewness (Eq. 6-5) is calculated by summing up Col. 5 and multiplying the sum by $n/[(n-1)(n-2)] = 11/90$. This results in $a = 6,717,359,675$ m⁹/s³. The skew coefficient (Eq. 6-6) is equal to the skewness divided by the cube of the standard deviation. This results in $C_s = 0.914$.

TABLE 6-1   CALCULATION OF MEAN, STANDARD DEVIATION, AND SKEW COEFFICIENT: EXAMPLE 6-1

(1)	(2)	(3)	(4)	(5)
Year	Peak Flow (m³/s)	$(x_i - \bar{x})$ (m³/s)	$(x_i - \bar{x})^2$ (m⁶/s²)	$(x_i - \bar{x})^3$ (m⁹/s³)
1	4,580	−591	349,281	−206,425,071
2	3,490	−1,681	2,825,761	−4,750,104,241
3	7,260	2,089	4,363,921	9,116,230,969
4	9,350	4,179	17,464,041	72,982,227,340
5	2,510	−2,661	7,080,921	−18,842,330,780
6	3,720	−1,451	2,105,401	−3,054,936,851
7	4,070	−1,101	1,212,201	−1,334,633,301
8	5,400	229	52,441	12,008,989
9	6,220	1,049	1,100,401	1,154,320,649
10	4,350	−821	674,041	−553,387,661
11	5,930	759	576,081	437,245,479
Sum	56,880		37,804,491	54,960,215,521

Continuous Probability Distributions

A continuous probability distribution is referred to as a probability density function (PDF). A PDF is an equation relating probability, random variable, and parameters of the distribution. Selected PDFs useful in engineering hydrology are described in this section.

Normal Distribution. The *normal distribution* is a symmetrical, bell-shaped PDF also known as the Gaussian distribution, or the natural law of errors. It has two parameters: the mean, μ, and the standard deviation, σ, of the population. In practical applications, the mean $\bar{x}$ and the standard deviation s derived from sample data are substituted for μ and σ. The PDF of the normal distribution is

$$f(x) = \frac{1}{\sigma\sqrt{2\pi}} e^{-(x-\mu)^2/(2\sigma^2)} \tag{6-7}$$

in which x is the random variable and $f(x)$ is the continuous probability.

By means of the transformation

$$z = \frac{x - \mu}{\sigma} \tag{6-8}$$

the normal distribution can be converted into a one-parameter distribution, as follows:

$$f(z) = \frac{1}{\sqrt{2\pi}} e^{-z^2/2} \tag{6-9}$$

in which z is the standard unit, which is normally distributed with zero mean and unit standard deviation.

From Eq. 6-8,

$$x = \mu + z\sigma \tag{6-10}$$

in which z, the standard unit, is the *frequency factor* of the normal distribution. In general, the frequency factor of a statistical distribution is referred to as K.

A cumulative density function (CDF) can be derived by integrating the probability density function. From Eq. 6-9, integration leads to

$$F(z) = \frac{1}{\sqrt{2\pi}} \int_{-\infty}^{z} e^{-u^2/2} \, du \tag{6-11}$$

in which $F(z)$ denotes cumulative probability and u is a dummy variable of integration. The distribution is symmetrical with respect to the origin; therefore, only half of the distribution needs to be evaluated. Table A-5 (Appendix A) shows values of $F(z)$ versus z, in which $F(z)$ is integrated from the origin to z.

Example 6-2.

The annual maximum flows of a certain stream have been found to be normally distributed, with mean 90 m³/s and standard deviation 30 m³/s. Calculate the probability that a flow larger that 150 m³/s will occur.

To enter Table A-5, it is necessary to calculate the standard unit. For a flow of 150 m^3/s, the standard unit (Eq. 6-8) is: $z = (150 - 90)/30 = 2$. This means that the flow of 150 m^3/s is located two standard deviations to the right of the mean (had z been negative, the flow would have been located to the left of the mean). In Table A-5, for $z = 2$, $F(z) = 0.4772$. This value is the cumulative probability measured from $z = 0$ to $z = 2$, i.e., from the mean (90 m^3/s) to the value being considered (150 m^3/s). Because the normal distribution is symmetrical with respect to the origin, the cumulative probability measured from $z = -\infty$ to $z = 0$, is 0.5. Therefore, the cumulative probability measured from $z = -\infty$ to $z = 2$, is $F(z) = 0.5 + 0.4772 = 0.9772$. This is the probability that the flow is less than 150 m^3/s. To find the probability that the flow is larger than 150 m^3/s, the complementary cumulative probability is calculated: $G(z) = 1 - F(z) = 0.0228$. Therefore, there is a $(0.0228 \times 100) = 2.28\%$ chance that the annual maximum flow for the given stream will be larger than 150 m^3/s.

Lognormal Distribution.   For certain natural phenomena, values of random variables do not follow a normal distribution, but their logarithms do. In this case, a suitable PDF can be obtained by substituting y for x in the equation for the normal distribution, Eq. 6-7, in which $y = \ln x$. The parameters of the lognormal distribution are the mean and standard deviation of y: μ_y and σ_y.

Gamma Distribution.   The gamma distribution is used in many applications of engineering hydrology. The PDF of the gamma distribution is the following:

$$f(x) = \frac{x^{\gamma-1} e^{-x/\beta}}{\beta^\gamma \Gamma(\gamma)} \tag{6-12}$$

with parameters β and γ and valid for $\gamma > 0$.

The mean of the gamma distribution is $\beta\gamma$, the variance is $\beta^2\gamma$, and the skewness is $2/(\gamma)^{1/2}$. The term $\Gamma(\gamma) = (\gamma - 1)!$ is an important definite integral referred to as the *gamma function*, defined as follows:

$$\Gamma(\gamma) = \int_0^\infty x^{\gamma-1} e^{-x} \, dx \tag{6-13}$$

Pearson Distributions.   Pearson [23] has derived a series of probability functions to fit virtually any distribution. These functions have been widely used in practical statistics to define the shape of many distribution curves. The general PDF of the Pearson distributions is the following [5]:

$$f(x) = e^{\int_{-\infty}^{x} [(a + x)/(b_0 + b_1 x + b_2 x^2)] \, dx} \tag{6-14}$$

in which a, b_0, b_1, and b_2 are constants. The criterion for determining the type of distribution is κ, defined as follows:

$$\kappa = \frac{\beta_1(\beta_2 + 3)^2}{4(4\beta_2 - 3\beta_1)(2\beta_2 - 3\beta_1 - 6)} \tag{6-15}$$

in which $\beta_1 = \mu_3^2/\mu_2^3$ and $\beta_2 = \mu_4/\mu_2^2$, with μ_2, μ_3, and μ_4 being the second, third, and fourth moments about the mean. With $\mu_3 = 0$ (i.e., zero skewness), $\beta_1 = 0$, $\kappa = 0$, and the Pearson distribution reduces to the normal distribution.

The Pearson Type III distribution has been widely used in flood frequency analysis. In the Pearson Type III distribution, $\kappa = \infty$, which implies that $2\beta_2 = (3\beta_1 + 6)$. This is a three-parameter skewed distribution with the following PDF:

$$f(x) = \frac{(x - x_o)^{\gamma-1} e^{-(x-x_o)/\beta}}{\beta^\gamma \Gamma(\gamma)} \tag{6-16}$$

and parameters β, γ, and x_o. For $x_o = 0$, the Pearson Type III distribution reduces to the gamma distribution (Eq. 6-12). For $\gamma = 1$, the Pearson Type III distribution reduces to the exponential distribution, with the following PDF:

$$f(x) = \left(\frac{1}{\beta}\right) e^{-(x-x_o)/\beta} \tag{6-17}$$

The mean of the Pearson Type III distribution is $x_o + \beta\gamma$, the variance is $\beta^2\gamma$, and the skewness is $2/(\gamma)^{1/2}$.

Extreme Value Distributions. The extreme value distributions Types I, II, and III are based on the theory of extreme values. Frechet (on Type II) in 1927 [9] and Fisher and Tippett (on Types I and III) in 1928 [7] independently studied the statistical distribution of extreme values. Extreme value theory implies that if a random variable Q is the maximum in a sample of size n from some population of x values, then, provided n is sufficiently large, the distribution of Q is one of three asymptotic types (I, II, or III), depending on the distribution of x.

The extreme value distributions can be combined into one and expressed as a general extreme value (GEV) distribution [22]. The cumulative density function of the GEV distribution is:

$$F(x) = e^{-[1-k(x-u)/\alpha]^{1/k}} \tag{6-18}$$

in which k, u and α are parameters. The parameter k defines the type of distribution, u is a location parameter, and α is a scale parameter. For $k = 0$, the GEV distribution reduces to the extreme value Type I (EV1), or Gumbel, distribution. For $k < 0$, the GEV distribution is the extreme value Type II (EV2), or Frechet, distribution. For $k > 0$, the GEV distribution is the extreme value Type III (EV3), or Weibull, distribution. The GEV distribution is useful in applications where an extreme value distribution is being considered but its type is not known a priori.

Gumbel [12, 13, 14] has fitted the extreme value Type I distribution to long records of river flows from many countries. The cumulative density function (CDF) of the Gumbel distribution is the following double exponential function:

$$F(x) = e^{-e^{-y}} \tag{6-19}$$

in which $y = (x - u)/\alpha$ is the Gumbel (reduced) variate.

The mean $\bar{y}_n$ and standard deviation σ_n of the Gumbel variate are functions of record length n. Values of $\bar{y}_n$ and σ_n as a function of n are given in Table A-8 (Appendix A). When the record length approaches ∞, the mean $\bar{y}_n$ approaches the value of the Euler constant (0.5772) [25], and the standard deviation σ_n approaches the value $\pi/\sqrt{6}$. The skew coefficient of the Gumbel distribution is 1.14.

The extreme value Type II distribution is also known as the log Gumbel. Its cumulative density function is

$$F(x) = e^{-y^{1/k}} \qquad (6\text{-}20)$$

for $k < 0$.

The extreme value Type III distribution has the same CDF as the Type II, but in this case $k > 0$. As k approaches 0, the EV2 and EV3 distributions converge to the EV1 distribution.

6.2 FLOOD FREQUENCY ANALYSIS

Flood frequency analysis refers to the application of frequency analysis to study the occurrence of floods. Historically, many probability distributions have been used for this purpose. The normal distribution was first used by Horton [18] in 1913, and shortly thereafter by Fuller [10]. Hazen [16] used the lognormal distribution to reduce skewness, whereas Foster [8] preferred to use the skewed Pearson distributions.

The logarithmic version of the Pearson Type III distribution, i.e., the log Pearson III, has been endorsed by the U.S. Interagency Advisory Committee on Water Data for general use in the United States [27]. The Gumbel distribution (extreme value Type I, or EV1) is also widely used in the United States and throughout the world. The log Pearson III and Gumbel methods are described in this section.

Selection of Data Series

The complete record of streamflows at a given gaging station is called the *complete duration series*. To perform a flood frequency analysis, it is necessary to select a *flood series*, i.e., a sample of flood events extracted from the complete duration series.

There are two types of flood series: (1) the partial duration series and (2) the extreme value series. The partial duration (or peaks-over-a-threshold (POT) [22]) series consists of floods whose magnitude is greater than a certain base value. When the base value is such that the number of events in the series is equal to the number of years of record, the series is called an *annual exceedence* series.

In the extreme value series, every year of record contributes one value to the extreme value series, either the maximum value (as in the case of flood frequency analysis) or the minimum value (as in the case of low-flow frequency analysis). The former is the *annual maxima* series; the latter is the *annual minima* series.

The annual exceedence series takes into account all extreme events above a certain base value, regardless of when they occurred. However, the annual maxima series considers only one extreme event per yearly period. The difference between the two series is likely to be more marked for short records in which the second largest annual events may strongly influence the character of the annual exceedence series. In practice, the annual exceedence series is used for frequency analyses involving short return periods, ranging from 2 to 10 y. For longer return periods the difference between annual exceedence and annual maxima series is small. The annual maxima series is used for return periods ranging from 10 to 100 y and more.

Return Period, Frequency, and Risk

The time elapsed between successive peak flows exceeding a certain flow Q is a random variable whose mean value is called the *return period* T (or recurrence interval) of the flow Q. The relationship between probability and return period is the following:

$$P(Q) = \frac{1}{T} \qquad (6\text{-}21)$$

in which $P(Q)$ is the *probability of exceedence* of Q, or frequency. The terms frequency and return period are often used interchangeably, although strictly speaking, frequency is the reciprocal of return period. A frequency of $1/T$, or one in T years, corresponds to a return period of T years.

The *probability of nonexceedence* $P(\bar{Q})$ is the *complementary probability* of the probability of exceedence $P(Q)$, defined as

$$P(\bar{Q}) = 1 - P(Q) = 1 - \frac{1}{T} \qquad (6\text{-}22)$$

The probability of nonexceedence in n successive years is

$$P(\bar{Q}) = \left(1 - \frac{1}{T}\right)^n \qquad (6\text{-}23)$$

Therefore, the probability, or *risk*, that Q will occur at least once in n successive years is

$$R = 1 - P(\bar{Q}) = 1 - \left(1 - \frac{1}{T}\right)^n \qquad (6\text{-}24)$$

Plotting Positions

Frequency distributions are plotted using probability papers. One of the scales on a probability paper is a probability scale; the other is either an arithmetic or logarithmic scale. Normal and extreme value probability distributions are most often used in probability papers.

An *arithmetic probability* paper has a normal probability scale and an arithmetic scale. This type of paper is used for plotting normal and Pearson distributions. A *log probability* paper has a normal probability scale and a logarithmic scale and is used for plotting lognormal and log Pearson distributions. An *extreme value* probability paper has an extreme value scale and an arithmetic scale and is used for plotting extreme value distributions.

Data fitting a normal distribution plot as a straight line on arithmetic probability paper. Likewise, data fitting a lognormal distribution plot as a straight line on log probability paper, and data fitting the Gumbel distribution plot as a straight line on extreme value probability paper.

For plotting purposes, the probability of an individual event can be obtained directly from the flood series. For a series of n annual maxima, the following ratio holds:

$$\frac{\bar{x}}{N} = \frac{m}{n+1} \tag{6-25}$$

in which $\bar{x}$ = mean number of exceedences; N = number of trials; n = number of values in the series; and m = the rank of descending values, with largest equal to 1.

For example, if $n = 79$, the second largest value in the series ($m = 2$) will be exceeded twice on the average ($\bar{x} = 2$) in 80 trials ($N = 80$). Likewise, the largest value in the series ($m = 1$) will be exceeded once on the average ($\bar{x} = 1$) after 80 trials ($N = 80$). Since return period T is associated with $\bar{x} = 1$, Eq. 6-25 can be expressed as follows:

$$\frac{1}{T} = P = \frac{m}{n+1} \tag{6-26}$$

in which P = exceedence probability.

Equation 6-26 is known as the *Weibull plotting position* formula. This equation is commonly used in hydrologic applications, particularly for computing plotting positions for unspecified distributions [1]. A general plotting position formula is of the following form [11]:

$$\frac{1}{T} = P = \frac{m - a}{n + 1 - 2a} \tag{6-27}$$

in which a = parameter. Cunnane [6] performed a detailed study of the accuracy of different plotting position formulas and concluded that the Blom formula [3], with $a = 0.375$ in Eq. 6-27, is most appropriate for the normal distribution, whereas the Gringorten formula, with $a = 0.44$, should be used in connection with the Gumbel distribution. According to Cunnane, the Weibull formula, for which $a = 0$, is most appropriate for a uniform distribution.

In computing plotting positions, when the ranking of values is in descending order (from highest to lowest), P is the probability of exceedence, or the probability of a value being *greater than or equal to* the ranked value. When the ranking of values is in ascending order (from lowest to highest), P is the probability of nonexceedence, or the probability of a value being *less than or equal to* the ranked value. The computation of plotting positions is illustrated by the following example.

Example 6-3.

Use Eq. 6-26 to calculate the plotting positions for the flood series (annual maxima) shown in Table 6-2, Col. 2.

The solution is shown in Table 6-2, Cols. 3–5. Column 3 shows the ranked values, from highest to lowest. Column 4 shows the rank of each value, from 1 to 16 ($n = 16$), with the highest value ranked as 1 and the lowest value ranked as 16. Column 5 shows the probability calculated by Eq. 6-26 (expressed in percent). Because the ranking was done in descending order, Col. 5 shows the probability that a value of flood discharge will be greater than or equal to the ranked value. To illustrate, there is a 5.88% probability that a value of flood discharge will be greater than or equal to 3320 m³/s. Conversely, there is a 94.12% probability that the value of flood discharge will be greater than or equal to 690 m³/s. Column 6 shows the return period calculated by Eq. 6-26.

TABLE 6-2 COMPUTATION OF PLOTTING POSITIONS: EXAMPLE 6-3

(1)	(2)	(3)	(4)	(5)	(6)
Year	Annual Flood (m³/s)	Ranked Values (m³/s)	Rank	Probability (percent)	Return Period (y)
1972	2520	3320	1	5.88	17.00
1973	1850	3170	2	11.76	8.50
1974	750	2520	3	17.65	5.67
1975	1100	2160	4	23.53	4.25
1976	1380	1950	5	29.41	3.40
1977	1910	1910	6	35.29	2.83
1978	3170	1850	7	41.18	2.43
1979	1200	1730	8	47.06	2.13
1980	820	1480	9	52.94	1.89
1981	690	1380	10	58.82	1.70
1982	1240	1240	11	64.71	1.55
1983	1730	1200	12	70.59	1.42
1984	1950	1100	13	76.47	1.31
1985	2160	820	14	82.35	1.21
1986	3320	750	15	88.24	1.13
1987	1480	690	16	94.12	1.06

Curve Fitting

Once the data have been plotted on probability paper, the next step is to fit a curve through the plotted points. Curve fitting can be accomplished by any of the following methods: (1) graphical, (2) least square, (3) moments, and (4) maximum likelihood. The graphical method consists of fitting a function visually to the data. This method, however, has the disadvantage that the results are highly dependent on the skills of the person doing the fitting. A more consistent procedure is to use either the least square, moments, or maximum likelihood methods.

In the least square method, the sum of the squares of the differences between observed data and fitted values is minimized. The minimization condition leads to a set of m normal equations, where m is the number of parameters to be estimated. The simultaneous solution of the normal equations leads to the parameters describing the fitting (Chapter 7).

To apply the method of moments, it is first necessary to select a distribution; then, the moments of the distribution are calculated based on the data. The method provides an exact theoretical fitting, but the accuracy is substantially affected by errors in the tail of the distribution (i.e., events of long return period). A disadvantage of the method is the uncertainty regarding the adequacy of the chosen probability distribution.

In the method of maximum likelihood, the distribution parameters are estimated in such a way that the product of probabilities (i.e., the joint probability, or likelihood) is maximized. This is obtained in a similar manner to the least square method by partially differentiating the likelihood with respect to each of the parameters and equating the result to zero.

The four fitting methods can be rated in ascending order of effectiveness: graphical, least square, moments, and maximum likelihood. The latter, however, is somewhat more difficult to apply [5, 20]. In practice, the method of moments is the most commonly used curve-fitting method (see, for instance, the log Pearson III and Gumbel methods described later in this section).

Frequency Factors

Any value of a random variable may be represented in the following form:

$$x = \bar{x} + \Delta x \qquad (6\text{-}28)$$

in which x = value of random variable; $\bar{x}$ = mean of the distribution, and Δx = departure from the mean, a function of return period and statistical properties of the distribution. This departure from the mean can be expressed in terms of the product of the standard deviation s and a frequency factor K such that $\Delta x = Ks$. The frequency factor is a function of return period and probability distribution to be used in the analysis. Therefore, Eq. 6-28 can be written in the following form:

$$x = \bar{x} + Ks \qquad (6\text{-}29)$$

or, alternatively,

$$\frac{x}{\bar{x}} = 1 + KC_v \qquad (6\text{-}30)$$

in which C_v = variance coefficient.

Equation 6-29 was proposed by Chow [4] as a general equation for hydrologic frequency analysis. For any probability distribution, a relationship can be determined between frequency factor and return period. This relationship can be expressed in analytical terms, in the form of tables, or by K-T curves. In using the procedure, the statistical parameters are first determined from the analysis of the flood series. For a given return period, the frequency factor is determined from the curves or tables and the flood magnitude computed by Eq. 6-29.

Log Pearson III Method

The log Pearson III method of flood frequency analysis is described in Bulletin 17B: *Guidelines for Determining Flood Flow Frequency*, published by the U.S. Interagency Advisory Committee on Water Data, Reston, Virginia [27].

Methodology. To apply the method, the following steps are necessary:

1. Assemble the annual flood series x_i.
2. Calculate the logarithms of the annual flood series:

$$y_i = \log x_i \qquad (6\text{-}31)$$

3. Calculate the mean $\bar{y}$, standard deviation s_y, and skew coefficient C_{sy} of the logarithms y_i.

4. Calculate the logarithms of the flood discharges, $\log Q_j$, for each of several chosen probability levels P_j, using the following frequency formula:

$$\log Q_j = \bar{y} + K_j s_y \qquad (6\text{-}32)$$

in which K_j is the frequency factor, a function of the probability P_j and the skew coefficient C_{sy}. Table A-6 (Appendix A) shows frequency factors K for 10 selected probability levels in the range 0.5 to 95 percent (and corresponding return periods in the range 200 to 1.05 y) and skew coefficients in the range -3.0 to 3.0.

5. Calculate the flood discharges Q_j for each P_j probability level (or return period T_j) by taking the antilogarithms of the $\log Q_j$ values.

6. Plot the flood discharges Q_j against probabilities P_j on log probability paper, with discharges in the log scale and probabilities in the probability scale. The log Pearson III fit to the data is obtained by linking the points with a smooth curve. For $C_{sy} = 0$, the curve reduces to a straight line.

The procedure is illustrated by the following example.

Example 6-4.

Apply the log Pearson III method to the flood series of Example 6-3. Plot the results on log probability paper along with the plotting positions calculated in Example 6-3.

The discharge values, Table 6-2, Col. 2, are converted to logarithms, and the mean, standard deviation, and skew coefficient of the logarithms calculated. This results in $\bar{y} = 3.187$, $s_y = 0.207$, and $C_{sy} = -0.116$. The computations are summarized in Table 6-3. Column 1 shows selected return periods, and Col. 2 shows the associated probabilities in percent (exceedence probability). Column 3 shows the frequency factors K for $C_{sy} = -0.116$ and for each return period or exceedence probability. Values in Col. 3 are obtained from Table A-6 by linear interpolation. Column 4 shows the logarithms of the flood discharges calculated by Eq. 6-32, and Col. 5 shows the flood discharges. The flood

TABLE 6-3 LOG PEARSON III METHOD: EXAMPLE 6-4

(1)	(2)	(3)	(4)	(5)
Return period T (y)	Probability P (percent)	Frequency factor K (for $C_{sy} = -0.116$)	$y_i = \log Q$	$x_i = Q$ (m³/s)
1.05	95	-1.677	2.840	692
1.11	90	-1.293	2.919	830
1.25	80	-0.835	3.014	1033
2	50	0.019	3.191	1552
5	20	0.847	3.362	2301
10	10	1.268	3.449	2812
25	4	1.710	3.541	3475
50	2	1.991	3.599	3972
100	1	2.240	3.651	4477
200	0.5	2.467	3.698	4989

discharges are plotted against corresponding probabilities, as shown by the solid line of Fig. 6-4, to obtain the log Pearson III fit to the data. The plotting positions calculated in Example 6-3 are also shown for comparison.

Regional Skew Characteristics. The skew coefficient of the flood series (i.e. the station skew) is sensitive to extreme events. The overall accuracy of the method is improved by using a weighted value of skew in lieu of the station skew. First, a value of regional skew is obtained, and the weighted skew is calculated by weighing station and regional skews in inverse proportion to their mean square errors (MSE). The formula for weighted skew is the following:

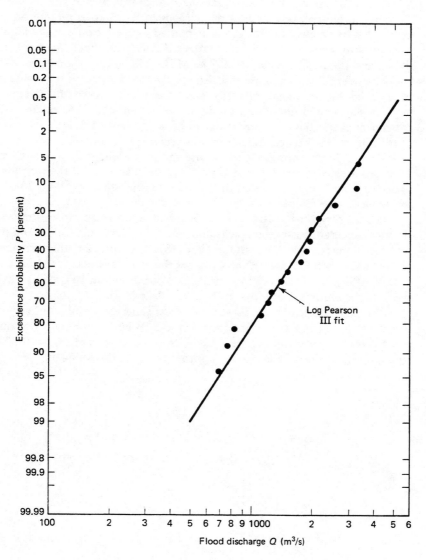

Figure 6-4 Log Pearson III fit: Example 6-4.

$$C_{sw} = \frac{(\text{MSE})_{sr}\, C_{sy} + (\text{MSE})_{sy}\, C_{sr}}{(\text{MSE})_{sr} + (\text{MSE})_{sy}} \tag{6-33}$$

in which C_{sw} = weighted skew; C_{sy} = station skew; C_{sr} = regional skew; $(\text{MSE})_{sy}$ = mean square error of the station skew; and $(\text{MSE})_{sr}$ = mean square error of the regional skew.

To develop a value of regional skew, it is necessary to assemble data from at least 40 stations or, alternatively, all stations within a 160-km radius. The stations should have at least 25 y of record. In certain cases, the paucity of data may require a relaxation of these criteria. The procedure includes analysis by three methods: (1) skew isolines map, (2) skew prediction equation, and (3) statistics of station skews.

To develop a skew isolines map, each station skew is plotted on a map at the centroid of its catchment area, and the plotted data are examined to identify any geographic or topographic trends. If a pattern is evident, isolines (lines of equal skew) are drawn and the MSE is computed. The MSE is the mean of the square of the differences between observed skews and isoline skews. If no pattern is evident, an isoline map cannot be developed, and this method is not considered further.

In the second method, a prediction equation is used to relate station skew to catchment properties and climatological variables. The MSE is the mean of the square of the differences between observed and predicted skews.

In the third method, the mean and variance of the station skews are calculated. In some cases, the variability of runoff may be such that all the stations may not be hydrologically homogeneous. If this is the case, the values of about 20 stations can be used to calculate the mean and variance of the data.

Of the three methods, the one providing the most accurate estimate of skew coefficient is selected. First a comparison of the MSEs from the isolines map and prediction equations is made. Then the smaller MSE is compared to the variance of the data. If the smaller MSE is significantly smaller than the variance, it should be used in Eq. 6-33 as MSE_{sr}. If this is not the case, the variance should be used as MSE_{sr}, with the mean of the station skews used as regional skew (C_{sr}).

In the absence of regional skew studies, generalized values of regional skew for use in Eq. 6-33 can be obtained from Fig. 6-5. When regional skew is obtained from this figure, the mean square error of the regional skew is $\text{MSE}_{sr} = 0.302$. The mean square error of the station skew is approximated by the following formula:

$$(\text{MSE})_{sy} = 10^{A - B \log(n/10)} \tag{6-34}$$

in which

$$A = -0.33 + 0.08G, \quad \text{for } G < 0.9 \tag{6-34a}$$

$$A = 0.52 + 0.30G, \quad \text{for } G \geq 0.9 \tag{6-34b}$$

$$B = 0.94 - 0.26G, \quad \text{for } G < 1.5 \tag{6-34c}$$

$$B = 0.55, \quad \text{for } G \geq 1.5 \tag{6-34d}$$

with G = absolute value of the station skew and n = record length in years.

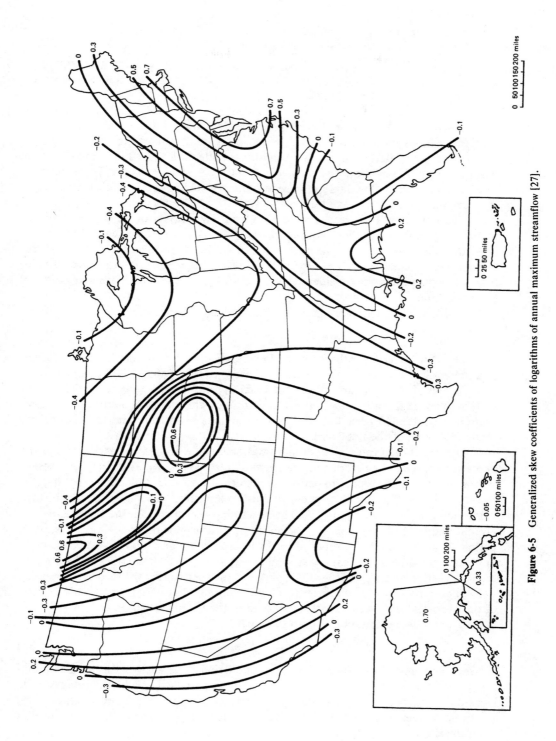

Figure 6-5 Generalized skew coefficients of logarithms of annual maximum streamflow [27].

Example 6-5.

A station in San Diego, California, has flood records for 34 y, with station skew $C_{sy} = -0.1$. Calculate a weighted skew following Eq. 6-33 and Fig. 6-5.

From Fig. 6-5, the generalized value of regional skew is $C_{sr} = -0.3$. The MSE of the station skew is calculated by Eq. 6-34, with $G = 0.1$: $(MSE)_{sy} = 0.156$. Therefore, the weighted skew is (Eq. 6-33): $C_{sw} = -0.168$.

Treatment of Outliers. *Outliers* are data points that depart significantly from the overall trend of the data. The treatment of these outliers (i.e., their retention, modification, or deletion) may have a significant effect on the value of the statistical parameters computed from the data, particularly for small samples. Procedures for treatment of outliers invariably require judgment involving mathematical and hydrologic considerations.

The detection and treatment of high and low outliers in the log Pearson III method is performed in the following way [27]. For station skew greater than $+0.4$, tests for high outliers are considered first. For station skew less than -0.4, tests for low outliers are considered first. For station skew in the range -0.4 to $+0.4$, tests for high and low outliers are considered simultaneously, without eliminating any outliers from the data.

The following equation is used to detect high outliers:

$$y_H = \bar{y} + K_n s_y \tag{6-35}$$

in which y_H = high outlier threshold (in log units); and K_n = outlier frequency factor, a function of record length n. Values of K_n are given in Table A-7 (Appendix A).

Values of y_i (logarithms of the flood series) greater than y_H are considered to be high outliers. If there is sufficient evidence to indicate that a high outlier is a maximum in an extended period of time, it is treated as historical data. Otherwise, it is retained as part of the flood series.

Historical data refers to flood information outside of the flood series, which can be used to extend the record to a period much longer than that of the flood series. Historical knowledge is used to define the historical period H, which is longer than the record period n. The number z of events that are known to be the largest in the historical period are given a weight of 1. The remaining n events from the flood series are given a weight of $(H - z)/n$. For instance, for a record length $n = 44$ y, a historical period $H = 77$ y, and a number of peaks in the historical period $z = 3$, the weight applied to the three historical peaks would be 1, and the weight applied to the remaining flood series would be $(77 - 3)/44 = 1.68$. In other words, the record is extended to 77 y, and the 44 y of flood series (excluding outliers that have been considered part of the historical data) represent 74 y of data in the historical period of 77 y [27].

The following equation is used to detect low outliers:

$$y_L = \bar{y} - K_n s_y \tag{6-36}$$

in which y_L = low outlier threshold (in log units) and other terms are as defined previously. If an adjustment for historical data has been previously made, the values on the right-hand side of Eq. 6-36 are those previously used in the historically weighted computation. Values of y_i smaller than y_L are considered to be low outliers and deleted from the flood series [27].

　　　　　　　　　　　　　　　　　　　　　Frequency Analysis　　　Chap. 6

Complements to Flood Frequency Estimates.   The accuracy of flood estimates based on frequency analysis deteriorates for values of probability much greater than the record length. This is due to sampling error and to the fact that the underlying distribution is not known with certainty. Alternative procedures that complement the information provided by flood frequency analysis are recommended. These procedures include flood estimates from precipitation data (e.g., unit hydrograph, Chapter 5) and comparisons with catchments of similar hydrologic characteristics (regional analysis, Chapter 7). Table 6-4 shows the relationship between the various types of analysis used in flood frequency studies.

Gumbel's Extreme Value Type I Method

The extreme value Type I distribution, also known as the Gumbel method [15], or EV1, has been widely used in the United States and throughout the world. The method is a special case of the three-parameter GEV distribution described in the British *Flood Studies Report* [22].

The cumulative density function $F(x)$ of the Gumbel method is the double exponential, Eq. 6-19, repeated here:

$$F(x) = e^{-e^{-y}} \tag{6-19}$$

in which $F(x)$ is the probability of nonexceedence. In flood frequency analysis, the probability of interest is the *probability of exceedence*, i.e. the complementary probability to $F(x)$:

$$G(x) = 1 - F(x) \tag{6-37}$$

The return period T is the reciprocal of the probability of exceedence. Therefore,

$$\frac{1}{T} = 1 - e^{-e^{-y}} \tag{6-38}$$

From Eq. 6-38:

$$y = -\ln \ln \frac{T}{T-1} \tag{6-39}$$

TABLE 6-4   TYPES OF ANALYSES USED IN FLOOD
FREQUENCY STUDIES [27]

Description	Record Length n (y)		
	10–24	25–49	50 or more
Statistical analysis of flood frequency	X	X	X
Comparisons with similar catchments	X	X	—
Flood estimates from precipitation	X	—	—

In the Gumbel method, values of flood discharge are obtained from the frequency formula, Eq. 6-29:

$$x = \bar{x} + Ks \qquad (6\text{-}29)$$

The frequency factor K is evaluated with the frequency formula:

$$y = \bar{y}_n + K\sigma_n \qquad (6\text{-}40)$$

in which y = Gumbel (reduced) variate, a function of return period (Eq.6-39); and $\bar{y}_n$ and σ_n are the mean and standard deviation of the Gumbel variate, respectively. These values are a function of record length n (see Table A-8, Appendix A).

In Eq. 6-29, for $K = 0$, x is equal to the mean annual flood $\bar{x}$. Likewise, in Eq. 6-40, for $K = 0$, the Gumbel variate y is equal to its mean $\bar{y}_n$. The limiting value of $\bar{y}_n$, for n approaching ∞ is the Euler constant, 0.5772 [25]. In Eq. 6-38, for $y = 0.5772$: $T = 2.33$ years. Therefore, the return period of 2.33 y is taken as the return period of the mean annual flood.

From Eqs. 6-29 and 6-40,

$$x = \bar{x} + \frac{y - \bar{y}_n}{\sigma_n} s \qquad (6\text{-}41)$$

and with Eq. 6-39.

$$x = \bar{x} - \frac{\ln \ln \dfrac{T}{T-1} + \bar{y}_n}{\sigma_n} s \qquad (6\text{-}42)$$

The following steps are necessary to apply the Gumbel method:

1. Assemble the flood series.
2. Calculate the mean $\bar{x}$ and standard deviation s of the flood series.
3. Use Table A-8 to determine the mean $\bar{y}_n$ and standard deviation σ_n of the Gumbel variate as a function of record length n.
4. Select several return periods T_j and associated exceedence probabilities P_j.
5. Calculate the Gumbel variates y_j corresponding to the return periods T_j by using Eq. 6-39, and calculate the flood discharge $Q_j = x_j$ for each Gumbel variate (and associated return period) using Eq. 6-41. Alternatively, the flood discharges can be calculated directly for each return period by using Eq. 6-42.

Values of Q are plotted against y or T (or P) on Gumbel probability paper, and a straight line is drawn through the points. Gumbel probability paper has an arithmetic scale of Gumbel variate y in the abscissas and an arithmetic scale of flood discharge Q in the ordinates. To facilitate the reading of frequencies and probabilities, Eq. 6-38 can be used to superimpose a scale of return period T (or probability P) on the arithmetic scale of Gumbel variate y.

Example 6-6.

Apply the Gumbel method to the flood series of Example 6-3. Plot the results on Gumbel paper along with the plotting positions calculated in Example 6-3.

The mean and standard deviation of the flood series are: $\bar{x} = 1704$ m³/s and $s = 795$ m³/s. From Table A-8, for $n = 16$, the mean and standard deviation of the Gumbel variate are $\bar{y} = 0.5157$ and $\sigma_n = 1.0316$. The results are shown in Table 6-5. Columns 1 and 2 show selected return periods T and associated exceedence probabilities (in percent). Column 3 shows the values of Gumbel variate calculated by Eq. 6-39. Column 4 shows the flood discharge Q calculated by Eq. 6-41 for each variate y, return period T, and associated exceedence probability P. The flood discharges define a straight line when plotted versus return period on Gumbel paper, as shown by the solid line of Fig. 6-6. Plotting positions calculated by Example 6-3 are shown for comparison purposes.

Modifications to the Gumbel Method. Since its inception in the 1940s, several modifications to the Gumbel method have been suggested. Gringorten [11] has shown that the Gumbel distribution does not follow the Weibull plotting rule, Eq. 6-26 (or Eq. 6-27 with $a = 0$). He recommended $a = 0.44$, which led to the Gringorten plotting position formula:

$$\frac{1}{T} = P = \frac{m - 0.44}{n + 0.12} \tag{6-43}$$

Lettenmaier and Burges [21] have suggested that better flood estimates are obtained by using the limiting values of mean and standard deviation of the Gumbel variate (i.e., those corresponding to $n = \infty$) in Eq. 6-40, instead of basing these values on the record length. In this case, $\bar{y}_n = 0.5772$, and $\sigma_n = \pi/\sqrt{6} = 1.2825$. Therefore, Eq. 6-41 reduces to

$$x = \bar{x} + (0.78y - 0.45)s \tag{6-44}$$

and Eq. 6-42 reduces to

$$x = \bar{x} - \left(0.78 \ln \ln \frac{T}{T - 1} + 0.45\right)s \tag{6-45}$$

TABLE 6-5 GUMBEL METHOD: EXAMPLE 6-6

(1)	(2)	(3)	(4)
Return period T (y)	Probabilty P (percent)	Gumbel variate y	Flood discharge Q (m³/s)
1.05	95	−1.113	449
1.11	90	−0.838	661
1.25	80	−0.476	940
2	50	0.367	1590
5	20	1.500	2462
10	10	2.250	3040
25	4	3.199	3772
50	2	3.902	4314
100	1	4.600	4851
200	0.5	5.296	5388

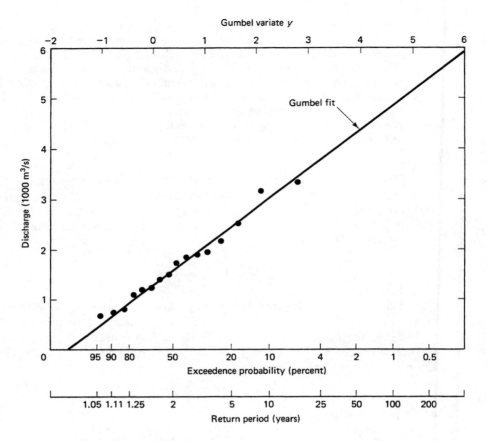

Figure 6-6 Flood-frequency analysis by Gumbel method: Example 6-6.

Lettenmaier and Burges [21] have also suggested that a biased variance estimate, using n as the divisor in Eq. 6-3, yields better estimates of extreme events that the usual unbiased estimate, that is, the divisor $n - 1$.

Comparison Between Flood Frequency Methods

In 1966, the Hydrology Subcommittee of the U.S. Water Resources Council began work on selecting a suitable method of flood frequency analysis that could be recommended for general use in the United States.

The committee tested the goodness of fit of six distributions: (1) lognormal, (2) log Pearson III, (3) Hazen, (4) gamma, (5) Gumbel (EV1) and (6) log Gumbel (EV2). The study included ten sets of records, the shortest of which was 40 y. The findings showed that the first three distributions had smaller average deviations that the last three. Since the Hazen distribution is a type of lognormal distribution and the lognormal is a special case of the log Pearson III, the Committee concluded that the latter was the most appropriate of the three, and hence recommended it for general use.

The same type of analysis was repeated for six sets of records in the United Kingdom, the shortest of which was 32 y [2]. The methods were: (1) gamma, (2) log gamma, (3) lognormal, (4) Gumbel (EV1), (5) GEV, (6) Pearson Type III, and (7) log Pearson III. At low return periods (T from 2 to 5 y), the GEV and Pearson Type III showed the smallest average deviations, whereas for return periods exceeding 10 y the log Pearson III method had the smallest average deviations.

Similar comparative studies were reported in the British *Flood Studies Report* [22]. The study concluded that the three-parameter distributions (GEV, Pearson Type III, and log Pearson III) provided a better fit than the two-parameter distributions (Gumbel, lognormal, gamma, log gamma). Based on mean absolute deviation criteria, the study rated the log Pearson III method better than the GEV and the latter better than the Pearson Type III. However, based on root mean square deviation, it rated the Pearson Type III better than both the log Pearson III and GEV distributions.

Although in general, the three-parameter methods seemed to fare better than the two-parameter methods, the latter should not be completely discarded. The British *Flood Studies Report* [22] observed that their use in connection with short record lengths often leads to results which are more sensible than those obtained by fitting three-parameter distributions. A three-parameter distribution fitted to a small sample may in some cases imply that there is an upper bound to the flood discharge equal to about twice the mean annual flood. While there may be an upper limit to flood magnitude, it is certainly higher than twice the mean annual flood.

6.3 LOW-FLOW FREQUENCY ANALYSIS

Whereas high flows lead to floods, sustained low flows can lead to droughts. A drought is defined as a lack of rainfall so great and continuing so long as to affect the plant and animal life of a region adversely and to deplete domestic and industrial water supplies, especially in those regions where rainfall is normally sufficient for such purposes [17].

In practice, a drought refers to a period of unusually low water supplies, regardless of the water demand. The regions most subject to droughts are those with the greatest variability in annual rainfall. Studies have shown that regions where the variance coefficient of annual rainfall exceeds 0.35 are more likely to have frequent droughts [5]. Low annual rainfall and high annual rainfall variability are typical of arid and semiarid regions. Therefore, these regions are more likely to be prone to droughts.

Studies of tree rings, which document long term trends of rainfall, show clear patterns of periods of wet and dry weather [26]. While there is no apparent explanation for the cycles of wet and dry weather, the dry years must be considered in planning water resource projects. Analysis of long records has shown that there is a tendency for dry years to group together. This indicates that the sequence of dry years is not random, with dry years tending to follow other dry years. It is therefore necessary to consider both the severity and duration of a drought period.

The severity of droughts can be established by measuring (1) the deficiency in rainfall and runoff, (2) the decline of soil moisture, and (3) the decrease in groundwater levels. Alternatively, low-flow-frequency analysis can be used in the assessment of the probability of occurrence of droughts of different durations.

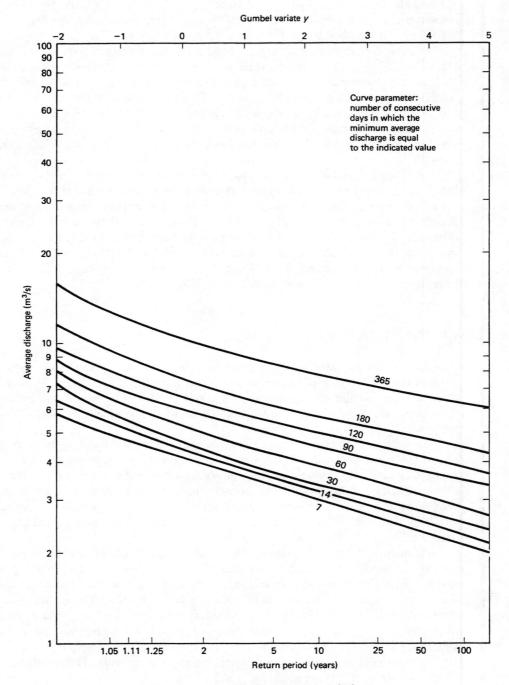

Figure 6-7 Low-flow frequency curves [24].

Frequency Analysis   Chap. 6

Methods of low-flow frequency analysis are based on an assumption of invariance of meteorological conditions. The absence of long records, however, imposes a stringent limitation on low-flow frequency analysis. When records of sufficient length are available, analysis begins with the identification of the low-flow series. Either the annual minima or the annual exceedence series are used. In a monthly analysis, the annual minima series is formed by the lowest monthly flow volumes in each year of record. If the annual exceedence method is chosen, the lowest monthly flow volumes in the record are selected, regardless of when they occurred. In the latter method, the number of values in the series need not be equal to the number of years of record.

A flow-duration curve can be used to give an indication of the severity of low flows. Such a curve, however, does not contain information on the sequence of low flows or the duration of possible droughts. The analysis is made more meaningful by abstracting the minimum flows over a period of several consecutive days. For instance, for each year, the 7-day period with minimum flow volume is abstracted, and the minimum flow is the average flow rate for that period. A frequency analysis on the low-flow series, using the Gumbel method, for instance, results in a function describing the probability of occurrence of low flows of a certain duration. The same analysis repeated for other durations leads to a family of curves depicting low-flow frequency, as shown in Fig. 6-7 [24].

In reservoir design, the assessment of low flows is aided by a flow-mass curve. The technique involves the determination of storage volumes required for all low-flow periods. Although it is practically impossible to provide sufficient storage to meet hydrologic risks of great rarity, common practice is to provide for a stated risk (i.e., a drought probability) and to add a suitable percent of the computed storage volume as reserve storage allowance. The variance coefficient of annual flows is used in determining the risk and storage allowance levels. Extraordinary drought levels are then met by cutting draft rates.

Regulated rivers may alter natural flow conditions to provide a minimum downstream flow for specific purposes. In this case, the reservoirs serve as the mechanism to diffuse the natural flow variability into downstream flows which can be made to be nearly constant in time. Regulation is necessary for downstream low flow maintenance, usually for the purpose of meeting agricultural, municipal and industrial water demands, minimum instream flows, navigation draft, and water pollution control requirements.

QUESTIONS

1. In statistical analysis, what are the measures of central tendency? Explain.
2. What is skewness? A distribution with a long tail on the right side has positive or negative skewness?
3. What are the parameters of the gamma distribution? How are the gamma and Pearson Type III distributions related?
4. What is the parameter that distinguishes the three extreme value distributions? What is the limiting value of the mean of the Gumbel variate?
5. What is the difference between the annual exceedence series and the annual maxima series? What is risk in the context of frequency analysis?

6. How is an extreme value probability paper constructed? What type of probability paper is used in the log Pearson Type III method?

7. What is the difference between the Weibull, Blom, and Gringorten plotting position formulas?

8. How is skewness variability accounted for in the log Pearson III method?

9. When are high outliers considered part of historical data? When is it necessary to perform a historically weighted computation?

10. Why are two-parameter distributions such as the Gumbel distribution appropiate for use in connection with short record lengths?

11. Compare floods and droughts from the standpoint of frequency analysis.

PROBLEMS

6-1. Develop a computer program to calculate the mean, standard deviation, and skew coefficient of a series of annual maximum flows. Test your computer program using the data of Example 6-1 in the text.

6-2. The annual maximum flows of a certain stream have been found to be normally distributed with mean 22,500 ft^3/s and standard deviation 7500 ft^3/s. Calculate the probability that a flow larger than 39,000 ft^3/s will occur.

6-3. The 10-y and 25-y floods of a certain stream are 73 and 84 m^3/s, respectively. Assuming a normal distribution, calculate the 50-y and 100-y floods.

6-4. The low flows of a certain stream have been shown to follow a normal distribution. The flows expected to be exceeded 95% and 90% of the time are 15 and 21 m^3/s, respectively. What flow can be expected to be exceeded 80% of the time?

6-5. A temporary cofferdam for a 5-y dam construction period is designed to pass the 25-y flood. What is the risk that the cofferdam may fail before the end of the construction period? What design return period is needed to reduce the risk to less than 10%?

6-6. Use the Weibull formula (Eq. 6-26) to calculate the plotting positions for the following series of annual maxima, in cubic feet per second: 1305, 3250, 4735, 5210, 4210, 2120, 2830, 3585, 7205, 1930, 2520, 3250, 5105, 4830, 2020, 2530, 3825, 3500, 2970, 1215.

6-7. Use the Gringorten formula to calculate the plotting positions for the following series of annual maxima, in cubic meters per second: 160, 350, 275, 482, 530, 390, 283, 195, 408, 307, 625, 513.

6-8. Modify the computer program of Problem 6-1 to calculate the mean, standard deviation, and skew coefficients of the logarithms of a series of annual maximum flows. Test your computer program using the results of Example 6-4 in the text.

6-9. Fit a log Pearson III curve to the data of Problem 6-6. Plot the calculated distribution on log probability paper, along with the Weibull plotting positions calculated in Problem 6-6.

6-10. Fit a Gumbel curve to the data of Problem 6-6. Plot the calculated distribution on Gumbel paper, along with the Weibull plotting positions calculated in Problem 6-6.

6-11. Develop a computer program to read a series of annual maxima, sort the data in descending order, and compute the corresponding plotting positions (percent chance and return period) by the Weibull and Gringorten formulas.

6-12. Given the following statistics of annual maxima for stream X: number of years $n = 35$; mean = 3545 ft^3/s; standard deviation = 1870 ft^3/s. Compute the 100-y flood by the Gumbel method.

6-13. Given the following statistics of annual maxima for river Y: number of years $n = 45$; mean = 2700 m³/s; standard deviation 1300 m³/s; mean of the logarithms = 3.1; standard deviation of the logarithms = 0.4; skew coefficient of the logarithms = −0.35. Compute the 100-y flood using the following probability distributions: (a) normal, (b) Gumbel, and (c) log Pearson III.

6-14. A station near Denver, Colorado, has flood records for 48 y, with station skew $C_{sy} = -0.18$. Calculate a weighted skew coefficient.

6-15. Determine if the value $Q = 13,800$ ft³/s is a high outlier in a 45-y flood series with the following statistics: mean of the logarithms = 3.572; standard deviation of the logarithms = 0.215.

6-16. Using the Lettenmaier and Burges modification to the Gumbel method, fit a Gumbel curve to the data of Example 6-6 in the text. Plot the calculated distribution on Gumbel paper, along with plotting positions calculated by the Gringorten formula.

6-17. Solve Example 6-4 in the text using the computer program EH600A (Log Pearson III method) included in Appendix D.

6-18. Solve Example 6-6 in the text using the computer program EH600B (Gumbel method) included in Appendix D.

6-19. Solve Problem 6-9 using the computer program EH600A (Log Pearson III method) included in Appendix D.

6-20. Solve Problem 6-10 using the computer program EH600B (Gumbel method) included in Appendix D.

REFERENCES

1. Benson, M. A. (1962). "Plotting Positions and Economics of Engineering Planning," *Journal of the Hydraulics Division*, ASCE, Vol. 88, November, pp. 57–71.
2. Benson, M. A. (1968). "Uniform Flood Frequency Estimating Methods for Federal Agencies," *Water Resources Research*, Vol. 4, pp. 891–908.
3. Blom, G. (1958). *Statistical Estimates and Transformed Beta Variables*. New York: John Wiley.
4. Chow, V. T. (1951). "A General Formula for Hydrologic Frequency Analysis," *Transactions, American Geophysical Union*, Vol. 32, pp. 231–237.
5. Chow, V. T. (1964). *Handbook of Applied Hydrology*. New York: McGraw-Hill.
6. Cunnane, C. (1978). "Unbiased Plotting Positions-A Review," *Journal of Hydrology*, Vol. 37, pp. 205–222.
7. Fisher, R. A., and L. H. C. Tippett. (1928). "Limiting Forms of a Frequency Distribution of the Smallest and Largest Member of a Sample," *Proceedings, Cambridge Philosophical Society*, Vol. 24, pp. 180–190.
8. Foster, H. A. (1924). "Theoretical Frequency Curves and their Application to Engineering Problems," *Transactions*, ASCE, Vol. 87, pp. 142–173.
9. Frechet, M. (1927). "Sur la loi de Probabilité de l'écart Maximum," (On the Probability Law of Maximum Error), *Annals of the Polish Mathematical Society*, (Cracow), Vol. 6, pp. 93–116.
10. Fuller, W. E. (1914). "Flood Flows," *Transactions*, ASCE, Vol. 77, pp. 564–617.
11. Gringorten, I. I. (1963). "A Plotting Rule for Extreme Probability Paper," *Journal of Geophysical Research*, Vol. 68, No. 3, February, pp. 813–814.
12. Gumbel, E. J. (1941). "Probability Interpretation of the Observed Return Periods of Floods," *Transactions, American Geophysical Union*, Vol. 21, pp. 836–850.

13. Gumbel, E. J. (1942). "Statistical Control Curves for Flood Discharges," *Transactions, American Geophysical Union,* Vol. 23, pp. 489–500.

14. Gumbel, E. J. (1943). "On the Plotting of Flood Discharges," *Transactions, American Geophysical Union,* Vol. 24, pp. 699–719.

15. Gumbel, E. J. (1958). *Statistics of Extremes.* Irvington, N.Y.: Columbia University Press.

16. Hazen, A. (1914). Discussion on "Flood Flows," by W. E. Fuller, *Transactions,* ASCE, Vol. 77, p. 628.

17. Havens, A. V. (1954). "Drought and Agriculture," *Weatherwise,* Vol. 7, pp. 51–55.

18. Horton, R. E. (1913). "Frequency of Recurrence of Hudson River Floods," *U.S. Weather Bureau Bulletin Z,* pp. 109–112.

19. Jenkinson, A. F. (1955). "The Frequency Distributions of the Annual Maximum (or Minimum) Values of Meteorological Elements," *Quarterly Journal of the Royal Meteorological Society,* , Vol. 87, p. 158.

20. Kite, G. W. (1977). *Frequency and Risk Analyses in Hydrology.* Fort Collins, Colorado: Water Resources Publications.

21. Lettenmaier, D. P., and S. J. Burges. (1982). "Gumbel's Extreme Value Distribution: A New Look," *Journal of the Hydraulics Division,* ASCE, Vol. 108, No. HY4, April, pp. 503–514.

22. Natural Environment Research Council. (1975). *Flood Studies Report,* Vol. 1 (of 5 volumes), London, England.

23. Pearson, K. (1930). "Tables for Statisticians and Biometricians," Part I, 3d. Ed., The Biometric Laboratory, University College. London: Cambridge University Press.

24. Riggs, H. C. (1972). "Low-Flow Investigations," *Techniques of Water Resources Investigations of the United States Geological Survey,* Book 4, Chapter B1.

25. Spiegel, M. *Mathematical Handbook of Formulas and Tables.* Schaum's Outline Series in Mathematics. New York: McGraw-Hill.

26. Troxell, H. C. (1937). "Water Resources of Southern California with Special Reference to the Drought of 1944–51," *U.S. Geological Survey Water Supply Paper* No. 1366.

27. U.S. Interagency Advisory Committee on Water Data, Hydrology Subcommittee. (1983). "Guidelines for Determining Flood Flow Frequency," *Bulletin No.* 17B, issued 1981, revised 1983, Reston, Virginia.

SUGGESTED READINGS

Gumbel, E. J. (1958). *Statistics of Extremes,* Irvington. N.Y.: Columbia University Press.

Natural Environment Research Council. (1975). *Flood Studies Report,* in 5 volumes, London, England.

Riggs, H. C. (1972). "Low-Flow Investigations," *Techniques of Water Resources Investigations of the United States Geological Survey,* Book 4, Chapter B1.

U.S. Interagency Advisory Committee on Water Data, Hydrology Subcommittee. (1983). "Guidelines for Determining Flood Flow Frequency," *Bulletin No.* 17B, issued 1981, revised 1983, Reston, Virginia.

REGIONAL ANALYSIS

In engineering hydrology, *regional analysis* encompasses the study of hydrologic phenomena with the aim of developing mathematical relations to be used in a regional context. Generally, mathematical relations are developed so that information from gaged or long-record catchments can be readily transferred to neighboring ungaged or short-record catchments of similar hydrologic characteristics. Other applications of regional analysis include regression techniques used to develop empirical (i.e., parametric) equations applicable within a broad geographical region. Regional analysis makes use of statistics and probability, including frequency analysis and joint probability distributions.

This chapter is divided into three sections. Section 7.1 describes joint probability distributions, including marginal distributions and conditional probability. Section 7.2 describes the techniques of regression analysis. Section 7.3 presents selected techniques for regional analysis of flood and rainfall characteristics.

7.1 JOINT PROBABILITY DISTRIBUTIONS

Probability distributions possessing one random variable (X) were discussed in Chapter 6. These are called *univariate* distributions. Probability distributions with two random variables, X and Y, are called *bivariate*, or joint distributions. A joint distribution expresses in mathematical terms the probability of occurrence of an outcome consisting of a pair of values of X and Y. In statistical notation, $P(X = x_i, Y = y_j)$ is the probability P that the random variables X and Y will take on the outcomes x_i and y_j simultaneously. A shorter notation is $P(x_i, y_j)$.

For $x_i(1, 2, \ldots, n)$, and $y_j(1, 2, \ldots, m)$, the sum of the probabilities of all possible outcomes is equal to unity:

$$\sum_{i=1}^{n} \sum_{j=1}^{m} P(x_i, y_j) = 1 \tag{7-1}$$

A classical example of joint probability is that of the outcome of the cast of two dice, say A and B. Intuitively, the probability of getting a 1 for A and a 6 for B is $P(A = 1, B = 6) = \frac{1}{36}$. In total, there are $6 \times 6 = 36$ possible outcomes, and each one of them has the same probability: $\frac{1}{36}$ (assuming, of course, that the dice are not loaded). This distribution is referred to as the bivariate uniform distribution because each outcome has a uniform and equal probability of occurrence. The sum of the probabilities of all possible outcomes is confirmed to be equal to 1.

Joint cumulative probabilities are defined in a similar way as for univariate probabilities:

$$F(x_k, y_l) = \sum_{i=1}^{k} \sum_{j=1}^{l} P(x_i, y_j) \tag{7-2}$$

in which $F(x_k, y_l)$ is the joint cumulative probability. Continuing with the example of the two dice, the probability of A being 5 or less and B being 3 or less is the sum of all the individual probabilities, for all combinations of i and j, as i varies from 1 to 5, and as j varies from 1 to 3; i.e., $5 \times 3 = 15$ possible combinations, resulting in a probability equal to $15 \times (\frac{1}{36}) = \frac{15}{36}$.

Marginal Probability Distributions

Marginal probability distributions are obtained by summing up $P(x_i, y_j)$ over all values of one of the variables, for instance, X. The resulting (marginal) distribution is the probability distribution of the other variable, in this case Y without regard to X. Marginal distributions are univariate distributions obtained from bivariate distributions. In statistical notation, the marginal probability distribution of X is

$$P(x_i) = \sum_{j=1}^{m} P(x_i, y_j) \tag{7-3}$$

Likewise, the marginal distribution of Y is:

$$P(y_j) = \sum_{i=1}^{n} P(x_i, y_j) \tag{7-4}$$

The example of the two dice A and B can be used to illustrate the concept of marginal probability. Intuitively, the probability of A being equal to 1, regardless of the value of B, is $6 \times (\frac{1}{36}) = \frac{1}{6}$. Likewise, the probability of B being equal to 4, regardless of the value of A, is also $\frac{1}{6}$. Notice that the joint probabilities $(\frac{1}{36})$ of each one of all 6 possible outcomes have been summed in order to calculate the marginal probability.

Marginal cumulative probability distributions are obtained by combining the concepts of marginal and cumulative distributions. In statistical notation, the marginal cumulative probability distribution of X is

$$F(x_k) = \sum_{i=1}^{k} \sum_{j=1}^{m} P(x_i, y_j)$$ (7-5)

Likewise, the marginal cumulative probability distribution of Y is

$$F(y_l) = \sum_{i=1}^{n} \sum_{j=1}^{l} P(x_i, y_j)$$ (7-6)

The example of the two dice A and B is again used to illustrate the concept of marginal cumulative probability. The probability of $A \leq 2$, regardless of the value of B, is $2 \times 6 \times (\frac{1}{36}) = \frac{1}{3}$. Likewise, the probability of $B \leq 5$, regardless of the value of A, is $5 \times 6 \times (\frac{1}{36}) = \frac{5}{6}$. To calculate the marginal cumulative probabilities, the concepts of marginal and cumulative distributions have been combined.

Conditional Probability

The concept of conditional probability is useful in regression analysis and other hydrologic applications. The conditional probability is the ratio of joint and marginal probabilities. In statistical notation,

$$P(x|y) = \frac{P(x, y)}{P(y)}$$ (7-7)

in which $P(x|y)$ is the conditional probability of x, given y. The conditional probability of y, given x, is:

$$P(y|x) = \frac{P(x, y)}{P(x)}$$ (7-8)

From Eqs. 7-7 and 7-8, it follows that joint probability is the product of conditional and marginal probabilities.

Joint probability distributions can be expressed as continuous functions. In this case they are called joint density functions, with the notation $f(x, y)$. For the conditional density function, the notation is $f(x|y)$, or alternatively, $f(y|x)$.

As with univariate distributions, the moments provide descriptions of the properties of joint distributions. For continuous functions, the joint moment of order r and s about the origin (indicated with $'$) is defined as follows:

$$\mu'_{r,s} = \int_{-\infty}^{\infty} \int_{-\infty}^{\infty} x^r y^s f(x, y) \, dy \, dx$$ (7-9)

With $r = 1$ and $s = 0$, Eq. 7-9 reduces to the mean of x:

$$\mu'_{1,0} = \int_{-\infty}^{\infty} x \left[\int_{-\infty}^{\infty} f(x, y) \, dy \right] dx$$ (7-10)

with the expression between brackets being the marginal PDF of x, or $f(x)$. Therefore, the expression for the mean of x is

$$\mu'_{1.0} = \mu_x = \int_{-\infty}^{\infty} x f(x) \, dx \qquad (7\text{-}11)$$

Similar equations hold for y.

The second moments are usually written about the mean:

$$\mu_{r.s} = \int_{-\infty}^{\infty} \int_{-\infty}^{\infty} (x - \mu_x)^r (y - \mu_y)^s f(x, y) \, dy \, dx \qquad (7\text{-}12)$$

For $r = 2$ and $s = 0$, Eq. 7-12 reduces to the variance of x. Likewise, for $r = 0$ and $s = 2$, Eq. 7-12 reduces to the variance of y. A third type of second moment, i.e., the covariance, arises for $r = 1$ and $s = 1$:

$$\sigma_{x.y} = \int_{-\infty}^{\infty} \int_{-\infty}^{\infty} (x - \mu_x)(y - \mu_y) f(x, y) \, dy \, dx \qquad (7\text{-}13)$$

in which $\sigma_{x.y}$ is the covariance.

The correlation coefficient is a dimensionless value relating the covariance $\sigma_{x.y}$ and standard deviations σ_x and σ_y:

$$\rho_{x.y} = \frac{\sigma_{x.y}}{\sigma_x \sigma_y} \qquad (7\text{-}14)$$

in which $\rho_{x.y}$ is the correlation coefficient based on population data. The sample correlation coefficient is

$$r_{x.y} = \frac{s_{x.y}}{s_x s_y} \qquad (7\text{-}15)$$

The calculation of sample correlation coefficient $r_{x.y}$, including sample covariance $s_{x.y}$, is illustrated by Example 7-1. The correlation coefficient is a measure of the linear dependence between x and y. It varies in the range of -1 to $+1$. A value of ρ or r close to or equal to 1 indicates a strong linear relationship between the variables, with large values of x associated with large values of y, and small values of x, with small values of y. A value of ρ or r close to or equal to -1 indicates a correlation such that large values of x are associated with small values of y and vice versa. A value of $\rho = 0$ or $r = 0$, i.e. a zero covariance, indicates the lack of linear dependence between x and y.

Example 7-1.

The monthly flows of the north and south fork tributaries of a certain stream have the following joint probability distribution $f(x, y)$ (expressed as mean value in each class):

North fork, x (hm^3)		100	200	300	400
South fork, y (hm^3)					
100		0.14	0.03	0.00	0.00
200		0.02	0.18	0.11	0.00
300		0.00	0.09	0.23	0.02
400		0.00	0.00	0.03	0.15

Calculate the marginal distributions, means, variances, standard deviations, covariance, and correlation coefficient for this joint distribution.

The north fork marginal distribution, $f(x)$, is obtained by summing up the joint probabilities across y. Therefore,

x (hm^3)	100	200	300	400
$f(x)$	0.16	0.30	0.37	0.17

Likewise, the south fork marginal distribution, $f(y)$, is obtained by summing up the joint probabilities across x:

y (hm^3)	100	200	300	400
$f(y)$	0.17	0.31	0.34	0.18

The means are the first moments of the marginal distributions with respect to the origin:

$$\bar{x} = (100 \times 0.16) + (200 \times 0.30) + (300 \times 0.37) + (400 \times 0.17) = 255 \text{ hm}^3$$

$$\bar{y} = (100 \times 0.17) + (200 \times 0.31) + (300 \times 0.34) + (400 \times 0.18) = 253 \text{ hm}^3$$

The variances are the second moments of the marginal distributions with respect to the means:

$$s_x^2 = \Sigma (x - \bar{x})^2 f(x) =$$

$$s_x^2 = (100 - 255)^2 \times 0.16 + (200 - 255)^2 \times 0.30 + (300 - 255)^2 \times 0.37$$
$$+ (400 - 255)^2 \times 0.17 = 9075 \text{ hm}^6$$

$$s_x = 95.26 \text{ hm}^3.$$

Likewise, for y:

$$s_y^2 = 9491 \text{ hm}^6 \text{ and } s_y = 97.42 \text{ hm}^3$$

The covariance is the second moment of the joint distribution:

$$
\begin{aligned}
s_{x.y} &= \Sigma (x - \bar{x})(y - \bar{y}) f(x, y) \\
&= [(100 - 255) \times (100 - 253) \times 0.14] \\
&\quad + [(200 - 255) \times (100 - 253) \times 0.03] \\
&\quad + [(100 - 255) \times (200 - 253) \times 0.02] \\
&\quad + [(200 - 255) \times (200 - 253) \times 0.18] \\
&\quad + [(300 - 255) \times (200 - 253) \times 0.11] \\
&\quad + [(200 - 255) \times (300 - 253) \times 0.09] \\
&\quad + [(300 - 255) \times (300 - 253) \times 0.23] \\
&\quad + [(400 - 255) \times (300 - 253) \times 0.02] \\
&\quad + [(300 - 255) \times (400 - 253) \times 0.03] \\
&\quad + [(400 - 255) \times (400 - 253) \times 0.15] = 7785 \text{ hm}^6.
\end{aligned}
$$

The correlation coefficient is $r_{x.y} = s_{x.y}/(s_x s_y) = 0.839$.

Bivariate Normal Distribution

Among the many joint probability distributions, the bivariate normal distribution is important in hydrology because it is the foundation of regression theory. The bivariate normal probability distribution is [10]:

$$f(x, y) = Ke^M \qquad (7\text{-}16)$$

in which x and y are the random variables, and K and M are coefficient and exponent, respectively, defined as follows:

$$K = \frac{1}{2\pi \sigma_x \sigma_y (1 - \rho^2)^{1/2}} \qquad (7\text{-}17)$$

$$M = -\frac{1}{2(1 - \rho^2)} \left[\left(\frac{x - \mu_x}{\sigma_x}\right)^2 - 2\rho \left(\frac{x - \mu_x}{\sigma_x}\right) \left(\frac{y - \mu_y}{\sigma_y}\right) + \left(\frac{y - \mu_y}{\sigma_y}\right)^2 \right] \qquad (7\text{-}18)$$

The distribution has five parameters: the means μ_x and μ_y, the standard deviations σ_x and σ_y, and the correlation coefficient ρ.

Following Eq. 7-8, the conditional distribution is obtained by dividing the bivariate normal (Eq. 7-16) by the univariate normal (Eq. 6-7), to yield

$$f(y|x) = \frac{f(x, y)}{f(x)} = K'e^{M'} \qquad (7\text{-}19)$$

in which K' and M' are coefficient and exponent, respectively, defined as follows:

$$K' = \frac{1}{\sigma_y [2\pi (1 - \rho^2)]^{1/2}} \qquad (7\text{-}20)$$

$$M' = -\frac{1}{2\sigma_y^2(1 - \rho^2)} \left[(y - \mu_y) - \rho \frac{\sigma_y}{\sigma_x} (x - \mu_x) \right]^2 \qquad (7\text{-}21)$$

By inspection of Eqs. 7-20 and 7-21, and comparison with Eq. 6-7, it is concluded that the conditional distribution is also normal, with mean and variance:

$$\mu_{y|x} = \mu_y + \rho \frac{\sigma_y}{\sigma_x}(x - \mu_x) \qquad (7\text{-}22)$$

$$\sigma_e^2 = \sigma_y^2(1 - \rho^2) \qquad (7\text{-}23)$$

Equations 7-22 and 7-23 are useful in regression analysis. Equation 7-22 expresses the linear dependence between x and y. The slope of the regression line is $\rho\sigma_y/\sigma_x$. Likewise, ρ is the fraction of the original variance explained or removed by the regression. For $\rho = 1$, all the variance is removed, and $\sigma_e = 0$. For $\rho = 0$, the variance remains $\sigma_e = \sigma_y$.

7.2 REGRESSION ANALYSIS

A fundamental tool of regional analysis is the equation relating two or more hydrologic variables. The variable for which values are given is called the *predictor* variable. The variable for which values must be estimated is called the *criterion* variable [7]. The equation relating criterion variable to one or more predictor variables is called the prediction equation.

The objective of regression analysis is to evaluate the parameters of the prediction equation relating the criterion variable to one or more predictor variables. The predictor variables are those whose variation is believed to cause or agree with variation in the criterion variable.

Correlation provides a measure of the goodness of fit of the regression. Therefore, while regression provides the parameters of the prediction equation, correlation describes its quality. The distinction between correlation and regression is necessary because the predictor and criterion variables cannot be switched unless the correlation coefficient is equal to 1. Stated in other terms, if a criterion variable Y is regressed on a predictor variable X, the regression parameters cannot be used to express X as a function of Y, unless the correlation coefficient is 1. In hydrologic modeling, regression analysis is useful in model calibration; correlation is useful in model formulation and verification.

The principle of least squares is used in regression analysis as a means of obtaining the best estimates of the parameters of the prediction equation. The principle is based on the minimization of the sum of the squares of the differences between observed and predicted values. The procedure can be used to regress one criterion variable on one or more predictor variables.

One-Predictor-Variable Regression

Assume a predictor variable x, a criterion variable y, and a set on n paired observations of x and y. In the simplest linear case, the line to be fitted has the following form:

$$y' = \alpha + \beta x \tag{7-24}$$

in which y' is an estimate of y and α and β are parameters to be determined by regression.

In the least squares procedure, values of the intercept α and slope β are sought such that y' is the best estimate of y. For this purpose, the sum of the squares of the differences between y and y' are minimized.

$$\Sigma (y - y')^2 = \Sigma [y - (\alpha + \beta x)]^2 \tag{7-25}$$

in which the symbol Σ indicates the sum of all values from $i = 1$ to $i = n$.

Setting the partial derivatives equal to zero,

$$\frac{\partial}{\partial \alpha} \{\Sigma [y - (\alpha + \beta x)]^2\} = 0 \tag{7-26}$$

and:

$$\frac{\partial}{\partial \beta} \{\Sigma [y - (\alpha + \beta x)]^2\} = 0 \tag{7-27}$$

This leads to the normal equations

$$\Sigma y - n\alpha - \beta \Sigma x = 0 \tag{7-28}$$

and:

$$\Sigma xy - \alpha \Sigma x - \beta \Sigma x^2 = 0 \tag{7-29}$$

Solving Eqs. 7-28 and 7-29 simultaneously gives

$$\beta = \frac{\Sigma xy - \dfrac{\Sigma x \, \Sigma y}{n}}{\Sigma x^2 - \dfrac{(\Sigma x)^2}{n}} \tag{7-30}$$

$$\alpha = \frac{\Sigma y - \beta \, \Sigma x}{n} \tag{7-31}$$

Since the slope of the regression line is $\beta = \rho \, \sigma_y/\sigma_x$, the estimate from sample data is $\beta = rs_y/s_x$. Therefore, the correlation coefficient is

$$r = \beta \frac{s_x}{s_y} \tag{7-32}$$

The standard error of estimate of the correlation is the square root of the variance of the conditional distribution:

$$s_e = \left[\frac{1}{n-2} \Sigma (y - y')^2 \right]^{1/2} \tag{7-33}$$

in which $n - 2$ is the number of *degrees of freedom*, i.e., the sample size minus the number of unknowns.

Alternatively, the standard error of estimate can be estimated from the variance of the conditional distribution, Eq. 7-23. For calculations based on sample data, the standard error of estimate is

$$s_e = s_y \left[\frac{n-1}{n-2} (1 - r^2) \right]^{1/2} \tag{7-34}$$

Equations 7-30 and 7-31 can also be used to fit power functions of the type $y = ax^b$. First, this equation is linearized by taking the logarithms: $\log y = \log a + b \log x$. With $u = \log x$, and $v = \log y$, this equation is: $v = \log a + bu$. The variables u and v are used in Eqs. 7-30 and 7-31 instead of x and y, respectively. Then $\alpha = \log a$, and $\beta = b$, and the regression equation is $y = 10^\alpha x^\beta$.

Example 7-2.

Find the regression equation linking the low flows (annual minima series) of streams X and Y shown in Cols. 2 and 3 of Table 7-1. Calculate the regression parameters α and β, the correlation coefficient, and the standard error of estimate.

Summing up the values of Cols. 2 and 3, and dividing by $n = 15$, the means are obtained: $\bar{x} = 72$ m³/s and $\bar{y} = 77$ m³/s. Columns 4 and 5 show the square of the deviations from the means. Summing up Cols. 4 and 5, dividing the sums by $(n - 1) = 14$, and taking the square roots, the standard deviations $s_x = 29.6$ m³/s and $s_y = 26.6$ m³/s are obtained. Column 6 shows the x^2 values, and Col. 7, the xy values. The sum of these values is $\Sigma x^2 = 90,000$ and $\Sigma xy = 93,056$. Using Eq. 7-30, $\beta = [93,056 - (1,080 \times 1,155)/15]/[90,000 - (1,080 \times 1,080)/15] = 0.8085$. Using Eq. 7-31, $\alpha = [1155 - (0.8085 \times$

TABLE 7-1 ONE-PREDICTOR-VARIABLE REGRESSION: EXAMPLE 7-2

(1)	(2)	(3)	(4)	(5)	(6)	(7)
Year	x (m³/s)	y (m³/s)	$(x - \bar{x})^2$	$(y - \bar{y})^2$	x^2	xy
1973	110	89	1,444	144	12,100	9,790
1974	42	51	900	676	1,764	2,142
1975	75	72	9	25	5,625	5,400
1976	120	112	2,304	1,225	14,400	13,440
1977	89	70	289	49	7,921	6,230
1978	32	45	1,600	1,024	1,024	1,440
1979	37	42	1,225	1,225	1,369	1,554
1980	56	59	256	324	3,136	3,304
1981	82	100	100	529	6,724	8,200
1982	90	92	324	225	8,100	8,280
1983	50	70	484	49	2,500	3,500
1984	30	42	1,764	1,225	900	1,260
1985	81	92	81	225	6,561	7,452
1986	110	130	1,444	2,809	12,100	14,300
1987	76	89	16	144	5,776	6,764
Sum	1,080	1,155	12,240	9,898	90,000	93,056

1080)]/15 = 18.788. Using Eq. 7-32, the correlation coefficient is $r = 0.8085 \times 29.6/26.6 = 0.899$. Using Eq. 7-34, the standard error of estimate is $s_e = 26.6 \times [(14/13)(1 - 0.899^2)]^{1/2} = 12.09$ m³/s. The data and regression line are plotted in Fig. 7-1.

Multiple Regression

The extension of the least squares technique to more than one predictor variable is referred to as multiple regression. In the case of two predictor variables, x_1 and x_2, with criterion variable y and a set of n observations of y, x_1, and x_2, the line to be fitted is:

$$y' = \alpha + \beta_1 x_1 + \beta_2 x_2 \tag{7-35}$$

in which x_1 and x_2 are measured values and y' is an estimate of y.

As with the two-variable case, values of the intercept α and slopes β_1 and β_2 are sought such that y' is the best estimate of y. For this purpose, the sum of the squares of the differences between y and y' are minimized.

$$\Sigma (y - y')^2 = \Sigma [y - (\alpha + \beta_1 x_1 + \beta_2 x_2)]^2 \tag{7-36}$$

Setting the partial derivatives with respect to α, β_1, and β_2 equal to zero leads to the normal equations

$$\Sigma y - n\alpha - \beta_1 \Sigma x_1 - \beta_2 \Sigma x_2 = 0 \tag{7-37}$$

$$\Sigma yx_1 - \alpha \Sigma x_1 - \beta_1 \Sigma x_1^2 - \beta_2 \Sigma x_1 x_2 = 0 \tag{7-38}$$

$$\Sigma yx_2 - \alpha \Sigma x_2 - \beta_2 \Sigma x_2^2 - \beta_1 \Sigma x_1 x_2 = 0 \tag{7-39}$$

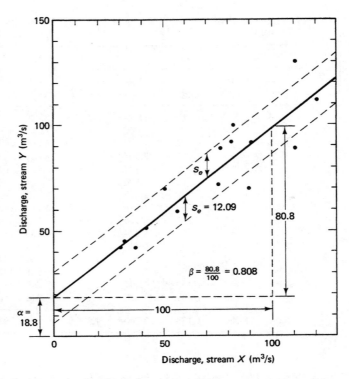

Figure 7-1 X-Y (One-predictor-variable) regression: Example 7-2.

Solving Eqs. 7-37 to 7-39 simultaneously:

$$\beta_1 = \frac{(n\Sigma yx_2 - \Sigma y\Sigma x_2)(n\Sigma x_1x_2 - \Sigma x_1\Sigma x_2) - [n\Sigma x_2^2 - (\Sigma x_2)^2][n\Sigma yx_1 - \Sigma y\Sigma x_1]}{(n\Sigma x_1x_2 - \Sigma x_1\Sigma x_2)^2 - [n\Sigma x_1^2 - (\Sigma x_1)^2][n\Sigma x_2^2 - (\Sigma x_2)^2]}$$

(7-40)

$$\beta_2 = \frac{(n\Sigma yx_1 - \Sigma y \Sigma x_1) - \beta_1[n\Sigma x_1^2 - (\Sigma x_1)^2]}{n\Sigma x_1x_2 - \Sigma x_1 \Sigma x_2}$$

(7-41)

$$\alpha = \frac{\Sigma y - \beta_1 \Sigma x_1 - \beta_2 \Sigma x_2}{n}$$

(7-42)

As in the case of the one-predictor-variable regression, the standard error of estimate of the correlation is the square root of the variance of the conditional distribution:

$$s_e = \left[\frac{1}{n-3}\Sigma(y-y')^2\right]^{1/2}$$

(7-43)

in which $n - 3$ is the number of degrees of freedom.

Alternatively, the standard error of estimate can be estimated from the variance of the conditional distribution. For calculations based on sample data, the standard error of estimate is

$$s_e = s_y \left[\frac{n-1}{n-3} (1 - R^2) \right]^{1/2} \tag{7-44}$$

in which R = multiple regression coefficient, or coefficient of determination [7].

Equations 7-40 to 7-42 can also be used to fit equations of the type:

$$y = ax_1^{b_1} x_2^{b_2} \tag{7-45}$$

First, this equation is linearized by taking the logarithms:

$$\log y = \log a + b_1 \log x_1 + b_2 \log x_2 \tag{7-46}$$

With $u = \log x_1$, $v = \log x_2$, and $w = \log y$, this equation is $w = \log a + bu + cv$. The variables u, v, and w are used in Eqs. 7-40 to 7-42 instead of x_1, x_2, and y, respectively. Then $\alpha = \log a$; $\beta_1 = b_1$, $\beta_2 = b_2$, and the regression equation is

$$y = 10^\alpha x_1^{\beta_1} x_2^{\beta_2} \tag{7-47}$$

Multiple regression analysis involving more than two predictor variables is based on the same least squares principle as in the cases shown here. Library programs are usually available to perform the large amount of computations involved. In general, regression calculations are more efficiently performed with the aid of a computer.

7.3 REGIONAL ANALYSIS OF FLOOD AND RAINFALL CHARACTERISTICS

Peak Flow Based on Catchment Area

The earliest approach to regionalization of hydrologic properties was to assume that peak flow is related to catchment area and to perform a regression to determine the parameters. The equation is of the following form:

$$Q_p = cA^m \tag{7-48}$$

in which Q_p = peak flow; A = catchment area; and c and m are regression parameters. In nature, as catchment area increases, the spatially averaged rainfall intensity decreases, and consequently peak flow does not increase as fast as catchment area. Therefore, the exponent m in Eq. 7-48 is always less than 1, usually in the range 0.4 to 0.9 [5, 9]. Practical examples of the use of this method are given in Section 14.6.

Other formulas relating peak flow to catchment area are the following:

$$Q_p = cA^{nA^{-m}} \tag{7-49}$$

$$Q_p = cA^{(a - b \, \log A)} \tag{7-50}$$

$$Q_p = \frac{cA}{(a + bA)^m} + dA \tag{7-51}$$

in which a, b, c, d, m, and n are parameters determined from statistical analysis of measured data and are applicable on a regional basis, i.e., for neighboring watersheds of similar physiographic, vegetative, and land-use patterns.

The Creager curves (Fig. 2-33) are an example of Eq. 7-49 [3]. Equation 7-50 has been used in regional flood studies in the Southwest [2, 6, 8], whereas Eq. 7-51 appears to be typical of European practice [5]. In principle, none of these equations accounts explicitly for flood frequency, being limited to providing a *maximum* flow. The effect of flood frequency, however, can be accounted for by varying the parameters (see also Section 14.6).

Index-Flood Method

The index-flood method is used to determine the magnitude and frequency of peak flows for catchments of any size, whether gaged or ungaged, located within a hydrologically homogeneous region (i.e., a region with similar hydrologic characteristics) [1, 4].

The application of the index-flood method consists of developing two curves. The first curve depicts the mean annual flood (i.e., that corresponding to the 2.33-y frequency) versus catchment area. The second curve shows peak flow ratio versus frequency. The peak flow ratio is the ratio of peak flow for a given frequency to the mean annual flood. Using these two curves, a flood-frequency curve can be developed for any catchment in the region.

The procedure consists of (1) measuring the catchment area, (2) using the first curve to obtain the mean annual flood, (3) using the second curve to obtain peak flow ratios for selected frequencies, (4) calculating the peak flows for each frequency, and (5) plotting peak flows versus frequencies.

Mean Annual Flood. The magnitude of the mean annual flood is a function of several physiographic and meteorologic factors. The physiographic factors that may influence the mean annual flood are (1) drainage area, (2) channel storage, (3) artificial or natural storage in lakes and ponds, (4) catchment slope, (5) land slope, (6) stream density and pattern, (7) mean elevation, (8) catchment shape, (9) orographic position, (10) underlying geology, (11) soil cover, and (12) vegetative and land-use patterns. The meteorologic factors include (1) regional climatic characteristics, (2) rainfall intensities, (3) storm direction, pattern and volume, (4) effect of snowmelt, and others.

Of the above factors, drainage area is the most important and the one most readily available. Measuring the other factors is usually more difficult. For instance, channel storage has an important effect but cannot be measured directly. For practical use, a regression of mean annual flood on catchment area is usually sufficient. Alternatively, equations relating mean annual flood to catchment characteristics other than area can be determined by using multiple regression techniques.

Regional Frequency Curve. The procedure to develop a regional frequency curve by the index-flood method consists of the following steps:

1. Assemble the records (annual exceedence or annual maxima series) of several stations (usually 10 to 15), each having more than 5 y of record.

2. Select a time base common to all the stations (common base period of analysis) in order to eliminate the effect of variability with time.

3. For each ith station, rank the records in descending order and compute return periods using a plotting position formula such as Weibull's (Eq. 6-26).

4. For each ith station, plot the annual flows versus return periods on extreme value probability paper and fit a line visually to determine the frequency curve.

5. For each ith station, determine the mean annual flood, that is, the peak flow corresponding to the 2.33-y frequency.

6. Choose several frequencies, and for each ith station and jth frequency calculate the peak flow ratio, i.e., the ratio of peak flow for the jth frequency to the mean annual flood.

7. For each jth frequency, determine the median value of peak flow ratios for all stations, that is, the median peak flow ratio.

8. Plot median peak flow ratios versus frequencies on extreme value probability paper and draw a line of best fit to obtain a regional flood frequency curve for the given data.

Test of Hydrologic Homogeneity. The index-flood method includes a test of regional hydrologic homogeneity. Any station not passing this test should be excluded from the set. The test procedure consists of the following steps [4]:

1. For each ith station, use its frequency curve to determine the 2.33-y and the 10-y floods.

2. For each ith station, calculate the 10-y peak flow ratio, i.e., the ratio of the 10-y flood to the 2.33-y flood.

3. Calculate the average of the 10-y peak flow ratios for all stations.

4. For each ith station, multiply the 2.33-y flood by the average 10-y peak flow ratio to obtain an adjusted 10-y peak flow.

5. For each ith station, use its frequency curve to determine the return period T_i for the adjusted 10-y peak flow.

6. For each ith station, plot the return period T_i versus the length of record n, in years, in Fig. 7-2. Points located within the confidence limits (solid lines) are considered to be hydrologically homogeneous. Points lying outside of the solid lines should not be used in the calculation of the median peak flow ratio (step 7 of the index-flood method).

Limitations of the Index-Flood Method. Benson [1] has noted the following limitations of the index-flood method:

1. The mean annual flood for stations with short periods of record may not be typical, which means that the peak flow ratios of different return periods may vary widely among stations.

2. The homogeneity test is used to determine whether the differences in the frequency curves are greater than those that could be attributed to chance alone. The index-flood test uses the 10-y flow ratio because of the lack of sufficient data

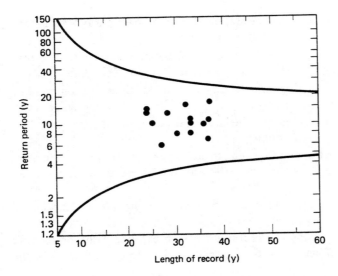

Figure 7-2 Homogeneity test chart for index-flood method [4].

Return period (y) — vertical axis
Length of record (y) — horizontal axis

to define the frequency curve adequately at longer return periods. Studies have shown that although homogeneity may be assumed on the basis of the 10-y peak flow ratio, the individual frequency curves may show wide and sometimes systematic differences at longer return periods.

3. The method combines frequency curves for all catchment sizes, excluding only the largest. At the 10-y peak flow ratio level, the effect of catchment size is small and can be neglected. Studies have shown that the peak flow ratios tend to vary inversely with catchment size. In general, the larger the catchment, the flatter the frequency curve and the lower the peak flow ratios. The effect of catchment size is particularly marked for floods of long return period.

Example 7-3.

Use the $Q_i/Q_{2.33}$ data for the five stations shown in Table 7-2 to develop a regional flood frequency curve by the index-flood method. Assuming $Q_{2.33} = 2.5A^{0.6}$, in which $Q_{2.33}$ is in cubic meters per second and catchment area A is in square kilometers, calculate the 50-y flood for a 150-km^2 catchment based on the regionally developed curve.

The median values are shown at the bottom of each column. These values are plotted against the return period, as shown in Fig. 7-3. The fitted line is the regional flood-frequency curve. For a 150-km^2 catchment, the mean annual flood is: 50.5 m^3/s. From Fig. 7-3, the peak flood ratio for the 50-y return period is 2.62. Therefore, the 50-y flood for this catchment is 132 m^3/s.

Rainfall Intensity-Duration-Frequency

Curves showing the relationship between intensity, duration, and frequency of rainfall (IDF curves) are required for peak flow computations in small catchments (see rational method, Chapter 4). These curves can be developed using either (a) depth-duration-frequency data provided by the National Weather Service or (b) regional or local rainfall intensity-duration data. The latter procedure is illustrated by the following example.

TABLE 7-2 INDEX-FLOOD METHOD: EXAMPLE 7-3

(1)	(2)	(3)	(4)	(5)	(6)	(7)	(8)
	$Q_j/Q_{2.33}$ for jth Return Period (years)						
Station i	1.11	1.25	2	5	10	25	50
1	0.32	0.49	0.90	1.45	1.82	2.28	2.62
2	0.35	0.51	0.92	1.44	1.79	2.23	2.56
3	0.39	0.55	0.92	1.40	1.73	2.14	2.44
4	0.27	0.45	0.90	1.50	1.88	2.38	2.74
5	0.31	0.50	0.91	1.46	1.84	2.32	2.68
Median	0.32	0.50	0.91	1.45	1.82	2.28	2.62

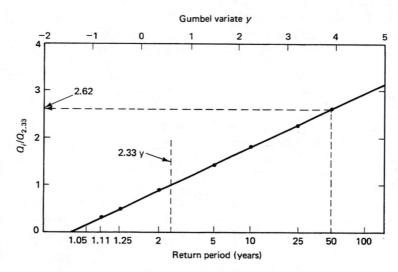

Figure 7-3 Index-flood method: Example 7-3.

Example 7-4.

Determine the equation relating rainfall intensity and duration for the following 10-y frequency rainfall data.

Rainfall duration t_r (min)	5	10	15	30	60	120	180
Rainfall intensity i (cm/h)	8	5	4	2.5	1.5	1.0	0.8

The data suggest that the relation is of hyperbolic type, with greater intensities associated with shorter durations. Therefore, an equation of the type of Eq. 2-6 is applicable:

$$i = \frac{a}{t_r + b} \qquad (7\text{-}52)$$

in which a and b are constants to be determined by regression analysis. This equation can be linearized in the following way:

$$\frac{1}{i} = \frac{t_r}{a} + \frac{b}{a} \tag{7-53}$$

With $y = 1/i$, $x = t_r$, $\alpha = b/a$, and $\beta = 1/a$, the application of the regression formulas (Eqs. 7-30 and 7-31) to the data leads to $1/i = 0.006422t_r + 0.1706$, in which $\alpha = 0.1706$ and $\beta = 0.006422$. Therefore, $a = 155.7$ and $b = 26.56$. The regression equation is $i = 155.7/(t_r + 26.56)$. The data and regression line are shown in Fig. 7-4.

QUESTIONS

1. What is a joint probability? What is a marginal probability?
2. What is a joint density function? Give an example.
3. What is a conditional probability? How is it used in regression analysis?
4. Define covariance.
5. What is a correlation coefficient?
6. What is the difference between correlation and regression?
7. Describe briefly the index-flood method for regional analysis of flood frequency.

PROBLEMS

7-1. Computer program EH700A (Appendix D) calculates the correlation coefficient of the joint probability distribution of monthly (or seasonal) runoff volumes of streams X and Y. Test this program using Example 7-1 in the text.

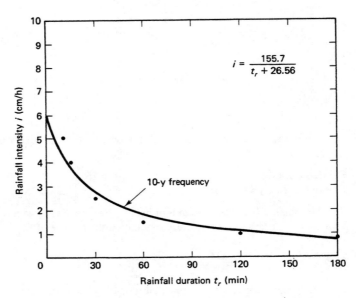

Figure 7-4 Fitting intensity-duration-frequency curve: Example 7-4.

7-2. Using computer program EH700A (Appendix D), calculate the correlation coefficient of the following joint distribution of quarterly flows (expressed as mean values in each class) in streams A and B:

Stream A (ac-ft)	1000	2000	3000	4000	5000
Stream B (ac-ft)					
1000	0.07	0.03	0.02	0.00	0.00
2000	0.03	0.08	0.04	0.03	0.00
3000	0.02	0.04	0.08	0.05	0.02
4000	0.00	0.04	0.08	0.11	0.06
5000	0.00	0.00	0.03	0.08	0.09

7-3. Develop a computer program to calculate the regression constants, correlation coefficient, and standard error of estimate of a series of paired flow values X and Y. Test your program using the data of Example 7-2 in the text.

7-4. Using the computer program developed in Problem 7-3, calculate the regression constants, correlation coefficient, and standard error of estimate for the following paired low-flow series (annual minima):

Stream X Flow (m^3/s)	Stream Y Flow (m^3/s)
50	65
66	76
32	45
78	95
12	18
34	50
23	31
50	64
43	67
89	99
76	89
22	33

7-5. Modify the computer program developed in Problem 7-3 to calculate the regression constants to fit a power function of the following form (Eq. 7-48):

$$Q_p = cA^m$$

in which Q_p = peak discharge; A = drainage area; c and m are coefficient and exponent, respectively.

7-6. Using the computer program developed in Problem 7-5, fit a power function to the following data:

Peak Discharge (m^3/s)	Drainage Area (km^2)
124	25
254	46
378	78

101	22
678	99
540	89
490	83
267	52
350	73

7-7. Program EH700B (Appendix D) solves the two-predictor-variable nonlinear regression problem (Eq. 7-45). Use this program to determine the regression constants for the following data set:

Y Time of Concentration (min)	X_1 Hydraulic Length (m)	X_2 Catchment Slope (m/m)
89	3245	0.008
75	2567	0.011
57	2783	0.009
34	1234	0.015
101	5345	0.006
121	5329	0.007
68	3002	0.008
79	2976	0.010
25	1034	0.018
59	2984	0.010
96	3892	0.007
12	534	0.020

7-8. Modify program EH700B to solve the two-predictor-variable linear regression problem of Eq. 7-35. Then fit a regression line to the data of Problem 7-7.

7-9. The median $Q_i/Q_{2.33}$ ratios (i = frequency) for 10 stations have been found to be 1.95 for the 10-y frequency and 2.45 for the 50-y frequency. Use the index-flood method to calculate the 25-y flood for a point in a stream having a 340-km^2 catchment and a mean annual flood given by the following formula:

$$Q_{2.33} = 3.93A^{0.75}$$

in which Q = flood discharge in cubic meters per second and A = drainage area in square kilometers.

7-10. Modify the computer program developed in Problem 7-3 to calculate the regression constants and correlation coefficient to fit intensity-duration-frequency rainfall data. Test your computer program using the data of Example 7-4 in the text.

7-11. Using the computer program developed in Problem 7-10, calculate the regression constants a and b (Eq. 7-52) and correlation coefficient for the following 25-y frequency rainfall data:

Duration (min)	5	10	15	30	60	120	180
Intensity (mm/h)	15.5	7.5	6.5	4.5	3.5	2.5	1.5

REFERENCES

1. Benson, M. A. (1962). "Evolution of Methods for Evaluating the Occurrence of Floods," *U.S. Geological Survey Water Supply Paper* No. 1580-A.

2. Boughton, W. C., and K. G. Renard. (1984). "Flood Frequency Characteristics of Some Arizona Watersheds," *Water Resources Bulletin,* Vol. 20, No. 5, October, pp. 761–769.

3. Creager, W. P., J. D. Justin, and J. Hinds. (1945). *Engineering for Dams,* Vol. 1. New York: John Wiley.

4. Dalrymple, T. (1960). "Flood Frequency Analyses," *U.S. Geological Survey Water Supply Paper* No. 1543A.

5. Hall, M. J. (1984). *Urban Hydrology.* London: Elsevier Applied Science Publishers.

6. Malvick, A. J. (1980). "A Magnitude-Frequency-Area Relation for Floods in Arizona," *Research Report* No. 2, College of Engineering, University of Arizona, Tucson.

7. McCuen, R. H. (1985). *Statistical Methods for Engineers.* Englewood Cliffs, N.J.: Prentice-Hall.

8. Reich, B. M., H. B. Osborn, and M. C. Baker. (1979). "Tests on Arizona New Flood Estimates," in *Hydrology and Water Resources in Arizona and the Southwest,* University of Arizona, Tucson, Vol. 9.

9. Roeske, R. H. (1978). "Methods for Estimating the Magnitude and Frequency of Floods in Arizona," Final Report, ADOT-RS-15-121, U.S. Geological Survey, Tucson, Arizona.

10. Viessman, W. Jr., J. W. Knapp, G. L. Lewis, and T. E. Harbaugh, *Introduction to Hydrology.* 2d. ed. New York: Harper & Row.

RESERVOIR ROUTING

In many applications of engineering hydrology, it is necessary to calculate the variation of flows in time and space. These applications include reservoir design, design of flood-control structures, flood forecasting, and water resources planning and analysis.

A reservoir is a natural or artificial feature designed to store incoming water and release it at regulated rates. Surface water reservoirs should be distinguished from natural groundwater reservoirs which store groundwater. Surface water reservoirs store water for diverse uses, including hydropower generation, municipal and industrial water supply, flood control, irrigation, navigation, fish and wildlife management, water quality, and recreation.

Reservoir routing uses mathematical relations to calculate outflow from a reservoir once inflow, initial conditions, reservoir characteristics, and operational rules are known. The classical approach to reservoir routing is based on the storage concept (Chapter 4). Reservoir-routing techniques based on the storage concept are referred to as *hydrologic* reservoir-routing methods, or storage routing methods, to distinguish them from *hydraulic* routing methods. The latter use principles of mass and momentum conservation to obtain detailed solutions for discharges and stages throughout the reservoir [2]. In practice, however, most applications of reservoir routing have used the storage concept.

Reservoirs can be of widely differing sizes; they can range from small detention ponds designed to diffuse flood flows from developed urban sites to very large reservoirs occupying substantial segments of large rivers. For a single reservoir, inflow is dependent on the natural upstream flows. Outflow, however, may be either (1) uncontrolled, (2) controlled, or (3) a combination of controlled and uncontrolled. Uncontrolled outflow is not subject to operator intervention, for example, outflow through an ungated spillway. On the other hand, controlled outflow is subject to operator intervention, e.g., a gated spillway.

Detention ponds and small flood-retention reservoirs are typical examples of reservoirs with uncontrolled outflow. In these cases, an ungated spillway (or a gated spillway that is fully open during the flood season) serves as outflow structure. Therefore, outflow from the reservoir is solely a function of reservoir stage.

There are two types of reservoir routing with uncontrolled outflow: (1) simulated and (2) actual. Simulated reservoir routing uses mathematical relations to mimic natural diffusion processes existing in nature. An example of simulated reservoir routing is the *linear reservoir* method, which is extensively used in catchment routing (Chapter 10).

Actual reservoir routing refers to the routing through a planned or existing reservoir, either for design or operational purposes. In this case, the outflow characteristics are determined by the geometric properties of the reservoir and the hydraulic properties of the outflow structure(s). The most widely used method of actual reservoir routing with uncontrolled outflow is the *storage indication* method.

In a reservoir with controlled outflow, gates are used for the purpose of regulating flow through the outlet structure(s). The gates are operated following established rules. These rules determine the relation between inflow, outflow, and storage, taking into account the daily, monthly, or seasonal downstream water demands. Most large reservoirs operate with controlled outflow conditions.

In certain cases, outflow may be a combination of controlled and uncontrolled modes—for instance, when the reservoir features a combined regulated outflow and emergency spillway that operates only above a certain pool level. Flow through an emergency spillway is usually of the uncontrolled type, outflow being governed solely by the hydraulic properties of the spillway, without the need for operator intervention.

This chapter is divided into four sections. Section 8.1 discusses general concepts of storage routing. Section 8.2 discusses linear reservoirs and their use in simulated reservoir routing. Section 8.3 describes the storage indication method and its use in actual reservoir routing with uncontrolled outflow. Section 8.4 discusses reservoir routing with controlled outflow.

8.1 STORAGE ROUTING

The storage concept is well established in flow-routing theory and practice. Storage routing is used not only in reservoir routing but also in stream channel and catchment routing (Chapters 9 and 10). Techniques for storage routing are invariably based on the differential equation of water storage. This equation is founded on the principle of mass conservation, which states that the change in flow per unit length in a control volume is balanced by the change in flow area per unit time. In partial differential form:

$$\frac{\partial Q}{\partial x} + \frac{\partial A}{\partial t} = 0 \tag{8-1}$$

in which Q = flow rate, A = flow area, x = space (length), and t = time.

The differential equation of storage is obtained by lumping spatial variations. For this purpose, Eq. 8-1 is expressed in finite increments:

$$\frac{\Delta Q}{\Delta x} + \frac{\Delta A}{\Delta t} = 0 \qquad (8\text{-}2)$$

With $\Delta Q = O - I$, in which O = outflow and I = inflow; and $\Delta S = \Delta A\, \Delta x$, in which ΔS = change in storage volume, Eq. 8-2 reduces to:

$$I - O = \frac{\Delta S}{\Delta t} \qquad (8\text{-}3)$$

in which inflow, outflow, and rate of change of storage are expressed in $L^3 T^{-1}$ units. Furthermore, Eq. 8-3 can be expressed in differential form, leading to the differential equation of storage:

$$I - O = \frac{dS}{dt} \qquad (8\text{-}4)$$

Equation 8-4 implies that any difference between inflow and outflow is balanced by a change of storage in time (Fig. 8-1). In a typical reservoir routing application, the inflow hydrograph (upstream boundary condition), initial outflow and storage (initial conditions), and reservoir physical and operational characteristics are known. Thus, the objective is to calculate the outflow hydrograph for the given initial condition, upstream boundary condition, reservoir characteristics, and operational rules.

Storage-Outflow Relations

Unlike in an ideal channel for which storage is a function of both inflow and outflow, in an ideal reservoir storage is a function only of outflow (Section 4.2). The relationship between storage and outflow can be expressed in the following general form:

$$S = f(O) \qquad (8\text{-}5)$$

A common relationship between outflow and storage is the following power function:

$$S = KO^n \qquad (8\text{-}6)$$

in which K = storage coefficient and n = exponent. For $n = 1$, Eq. 8-6 reduces to the linear form

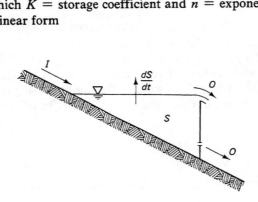

Figure 8-1 Inflow, outflow, and change of storage in a reservoir.

$$S = KO \qquad (8\text{-}7)$$

in which K is a proportionality constant or linear storage coefficient, which has units of time (T).

Real reservoirs usually have a nonlinear storage-outflow relationship; therefore, Eq. 8-6 is applicable to planned or existing reservoirs. Exceptions are the cases where the storage-outflow relation is indeed linear, as in the case of the proportional weir. The latter is used in connection with irrigation diversions or measurement of very small flows.

Simulated reservoirs are usually of the linear type (Eq. 8-7), although nonlinear reservoirs have also been used in simulation. Several linear reservoirs in series lead to a *cascade of linear reservoirs*, a mathematical procedure that is useful in catchment routing (Chapter 10). For linear reservoirs, the constant K is the linear storage coefficient. Increasing the value of K increases the amount of storage simulated by the system. In other words, greater values of K result in increased outflow hydrograph diffusion.

For routing in actual reservoirs, the nonlinear properties of the storage-outflow relation must be determined in advance. Outflow from an actual reservoir will depend on whether the flow is discharged through either closed conduit(s), overflow spillway(s), or a combination of the two. A general hydraulic outflow formula is the following:

$$O = C_d Z H^y \qquad (8\text{-}8)$$

in which O = outflow; C_d = discharge coefficient; Z = variable representing either cross-sectional area (for a free-outlet closed conduit) or length of spillway crest (for a free-surface overflow spillway); H = hydraulic head, either above outlet elevation (for a closed conduit) or above spillway crest (for an overflow spillway); and y = exponent of the rating.

Theoretical values of discharge coefficient C_d and rating exponent y are determined using hydraulic principles. For the free-outlet closed conduit, the conservation of energy between reservoir pool and outlet elevations—neglecting entrance and friction losses—leads to:

$$H = \frac{V^2}{2g} \qquad (8\text{-}9)$$

in which V = mean velocity, and g = gravitational acceleration. Therefore, the outflow is

$$O = (2gH)^{1/2} Z \qquad (8\text{-}10)$$

Comparing Eq. 8-10 with Eq. 8-8, it follows that $y = \frac{1}{2}$, with $C_d = 4.43$ in SI units and $C_d = 8.02$ in U.S. customary units. In practice, these theoretical values of discharge coefficient are reduced by about 30 percent to account for flow contraction and entrance and friction losses.

For an ungated overflow spillway, the critical flow condition in the vicinity of the crest leads to:

$$O = [g(\tfrac{2}{3})H]^{1/2}[(\tfrac{2}{3})H]Z \qquad (8\text{-}11)$$

which reduces to:

$$O = (\tfrac{2}{3})[(\tfrac{2}{3})g]^{1/2}ZH^{3/2} \qquad (8\text{-}12)$$

Comparing Eq. 8-12 with Eq. 8-8, it follows that $y = \tfrac{3}{2}$. In SI units, $C_d = 1.70$; in U.S. customary units, $C_d = 3.09$. In practice, the discharge coefficient of an overflow spillway (ogee-shaped crest) is not constant, varying with hydraulic head approximately between 95 percent and 130 percent of the theoretical value given by Eq. 8-12.

In the proportional weir, the cross-sectional flow area grows in proportion to the half-power of the hydraulic head. Therefore, outflow is linearly related to hydraulic head, and a spillway rating based on Eq. 8-7 is applicable.

8.2 LINEAR RESERVOIR ROUTING

Equation 8-4 can be solved by analytical or numerical means. The numerical approach is usually preferred because it can account for an arbitrary inflow hydrograph and because it lends itself readily to computer solution. The solution is accomplished by discretizing Eq. 8-4 on the xt plane (Fig. 8-2). The xt plane is a graph showing the values of a certain variable in discrete points in time and space.

Figure 8-2 shows two consecutive time levels, 1 and 2, separated between them an interval Δt, and two spatial locations depicting inflow and outflow, with the reservoir located between them. The discretization of Eq. 8-4 on the xt plane leads to:

$$\frac{I_1 + I_2}{2} - \frac{O_1 + O_2}{2} = \frac{S_2 - S_1}{\Delta t} \qquad (8\text{-}13)$$

in which I_1 = inflow at time level 1; I_2 = inflow at time level 2; O_1 = outflow at time level 1; O_2 = outflow at time level 2; S_1 = storage at time level 1; S_2 = storage at time level 2; and Δt = time interval. Equation 8-13 states that between two time levels 1 and 2 separated by a time interval Δt, average inflow minus average outflow is equal to change in storage.

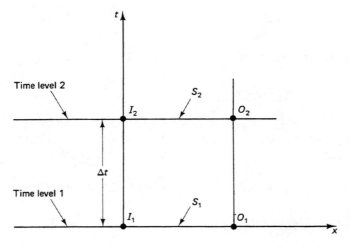

Figure 8-2 Discretization of storage equation in xt plane.

For linear reservoirs, Eq. 8-7 is the relation between storage and outflow. Therefore

$$S_1 = KO_1 \tag{8-14a}$$

and

$$S_2 = KO_2 \tag{8-14b}$$

in which K is the storage constant.

Substituting Eqs. 8-14 into 8-13, and solving for O_2:

$$O_2 = C_0 I_2 + C_1 I_1 + C_2 O_1 \tag{8-15}$$

in which C_0, C_1 and C_2 are routing coefficients defined as follows:

$$C_0 = \frac{\Delta t/K}{2 + (\Delta t/K)} \tag{8-16}$$

$$C_1 = C_0 \tag{8-17}$$

$$C_2 = \frac{2 - (\Delta t/K)}{2 + (\Delta t/K)} \tag{8-18}$$

Since $C_0 + C_1 + C_2 = 1$, the routing coefficients are interpreted as weighting coefficients. These routing coefficients are a function of $\Delta t/K$, the ratio of time interval to storage constant. Values of the routing coefficients as a function of $\Delta t/K$ are given in Table 8-1.

The linear reservoir routing procedure is illustrated by the following example.

TABLE 8-1 LINEAR RESERVOIR-ROUTING COEFFICIENTS

(1)	(2)	(3)	(4)
$\Delta t/K$	C_0	C_1	C_2
$\frac{1}{8}$	$\frac{1}{17}$	$\frac{1}{17}$	$\frac{15}{17}$
$\frac{1}{4}$	$\frac{1}{9}$	$\frac{1}{9}$	$\frac{7}{9}$
$\frac{1}{2}$	$\frac{1}{5}$	$\frac{1}{5}$	$\frac{3}{5}$
$\frac{3}{4}$	$\frac{3}{11}$	$\frac{3}{11}$	$\frac{5}{11}$
1	$\frac{1}{3}$	$\frac{1}{3}$	$\frac{1}{3}$
$\frac{5}{4}$	$\frac{5}{13}$	$\frac{5}{13}$	$\frac{3}{13}$
$\frac{3}{2}$	$\frac{3}{7}$	$\frac{3}{7}$	$\frac{1}{7}$
$\frac{7}{4}$	$\frac{7}{15}$	$\frac{7}{15}$	$\frac{1}{15}$
2	$\frac{1}{2}$	$\frac{1}{2}$	0
4	$\frac{2}{3}$	$\frac{2}{3}$	$-\frac{1}{3}$
6	$\frac{3}{4}$	$\frac{3}{4}$	$-\frac{1}{2}$
8	$\frac{4}{5}$	$\frac{4}{5}$	$-\frac{3}{5}$

Example 8-1.

A linear reservoir has a storage constant $K = 2$ h, and it is initially at equilibrium with inflow and outflow equal to 100 m^3/s. Route the following inflow hydrograph through the reservoir.

Time (h)	0	1	2	3	4	5	6	7	8	9	10	11	12	13	14	15
Inflow (m^3/s)	100	150	250	400	800	1000	900	700	550	400	300	250	200	150	120	100

First it is necessary to select an appropriate time interval. An examination of the inflow hydrograph reveals that the time-to-peak is $t_p = 5$ h. A rule-of-thumb for adequate temporal resolution is to make the ratio $t_p/\Delta t$ at least equal to 5. Setting $\Delta t = 1$ h assures that $t_p/\Delta t = 5$. With $\Delta t = 1$ h, the ratio $\Delta t/K = \frac{1}{2}$. From Eqs. 8-16 to 8-18, or Table 8-1, $C_0 = C_1 = \frac{1}{5}$, and $C_2 = \frac{3}{5}$. The routing calculations are shown in Table 8-2. Column 1 shows the time and Col. 2 shows the inflow hydrograph ordinates. Columns 3 to 6 are calculated by the recursive application of Eq. 8-15 between two successive time levels. Columns 3 to 5 are the partial flows and Col. 6 is the sum of the partial flows at each time level. The recursive procedure continues until the calculated outflow (Col. 6) is within 5 percent of baseflow (100 m^3/s). Plotted inflow and outflow hydrographs (Cols. 2 and 6) are shown in Fig. 8-3. The calculated peak outflow (758 m^3/s) occurs at $t = 7$ h. However, the shape of the outflow hydrograph reveals that the true peak outflow occurs some-

TABLE 8-2 LINEAR RESERVOIR ROUTING: EXAMPLE 8-1

(1) Time (h)	(2) Inflow (m^3/s)	(3) $C_0 I_2$	(4) $C_1 I_1$	(5) $C_2 O_1$	(6) Outflow (m^3/s)
		Partial Flows (m^3/s)			
0	100	—	—	—	100.0
1	150	30	20	60	110.0
2	250	50	30	66	146.0
3	400	80	50	87.6	217.6
4	800	160	80	130.6	370.6
5	1000	200	160	222.4	582.4
6	900	180	200	349.4	729.4
7	700	140	180	437.6	757.6
8	550	110	140	454.6	704.6
9	400	80	110	422.8	612.8
10	300	60	80	367.7	507.7
11	250	50	60	304.6	414.6
12	200	40	50	248.8	338.8
13	150	30	40	203.3	273.3
14	120	24	30	164.0	218.0
15	100	20	24	131.8	174.8
16	100	20	20	104.9	144.9
17	100	20	20	86.9	126.9
18	100	20	20	76.1	116.1
19	100	20	20	69.7	109.7
20	100	20	20	65.8	105.8
21	100	20	20	63.5	103.5

Reservoir Routing   Chap. 8

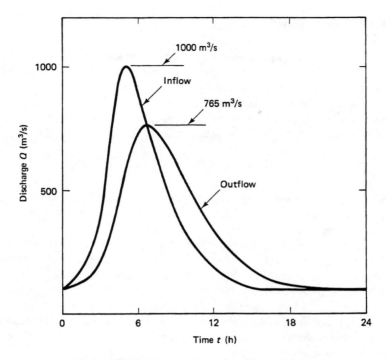

Figure 8-3 Linear reservoir routing: Example 8-1.

where between 6 and 7 h. The true peak outflow is approximated graphically at 765 m^3/s, occurring at about 6.6 h. The peak outflow is substantially less than the peak inflow (1000 m^3/s), showing the attenuating effect of the reservoir. Also, the time elapsed between the occurrences of peak inflow and peak outflow (1.6 h) is approximately equal to the storage constant.

The reservoir exerts a diffusive action on the flow, with the net result that peak flow is attenuated and time base is increased. In the linear reservoir case, the amount of attenuation is a function of $\Delta t/K$. The smaller this ratio, the greater the amount of attenuation exerted by the reservoir. Conversely, large values of $\Delta t/K$ cause less attenuation. Values of $\Delta t/K$ greater than 2 (see Table 8-1) can lead to negative attenuation. This amounts to amplification; therefore, values of $\Delta t/K$ greater than 2 are not used in reservoir routing.

A distinct characteristic of reservoir routing is the occurrence of peak outflow at the time when inflow equals outflow (see Fig. 8-3). Since outflow is proportional to storage (Eq. 8-7), peak outflow corresponds to maximum storage. Since storage ceases to increase when outflow equals inflow, maximum storage and peak outflow must occur at the time when inflow and outflow coincide.

Another characteristic of reservoir routing is the immediate outflow response, with no apparent lag between the start of inflow and the start of outflow (see Fig. 8-3). From a mathematical standpoint, this property is attributed to the infinite propagation velocity of surface waves in an ideal reservoir.

8.3 STORAGE INDICATION METHOD

The storage indication method is also known as the *modified Puls* method [1]. It is used to route streamflows through actual reservoirs, for which the relationship between outflow and storage is usually of a nonlinear nature.

The method is based on the differential equation of storage, Eq. 8-4. The discretization of this equation on the *xt* plane (Fig. 8-2) leads to Eq. 8-13. In the storage indication method, Eq. 8-13 is transformed to its equivalent form:

$$\frac{2S_2}{\Delta t} + O_2 = I_1 + I_2 + \frac{2 S_1}{\Delta t} - O_1 \tag{8-19}$$

in which the unknown values (S_2 and O_2) are on the left side of the equation and the known values (inflows, initial outflow and storage) are on the right side. The left side of Eq. 8-19 is known as the *storage indication quantity*.

In the storage indication method, it is first necessary to assemble geometric and hydraulic reservoir data in suitable form. For this purpose, the following curves (or tables) are prepared: (1) elevation-storage, (2) elevation-outflow, (3) storage-outflow, and (4) storage indication-outflow. For computer applications, these curves are replaced by tables of elevation-outflow-storage-storage indication quantities.

The elevation-storage relation is determined based on topographic information. The minimum elevation is that for which storage is zero, and the maximum elevation is the minimum elevation of the dam crest.

The elevation-outflow relation is determined based on the hydraulic properties of the outlet works, either closed conduit, overflow spillway, or a combination of the two. In the typical application, the reservoir pool elevation provides a head over the outlet or spillway crest, and the outflow can be calculated using an equation such as Eq. 8-8. When routing floods through emergency spillways, storage is alternatively expressed in terms of *surcharge storage*, i.e., the storage above a certain level, usually the emergency spillway crest elevation (see Section 14.3).

Elevation-storage and elevation-outflow relations lead to the storage-outflow relation. In turn, the storage-outflow relation is used to develop the storage indication-outflow relation. The storage indication variable is the left side of Eq. 8-19. In general, the storage indication quantity is $[(2S/\Delta t) + O]$, with S = storage, O = outflow, and Δt = time interval. To develop the storage indication-outflow relation it is first necessary to select a time interval such that the resulting linearization of the inflow hydrograph remains a close approximation of the actual nonlinear shape of the hydrograph. For smoothly rising hydrographs, a minimum value of $t_p/\Delta t = 5$ is recommended, in which t_p is the time-to-peak of the inflow hydrograph. In practice, a computer-aided calculation would normally use a much greater ratio, say 10 to 20.

Once the data has been prepared, Eq. 8-19 is used to perform the reservoir routing. The procedure consists of the following steps:

1. Set the counter at $n = 1$ to start.
2. Use Eq. 8-19 to calculate the storage indication quantity $[(2S_{n+1}/\Delta t) + O_{n+1}]$ at time level $n + 1$.
3. Use the storage indication quantity versus outflow relation to determine the outflow O_{n+1} at time level $n + 1$.

4. Use the storage indication quantity and outflow at time level $n + 1$ to calculate $[(2S_{n+1}/\Delta t) - O_{n+1}] = [(2S_{n+1}/\Delta t) + O_{n+1}] - 2(O_{n+1})$.

5. Increment the counter by 1, go back to step 2 and repeat. The recursive procedure is terminated either when the inflow ceases or when the outflow hydrograph has substantially receded to baseflow discharge.

The procedure is illustrated in Example 8-2 using the same data as in Example 8-1, confirming that the storage indication method is applicable to linear reservoir data. Example 8-3 illustrates the application of the storage indication method to an actual reservoir featuring a nonlinear storage-outflow relation.

Example 8-2.

Use the data in Example 8-1 to perform a reservoir routing by the storage indication method.

Since $K = 2$ h and the reservoir is linear, the outflow-storage relation is the following:

$$S = 2(O) \tag{8-20}$$

in which outflow O is in cubic meters per second and storage S is in (cubic meters per second)-hour for computational convenience. Selecting $\Delta t = 1$ h as in the previous example, the storage indication variable is $[(2S/\Delta t) + O] = 5(O)$, from which

$$O = \frac{(2S/\Delta t) + O}{5} \tag{8-21}$$

The calculations are shown in Table 8-3. At $t = 0$, the counter is set at $n = 1$, the outflow is 100 m³/s (baseflow), and the storage indication (Eq. 8-21) is: $100 \times 5 = 500$ m³/s. Therefore, Col. 3 is: $500 - (2 \times 100) = 300$ m³/s. For $n = 1$, between $t = 0$ and $t = 1$, Eq. 8-19 is used to calculate the storage indication at $t = 1$: $300 + 100 + 150 = 550$ m³/s. The outflow at $t = 1$ (Eq. 8-21) is: $550/5 = 110$ m³/s. Column 3 is $550 - (2 \times 110) = 330$ m³/s. The counter is incremented by 1 and the recursive procedure is continued until the outflow hydrograph ordinate (Col. 5) is within 5% of baseflow discharge (100 m³/s). The results of Table 8-2, Col. 5 are confirmed to be the same as those of Table 8-1, Col. 6.

Example 8-3.

The design of an emergency spillway calls for a broad-crested weir of width $Z = 10.0$ m; rating coefficient $C_d = 1.70$; and exponent $y = 1.5$. The spillway crest is at elevation 1070 m. Above this level, the reservoir walls can be considered to be vertical, with a surface area of 100 ha. The dam crest is at elevation 1076 m. Baseflow is 17 m³/s, and initially the reservoir level is at elevation 1071 m. Route the following design hydrograph through the reservoir.

Time (h)	1	2	3	4	5	6	7	8
Inflow (m³/s)	20	50	100	130	150	140	110	90

Time (h)	9	10	11	12	13	14	15	16
Inflow (m³/s)	70	50	30	20	17	17	17	17

What is the maximum pool elevation reached?

The calculations of the storage indication function are shown in Table 8-4. Column 1 shows water surface elevations, from 1070 to 1076. Column 2 shows the head above spillway crest. Column 3 shows the outflows, calculated by the following formula:

TABLE 8-3 STORAGE INDICATION METHOD: EXAMPLE 8-2

(1) Time (h)	(2) Inflow (m³/s)	(3) $[(2S/\Delta t) - O]$ (m³/s)	(4) $[(2S/\Delta t) + O]$ (m³/s)	(5) Outflow (m³/s)
0	100	300.0	500.0	100.0
1	150	330.0	550.0	110.0
2	250	438.0	730.0	146.0
3	400	652.8	1088.0	217.6
4	800	1111.6	1852.8	370.6
5	1000	1747.0	2911.6	582.3
6	900	2188.2	3647.0	729.4
7	700	2273.0	3788.2	757.6
8	550	2113.8	3523.0	704.6
9	400	1838.2	3063.8	612.8
10	300	1523.0	2538.2	507.6
11	250	1243.8	2073.0	414.6
12	200	1016.2	1693.8	338.8
13	150	819.8	1366.2	273.2
14	120	653.8	1089.8	218.0
15	100	524.2	873.8	174.8
16	100	434.6	724.2	144.8
17	100	380.8	634.6	126.9
18	100	348.4	580.8	116.2
19	100	329.0	548.4	109.7
20	100	317.4	529.0	105.8
21	100	310.4	517.4	103.5

TABLE 8-4 STORAGE INDICATION METHOD: EXAMPLE 8-3
STORAGE INDICATION VERSUS OUTFLOW RELATION

(1) Elevation (m)	(2) Head (m)	(3) Outflow (m³/s)	(4) Storage (m³)	(5) Storage (m³/s)-h	(6) $[(2S/\Delta t) + O]$ (m³/s)
1070	0	0	0	0	0
1071	1	17.00	1000,000	277.78	572.56
1072	2	48.08	2000,000	555.55	1159.18
1073	3	88.33	3000,000	833.33	1754.99
1074	4	136.00	4000,000	1111.11	2358.22
1075	5	190.07	5000,000	1388.89	2967.85
1076	6	249.85	6000,000	1666.66	3583.17

$$O = C_d Z H^y = 1.70(10.0)H^{1.5} \tag{8-22}$$

Column 4 shows the storage volumes in cubic meters above spillway crest elevation (i.e., surcharge storage), calculated as the product of reservoir surface area (100 ha) times head above spillway crest. Column 5 shows storage volumes in (cubic meters per second)-hour. A time interval $\Delta t = 1$ h is appropriate for this example. Column 6 shows the storage indication quantities $[(2S/\Delta t) + O]$, in m³/s. Fig. 8-4 shows the storage indication versus outflow relation. The routing is summarized in Table 8-5. Column 1 shows

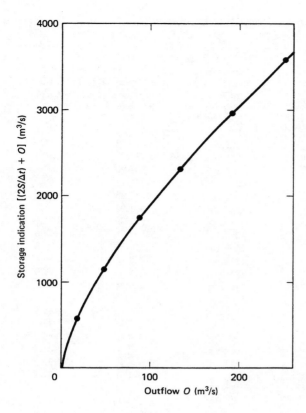

Figure 8-4 Storage indication function: Example 8-3.

time; Col. 2 shows the inflow hydrograph; Col. 3 shows $[(2S/\Delta t) - O]$; Col. 4 shows the storage indication quantities $[(2S/\Delta t) + O]$; Col. 5 shows the calculated outflow. The recursive procedure is the same as in the previous example. The initial outflow is 17 m³/s; the initial storage indication value is 572.56 m³/s; the initial value of Col. 3 is 538.56 m³/s. The next storage indication value is 17 + 20 + 538.56 = 575.56 m³/s, which through Fig. 8-4 leads to an outflow of 17.1 m³/s. The recursive procedure continues until the outflow has substantially reached baseflow conditions. To calculate the maximum pool elevation, use Eq. 8-22 and solve for H with the peak outflow value of 72.5 m³/s. This results in a maximum head of 2.63 m above spillway crest. Therefore, the maximum pool elevation is 1070.0 + 2.63 = 1072.63 m.

8.4 RESERVOIR ROUTING WITH CONTROLLED OUTFLOW

Most large reservoirs have some type of outflow control, wherein the amount of outflow is regulated by gated spillways. In this case, the prescribed outflow is determined by both hydraulic conditions and operational rules. Operational rules take into account the various uses of water. For instance, a multipurpose reservoir may be designed for hydropower generation, flood control, irrigation, and navigation. For hydropower generation, reservoir pool level is kept within a narrow range, usually close to the optimum operating level of the installation. On the other hand, flood-control operation may require that a certain storage volume be kept empty during the flood

TABLE 8-5 STORAGE INDICATION METHOD, EXAMPLE 8-3
ROUTING OF INFLOW HYDROGRAPH

(1)	(2)	(3)	(4)	(5)
Time (h)	Inflow (m^3/s)	$[(2S/\Delta t) - O]$ (m^3/s)	$[(2S/\Delta t) + O]$ (m^3/s)	Outflow (m^3/s)
0	17	538.56	572.56	17.0
1	20	541.36	575.56	17.1
2	50	573.96	611.36	18.7
3	100	675.96	723.96	24.0
4	130	839.16	905.96	33.4
5	150	1027.96	1119.16	45.6
6	140	1201.96	1317.96	58.0
7	110	1318.36	1451.96	66.8
8	90	1375.76	1518.36	71.3
9	70	1390.76	1535.76	72.5
10	50	1369.16	1510.76	70.8
11	30	1315.96	1449.16	66.6
12	20	1243.96	1365.96	61.0
13	17	1169.76	1280.96	55.6
14	17	1102.36	1203.76	50.7
15	17	1043.16	1136.36	46.6
16	17	990.96	1077.16	43.1
17	17	944.96	1024.16	40.0
18	17	904.96	978.96	37.4
19	17	867.76	938.16	35.2
20	17	835.56	901.76	33.1
21	17	806.96	869.56	31.3
22	17	781.16	840.96	29.9
23	17	757.96	815.16	28.6
24	17	737.56	791.96	27.4

season in order to receive and attenuate the incoming floods. Flood-control operations also require that the reservoir releases be kept below a certain maximum, usually taken as the flow corresponding to bank-full stage. Irrigation requirements may vary from month to month depending on the consumptive needs and crop patterns. For navigation purposes, outflow should be a nearly constant value that will ensure a minimum draft downstream of the reservoir.

Reservoir operational rules are designed to take into account the various water demands. These are often conflicting and, therefore, compromises must be reached. Multipurpose reservoirs allocate reservoir volumes to the different uses. In this way, operational rules may be developed to take into account the requirements of each use. In general, outflow from a reservoir with gated outlets is determined by prescribed operational policies. In turn, the latter are based on the current level of storage, incoming flow, and downstream flow requirements.

The differential equation of storage can be used to route flows through reservoirs with controlled outflow. In general, the outflow can be either (1) uncontrolled (ungated), (2) controlled (gated), or (3) a combination of controlled and uncontrolled. The discretized equation, including controlled outflow, is

$$\frac{I_1 + I_2}{2} - \frac{O_1 + O_2}{2} - \bar{O}_r = \frac{S_2 - S_1}{\Delta t} \tag{8-23}$$

in which $\bar{O}_r$ is the mean regulated outflow during the time interval Δt. Equation 8-23 can be expressed in storage indication form:

$$\frac{2S_2}{\Delta t} + O_2 = I_1 + I_2 + \frac{2S_1}{\Delta t} - O_1 - 2\bar{O}_r \tag{8-24}$$

With $\bar{O}_r$ known, the solution proceeds in the same way as with the uncontrolled outflow case. In the case where all the outflow is controlled, Eq. 8-23 reduces to:

$$S_2 = S_1 + \frac{\Delta t}{2}(I_1 + I_2) - (\Delta t)\bar{O}_r \tag{8-25}$$

by which the storage volume can be updated based on average inflows and mean regulated outflow. Other requirements, such as estimates of reservoir evaporation where warranted (i.e., in semiarid and arid regions) may be implemented to properly account for the storage volumes.

Rating of Gated Spillways

A typical rating of a gated spillway is shown in Fig. 8-5 [4]. Outflow (abscissas) is a function of reservoir pool level (ordinates) and gate opening. Each curve represents a different gate opening. Also shown is the spillway rating when all gates are fully open.

QUESTIONS

1. What is reservoir routing? What is a linear reservoir?
2. What is the differential equation of storage? What principle is it based on?
3. What is the xt plane? Above what value of the storage constant will one of the routing coefficients be negative?
4. Explain why in reservoir routing with uncontrolled outflow, the peak outflow occurs when inflow and outflow coincide.
5. What is the storage indication quantity? What is an appropriate value of time interval to choose in reservoir routing?
6. What is surcharge storage?
7. Name three applications of reservoir routing.

PROBLEMS

8-1. Route the following inflow hydrograph through a linear reservoir:

Time (h)	0	1	2	3	4	5	6	7	8	9	10
Inflow (m^3/s)	0	10	20	30	40	50	40	30	20	10	0

Assume baseflow $= 0$ m^3/s, $K = 3$ h, $\Delta t = 1$ h.

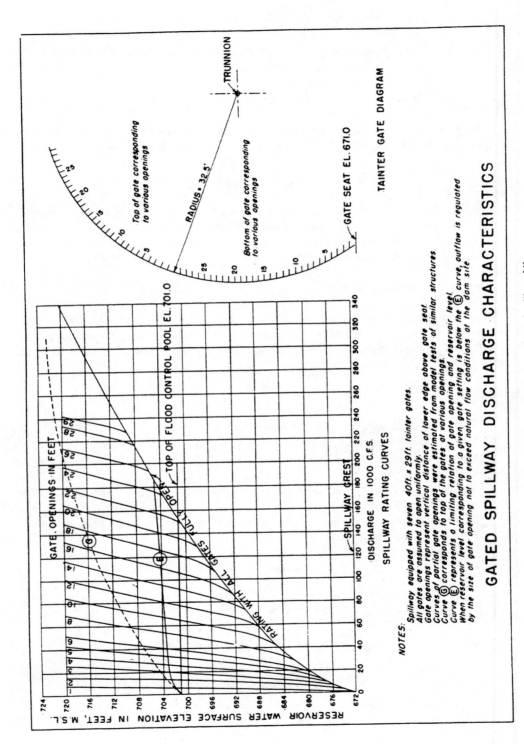

Figure 8-5 Example of rating of gated spillway [4].

NOTES: Spillway equipped with seven 40 ft. x 29 ft. tainter gates.
All gates are assumed to open uniformly.
Gate openings represent vertical distance of lower edge above gate seat.
Curves of partial gate openings were estimated from model tests of similar structures.
Curve ⓖ corresponds to top of the gates at various openings.
Curve ⓔ represents a limiting relation of gate opening and reservoir level.
When reservoir level corresponding to a given gate setting is below the ⓔ curve, outflow is regulated by the size of gate opening not to exceed natural flow conditions at the dam site.

GATED SPILLWAY DISCHARGE CHARACTERISTICS

8-2. Route the following triangular inflow hydrograph through a linear reservoir: peak inflow = 120 m³/s, baseflow = 0 m³/s, time-to-peak = 6 h, time base = 16 h, storage constant $K = 2$ h, and time interval $\Delta t = 1$ h.

8-3. Route the following inflow hydrograph through a linear reservoir.

Time (h)	0	1	2	3	4	5	6	7	8	9	10	11	12
Inflow (m³/s)	10	20	50	80	90	100	90	60	50	40	30	20	10

Assume baseflow = 10 m³/sec, $K = 4$ h, $\Delta t = 1$ h.

8-4. Develop a computer program to route a triangular inflow hydrograph through a linear reservoir. Inputs to the program are the following: peak inflow, baseflow, time-to-peak, time base, storage constant, and time interval. Test your program using Problem 8-2.

8-5. Use the computer program developed in Problem 8-4 to route the following inflow hydrograph: peak inflow = 750 m³/s, baseflow = 50 m³/s, time-to-peak = 3 h, time base = 8 h, storage constant = 1.5 h, time interval = 0.5 h.

8-6. Develop a computer program to route an inflow hydrograph of arbitrary shape through a linear reservoir. Inputs to the program are the following: inflow hydrograph ordinates, baseflow, reservoir storage constant, and time interval. Test your program using Problem 8-3.

8-7. Use the computer program developed in Problem 8-6 to study the sensitivity of the outflow hydrograph to the chosen value of storage constant K. Use the following inflow hydrograph:

Time (h)	0	1	2	3	4	5	6	7	8	9	10	11	12
Inflow (m³/s)	0	10	30	50	80	100	90	60	40	30	20	10	0

Assume baseflow = 0 m³/s and $\Delta t = 1$ h. Report calculated peak outflow and time-to-peak for (a) $K = 1$ h, (b) $K = 2$ h, (c) $K = 3$ h, and (d) $K = 4$ h.

8-8. Solve Problem 8-1 by the storage indication method.

8-9. Solve Problem 8-2 by the storage indication method.

8-10. Solve Problem 8-3 by the storage indication method.

8-11. Use the data of Example 8-3 in the text to test program EH800 included in Appendix D. Then route the same inflow hydrograph through the same reservoir and spillway, but change the rating exponent to $y = 1.6$. Report peak outflow, time-to-peak, and maximum pool elevation.

8-12. Use program EH800 (Appendix D) to solve the following reservoir routing problem: emergency spillway width = 15 m, rating coefficient = 1.952, rating exponent = 1.55, emergency spillway crest elevation = 730 m, dam crest elevation = 735 m, initial pool elevation = 730.5 m, baseflow = 10 m³/s. At spillway crest elevation, the reservoir storage is 3 hm³, increasing linearly to 4 hm³ at dam crest elevation. The inflow hydrograph is the following:

Time (h)	0	1	2	3	4	5	6	7	8	9	10	11	12
Inflow (m³/s)	10	30	70	150	210	250	170	110	70	50	30	20	10

Set Δt equal to: (a) 1 h; and (b) 0.5 h. Report peak outflow, time-to-peak, and maximum pool elevation. Explain any difference in results between the runs for $\Delta t = 1$ h and $\Delta t = 0.5$ h.

8-13. Using the data of Problem 8-12, modify the volumetric characteristics of the reservoir to the following: storage at spillway crest elevation, 6 hm³; storage at dam crest elevation, 8 hm³. Run EH800 using $\Delta t = 0.5$ h. Compare with the results of Problem 8-12, explaining the differences.

8-14. Use program EH800 (Appendix D) to solve the following reservoir routing problem: emergency spillway width = 100 ft, rating coefficient = 3.0, rating exponent = 1.55, emergency spillway crest elevation = 1572 ft, dam crest elevation = 1585 ft, initial pool elevation = 1573 ft, baseflow = 300 ft^3/s. At spillway crest elevation, the reservoir storage is 2700 ac-ft, increasing linearly to 4000 ac-ft at dam crest elevation. The inflow hydrograph is the following:

Time (h)	0	1	2	3	4	5	6
Inflow (ft^3/s)	300	800	2000	6000	8000	9000	7000

Time (h)		7	8	9	10	11	12
Inflow (ft^3/s)		5000	3000	2000	1000	500	300

Assume $\Delta t = 1$ h. Report peak outflow, time-to-peak, and maximum pool elevation.

8-15. Determine the actual freeboard for the following dam, reservoir, and flood conditions: dam crest elevation = 125 m; emergency spillway crest elevation = 120 m; coefficient of spillway rating = 1.7; exponent of spillway rating = 1.5; width of emergency spillway (rectangular cross section) = 18 m.
Elevation-storage relation:

Elevation (m)	120	121	122	123	124	125
Storage (hm^3)	3.00	3.05	3.15	3.35	3.75	4.25

Inflow hydrograph to reservoir:

Time (h)	0	1	2	3	4	5	6	7	8	9	10
Inflow (m^3/s)	0	10	15	30	55	85	105	125	150	135	110

Time (h)	11	12	13	14	15	16	17	18	19	20	21
Inflow (m^3/s)	95	72	55	38	29	14	9	7	2	1	0

Assume the initial reservoir pool level at spillway crest. Use program EH800 included in Appendix D.

8-16. Design the emergency spillway width (rectangular cross section) for the following dam, reservoir, and flood conditions: dam crest elevation = 483 m; emergency spillway crest elevation = 475 m; coefficient of spillway rating = 1.7; exponent of spillway rating = 1.5. Elevation-storage relation:

Elevation (m)	475	477	479	481	483
Storage (hm^3)	5.1	5.3	5.6	6.4	7.6

Inflow hydrograph to reservoir:

Time (h)	0	1	2	3	4	5	6	7	8	9	10	11	12
Inflow (m^3/s)	0	10	30	50	90	150	250	350	280	210	190	170	130

Time (h)	13	14	15	16	17	18	19	20	21	22	23	24
Inflow (m^3/s)	100	90	75	50	40	30	15	10	5	2	1	0

Assume design freeboard = 3 m and initial reservoir pool level at spillway crest. Use program EH800 included in Appendix D.

REFERENCES

1. Chow, V. T. (1964). *Handbook of Applied Hydrology.* New York: McGraw-Hill.

2. Garrison, J. M., J. Granju, and J. T. Price. (1969). "Unsteady Flow Simulation in Rivers and Reservoirs," *Journal of the Hydraulics Division,* ASCE, Vol. 95, No. HY9, September, pp. 1559–1576.

3. Laurenson, E. M. (1961). "Hydrograph Synthesis by Runoff Routing," *Technical Report No.* 66, Water Research Laboratory, University of New South Wales, Kensington, New South Wales, Australia.

4. U.S. Army Corps of Engineers. (1959). "Reservoir Regulation," EM 1110-2-3600, Engineering and Design, Office of the Chief of Engineers, Washington, D.C., May 25, with changes of 26 December 1962.

STREAM CHANNEL ROUTING

Stream channel routing uses mathematical relations to calculate outflow from a stream channel once inflow, lateral contributions, and channel characteristics are known.

Stream channel routing usually implies open channel flow conditions, although there are exceptions, such as storm sewer flow, for which mixed open channel-closed conduit flow conditions may prevail. In this chapter, stream channel routing refers to unsteady flow calculations in streams and rivers. Channel *reach* refers to a specific length of stream channel possessing certain translation and storage properties. The hydrograph at the upstream end of the reach is the inflow hydrograph; the hydrograph at the downstream end is the outflow hydrograph. Lateral contributions consist of point tributary inflows and/or distributed inflows (i.e. interflow and groundwater flow).

The terms stream channel routing and *flood routing* are often used interchangeably. This is attributed to the fact that most stream channel-routing applications are in flood flow analysis, flood control design, or flood forecasting.

Two general approaches to stream channel routing are recognized: (1) hydrologic and (2) hydraulic. As in the case of reservoir routing, hydrologic stream channel routing is based on the storage concept. Conversely, hydraulic channel routing is based on the principles of mass and momentum conservation. Hydraulic routing techniques are of three types: (1) kinematic wave, (2) diffusion wave, and (3) dynamic wave. The dynamic wave is the most complete model of unsteady open channel flow. Kinematic and diffusion waves are convenient and practical approximations to the dynamic wave.

An alternate approach to hydrologic and hydraulic routing has emerged in recent years. This approach is similar in nature to the hydrologic routing methods yet contains sufficient physical information to compare favorably with the more complex hydraulic routing techniques. This *hybrid* approach is the basis of the Muskingum-Cunge method of flood routing.

At the outset of the study of stream channel routing, it is necessary to introduce a few basic modeling concepts. A typical hydrologic model consists of system, input, and output. In surface water hydrology, the system is usually a catchment, a reservoir, or a stream channel. In the case of a catchment, the input is a storm hyetograph. For reservoirs and stream channels, the input is an inflow hydrograph. For all three cases, catchments, reservoirs, and channels, the output is an outflow hydrograph.

In general, modeling problems are classified into three types: (1) prediction, (2) calibration, and (3) inversion. In the prediction problem, input and system are known and described by properties or parameters, and the task is to calculate the output based on the knowledge of system and input. For instance, with known inflow hydrograph, lateral contributions, and channel reach parameters, the outflow hydrograph from a stream channel can be computed using routing techniques.

In the calibration problem, input and output are known, and the objective is to determine the properties or parameters describing the system. In the case of a stream channel, with known upstream inflow, lateral contributions, and outflow hydrograph, the routing parameters are calculated by a calibration procedure.

The inversion problem is the third type of modeling problem. In this case, system and output are known, and the task is to calculate the inflow or inflows. This is accomplished by reversing the routing process in a technique known as *inverse* channel routing. For instance, with known upstream inflow, outflow, and channel reach parameters, the lateral contributions can be calculated by inverse routing.

The prediction problem is the more common type of modeling application. However, a calibration is usually required in advance of the prediction. Model verification is the process of testing the model with actual data to establish its predictive accuracy. To calibrate and verify a model, it is usually necessary to assemble two different data sets. The first set is used in model calibration, and the second set is used in model verification. A close agreement between calculated and measured data is an indication that the model has been verified. A detailed discussion of these subjects is given in Chapter 13.

This chapter is divided into five sections. Section 9.1 describes the Muskingum method, the most widely used method of hydrologic stream channel routing. Sections 9.2 and 9.3 discuss simplified hydraulic routing techniques: kinematic and diffusion waves, respectively. Section 9.4 describes the Muskingum-Cunge method. Section 9.5 introduces the subject of dynamic wave routing, the most complete hydraulic routing technique.

9.1 MUSKINGUM METHOD

The Muskingum method of flood routing was developed in the 1930s in connection with the design of flood protection schemes in the Muskingum River Basin, Ohio [11].

It is the most widely used method of hydrologic stream channel routing, with numerous applications in the United States and throughout the world.

The Muskingum method is based on the differential equation of storage, Eq. 8-4, reproduced here:

$$I - O = \frac{dS}{dt} \tag{8-4}$$

In an ideal channel, storage is a function of inflow and outflow. This is in constrast with ideal reservoirs, in which storage is solely a function of outflow (see Eqs. 8-5 to 8-7). In the Muskingum method, storage is a linear function of inflow and outflow:

$$S = K[XI + (1 - X)O] \tag{9-1}$$

in which S = storage volume; I = inflow; O = outflow; K = a time constant or storage coefficient; and X = a dimensionless weighting factor. With inflow and outflow in cubic meters per second, K in hours, storage volume is in (cubic meters per second)-hour. Alternatively, K could be expressed in seconds, in which case storage volume is in cubic meters.

Equation 9-1 was developed in 1938 [11] and has been widely used since then. It is essentially a generalization of the linear reservoir concept (Eq. 8-7). In fact, for $X = 0$, Eq. 9-1 reduces to Eq. 8-7. In other words, linear reservoir routing is a special case of Muskingum channel routing for which $X = 0$.

To derive the Muskingum routing equation, Eq. 8-4 is discretized on the xt plane (Fig. 8-2), to yield Eq. 8-13, repeated here:

$$\frac{I_1 + I_2}{2} - \frac{O_1 + O_2}{2} = \frac{S_2 - S_1}{\Delta t} \tag{8-13}$$

Equation 9-1 is expressed at time levels 1 and 2:

$$S_1 = K[XI_1 + (1 - X)O_1] \tag{9-2}$$

$$S_2 = K[XI_2 + (1 - X)O_2] \tag{9-3}$$

Substituting Eqs. 9-2 to 9-3 into Eq. 8-13 and solving for O_2 yields Eq. 8-15, repeated here:

$$O_2 = C_0 I_2 + C_1 I_1 + C_2 O_1 \tag{8-15}$$

in which C_0, C_1 and C_2 are routing coefficients defined in terms of Δt, K, and X as follows:

$$C_0 = \frac{(\Delta t/K) - 2X}{2(1 - X) + (\Delta t/K)} \tag{9-4}$$

$$C_1 = \frac{(\Delta t/K) + 2X}{2(1 - X) + (\Delta t/K)} \tag{9-5}$$

$$C_2 = \frac{2(1 - X) - (\Delta t/K)}{2(1 - X) + (\Delta t/K)} \tag{9-6}$$

Since $(C_0 + C_1 + C_2) = 1$, the routing coefficients can be interpreted as weighting coefficients. For $X = 0$, Eqs. 9-4, 9-5, and 9-6 reduce to Eqs. 8-16, 8-17, and 8-18, respectively.

Given an inflow hydrograph, an initial flow condition, a chosen time interval Δt, and routing parameters K and X, the routing coefficients can be calculated with Eqs. 9-4 to 9-6 and the outflow hydrograph, with Eq. 8-15. The routing parameters K and X are related to flow and channel characteristics, K being interpreted as the *travel time* of the flood wave from upstream end to downstream end of the channel reach. Therefore, K accounts for the translation (or concentration) portion of the routing (Fig. 9-1).

The parameter X accounts for the storage portion of the routing. For a given flood event, there is a value of X for which the storage in the calculated outflow hydrograph matches that of the measured outflow hydrograph. The effect of storage is to reduce the peak flow and spread the hydrograph in time (Fig. 9-1). Therefore, it is often used interchangeably with the terms *diffusion* and peak attenuation.

The routing parameter K is a function of channel reach length and flood wave speed; parameter X is a function of the flow and channel characteristics that cause runoff diffusion. In the Muskingum method, X is interpreted as a weighting factor and restricted in the range 0.0 to 0.5. Values of X greater than 0.5 produce hydrograph amplification (i.e., negative diffusion), which does not correspond with reality. With $K = \Delta t$ and $X = 0.5$, flow conditions are such that the outflow hydrograph retains the same shape as the inflow hydrograph, but it is translated downstream a time equal to K. For $X = 0$, Muskingum routing reduces to linear reservoir routing (Section 8.2).

In the Muskingum method, parameters K and X are determined by calibration using streamflow records. Simultaneous inflow-outflow discharge measurements for a given channel reach are coupled with a trial-and-error procedure, leading to the deter-

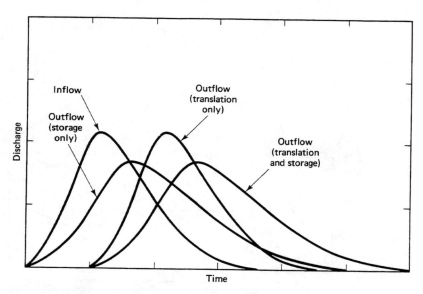

Figure 9-1 Translation and storage processes in stream channel routing.

mination of K and X (see Example 9-2). The procedure is time-consuming and lacks predictive capability. Values of K and X determined in this way are valid only for the given reach and flood event used in the calibration. Extrapolation to other reaches or to other flood events (of different magnitude) within the same reach is usually unwarranted.

When sufficient data are available, a calibration can be performed for several flood events, each of different magnitude, to cover a wide range of flood levels. In this way, the variation of K and X as a function of flood level can be ascertained. In practice, K is more sensitive to flood level than X. A sketch of the variation of K with stage and discharge is shown in Fig. 9-2.

Example 9-1.

An inflow hydrograph to a channel reach is shown in Col. 2 of Table 9-1. Assume baseflow is 352 m³/s. Using the Muskingum method, route this hydrograph through a channel reach with $K = 2$ d and $X = 0.1$ to calculate an outflow hydrograph.

First, it is necessary to select a time interval Δt. In this case, it is convenient to choose $\Delta t = 1$ d. As with reservoir routing, the ratio of time-to-peak to time interval ($t_p/\Delta t$) should be greater than or equal to 5. In addition, the chosen time interval should be such that the routing coefficients remain positive. With $\Delta t = 1$ d, $K = 2$ d, and $X = 0.1$, the routing coefficients (Eqs. 9-4 to 9-6) are: $C_0 = 0.1304$; $C_1 = 0.3044$; and $C_2 = 0.5652$. It is verified that $C_0 + C_1 + C_2 = 1$. The routing calculations are shown in Table 9-1. Column 1 shows the time in days, and Col. 2 shows the inflow hydrograph ordinates in cubic meters per second. Columns 3-5 show the partial flows. Following Eq. 8-15, Cols. 3-5 are summed to obtain Col. 6, the outflow hydrograph ordinates in cubic meters per second. To explain the procedure briefly, the outflow at the start (day 0) is assumed to be equal to the inflow at the start: 352 m³/s. The inflow at day 1 multiplied by C_0 is entered in Col. 3, day 1: 76.6 m³/s. The inflow at day 0 multiplied by C_1 is entered in Col. 4, day

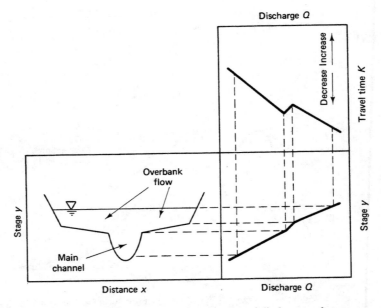

Figure 9-2 Sketch of travel time as a function of discharge and stage.

Stream Channel Routing   Chap. 9

TABLE 9-1 CHANNEL ROUTING BY THE MUSKINGUM METHOD: EXAMPLE 9-1

(1)	(2)	(3)	(4)	(5)	(6)
		Partial Flows (m³/s)			
Time (d)	Inflow (m³/s)	$C_0 I_2$	$C_1 I_1$	$C_2 O_1$	Outflow (m³/s)
0	352.0	—	—	—	352.0
1	587.0	76.6	107.1	199.0	382.7
2	1353.0	176.5	178.6	216.3	571.4
3	2725.0	355.4	411.8	323.0	1090.2
4	4408.5	575.0	829.4	616.2	2020.6
5	5987.0	780.9	1341.7	1142.1	3264.7
6	6704.0	874.4	1822.1	1845.3	4541.8
7	6951.0	906.7	2040.3	2567.1	5514.1
8	6839.0	892.0	2115.5	3116.7	6124.2
9	6207.0	809.6	2081.5	3461.5	6352.6
10	5346.0	697.3	1889.1	3590.6	6177.0
11	4560.0	594.8	1627.0	3491.4	5713.2
12	3861.5	503.7	1387.8	3229.2	5120.7
13	3007.0	392.2	1175.2	2894.3	4461.7
14	2357.5	307.5	915.2	2521.8	3744.5
15	1779.0	232.0	717.5	2116.5	3066.0
16	1405.0	183.3	541.4	1733.0	2457.7
17	1123.0	146.5	427.6	1389.1	1963.2
18	952.5	124.2	341.8	1109.6	1575.6
19	730.0	95.2	289.9	890.6	1275.7
20	605.0	78.9	222.2	721.0	1022.1
21	514.0	67.1	184.1	577.7	828.9
22	422.0	55.1	156.4	468.5	680.0
23	352.0	45.9	128.4	384.4	558.7
24	352.0	45.9	107.1	315.8	468.8
25	352.0	45.9	107.1	265.0	418.0

1: 107.1 m³/s. The outflow at day 0 multiplied by C_2 is entered in Col. 5, day 1: 199 m³/s. Columns 3–5 of day 1 are summed to obtain Col. 6 of day 1: $76.6 + 107.1 + 199.0 = 382.7$ m³/s. The calculations proceed in a recursive manner until all outflows in Col. 6 have been calculated. Inflow and outflow hydrographs are plotted in Fig. 9-3. The outflow peak is 6352.6 m³/s, which shows that the inflow peak, 6951 m³/s, has attenuated to about 91% of its initial value. The peak outflow occurs at day 9, 2 d after the peak inflow, which occurs at day 7. The time elapsed between the occurrence of peak inflow and peak outflow is generally equal to K, the travel time.

Unlike reservoir routing, stream channel–routing calculations exhibit a definite (time) lag between inflow and outflow. Furthermore, in the general case ($X \neq 0$), maximum outflow does not occur at the time when inflow and outflow coincide.

Example 9-1 has illustrated the predictive stage of the Muskingum method, in which the routing parameters are known in advance of the routing. If the parameters are not known, it is first necessary to perform a calibration. The trial-and-error procedure to calibrate the routing parameters is illustrated by the following example.

Example 9-2.

Use the outflow hydrograph calculated in the previous example together with the given

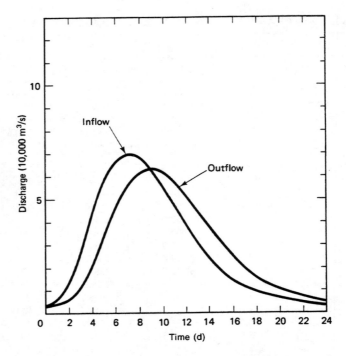

Figure 9-3 Stream channel routing by Muskingum method: Example 9-1.

inflow hydrograph to calibrate the Muskingum method, that is, to find the routing parameters K and X.

The procedure is summarized in Table 9-2. Column 1 shows the time in days. Cols. 2 shows the inflow hydrograph in cubic meters per second; Col. 3 shows the outflow hydrograph in cubic meters per second; Column 4 shows the channel storage in (cubic meters per second)-days. Channel storage at the start is assumed to be 0, and this value is entered in Col. 4, day 0. Channel storage is calculated by solving Eq. 8-13 for S_2:

$$S_2 = S_1 + (\Delta t/2)(I_1 + I_2 - O_1 - O_2) \tag{9-7}$$

Several values of X are tried, within the range 0.0 to 0.5, for example, 0.1, 0.2 and 0.3. For each trial value of X, the weighted flows $[XI + (1 - X)O]$ are calculated, as shown in Cols. 5-7. Each of the weighted flows is plotted against channel storage (Col. 4), as shown in Fig. 9-4. The value of X for which the storage versus weighted flow data plots closest to a line is taken as the correct value of X. In this case, Fig. 9-4(a): $X = 0.1$ is chosen. Following Eq. 9-1, the value of K is obtained from Fig. 9-4(a) by calculating the slope of the storage vs weighted outflow curve. In this case, the value of $K = [2000\,(\text{m}^3/\text{s})\text{-d}]/(1000\,\text{m}^3/\text{s}) = 2$ d. Thus, it is shown that $K = 2$ days and $X = 0.1$ are the Muskingum routing parameters for the given inflow and outflow hydrographs.

The estimation of routing parameters is crucial to the application of the Muskingum method. The parameters are not constant, tending to vary with flow rate. If the routing parameters can be related to flow and channel characteristics, the need for trial-and-error calibration would be eliminated. Parameter K could be related to

TABLE 9-2 CALIBRATION OF MUSKINGUM ROUTING PARAMETERS: EXAMPLE 9-2

(1)	(2)	(3)	(4)	(5)	(6)	(7)
				Weighted Flow (m³/s)		
Time (d)	Inflow (m³/s)	Outflow (m³/s)	Storage (m³/s)-d	$X = 0.1$	$X = 0.2$	$X = 0.3$
0	352.0	352.0	0	—	—	—
1	587.0	382.7	102.2	403.0	423.5	443.9
2	1,353.0	571.4	595.2	649.6	727.7	805.9
3	2,725.0	1,090.2	1,803.4	1,253.7	1,417.2	1,580.6
4	4,408.5	2,020.6	3,814.7	2,259.4	2,498.2	2,737.0
5	5,987.0	3,264.7	6,369.8	3,536.9	3,809.2	4,081.4
6	6,704.0	4,541.8	8,812.1	4,758.0	4,974.2	5,190.5
7	6,951.0	5,514.1	10,611.6	5,657.8	5,801.5	5,945.2
8	6,839.0	6,124.2	11,687.5	6,195.7	6,267.2	6,338.6
9	6,207.0	6,352.6	11,972.1	6,338.0	6.323.5	6,308.9
10	5,346.0	6,177.0	11,483.8	6,093.9	6,010.8	5,927.7
11	4,560.0	5,713.2	10,491.7	5,597.9	5,482.6	5,367.2
12	3,861.5	5,120.7	9,285.5	4,994.8	4,868.9	4,742.9
13	3,007.0	4,461.7	7,928.5	4,316.2	4,170.8	4,025.3
14	2,357.5	3,744.5	6,507.7	3,605.8	3,467.1	3,328.4
15	1,779.0	3,066.0	5,170.7	2,937.3	2,808.6	2,679.9
16	1,405.0	2,457.7	4,000.8	2,352.4	2,247.2	2,141.9
17	1,123.0	1,963.2	3,054.4	1,879.2	1,795.2	1,711.1
18	952.5	1,575.6	2,322.7	1,513.4	1,451.1	1,388.7
19	730.0	1,275.7	1,738.2	1,221.1	1,166.6	1,112.0
20	605.0	1,022.1	1,256.8	980.4	938.7	897.0
21	514.0	828.9	890.8	797.4	765.9	734.4
22	422.0	680.0	604.4	654.2	628.4	602.6
23	352.0	558.7	372.0	537.9	517.3	496.6
24	352.0	468.8	210.3	457.1	445.4	433.8
25	352.0	418.0	118.9	411.4	404.8	398.2

reach length and flood wave velocity, whereas X could be related to the diffusivity characteristics of flow and channel. These propositions are the basis of the Muskingum-Cunge method (Section 9.4).

9.2 KINEMATIC WAVES

Three types of unsteady open channel flow waves are commonly used in engineering hydrology: (1) kinematic, (2) diffusion, and (3) dynamic waves. Kinematic waves are the simplest type of wave, and dynamic waves are the most complex. Diffusion waves lie somewhere in between kinematic and dynamic waves. Kinematic waves are discussed in this section, and diffusion waves are discussed in Section 9.3. An introduction to dynamic waves is given in Section 9.5.

Kinematic Wave Equation

The derivation of the kinematic wave equation is based on the principle of mass conservation within a control volume. This principle states that the difference between

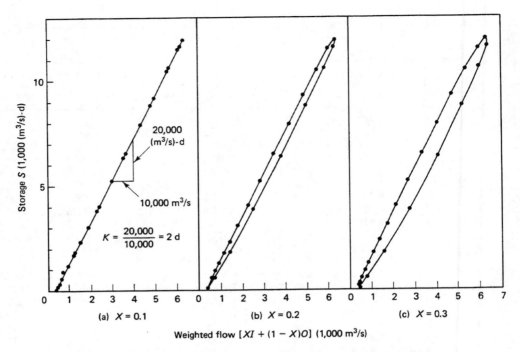

Figure 9-4 Calibration of Muskingum routing parameters: Example 9-2.

outflow and inflow within one time interval is balanced by a corresponding change in volume. In terms of finite intervals (i.e., finite differences) it is:

$$(Q_2 - Q_1)\Delta t + (A_2 - A_1)\Delta x = 0 \tag{9-8}$$

in which Q = flow; A = flow area; Δt = time interval; and Δx = space interval. In differential form, Eq. 9-8 can be written as:

$$\frac{\partial Q}{\partial x} + \frac{\partial A}{\partial t} = 0 \tag{9-9}$$

which is the equation of conservation of mass, or equation of continuity.

The equation of conservation of momentum (Eq. 4-22) contains local inertia, convective inertia, pressure gradient (due to flow depth gradient), friction (friction slope), gravity (bed slope), and a momentum source term (Section 4.2). In deriving the kinematic wave equation, a statement of uniform flow is used in lieu of conservation of momentum. Since uniform flow is strictly a balance of friction and gravity, it follows that local and convective inertia, pressure gradient, and momentum source terms are excluded from the formulation of kinematic waves. In other words, a kinematic wave is a simplified wave that does not include these terms or processes. As shown later in this section, this simplification imposes limits to the applicability of kinematic waves.

Uniform flow in open channels is described by the Manning or Chezy formulas (Section 2.4). The Manning equation is:

$$Q = \frac{1}{n} AR^{2/3}S_f^{1/2} \tag{9-10}$$

in which R is the hydraulic radius in meters, S_f is the friction slope in meters per meter, and n is the Manning friction coefficient.

The Chezy equation is:

$$Q = CAR^{1/2}S_f^{1/2} \qquad (9\text{-}11)$$

in which C = Chezy coefficient. Notice that in unsteady flow, friction slope is used in Eqs. 9-10 and 9-11 in lieu of channel slope.

The hydraulic radius is $R = A/P$, in which P is the wetted perimeter. Substituting this into Eq. 9-10, leads to:

$$Q = \frac{1}{n} \frac{S_f^{1/2}}{P^{2/3}} A^{5/3} \qquad (9\text{-}12)$$

Assume for the sake of simplicity that n, S_f, and P are constant. This may be the case of a wide channel in which P can be assumed to be essentially independent of A. Equation 9-12 can then be written as:

$$Q = \alpha A^\beta \qquad (9\text{-}13)$$

in which α and β are parameters of the discharge-area rating (see rating curve, Section 2.4), defined as follows:

$$\alpha = \frac{1}{n} \frac{S_f^{1/2}}{P^{2/3}} \qquad (9\text{-}14)$$

$$\beta = \frac{5}{3} \qquad (9\text{-}15)$$

In Eq. 9-13, differentiating Q with respect to A leads to

$$\frac{dQ}{dA} = \beta \frac{Q}{A} = \beta V \qquad (9\text{-}16)$$

in which V is the mean flow velocity.

Multiplying Eqs. 9-9 and 9-16 and applying the chain rule, the kinematic wave equation is obtained:

$$\frac{\partial Q}{\partial t} + \left(\frac{dQ}{dA}\right)\frac{\partial Q}{\partial x} = 0 \qquad (9\text{-}17)$$

or, alternatively

$$\frac{\partial Q}{\partial t} + (\beta V)\frac{\partial Q}{\partial x} = 0 \qquad (9\text{-}18)$$

Equation 9-17 (or 9-18) describes the movement of waves which are kinematic in nature. These are referred to as kinematic waves, i.e., waves for which inertia and pressure (flow depth) gradient have been neglected. Equation 9-17 is a first order partial differential equation. Therefore, kinematic waves travel with wave celerity dQ/dA (or βV) and do not attenuate. Wave attenuation can only be described by a second-order partial differential equation.

The absence of wave attenuation can be further explained by resorting to a mathematical argument. Since dQ/dA is the celerity of the unsteady (i.e., wavelike) Q, it can be replaced by dx/dt. Therefore, in Eq. 9-17:

$$\frac{\partial Q}{\partial t} + \left(\frac{dx}{dt}\right)\frac{\partial Q}{\partial x} = 0 \qquad (9\text{-}19)$$

which is equal to the total derivative dQ/dt. Since the right side of Eq. 9-19 is zero, it follows that Q remains constant in time for waves traveling with celerity dQ/dA.

Discretization of Kinematic Wave Equation

Equation 9-18 (or 9-17) is a nonlinear first-order partial differential equation describing the change of discharge Q in time and space. It is nonlinear because the wave celerity βV (or dQ/dA) varies with discharge. The nonlinearity, however, is usually mild, and therefore, Eq. 9-18 can also be solved in a linear mode by considering the wave celerity to be constant.

The solution of Eq. 9-18 can be obtained by analytical or numerical methods. The simplest kinematic wave solution is a linear numerical solution. For this purpose, it is necessary to select a numerical scheme with which to discretize Eq. 9-18 on the xt plane (Fig. 9-5). A review of basic concepts of numerical analysis is necessary before discussing numerical schemes.

Order of Accuracy of Numerical Schemes.

The *order of accuracy* of a numerical scheme measures the ability of the scheme to reproduce (i.e., recreate) the terms of the differential equation. In general, the higher the order of accuracy of a scheme, the better it is able to reproduce the terms of the differential equation. *Forward* and *backward* finite differences have first-order accuracy, i.e., discretization errors of first order. *Central* differences have second-order accuracy, with discretization errors of second order. In connection with the numerical solution of Eq. 9-18,

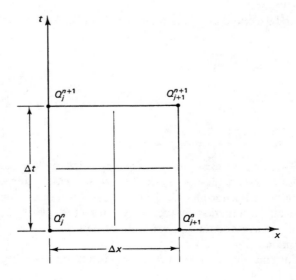

Figure 9-5 Space-time discretization of kinematic wave equation.

first-order schemes create *numerical diffusion* and *numerical dispersion*, while second-order schemes create only numerical dispersion. A third-order scheme creates neither numerical diffusion nor dispersion. Numerical diffusion and/or dispersion are caused by the finite grid size and are not necessarily related to the physical problem.

Second-order-accurate Numerical Scheme. The discretization of Eq. 9-18 following a linear second-order-accurate scheme (i.e., using central differences in space and time) leads to (Fig. 9-5):

$$\frac{\dfrac{Q_{j+1}^{n+1} + Q_j^{n+1}}{2} - \dfrac{Q_{j+1}^n + Q_j^n}{2}}{\Delta t} + \beta V \frac{\dfrac{Q_{j+1}^n + Q_{j+1}^{n+1}}{2} - \dfrac{Q_j^n + Q_j^{n+1}}{2}}{\Delta x} = 0 \quad (9\text{-}20)$$

in which βV has been held constant (linear mode), leading to:

$$Q_{j+1}^{n+1} = C_0 Q_j^{n+1} + C_1 Q_j^n + C_2 Q_{j+1}^n \quad (9\text{-}21)$$

in which

$$C_0 = \frac{C - 1}{1 + C} \quad (9\text{-}22)$$

$$C_1 = 1 \quad (9\text{-}23)$$

$$C_2 = \frac{1 - C}{1 + C} \quad (9\text{-}24)$$

and C is the Courant number, defined as follows:

$$C = \beta V \frac{\Delta t}{\Delta x} \quad (9\text{-}25)$$

Notice that Courant number is the ratio of physical wave celerity βV to *grid celerity* $\Delta x / \Delta t$. The Courant number is a fundamental concept in the numerical solution of *hyperbolic* partial differential equations.

Example 9-3.

Use Eq. 9-21 with the routing coefficients of Eqs. 9-22 to 9-24 (linear kinematic wave numerical solution using central differences in space and time) to route the following triangular flood wave. Consider the following three cases: (1) $V = 1.2$ m/s and $\Delta x = 7200$ m; (2) $V = 1.2$ m/s and $\Delta x = 4800$ m; and (3) $V = 0.8$ m/s and $\Delta x = 4800$ m. Use $\beta = 5/3$, and $\Delta t = 1$ h.

Time (h)	0	1	2	3	4	5	6	7	8	9	10
Inflow (m³/s)	0	30	60	90	120	150	120	90	60	30	0

1. Using Eq. 9-25: $C = 1$. Using Eqs. 9-22 to 9-24: $C_0 = 0$; $C_1 = 1$; $C_2 = 0$. The routing by Eq. 9-21 shown in Table 9-3 depicts the pure translation of the hydrograph a time equal to Δt. In other words, for $\beta V = \Delta x / \Delta t$ (i.e., $C = 1$), the central difference scheme is of third order, and the numerical solution is exactly equal to the analytical solution.
2. Using Eq. 9-25: $C = 1.5$. Using Eqs. 9-22 to 9-24: $C_0 = 0.2$; $C_1 = 1.0$; $C_2 = -0.2$. The routing by Eq. 9-21 shown in Table 9-4 depicts the translation of the hydrograph a time approximately equal to Δt, but it also shows a small amount of numerical dispersion because βV is not equal to $\Delta x / \Delta t$. The dispersion, including the notorious negative out-

TABLE 9-3 KINEMATIC WAVE ROUTING: PURE TRANSLATION
EXAMPLE 9-3, PART 1

(1)	(2)	(3)	(4)	(5)	(6)
		Partial Flows (m³/s)			
Time (h)	Inflow (m³/s)	$C_0 I_2$	$C_1 I_1$	$C_2 O_1$	Outflow (m³/s)
0	0	—	—	—	0
1	30	0	0	0	0
2	60	0	30	0	30
3	90	0	60	0	60
4	120	0	90	0	90
5	150	0	120	0	120
6	120	0	150	0	150
7	90	0	120	0	120
8	60	0	90	0	90
9	30	0	60	0	60
10	0	0	30	0	30
11	0	0	0	0	0

TABLE 9-4 KINEMATIC WAVE ROUTING: TRANSLATION AND DISPERSION
EXAMPLE 9-3, PART 2

(1)	(2)	(3)	(4)	(5)	(6)
		Partial Flows (m³/s)			
Time (h)	Inflow (m³/s)	$C_0 I_2$	$C_1 I_1$	$C_2 O_1$	Outflow (m³/s)
0	0	—	—	—	0
1	30	6	0	0	6
2	60	12	30	−1.20	40.80
3	90	18	60	−8.16	69.84
4	120	24	90	−13.97	100.03
5	150	30	120	−20.91	129.99
6	120	24	150	−26.00	148.00
7	90	18	120	−29.60	108.40
8	60	12	90	−21.68	80.32
9	30	6	60	−16.06	49.94
10	0	0	30	−9.99	20.01
11	0	0	0	−4.00	−4.00
12	0	0	0	0.80	0.80
13	0	0	0	−0.16	−0.16

flows at the trailing end of the hydrograph, are caused by errors associated with the scheme's second-order accuracy.

3. Using Eq. 9-25, $C = 1$. Therefore, the solution is the same as in the first case, exhibiting pure hydrograph translation.

The three cases of Example 9-3 illustrate the properties of kinematic waves. The second-order-accurate scheme has no numerical diffusion. In addition, for Courant number $C = 1$, i.e., the wave celerity βV equal to the grid celerity $\Delta x/\Delta t$, the scheme has no numerical dispersion, with the hydrograph being translated downstream without change in shape. In other words, the numerical solution by Eqs. 9-21 to 9-25 is exact only for Courant number $C = 1$. For any other value of C, the numerical solution exhibits a small but perceptible amount of numerical dispersion.

First-order-accurate Numerical Scheme. The numerical solution of Eq. 9-18 can also be attempted using a first-order-accurate scheme, i.e., one featuring forward or backward finite differences. The discretization of Eq. 9-18 in a linear mode, using backward differences in both space and time yields (Fig. 9-5):

$$\frac{Q_{j+1}^{n+1} - Q_{j+1}^n}{\Delta t} + \beta V \frac{Q_{j+1}^{n+1} - Q_j^{n+1}}{\Delta x} = 0 \tag{9-26}$$

from which

$$Q_{j+1}^{n+1} = C_0 \, Q_j^{n+1} + C_2 \, Q_{j+1}^n \tag{9-27}$$

in which

$$C_0 = \frac{C}{1 + C} \tag{9-28}$$

$$C_2 = \frac{1}{1 + C} \tag{9-29}$$

and C is the Courant number defined by Eq. 9-25.

Example 9-4.

Use Eq. 9-27 with the coefficients calculated by Eq. 9-28 and 9-29 to route the same inflow hydrograph as in the previous example. Use $V = 1.2$ m/s; $\Delta x = 7200$ m; $\beta = \frac{5}{3}$; and $\Delta t = 1$ h.

Using Eq. 9-25, $C = 1$. Therefore, $C_0 = 0.5$, and $C_2 = 0.5$. The routing using Eq. 9-27 is shown in Table 9-5. It is observed that off-centering the derivatives by using backward differences has caused a significant amount of numerical diffusion, with peak outflow of 120.93 m³/s as compared to peak inflow of 150 m³/s. The conclusion is that different schemes for solving Eq. 9-18 lead to different answers, depending on the time and space intervals, Courant number, order of accuracy of the scheme, and associated numerical diffusion and/or dispersion.

Convex Method. The convex method of stream channel routing belongs to the family of linear kinematic wave methods. Until recently (1982), it was part of the SCS TR-20 model for hydrologic simulation (Chapter 13). The routing equation for the convex method is obtained by discretizing Eq. 9-18 in a linear mode using a forward-in-time, backward-in-space finite difference scheme, to yield (Fig. 9-5):

$$\frac{Q_{j+1}^{n+1} - Q_{j+1}^n}{\Delta t} + \beta V \frac{Q_{j+1}^n - Q_j^n}{\Delta x} = 0 \tag{9-30}$$

from which

TABLE 9-5 KINEMATIC WAVE ROUTING: TRANSLATION AND DIFFUSION
EXAMPLE 9-4

(1)	(2)	(3)	(4)	(5)	(6)
		Partial Flows (m³/s)			Outflow
Time (h)	Inflow (m³/s)	$C_0 I_2$	$C_1 I_1$	$C_2 O_1$	(m³/s)
0	0	—	—	—	0
1	30	15	—	0	15.00
2	60	30	—	7.5	37.50
3	90	45	—	18.75	63.75
4	120	60	—	31.87	91.87
5	150	75	—	45.93	120.93
6	120	60	—	60.46	120.46
7	90	45	—	60.23	105.23
8	60	30	—	52.62	82.62
9	30	15	—	41.31	56.31
10	0	0	—	28.15	28.15
11	0	0	—	14.08	14.08
12	0	0	—	7.04	7.04
13	0	0	—	3.52	3.52
14	0	0	—	1.76	1.76
15	0	0	—	0.88	0.88

$$Q_{j+1}^{n+1} = C_1 \, Q_j^n + C_2 \, Q_{j+1}^n \qquad (9\text{-}31)$$

in which

$$C_1 = C \qquad (9\text{-}32)$$

$$C_2 = 1 - C \qquad (9\text{-}33)$$

and C is the Courant number (Eq. 9-25), restricted to values less than or equal to 1 for numerical stability reasons. In the convex method, C is regarded as an empirical routing coefficient. The following example illustrates the application of the convex method.

Example 9-5.

Use Eq. 9-31 (the convex method) to route the same inflow hydrograph as in Example 9-3. Assume $C = \frac{2}{3}$.

The routing coefficients are $C_1 = C = \frac{2}{3}$; and $C_2 = 1 - C = \frac{1}{3}$. The routing is shown in Table 9-6. The convex method leads to a significant amount of diffusion, with peak outflow of 135.06 m³/s as compared to peak inflow of 150 m³/s. The calculated diffusion amount is a function of C, with practical values of C being restricted in the range 0.5 to 0.9. For $C = 1$, the hydrograph is translated with no diffusion or dispersion, as in the first and third parts of Example 9-3. Values of C greater than 1 render the calculation unstable (large negative values of discharge) and are therefore not recommended. It should be noted that the instability of the convex method for C values greater than 1 has a parallel in the instability of the Muskingum method for X values greater than 0.5.

The convex method is relatively simple, but the solution is dependent on the routing parameter C. The latter could be interpreted as a Courant number and related

TABLE 9-6 KINEMATIC WAVE ROUTING: CONVEX METHOD
EXAMPLE 9-5

(1)	(2)	(3)	(4)	(5)	(6)
Time (h)	Inflow (m^3/s)	Partial Flows (m^3/s)			Outflow (m^3/s)
		$C_0 I_2$	$C_1 I_1$	$C_2 O_1$	
0	0	—	—	—	0.00
1	30	—	0	0	0.00
2	60	—	20	0	20.00
3	90	—	40	6.67	46.67
4	120	—	60	15.56	75.56
5	150	—	80	25.19	105.19
6	120	—	100	35.06	135.06
7	90	—	80	45.02	125.02
8	60	—	60	41.67	101.67
9	30	—	40	33.89	73.89
10	0	—	20	24.63	44.63
11	0	—	0	14.88	14.88
12	0	—	0	4.96	4.96
13	0	—	0	1.65	1.65
14	0	—	0	0.55	0.55

to kinematic wave celerity and grid size, as in Eq. 9-25. However, for values of C other than 1, the amount of diffusion introduced in the numerical problem is unrelated to the true diffusion, if any, of the physical problem. Therefore, the convex method (as well as all kinematic wave methods featuring uncontrolled amounts of numerical diffusion) is regarded as a somewhat crude approach to stream channel routing.

Kinematic Wave Celerity

The kinematic wave celerity is dQ/dA, or βV. A value of $\beta = \frac{5}{3}$ was derived for the case of a wide channel (for which the wetted perimeter P is independent of flow area A) governed by Manning friction.

The kinematic wave celerity is also known as the Kleitz-Seddon or Seddon, law [8, 18]. In 1900, Seddon published a paper in which he studied the nature of unsteady flow movement in rivers and concluded that the celerity of long disturbances was equal to dQ/dA. In view of the fact that $dA = T\,dy$, in which T is the channel top width and y is the stage or water surface elevation, the Seddon law can be expressed in practical terms as follows:

$$c = \frac{1}{T} \frac{dQ}{dy} \qquad (9\text{-}34)$$

in which c = kinematic wave celerity.

From Eq. 9-34 it is concluded that kinematic wave celerity is a function of the slope of the discharge-stage rating (Q versus y). This slope is likely to vary with stage; therefore, kinematic wave celerity is not constant but varies with stage and flow level. If $c = \beta V$ is a function of Q, then Eq. 9-18 is a nonlinear equation requiring an iterative solution. Nonlinear kinematic wave solutions account for the variation of kinematic wave celerity with stage and flow level. The simpler linear solutions, as in Examples 9-3 and 9-4, assume a constant value of kinematic wave celerity βV. Notice that there is a striking similarity between the linear kinematic wave solutions and the Muskingum method. This subject is further examined in Section 9.4.

Theoretical β values other than $\frac{5}{3}$ can be obtained for other friction formulations and cross-sectional shapes. For turbulent flow governed by Manning friction, β has an upper limit of $\frac{5}{3}$ but is usually not less than 1. For laminar flow in wide channels, β is equal to 3; for mixed or transitional flow—between laminar and turbulent Manning—it ranges from $\frac{5}{3}$ to 3. For flow in a wide channel described by the Chezy formula, $\beta = \frac{3}{2}$ (Section 4.2). The calculation of β as a function of frictional type and cross-sectional shape is illustrated by the following example.

Example 9-6.

Calculate the β value for a triangular channel (see Fig. 9-6) with Manning friction.

Equation 9-10 is the Manning equation. Substituting $R = A/P$ leads to Eq. 9-12. Since P is a function of A, Eq. 9-12 can be written as follows:

$$Q = K_1 \frac{A^{5/3}}{P^{2/3}} \tag{9-35}$$

in which K_1 is a constant containing n and S_f. The latter have been assumed to be independent of either A or P. For the triangular-shaped channel of Fig. 9-6, the top width is proportional to the flow depth, say $T = Kd$, in which T is the top width, d is the flow depth, and K is a proportionality constant. The flow area is

$$A = K \frac{d^2}{2} \tag{9-36}$$

and the wetted perimeter is

$$P = 2d\left[1 + \frac{K^2}{4}\right]^{1/2} \tag{9-37}$$

Eliminating d from Eqs. 9-36 and 9-37:

$$P = \frac{2(2^{1/2})A^{1/2}}{K^{1/2}}\left[1 + \frac{K^2}{4}\right]^{1/2} \tag{9-38}$$

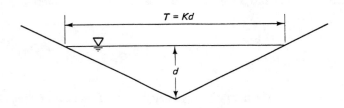

Flow area $A = \dfrac{Kd^2}{2}$

Wetted perimeter $P = 2d\left(1 + \dfrac{K^2}{4}\right)^{1/2}$

Figure 9-6 Properties of triangular channel cross section.

from which

$$P^{2/3} = K_2 A^{1/3} \tag{9-39}$$

in which K_2 is a constant containing K. Substituting Eq. 9-39 into Eq. 9-35 leads to:

$$Q = K_3 A^{4/3} \tag{9-40}$$

in which K_3 is a constant containing K_1 and K_2. From Eq. 9-40:

$$\frac{dQ}{dA} = \frac{4}{3} \frac{Q}{A} \tag{9-41}$$

and the value of β for a triangular channel with Manning friction is $\beta = \frac{4}{3}$.

Kinematic Waves with Lateral Inflow

Practical applications of stream channel routing often require the specification of lateral inflows. The latter could be either concentrated, as in the case of tributary inflow at a point along the channel reach, or distributed along the channel, as with groundwater exfiltration (for effluent streams) or infiltration (for influent streams). As with Eq. 9-9, a mass balance leads to:

$$\frac{\partial Q}{\partial x} + \frac{\partial A}{\partial t} = q_L \tag{9-42}$$

which, unlike Eq. 9-9, includes the source term q_L, the lateral flow per unit channel length. For Q given in cubic meters per second and x in meters, q_L is given in cubic meters per second per meter $[L^2 T^{-1}]$. Multiplying Eq. 9-42 by dQ/dA (or βV), as with Eq. 9-17 (or Eq. 9-18), leads to:

$$\frac{\partial Q}{\partial t} + (\beta V) \frac{\partial Q}{\partial x} = (\beta V) q_L \tag{9-43}$$

which is the kinematic wave equation with lateral inflow (or outflow). For q_L positive, there is lateral inflow (e.g., tributary flow); for q_L negative, there is lateral outflow (e.g., channel transmission losses).

Applicability of Kinematic Waves

The kinematic wave celerity is a fundamental streamflow property. Flood waves which approximate kinematic waves travel with the kinematic wave celerity ($c = \beta V$) and are subject to very little or no attenuation.

In practice, flood waves are kinematic if they are of long duration or travel on a channel of steep slope. Criteria for the applicability of kinematic waves to overland flow [19] (Section 4.2) and stream channel flow [14] have been developed. The stream channel criterion states that in order for a wave to be kinematic, it should satisfy the following dimensionless inequality:

$$\frac{t_r S_o V_o}{d_o} \geq N \tag{9-44}$$

in which t_r is the time-of-rise of the inflow hydrograph, S_o is the bottom slope, V_o is the average velocity, and d_o is the average flow depth. For 95 percent accuracy in one period of translation [14], a value of $N = 85$ is indicated.

Example 9-7.

Use the kinematic wave criterion (Eq. 9-44) to determine whether a flood wave with the following characteristics is a kinematic wave: time-of-rise $t_r = 12$ h; bottom slope $S_o = 0.001$; average velocity $V_o = 2$ m/s; average flow depth $d_o = 2$ m.

For the given channel and flow characteristics, the left side of Eq. 9-44 is equal to 43.2, which is less than 85. For values greater than 85, the wave would be kinematic—therefore, subject to negligible diffusion. Since the value is 43.2, this wave is not kinematic and is likely to experience a significant amount of diffusion. If this wave is routed as a kinematic wave with zero diffusion and dispersion, as in Example 9-3 (Part 1), the peak outflow would be much larger than in reality. If this wave is routed as a kinematic wave with diffusion or dispersion, as in Examples 9-3 (Part 2) and 9-4, it is likely that the amount of numerical diffusion and/or dispersion would be different from the actual amount of physical diffusion. It should be noted that had the bottom slope been $S_o = 0.01$, the left side of Eq. 9-44 would be 432, satisfying the kinematic wave criterion. Therefore, it is concluded that the steeper the channel slope, the more kinematic the flow is.

9.3 DIFFUSION WAVES

In Section 9.1, the Muskingum method was used to calculate unsteady flows in a hydrologic sense. In Section 9.2, the principle of mass conservation was coupled with a uniform flow formula to derive the kinematic wave equation. Solutions to this equation have been widely used in engineering hydrology, particularly for overland flow and other routing applications involving steep slopes or slow-rising hydrographs.

The Muskingum method and linear kinematic wave solutions show striking similarities. Both methods have the same type of routing equation. The Muskingum method, however, can calculate hydrograph diffusion, whereas the kinematic wave can do so only by the introduction of numerical diffusion. The latter is dependent on the grid size and type of numerical scheme.

Kinematic wave theory can be enhanced by allowing a small amount of physical diffusion in its formulation [10]. In this way, an improved type of kinematic wave can be formulated, a kinematic-with-diffusion wave, for short, a diffusion wave. A definite advantage of the diffusion wave is that it includes the diffusion which is present in most natural unsteady open channel flows.

Diffusion Wave Equation

In Section 9.2, the kinematic wave equation was derived by using a statement of steady uniform flow (i.e., friction slope is equal to bottom slope) in lieu of momentum conservation. In deriving the diffusion wave, a statement of steady *nonuniform* flow (i.e., friction slope is equal to water surface slope) is used instead (Fig. 9-7). This leads to

$$Q = \frac{1}{n} A R^{2/3} \left(S_o - \frac{dy}{dx} \right)^{1/2} \tag{9-45}$$

in which the term $S_o - (dy/dx)$ is the water surface slope. The difference between

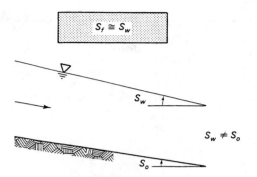

Figure 9-7 Diffusion wave assumption.

kinematic and diffusion waves is in the term dy/dx. From a physical standpoint, the term dy/dx accounts for the natural diffusion processes present in unsteady open channel flow phenomena.

To derive the diffusion wave equation, Eq. 9-45 is expressed in a slightly different form:

$$mQ^2 = S_o - \frac{dy}{dx} \tag{9-46}$$

in which m is the reciprocal of the square of the channel conveyance K, defined as

$$K = \frac{1}{n} AR^{2/3} \tag{9-47}$$

With $dA = T\,dy$, ($T =$ top width), Eq. 9-46 changes to:

$$\left(\frac{1}{T}\right)\frac{dA}{dx} + mQ^2 - S_o = 0 \tag{9-48}$$

Equations 9-9 and 9-48 constitute a set of two partial differential equations describing diffusion waves. These equations can be combined into one equation with Q as dependent variable. However, it is first necessary to linearize the equations around *reference flow* values. For simplicity, a constant top width is assumed (i.e., a wide channel assumption).

The linearization of Eqs. 9-9 and 9-48 is accomplished by small perturbation theory [4]. This procedure, while heuristic, has seemed to work well in a number of applications. The variables Q, A, and m can be expressed in terms of the sum of a reference value (with subscript o) and a small perturbation to the reference value (with superscript $'$): $Q = Q_o + Q'$; $A = A_o + A'$; $m = m_o + m'$. Substituting these into Eqs. 9-9 and 9-48, neglecting squared perturbations, and subtracting the reference flow leads to:

$$\frac{\partial Q'}{\partial x} + \frac{\partial A'}{\partial t} = 0 \tag{9-49}$$

and

$$\frac{1}{T}\frac{\partial A'}{\partial x} + Q_o^2 m' + 2m_o Q_o Q' = 0 \tag{9-50}$$

Differentiating Eq. 9-49 with respect to x and Eq. 9-50 with respect to t gives

$$\frac{\partial^2 Q'}{\partial x^2} + \frac{\partial^2 A'}{\partial x \, \partial t} = 0 \tag{9-51}$$

$$\frac{1}{T} \frac{\partial^2 A'}{\partial x \, \partial t} + Q_o^2 \frac{\partial m'}{\partial t} + 2m_o Q_o \frac{\partial Q'}{\partial t} = 0 \tag{9-52}$$

Using the chain rule and Eq. 9-49 yields

$$\frac{\partial m'}{\partial t} = \frac{\partial m'}{\partial A'} \frac{\partial A'}{\partial t} = -\frac{\partial m'}{\partial A'} \frac{\partial Q'}{\partial x} \tag{9-53}$$

Combining Eq. 9-52 with Eq. 9-53,

$$\frac{1}{T} \frac{\partial^2 A'}{\partial x \, \partial t} - Q_o^2 \frac{\partial m'}{\partial A'} \frac{\partial Q'}{\partial x} + 2m_o Q_o \frac{\partial Q'}{\partial t} = 0 \tag{9-54}$$

Combining Eqs. 9-51 and 9-54 and rearranging gives

$$\frac{\partial Q'}{\partial t} - \left[\frac{Q_o}{2m_o}\right] \frac{\partial m'}{\partial A'} \frac{\partial Q'}{\partial x} = \left[\frac{1}{2Tm_o Q_o}\right] \frac{\partial^2 Q'}{\partial x^2} \tag{9-55}$$

Since by definition: $mQ^2 = S_f$, it follows that

$$\frac{\partial Q'}{\partial m'} = \frac{\partial Q}{\partial m} = -\frac{Q_o}{2m_o} \tag{9-56}$$

and also

$$m_o Q_o = \frac{S_o}{Q_o} \tag{9-57}$$

Substituting Eqs. 9-56 and 9-57 into Eq. 9-55, using the chain rule, and dropping the superscripts for simplicity, the following equation is obtained:

$$\frac{\partial Q}{\partial t} + \left[\frac{\partial Q}{\partial A}\right] \frac{\partial Q}{\partial x} = \left[\frac{Q_o}{2TS_o}\right] \frac{\partial^2 Q}{\partial x^2} \tag{9-58}$$

The left side of Eq. 9-58 is recognized as the kinematic wave equation, with $\partial Q/\partial A$ as the kinematic wave celerity. The right side is a second-order (partial differential) term that accounts for the physical diffusion effect. The coefficient of the second-order term has the units of diffusivity $[L^2 T^{-1}]$, being referred to as the *hydraulic diffusivity,* or channel diffusivity.

The hydraulic diffusivity is a characteristic of the flow and channel, defined as:

$$\nu_h = \frac{Q_o}{2TS_o} = \frac{q_o}{2S_o} \tag{9-59}$$

in which $q_o = Q_o/T$ is the reference flow per unit of channel width. From Eq. 9-59, it is concluded that hydraulic diffusivity is small for steep bottom slopes (e.g., those of small mountain streams), and large for mild bottom slopes (e.g., tidal rivers).

Equation 9-58 describes the movement of flood waves in a better way than Eq. 9-

17 or 9-18. It falls short from describing the full momentum effects, but it does physically account for peak flow attenuation.

Equation 9-58 is a second-order *parabolic* partial differential equation. It can be solved analytically, leading to Hayami's *diffusion analogy* solution for flood waves [7], or numerically with the aid of a numerical scheme for parabolic equations such as the Crank-Nicolson scheme [3]. An alternate approach is to match the hydraulic diffusivity with the numerical diffusion coefficient of the Muskingum scheme. This approach is the basis of the Muskingum-Cunge method [4, 12] (Section 9.4).

Applicability of Diffusion Waves

Most flood waves have a small amount of physical diffusion; therefore, they are better approximated by the diffusion wave rather than by the kinematic wave. For this reason, diffusion waves apply to a much wider range of practical problems than kinematic waves. Where the diffusion wave fails, only the dynamic wave can properly describe the translation and diffusion of flood waves. The dynamic wave, however, is very strongly diffusive, especially for flows well in the subcritical regime [14]. In practice, most flood flows are only mildly diffusive, and therefore, are subject to modeling with the diffusion wave.

To determine if a wave is a diffusion wave, it should satisfy the following dimensionless inequality [14]:

$$t_r S_o \left[\frac{g}{d_o} \right]^{1/2} \geq M \qquad (9\text{-}60)$$

in which t_r is the time-of-rise of the inflow hydrograph, S_o is the bottom slope, d_o is the average flow depth, and g is the gravitational acceleration. The greater the left side of this inequality, the more likely it is that the wave is a diffusion wave. In practice, a value of $M = 15$ is recommended for general use.

Example 9-8.

Use the criterion of Eq. 9-60 to determine whether the flood wave of Example 9-7 can be considered a diffusion wave.

For $t_r = 12$ h, $S_o = 0.001$, and $d_o = 2$ m, the left side of Eq. 9-60 is 95.7, which is greater than 15. In the previous example, this wave was shown not to satisfy the kinematic wave criterion. This example shows, however, that this wave is a diffusion wave. Had Eq. 9-60 not been satisfied, the flood wave would have been properly a dynamic wave, subject only to dynamic wave routing. Dynamic wave routing takes into account the complete momentum equation, including the inertia terms (local and convective) that were neglected in the formulation of kinematic and diffusion waves. Section 9.5 contains a brief introduction to dynamic waves.

9.4 MUSKINGUM-CUNGE METHOD

The Muskingum method can calculate runoff diffusion, ostensibly by varying the parameter X. A numerical solution of the linear kinematic wave equation using a third-order-accurate scheme ($C = 1$) leads to pure flood hydrograph translation (see Example 9-3, Part 1). Other numerical solutions to the linear kinematic wave equation invariably produce a certain amount of numerical diffusion and/or dispersion (See Example 9-3, Part 2). The Muskingum and linear kinematic wave routing equations are strikingly similar. Furthermore, unlike the kinematic wave equation, the diffusion wave equation does have the capability to describe physical diffusion.

From these propositions, Cunge [4] concluded that the Muskingum method is a linear kinematic wave solution and that the flood wave attenuation shown by the calculation is due to the numerical diffusion of the scheme itself. To prove this assertion, the kinematic wave equation (Eq. 9-18) is discretized on the xt plane (Fig. 9-8) in a way that parallels the Muskingum method, centering the spatial derivative and off-centering the temporal derivative by means of a weighting factor X:

$$\frac{X(Q_j^{n+1} - Q_j^n) + (1 - X)(Q_{j+1}^{n+1} - Q_{j+1}^n)}{\Delta t} + c \frac{(Q_{j+1}^n - Q_j^n) + (Q_{j+1}^{n+1} - Q_j^{n+1})}{2\Delta x} = 0$$

(9-61)

in which $c = \beta V$ is the kinematic wave celerity.

Solving Eq. 9-61 for the unknown discharge leads to the following routing equation:

$$Q_{j+1}^{n+1} = C_0 Q_j^{n+1} + C_1 Q_j^n + C_2 Q_{j+1}^n$$

(9-62)

The routing coefficients are

$$C_0 = \frac{c(\Delta t/\Delta x) - 2X}{2(1 - X) + c(\Delta t/\Delta x)}$$

(9-63)

$$C_1 = \frac{c(\Delta t/\Delta x) + 2X}{2(1 - X) + c(\Delta t/\Delta x)}$$

(9-64)

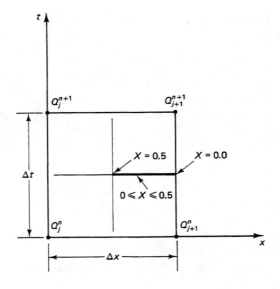

Figure 9-8 Space-time discretization of kinematic wave equation paralleling Muskingum method.

Stream Channel Routing   Chap. 9

$$C_2 = \frac{2(1-X) - c\,(\Delta t/\Delta x)}{2(1-X) + c\,(\Delta t/\Delta x)} \tag{9-65}$$

By defining

$$K = \frac{\Delta x}{c} \tag{9-66}$$

it is seen that the two sets of Eqs. 9-63 to 9-65 and Eqs. 9-4 to 9-6 are the same.

Equation 9-66 confirms that K is in fact the flood-wave travel time, i.e., the time it takes a given discharge to travel the reach length Δx with the kinematic wave celerity c. In a linear mode, c is constant and equal to a reference value; in a nonlinear mode, it varies with discharge.

It can be seen that for $X = 0.5$, Eqs. 9-63 to 9-65 reduce to the routing coefficients of the linear second-order-accurate kinematic wave solution, Eqs. 9-22 to 9-24. For $X = 0.5$ and $C = 1$ ($C = c\,\Delta t/\Delta x = \beta V \Delta t/\Delta x$, the Courant number, Eq. 9-25) the routing equation is third-order accurate, i.e., the numerical solution is equal to the analytical solution of the kinematic wave equation. For $X = 0.5$ and $C \neq 1$, it is second-order accurate, exhibiting only numerical dispersion. For $X < 0.5$ and $C \neq 1$, it is first-order accurate, exhibiting both numerical diffusion and dispersion. For $X < 0.5$ and $C = 1$, it is first-order accurate, exhibiting only numerical diffusion. These relations are summarized in Table 9-7.

In practice, the numerical diffusion can be used to simulate the physical diffusion of the actual flood wave. By expanding the discrete function $Q(j\Delta x, n\Delta t)$ in Taylor series about grid point $(j\Delta x, n\Delta t)$, the numerical diffusion coefficient of the Muskingum scheme is derived (see Appendix B):

$$\nu_n = c\,\Delta x\left(\frac{1}{2} - X\right) \tag{9-67}$$

in which ν_n is the numerical diffusion coefficient of the Muskingum scheme. This equation reveals the following: (1) for $X = 0.5$ there is no numerical diffusion, although there is numerical dispersion for $C \neq 1$; (2) for $X > 0.5$, the numerical diffusion coefficient is negative, i.e., numerical amplification, which explains the behavior of the Muskingum method for this range of X values; (3) for $\Delta x = 0$, the numerical diffusion coefficient is zero, clearly the trivial case.

A predictive equation for X can be obtained by matching the hydraulic diffusivity ν_h (Eq. 9-59) with the numerical diffusion coefficient of the Muskingum scheme ν_n (Eq. 9-67). This leads to the following expression for X:

TABLE 9-7 NUMERICAL PROPERTIES OF MUSKINGUM-CUNGE METHOD

Parameter X	Parameter C	Order of Accuracy	Numerical Diffusion	Numerical Dispersion
0.5	1	Third	No	No
0.5	$\neq 1$	Second	No	Yes
< 0.5	$\neq 1$	First	Yes	Yes
< 0.5	1	First	Yes	No

$$X = \frac{1}{2}\left(1 - \frac{q_o}{S_o \, c \, \Delta x}\right) \qquad (9\text{-}68)$$

With X calculated by Eq. 9-68, the Muskingum method is referred to as Muskingum-Cunge method [12]. Using Eq. 9-68, the routing parameter X can be calculated as a function of the following numerical and physical properties: (1) reach length Δx, (2) reference discharge per unit width q_o, (3) kinematic wave celerity c, and (4) bottom slope S_o.

It should be noted that Eq. 9-68 was derived by matching physical and numerical diffusion (i.e., second-order processes), and does not account for dispersion (a third-order process). Therefore, in order to simulate wave diffusion properly with the Muskingum-Cunge method, it is necessary to optimize numerical diffusion (with Eq. 9-68) while minimizing numerical dispersion (by keeping the value of C as close to 1 as practicable).

A unique feature of the Muskingum-Cunge method is the grid independence of the calculated outflow hydrograph, which sets it apart from other linear kinematic wave solutions featuring uncontrolled numerical diffusion and dispersion (e.g., the convex method). If numerical dispersion is minimized, the calculated outflow at the downstream end of a channel reach will be essentially the same, regardless of how many subreaches are used in the computation. This is because X is a function of Δx, and the routing coefficients C_0, C_1, and C_2 vary with reach length.

An improved version of the Muskingum-Cunge method is due to Ponce and Yevjevich [15]. The C value is the Courant number, i.e., the ratio of wave celerity c to grid celerity $\Delta x / \Delta t$:

$$C = c \, \frac{\Delta t}{\Delta x} \qquad (9\text{-}69)$$

The grid diffusivity is defined as the numerical diffusivity for the case of $X = 0$. From Eq. 9-67, the grid diffusivity is

$$\nu_g = \frac{c \, \Delta x}{2} \qquad (9\text{-}70)$$

The cell Reynolds number [17] is defined as the ratio of hydraulic diffusivity (Eq. 9-59) to grid diffusivity (Eq. 9-70). This leads to

$$D = \frac{q_o}{S_o \, c \, \Delta x} \qquad (9\text{-}71)$$

in which D = cell Reynolds number. Therefore

$$X = \tfrac{1}{2}(1 - D) \qquad (9\text{-}72)$$

Equations 9-71 and 9-72 imply that for very small values of Δx, D may be greater than 1, leading to negative values of X. In fact, for the *characteristic reach length*

$$\Delta x_c = \frac{q_o}{S_o c} \qquad (9\text{-}73)$$

the cell Reynolds number is $D = 1$, and $X = 0$. Therefore, in the Muskingum-Cunge method, reach lengths shorter than the characteristic reach length result in negative values of X. This should be contrasted with the classical Muskingum method (Section 9.1), in which X is restricted in the range 0.0–0.5. In the classical Muskingum, X is interpreted as a weighting factor. As shown by Eqs. 9-71 and 9-72, nonnegative values of X are associated with long reaches, typical of the manual computation used in the development and early application of the Muskingum method.

In the Muskingum-Cunge method, however, X is interpreted in a moment-matching sense [2] or diffusion-matching factor. Therefore, negative values of X are entirely possible. This feature allows the use of shorter reaches than would otherwise be possible if X were restricted to nonnegative values.

The substitution of Eqs. 9-69 and 9-72 into Eqs. 9-63 to 9-65 leads to routing coefficients expressed in terms of Courant and cell Reynolds numbers:

$$C_0 = \frac{-1 + C + D}{1 + C + D} \tag{9-74}$$

$$C_1 = \frac{1 + C - D}{1 + C + D} \tag{9-75}$$

$$C_2 = \frac{1 - C + D}{1 + C + D} \tag{9-76}$$

The calculation of routing parameters C and D, Eqs. 9-69 and 9-71, can be performed in several ways. The wave celerity can be calculated with either Eq. 9-16 or Eq. 9-34. With Eq. 9-16, $c = \beta V$; with Eq. 9-34, $c = (1/T) \, dQ/dy$. Theoretically, these two equations are the same. For practical applications, if a stage-discharge rating and cross-sectional geometry are available (i.e., stage-discharge-top width tables), Eq. 9-34 is preferred over Eq. 9-16 because it accounts directly for cross-sectional shape. In the absence of a stage-discharge rating and cross-sectional data, Eq. 9-16 can be used to estimate flood wave celerity.

With the aid of Eqs. 9-69 and 9-71, the routing parameters can be based on flow characteristics. The calculations can proceed in a linear or nonlinear mode. In the linear mode, the routing parameters are based on reference flow values and kept constant throughout the computation in time. The choice of reference flow has a bearing on the calculated results [2, 15], although the overall effect is likely to be small. For practical applications, either an average or peak flow value can be used as reference flow. The peak flow value has the advantage that it can be readily ascertained, although a better approximation may be obtained by using an average value [15]. The linear mode of computation is referred to as the constant-parameter Muskingum-Cunge method to distinguish it from the variable-parameter Muskingum-Cunge method, in which the routing parameters are allowed to vary with the flow. The constant-parameter method resembles the Muskingum method, with the difference that the routing parameters are based on measurable flow and channel characteristics instead of historical streamflow data.

Example 9-9.

Use the constant-parameter Muskingum-Cunge method to route a flood wave with the following flood and channel characteristics: peak flow $Q_p = 1000 \text{ m}^3/\text{s}$; baseflow $Q_b = 0$

m^3/s; channel bottom slope $S_o = 0.000868$; flow area at peak discharge $A_p = 400 \ m^2$; top width at peak discharge $T_p = 100 \ m$; rating exponent $\beta = 1.6$; reach length $\Delta x = 14.4 \ km$; time interval $\Delta t = 1 \ h$.

Time (h)	0	1	2	3	4	5	6	7	8	9	10
Flow (m^3/s)	0	200	400	600	800	1000	800	600	400	200	0

The mean velocity (based on the peak discharge) is $V = Q_p/A_p = 2.5 \ m/s$. The wave celerity is $c = \beta V = 4 \ m/s$. The flow per unit width (based on the peak discharge) is $q_o = Q_p/T_p = 10 \ m^2/s$. The Courant number (Eq. 9-69) is $C = 1$. The cell Reynolds number (Eq. 9-71) is $D = 0.2$. The routing coefficients (Eqs. 9-74 to 9-76) are $C_0 = 0.091$; $C_1 = 0.818$; and $C_2 = 0.091$. It is confirmed that the sum of routing coefficients is equal to 1. The routing calculations are shown in Table 9-8.

Resolution Requirements

When using the Muskingum-Cunge method, care should be taken to ensure that the values of Δx and Δt are sufficiently small to approximate closely the actual shape of the hydrograph. For smoothly rising hydrographs, a minimum value of $t_p/\Delta t = 5$ is recommended. This requirement usually results in the hydrograph time base being resolved into at least 15 to 25 discrete points, considered adequate for Muskingum routing.

Unlike temporal resolution, there is no definite criteria for spatial resolution. A criterion borne out by experience is based on the fact that Courant and cell Reynolds numbers are inversely related to reach length Δx. Therefore, to keep Δx sufficiently small, Courant and cell Reynolds numbers should be kept sufficiently large. This leads to the practical criterion [16]:

TABLE 9-8 CHANNEL ROUTING BY MUSKINGUM-CUNGE METHOD: EXAMPLE 9-9

(1)	(2)	(3)	(4)	(5)	(6)
		Partial Flows			
Time (h)	Inflow (m^3/s)	$C_0 I_2$	$C_1 I_1$	$C_2 O_1$	Outflow (m^3/s)
0	0	—	—	—	0.0
1	200	18.2	0.0	0.0	18.20
2	400	36.4	163.6	1.66	201.66
3	600	54.6	327.2	18.35	400.15
4	800	72.8	490.8	36.41	600.01
5	1000	91.0	654.4	54.60	800.00
6	800	72.8	818.0	72.80	963.60
7	600	54.6	654.4	87.69	796.69
8	400	36.4	490.8	72.50	599.70
9	200	18.2	327.2	54.57	399.97
10	0	0.0	163.6	36.40	200.00
11	0	0.0	0.0	18.20	18.20
12	0	0.0	0.0	1.66	1.66
13	0	0.0	0.0	0.16	0.16

$$C + D \geq 1 \qquad (9\text{-}77)$$

which can be written as: $-1 + C + D \geq 0$. This confirms the necessity of avoiding negative values of C_0 in Muskingum-Cunge routing (See Eq. 9-74). Experience has shown that negative values of either C_1 or C_2 do not adversely affect the method's overall accuracy [16].

Notwithstanding Eq. 9-77, the Muskingum-Cunge method works best when the numerical dispersion is minimized, that is, when C is kept close to 1. Values of C substantially different from 1 are likely to cause the notorious dips, or negative outflows, in portions of the calculated hydrograph. This computational anomaly is attributed to excessive numerical dispersion and should be avoided.

Nonlinear Muskingum-Cunge Method

The kinematic wave equation, Eq. 9-18, is nonlinear because the kinematic wave celerity varies with discharge. The nonlinearity is mild, among other things because the wave celerity variation is usually restricted within a narrow range. However, in certain cases it may be necessary to account for this nonlinearity. This can be done in two ways: (1) during the discretization, by allowing the wave celerity to vary, resulting in a nonlinear numerical scheme to be solved by iterative means; and (2) after the discretization, by varying the routing parameters, as in the variable-parameter Muskingum-Cunge method [15]. The latter appproach is particularly useful if the overall nonlinear effect is small, which is often the case.

In the variable-parameter method, the routing parameters are allowed to vary with the flow. The values of C and D are based on local q_o and c values instead of peak flow or other reference value as in the constant-parameter method. To vary the routing parameters, the most expedient way is to obtain an average value of q_o and c for each computational cell. This can be achieved with a direct three-point average of the values at the known grid points (See Fig. 9-8), or by an iterative four-point average, which includes the unknown grid point. To improve the convergence of the iterative four-point procedure, the three-point average can be used as the first guess of the iteration. Once q_o and c have been determined for each computational cell, the Courant and cell Reynolds numbers are calculated by Eqs. 9-69 and 9-71. The value of bottom slope S_o remains unchanged within each computational cell.

The variable parameter Muskingum-Cunge method represents a small yet sometimes perceptible improvement over the constant parameter method. The differences are likely to be more marked for very long reaches and/or wide variations in flow levels. Flood hydrographs calculated with variable parameters show a certain amount of distortion, either wave steepening in the case of flows contained inbank or wave attenuation in the case of typical overbank flows. This is a physical manifestation of the nonlinear effect, i.e., different flow levels traveling with different celerities. On the other hand, flood hydrographs calculated using constant parameters do not show such wave distortion.

Assessment of Muskingum-Cunge Method

The Muskingum-Cunge method is a physically based alternative to the Muskingum method. Unlike the Muskingum method where the parameters are calibrated using streamflow data, in the Muskingum-Cunge method the parameters are calculated based on flow and channel characteristics. This makes possible channel routing without the need for time-consuming and cumbersome parameter calibration. More importantly, it makes possible extensive channel routing in ungaged streams with a reasonable expectation of accuracy. With the variable-parameter feature, nonlinear properties of flood waves (which could otherwise only be obtained by more elaborate numerical procedures) can be described within the context of the Muskingum formulation.

Like the Muskingum method, the Muskingum-Cunge method is limited to diffusion waves. Furthermore, the Muskingum-Cunge method is based on a single-valued rating and does not take into account strong flow nonuniformity or unsteady flows exhibiting substantial loops in discharge-stage rating (i.e., dynamic waves). Thus, the Muskingum-Cunge method is suited for channel routing in natural streams without significant backwater effects and for unsteady flows that classify under the diffusion wave criterion (Eq. 9-60).

An important difference between the Muskingum and Muskingum-Cunge methods should be noted. The Muskingum method is based on the storage concept (Chapter 4) and, therefore, the parameters K and X are reach averages. The Muskingum-Cunge method, however, is kinematic in nature, with the parameters C and D being based on values evaluated at channel cross sections rather than being reach averages. Therefore, for the Muskingum-Cunge method to improve on the Muskingum method, it is necessary that the routing parameters evaluated at channel cross sections be representative of the channel reach under consideration.

Historically, the Muskingum method has been calibrated using streamflow data. On the contrary, the Muskingum-Cunge method relies on physical characteristics such as rating curves, cross-sectional data and channel slope. The different data requirements reflect the different theoretical bases of the methods, i.e., storage concept in the Muskingum method, and kinematic wave theory in the Muskingum-Cunge method.

9.5 INTRODUCTION TO DYNAMIC WAVES

In Section 9.2, kinematic waves were formulated by simplifying the momentum conservation principle to a statement of steady uniform flow. In Section 9.3, diffusion waves were formulated by simplifying the momentum principle to a statement of steady nonuniform flow. These two waves, in particular the diffusion wave, have been extensively used in stream channel routing applications. The Muskingum and Muskingum-Cunge methods are examples of calculations using the concept of diffusion wave.

A third type of open channel flow wave, the dynamic wave, is formulated by taking into account the complete momentum principle, including its inertial components. As such, the dynamic wave contains more physical information than either kinematic or diffusion waves. Dynamic wave solutions, however, are more complicated than either kinematic or diffusion wave solutions.

In a dynamic wave solution, the equations of mass and momentum conservation are solved by a numerical procedure, either the method of finite differences, the method of characteristics, or the finite element method. In the method of finite differences, the partial differential equations are discretized following a chosen numerical scheme [9]. The method of characteristics is based on the conversion of the set of partial differential equations into a related set of ordinary differential equations, and the solution along a *characteristic grid,* i.e. a grid that follows characteristic directions. The method of finite elements solves a set of integral equations over a chosen grid of finite elements.

In the past two decades, the method of finite differences has come to be regarded as the most expedient way of obtaining a dynamic wave solution for practical applications [6, 9]. Among several numerical schemes that have been used in connection with the dynamic wave, the Preissmann scheme is perhaps the most popular. This is a four-point scheme, centered in the temporal derivatives and slightly off-centered in the spatial derivatives. The off-centering in the spatial derivatives introduces a small amount of numerical diffusion necessary to control the numerical stability of the non-linear scheme. This produces a workable yet sufficiently accurate scheme.

The stream channel is divided into several reaches for computational purposes. The application of the Preissmann scheme to the governing equations for the various reaches results in a matrix solution requiring a *double sweep* algorithm, i.e., one that accounts only for the nonzero entries of the coefficient matrix, which are located within a narrow band surrounding the main diagonal. This technique leads to a considerable savings in storage and execution time. With the appropriate upstream and downstream boundary conditions, the solution of the set of hyperbolic equations marches in time until a specified number of time intervals is completed.

In practice, a dynamic wave solution represents an order of magnitude increase in complexity and associated data requirements when compared to either kinematic or diffusion wave solutions. Its use is recommended in situations where neither kinematic or diffusion wave solutions are likely to represent adequately the physical phenomena. In particular, dynamic wave solutions are applicable to flow over very flat slopes, flow into large reservoirs, strong backwater conditions and flow reversals. In general, the dynamic wave is recommended for cases warranting a precise determination of the unsteady variation of river stages.

Relevance of Dynamic Waves to Engineering Hydrology

Dynamic wave solutions are often referred to as *hydraulic river routing* . As such, they have the capability to calculate unsteady discharges and stages when presented with the appropriate geometric channel data and initial and boundary conditions. Their relevance to engineering hydrology is examined here by comparing them to kinematic and diffusion wave solutions.

Kinematic waves calculate unsteady discharges; the corresponding stages are subsequently obtained from the appropriate rating curves. Usually, equilibrium (steady, uniform) rating curves are used for this purpose. Diffusion waves may or may not use equilibrium rating curves to calculate stages. Some methods, e.g., Muskingum-Cunge, use equilibrium ratings, but more elaborate diffusion wave solutions may not.

Dynamic waves rely on the physics of the phenomena as built into the governing equations to generate their own unsteady rating. A looped rating curve is produced at every cross section, as shown in Fig. 9-9. For any given stage, the discharge is higher in the rising limb of the hydrograph and lower in the receding limb. This loop is due to hydrodynamic reasons and should not be confused with other loops, which may be due to erosion, sedimentation, or changes in bed configuration (Chapter 15).

The width of the loop is a measure of the flow unsteadiness, with wider loops corresponding to highly unsteady flow, i.e, dynamic wave flow. If the loop is narrow, it implies that the flow is mildly unsteady, perhaps a diffusion wave. If the loop is practically nonexistent, the flow can be approximated as kinematic flow. In fact, the basic assumption of kinematic flow is that momentum can be simulated as steady uniform flow, i.e., that the rating curve is single-valued.

The preceding observations lead to the conclusion that the relevance of dynamic waves in engineering hydrology is directly related to the flow unsteadiness and the associated loop in the rating curve. For highly unsteady flows such as dam-break flood waves, it may well be the only way to account properly for the looped rating. For other less unsteady flows, kinematic and diffusion waves are a viable alternative, provided their applicability can be clearly demonstrated (Eqs. 9-44 and 9-60).

Diffusion Wave Solution with Dynamic Component

A simplified approach to dynamic wave routing is that of the diffusion wave with dynamic component [2]. In this approach, the complete governing equations, including inertia terms, are linearized in a similar way as with diffusion waves. This leads to a diffusion equation similar to Eq. 9-58, but with a modified hydraulic diffusivity. The equation is [5]:

$$\frac{\partial Q}{\partial t} + \left[\frac{\partial Q}{\partial A}\right]\frac{\partial Q}{\partial x} = \left[\frac{Q_o}{2TS_o}\right][1 - (\beta - 1)^2 F_o^2]\frac{\partial^2 Q}{\partial x^2} \qquad (9\text{-}78)$$

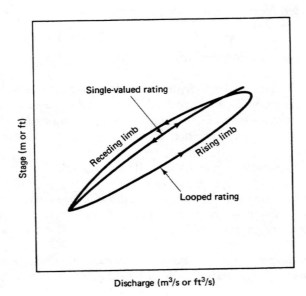

Stage (m or ft)

Single-valued rating

Receding limb

Rising limb

Looped rating

Discharge (m³/s or ft³/s)

Figure 9-9 Sketch of the looped rating of dynamic waves.

in which the hydraulic diffusivity is also a function of the rating curve parameter β and the Froude number, defined as:

$$F_o = \frac{V_o}{(gd_o)^{1/2}} \tag{9-79}$$

with g = gravitational acceleration and d_o = reference flow depth.

Equation 9-78 provides an enhanced predictive capability for the simulation of diffusion waves including a dynamic component. For instance, for $\beta = 1.5$ (i.e., Chezy friction in wide channels) and $F_o = 2$, the hydraulic diffusivity in Eq. 9-78 vanishes, which is in agreement with physical reality [10, 13]. On the other hand, the hydraulic diffusivity of the diffusion wave (Eq. 9-58) is independent of the Froude number. Therefore, Eq. 9-78 is a somewhat better model than Eq. 9-58, especially for Froude numbers in the supercritical regime. Most natural flows, however, are in the range well below critical, with Eq. 9-58 remaining a practical model of unsteady open channel flow phenomena.

QUESTIONS

1. What is routing? What types of waves are used in describing unsteady open channel flow processes?
2. What is model calibration? What is model verification?
3. In the Muskingum method, what does the parameter K represent? What does the parameter X represent?
4. How does channel routing differ from reservoir routing? What differences are to be noted in the routed hydrographs?
5. What is the kinematic wave celerity? What is the practical range of turbulent flow values of β, the rating constant used in the kinematic wave celerity?
6. What is the order of accuracy of a numerical scheme? What is the difference between numerical diffusion and numerical dispersion in connection with kinematic wave solutions?
7. What is a linear model in the context of kinematic wave routing? What is a nonlinear model?
8. Why are the results of convex routing dependent on the grid size?
9. What is a diffusion wave? How does it differ from a kinematic wave?
10. What is hydraulic diffusivity? Why is it important in flood routing?
11. What values of parameters X and C optimize numerical diffusion and minimize numerical dispersion in the Muskingum-Cunge method?
12. Why are negative values of X entirely possible in Muskingum-Cunge routing? Why are values of X in excess of 0.5 unfeasible?
13. What is the Courant number? What is the cell Reynolds number?
14. Describe the difference between linear and nonlinear solutions to channel routing problems.
15. What is a dynamic wave? How does it differ from the diffusion and kinematic waves?
16. How does the method of finite differences differs from the method of characteristics? What is a double sweep algorithm?

17. Discuss the influence of the loop in the rating in determining whether an open channel flow wave is dynamic in nature.

18. What is the effect of the inclusion of a dynamic component in diffusion-wave modeling?

PROBLEMS

9-1. Given the following inflow hydrograph to a certain stream channel reach, calculate the outflow by the Muskingum method.

Time (h)	0	1	2	3	4	5	6	7	8	9	10	11	12
Inflow (m³/s)	10	20	40	80	120	150	120	60	50	40	30	20	10

Assume baseflow 10 m³/s, $K = 1$ h, $X = 0.2$, $\Delta t = 1$ h.

9-2. Given the following inflow hydrograph to a certain stream channel reach, calculate the outflow by the Muskingum method.

Time (h)	0	3	6	9	12	15	18	21	24	27	30	33	36
Inflow (m³/s)	100	120	150	200	250	275	250	210	180	150	120	110	100

Assume baseflow 100 m³/s, $K = 2.4$ h, $X = 0.1$, $\Delta t = 3$ h.

9-3. Given the following inflow and outflow hydrographs for a certain stream channel reach, calculate the Muskingum parameters K and X.

Time (h)	0	1	2	3	4	5
Inflow (ft³/s)	2520	3870	4560	6795	8975	9320
Outflow (ft³/s)	2520	2643	3598	4500	6367	8295

Time (h)	6	7	8	9	10	11
Inflow (ft³/s)	7780	6520	5340	4105	3210	2520
Outflow (ft³/s)	8900	7971	6808	5628	4439	3482

Time (h)	12	13	14	15	16	17
Inflow (ft³/s)	2520	2520	2520	2520	2520	2520
Outflow (ft³/s)	2782	2592	2540	2525	2521	2520

9-4. Develop a computer program to solve the Muskingum method of stream channel routing, given the following data: (1) an inflow hydrograph of arbitrary shape, (2) baseflow, (3) storage constant K, (4) weighting factor X, and (5) time interval Δt. Test your program with Example 9-1 in the text.

9-5. Given the following inflow hydrograph, use the computer program developed in Problem 9-4 to calculate the outflow hydrograph by the Muskingum method.

Time (h)	0.00	0.25	0.5	0.75	1.0	1.25
Inflow (m³/s)	0	1	2	4	8	10

Time (h)	1.50	1.75	2.0	2.25	2.5	2.75
Inflow (m³/s)	8	6	4	2	1	0

Assume baseflow 0 m³/s, $K = 0.4$ h, $X = 0.15$, and $\Delta t = 0.25$ h.

9-6. Develop a computer program to estimate the parameters of the Muskingum method, given a matching set of inflow and outflow hydrographs for a certain channel reach. A suggested algorithm is to search for the value of X that minimizes the root mean square (RMS) of the differences between predicted and measured storage. For this purpose, several values of X (between the range 0.0 to 0.5) are tried. For each trial value, a regression line is fitted to the (measured) storage (calculated using Eq. 9-7) versus weighted flow data, with weighted flow in the abscissas and measured storage in the ordinates. The differences between measured storage and predicted storage, i.e., storage predicted by the regression, are calculated. The RMS is evaluated by the following formula:

$$\text{RMS} = \left[\frac{1}{n-1} \sum_{n} (S - S')^2 \right]^{1/2} \tag{9-80}$$

in which S = measured storage, S' = predicted storage, and n = number of values. The X corresponding to the minimum RMS value is the estimated X. The Muskingum parameter K is the slope of the regression line corresponding to the chosen X value. Use Example 9-2 in the text to test your program.

9-7. Use the data of Problem 9-3 to test further the computer program developed in Problem 9-6.

9-8. Route the following flood wave using a linear forward-in-time/backward-in-space numerical scheme of the kinematic wave equation (similar to the convex method).

Time (min)	0	10	20	30	40	50	60	70	80	90	100
Inflow (m^3/s)	0	1	2	4	8	10	8	4	2	1	0

Assume base flow 0 m^3/s, $V = 1$ m/s, $\beta = 1.5$, $\Delta x = 1200$ m, and $\Delta t = 10$ min.

9-9. Derive the routing coefficients for a linear forward-in-space/backward-in-time numerical scheme of the kinematic wave equation.

9-10. Use the routing coefficients derived in Problem 9-9 to route the inflow hydrograph of Problem 9-8. Assume a reach length $\Delta x = 800$ m.

9-11. Calculate the β value for a triangular channel with Chezy friction.

9-12. A large river of nearly constant width $B = 900$ m is seen to be rising at the rate of 10 cm/h. At the observation point, a stage measurement indicates that the current value of discharge is 2200 m^3/s. What is a rough estimate of the discharge at a point 5 km upstream?

9-13. Solve Problem 9-12 if the tributary contribution between the two points is estimated to be constant and equal to 225 m^3/s.

9-14. Determine if a flood wave with the following characteristics is a kinematic wave: time-of-rise $t_r = 6$ h, bottom slope $S_o = 0.015$, average flow velocity $V_o = 1.5$ m/s, and average flow depth $d_o = 3$ m.

9-15. Determine if a flood wave with the following characteristics is a diffusion wave: time-of-rise $t_r = 6$ h, bottom slope $S_o = 0.005$, and average flow depth $d_o = 3$ m.

9-16. Program EH900 included in Appendix D solves the Muskingum-Cunge method of flood routing for a triangular inflow hydrograph, with routing parameters based on peak flow. Test program EH900 using Example 9-9 in the text. Then run program EH900 using the following data: peak discharge = 550 m^3/s, baseflow = 50 m^3/s, time-to-peak = 5 h, time base = 15 h, channel bed slope = 0.0008, flow area corresponding to the peak discharge = 200 m^2, channel top width corresponding to the peak discharge = 50 m, rating exponent $\beta = 1.65$, reach length = 15 km, time interval $\Delta t = 1$ h. Report peak outflow and time-to-peak.

9-17. Modify Program EH900 to solve the Muskingum-Cunge method using an inflow hydrograph of arbitrary shape. Test your program using the inflow hydrograph of Example 9-9 in the text.

9-18. Given the following inflow hydrograph to a stream channel reach, use the computer program developed in Problem 9-17 to calculate the outflow hydrograph.

Time (h)	0	1	2	3	4	5	6	7	8	9	10	11	12	13
Inflow (m³/s)	10	20	40	80	160	320	400	320	240	160	80	40	20	10

Assume baseflow $= 10$ m³/s, channel bed slope $= 0.001$, flow area corresponding to the peak discharge $= 800$ m², channel top width corresponding to the peak discharge $= 35$ m, rating exponent $\beta = 1.6$, reach length $= 3$ km, and time interval $\Delta t = 1$ h.

9-19. Given the following inflow hydrograph to a channel reach, calculate the outflow hydrographs for (a) a reach length of 4 km, and (b) for a reach length of 5 km.

Time (h)	0	1	2	3	4	5	6	7	8	9	10	11	12
Inflow (m³/s)	5	8	12	20	28	33	29	22	19	13	8	6	5

Assume channel bed slope $= 0.0015$, flow area corresponding to the peak discharge $= 42$ m², channel top width corresponding to the peak discharge $= 18$ m, rating exponent $\beta = 1.5$.

9-20. Calculate the hydraulic diffusivity for the following flow conditions: channel bed slope $= 0.002$, mean flow depth 4 m, mean flow velocity 2 m/s, and rating exponent $\beta = 1.6$. Compare the two cases: (a) without inertia and (b) with inertia.

REFERENCES

1. Abbott, M. A. (1975). "Method of Characteristics," in *Unsteady Flow in Open Channels*, Vol. 1, K. Mahmood and V. Yevjevich, editors. Fort Collins, Colorado: Water Resources Publications.

2. Agricultural Research Service, U.S. Department of Agriculture. (1973). "Linear Theory of Hydrologic Systems," *Technical Bulletin No.* 1468, (J. C. I. Dooge, author). Washington, D.C.

3. Crandall, S. H. (1956). *Engineering Analysis*. Engineering Society Monographs. New York: McGraw-Hill.

4. Cunge, J. A. (1969). "On the Subject of a Flood Propagation Computation Method (Muskingum Method)," *Journal of Hydraulic Research*, Vol. 7, No. 2, pp. 205–230.

5. Dooge, J. C. I., W. B. Strupczewski, W. B. and J. J. Napiorkowski. (1982). "Hydrodynamic Derivation of Storage Parameters of the Muskingum Model," *Journal of Hydrology*, Vol. 54, pp. 371–387.

6. Fread, D. L. (1985). "Channel Routing," in *Hydrological Forecasting*. M. G. Anderson and T. P. Burt, editors. New York: John Wiley.

7. Hayami, S. (1951). "On the Propagation of Flood Waves," *Bulletin of the Disaster Prevention Research Institute*, Kyoto University, Kyoto, Japan, No. 1, December.

8. Kleitz, M. (1877). "Note sur la Théorie du Mouvement non Permanent des Liquides et sur application à la Propagation del Crues des Rivières, (Note on the Theory of Unsteady Flow of Liquids and on Application to Flood Propagation in Rivers)," *Annales des Ponts et Chaussées*, Ser. 5, Vol. 16, 2e semestre, pp. 133–196.

9. Liggett, J. A., and J. A. Cunge. (1975). "Numerical Methods of Solution of the Unsteady Flow Equations," in *Unsteady Flow in Open Channels*, Vol. 1, K. Mahmood and V. Yevjevich, editors. Fort Collins, Colorado: Water Resources Publications.

10. Lighthill, M. J., and G. B. Whitham. (1955). "On Kinematic Waves. I. Flood Movement in Long Rivers," *Proceedings of the Royal Society of London*, Vol. A229, May, pp. 281–316.

11. McCarthy, G. T. (1938). "The Unit Hydrograph and Flood Routing," unpublished manuscript, presented at a Conference of the North Atlantic Division, U.S. Army Corps of Engineers, June 24.

12. Natural Environment Research Council. (1975). *Flood Studies Report, Vol. 3: Flood Routing,* London, England.
13. Ponce, V. M., and D. B. Simons. (1977). "Shallow Wave Propagation in Open Channel Flow," *Journal of the Hydraulics Division,* ASCE, Vol. 103, No. HY12, December, pp. 1461–1476.
14. Ponce, V. M., R. M. Li, and D. B. Simons. (1978). "Applicability of Kinematic and Diffusion Models," *Journal of the Hydraulics Division,* ASCE, Vol. 104, No. HY3, March, pp. 353–360.
15. Ponce, V. M., and V. Yevjevich. (1978). "Muskingum-Cunge Method with Variable Parameters," *Journal of the Hydraulics Division,* ASCE, Vol. 104, No. HY12, December, pp. 1663–1667.
16. Ponce, V. M., and F. D. Theurer. (1982). "Accuracy Criteria in Diffusion Routing," *Journal of the Hydraulics Division,* ASCE, Vol. 108, No. HY6, June, pp. 747–757.
17. Roache, P. (1972). *Computational Fluid Dynamics.* Hermosa Publishers. New Mexico: Albuquerque.
18. Seddon, J. A. (1900). "River Hydraulics," *Transactions,* ASCE, Vol. 43, pp. 179–229.
19. Woolhiser, M. H., and Liggett, J. A. (1967). "Unsteady One-Dimensional Flow Over a Plane: The Rising Hydrograph," *Water Resources Research,* Vol. 3, No. 3, pp. 753–771.

SUGGESTED READINGS

Agricultural Research Service, U.S. Department of Agriculture. (1973). "Linear Theory of Hydrologic Systems," *Technical Bulletin No.* 1468, (J. C. I. Dooge, author). Washington, D.C.

Cunge, J. A. (1969). "On the Subject of a Flood Propagation Computation Method (Muskingum Method)," *Journal of Hydraulic Research,* Vol. 7, No. 2, 1969, pp. 205–230.

Fread, D. L. (1985). "Channel Routing," in *Hydrological Forecasting.* M. G. Anderson and T. P. Burt, editors. New York: John Wiley.

Lighthill, M. J., and Whitham, G. B. (1955). "On Kinematic Waves. I. Flood Movement in Long Rivers," *Proceedings of the Royal Society of London,* Vol. A229, May, 1955, pp. 281–316.

Natural Environment Research Council. (1975). *Flood Studies Report, Vol. 5: Flood Routing,* London, England, 1975.

Ponce, V. M., Li, R. M., and Simons, D. B., (1978). "Applicability of Kinematic and Diffusion Models," *Journal of the Hydraulics Division,* ASCE, Vol. 104, No. HY3, March, 1978, pp. 353–360.

CATCHMENT ROUTING

Catchment routing refers to the calculation of flows in time and space within a catchment. The objective of catchment routing is to transform effective rainfall into streamflow. This is accomplished either in a lumped mode (e.g., time-area method) or in a distributed mode (e.g., kinematic wave method).

Methods for catchment routing are similar to those of reservoir and stream channel routing. In fact, many techniques used in reservoir and channel routing are also applicable to catchment routing. For instance, the concept of linear reservoir is used in both reservoir and catchment routing. Kinematic wave techniques were originally developed for river routing [9], but later were applied to catchment routing [19, 22].

Methods for catchment routing are of two types: (1) hydrologic and (2) hydraulic. Hydrologic methods are based on the storage concept and are spatially lumped to provide a runoff hydrograph at the catchment outlet. Examples of hydrologic catchment routing methods are the time-area method and the cascade of linear reservoirs. Hydraulic methods use kinematic or diffusion waves to simulate surface runoff within a catchment in a distributed context. Unlike hydrologic methods, hydraulic methods can provide runoff hydrographs inside the catchment.

Catchment routing models can use parametric, conceptual, and/or deterministic components. For instance, the hydrograph obtained by the time-area method can be routed through a linear reservoir using a storage constant derived by empirical (i.e., parametric) means. The cascade of linear reservoirs is a typical example of a conceptual model used in catchment routing. Kinematic and diffusion models are examples of deterministic methods used in catchment routing.

The concepts of *translation* and *storage* are central to the study of flow routing, whether in catchments, reservoirs, or stream channels. They are particularly important in catchment routing because they can be studied separately, unlike in reservoir and channel routing. Translation may be interpreted as the movement of water in a direction parallel to the channel bottom. Storage may be interpreted as the movement of water in a direction perpendicular to the channel bottom. Translation is synonymous with runoff concentration; storage is synonymous with runoff diffusion.

In reservoir routing, storage is the primary mechanism, with translation almost nonexistent. In stream channel routing, the situation is reversed, with translation being the predominant mechanism and storage playing only a minor role. This is the reason why kinematic and diffusion waves are useful models of stream channel routing. In catchment routing, translation and storage are about equally important, and, therefore, they are often accounted for separately. The translation effect can be related to runoff concentration, whereas the storage effect can be simulated with linear reservoirs.

This chapter is divided into six sections. Section 10.1 describes the time-area method of hydrologic catchment routing. Section 10.2 describes the Clark unit hydrograph, a procedure closely related to the time-area method. Section 10.3 deals with the cascade of linear reservoirs, a widely accepted method of hydrologic catchment routing. Sections 10.4 and 10.5 describe two hydraulic methods of catchment routing, based on kinematic and diffusion waves, respectively. Section 10.6 contains a discussion of the capabilities and limitations of catchment routing techniques.

10.1 TIME-AREA METHOD

The time-area method of hydrologic catchment routing transforms an effective storm hyetograph into a runoff hydrograph. The method accounts for translation only and does not include storage. Therefore, hydrographs calculated with the time-area method show a lack of diffusion, resulting in higher peaks than those that would have been obtained if storage had been taken into account. If necessary, the required amount of storage can be incorporated by routing the hydrograph obtained by the time-area method through a linear reservoir. The required amount of storage is determined by calibrating the linear reservoir storage constant K with measured data. Alternatively, suitable values of K can be estimated based on regionally derived formulas.

The time-area method is essentially an extension of the runoff concentration principle used in the rational method (Chapter 4). Unlike the rational method, however, the time-area method can account for the temporal variation of rainfall intensity. Therefore, the applicability of the time-area method is extended to midsize catchments.

The time-area method is based on the concept of *time-area histogram*, i.e., a histogram of contributing catchment subareas. To develop a time-area histogram, the catchment's time of concentration is divided into a number of equal time intervals. Cumulative time at the end of each time interval is used to divide the catchment into zones delimited by *isochrone lines*, i.e., the loci of points of equal travel time to the catchment outlet, as shown in Fig. 10-1(a). For any point inside the catchment, the travel time refers to the time that it would take a parcel of water to travel from that

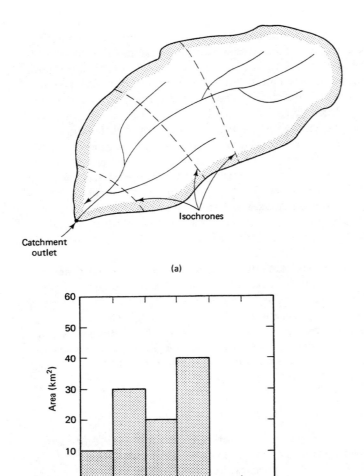

Catchment
outlet

Isochrones

(a)

(b)

Figure 10-1 Time-area method: (a) isochrone delineation; (b) time-area histogram.

point to the outlet. The catchment subareas delimited by the isochrones are measured and plotted in histogram form as shown in Fig. 10-1(b).

The time interval of the effective rainfall hyetograph should be equal to the time interval of the time-area histogram. The rationale of the time-area method is that, according to the runoff concentration principle (Section 2.4), the partial flow at the end of each time interval is equal to the product of effective rainfall times contributing subarea. The lagging and summation of the partial flows results in a runoff hydrograph for the given effective rainfall hyetograph and time-area histogram. While the time-area method accounts for runoff concentration only, it has the advantage that the catchment shape is reflected in the time-area histogram and, therefore, in the runoff hydrograph. The procedure is illustrated by the following example.

Catchment Routing   Chap. 10

Example 10-1.

A 100-km² catchment has a 4-h concentration time, with isochrones at 1-h intervals resulting in the time-area histogram shown in Fig. 10-1(b). A 6-h storm has the following effective rainfall hyetograph (Fig. 10-2):

Time (h)	0	1	2	3	4	5	6
Effective rainfall (cm/h)		0.5	1.0	2.0	1.5	1.0	0.5

Use the time-area method to calculate the outflow hydrograph.

The routing is shown in Table 10-1. Column 1 shows time in hours. The flows shown in Cols. 2-7 were obtained by multiplying effective rainfall intensity times the contributing partial area. For instance, Col. 2 shows the contribution of the first effective rainfall interval (0.5 cm/h) on each of the subareas (10, 30, 20, and 40 km²). At $t = 1$ h, the partial flow due to the first effective rainfall interval is 0.5 cm/h × 10 km² = 5 km²-cm/h (i.e., the flow contributed by the subarea enclosed within the catchment outlet and the first isochrone takes 1 h to concentrate). Likewise, at $t = 2$ h, the partial flow due to the first effective rainfall interval is 0.5 cm/hr × 30 km² = 15 km²-cm/h (i.e., the flow contributed by the subarea enclosed within the first and second isochrones takes 2 h to concentrate at the catchment outlet). The remaining values in Col. 2 (10 and 20) are calculated in a similar way. Finally, at $t = 5$ h, the flow is zero because it takes a full time interval (in the absence of runoff diffusion) for the last concentrated partial flow to recede back to zero. Columns 2 to 7 show the partial flows contributed by the six effective rainfall intervals, each appropriately lagged a time interval (because the contribution of the second rainfall interval starts at $t = 1$, the third rainfall interval at $t = 2$, and so on). The sum of these partial flows, shown in Col. 8, is the catchment outflow hydrograph. In Col. 9, the hydrograph of Col. 8 is expressed in cubic meters per second. The time base of the outflow hydrograph is 10 h, which is equal to the concentration time (4 h) plus the effective rainfall duration (6 h). To verify the accuracy of the computations, the sum of Col. 8 is 650 km²-cm/h, which represents 6.5 cm of effective rainfall depth uniformly distributed over the entire catchment area (100 km²). This value (6.5 cm) agrees with the total amount of effective rainfall.

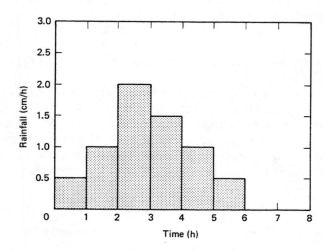

Figure 10-2 Effective rainfall hyetograph: Example 10-1.

TABLE 10-1 TIME-AREA METHOD OF CATCHMENT ROUTING: EXAMPLE 10-1

(1)	(2)	(3)	(4)	(5)	(6)	(7)	(8)	(9)
	Partial Flows and Sum (km²-cm/h)							
Time (h)	0.5 cm/h	1.0 cm/h	2.0 cm/h	1.5 cm/h	1.0 cm/h	0.5 cm/h	Sum	Flow (m³/s)
0	0						0	0.0
1	5	0					5	13.9
2	15	10	0				25	69.4
3	10	30	20	0			60	166.7
4	20	20	60	15	0		115	319.4
5	0	40	40	45	10	0	135	375.0
6		0	80	30	30	5	145	402.8
7			0	60	20	15	95	263.9
8				0	40	10	50	138.9
9					0	20	20	55.6
10						0	0	0.0
Total							650	

It is readily seen that the time-area method and the rational method share a common theoretical basis. However, since the time-area method uses effective rainfall and does not rely on runoff coefficients, it can account only for runoff concentration, with no provision for runoff diffusion. Diffusion can be provided by routing the hydrograph calculated by the time-area method through a linear reservoir with an appropriate storage constant.

The time-area method leads to an alternate way of calculating concentration time. Provided there is no runoff diffusion, as would be the case of a hydrograph calculated by the time-area method, concentration time can be calculated as the difference between hydrograph time base and effective rainfall duration. Intuitively, as rainfall ceases, the farthest parcels of water concentrate at the catchment outlet at a time equal to the concentration time. Therefore,

$$t_c = T_b - t_r \qquad (10\text{-}1)$$

in which t_c = concentration time, T_b = time base of the translated-only hydrograph, and t_r = effective rainfall duration.

Equation 10-1 can also be expressed in a slightly different form. Assuming that the point of inflection (i.e., the point of zero curvature) on the receding limb of a measured (i.e., translated and diffused) hydrograph coincides with the end of the translated-only hydrograph, the time to point of inflection of the measured hydrograph can be used in Eq. 10-1 in lieu of time base. Therefore, concentration time can be defined as the difference between the time to point of inflection and the effective rainfall duration (see Fig. 10-3):

$$t_c = t_i - t_r \qquad (10\text{-}2)$$

in which t_i = time to point of inflection on the receding limb of a measured hydrograph. The advantage of Eq. 10-2 over Eq. 10-1 is that, unlike the time base of the

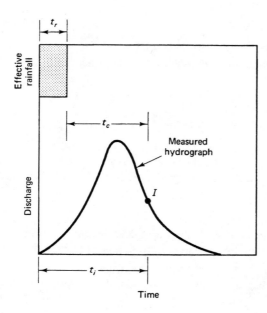

Figure 10-3 Alternate definition of concentration time.

translated-only hydrograph, the point of inflection on the receding limb of a measured hydrograph can be readily ascertained.

10.2 CLARK UNIT HYDROGRAPH

The procedure to derive a Clark unit hydrograph parallels that of the time-area method [2]. First, it is necessary to determine the catchment isochrones. In the Clark method, however, a *unit* effective rainfall is used in lieu of the effective storm hyetograph used in the time-area method. This leads to an outflow hydrograph corresponding to a unit runoff depth, that is, a unit hydrograph. Since the unit hydrograph calculated in this way lacks runoff diffusion, Clark suggested that it be routed through a linear reservoir. As with the time-area method, an estimate of the linear reservoir storage constant is required. This can be obtained either from the tail of a measured hydrograph or by using a regionally derived formula. In the latter case, the Clark unit hydrograph can be properly regarded as a synthetic unit hydrograph.

Like the time-area method, the Clark unit hydrograph method has the advantage that the catchment's properties (shape, hydraulic length, surface roughness, and so on) are reflected in the time-area histogram and, therefore, on the shape of the unit hydrograph. This feature has contributed to the popularity of the Clark unit hydrograph in engineering practice [7].

When using the Clark or time-area methods, the storage constant can be estimated from the tail of a measured hydrograph. For this purpose, the differential equation of storage (Eq. 8-4) is evaluated at a time for which inflow equals zero ($I = 0$), i.e., past the end of the translated-only hydrograph. Alternatively, it can be evaluated at the point of inflection on the receding limb of a measured hydrograph (Fig. 10-3). This leads to:

$$-O = \frac{dS}{dt} \tag{10-3}$$

and since $S = KO$, the following expression for K is obtained:

$$K = -\frac{O}{\dfrac{dO}{dt}} \tag{10-4}$$

in which O and dO/dt are evaluated past the end of the translated-only hydrograph or at (the time to) the point of inflection on the receding limb of a measured hydrograph.

The derivation of the Clark unit hydrograph is illustrated by the following example.

Example 10-2.

Use the Clark method to derive a 2-h unit hydrograph for the catchment of Example 10-1. To provide storage, route the translated-only hydrograph through a linear reservoir of storage constant $K = 2$ h. Use $\Delta t = 1$ h.

The 2-h unit hydrograph has an effective rainfall intensity of 0.5 cm/h (i.e., 1-cm depth over a 2-h duration). The calculations are shown in Table 10-2. Column 1 shows time in

TABLE 10-2   DERIVATION OF CLARK UNIT HYDROGRAPH: EXAMPLE 10-2

(1)	(2)	(3)	(4)	(5)	(6)	(7)	(8)	(9)
			Partial Flows and Sum (km^2-cm/h)					
Time (h)	0.5 cm/h	0.5 cm/h	Sum	$C_0 I_2$	$C_1 I_1$	$C_2 O_1$	Sum	Flow (m^3/s)
0	0		0	—	—	—	0.00	0.00
1	5	0	5	1	0	0	1.00	2.78
2	15	5	20	4	1	0.60	5.60	15.55
3	10	15	25	5	4	3.36	12.36	34.33
4	20	10	30	6	5	7.42	18.42	51.17
5	0	20	20	4	6	11.05	21.05	58.47
6	0	0	0	0	4	12.63	16.63	46.19
7	0	0	0	0	0	9.98	9.98	27.72
8	0	0	0	0	0	5.99	5.99	16.64
9						3.59	3.59	9.97
10						2.15	2.15	5.97
11						1.29	1.29	3.58
12						0.78	0.78	2.17
13						0.46	0.46	1.28
14						0.28	0.28	0.78
15						0.17	0.17	0.47
16						0.10	0.10	0.28
17						0.06	0.06	0.17
18						0.04	0.04	0.11
19						0.02	0.02	0.06
20						0.01	0.01	0.03
Total			100				99.98	

hours. Column 2 shows the contribution of the first hour, with 0.5 cm/h of effective rainfall. The procedure is the same as in Table 10-1, Col. 2. Column 3 shows the contribution of the second hour, with 0.5 cm/h of effective rainfall. Again, the procedure is the same as in Table 10-1, Col. 2, but the partial flows are lagged 1 h. The translated-only unit hydrograph shown in Col. 4 is the sum of the partial flows (Cols. 2 and 3). The translated-only unit hydrograph (Col. 4) is the inflow to the linear reservoir. With $\Delta t/K = \frac{1}{2}$, the routing coefficients (Table 8-1) are $C_0 = \frac{1}{5}$, $C_1 = \frac{1}{5}$, and $C_2 = \frac{3}{5}$. The partial flows of the linear reservoir routing are shown in Cols. 5 to 7, and the translated-and-diffused unit hydrograph (in km²-cm/h) shown in Col. 8 is the sum of Cols. 5 to 7 (See Example 8-1 for details of the linear reservoir-routing procedure). Column 9 shows the translated-and-diffused Clark unit hydrograph in cubic meters per second. The sum of the ordinates of the translated-only hydrograph (Col. 4) is 100, which amounts to 1 cm of effective rainfall depth uniformly distributed over 100 km² of catchment area. Likewise, the sum of the ordinates of the translated-and-diffused hydrograph (Col. 8) is 99.98, which verifies not only that the calculated hydrograph is a unit hydrograph but also that the calculation is mass (i.e., volume) conservative. Notice that the peak of the translated-only unit hydrograph (Col. 4) is 30 km²-cm/h, whereas the peak of the translated-and-diffused unit hydrograph (Col. 8) is 21.05 km²-cm/h. Also, notice that the time base of the translated-only unit hydrograph ends sharply at 6 h, whereas the time base of the translated and diffused unit hydrograph is much longer, with the receding limb of the unit hydrograph gradually approaching zero. This reveals the substantial amount of runoff diffusion provided by the linear reservoir.

By using Eq. 10-4, the linear reservoir storage constant can be calculated directly from the tail of a measured hydrograph. To illustrate the procedure, in Table 10-2, Col. 9, the two lines for $t = 6$ h and $t = 7$ h show zero outflow in the translated-only unit hydrograph (Col. 4), that is, *zero inflow* to the linear reservoir. Therefore, Eq. 10-4 can be applied between $t = 6$ h and $t = 7$ h. The average outflow (Col. 9) is $(46.19 + 27.72)/2 = 36.955$ m³/s and the rate of change of outflow is $(27.72 - 46.19)/(1 \text{ h}) = -18.47$ (m³/s)/h. Therefore, the storage constant (Eq. 10-4) is: $K = -(36.955)/(-18.47) = 2$ h. Likewise, between $t = 7$ h and $t = 8$ h: $K = -[(27.72 + 16.64)/2]/[(16.64 - 27.72)/(1)] = 2$ h. In other words, Eq. 10-4 applies at the tail of the outflow hydrograph, after the translated-only (inflow) hydrograph has receded back to zero. When using the Clark (or time-area) method, the time base of the translated-only hydrograph is equal to the sum of concentration time plus the unit hydrograph (or effective storm) duration (See Eq. 10-1).

With the help of regional analysis (Chapter 7), the Clark parameters (concentration time and linear reservoir storage constant) can be estimated based on catchment characteristics. This effectively qualifies the Clark unit hydrograph as a synthetic unit hydrograph. The Eaton [4], O'Kelly [11], and Cordery [3] models are examples of this approach. See Singh [16] for a recent review.

10.3 CASCADE OF LINEAR RESERVOIRS

As seen in Section 8.2, a linear reservoir has a diffusion effect on the inflow hydrograph. If an inflow hydrograph is routed through a linear reservoir, the outflow hydrograph has a reduced peak and an increased time base. This increase in time base causes a difference in the relative timing of inflow and outflow hydrographs, referred to as the *lag*. The amount of diffusion (and associated lag) is a function of the ratio $\Delta t/K$, a larger diffusion effect corresponding to smaller values of $\Delta t/K$.

The cascade of linear reservoirs is a widely used method of hydrologic catchment routing. As its name implies, the method is based on the connection of several linear reservoirs in series. For N such reservoirs, the outflow from the first would be taken as inflow to the second, the outflow from the second as inflow to the third, and so on, until the outflow from the $(N - 1)$th reservoir is taken as inflow to the Nth reservoir. The outflow from the Nth reservoir is taken as the outflow from the cascade of linear reservoirs. Admittedly, the cascade of reservoirs to simulate catchment response is an abstract concept. Nevertheless, it has proven to be quite useful in practice.

Each reservoir in the series provides a certain amount of diffusion and associated lag. For a given set of parameters $\Delta t/K$ and N, the outflow from the last reservoir is a function of the inflow to the first reservoir. In this way, a one-parameter linear reservoir method ($\Delta t/K$) is extended to a two-parameter catchment routing method. Moreover, the basic routing formula (Eq. 8-15) and routing coefficients (Eqs. 8-16 to 8-18) remain essentially the same.

The addition of the second parameter (N) provides considerable flexibility in simulating a wide range of diffusion and associated lag effects. However, the conceptual basis of the method restricts its general use, since no relation between either of the parameters to the physical problem can be readily envisaged. Notwithstanding this apparent limitation, the method has been widely used in catchment simulation, primarily in applications involving large gaged river basins. Rainfall-runoff data can be used to calibrate the method, i.e., to determine a set of parameters $\Delta t/K$ and N that produces the best fit to the measured data.

The analytical version of the cascade of linear reservoirs is referred to as the Nash model [10]. The numerical version is featured in several hydrologic simulation models developed in the United States and other countries. Notable among them is the SSARR model (Chapter 13), which uses it in its watershed and stream channel routing modules [18]. To derive the routing equation for the method of cascade of linear reservoirs, Eq. 8-15 is reproduced here in a slightly different form:

$$Q_{j+1}^{n+1} = C_0 \, Q_j^{n+1} + C_1 \, Q_j^n + C_2 \, Q_{j+1}^n \tag{10-5}$$

in which Q represents discharge, whether inflow or outflow and j and n are space and time indexes, respectively (Fig. 10-4).

As with Eq. 8-15, the routing coefficients C_0, C_1 and C_2 are a function of the dimensionless ratio $\Delta t/K$. This ratio is properly a Courant number ($C = \Delta t/K$). In terms of Courant number, Eqs. 8-16 to 8-18 are expressed as follows:

$$C_0 = \frac{C}{2 + C} \tag{10-6}$$

$$C_1 = C_0 \tag{10-7}$$

$$C_2 = \frac{2 - C}{2 + C} \tag{10-8}$$

For application to catchment routing, it is convenient to define the average inflow as follows:

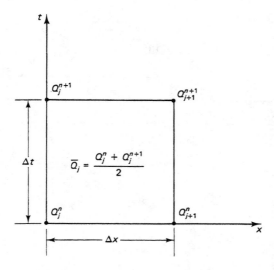

Figure 10-4 Space-time discretization in the method of cascade of linear reservoirs.

$$\bar{Q}_j = \frac{Q_j^n + Q_j^{n+1}}{2} \qquad (10\text{-}9)$$

Substituting Eq. 10-6 to 10-9 into 10-5 gives

$$Q_{j+1}^{n+1} = 2C_1\bar{Q}_j + C_2 Q_{j+1}^n \qquad (10\text{-}10)$$

or, alternatively:

$$Q_{j+1}^{n+1} = \frac{2C}{2+C}[\bar{Q}_j - Q_{j+1}^n] + Q_{j+1}^n \qquad (10\text{-}11)$$

which is the routing equation used in the SSARR model [18]. Equations 10-10 and 10-11 are in a form convenient for catchment routing because the inflow is usually a rainfall hyetograph, that is, a constant average value per time interval.

Smaller values of C lead to greater amounts of runoff diffusion. For values of C greater than 2, the behavior of Eq. 10-10 (or Eq. 10-11) is highly dependent on the type of input. For instance, in the case of a unit impulse (rainfall duration equal to the time interval), Eq. 10-10 (or Eq. 10-11) results in negative outflow values (numerical instability). For this reason, Eq. 10-10 is restricted in practice to $C \leq 2$.

The method of cascade of linear reservoirs is illustrated by the following example.

Example 10-3.

Use the method of cascade of linear reservoirs to route the following effective storm hyetograph for a 1000 km² basin. Use $N = 3$, $\Delta t = 6$ h and $K = 12$ h.

Time (h)	0	6	12	18	24
Effective rainfall (cm/h)		0.2	1.0	0.8	0.4
Effective rainfall (cm)		1.2	6.0	4.8	2.4

The Courant number is $C = \Delta t/K = \frac{6}{12} = \frac{1}{2}$, which results in $2C_1 = \frac{2}{5}$, and $C_2 = \frac{3}{5}$. The computations are shown in Table 10-3. Column 1 shows time in hours. Column 2 shows the inflow to the first reservoir (in km²-cm/h) calculated by multiplying each one of

TABLE 10-3 CASCADE OF LINEAR RESERVOIRS: EXAMPLE 10-3

(1)	(2)	(3)	(4)	(5)	(6)	(7)
	Flow (km²-cm/h)					
	N = 1		N = 2		N = 3	
Time (h)	Inflow	Outflow	Inflow	Outflow	Inflow	Outflow
0	200	0.00	40.00	0.00	8.00	0.00
6	1000	80.00	264.00	16.00	65.60	3.20
12	800	448.00	518.40	115.20	195.84	28.16
18	400	588.80	551.04	276.48	331.39	95.23
24	0	513.28	411.62	386.30	391.36	189.69
30	0	309.97	247.38	396.43	366.62	270.36
36	0	184.78	147.83	336.81	299.01	308.86
42	0	110.87	88.69	261.22	226.71	304.92
48	0	66.52	53.21	192.21	164.41	273.64
54	0	39.91	31.93	136.61	115.68	229.95
60	0	23.95	19.16	94.74	79.63	184.24
66	0	14.37	11.50	64.51	53.91	142.40
72	0	8.62	6.89	43.31	36.02	107.00
78	0	5.17	4.14	28.74	23.82	78.61
84	0	3.10	2.48	18.90	15.62	56.69
90	0	1.86	1.49	12.33	10.16	40.26
96	0	1.12	0.89	7.99	6.57	28.22
102	0	0.67	0.53	5.15	4.22	19.56
108	0	0.40	0.32	3.30	2.70	13.42
114	0	0.24	0.19	2.11	1.73	9.30
120	0	0.14		1.34		6.17
	2400	2401.77		2399.68		2389.71

the effective rainfall intensities (0.2, 1.0, 0.8, and 0.4 cm/h) times the basin area (1000 km²). Column 3 is the outflow from the first reservoir. Columns 4 and 5 are the inflow and outflow for the second reservoir. Columns 6 and 7 are the inflow and outflow for the third reservoir. (All values shown are in km²-cm/h; to convert to m³/s, multiply by 2.78). To illustrate the calculations for the first reservoir, following Eq. 10-10, $\frac{2}{5}$ of the average inflow for the first time interval [$(\frac{2}{5}) \times 200$ km²-cm/h] plus $\frac{3}{5}$ of the outflow at time $t = 0$ h [$(\frac{3}{5}) \times 0$ km²-cm/h] is equal to the outflow at 6 h: 80 km²-cm/h. Likewise, $\frac{2}{5}$ of the average inflow for the second time interval [$(\frac{2}{5}) \times 1000$ km²-cm/hr] plus $\frac{3}{5}$ of the outflow at time $t = 6$ h [$(\frac{3}{5}) \times 80$ km²-cm/h] is equal to the outflow at 12 h: 448 km²-cm/h, and so on. The (average) inflow to the second reservoir (Col. 4) is the average outflow from the first reservoir (Col. 3). For instance, for the first time interval, 40 km²-cm/h is the average of 0 and 80 km²-cm/h. The calculations proceed in a recursive fashion until the routing through the three linear reservoirs has been completed. Notice that the sum of Cols. 3, 5 and 7 is approximately the same: 2400 km²-cm/hr. Since the time interval is 6 h, this is equivalent to 2400 × 6/1000 = 14.4 cm of effective rainfall depth uniformly distributed over 1000 km² of basin area. Also notice that the peak outflow from the first reservoir is 588.8 km²-cm/h, and it occurs at 18 h; the peak outflow from the second reservoir is 396.43 km²-cm/h, occurring at 30 h; and the peak outflow from the third reservoir is 308.86 km²-cm/h, occurring at 36 h. This shows that the effect of the cascade is to produce a certain amount of runoff diffusion at every step, with a corresponding increase in the lag of catchment response.

The cascade of linear reservoirs provides a convenient mechanism for simulating a wide range of catchment routing problems. Furthermore, the method can be applied to each runoff component (surface runoff, subsurface runoff, and baseflow) separately, and the catchment response can be taken as the sum of the responses of the individual components. For instance, assume that a certain basin has 10 cm of runoff, of which 7 cm are surface runoff, 2 cm are subsurface runoff, and 1 cm is baseflow. Since surface runoff is the less diffused process, it can be simulated with a high Courant number, say $C = 1$, and a small number of reservoirs, say $N = 3$. Subsurface runoff is much more diffused than surface runoff; therefore, it can be simulated with $C = 0.4$ and $N = 5$. Baseflow, being very diffused, can be simulated with $C = 0.1$, and $N = 7$. In practice, the parameters C and N are determined by extensive calibration. In this sense, the cascade of linear reservoirs remains essentially a conceptual model [15].

10.4 CATCHMENT ROUTING WITH KINEMATIC WAVES

Hydraulic catchment routing using kinematic waves was introduced by Wooding in 1965 [19, 20, 21]. Since then, the kinematic wave approach has been widely used in deterministic catchment modeling. The approach can be either lumped or distributed, depending on whether the parameters are kept constant or allowed to vary in space. Analytical solutions are suited to lumped modeling, whereas numerical solutions are more appropriate for distributed modeling.

Wooding used an open-book geometric configuration (Fig. 4-15) to represent the catchment-stream problem physically. As its name implies, an open-book configuration consists of two rectangular catchments separated by a stream and draining laterally into it; in turn the streamflow drains out of the catchment outlet. Wooding used analytical solutions of kinematic waves and the method of characteristics to formulate his method. Since diffusion is absent from these solutions, the method is strictly applicable only to kinematic waves. Criteria for the applicability of kinematic waves have been developed by Woolhiser and Liggett [22] (Eq. 4-51) for overland flow, and by Ponce et al. [12] for stream channel flow (Eq. 9-44).

Kinematic catchment routing models can be approached in a variety of ways. Methods can be either (1) analytical or numerical, (2) lumped or distributed, (3) linear or nonlinear, or (4) single plane, two-plane, or a cascade of planes [7, 8]. Analytical models take advantage of the nondiffusive properties of kinematic waves, whereas numerical models are usually based on the method of finite differences or the method of characteristics. Linear models assume a constant wave celerity, but nonlinear models relax this restriction. The feature of variable wave celerity often renders the nonlinear models impractical because of wave steepening and associated *kinematic shock* development [8, 13]. Single- and two-plane models are used in hydrologic engineering practice [7].

The application of kinematic wave modeling to catchment routing is illustrated here with an example of a two-plane finite difference numerical model. The model could be either lumped or distributed, depending on whether the inputs and parameters are allowed to vary in space or not. For simplicity, this example considers constant input (i.e., constant effective rainfall) and constant parameters (i.e., a linear mode of computation). In practice, a computer-aided solution may relax this restriction.

Two-Plane Linear Kinematic Catchment Routing Model

Assume a catchment configured as two rectangular planes adjacent to each other, draining laterally into a stream channel located between them. Each of the planes is 100 m long by 200 m wide, and the channel is 200 m long (Fig. 10-5). The bottom friction in the planes and channel is such that the average velocity in the planes is 0.0417 m/s and the average velocity in the channel is 0.3 m/s. It is desired to obtain the hydrograph at the catchment outlet resulting from an effective rainfall of 9 cm/h lasting 20 min.

Calculation of Flow Parameters.   Since the model is linear, it is first necessary to calculate the flow parameters on which to base the calculation of the routing parameters and coefficients.

The flow per unit width in the midlength of each plane is equal to the effective rainfall intensity times the contributing area (50 m $\times$ 1 m):

$$q_p = \frac{9 \text{ cm/h} \times 50 \text{ m} \times 0.01 \text{ m/cm}}{3600 \text{ s/h}} = 0.00125 \text{ m}^2/\text{s} \qquad (10\text{-}12)$$

Since the average velocity in the planes is $v_p = 0.0417$ m/s, the average flow depth in the planes is $d_p = q_p/v_p = 0.03$ m. Laminar flow in the planes is assumed, with discharge-depth rating exponent $\beta_p = 3$. Therefore, the wave celerity in the planes is $c_p = \beta_p v_p = 0.125$ m/s.

The flow in the midlength of the channel is equal to the effective rainfall intensity times the contributing area (2 planes $\times$ 100 m $\times$ 100 m):

$$Q_c = \frac{9 \text{ cm/h} \times 2 \times 100 \text{ m} \times 100 \text{ m} \times 0.01 \text{ m/cm}}{3600 \text{ s/h}} = 0.5 \text{ m}^3/\text{s} \qquad (10\text{-}13)$$

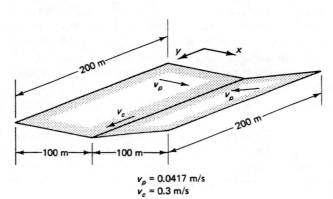

$v_p = 0.0417$ m/s
$v_c = 0.3$ m/s

Figure 10-5 Two-plane linear kinematic catchment routing model.

Catchment Routing   Chap. 10

Assume a channel top width $T_c = 5$ m. Therefore, the flow per unit width in the channel is $q_c = Q_c/T_c = 0.1$ m²/s. Since the average velocity in the channel is $v_c = 0.3$ m/s, the average flow depth in the channel (at midlength) is $d_c = q_c/v_c = 0.333$ m. A wide channel and turbulent Manning friction is assumed, with discharge-area rating exponent $\beta_c = 1.67$. Therefore, the wave celerity in the channel is $c_c = \beta_c v_c = 0.5$ m/s.

The concentration time is equal to the travel time in the planes plus the travel time in the channel. The travel time in the planes is (100 m)/(0.125 m/s) = 800 s. The travel time in the channel is (200 m)/(0.5 m/s) = 400 s. Therefore, the concentration time is 800 + 400 = 1200 s, which is equal to the effective rainfall duration. This assures concentrated flow at the catchment outlet.

The maximum possible (i.e., equilibrium) peak flow is equal to the product of rainfall intensity and catchment area:

$$Q_e = \frac{9 \text{ cm/h} \times 2 \times 100 \text{ m} \times 200 \text{ m} \times 0.01 \text{ m/cm}}{3600 \text{ s/h}} = 1 \text{ m}^3/\text{s} \tag{10-14}$$

The total volume of runoff is

$$V_r = \frac{9 \text{ cm/h} \times 20 \text{ min} \times 2 \times 100 \text{ m} \times 200 \text{ m}}{100 \text{ cm/m} \times 60 \text{ min/h}} = 1200 \text{ m}^3 \tag{10-15}$$

Selection of Discrete Intervals. For simplicity, a space interval $\Delta x = 100$ m is chosen for the planes. This amounts to one spatial increment in the planes. In an actual application using a computer, a smaller value of Δx would be indicated. The time interval is chosen as $\Delta t = 10$ min. This leads to a Courant number in the planes $C_p = c_p(\Delta t/\Delta x) = 0.75$. In the case of the channel, a space interval $\Delta y = 200$ m is chosen, that is, one spatial increment in the channel. This leads to a Courant number in the channel $C_c = c_c(\Delta t/\Delta y) = 1.5$.

Selection of Routing Scheme. There are many possible choices for routing scheme. Either first- or second-order schemes may be used (Section 9.2). In practice, first-order schemes are preferred because they are *more stable* than second-order schemes (compare the results of a first order scheme, Table 9-5, with those of a second order scheme, Table 9-4).

Two first-order schemes are chosen here: (1) Scheme I, forward-in-time, backward-in-space, stable for Courant numbers less than or equal to 1 (similar to the convex method, see Example 9-5), and (2) Scheme II, forward-in-space, backward-in-time, stable for Courant numbers greater than or equal to 1 (exact opposite of the convex method). The use of these two schemes guarantees that the solution will remain stable because scheme I is used for Courant numbers less than or equal to 1, whereas scheme II is used for Courant numbers greater than 1 [7]. In the present application, scheme I is used for routing in the planes ($C_p = 0.75$), and scheme II is used for routing in the channel ($C_c = 1.5$).

Lateral inflows are an integral part of catchment routing. For routing in the planes, lateral inflow is the effective rainfall; for channel routing, lateral inflow is the lateral contribution from the planes. Therefore, it is necessary to discretize the kinematic wave equation with lateral inflow, Eq. 9-43.

The discretization of Eq. 9-43 in a forward-in-time, backward-in-space linear scheme (Fig. 10-6(a)) leads to:

$$Q_{j+1}^{n+1} = C_1 Q_j^n + C_2 Q_{j+1}^n + C_3 Q_L \tag{10-16}$$

in which $C_1 = C$; $C_2 = 1 - C$; and $C_3 = C$, with C being the Courant number ($C = \beta v \, \Delta t/\Delta s$), with Δs either Δx (planes) or Δy (channel). The term Q_L is the lateral inflow in cubic meters per second. For routing in the planes, the lateral inflow is equal to the effective rainfall (centimeters per hour) times the applicable area (square meters). For channel routing, the lateral inflow is the average distributed lateral inflow (cubic meters per second per meter) multiplied by the channel length (meters).

The discretization of Eq. 9-43 in a forward-in-space, backward-in-time linear scheme (Fig. 10-6(b)) leads to:

$$Q_{j+1}^{n+1} = C_0 Q_j^{n+1} + C_1 Q_j^n + C_3 Q_L \tag{10-17}$$

in which $C_0 = (C - 1)/C$; $C_1 = 1/C$; and $C_3 = 1$.

The catchment routing is shown in Table 10-4. Column 1 shows time in minutes. Columns 2–4 show the plane routing, and Cols. 5–7 show the channel routing. Column 2 shows the lateral inflow to each plane:

$$Q_{Lp} = \frac{9 \text{ cm/h} \times 100 \text{ m} \times 200 \text{ m} \times 0.01 \text{ m/cm}}{3600 \text{ s/h}} = 0.5 \text{ m}^3/\text{s} \tag{10-18}$$

Notice that the lateral inflow is an average value for a time interval, and it lasts 20 min (i.e., the effective rainfall duration). Column 3 shows the upstream inflow to the plane, that is, zero (this example does not consider upstream inflow to the planes). Column 4 is obtained by routing with Eq. 10-16, with Courant number in the planes $C_p = 0.75$. Column 4 is the outflow hydrograph from each plane. The sum of Col. 2 is 1.0; likewise, the sum of Col. 4 is 0.9999, which confirms that the volume under the

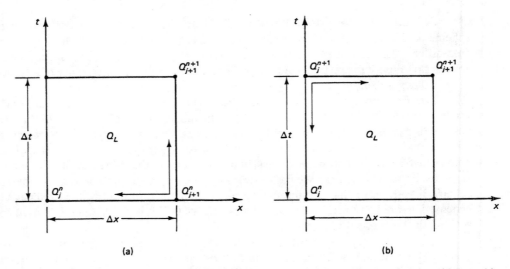

(a)   (b)

Figure 10-6 Space-time discretization of first-order schemes of kinematic wave equation with lateral inflow: (a) forward-in-time, backward-in-space; (b) forward-in-space, backward-in-time.

TABLE 10-4 TWO-PLANE CATCHMENT ROUTING WITH KINEMATIC WAVE TECHNIQUE

(1)	(2)	(3)	(4)	(5)	(6)	(7)
Time (min)	Flow (m³/s)					
	Q_{Lp}	I_p	O_p	Q_{Lc}	I_c	O_c
0		0	0		0	0
	0.5			0.3750		
10		0	0.3750		0	0.3750
	0.5			0.8437		
20		0	0.4687		0	0.8437
	0			0.5859		
30		0	0.1172		0	0.5859
	0			0.1465		
40		0	0.0293		0	0.1465
	0			0.0366		
50		0	0.0073		0	0.0366
	0			0.0091		
60		0	0.0018		0	0.0091
	0			0.0023		
70		0	0.0005		0	0.0023
	0			0.0006		
80		0	0.0001		0	0.0006
	0			0.0001		
90		0	0		0	0.0001
Sum	1.00		0.9999			1.9998

outflow hydrograph from each plane is 1 m³/s × 10 min × 60 s/min = 600 m³. Column 5 shows the average lateral inflow to the channel, obtained by multiplying by 2 the average lateral inflow from each plane (Col. 4) (to account for two planes of the same dimensions). Column 6 is the upstream inflow to the channel, that is, zero (this example does not consider upstream inflow to the channel). Column 7 is obtained by routing with Eq. 10-17, with Courant number in the channel $C_c = 1.5$. Column 7 is the outflow hydrograph from the catchment. The sum of Col. 7 is 1.9998, which confirms that the total runoff volume is 1.9998 m³/s × 10 min × 60 s/min = 1200 m³.

Assessment of Kinematic Wave Method. The calculated outflow hydrograph peak is 0.8437 m³/s, and it occurs at 20 min. This value is less than the maximum peak flow, Eq. 10-14: 1.0 m³/s. Since the rainfall duration is equal to the concentration time, this implies that the hydrograph has undergone a certain amount of runoff diffusion. This diffusion is really numerical diffusion, due primarily to the coarse grid size and secondarily to the Courant numbers (of planes and channel) being different than 1.

To prove this assertion, it is necessary to reduce the grid size and test the *convergence* of the kinematic wave schemes, Eqs. 10-16 and 10-17. Convergence refers to the ability of the numerical scheme to approach the analytical solution as the grid is refined. Due to the large number of calculations involved, the procedure is better accomplished with the aid of a computer program. Table 10-5 shows the results obtained

TABLE 10-5 KINEMATIC WAVE CATCHMENT ROUTING: EFFECT OF GRID RESOLUTION

Number of Increments	Δx (m)	Δy (m)	Δt (s)	Peak Flow (m^3/s)
1	100	200	600	0.8437
2	50	100	300	0.9063
4	25	50	150	0.9490
8	12.5	25	75	0.9776
16	6.25	12.5	37.5	0.9899

by successive grid refinement, using program EH1000B included in Appendix D. It is seen that the results are a function of grid size and that the peak flow value converges to the maximum possible value (1.0 m^3s) as the grid is refined.

It is concluded that a kinematic wave numerical solution for catchment routing is grid dependent. If necessary, numerical diffusion can be eliminated by successive grid refinement (while keeping the Courant number as close to 1 as possible). However, in this case the calculated hydrograph would be translated-only, with no diffusion. This may be adequate for catchments with negligible runoff diffusion (e.g., small catchments with slopes on the order of 1%), but is generally not adequate for catchments showing substantial amounts of runoff diffusion (e.g., midsize catchments). For the latter, the diffusion wave technique may be used as a viable alternative to the kinematic wave.

10.5 CATCHMENT ROUTING WITH DIFFUSION WAVES

Catchment routing with diffusion waves is applicable to cases where both translation and diffusion are important, that is, for routing in midsize catchments where catchment slope is such that the kinematic wave criterion is not satisfied. Although the concept of diffusion waves and catchment routing dates back to the work of Dooge [1], actual numerical applications have only recently been attempted [6, 14]. Diffusion wave routing can provide grid-independent results, and is therefore regarded as an improvement over grid-dependent techniques.

The diffusion wave catchment routing approach is illustrated here by using the same example as in the previous section. The Muskingum-Cunge method (Chapter 9) is used as the routing scheme of the diffusion wave method [14].

Two-Plane Linear Diffusion Catchment Routing Model

This example is similar to that of the previous section. Assume a catchment configured as two rectangular planes adjacent to each other, draining laterally into a stream channel located between them. Each of the planes is 100 m long by 200 m wide, and the channel is 200 m long (Fig. 10-5). The slopes of planes and channel (in the direction of the flow) are $S_{op} = 0.01$; and $S_{oc} = 0.01$, respectively. The bottom friction in the planes and channel is such that the average velocity in the planes is 0.0417 m/s, and the average velocity in the channel is 0.3 m/s. It is desired to calculate the runoff hydrograph at the catchment outlet resulting from an effective rainfall of 9 cm/h lasting 20 min.

Calculation of Flow Parameters. Since the model is linear, it is first necessary to calculate the flow parameters on which to base the calculation of the routing parameters and coefficients. As described in the preceding section, the flow per unit width in the midlength of each plane is equal to $q_p = 0.00125$ m^2/s. Since the average velocity in the planes is $v_p = 0.0417$ m/s, the average flow depth in the planes is $d_p = q_p/v_p = 0.03$ m. Laminar flow in the planes is assumed, with a discharge-depth rating exponent $\beta_p = 3$. Therefore, the wave celerity in the planes is $c_p = \beta_p v_p = 0.125$ m/s. The flow in the midlength of the channel is equal to 0.5 m^3/s. Assume a channel top width $T_c = 5$ m. Therefore, the flow per unit width in the channel is $q_c = Q_c/T_c = 0.1$ m^2/s. Since the average velocity in the channel is $v_c = 0.3$ m/s, the average flow depth in the channel (at midlength) is $d_c = q_c/v_c = 0.333$ m. A wide channel with turbulent Manning friction is assumed, with a discharge-area rating exponent $\beta_c = 1.67$. Therefore, the wave celerity in the channel is $c_c = \beta_c v_c = 0.5$ m/s. The concentration time is equal to 1200 s, which is equal to the rainfall duration. The maximum possible peak flow is 1.0 m^3/s. The total volume of runoff is 1200 m^3.

Selection of Discrete Intervals. For simplicity, a space interval of $\Delta x = 100$ m is chosen for the planes, and $\Delta y = 200$ m for the channel. This amounts to one spatial increment in planes and channel. The time interval is chosen as $\Delta t = 10$ minutes. In an actual computer application, a finer grid size would be indicated.

Selection of Routing Scheme. The chosen routing scheme is the Muskingum-Cunge method, Eq. 9-62, with the routing coefficients calculated by Eqs. 9-74 to 9-76 but modified with the addition of lateral inflow. For this purpose, Eq. 9-43 is discretized in the same way as Eq. 9-61, leading to

$$Q_{j+1}^{n+1} = C_0 Q_j^{n+1} + C_1 Q_j^n + C_2 Q_{j+1}^n + C_3 Q_L \tag{10-19}$$

which has the same meaning as Eq. 9-62 with Eqs. 9-74 to 9-76, except for the addition of the lateral inflow term, with routing coefficient:

$$C_3 = \frac{2C}{1 + C + D} \tag{10-20}$$

The term Q_L is the lateral inflow in cubic meters per second. For overland flow routing, the lateral inflow is equal to the effective rainfall (centimeters per hour) times the applicable area (square meters). For channel routing, the lateral inflow is the average distributed lateral inflow (cubic meters per second per meter) multiplied by the channel length (meters).

The grid size and physical parameters allow the calculation of the routing parameters. The Courant numbers in the planes and channel (Eq. 9-69) are $C_p = 0.75$ and $C_c = 1.5$, respectively. The cell Reynolds numbers in the planes and channel (Eq. 9-71) are $D_p = 0.01$ and $D_c = 0.1$, respectively.

With Eqs. 9-74 to 9-76 and 10-20, the routing coefficients in the planes are $C_0 = -0.136$, $C_1 = 0.988$, $C_2 = 0.148$, and $C_3 = 0.852$. Likewise, the routing coefficients in the channel are $C_0 = 0.231$, $C_1 = 0.923$, $C_2 = -0.154$, and $C_3 = 1.154$.

TABLE 10-6 TWO-PLANE CATCHMENT ROUTING WITH DIFFUSION WAVE TECHNIQUE

(1)	(2)	(3)	(4)	(5)	(6)	(7)
Time (min)	Flow (m³/s)					
	Q_{Lp}	I_p	O_p	Q_{Lc}	I_c	O_c
0		0	0		0	0
	0.5			0.4260		
10		0	0.4260		0	0.4916
	0.5			0.9150		
20		0	0.4890		0	0.9802
	0			0.5614		
30		0	0.0724		0	0.4969
	0			0.0831		
40		0	0.0107		0	0.0194
	0			0.0123		
50		0	0.0016		0	0.0112
	0			0.0018		
60		0	0.0002		0	0.0004
	0			0.0002		
70		0	0		0	0.0002
Sum	1.00		0.9999			1.9999

The catchment routing is shown in Table 10-6. Column 1 shows time in hours. Columns 2–4 show the plane routing and Cols. 5–7 show the channel routing. As in the previous example, the lateral inflow to the plane is 0.5 m³/s (Eq. 10-18). Column 3 shows the upstream inflow to each plane, that is, zero (this example does not consider upstream inflow to the planes). Column 4 is the outflow from each one of the planes, obtained by routing with Eq. 10-19. Column 5 shows the average lateral inflow to the channel, obtained by multipying by 2 the average inflow from each plane (Col. 4) (two planes, each of the same dimensions). Column 6 is the upstream inflow to the channel, that is, zero (this example does not consider upstream inflow to the channel). Column 7 is the outflow from the catchment, obtained by routing with Eq. 10-19. As in the previous example, the sums of Cols. 2, 4 and 7 confirm that the runoff volumes are appropriately conserved.

Assessment of the Diffusion Wave Method. The calculated outflow hydrograph peak is 0.9802 m³/s, and it occurs at 20 min. This value is very close to the maximum possible peak flow (1.0 m³/s), revealing that the amount of physical diffusion for this particular example is relatively small.

To study the effect of grid size on the hydrograph calculated by the diffusion wave method, a test similar to that of the previous section is performed with the aid of a computer program. Table 10-7 shows the results obtained by successive grid refinement, using program EH1000C included in Appendix D. It is shown that the results are essentially independent of grid size. A coarse grid (one increment in space and time) results in a peak flow of 0.9804, whereas a fine grid (16 increments in space and time) results in a value of 0.9845. However, it should be noted that the coarse-grid

TABLE 10-7 DIFFUSION WAVE CATCHMENT ROUTING: EFFECT OF GRID RESOLUTION

Number of Increments	Δx (m)	Δy (m)	Δt (s)	Peak Flow (m³/s)
1	100	200	600	0.9804
2	50	100	300	0.9716
4	25	50	150	0.9766
8	12.5	25	75	0.9814
16	6.25	12.5	37.5	0.9845

solution (one increment in space and time) exhibits a small but perceptible amount of numerical dispersion, as demonstrated by its peak (0.9804) being somewhat greater than the peak obtained with two increments (0.9716). This is caused by the negative C_0 in the planes ($C_0 = -0.316$).

Unlike in the kinematic wave method, in the diffusion wave technique the numerical diffusion is matched to the physical problem. As in the case of stream channel routing (Section 10.4), this procedure works best when numerical dispersion is minimized, that is, when Courant numbers are kept reasonably close to 1. In practice, substantial deviations from this condition may lead to increased numerical dispersion and associated numerical instability. As with other finite difference schemes, a test of grid independence similar to that shown in Table 10-7 is necessary to verify the convergence of the numerical scheme.

10.6 ASSESSMENT OF CATCHMENT-ROUTING TECHNIQUES

Catchment-routing techniques have evolved from the simple time-area methods to the more elaborate physically based kinematic and diffusion wave techniques. The variety of existing methods and techniques reflects the fact that no one method is applicable to all cases. Surface runoff in catchments is a complex phenomenon, and research continues to unveil improved ways of solving the problem.

In nature, catchments can be either small, midsize, or large. Runoff processes are nonlinear and distributed—nonlinear in the sense that the parameters do not increase in the same proportion as the flow and distributed in the sense that the parameters vary within the catchment. Hydrographs originating in catchment runoff are generally concentrated, diffused, and dispersed. From the mathematical standpoint, concentration is a first-order process, diffusion a second-order process, and dispersion a third-order process [5]. Being a first-order process, runoff concentration is the primary mechanism; it is also referred to as translation or wave travel. For certain applications, diffusion also plays an important role—for instance, in the modeling of runoff response in catchments with mild slopes. Early description of the diffusion mechanism referred to it as storage. The usage is so widespread that it has been preserved in this book. In a routing context, diffusion and (longitudinal channel) storage have essentially the same meaning. Dispersion is a third order process; therefore, it is usually much smaller than runoff concentration and diffusion. Physical dispersion translates into wave steepening, a concept similar to that of skewness, or the third moment of a statistical distribution (Chapter 6).

The time-area method can calculate concentration but it cannot account for diffusion or dispersion. Therefore, the time-area method should be limited to small and midsize catchments where translation is by far the predominant mechanism. When used indiscriminately, the time-area method always overestimates the peak of the outflow hydrograph. If necessary, diffusion can be added by routing the time-area hydrograph through a linear reservoir. However, the storage constant would have to be determined either from measured data or by synthetic means. While the time-area method is a lumped method, it can be used as a component of larger network models which have a distributed structure [17].

In principle, the method of cascade of linear reservoirs accounts for runoff diffusion only. However, the connection of several linear reservoirs in series provides enough diffusion so that translation is actually being simulated by means of diffusion. The method is linear insofar as the routing parameters are determined by calibration. Nevertheless, by successive calibrations at different flow levels, from low to high, the method can be made to reflect the nonlinearity actually existing in nature. The method is lumped, but it can be used as a component of larger network models that have a distributed structure. The method has been successfully applied to very large basins (in excess of 10,000 km^2), which exhibit substantial amounts of runoff diffusion. For such large basins, the distributed models of the kinematic and diffusion type would be impractical due to the prohibitive amount of data required to properly specify the spatial diversity.

The kinematic wave model provides translation and diffusion, the latter, however, due only to the finite grid size. The method can be linear or nonlinear, and lumped or distributed, depending on the numerical scheme and input data. The method is applicable to small catchments with steep slopes where diffusion is small and can be controlled by grid refinement. Theoretically, the method could also be applicable to midsize catchments, as long as physical diffusion remains small. In practice, the larger the catchment, the more unlikely it is that physical diffusion is negligible. The distributed nature of kinematic wave models results in substantial data needs; the use of average parameters would render the model lumped, with the consequent loss of detail. Another important consideration in kinematic wave models is the validity of the geometric configuration. For instance, two-plane descriptions are adequate as long as the catchment geometry fits the two-plane model configuration. Otherwise, a certain amount of lumping would be introduced by the model's inability to properly account for the physical detail. In practice, the larger the catchment, the more difficult it is to fit catchments within two-plane descriptions. Multiple-plane descriptions are possible but invariably lead to additional complexity [8].

The diffusion wave technique (Muskingum-Cunge scheme) provides translation and diffusion, and, unlike the kinematic wave model, its solution is generally independent of grid size. Therefore, it is applicable to catchments with substantial amounts of physical diffusion, either small catchments of mild slope or midsize catchments with average slopes. The method is linear or nonlinear and lumped or distributed, depending on the numerical scheme and input data. As with kinematic models, the validity of the geometric configuration is important in diffusion wave catchment models. Unless the model's geometric abstraction is a reasonable representation of the catchment's actual geometry, the degree of lumping introduced would tend to mask the distributed nature of the model.

The preceding comments have referred specifically to catchment routing, i.e., the conversion of effective rainfall into runoff. In practice, comparative evaluations of the performance of catchment-routing methods are hampered by the fact that it is seldom possible to determine effective rainfall with any degree of certainty. Hydrologic abstractions are time-variant, distributed, and nonlinear. Therefore, a proper estimation of hydrologic abstractions is crucial to the performance evaluation of catchment-routing models.

From this discussion, it can be concluded that no one method or model is suitable for all applications. All have strengths and weaknesses; they are either simple or complex and suffer from lack of detail or require a substantial amount of data for their successful operation. In practice, the choice of catchment-routing method remains one of individual preference and experience.

QUESTIONS

1. What is the difference between hydrologic and hydraulic methods of catchment routing?
2. What are catchment isochrones? How are they determined?
3. How is concentration time defined when using hydrographs generated by the time-area method?
4. When can the Clark unit hydrograph be considered synthetic? Explain.
5. What is the difference between translation and diffusion? How are the time-area and rational methods related?
6. How can the linear reservoir storage coefficient be determined for runoff data?
7. What is the principle behind the method of cascade of linear reservoirs used in catchment routing?
8. Why is it necessary to use two schemes in catchment routing using first-order kinematic wave techniques?
9. Why is a kinematic wave solution using numerical techniques usually grid dependent? Why is the diffusion wave solution grid independent?

PROBLEMS

10-1. A 45-km^2 catchment has a 6-h concentration time with isochrones at 2-h intervals, resulting in the following time-area histogram:

Time (h)	0		2		4		6
Area (km^2)		9		21		15	

Use the time-area method to calculate the outflow hydrograph from the following effective storm pattern:

Time (h)	0		2		4		6		8		10		12
Effective rainfall (cm/h)		0.5		1.0		2.0		3.0		1.0		0.5	

10-2. A 68-km² catchment has a 4-h concentration time, with isochrones at 1-h intervals resulting in the following time-area histogram:

Time (h)	0	1	2	3	4
Area (km²)		12	19	26	11

Use the time-area method to calculate the outflow hydrograph from the following storm pattern:

Time (h)	0	1	2	3	4	5	6
Total rainfall (cm/h)		1	2	4	3	2	1

The runoff curve number is CN = 75.

10-3. Develop an interactive computer program to solve the time-area method of catchment routing. Input should be the following: (1) time interval, (2) number of isochrones, (3) the subareas corresponding to each isochrone, (4) number of effective rainfall increments, and (5) effective rainfall intensities corresponding to each increment of time. Test your program using Example 10-1 in the text.

10-4. Modify the computer program developed in Problem 10-3 to route the hydrograph calculated by the time-area method through a linear reservoir of storage constant K. Test your program using Example 10-2 in the text.

10-5. A 82-km² catchment has the following characteristics:

Time (h)	0	1	2	3	4	5	6
Area (km²)		14	22	29	17	0	0
Effective rainfall (cm/h)		0.8	1.2	1.3	2.1	0.7	0.5

Use the computer program developed in Problem 10-3 to calculate the outflow hydrograph by the time-area method.

10-6. Use the computer program developed in Problem 10-4 to route the result of Problem 10-5 through a linear reservoir. Compare results with the following storage constants: (1) $K = 1$ h, (2) $K = 2$ h, and (3) $K = 3$ h.

10-7. Use the Clark method to derive a 3-h unit hydrograph for a catchment with the following time-area diagram:

Time (h)	0	3	6	9	12
Area (km²)		57	72	39	15

Use $\Delta t = 3$ h and $K = 3$ h.

10-8. Use the Clark method to derive a 1-h unit hydrograph for a catchment with the following time-area diagram:

Time (h)	0	1	2	3	4	5	6
Area (km²)		12	20	42	66	30	16

Use $\Delta t = 1$ h and $K = 2$ h.

10-9. Use the Clark method to derive a 2-h unit hydrograph for a catchment with the following time-area diagram:

Time (h)	0	1	2	3	4	5	6
Area (km²)		10	20	30	20	12	8

Use $\Delta t = 1$ h and $K = 4$ h.

10-10. The 2-h unit hydrograph for a 92-km² catchment is the following:

Time (h)	0	2	4	6	8	10	12
Flow (m³/s)	0	2.778	8.611	14.333	19.433	21.938	20.942
Time (h)	14	16	18	20	22	24	26
Flow (m³/s)	15.897	9.538	5.722	3.433	2.061	1.236	0.742
Time (h)	28	30					
Flow (m³/s)	0.444	0.267					

Given the following time-area diagram, what is the linear reservoir storage constant in the Clark method?

Time (h)	0	2	4	6	8	10	12
Area (km²)		10	15	18	21	16	12

10-11. A 1-h unit hydrograph derived from measured data has the following ordinates:

Time (h)	0	1	2	3	4	5	6	7
Flow (m³/s)	0	7	22	48	60	90	74	47
Time (h)		8	9	10	11	12	13	14
Flow (m³/s)		28	17	10	6	4	3	2

Assuming a time of concentration $t_c = 6$ h, calculate the linear reservoir storage constant in the Clark method.

10-12. Use Example 10-3 in the text (cascade of linear reservoirs) to test program EH1000A included in Appendix D. Then route the effective rainfall hyetograph of Example 10-3 (with time interval $\Delta t = 6$ h) using (a) $K = 12$ h and $N = 4$ and (b) $K = 18$ h and $N = 3$.

10-13. Use program EH1000A to route the following storm hyetograph through a 535-km² catchment.

Time (h)	0	6	12	18	24
Effective rainfall depth (cm)		1.0	1.5	2.5	1.2

Set $K = 12$ h and $N = 3$. Report peak outflow and time-to-peak.

10-14. Using the method of cascade of linear reservoirs, derive a 3-h unit hydrograph for a 432-km² basin. Assume $\Delta t = 3$ h, $K = 9$ h and two reservoirs ($N = 2$). Verify the results with program EH1000A.

10-15. Use program EH1000A to derive a 6-h unit hydrograph for a 1235-km² basin. Assume $\Delta t = 6$ h, $K = 6$ h and $N = 4$.

10-16. Derive Eqs. 10-16 and 10-17.

10-17. Test program EH1000B included in Appendix D (two-plane linear kinematic catchment routing model) using the illustrative problem of Section 10.4.

10-18. Use program EH1000B to solve the following catchment routing problem: plane length, 840 m; number of plane increments, 4; channel length, 1250 m; number of channel increments, 6; total simulation time, 1800 min; number of time intervals, 180; effective rainfall intensity, 5 cm/h; effective rainfall duration, 180 min; average wave celerity in the planes, 0.1 m/s; average wave celerity in the channel, 0.5 m/s. Report peak outflow, time-to-peak, and runoff volume.

10-19. Derive Eq. 10-19.

10-20. Test program EH1000C included in Appendix D (two-plane linear diffusion catchment routing model) using the illustrative problem of Section 10.5.

10-21. Use program EH1000C to solve the catchment-routing problem of Problem 10-18. Use the following additional data: slope of planes, 0.01; slope of channel, 0.0075; average flow per unit width in the planes, 0.001 m³/s/m; average flow per unit width in the channel, 0.1 m³/s/m. Report peak outflow, time-to-peak, and runoff volume.

REFERENCES

1. Agricultural Research Service, U.S. Department of Agriculture. (1973). "Linear Theory of Hydrologic Systems," *Technical Bulletin No.* 1468, (J. C. I. Dooge, author), Washington, D.C.

2. Clark, C. O. (1945). "Storage and the Unit Hydrograph," *Transactions,* ASCE, Vol. 110, pp. 1416–1446.

3. Cordery, I. (1968). "Synthetic Unit Graphs for Small Catchments in Eastern New South Wales," *Civil Engineering Transactions of the Institution of Engineers,* Australia, Vol. 10, pp. 47–58.

4. Eaton, T. D. (1954). "The Derivation and Synthesis of Unit Hydrographs When Rainfall Records Are Inadequate," *Institution of Engineers,* Australia, Vol. 26, pp. 239–246.

5. Ferrick, M. G., J. Bilmes, and S. E. Long. (1983). "Analysis of a Diffusion Wave Flow Routing Model with Application to Flow in Tailwaters," U.S. Army Corps of Engineers, Cold Regions Research and Engineering Laboratory, CRREL Report 83-7, March.

6. Hromadka, T. V., R. H. McCuen, and C. C. Yen. (1987). "Comparison of Overland Flow Hydrograph Models," *Journal of Hydraulic Engineering,* ASCE, Vol. 113, No. 11, November, pp. 1422–1440.

7. Hydrologic Engineering Center, U.S. Army Corps of Engineers. (1985). "HEC-1, Flood Hydrograph Package, Users Manual," September, 1981, revised January 1985.

8. Kibler, D. F., and D. A. Woolhiser. (1970). "The Kinematic Cascade as a Hydrologic Model," Colorado State University *Hydrology Paper No.* 39, Fort Collins, Colorado.

9. Lighthill, M. J., and G. B. Whitham. (1955). "On Kinematic Waves. I. Flood Movement in Long Rivers," *Proceedings, Royal Society,* London, Vol. A229, May, pp. 281–316.

10. Nash, J. E. (1958). "The Form of the Instantaneous Unit Hydrograph," International Association for Scientific Hydrology, *Publication No.* 42, No. 3, pp. 114–118.

11. O'Kelly, J. J. (1955). "The Employment of Unit Hydrograph to Determine the Flows of Irish Arterial Drainage Channels," *Proceedings of the Institution of Civil Engineers,* Dublin, Ireland, Vol. 4, No. 3, pp. 365–412.

12. Ponce, V. M., R. M. Li, and D. B. Simons. (1978). "Applicability of Kinematic and Diffusion Models," *Journal of the Hydraulics Division,* ASCE, Vol. 104, No. HY3, March, pp. 353–360.

13. Ponce, V. M., and D. Windingland. (1985). "Kinematic Shock: Sensitivity Analysis," *Journal of Hydraulic Engineering,* ASCE. Vol. 111, No. 4, April, pp. 600–611.

14. Ponce, V. M. (1986). "Diffusion Wave Modeling of Catchment Dynamics," *Journal of Hydraulic Engineering,* ASCE, Vol. 112, No. 8, August, pp. 716–727.

15. Rao, R. A., J. W. Delleur, and B. S. Sarma. (1972). "Conceptual Hydrologic Models for Urbanizing Basins," *Journal of the Hydraulics Division,* ASCE, Vol. 98, No. HY7, July, pp. 1205–1220.

16. Singh, V. P. (1988). *Hydrologic Systems, Rainfall-Runoff Modeling.* Vol. 1. Englewood Cliffs, N.J.: Prentice-Hall.

17. Stall, J. B., and Terstriep, M. L. (1972). "Storm-Sewer Design: An Evaluation of the RRL Method," *EPA Technology Series* EPA-R2-72-068, October.

18. U.S. Army Engineer Division, North Pacific Division. (1986). "Program Description and User Manual for SSARR Model, Streamflow Synthesis and Reservoir Regulation," Portland, Oregon, Draft, April.

19. Wooding, R. A. (1965). "A Hydraulic Model for the Catchment-stream Problem, I. Kinematic Wave Theory," *Journal of Hydrology,* Vol. 3, pp. 254–267.

20. Wooding, R. A. (1965). "A Hydraulic Model for the Catchment-stream Problem, II. Numerical Solutions," *Journal of Hydrology,* Vol. 3, 1965, pp. 268-282.
21. Wooding, R. A. (1966). "A Hydraulic Model for the Catchment-stream Problem, III. Comparison with Runoff Observations," *Journal of Hydrology,* Vol. 4, pp. 21-37.
22. Woolhiser, D. A., and Liggett, J. A. (1967). "Unsteady One-dimensional Flow Over a Plane: The Rising Hydrograph," *Water Resources Research,* Vol. 3, No. 3, pp. 753-771.

SUGGESTED READINGS

Agricultural Research Service, U.S. Department of Agriculture. (1973). "Linear Theory of Hydrologic Systems," *Technical Bulletin No.* 1468, (J. C. I. Dooge, author), Washington, D.C.

Clark, C. O. (1945). "Storage and the Unit Hydrograph," *Transactions,* ASCE, Vol. 110, 1945, pp. 1416-1446.

U.S. Army Engineer Division, North Pacific Division. (1986). "Program Description and User Manual for SSARR Model, Streamflow Synthesis and Reservoir Regulation," Portland, Oregon, Draft, April.

Wooding, R. A. (1965). "A Hydraulic Model for the Catchment-stream Problem, I. Kinematic Wave Theory," *Journal of Hydrology,* Vol. 3, 1965, pp. 254-267.

SUBSURFACE WATER

The study of surface water is incomplete without the knowledge of its interaction with *subsurface water*. Subsurface water comprises all water either in storage or flowing below the ground surface. There are two types of subsurface water: (1) interflow, and (2) groundwater flow. Interflow takes place in the unsaturated zone, close to the ground surface. Groundwater flow takes place in the saturated zone, which may be either close to the ground surface or deep in underground waterbearing formations. The surface separating the unsaturated and saturated zones is referred to as the *groundwater table*, or *water table*.

In Chapter 2, the following three components of runoff were identified: (1) surface runoff, (2) runoff contributed by interflow, and (3) runoff contributed by groundwater, i.e., baseflow. These components depict the path of runoff. At any one time, runoff consists of a combination of the three. Generally, during wet-weather periods, surface runoff and interflow are the primary contributors to runoff. Conversely, during dry-weather periods, baseflow is the major—if not the only—contributor to runoff.

Traditionally, surface runoff has been regarded as the single most important component of flood flows. This approach is embodied in the concept of overland flow, or *Hortonian* flow, after Horton, who pioneered the theory of infiltration capacity [12]. As shown in Chapters 4 and 10, overland flow can be used to simulate runoff response.

Notwithstanding the classical Hortonian approach, recent theories of *hillslope hydrology* have emphasized the role of interflow and the timing—rather than the path—of runoff. Two runoff components are recognized under this framework: (1) *quickflow*, consisting of overland flow, fast interflow, and rain falling directly on the channel network, and (2) baseflow, consisting of slow interflow and groundwater flow [11].

This chapter is divided into five sections. Section 11.1 describes general properties of subsurface water, and Section 11.2 describes physical properties. Section 11.3 describes the equations of groundwater flow, and Section 11.4 deals with well hydraulics. Section 11.5 discusses surface-subsurface flow interaction, including concepts of hillslope hydrology and baseflow recession.

11.1 GENERAL PROPERTIES OF SUBSURFACE WATER

Subsurface water occurs by infiltration of rainfall and/or snowmelt into the ground. Once the water has infiltrated, it can follow one of two paths: (1) move in a general lateral direction within the unsaturated zone close to the ground surface or (2) move in a general downward direction and join the saturated zone. Interflow is flow in the unsaturated zone; groundwater flow is flow in the saturated zone.

The earth's crust is composed of soils and rocks containing pores (i.e., voids) that can hold air and water. The various types of soils and rocks have different relative amounts of pore space and, consequently, can hold different amounts of air and water. In subsurface water evaluations, the earth's crust is divided into two zones: (1) the unsaturated zone, where the pores are filled with both air and water, and (2) the saturated zone, where the pores are filled only with water. The boundary between the unsaturated and saturated zones is the water table.

The distance from the ground surface to the water table varies from place to place. In some places it may be less than 1 m, whereas in others it may be more than 100 m. In general, the water table is not flat, tending to follow the surface topography in a subdued way, deeper beneath the hills and shallower beneath the valleys. In certain cases it may even coincide with the ground surface, as with ponds and marshes, or lie slightly above it, as in the typical exfiltration to perennial streams and rivers.

Extent of Groundwater Resources

Although only a fraction of precipitation infiltrates into the ground, the total amount of subsurface water is far greater than the total amount of land surface water. This is because groundwater flow is characteristically a very slow process, whereas land surface water moves at comparatively faster speeds. The average residence time of surface water (i.e., the time elapsed while flowing on the earth's surface) is estimated at 2 wk. On the other hand, the average residence time of subsurface water has been estimated to vary between 2 wk to 10,000 y [17]. Both surface and subsurface water are driven by the force of gravity in their unrelenting movement toward the sea.

To place the relationship between surface and subsurface water amounts in the proper perspective, it is necessary to examine the world's water balance. Studies have shown that about 94 percent of all the world's water is seawater. Of the remaining 6 percent, one-third occurs in solid form in glaciers and polar ice caps, and two-thirds is fresh water, which includes surface and subsurface water. Of the total amount of fresh water, more than 99 percent is groundwater. The water stored in lakes, reservoirs, streams, rivers, the unsaturated zone below the ground surface, and in vapor form in the atmosphere accounts for only a small fraction of the total amount of fresh water [8].

About half of the groundwater is contained within 800 m of the earth's surface [18]. Not all can be used, either because of its salinity or because of the great depths at which it occurs. The distribution of groundwater varies throughout the land areas of the world. Where it does occur, it can been used to supplement surface water supplies. Furthermore, in regions with little or no surface water resources, groundwater is often the only source of fresh water.

The feasibility of mining or extracting water from a groundwater reservoir is determined by the following three properties: (1) porosity, (2) permeability, and (3) replenishment. *Porosity* is the ratio of void volume to total volume of soil or rock. It is interpreted as a measure of the ability of the soil deposit or rock formation to hold water in sufficiently large quantities.

Permeability describes the rate at which water can pass through a soil deposit or rock formation. Permeable materials are those that allow water to pass through them easily. Conversely, impermeable materials are those that allow water to pass through them only with difficulty or not at all. The permeability value is a function of the size of pores or voids and the degree to which they are interconnected.

Replenishment relates to the size and extent of a groundwater reservoir and its connection to other neighboring groundwater resources in the region. Replenishment is largely controlled by nature, with little or no human intervention.

Aquifers

An *aquifer* is a saturated permeable geologic formation which can yield significant quantities of water to wells and springs. By contrast, an *aquiclude* is a saturated geologic formation that is incapable of transmitting significant amounts of water under ordinary circumstances.

The term *aquitard* describes the less permeable beds in a stratigraphic sequence. These beds may be permeable enough to transmit water in significant quantities, but not sufficient to justify the cost of drilling wells to exploit the groundwater resource. Most geologic formations are classified as either aquifers or aquitards, with very few formations fitting the definition of an aquiclude.

Aquifer Types. Aquifers can be of two types: (1) unconfined and (2) confined. An unconfined aquifer, or water table aquifer, is an aquifer in which the water table constitutes its upper boundary. A confined aquifer is an aquifer that is confined between two relatively impermeable layers or aquitards. Unconfined aquifers occur near the ground surface; confined aquifers occur at substantial depths below the ground surface. Figure 11-1 shows typical configurations of confined and unconfined aquifers.

The water level in an unconfined aquifer rests at the water table. In a confined aquifer, the water level in a well may rise above the top of the aquifer. If this is the case, the well is referred to as an *artesian well,* and the aquifer is said to exist under artesian conditions. In some cases, the water level may flow above the ground surface, in which case the aquifer is known as a *flowing artesian well,* and the aquifer is said to exist under flowing artesian conditions.

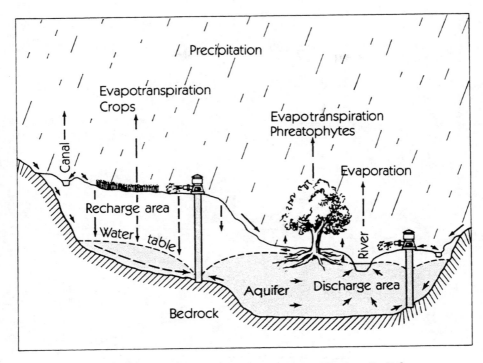

Figure 11-1(a)   Groundwater flow through an unconfined aquifer [24].

The water level in wells located in a confined aquifer defines an imaginary surface referred to as the potentiometric surface [8]. Several wells can help establish a potentiometric contour map, a map depicting lines of equal hydraulic head in the aquifer. A potentiometric map provides an indication of the direction of groundwater flow in an aquifer.

A *perched aquifer* is a special case of unconfined aquifer. A perched aquifer forms on top of an impermeable layer located well above the water table. Infiltrating water is held on top of this impermeable layer to form a saturated lens, usually of limited extent and not connected to the main water table. The water table of a perched aquifer is referred to as a *perched water table*.

Recharge and Discharge.   Typically, groundwater flows from a recharge area, through a groundwater reservoir, to a discharge area. The recharge area is an area of replenishment with infiltrated water. The groundwater reservoir is the main body of the aquifer. The discharge area is the area where the infiltrated water returns back to the surface.

In humid and subhumid climates, aquifer recharge usually takes place in upland slopes, with aquifer discharge occurring in the valleys, where the water table is shallow enough to be intercepted by streams and rivers. In arid and semiarid regions, however, the situation may be quite different. In this case, the water table in the valleys is usually much deeper, with aquifer recharge taking place primarily by channel transmission losses in streams and rivers.

Discharge from an unconfined aquifer is accomplished in three ways. First, if

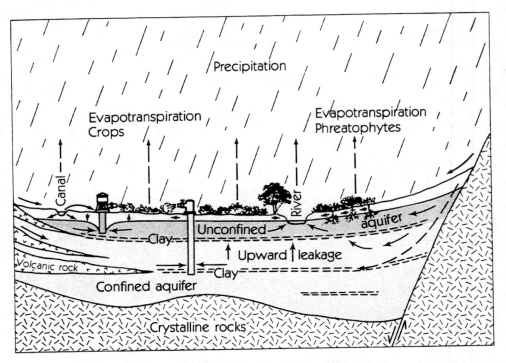

Figure 11-1(b) Groundwater flow through an unconfined aquifer, a confined aquifer, and a poorly permeable clay layer separating them [24].

the water table is close to the ground surface, water may be discharged from the aquifer either by vapor diffusion upward through the soil or through evapotranspiration by vegetation. Second, if the water table is intersected by a stream, discharge is accomplished by exfiltration. Third, an aquifer can be discharged by human-induced means, i.e., by pumping through a well, either for agricultural, municipal, or industrial uses.

Discharge from a confined aquifer is accomplished in two ways: first, by fast seepage through a permeable path in the overlying impermeable material, or by slow seepage through aquitards; and second, by human-induced means, i.e., by well pumping as with unconfined aquifers [24].

11.2 PHYSICAL PROPERTIES OF SUBSURFACE WATER

In an unconfined aquifer, the water table is the surface at which the water pressure is exactly equal to atmospheric pressure. The soil or rock below the water table is generally considered to be saturated with water. Indeed, the water table is the upper limit of a zone of saturation, or saturated zone.

The capillary fringe is located immediately above the water table. Water is held in this fringe by capillarity, at moisture levels close to saturation. However, the capillary fringe differs from the saturated zone in that a well will fill with water only to the base of the capillary fringe, i.e., the water table. Water in the capillary fringe is referred to as *capillary water* to distinguish it from the water in the saturated zone, or groundwater proper.

The thickness of the capillary fringe varies from one rock formation to another—depending on the size of the pores—from a few millimeters to several meters. Due to natural irregularities, the top of the capillary fringe is likely to be an irregular surface, with the moisture likely to decrease gradually in direction away from the water table.

Lowering of the water table by drainage, pumping, or other means will result in a lowering of the capillary fringe. However, all water cannot be drained out of the soil or rock formation. Surface tension and molecular effects are responsible for a certain amount of water being retained in the pores against the action of gravity.

Specific Yield

The total amount of water in an aquifer of area A and thickness b is

$$V = Abn \tag{11-1}$$

in which V = total volume of water, A = surface area of the aquifer, b = aquifer thickness, and n = porosity. However, the total amount of water that will drain freely from an aquifer is

$$V_w = AbS_y \tag{11-2}$$

in which V_w = volume of free-draining water and S_y = specific yield, the ratio of free-draining water volume to aquifer volume. Since a certain amount of water is always retained in the pore volume, the specific yield of an aquifer is always less than its porosity.

Specific retention is the ratio of volume of retained water to volume of aquifer. Therefore, the sum of specific yield and specific retention is equal to the porosity. In coarse-grained rocks with large pores, specific yield will be almost equal to the porosity, with specific retention reduced to a minimum. Conversely, in fine-grained rocks, specific retention can approximate the value of porosity, with specific yield being close to zero.

Above the capillary fringe is the intermediate zone, which may range in thickness from zero to more than 100 m. The water within this zone is referred to as *pellicular water,* with the water content generally close to specific retention. Above the intermediate zone is the upper portion of the earth's crust, referred to as the soil. The capillary fringe, the intermediate zone, and the soil are all part of the unsaturated zone. Interflow takes place in the unsaturated zone.

Soil Moisture Levels

Water enters the unsaturated zone by infiltration, where it is held in thin films around the solid particles or in the pore space between them. Within a given volume, the degree of saturation is a function of the relative amount of pore space that is occupied by water. The degree of saturation is a measure of the prevailing soil-moisture condition.

Field Capacity. Assume that a heavy rainfall causes significant amounts of water to infiltrate into the ground. If this situation persists for a sufficiently long period, the voids will eventually fill with water, i.e., a saturated condition will be attained.

During and immediately following rainfall, water will drain freely out of the soil under the effect of gravity, a characteristically slow process that can last anywhere from a few hours to several days. The total amount of water drained in this way represents the specific yield, with the remaining water constituting the specific retention. The moisture level equivalent to specific retention is the *field capacity*, i.e., the amount of water that can be held in the soil against the action of gravity.

Following a rainy period, evaporation from soil surfaces and evapotranspiration from vegetation act together to reduce the soil moisture below field capacity, and a soil-moisture deficit is gradually developed. The soil-moisture deficit is usually expressed in terms of the rainfall depth necessary to restore the soil moisture to field capacity.

Permanent Wilting Point. At or near field capacity, the water is largely held in thin films around the solid particles, with substantial amounts of air space between them. At the air-water interface, the difference in pressures—atmospheric in the air spaces and less than atmospheric within the thin films—results in a net suction effect. This suction effect, referred to as *soil-water suction*, keeps the water films adhered to the soil particles against the action of gravity.

With continuing absence of rainfall, increasing amounts of soil water are abstracted by evaporation, evapotranspiration, and vapor diffusion. This causes the water films surrounding the solid particles to become thinner and the soil-water suction to increase. Consequently, it becomes increasingly difficult for plants to extract water from the soil. Faced with this condition, the plants respond by using less water and reducing their growth rate. Plants that are deprived of normal amounts of water for an extended period of time begin to wilt. If water does not become available within a reasonable period, the *permanent wilting point* is reached and the plants die.

Flow Through Porous Media

The soil or rock deposits through which water flows can be regarded as porous media. The fundamental law governing flow through porous media is Darcy's law.

Darcy's Law. In 1856, Darcy pioneered the analysis of flow of water through sands [6]. His experiments led to the formulation of the empirical equation of flow through porous media bearing his name.

With reference to Fig. 11-2, Darcy's law states that the flow rate Q is directly proportional to the cross-sectional flow area A and hydraulic drop Δh and inversely proportional to the length Δl:

$$Q = KA \frac{\Delta h}{\Delta l} \tag{11-3}$$

in which K is a proportionality constant known as the *hydraulic conductivity*. The quantity $\Delta h/\Delta l$ is a dimensionless ratio referred to as *hydraulic gradient i* :

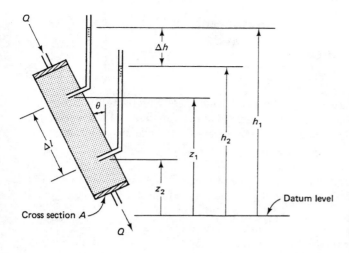

Cross section A

Datum level

Figure 11-2 Experimental setup for the illustration of Darcy's law [8].

$$i = \frac{\Delta h}{\Delta l} \tag{11-4}$$

Therefore,

$$Q = KiA \tag{11-5}$$

and

$$v = \frac{Q}{A} = Ki \tag{11-6}$$

is the *specific discharge,* or discharge per unit area, with the dimensions of velocity. The specific discharge, also known as Darcy velocity or Darcy flux, is a macroscopic concept that can be readily measured. It must be clearly differentiated from the microscopic velocities associated with the actual paths of the water as it winds its way through porous media. These velocities are real but—for all practical purposes—intractable.

Darcy's law is valid for groundwater flow in any direction. Given a constant hydraulic conductivity and hydraulic gradient, the specific discharge is independent of the angle θ shown in Fig. 11-2. This holds even for values of θ greater than 90°, when the flow is being forced up through the cylinder against gravity.

Hydraulic Conductivity. The hydraulic conductivity, which according to Eq. 11-6 has the dimensions of velocity $[LT^{-1}]$, is a function of the physical properties of fluid and porous media. To illustrate, consider the experimental setup of Fig. 11-2. Assume two experiments, each consisting of two tests, with Δh and Δl being held constant. In the first experiment, the fluid is the same (e.g., water), and two types of porous media are tested: (1) a coarse sand and (2) a fine sand. The specific discharge of the second test (fine sand) will be smaller than that of the first test (coarse sand). In the second experiment, the porous medium is the same (e.g., coarse sand), and two fluids of different viscosity are tested: (1) water and (2) oil. The specific discharge of the second test (oil) will be smaller than that of the first test (water).

An expression for hydraulic conductivity in terms of fluid and porous media properties is [13]:

$$K = \frac{Cd^2\rho g}{\mu} \tag{11-7}$$

in which d = mean grain diameter, ρ = fluid density, μ = fluid absolute viscosity, g = gravitational acceleration, and C = a dimensionless constant, a function of porous media properties other than mean grain diameter—grain size distribution, sphericity and roundness of the particles, and the nature of their packing.

Intrinsic Permeability. In Eq. 11-7, the product Cd^2 is a function only of the porous media, whereas ρ and μ are functions of the fluid. Therefore,

$$k = Cd^2 \tag{11-8}$$

in which k is the specific, or *intrinsic*, permeability. From Eqs. 11-7 and 11-8,

$$K = \frac{k\rho g}{\mu} \tag{11-9}$$

Intrinsic permeability is expressed in square centimeters or, alternatively, in *darcys*. One darcy unit is defined as the intrinsic permeability that produces a specific discharge of 1 cm/s, using a fluid of absolute viscosity equal to 1 centipoise, under a hydraulic gradient that results in the term $\rho g i$ being equal to 1 atm/cm. One darcy is approximately equal to 10^{-8} cm^2.

Equation 11-9 establishes a clear distinction between hydraulic conductivity K and intrinsic permeability k. However, it should be noted that hydraulic conductivity is often referred to as *coefficient of permeability*, whereas intrinsic permeability is commonly referred to as *permeability*. The different units, centimeters per second in the case of hydraulic conductivity and square centimeters (or darcys) in the case of intrinsic permeability, can be used to avoid confusion.

Variability of Hydraulic Conductivity. Soil and rock formations in which hydraulic conductivity is independent of position are said to be *homogeneous*. Conversely, formations in which hydraulic conductivity is a function of position are said to be *heterogeneous*.

If at a given point the hydraulic conductivity is independent of the direction of measurement, the formation is said to be *isotropic* at that point. On the other hand, if the hydraulic conductivity varies with the direction of measurement, the formation is said to be *anisotropic* at that point. The four possible combinations of homogeneity, heterogeneity, isotropy, and anisotropy are shown in Fig. 11-3.

Most sedimentary rocks are anisotropic with respect to hydraulic conductivity because the individual particles are not spherical, but they tend to be elongated in a direction parallel to the bedding. This causes the hydraulic conductivity to be greater in a direction parallel to the bedding than in a direction perpendicular to it.

Measured hydraulic conductivities vary by several orders of magnitude. A formation with hydraulic conductivity of 1 m/d would generally be regarded as permeable and likely to be a good aquifer. Conversely, a formation with hydraulic conductivity of less than 10^{-3} m/d would generally be regarded as impermeable and, therefore,

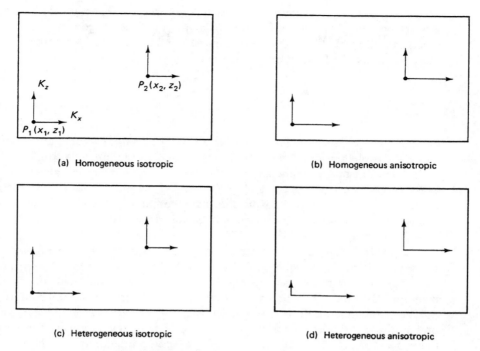

Figure 11-3 Possible combinations of homogeneity, heterogeneity, isotropy, and anisotropy in porous media [8].

(a) Homogeneous isotropic

(b) Homogeneous anisotropic

(c) Heterogeneous isotropic

(d) Heterogeneous anisotropic

not a good aquifer. However, depending on the intended application, comparisons of hydraulic conductivity (or permeability) are relative. For instance, a formation that is too impermeable for use as an aquifer may at the same time be too permeable for use as a water barrier in the foundation and core of an earth dam.

Compressibility. Both fluid (water) compressibility and medium (aquifer) compressibility affect groundwater flow.

Assume that an increase in pressure leads to a decrease in volume of a given mass of water. The water compressibility is defined as

$$\beta = \frac{-\dfrac{\Delta V_w}{V_w}}{\Delta p} \tag{11-10}$$

in which β = water compressibility, V_w = volume of a given mass of water, and ΔV_w = change in volume caused by a Δp change in pressure. For practical applications, β can be considered to be a constant, since it changes very little over the range of fluid pressures normally encountered in practice. The accepted value of β is 4.4×10^{-10} m²/N (Pa⁻¹).

The compressibility of porous media is quite different from the compressibility of water. Assuming that the compressibility of individual soil grains is negligible, porous media can only be compressed by rearrangement of the soil particles into a tighter packing. The total stress on the media is borne partly by the solid particles forming the granular skeleton and partly by the fluid in pore spaces. The portion of total stress that is borne by solid particles is called the *effective stress*. Rearrangement of the soil particles is caused by changes in the effective stress and not by changes in total stress. The compressibility of porous media is defined as

$$\alpha = \frac{-\dfrac{\Delta V_T}{V_T}}{\Delta \sigma_e}$$

(11-11)

in which α = compressibility of porous media, V_T = total volume of the soil mass, and ΔV_T = change in volume caused by a $\Delta \sigma_e$ change in effective stress.

The value of α is a function of soil type. Clays have typical α values in the range of 10^{-6} to 10^{-8} m²/N; sands, 10^{-7} to 10^{-9} m²/N; gravel, 10^{-8} to 10^{-10} m²/N; sound rock, 10^{-9} to 10^{-11} m²/N [8]. Note that water compressibility is of the same order of magnitude as that of the less compressible geologic materials.

Specific Storage, Transmissivity, and Storativity. Six basic physical properties of fluid (i.e., water) and porous media (soil and rock formation) are needed for the description of saturated groundwater flow; three for the fluid and three for the medium. The water properties are density ρ, absolute viscosity μ, and compressibility β. The porous media properties are porosity n, intrinsic permeability k, and compressibility α. All other parameters describing hydrogeologic properties are based on these; see, for example, the definition of hydraulic conductivity, Eq. 11-9.

Specific storage of a confined aquifer is defined as the volume of water released per unit volume of aquifer per unit decrease in hydraulic head. Compressibility considerations lead to the following expression for specific storage [8]:

$$S_s = \rho g (\alpha + n\beta)$$

(11-12)

in which S_s = specific storage in L^{-1} units.

The *transmissivity* of a confined aquifer is defined as follows:

$$T = Kb$$

(11-13)

in which T = transmissivity [$L^2 T^{-1}$ units], K = hydraulic conductivity [LT^{-1} units], and b = aquifer thickness [L units]. In SI units, the transmissivity is given in square meters per second. A transmissivity greater than 0.015 m²/s is indicative of a good aquifer, suitable for water well exploitation.

The *storativity* of a confined aquifer is defined as

$$S = S_s b$$

(11-14)

in which S = storativity, a dimensionless value. Typical values of storativity vary in the range 0.005 to 0.00005. Given the definition of specific storage, large decreases in hydraulic head over extensive formations are required in order for a confined aquifer to yield substantial amounts of water.

In unconfined aquifers, the concept of specific yield is equivalent to the storativity of confined aquifers. Typical values of specific yield vary in the range 0.01 to 0.30. The higher values of specific yield—as compared to storativity—reflect the fact that releases from an unconfined aquifer represent an actual dewatering of the pore spaces. On the other hand, releases from confined aquifers represent only the secondary effect of aquifer compaction caused by changes in fluid pressure. The favorable yield properties of unconfined aquifers make them more suited to well exploitation.

11.3 EQUATIONS OF GROUNDWATER FLOW

Depending on whether the flow is steady or unsteady, and saturated or unsaturated, the equations of groundwater flow can be formulated in one of the following three ways: (1) steady-state saturated flow, (2) transient (i.e., unsteady) saturated flow, and (3) transient unsaturated flow. Equations for steady-state and transient saturated flow are described here. See [8] for details on transient unsaturated flow through porous media.

Steady-State Saturated Flow

The law of conservation of mass for steady-state flow through saturated porous media requires that the net fluid mass flux through a control volume be equal to zero, i.e., the inflow must equal the outflow. This leads to the equation of continuity:

$$\frac{\partial(\rho v_x)}{\partial x} + \frac{\partial(\rho v_y)}{\partial y} + \frac{\partial(\rho v_z)}{\partial z} = 0 \tag{11-15}$$

in which the quantities v are specific discharges in three orthogonal directions x, y, and z, respectively. Assuming fluid incompressibility, $\rho(x, y, z)$ is constant, and consequently it can be eliminated from Eq. 11-15. Substitution of Darcy's law in Eq. 11-15 yields:

$$\frac{\partial(K_x i_x)}{\partial x} + \frac{\partial(K_y i_y)}{\partial y} + \frac{\partial(K_z i_z)}{\partial z} = 0 \tag{11-16}$$

in which the quantities K and i are hydraulic conductivities and hydraulic gradients, respectively.

For an isotropic medium, $K_x = K_y = K_z = K$. For a homogeneous medium, $K(x, y, z)$ is constant, and it can be eliminated from Eq. 11-16. Given $i_x = \partial h/\partial x$, $i_y = \partial h/\partial y$, and $i_z = \partial h/\partial z$, in which $h =$ hydraulic head, Eq. 11-16 reduces to:

$$\frac{\partial^2 h}{\partial x^2} + \frac{\partial^2 h}{\partial y^2} + \frac{\partial^2 h}{\partial z^2} = 0 \tag{11-17}$$

Equation 11-17 is the Laplace equation. The solution of this equation is a function $h(x, y, z)$ describing the value of hydraulic head at any point in a three-dimensional flow field. For two-dimensional flow, the third term on the left side of Eq. 11-17 cancels out, and the solution is a function $h(x, y)$.

Transient Saturated Flow

The law of conservation of mass for transient flow through saturated porous media requires that the net fluid mass flux through a control volume be equal to the time rate of change of fluid mass storage within the control volume. Therefore, the equation of continuity, Eq. 11-15, is modified to

$$\frac{\partial(\rho v_x)}{\partial x} + \frac{\partial(\rho v_y)}{\partial y} + \frac{\partial(\rho v_z)}{\partial z} = \frac{\partial(\rho n)}{\partial t} \qquad (11\text{-}18)$$

Assuming that the fluid is incompressible, $\rho(x, y, z)$ is constant, and it can be eliminated from Eq. 11-18. Substitution of Darcy's law in Eq. 11-18 yields

$$\frac{\partial(K_x i_x)}{\partial x} + \frac{\partial(K_y i_y)}{\partial y} + \frac{\partial(K_z i_z)}{\partial z} = \frac{\partial n}{\partial t} \qquad (11\text{-}19)$$

For an isotropic medium, $K_x = K_y = K_z = K$. For a homogeneous medium, $K(x, y, z)$ is constant. The time rate of change of porosity can be related to time rate of change of hydraulic head by the following:

$$\frac{\partial n}{\partial t} = S_s \frac{\partial h}{\partial t} \qquad (11\text{-}20)$$

Therefore, Eq. 11-19 reduces to [14, 15]:

$$\frac{\partial^2 h}{\partial x^2} + \frac{\partial^2 h}{\partial y^2} + \frac{\partial^2 h}{\partial z^2} = \frac{S_s}{K} \frac{\partial h}{\partial t} \qquad (11\text{-}21)$$

Equation 11-21 is a diffusion equation, with K/S_s being the *hydraulic diffusivity* of the aquifer [$L^2 T^{-1}$ units]. The solution of this equation is a function $h(x, y, z, t)$ describing the value of hydraulic head in three dimensions at any time. It requires the knowledge of S_s and K or, alternatively, the basic fluid and aquifer properties ρ, μ, α, n, k, and β.

For the special case of a horizontal confined aquifer of thickness b, the third term on the left side of Eq. 11-21 drops out, and with Eqs. 11-13 and 11-14:

$$\frac{\partial^2 h}{\partial x^2} + \frac{\partial^2 h}{\partial y^2} = \frac{S}{T} \frac{\partial h}{\partial t} \qquad (11\text{-}22)$$

The solution of this equation is a function $h(x, y, t)$ describing the value of hydraulic head in two dimensions at any time. It requires the knowledge of aquifer storativity S and transmissivity T. The ratio T/S is the hydraulic diffusivity of the aquifer.

11.4 WELL HYDRAULICS

Well hydraulics describes groundwater flow problems involving discharge to wells. There is a wealth of literature on the subject, including Bear [1], U.S. Bureau of Reclamation [23], and Walton [25]. The classical problem of radial flow to a well is described in this section.

Radial Flow to a Well

Consider an aquifer with the following characteristics: (1) horizontal, (2) confined between impermeable layers on top and bottom, (3) infinite in horizontal extent, (4) constant thickness, (5) homogeneous and isotropic. Furthermore, assume for simplicity: (1) a single pumping well, (2) a constant pumping rate, (3) negligible well diameter relative to the aquifer's horizontal dimensions, (4) well penetration through the entire aquifer depth, and (5) uniform initial hydraulic head throughout the aquifer. In horizontal and vertical coordinates, Eq. 11-22 is the governing equation for this problem, its solution being a function $h(x, y, t)$.

The idealized nature of this problem justifies the assumption of radial symmetry. A sketch of aquifer, pumping well, observational well, and potentiometric surface is shown in Fig. 11-4. The conversion of Eq. 11-22 into radial coordinates, using $r = (x^2 + y^2)^{1/2}$, leads to [15]:

$$\frac{\partial^2 h}{\partial r^2} + \frac{1}{r} \frac{\partial h}{\partial r} = \frac{S}{T} \frac{\partial h}{\partial t} \qquad (11\text{-}23)$$

in which r is the radial distance from the pumping well to the observational well.

The solution of Eq. 11-23 is a function $h(r, t)$ describing the potentiometric surface. The uniform initial hydraulic head is h_0. For convenience, solutions are often expressed in terms of *drawdown*, the difference between uniform initial hydraulic head and potentiometric surface:

$$Z = h_0 - h \qquad (11\text{-}24)$$

in which Z = drawdown, h_0 = uniform initial hydraulic head, and h = hydraulic head (i.e., potentiometric surface elevation).

The assumption of uniform initial hydraulic head throughout the aquifer leads to the following initial condition:

$$h(r, 0) = h_0 \qquad (11\text{-}25)$$

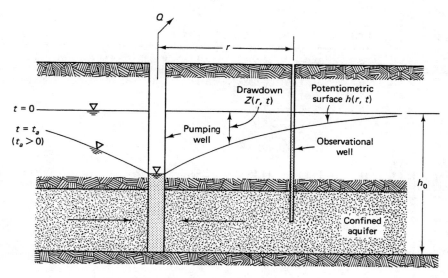

Figure 11-4 Sketch of radial flow to a pumping well.

for $r \geq 0$. Furthermore, assuming no change in hydraulic head at $r = \infty$, the following boundary condition is obtained:

$$h(\infty, t) = h_0 \tag{11-26}$$

for $t \geq 0$.

Theis Solution

The solution of Eq. 11-23 subject to the given initial and boundary conditions is due to Theis [21]:

$$Z = \frac{Q}{4\pi T} \int_u^\infty \frac{e^{-u}}{u} \, du \tag{11-27}$$

in which Q = (constant) pumping rate $[L^3 T^{-1}$ units], T = aquifer transmissivity $[L^2 T^{-1}$ units], and u is a dimensionless variable defined as

$$u = \frac{r^2 S}{4Tt} \tag{11-28}$$

with r = radial distance from pumping well to observational well, S = aquifer storativity (dimensionless), and t = time.

With u defined by Eq. 11-28, the integral in Eq. 11-27 is referred to as the *well function* $W(u)$, applicable to homogeneous isotropic confined aquifers with full-depth well penetration and at constant pumping rate. Values of $W(u)$ as a function of u are shown in Table 11-1. Equation 11-27 reduces to

$$Z = \frac{QW(u)}{4\pi T} \tag{11-29}$$

TABLE 11-1 VALUES OF $W(u)$ AS A FUNCTION OF u [26]

u	1.0	2.0	3.0	4.0	5.0	6.0	7.0	8.0	9.0
$\times 10^0$	0.219	0.049	0.013	0.0038	0.0011	0.00036	0.00012	0.000038	0.000012
$\times 10^{-1}$	1.82	1.22	0.91	0.70	0.56	0.45	0.37	0.31	0.26
$\times 10^{-2}$	4.04	3.35	2.96	2.68	2.47	2.30	2.15	2.03	1.92
$\times 10^{-3}$	6.33	5.64	5.23	4.95	4.73	4.54	4.39	4.26	4.14
$\times 10^{-4}$	8.63	7.94	7.53	7.25	7.02	6.84	6.69	6.55	6.44
$\times 10^{-5}$	10.94	10.24	9.84	9.55	9.33	9.14	8.99	8.86	8.74
$\times 10^{-6}$	13.24	12.55	12.14	11.85	11.63	11.45	11.29	11.16	11.04
$\times 10^{-7}$	15.54	14.85	14.44	14.15	13.93	13.75	13.60	13.46	13.34
$\times 10^{-8}$	17.84	17.15	16.74	16.46	16.23	16.05	15.90	15.76	15.65
$\times 10^{-9}$	20.15	19.45	19.05	18.76	18.54	18.35	18.20	18.07	17.95
$\times 10^{-10}$	22.45	21.76	21.35	21.06	20.84	20.66	20.50	20.37	20.25
$\times 10^{-11}$	24.75	24.06	23.65	23.36	23.14	22.96	22.81	22.67	22.55
$\times 10^{-12}$	27.05	26.36	25.96	25.67	25.44	25.26	25.11	24.97	24.86
$\times 10^{-13}$	29.36	28.66	28.26	27.97	27.75	27.56	27.41	27.28	27.16
$\times 10^{-14}$	31.66	30.97	30.56	30.27	30.05	29.87	29.71	29.58	29.46
$\times 10^{-15}$	33.96	33.27	32.86	32.58	32.35	32.17	32.02	31.88	31.76

Two types of problems can be solved by the Theis method: (1) prediction and (2) identification. In the prediction problem, the pumping rate Q, radial distance r, aquifer transmissivity T, and storativity S are known, and the variation of observational well drawdown Z with time t is sought. In the identification problem, the pumping rate Q, radial distance r, and observational well drawdown versus time data (Z versus t) are known, and the aquifer properties S and T are sought.

The prediction problem can be solved directly with Eqs. 11-28 and 11-29. For any assumed time t, u is calculated with Eq. 11-28 and used to obtain $W(u)$ from Table 11-1. Then, Z is calculated with Eq. 11-29 for the assumed time t and given radial distance r.

A variation of the prediction problem is that of the calculation of the *cone of depression*, or drawdown cone, as shown in Fig. 11-4. In this case, the pumping rate Q and aquifer properties (storativity S and transmissivity T) are known, and the variation of drawdown Z with radial distance r is sought at a fixed time t. For any assumed radial distance r, u is calculated with Eq. 11-28 and used to obtain $W(u)$ from Table 11-1. Then, Z is calculated with Eq. 11-29 for the assumed radial distance r and fixed time t.

The identification problem is solved by a graphical procedure. Eliminating T from Eqs. 11-28 and 11-29 leads to

$$\frac{Z}{t} = \frac{Q}{\pi r^2 S} \frac{W(u)}{(1/u)} \tag{11-30}$$

which indicates that the ratio Z/t is proportional to the ratio $W(u)/(1/u)$. The solution is accomplished by the following steps:

1. Using Table 11-1, plot the well function $W(u)$ versus $1/u$ on logarithmic paper, with $1/u$ in the abscissas and $W(u)$ in the ordinates.
2. Using the drawdown versus time data for the observational well, plot Z versus t on logarithmic paper, with t in the abscissas and Z in the ordinates.
3. Overlay both plots making sure that they coincide, as shown in Fig. 11-5.
4. Choose a pair of $W(u)$ and $1/u$ values on the first plot and a pair of matching Z and t values on the second plot. Any pair of $W(u)$ and $1/u$ is appropriate. For simplicity, the following pair can be chosen: $W(u) = 1$, and $1/u = 1$ (see Fig. 11-5). In this case, the matching pair of Z and t are Z_1 and t_1. Therefore, using Eqs. 11-28 and 11-29, the transmissivity is

$$T = \frac{Q}{4\pi Z_1} \tag{11-31}$$

and storativity is

$$S = \frac{4Tt_1}{r^2} \tag{11-32}$$

For a given radial distance, drawdown increases with increasing time. For a given time, drawdown decreases with increasing radial distance. For a given time and radial distance, drawdown is directly proportional to pumping rate. For a given time, distance, and pumping rate, drawdown is inversely proportional to transmissivity and

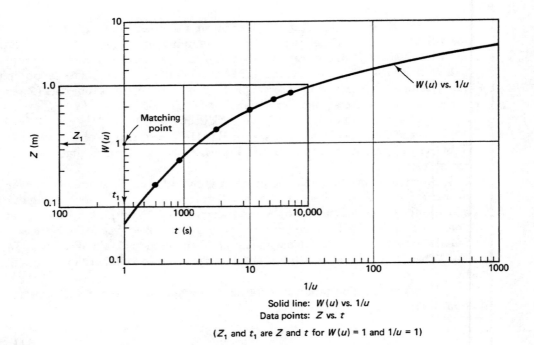

Figure 11-5 Graphical procedure for Theis solution for radial flow to a well [8].

storativity. Other conditions being constant, low transmissivity results in deep draw-down cones of limited extent, whereas high transmissivity results in shallow cones of wide extent. Likewise, an aquifer of low storativity has a comparatively deeper draw-down cone than an aquifer of high storativity [8].

For a detailed discussion of other types of radial flow to a well or wells, including unconfined aquifers, varying pumping rates, multiple-well configurations, and partially penetrating wells, see [8] and [25].

Example 11-1.

Assume a pumping rate $Q = 0.00314$ m^3/s, radial distance from well $r = 100$ m, aquifer transmissivity $T = 0.0025$ m^2/s, and aquifer storativity $S = 0.001$. Calculate the draw-down after an elapsed time of (a) 1000 s, (b) 10,000 s, and (c) 100,000 s.

For each time t, values of u are calculated with Eq. 11-28. Corresponding values of $W(u)$ and Z are obtained from Table 11-1 and Eq. 11-29, respectively. The results are summarized as follows:

Time t (s)	u	$W(u)$	Drawdown Z (m)
1,000	1.0	0.219	0.022
10,000	0.1	1.820	0.182
100,000	0.01	4.040	0.404

Example 11-2.

Assume a pumping rate $Q = 0.00314$ m^3/s, aquifer transmissivity $T = 0.0025$ m^2/s, aquifer storativity $S = 0.001$, elapsed time $t = 10,000$ s. Calculate the drawdown at the following radial distances from the well: (a) 10 m, (b) 100 m, and (c) 200 m.

For each radial distance, values of u are calculated with Eq. 11-28. Corresponding values of $W(u)$ and Z are obtained from Table 11-1 and Eq. 11-29, respectively. The results are summarized as follows:

Distance r (m)	u	$W(u)$	Drawdown Z (m)
10	0.001	6.33	0.633
100	0.1	1.82	0.182
200	0.4	0.70	0.070

11.5 SURFACE-SUBSURFACE FLOW INTERACTION

The interaction between surface and subsurface flow is central to the study of stream types and baseflow recession. Surface flow occurs in the form of overland flow in the catchments and, subsequently, streamflow in the channel network. Subsurface flow occurs as interflow in the unsaturated zone and as groundwater flow in the saturated zone.

The contribution of subsurface water to surface water is a function of stream type. Depending on whether streams serve as discharge or recharge areas for aquifers, they are classified as (1) *effluent,* or gaining streams, or (2) *influent,* or losing streams.

Effluent streams are discharge areas for aquifers. Usually, an aquifer is intersected by the effluent stream, discharging into it. This type of aquifer discharge is distributed along the stream length, in contrast to aquifer discharge that occurs at a point, i.e., a spring. The distributed discharge to effluent streams is referred to as exfiltration, or baseflow, and is typical of humid and subhumid environments where aquifers are at relatively shallow depth.

Influent streams are recharge areas for aquifers, largely due to the high permeability of their channel beds. This type of aquifer recharge—known as channel transmission loss(es) in stream channel routing—is distributed along the stream length and is typical of streams in arid and semiarid regions.

Depending on whether they flow all year, a few days of the year, or seasonally, streams can be further classified as (1) *perennial,* (2) *ephemeral,* or (3) *intermittent.* Perennial streams are those that flow throughout the year, during both wet and dry weather. Their dry weather flow is largely baseflow, i.e., discharge from groundwater reservoirs. Unlike perennial streams, ephemeral streams are those that flow only in wet weather, i.e., during and immediately following rainfall. In the southwestern United States, ephemeral streams are known as *arroyos,* washes, or dry washes, and flow approximately 30 d per year on the average. Typically, perennial streams are effluent, whereas ephemeral streams are influent. Furthermore, it is common for ephemeral streams to abstract most or all of their wet-weather flow through channel transmission losses. Intermittent streams are those possessing mixed characteristics, discharging either to or from groundwater reservoirs.

Hillslope Hydrology and Streamflow Generation

Hillslope hydrology refers to the hydrologic processes taking place on hillslopes. These processes are intrinsically related to streamflow generation. An important question is whether the preferred path of runoff is on the surface, as overland flow, or through the subsurface, by *subsurface stormflow,* or interflow.

Several theories of hillslope hydrology have been developed, together constituting a spectrum of plausible processes and models [5]. They range from classical Hortonian theory [12] to throughflow theory [16]. Hortonian theory emphasizes overland flow processes, whereas throughflow theory focuses on interflow processes. Hortonian theory is applicable to poorly vegetated slopes with relatively thin soil covers; throughflow theory is more apt to explain the processes that take place in heavily vegetated areas with thick permeable soil layers.

Between these two extremes lie a variety of models in which runoff is assumed to consist of a mix of Hortonian overland flow, and unsaturated and saturated throughflow. The *partial-area concept* [10, 11, 27] is a combination of throughflow in the upper hillslopes and overland flow in the lower hillslopes (Fig. 11-6). Based on recent field studies, Freeze and Cherry [8] have concluded that most overland flow hydrographs in humid, vegetated basins originate from not more than 10 percent of the catchment area, and then, are likely to prevail in only 10 to 30 percent of the storms.

Interest in the partial-area concept arose from the recognition that runoff estimates were improved by assuming that only a small portion of the catchment is able to contribute to runoff during the hydrograph peak. Betson [2], among others, has documented a substantial improvement in runoff predictions by using the partial-area concept. The size of the partial areas is a function of storm depth, rainfall intensity, and

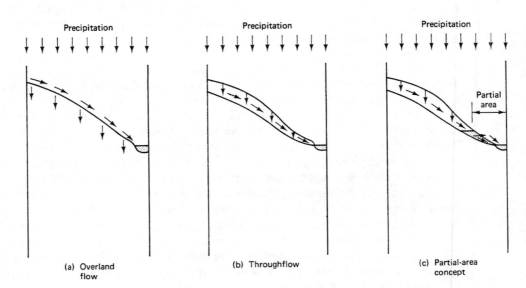

Figure 11-6 Sketch of overland flow, throughflow, and partial-area concepts in hillslope hydrology [5].

antecedent moisture conditions. As a percentage of the total catchment, partial areas in the Southern Appalachians were shown to vary between 5 percent for light-to-moderate storms and 40 percent for heavy storms [2]. With deforestation, the percentage can amount to 80 percent or more. Generally, partial areas may be expected to vary between 5 percent and 20 percent of the total basin [20].

Refinements of the partial-area concept have led to the *variable-source-area* model of hillslope hydrology [19, 22]. These variable-source areas are envisioned as comprising low-lying lands adjacent to streams and rivers and concentrated near watershed outlets [3, 7]. Variations in the extent of source areas are dictated by antecedent soil moisture conditions, soil-moisture storage capacity, and rainfall intensity. When the upper soil horizon is saturated, both throughflow and overland flow occur, with eventual exfiltration to stream banks. According to this scheme, the variable-source-area can be interpreted as an expanded stream system, which helps explain the growth in drainage density experienced by small watersheds under heavy rainfall [4].

The variable-source-area model is a dynamic version of the partial-area concept. This dynamism is manifested in soil moisture and runoff changes occurring annually, seasonally, between storms, and during storms. The source areas are largely responsible for overland flow, whereas the remainder of the watershed acts primarily as a reservoir to provide baseflow and to maintain saturation of the source areas [7].

The variable-source-area model visualizes the average storm hydrograph as consisting of (1) rain falling directly onto the channel, and (2) water transmitted rapidly through wet soil adjacent to the stream. The source areas shrink and expand as a function of rainfall amounts and antecedent soil-moisture conditions. In the absence of rainfall, streams are fed to a large extent by moisture moving downslope under unsaturated flow conditions. This downward migration favors the quick release of water from the source areas into the channels during storms. Typical storm hydrographs reflecting the variable-source-area concept are shown in Fig. 11-7.

Hydrograph Separation and Baseflow Recession

The variable-source-area concept has led to increased emphasis on the timing of runoff during and immediately following a storm. This justifies the separation of runoff into two components: (1) quickflow, consisting of fast interflow (subsurface stormflow), overland flow, and rain falling directly on the channels, and (2) baseflow, consisting of slow interflow and groundwater flow.

Hydrograph separation is an established procedure of hydrologic analysis (Section 5.2). It has an important role not only in flood hydrology but also in groundwater hydrology studies. In flood hydrology, it serves as a means to quantify the relationship between quickflow and baseflow. In groundwater hydrology, it provides information on the nature and behavior of local and regional groundwater regimes.

Baseflow Recession.   Consider the streamflow hydrograph shown in Fig. 11-8. Flow varies throughout the year, with peaks of approximately 100 m³/s and lows of about 1 m³/s. The smooth line is the baseflow hydrograph or *baseflow curve*, reflecting the seasonally varying groundwater contributions. The ragged line is the streamflow hydrograph, reflecting the relatively fast response typical of steep catchments with shallow soils of low permeability.

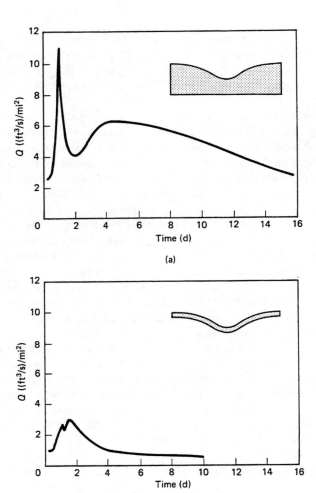

Figure 11-7 Typical storm hydrographs reflecting variable-source-area concept: (a) deep soil; (b) shallow soil [5].

The recession portion of the baseflow curve usually plots as a straight line (or a series of straight lines) on semilogarithmic paper, with discharge plotted on the log scale, as shown in Fig. 11-8. This leads to an equation of the form

$$Q = Q_o e^{-t/t_s} \tag{11-33}$$

in which Q = baseflow at any time t after the starting time t_o, Q_o = baseflow at t_o, and t_s = a recession constant known as the *time of storage,* defined as the time required for the flow Q to recede to $0.368 Q_o$ (i.e., for $t = t_s$, $Q = Q_o e^{-1}$).

Other forms of baseflow recession equations are

$$Q = Q_o e^{-\alpha t} \tag{11-34}$$

$$Q = Q_o 10^{-t/t_s} \tag{11-35}$$

$$Q = Q_o 10^{-\beta t} \tag{11-36}$$

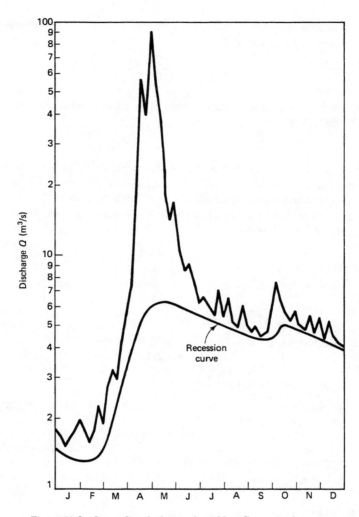

Figure 11-8 Streamflow hydrograph and baseflow-recession curve.

$$Q = Q_o K^t \qquad (11\text{-}37)$$

in which $\alpha = \beta = 1/t_s$ and $K =$ a recession constant. The values α and β are positive, and K is generally less than 1.0. Plotting Eqs. 11-33 to 11-37 on semilogarithmic paper, with discharge on the logarithmic scale, yields a straight line.

From Eqs. 11-33 and 11-37, the relationship between t_s and K is

$$t_s = -\frac{1}{\ln K} \qquad (11\text{-}38)$$

which leads to

$$Q = Q_o e^{(\ln K)t} \qquad (11\text{-}39)$$

The integration of Eq. 11-39 between a time corresponding to Q_o and a time corresponding to Q leads to:

$$S = \frac{Q - Q_o}{\ln K} \qquad (11\text{-}40)$$

in which S = storage volume released from groundwater [L^3 units]. Combining Eqs. 11-38 and 11-40:

$$S = (Q_o - Q)t_s \qquad (11\text{-}41)$$

The volume under the recession curve (from Q_o to Q) can be calculated by assuming that as $t \to \infty$, $Q \to 0$. Using Eq. 11-40, $S = -Q_o/(\ln K)$; using Eq. 11-41, $S = Q_o t_s$. Values of the recession constants can be obtained either from experience or through analysis of measured hydrographs. A historical perspective of baseflow recessions is given by Hall [9].

Example 11-3.

Given the following measured recession flows in a certain river:

Time (h)	0	3	6	9	12
Discharge (m³/s)	100.00	88.25	77.88	68.73	60.65

Calculate: (a) the time of storage t_s, (b) the recession constant α, (c) the recession constant K, (d) the recession volume released between $t = 0$ h and $t = 6$ h, (e) the recession volume released between $t = 3$ h and $t = 12$ h, and (f) the recession volume released between $t = 0$ h and $t = \infty$.

a. From Eq. 11-33, $t_s = t/(\ln Q_o - \ln Q) = 3$ h$/(4.605 - 4.480) = 24$ h.
b. $\alpha = 1/t_s = 0.04167$.
c. From Eq. 11-38, $K = e^{(-1/t_s)} = 0.9592$ h^{-1}.
d. From Eq. 11-41, $S = (100.00 - 77.88) \times 24 = 530.9$ (m³/s)-h $= 1{,}911{,}240$ m³.
e. From Eq. 11-41, $S = (88.25 - 60.65) \times 24 = 662.4$ (m³/s)-h $= 2{,}384{,}640$ m³.
f. $S = Q_o t_s = 100 \times 24 = 2400$ (m³/s)-h $= 8{,}640{,}000$ m³.

Example 11-4.

Given the following flows measured in the receding limb of a flood hydrograph:

Time (h)	0	1	2	3	4	5	6	7	8	9	10
Flow (cfs)	1100	466	231	144	110	97	90	87	85	83	81

(a) Calculate the time of storage for baseflow, (b) separate the hydrograph into baseflow and quickflow, and (c) calculate the volume released from baseflow in the elapsed 10-h period.

a. The last 3 h have an approximately constant time of storage. Therefore, the time of storage for baseflow is $t_s = -3$ h$/(\ln 81 - \ln 87) \cong 42$ h.
b. The calculation of baseflow values is based on Eq. 11-33:

$$Q_t = Q_{(t+\Delta t)} e^{(\Delta t/42)}$$

in which $\Delta t = 1$ h. To calculate the quickflow values, the baseflow values are subtracted from the total flows. The calculations are summarized in the following table.

Time (h)	Flows (ft³/s)	Baseflow (ft³/s)	Quickflow (ft³/s)
0	1100	101	999
1	466	99	367
2	231	97	134
3	144	95	49
4	110	93	17
5	97	91	6
6	90	89	1
7	87	87	0
8	85	85	0
9	83	83	0
10	81	81	0

c. Using Eq. 11-41, the volume released from baseflow is $S = (101 - 81)$ ft³/s $\times$ 42 h = 840 (ft³/s)-h = 69.4 ac-ft.

Bank Storage. A concept closely related to baseflow recession is that of bank storage. In the upland reaches, subsurface contributions to streamflow generally aid in the buildup of the flood wave. In the lower reaches, however, bank storage may contribute to flood wave attenuation. As shown in Fig. 11-9, a substantial increase in river stage generally results in net infiltration into the river banks, providing temporary bank storage. As the stage declines in the flood hydrograph recession, the flow reverses direction and exfiltration takes place. Where bank-storage effects are deemed to be significant, hydrograph separation must be performed with utmost care.

QUESTIONS

1. What is the limit between unsaturated and saturated zones in groundwater flow?

2. What factors determine the feasibility of mining a groundwater reservoir? Explain.

3. What is an aquifer? An aquiclude? An aquitard?

4. What are the major differences between confined and unconfined aquifers?

5. What is a potentiometric surface? A perched aquifer?

6. Explain the processes of recharge and discharge in unconfined and confined aquifers.

7. What is specific yield? Specific retention?

8. What is field capacity? Permanent wilting point?

9. What is specific discharge? How does hydraulic conductivity differ from intrinsic permeability?

10. Explain the differences between the compressibility of water and that of porous media.

11. Define specific storage, transmisivity and storativity. How does the concept of storativity differ from that of specific yield?

12. What fluid and aquifer properties are needed in the description of saturated groundwater flow?

13. What additional term is added to the equation of continuity in order to account for transient flow conditions?

14. What is hydraulic diffusivity in connection with groundwater flow?

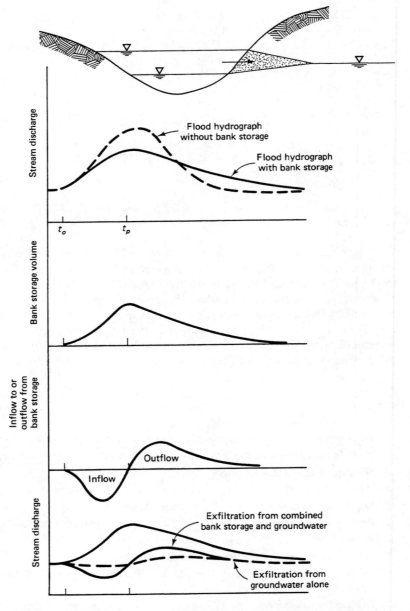

Figure 11-9 Effect of bank storage on flood hydrograph magnitude and shape [8].

15. What determines whether streams are classified as effluent or influent?

16. What theories of streamflow generation lie at the two extremes of the spectrum of hillslope hydrologic models?

17. What is the essential difference between the partial-area and the variable-source-area models of hillslope hydrology?

18. What runoff components are justified by current theories of hillslope hydrology? Explain.

PROBLEMS

11-1. What is the volume of free-draining water in an unconfined aquifer of surface area $A = 130 \text{ mi}^2$, thickness $b = 300$ ft, and specific yield $S_y = 0.05$?

11-2. What is the specific retention of an unconfined aquifer of surface area $A = 125 \text{ mi}^2$, thickness $b = 80$ ft, porosity $n = 0.30$, which can yield 1,536,000 ac-ft of free-draining water?

11-3. What is the flow through a porous media with cross-sectional area $A = 1250 \text{ m}^2$, hydraulic conductivity $K = 0.8 \times 10^{-4}$ m/s, under a hydraulic gradient $i = 0.01$?

11-4. Derive the exact equivalence between darcys and square centimeters.

11-5. What is the hydraulic conductivity of a porous medium with fluid density $\rho = 1 \text{ g/cm}^3$, intrinsic permeability $k = 5$ darcys, and fluid absolute viscosity $\mu = 1$ cp.

11-6. Show that specific storage S_s has units L^{-1}.

11-7. Calculate the storativity of a confined aquifer of thickness $b = 50$ m, porosity $n = 0.05$, and compressibility $\alpha = 1.0 \times 10^{-8} \text{ Pa}^{-1}$. Assume fluid density $\rho = 1 \text{ g/cm}^3$ and compressibility $\beta = 4.4 \times 10^{-10} \text{ Pa}^{-1}$.

11-8. Calculate the hydraulic diffusivity of an aquifer with the following properties: intrinsic permeability $k = 1$ darcy, fluid absolute viscosity $\mu = 1$ cp, compressibility $\alpha = 1.0 \times 10^{-8} \text{ Pa}^{-1}$, porosity $n = 0.08$. Assume fluid compressibility $\beta = 4.4 \times 10^{-10} \text{ Pa}^{-1}$.

11-9. Calculate the hydraulic diffusivity of an aquifer with storativity $S = 0.002$, hydraulic conductivity $K = 1.0 \times 10^{-5}$ cm/s, and thickness $b = 100$ m.

11-10. A well is pumped at the constant rate of $Q = 0.004 \text{ m}^3/\text{s}$ in an aquifer of transmissivity $T = 0.004 \text{ m}^2/\text{s}$ and storativity $S = 0.0005$. Calculate the drawdown 24 h after the start of pumping in an observational well located at a distance of 250 m from the pumped well.

11-11. A well is pumped at the constant rate of $Q = 0.008 \text{ m}^3/\text{s}$. A match of the well function ($W(u)$ versus $1/u$) with drawdown versus time data from an observational well located at a distance of 430 m from the pumped well has produced the following matching values: $W(u) = 1$, $u = 1$, $Z_1 = 0.21$ m, and $t_1 = 1.5$ h. Calculate the aquifer transmissivity and storativity.

11-12. A well is pumped at a constant rate $Q = 0.005 \text{ m}^3/\text{s}$ in an aquifer of transmissivity $T = 0.0015 \text{ m}^2/\text{s}$ and storativity $S = 0.0005$. Calculate the drawdown 24 h after the start of pumping in an observational well located 1000 m from the pumped well.

11-13. A well is pumped at a constant rate of 4 L/s. The drawdown in an observational well located at a distance of 150 m from the pumped well has been measured as follows:

Time (min)	0	10	15	30	60	90	120
Drawdown (m)	0	0.16	0.25	0.42	0.62	0.73	0.85

Calculate the transmissivity and storativity of the aquifer.

11-14. Given the following measured recession flows in a certain river:

Time (h)	0	12	24	36	48
Discharge (m^3/s)	1000	882	779	687	606

Calculate: (a) the time of storage t_s (b) the volume released from storage in the first 24 h, (c) the volume released from storage in the second 24 h, and (d) the total volume released from storage assuming that the flow eventually recedes back to zero.

11-15. Given the following measured recession flows in a certain stream:

Time (h)	0	1	2	3	4	5	6
Discharge (ft^3/s)	1000	920	846	779	716	659	606

Calculate: (a) the time of storage t_s, (b) the recession constant K, (c) the volume released from storage between 3 h and 6 h, using both t_s (Eq. 11-41) and K (Eq. 11-40), and (d) the total volume released from storage assuming that the flow eventually recedes back to zero, using both Eqs. 11-40 and 11-41.

11-16. Given the following flows measured in the receding limb of a flood hydrograph:

Time (h)	0	1	2	3	4	5	6	7	8	9	10
Flow (m^3/s)	32.0	25.2	20.5	17.9	16.1	14.5	13.5	12.9	12.6	12.3	12.0

Calculate: (a) the time of storage for baseflow, (b) the baseflow and quickflow components, (c) the volume released from baseflow in the elapsed 10-h period, and (d) the volume released from baseflow from $t = 0$ to $t = \infty$, assuming that the flow eventually recedes back to zero.

REFERENCES

1. Bear, J. (1979). *Hydraulics of Groundwater*. New York: McGraw-Hill.
2. Betson, R. P. (1964). "What is Watershed Runoff?" *Journal of Geophysical Research*, Vol. 69, pp. 1541–1551.
3. Betson, R. P., and J. B. Marius. (1969). "Source Areas of Storm Runoff," *Water Resources Research*, Vol. 5, pp. 574–582.
4. Carson, M. A., and E. A. Sutton. (1971). "The Hydrologic Response of the Eaton River Basin, Quebec," *Canadian Journal of Earth Science*, Vol. 8, pp. 102–115.
5. Chorley, R. J. (1978). "The Hillslope Hydrologic Cycle," in *Hillslope Hydrology*. M. J. Kirkby, editor. New York: John Wiley, pp. 1–42.
6. Darcy, H. (1856). "Les Fontaines Publiques de la Ville de Dijon," Paris: Victor Dalmont.
7. Dunne, T., and R. D. Black. (1970). "Partial Area Contributions to Storm Runoff in a Small New England Watershed," *Water Resources Research*, Vol. 6, pp. 1296–1311.
8. Freeze, R. A., and Cherry, J. A. (1979). *Groundwater*, Englewood Cliffs, N.J.: Prentice-Hall.
9. Hall, F. R. (1968). "Baseflow Recessions-A Review," *Water Resources Research*, Vol. 4, pp. 973–983.
10. Hewlett, J. D., and A. R. Hibbert. (1963). "Moisture and Energy Conditions Within a Sloping Soil Mass During Drainage," *Journal of Geophysical Research*, Vol. 68, pp. 1081–1087.
11. Hewlett, J. D., and A. R. Hibbert. (1967). "Factors Affecting the Response of Small Watersheds to Precipitation in Humid Areas," *Proceedings of the International Symposium on Forest Hydrology*, Pennsylvania State University, 1967, pp. 275–290.
12. Horton, R. E. (1933). "The Role of Infiltration in the Hydrologic Cycle," *Transactions, American Geophysical Union*, Vol. 14, pp. 446–460.
13. Hubbert, M. K. (1940). "The Theory of Groundwater Motion," *Journal of Geology*, Vol. 48, pp. 785–944.
14. Jacob, C. E. (1940). "On the Flow of Water in an Elastic Artesian Aquifer," *Transactions, American Geophysical Union*, Vol. 20, pp. 574–586.
15. Jacob, C. E. (1950). "Flow of Groundwater," in *Engineering Hydraulics*, Hunter Rouse, editor. New York: John Wiley, pp. 574–586.
16. Kirkby, M. J., and R. J. Chorley. (1967). "Throughflow, Overland Flow, and Erosion," *Bulletin of the International Association of Scientific Hydrology*, Vol. 12, pp. 5–21.

17. Nace, R. L. (1971). "Scientific Framework of World Water Balance," UNESCO *Technical Papers in Hydrology,* No. 7.

18. Price, M. (1985). *Introducing Groundwater.* London: Allen and Unwin.

19. Ragan, R. M. (1968). "An Experimental Investigation of Partial Area Contributions," *Proceedings of the General Assembly, International Association for Scientific Hydrology,* Berne, Publication 76, pp. 241-249.

20. Tennessee Valley Authority. (1966). "Cooperative Research Project in Western North Carolina, Annual Report, Water Year 1964-65," Knoxville, Tennessee.

21. Theis, C. V. (1935). "The Relation between the Lowering of the Piezometric Surface and the Rate of Duration of Discharge of a Well Using Groundwater Storage," *Transactions, American Geophysical Union,* pp. 519-524.

22. Tsukamoto, Y. (1963). "Storm Discharge from an Experimental Watershed," *Journal of the Japanese Society of Forestry,* Vol. 45, No. 6, pp. 186-190.

23. U.S. Dept. of Interior, Bureau of Reclamation. (1977). *Groundwater Manual.*

24. U.S. Water Resources Council, Hydrology Committee. (1980). "Essentials of Groundwater Hydrology Pertinent to Water Resources Planning," *Bulletin 16,* (revised).

25. Walton, W. C. (1970). *Groundwater Resource Evaluation.* New York: McGraw-Hill.

26. Wenzel, L. K. (1942). "Methods for Determining Permeability of Water-Bearing Materials," U.S. Geological Survey *Water Supply Paper* 887.

27. Whipkey, R. Z. (1965). "Subsurface Stormflow from Forested Slopes," *Bulletin of the International Association for Scientific Hydrology,* Vol. 10, No. 2, pp. 74-85.

SUGGESTED READINGS

Chorley, R. J. (1978). "The Hillslope Hydrologic Cycle," in *Hillslope Hydrology.* M. J. Kirkby, editor. New York: John Wiley, pp. 1-42.

Freeze, R. A., and Cherry, J. A. (1979). *Groundwater.* Englewood Cliffs, N.J.: Prentice-Hall.

Horton, R. E. (1933). "The Role of Infiltration in the Hydrologic Cycle," *Transactions, American Geophysical Union,* Vol. 14, pp. 446-460.

Hubbert, M. K. (1940). "The Theory of Groundwater Motion," *Journal of Geology,* Vol. 48, pp. 785-944.

SNOW HYDROLOGY

Snow hydrology studies the properties of snow, its formation, distribution, and measurement, including snowmelt and snowmelt-generated runoff. In many areas of the United States and the world, snow is a major contributor to streamflow, especially in regions of high altitude and/or temperate climate. For these regions, a knowledge of snowfall and snowmelt is necessary for the assessment of the seasonal variability of streamflow. In turn, snowfall and snowmelt are governed by meteorological principles. The atmosphere is the source of moisture for snowfall, and it regulates a basin's energy exchange, which determines snowmelt.

Principles of snow hydrology are useful in design and operation of engineering projects. Typical problems are the following:

1. The evaluation of the amount of water stored in the snowpack and its relation to the hydrologic balance of a catchment.
2. The evaluation of snowmelt rates, including the physical causes for snowmelt.
3. The evaluation of the effect of snowpack on runoff, both from snowmelt and rain on snow.

For project design, fixed sequences of meteorologic and hydrologic conditions are selected based on the overall functional requirements of the project. On the other hand, project operation requires evaluation of specific meteorologic and hydrologic conditions, and associated streamflow forecasts for both short- and long-term periods.

This chapter is divided into five sections. Section 12.1 describes the processes leading to snow formation and accumulation. Section 12.2 describes the processes responsible for melting of the snowpack. Section 12.3 deals with snowmelt indexes and their use in snowmelt computations. Section 12.4 discusses the effect of snowpack condition on runoff. Section 12.5 describes the synthesis of snowmelt hydrographs. See Section 3.2 for a description of snowpack measurements.

12.1 SNOW FORMATION AND ACCUMULATION

Snow Formation

The major atmospheric and environmental factors responsible for snow formation are (1) surface air temperature, (2) elevation, and (3) terrain features.

Temperature. Surface air temperature, measured at approximately 4-ft (1.2-m) depth, is considered to be a reliable indicator of the presence or absence of snow. A study of the effect of surface air temperature on snow formation was reported in *Snow Hydrology,* the U.S. Army Corps of Engineers Summary Report of Snow Investigations [10]. In this study, some 2400 occurrences of precipitation in Donner Summit, California (elevation 7200 ft), ranging from 29°F to 40°F (−2°C to 4°C) were analyzed to determine the effect of surface air temperature on the form of precipitation, either (a) rain, (b) snow, or (c) mixed rain and snow. The results, summarized in Table 12-1, show that precipitation occurs in the form of snow when the surface air temperature is approximately less than 35°F (1°C).

Elevation. The same study concluded that elevation is an important variable in snow formation. Precipitation data indicated that snowstorms accounted for approximately 95 percent of all precipitation events at 7000 ft (2100 m), 50 percent at 4000 ft (1200 m), and 1 percent at 1000 ft (300 m). While these data were obtained in the central Sierra Nevada of California, at 39°N latitude, they are believed to be representative of regions along the windward side of major mountains ranges in the United States and Canada. A tendency for an increase in snowstorms with an increase in latitude was also documented.

Terrain Features. In leveled (i.e., flat) terrain, the distribution of precipitation is a function solely of atmospheric variables. Conversely, in orographic regions,

TABLE 12-1 PERCENTAGES OF SNOW, RAIN, AND MIXED RAIN AND SNOW IN DONNER SUMMIT, CALIFORNIA [10]

Surface Air Temperature (°F)	29	30	31	32	33	34	35	36	37	38	39	40
Snow (%)	99	99	97	93	74	44	32	29	8	8	0	3
Rain (%)	0	1	2	3	12	31	51	57	81	90	100	97
Rain and snow (%)	1	0	1	4	14	25	17	14	11	2	0	0

precipitation is also a function of the character of the terrain. For a satisfactory evaluation of precipitation in orographic regions, the relationship of terrain to meteorologic conditions must be examined.

The effects of terrain on precipitation are classified as either (a) small scale or (b) large scale. Studies of small-scale orographic effects have shown that the water equivalent of the snowpack (Chapter 3) is a function of (1) elevation, (2) slope, (3) exposure, and (4) southern aspect. The California data in *Snow Hydrology* supported the following conclusions regarding water equivalent:

1. An increase of 1.0–2.5 in. (2.5–6.3 cm) for each 100 ft (30 m) increase in elevation.
2. A decrease of 0.2–0.5 in. (0.5–1.2 cm) for each 1 percent increase in slope.
3. A decrease of 0.5–0.75 in. (1.2–1.8 cm) for each 10° increase in *exposure sector*. An exposure sector is the sector of a circle of 0.5-mi radius, centered at the snow course (Section 3.2), within which there is no land higher than the points in the snow course.
4. An increase of 0.25–1.0 in. (0.6–2.5 cm) for each 10° deviation from southern aspect.

Spreen [9], using data from western Colorado, correlated average winter precipitation to large scale orographic features. He used elevation, slope, exposure, and orientation as independent variables and concluded that together they account for a high percentage of the precipitation variability in that area. Although general qualitative estimates can be deduced from these studies, values for individual basins or regions are likely to vary widely.

Snow Accumulation

Snow accumulation is a function of the following atmospheric and environmental conditions: (1) surface air temperature, (2) elevation, (3) slope and aspect of terrain, (4) wind, (5) energy and moisture transfer, and (6) vegetative cover.

Temperature. The temperature at the time of snowfall controls the dryness of snow and, therefore, its susceptibility to erosion by wind. On mountain slopes, an increase in snowcover is usually associated with a temperature decrease with an increase in elevation. Wet snow falls where temperatures are close to the melting point, usually in the proximity of large bodies of water. Conversely, dry snow is typical of the continental interiors, where colder temperatures prevail.

Elevation. In mountainous regions, elevation is considered to be the most important factor affecting snowcover distribution. Often a linear relationship between snow accumulation and elevation can be found for specific sites and elevation ranges. Since snow accumulation is a function not only of elevation but also of slope, exposure, aspect, and so on, these linear relationships tend to reflect local conditions.

Slope. Orographic precipitation rate is largely a function of terrain slope. If the air is saturated, the rate at which precipitation is produced is proportional to the rate of ascent of the air mass. Rhea and Grant [8] analyzed Colorado winter precipitation data and concluded that the long-term average of orographic precipitation at a point was strongly correlated with the topographic slope computed over the first 20 km upwind of the point.

Aspect. The importance of aspect on snow accumulation is shown by the large differences between snowcover amounts found on windward and leeward slopes of coastal mountain ranges. The influence of aspect is related to the direction of snowfall-producing masses, the frequency of snowfall, and the energy exchange processes influencing snowmelt. However, the effect of aspect on snow accumulation tends to be much less than that of elevation [7].

Wind. Wind is responsible for the movement of snow particles, changing their shape and physical properties and depositing them into drifts or banks of greater density than the parent material. A loose snow cover, with particles 1 to 2 mm in diameter is readily entrained by fairly light winds of about 10 km/h [4]. The formation of a glaze by the freezing of surface melt may inhibit transport by wind; however, very strong winds may move even large sheets of glazed snow. Erosion prevails at locations where the wind accelerates (at the crest of a ridge), and deposition occurs where the wind decelerates (along the edges of forests and cities). The rate of snow transport by wind is greatest over flat, extensive open areas and least in areas exhibiting great resistance to flow (forests and cities).

Energy and Moisture Transfer. During the winter months, energy and moisture transfers to and from the snowcover are responsible for changes in its state. Radiation fluxes are primarily responsible for changes in depth and density of the snowpack. The underlying surface, the physical properties of the snowcover, vegetation, buildings, roads, and other cultural features affect the net radiation flux reaching the snow, changing its erodibility, mass and state. The net radiation is a function of the snowcover's *albedo,* the ratio of reflected to incident shortwave radiation. Typical values of albedo for different snowcover surfaces vary from 0.8 for exposed surfaces to 0.12 under extensive coniferous forest cover.

Vegetative Cover. Vegetation influences the surface roughness and wind velocity, thereby affecting the erosional, transport, and depositional characteristics of the surface. When vegetation extends above the snowcover it affects the process of energy exchange and the amount of snow reaching the ground.

A forest provides a large intercepting and radiating biomass above the snowcover surface. Studies have revealed that more snow is usually found on forest openings than within the forest stands. In addition to modifying wind velocity and providing additional interception, a forest acts to modify the energy exchange processes that affect the snowcover's erodibility, mass, and state.

Distribution of Snowcover

The extent of snowcover is directly related to altitude, since temperatures near or below freezing affect both the frequency of snowfall and the probability of snowmelt. On a global basis, the duration of snowcover is longest near the poles and on high mountain ridges.

Snowcover may form and disappear several times within a season. At high latitudes, a long period of winter snowcover is virtually assured (exceeding 180 d in continental areas north of 60°N). At lower latitudes, snowcover may form briefly before melting. Due to the ephemeral nature of early and late seasonal snowstorms, it is often difficult to determine the length of the seasonal snowcover period.

Grasslands. Snowcover usually forms on the colder, continental grasslands of the Northern Hemisphere starting in November. In the southern Great Plains of North America, the snowcover becomes permanent in December or January. The characteristics of permanent snowcover are largely a function of air temperature. Seasonal snowcover periods vary from 120 to 160 d in the northern grasslands and 30 to 60 d in Oklahoma, to only a few days in Texas. The mean annual accumulated snowcover depth in grasslands is in the range of 20 to 50 cm, with a density of approximately 20 percent (200 kg/m^3) [4].

Mixed Forests. In the mixed forests of the Northern Hemisphere, snowcover usually forms in late November or December and recedes in two directions: from the south during February and from the north in late March or early April. Predictions of length of snowcover are unreliable because the cover does not remain on the ground for long periods. The average snowcover density is 20 percent, increasing progressively over the winter to 30 percent by late March.

Mountain Areas. Snow exists on most high mountain ridges every month of the year at elevations that vary with altitude and climate. The snowcover on rugged mountain terrain is highly variable due to its exposure to slides and wind action. Tundra conditions (i.e., those typical of forest-free arctic and subarctic regions) prevail at the higher elevations, with the snowcover undergoing severe erosion and wind packing resulting in the formation of slabs.

12.2 MELTING OF THE SNOWPACK

Snowmelt is the product of several heat transfer processes acting on the snowpack. Moreover, the quantity of snowmelt is a function of the condition of the snowpack itself. Therefore, snowmelt determinations are quite complex, and certain simplifying assumptions are necessary for practical applications.

The heat-transfer processes acting on the snowpack vary with time and location. Solar radiation, for example, is relatively important in the central plains of the United States but not in the Pacific Northwest. Solar radiation is also more important during the spring than during the winter, while its role diminishes with an increase in latitude. No single method for computing snowmelt is applicable for all regions and seasons. A thorough understanding of the snowmelt process is necessary to select the best method for a given location and time of the year.

Snowpack Energy Balance

The principal sources of heat energy involved in the melting of the snowpack are the following [12]:

1. Net shortwave (i.e., solar) radiation, H_s
2. Net long-wave (i.e., terrestrial) radiation, H_l
3. Convective heat transfer from atmosphere to snowpack, H_c
4. Heat transfer caused by condensation of water vapor onto snowpack, H_e
5. Heat transfer from rainwater to snowpack, H_p
6. Heat conduction from underlying ground to snowpack, H_g

Each of these items is a function of several factors. For instance, net shortwave radiation is the difference between incident and reflected solar radiation. Net long-wave radiation *loss* is the difference between the radiation emitted by the snowpack and the portion of it reflected back by the atmosphere.

Radiation melt is the snowmelt due to the combined effect of shortwave and long-wave radiation, i.e., all-wave radiation. *Convective melt* is the snowmelt caused by convective heat transfer from atmosphere to snowpack. *Condensation melt* is the snowmelt caused by condensation of water vapor onto snowpack. *Rain melt* is the snowmelt caused by heat transfer from rainwater to snowpack. *Ground melt* is the snowmelt caused by heat conduction from the ground to snowpack.

Snowmelt Heat Equivalent. Snowmelt heat equivalent is the total amount of heat energy involved in snowmelt. It is calculated as follows:

$$H_m = H_s + H_l + H_c + H_e + H_p + H_g \tag{12-1}$$

in which H_m = snowmelt heat equivalent. In Eq. 12-1, H_s is positive, H_l is usually negative in the open (i.e., long-wave radiation loss), H_c is usually positive, H_e may be either positive or negative, and H_p and H_g are almost always positive.

Thermal Quality of the Snowpack. The amount of snowmelt produced by a given amount of heat energy is a function of the *thermal quality* of the snowpack. Thermal quality is the ratio of the heat necessary to produce a given amount of water from the snowpack to the heat necessary to produce the same amount of water from pure ice, expressed as a percentage.

At temperatures below freezing the thermal quality of snow is greater than 100 percent. Conversely, the thermal quality of snow containing free water is less than 100 percent. A *ripe snowpack* is one that is at 0°C temperature and holds water only by adsorption and capillarity. The thermal quality of a ripe snowpack is approximately 97 percent [10].

Rate of Snowmelt. For a snowpack with a thermal quality of 100 percent, the latent heat of fusion is 80 cal/g. At 0°C, the density of water is approximately equal to 1 g/cm³. Therefore, for a snowpack of a thermal quality of 100 percent, the heat required to produce 1 cm of melt is 80 cal/cm², or 80 langleys. In general, for a snowpack with a thermal quality of B%, the heat required to produce 1 cm of melt is 80(B/100) ly = (B/1.25) ly. Therefore, the snowmelt rate can be calculated as follows:

$$M = \frac{1.25 \ H_m}{B} \qquad (12\text{-}2)$$

in which M = snowmelt rate in centimeters per day, H_m = snowmelt heat equivalent in langleys per day (ly/d), and B = thermal quality of the snowpack in percent. Similarly, the snowmelt rate associated with each of the heat energies of Eq. 12-1 can be calculated. For instance, $M_s = 1.25(H_s/B)$, in which H_s is the net shortwave radiation and M_s is the snowmelt rate due to shortwave radiation.

Solar Radiation

Only an infinitesimally small portion of all the radiant energy emitted by the sun reaches the earth, yet this small portion is the ultimate source of all the earth's energy. The amount of solar energy intercepted by the earth varies with the solar output and with the seasons. These variations, however, are quite small.

The *solar constant* is the intensity of solar radiation received on a unit area of a plane normal to the incident radiation at the outer limit of the earth's atmosphere, with the earth at its mean distance from the sun. The value of the solar constant is generally taken to be 1.94 ly/min, although variations in the range of 1.90-2.00 ly/min have been reported [10].

The amount of solar radiation incident on a horizontal surface is referred to as *insolation*. Daily insolation amounts received at the outer limit of the earth's atmosphere can be calculated from the solar constant for any given latitude and time of the year, as shown in Fig. 12-1. These insolation amounts are subject to reflection, scattering, and absorption by the atmosphere. In the absence of clouds, and barring unusual atmospheric conditions, the amounts reflected, scattered or absorbed are quite constant and relatively small. Variations in these amounts are caused primarily by variations in the content of water vapor and dust in the atmosphere.

The amount of insolation reaching the earth's surface is a function of the atmospheric transmission coefficient:

$$C_a = \frac{I_c}{I_o} \qquad (12\text{-}3)$$

in which C_a = atmospheric transmission coefficient; I_c = insolation reaching the earth's surface under clear sky; and I_o = insolation reaching the outer limit of the earth's atmosphere.

Atmospheric transmission coefficients include direct solar radiation and diffuse sky radiation, i.e., the scattered radiation that manages to reach the earth's surface. Therefore, an increase in diffuse sky radiation causes an increase in atmospheric transmission coefficients. The albedo of the earth's surface has a direct bearing on diffuse sky radiation. Other things being equal, the greater the albedo, the greater the diffuse sky radiation. The relatively high albedo of snow surfaces results in increased diffuse sky radiation. During the winter, the higher albedos associated with new-fallen snow (A = 0.8) cause a substantial increase in diffuse sky radiation. Dur-

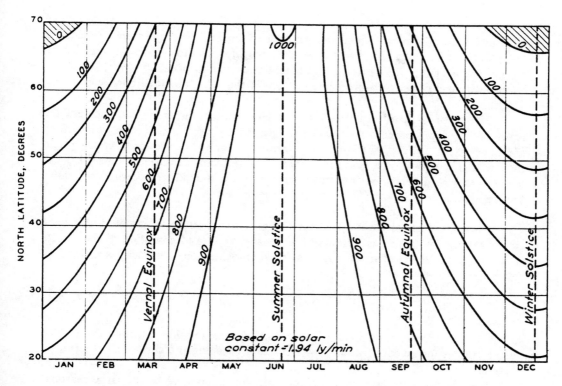

Figure 12-1 Insolation at the outer limit of the earth's atmosphere as a function of latitude and time of the year, in langleys per day [10].

ing the summer, the increase is attenuated by the lower albedos of the older snow (A = 0.4).

Effect of Clouds. By far the largest variations in the portion of solar radiation transmitted by the atmosphere are caused by clouds. The transmitted radiation varies with type, height, density, and cloud cover. The cloud cover coefficient is defined as follows:

$$C_c = \frac{I}{I_c} \tag{12-4}$$

in which C_c = cloud cover coefficient; I = insolation reaching the earth's surface under cloud cover; and I_c = insolation reaching the earth's surface under clear sky. The cloud cover coefficient can be related to cloud height and amount of cloud cover as follows [5]:

$$C_c = 1 - [0.82 - (0.024)Z]N \tag{12-5}$$

in which Z = cloud height in thousands of feet; and N = amount of cloud cover, the ratio of area of cloud cover to area of sky.

The effect of the reflectivity of the earth's surface on the amount of diffuse sky radiation is more pronounced for a cloudy sky than for a clear sky. Not only is the ratio of diffuse sky radiation to direct solar radiation increased by the presence of clouds, but also the diffuse sky radiation reflected by the snow surface is strongly rereflected by the clouds. Thus, for a given cloud height and amount of cloud cover, the cloud cover coefficient of snow-covered areas is greater than that of snow-free areas.

Effect of Slope. In the Northern Hemisphere, the radiation incident on south-facing slopes exceeds that incident on north-facing slopes. During the spring the slope effect is slight, but during the winter it is more pronounced. At any given instant, the radiation on a sloping surface (relative to the radiation on a horizontal surface) may be determined from the geometry of the individual situation (the slope and its aspect, and the solar altitude and azimuth) [10].

Effect of Forest Cover. The effect of forest cover on the amount of insolation reaching the ground is a function of density, type, and condition of the forest. To evaluate the effect of forest cover, a forest transmission coefficient is defined as follows:

$$C_f = \frac{I_f}{I_c} \tag{12-6}$$

in which C_f = forest transmission coefficient; I_f = insolation reaching the earth's surface under forest cover; and I_c = insolation reaching the earth's surface under clear sky.

For deciduous forests, the insolation amount is affected by the large seasonal variability in forest transmission coefficients. However, for coniferous forests, the variability is quite small throughout the year, and therefore average values are appropriate. For coniferous forests, transmission coefficients are inversely related to the density of forest canopy, i.e. the ratio of area covered by forest canopy to total area.

Measurement of Insolation. Insolation is measured with a *pyranometer*, an instrument consisting of a vacuum bulb, in the center of which is a disk having a white center and concentric black and white bands. An electromagnetic force is produced, which is proportional to the temperature difference between the rings and hence to the radiation incident upon them. The electromagnetic force is recorded with a potentiometer calibrated to measure radiation intensity. In the United States, insolation is measured in selected first-order stations operated by the National Weather Service. Insolation data may be acquired through the National Climatic Data Center, Asheville, North Carolina.

Albedo of a Snowpack. The albedo of a snowpack may vary widely, ranging from 0.4 for a ripe, granular snowpack, to 0.8 or more for new-fallen snow. The albedo is primarily a function of the condition of the surface layers of the snowpack. It is measured by means of two *pyranometers,* one measuring the insolation received and the other measuring the shortwave radiation reflected by the snowpack.

The variation of albedo with temperature index (summation of daily maximum temperatures since last snowfall) is shown in Fig. 12-2. Also, there is a tendency for a decrease in albedo from the accumulation season (new snow) to the melt season (older snow). Figure 12-3 shows the seasonal reduction in albedo.

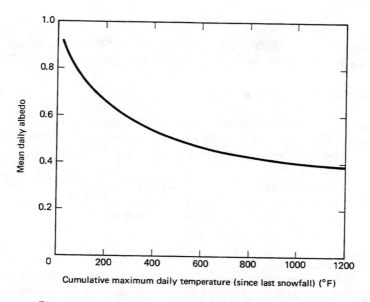

Figure 12-2 Variation of albedo with temperature index [10].

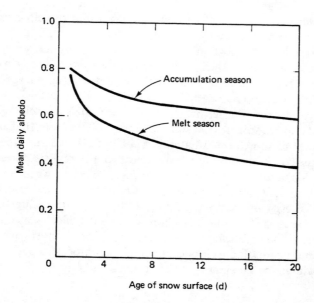

Figure 12-3 Seasonal reduction in albedo from accumulation to melt season [10].

Shortwave Radiation Melt. Following Eq. 12-2, the shortwave radiation melt is calculated as

$$M_s = \frac{1.25 H_s}{B} \qquad (12\text{-}7)$$

in which M_s = shortwave radiation melt rate in centimeters per day; H_s = net shortwave radiation in langleys per day; and B = thermal quality of the snowpack in percent. The net shortwave radiation is equal to:

Sec. 12.2   Melting of the Snowpack

$$H_s = I(1 - A) \qquad (12\text{-}8)$$

in which I = insolation reaching the earth's surface, in langleys per day, after appropriate correction for cloud cover, slope, and forest cover; and A = albedo.

Snowmelt Computations

Snowmelt computations are accomplished by disaggregating the snowmelt process into the following melt components: (1) all-wave radiation melt, (2) convection melt, (3) condensation melt, (4) rain melt, and (5) ground melt.

All-wave Radiation Melt. The effects of shortwave and long-wave radiation are usually combined into an all-wave radiation melt. During clear weather, the important variables in radiation melt are (1) insolation, (2) albedo, and (3) air temperature. The humidity of the air also affects the radiation melt; however, its effect is relatively minor compared to the other three variables. Figure 12-4 illustrates the daily radiation melt (inches per day) for the central Sierra of California as a function of albedo and air temperature, for (a) spring conditions, with insolation of 800 ly/d, and (b) winter conditions, with insolation of 400 ly/d. In Fig. 12-4(b), *negative melts* are shown as dashed lines.

Figure 12-5 illustrates the effect of clouds on daily radiation melt during (a) spring (May 20) and (b) winter (February 15). It should be noted that the effect of cloud height and cloud cover on radiation melt is less during the winter than during the spring. Also, notice the trend of the function representing radiation melt versus cloud cover varying from winter to spring. During the spring, they are inversely related, whereas during the winter they are directly related. The winter reversal is largely due to the increased role of long-wave radiation during this time of the year.

The forest canopy exerts a large influence on the combined radiation exchange between the snowpack and its environment. However, its effect differs from that of clouds, particularly with respect to shortwave radiation. While both clouds and trees restrict the transmission of insolation, clouds are highly reflective, whereas the forest canopy absorbs most of the insolation. This causes the forest canopy to warm up, releasing to the snowpack a portion of the incident shortwave radiation energy. Figure 12-6 illustrates the effect of forest canopy cover on daily radiation melt during (a) spring and (b) winter. The relations shown in this figure represent typical seasonal radiation snowmelt conditions in the middle latitudes under a coniferous forest cover. During the spring the maximum radiation melt occurs in the open (for zero percent canopy cover), while in the winter the maximum radiation melt occurs with 100 percent canopy cover.

Convection Melt. Unlike radiative heat transfer, convective heat transfer from atmosphere to snowpack cannot be measured directly. An empirical expression for convection melt rate is

$$M_c = k_c v_b (T_a - T_s) \qquad (12\text{-}9)$$

in which M_c = convection melt rate, T_a = mean air temperature, T_s = snow surface temperature, v_b = wind speed, and k_c = convective melt coefficient. With tempera-

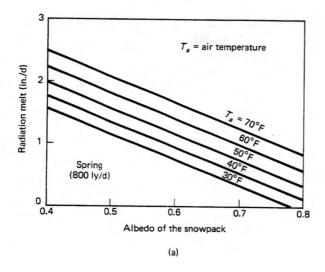

(a)

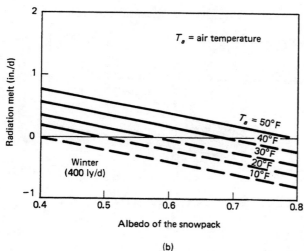

(b)

Figure 12-4 Variation of radiation melt with albedo and air temperature: (a) spring; (b) winter [10].

tures in degrees Celsius, wind speed at 15 m above ground level in kilometers per hour, and M_c in centimeters per day, the value of k_c is equal to 0.01137 [1].

Condensation Melt. The rate of condensation of water vapor onto the snow or ice surface may be important, particularly under conditions of rapid melt. At 0°C, the latent heat of vaporization (or condensation) is 597.3 cal/g. Compared with the latent heat of fusion (80 cal/g), this means that the volume of water available for runoff is approximately 7.5 times the volume of water actually condensed.

The amount of condensate can be related to vapor pressure and wind speed in the following way:

$$q_e = k_e v_b (e_a - e_s) \tag{12-10}$$

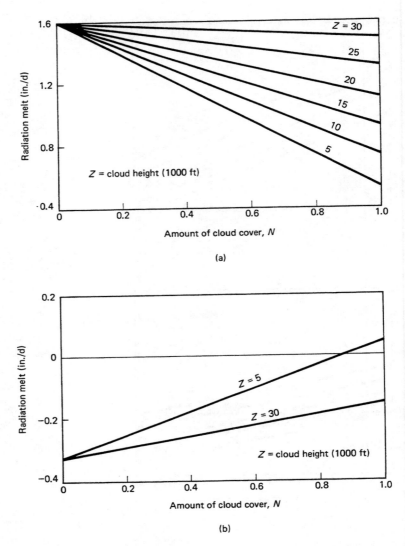

Figure 12-5 Variation of radiation melt with cloud height and cloud cover: (a) spring; (b) winter [10].

in which q_e = amount of condensate, e_a = vapor pressure of the air, e_s = vapor pressure of the snow surface, v_b = wind speed, and k_e = a coefficient.

For every unit of water vapor condensed, the additional heat of vaporization released is capable of melting 7.5 times this amount of snow. Therefore, the condensation melt equation becomes:

$$M_e = 8.5 \, k_e v_b (e_a - e_s) \tag{12-11}$$

in which M_e = condensation melt, which includes melt plus condensate.

Rain Melt. When rain falls on a snowpack, a certain amount of heat is transferred to the snow. For snowpacks at 0°C, this heat transfer produces snowmelt,

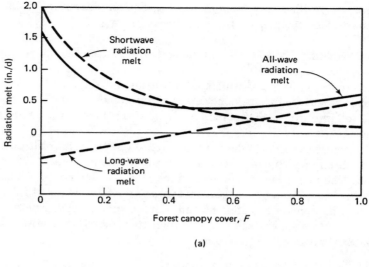

(a)

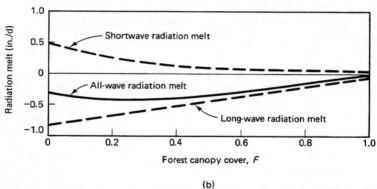

(b)

Figure 12-6 Variation of radiation melt with forest canopy cover: (a) spring; (b) winter [10].

whereas for colder snowpacks it causes a rise in snow temperature. The amount of heat released is directly proportional to the quantity of rainwater and to its temperature excess above that of the snowpack. Considering a melting snowpack, 1 cal of heat is available for every gram of rainwater and for every degree in excess of 0°C. This leads to

$$H_p = (T_r - T_s) P_r \tag{12-12}$$

in which H_p = heat released by rainwater in langleys, T_r = rainwater temperature in degrees Celsius, T_s = snowpack temperature in degrees Celsius, and P_r = rainwater depth in centimeters. Therefore, following Eq. 12-7, the rain melt is:

$$M_p = \frac{1.25(T_r - T_s) P_r}{B} \tag{12-13}$$

in which M_p = rain melt in centimeters, and B = thermal quality of the snowpack in percent. To illustrate, 1 cm of water at 5°C produces 0.0625 cm of rain melt on a snowpack of 100 percent thermal quality ($T_s = 0°C$).

Ground Melt. The conduction of heat from the underlying ground becomes important in snowmelt computations when the melt season as a whole is considered. This source of heat can cause melting during the winter and early spring when melt at the snow surface may be nonexistent. Ground melt is capable of priming the underlying soil prior to the melt season and may also help to ripen the snowpack, readying it for melt.

Measurements in the central Sierra of California indicate that ground melt rates vary throughout the melt season, with a tendency to increase as the season progresses [10]. The total ground melt during the month of January was measured as 0.28 cm; during May, 2.44 cm. The seasonal ground melt (for a total of 160 d) was measured at 8.28 cm, which translates into an average ground melt rate of 0.05 cm/d. These results, while not universally applicable, are generally indicative of the magnitude and seasonal variation of ground melt.

Total Melt. The various melt components—radiation, convection, condensation, rain and ground melts—can be summed up to obtain the variation of total melt with time of the year, as shown in Fig. 12-7 for the California data included in *Snow Hydrology*. The following conclusions can be drawn from Fig. 12-7:

1. The radiation melt (M_r) is negative during the winter and increasingly positive during the spring.
2. The sum of convection and condensation melts (M_{ce}) is very close to zero during the winter and increasingly positive during the spring.
3. The sum of ground melt (M_g) and rain melt (M_p) is very close to zero during both winter and spring.
4. The total melt ($M_r + M_{ce} + M_g + M_p$) is close to zero during the winter and increasingly positive during the spring.

12.3 SNOWMELT INDEXES

In hydrologic practice, an *index* is a readily measured meteorologic or hydrologic variable that is related to a physical process in need of monitoring and whose variability can be used as a measure of the variability of the physical process. The reliability of an index depends upon (1) its ability to depict the variability of the physical processes, (2) its spatial and temporal variability, (3) its random variability, and (4) the quality and representativeness of its measurements.

Prior to the publication of *Snow Hydrology*, temperature indexes were widely used to estimate snowmelt runoff. Temperature was used because it was generally regarded as the best index of the heat transfer processes associated with snowmelt and because it was—and in many cases will continue to be—the only reliable and regularly available meteorological data. More recently, however, substantial improvements in the understanding of snowmelt processes have led to the use of alternative snowmelt indexes.

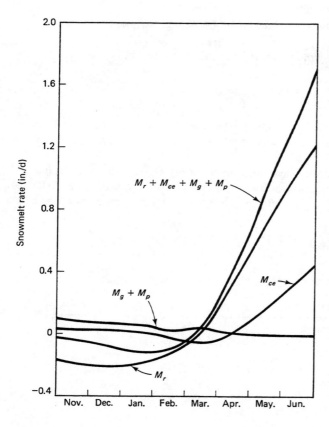

Figure 12-7 Variation of melt components and total melt with time of the year [10].

For instance, studies have shown that all-wave radiation is the controlling factor for snowmelt runoff from an open site. When all-wave radiation data is not available, estimates of shortwave and long-wave radiation can be summed up to obtain an estimate of all-wave radiation. *Duration-of-sunshine* data can be used as an index of shortwave radiation, with the appropriate albedo estimates to determine the net shortwave radiation. Air temperature data can be used as an index of long-wave radiation exchange. However, air temperature *alone* is generally a poor index of snowmelt runoff from an open site.

For forested areas, convection and condensation melts are by far the most important components in snowmelt runoff, largely because of the relation to the long-wave radiation exchange in the forest environment. The usual convection parameter is the product of daily maximum temperature and 12-h diurnal wind travel. The usual condensation parameter is the product of vapor pressure and 12-h diurnal wind travel. Therefore, accurate determinations of wind and vapor pressure are necessary for the evaluation of convection and condensation melt.

In the absence of wind and vapor pressure data, maximum daily air temperature can be used as an index to provide a fair estimate of daily melt for forested areas. Likewise, for heavily forested areas, either maximum daily or mean daily air temperature combined with vapor pressure are about equally effective as indexes of daily melt.

General Equation for Snowmelt Runoff in Terms of Indexes

The general equation for snowmelt runoff in terms of indexes is

$$M = a + \sum_{i=1}^{7} b_i X_i \qquad (12\text{-}14)$$

in which M = snowmelt runoff in centimeters per day (or inches per day), a = regression constant, b_i = regression constants, and X_i = snowmelt indexes described in Table 12-2.

Not all seven indexes are always significant for a particular basin. Typically, a regional snowmelt runoff equation is based on two or three indexes. For instance, a snowmelt runoff equation for the Boise River above Twin Springs, Idaho, is the following [10]:

$$M = -1.89 + 0.0245\,T + 0.00238\,G \qquad (12\text{-}15)$$

TABLE 12-2 VARIABLES IN GENERAL SNOWMELT RUNOFF EQUATION [10]

Variable	Description	Units	Index for
X_1	Daily total net shortwave radiation in the open	ly	Shortwave radiation
X_2	Daily long-wave radiation loss in the open, or daily total incident shortwave radiation in the open	ly	Long-wave radiation exchange in the open
X_3	Convection parameter, the production of maximum daily temperature and 12-h diurnal wind travel	°C-km (°F-mi)	Convection and long-wave radiation exchange in forest
X_4	Condensation parameter, the product of vapor pressure and 12-h diurnal wind travel	mb-km (mb-mi)	Condensation
X_5	12-h diurnal wind travel	km (mi)	Temperature and vapor pressure bases
X_6	Maximum daily temperature	°C (°F)	Convection and long-wave radiation exchange in forest
X_7	Vapor pressure, either maximum or at certain time of day	mb	Condensation

in which M = snowmelt runoff over snow-covered area in inches per day, T = maximum daily temperature at Boise, Idaho, and G = net all-wave radiation exchange in the open in langleys per day.

Temperature Indexes

In many areas where snowmelt is an important contributor to runoff, air temperature measurements are the only data available from which snowmelt can be computed. Moreover, air temperature is regarded as the best single snowmelt index for forested areas. For these reasons, temperature indexes are the most widely used method of computing snowmelt.

A commonly used temperature index is *degree-days* above a chosen temperature base, obtained by counting the number of degrees above the temperature base for each day in which the temperature remains above the temperature base. Usually, mean daily temperature (i.e., the mean of maximum daily and minimum daily temperatures) is used in calculating the number of degree-days. The freezing level (0°C) is normally chosen as the temperature base. For certain applications, however, maximum daily temperature may be used in lieu of mean daily temperature. Moreover, a temperature other than freezing is sometimes used as the temperature base.

A widely used snowmelt indicator is the *degree-day factor,* a unit melt rate defined as the number of centimeters (inches) of melt per degree-day. Its units are either centimeters per degree Celsius-day [cm/(°C-d)] or inches per degree Fahrenheit-day [in./(°F-d)], with 1 in./(°F-d) = 4.57 cm/(°C-d).

Point Melt Rates. Early investigations of temperature indexes focused on point melt rates. Horton [6] performed experiments wherein cylinders of snow were cut from the snowpack and melted under laboratory-controlled temperature conditions. He found the degree-day factor to be in the range 0.04-0.06 in./(°F-d). Clyde [2] performed similar experiments and determined average degree-day factors in the range 0.05-0.07 in/(°F-d). Church [3], using actual data from Soda Springs, California, obtained a degree-day factor of 0.051 in./(°F-d). In the *Snow Hydrology* report, values of degree-day factors ranging from 0.06 to 0.106 in./(°F-d) were obtained from measurements at three sites, with a mean of 0.085 in./(°F-d) [10].

Basinwide Snowmelt. Although point melt rates are usually restricted within a relatively narrow range, their extrapolation to basinwide snowmelt rates is complicated because of the spatial and seasonal variability of snow cover. During the snowmelt season, the progressive retreat of the snow line results in a gradual change in mean elevation of the snow-covered area. Furthermore, only a part of the snow-covered area may be contributing to snowmelt. For instance, in the Northern Hemisphere, the southerly exposed open areas tend to melt first, leaving the more sheltered areas to produce the last of the snowmelt. These complexities contribute to making the basinwide evaluation of snowmelt quite a difficult undertaking.

A complete water balance is required so that the sources of snowmelt runoff can be properly identified. Moreover, the areal extent of the snowpack needs to be determined. Extensive studies of basinwide snowmelt runoff reported in *Snow Hydrology* resulted in the basinwide degree-day factors shown in Table 12-3. The values shown

TABLE 12-3 DEGREE-DAY FACTORS FOR BASINWIDE SNOWMELT RUNOFF [10]

Basin	Degree-day factor based on mean temperature		Degree-day factor based on maximum temperature	
	April	May	April	May
Central Sierra Snow Laboratory	0.089	0.100	0.024	0.043
Upper Columbia Snow Laboratory	0.037	0.072	0.010	0.031
Willamette Basin Snow Laboratory	0.039	0.042	0.021	0.025

Note: Degree-day factors shown are given in in./(°F-d), with a temperature base of 32°F (0°C). To convert to SI units [cm/(°C-d)], multiply values shown by 4.57.

are means for the several years of record in the three testing sites. They reflect the general decrease in melt amounts with increasing forest cover, from the central Sierra in California to the Upper Columbia and Willamette basins in the Northwest. They also reflect the general increase in melt as the melt season progresses, from April to May. The April melt is shown to be generally less than the May melt because some of the April heat is consumed in ripening the snowpack rather than in melting it.

Generalized Basin Snowmelt Equations

A suitable mix of the relevant physical processes and statistical analyses was used in *Snow Hydrology* to develop generalized basin snowmelt equations. These equations are based on commonly available meteorological data, with appropriate simplifications to enhance their practicality.

Equations for Rain-free Periods. The generalized basin snowmelt equations for rain-free periods are as follows [10]:

a. For heavily forested areas:

$$M = 0.074(0.53T_a' + 0.47T_d') \tag{12-16a}$$

b. For forested areas:

$$M = k(0.0084v)(0.22T_a' + 0.78T_d') + 0.029T_a' \tag{12-16b}$$

c. For partly forested areas:

$$M = k'(1 - F)(0.004I_i)(1 - A) + k(0.0084v)(0.22T_a' + 0.78T_d') + F(0.029T_a') \tag{12-16c}$$

d. For open areas:

$$M = k'(0.00508I_i)(1 - A) + (1 - N)(0.0212T_a' - 0.84) + N(0.029T_c') + k(0.0084v)(0.22T_a' + 0.78T_d') \tag{12-16d}$$

in which

M = snowmelt rate in inches per day

T_a' = difference between the air temperature at a 10-ft height and the snow surface temperature in degrees Fahrenheit

T_d' = difference between the dew point temperature at a 10-ft height and the snow surface temperature in degrees Fahrenheit

v = wind speed at 50 ft above the snow in miles per hour

I_i = observed or estimated daily insolation in langleys

A = observed or estimated average snow surface albedo, expressed as a decimal fraction

k' = basin shortwave radiation melt factor, which depends on the average exposure of open areas compared to an unshielded horizontal surface; for south slopes, it varies between 1.1 in the summer solstice to 2.4 in the winter solstice

F = estimated average basin forest canopy cover, effective in shading the area from solar radiation, expressed as a decimal fraction

T_c' = difference between the cloud base temperature and the snow surface temperature in degrees Fahrenheit

N = estimated cloud cover, expressed as a decimal fraction

k = basin factor, a function of basin topographic characteristics and exposure to wind, varying from about 0.2 for densely forested areas to slightly over 1.0 for exposed ridges and mountain passes (for plain areas with no forest cover, $k = 1.0$).

Equations for Periods with Rainfall. The generalized basin snowmelt equations for periods with rainfall are as follows [10]:

a. For open or partly forested basins:

$$M = (0.029 + 0.0084kv + 0.007P_r)(T_a - 32) + 0.09 \qquad \text{(12-17a)}$$

b. For heavily forested areas:

$$M = (0.074 + 0.007P_r)(T_a - 32) + 0.05 \qquad \text{(12-17b)}$$

in which T_a = temperature of saturated air at a 10-ft height in degrees Fahrenheit; P_r = rainfall rate in inches per day; and k, v, and M have been previously defined (Eq. 12-16).

12.4 EFFECT OF SNOWPACK CONDITION ON RUNOFF

Character of the Snowpack

Snow is composed of ice crystals, which are formed in the atmosphere at temperatures below freezing by sublimation of water vapor on hygroscopic nuclei. There are many different types of snow crystals, depending on the shape of the nucleus, the rate of sublimation, and the temperature of the air. Due to the usual dendritic structure of snow crystals, new-fallen snow is generally of low density. With time, however, the snowpack density increases. The change from a loose, dry, subfreezing, low-density snowpack to a coarse, granular, moist, high-density snowpack is referred to as the *ripening*, or conditioning, of the snowpack. A ripe snowpack is said to be primed to produce runoff. Snowpack density can be used to characterize other physical properties of snowpack, including its thermal properties and its affinity for water.

Liquid Water in the Snowpack

Water exists in the snowpack in three forms:

1. *Hygroscopic water,* which is adsorbed as a thin film on the surfaces of snow crystals and is unavailable for runoff until the snow crystals have melted or changed in form.
2. *Capillary water,* held by surface tension in the capillary spaces around the snow particles, free to move under the influence of capillary forces but unavailable for runoff until the snow melts or the spacing between snow crystals changes.
3. *Gravitational water,* in transit through the snowpack under the influence of gravity, draining freely from the snowpack, and available for runoff.

Water moves within the snowpack in both vapor and liquid phases. The amounts of water vapor are usually small compared to the amounts of liquid water. The *liquid-water-holding capacity* of a snowpack is the sum of hygroscopic and capillary water, expressed in percent of snow by weight. Any water in excess of the liquid-water-holding capacity moves in a generally downward direction driven by the gravitational force.

The snowpack is said to be dry when its temperature is below freezing. At 0°C, the degree of wetness of the snowpack depends on its liquid-water-holding capacity and the availability of free water. Winter rains or melt may bring the snowpack to its liquid-water-holding capacity. Subsequent weather may change the character of the snowpack and, with it, its liquid-water-holding capacity.

A snowpack at 0°C has a liquid-water-holding capacity of approximately 2 to 5 percent, depending on (1) its density and depth, (2) the size, shape, and spacing of snow crystals, and (3) its structure. The liquid-water-holding capacity of a snowpack can be related to its density, but the relationship is not immediately apparent.

Metamorphism of the Snowpack

The change in character with time, i.e., the *metamorphism* of the snowpack, determines to a large extent the amount of snowmelt and runoff. Generally, for equal amounts of melt and hydrologic abstractions (losses), the generated runoff is a function of snowpack condition. For instance, an initially cold (i.e., subfreezing) snow will freeze a certain amount of liquid water entering it, raising the temperature of the snow to the melting point. Before releasing any free-draining water, the liquid-water-holding capacity of the snowpack needs to be satisfied. If, however, the entire snowpack is already conditioned to yield water, all water inflow will pass through the snowpack to the ground without depletion.

Factors Affecting the Metamorphism of Snow. Time is the principal factor in the metamorphism of snow. The physical processes contributing to the metamorphism of snow are (1) heat transfer at the snow surface by radiation, convection, and condensation, (2) percolation of melt or rainwater through the snowpack, (3) internal pressure due to the weight of the snow, (4) wind, (5) temperature and water vapor variation within the snowpack, and (6) heat transfer at the ground surface. These physical processes cause changes in the following properties of the snowpack: (1) density, (2) structure, (3) air, water, and heat permeability and diffusivity, (4) liquid-water-holding capacity, and (5) temperature.

Structure of the Snowpack. As each new layer of snow is deposited, its upper surface is subjected to the weathering effects of radiation, rain, and wind, its undersurface is subjected to ground heat, and its interior to the action of the percolating water and water vapor. This results in the stratification of the snowpack, that is, it becomes a layered structure reflecting individual snowstorm deposits. As the season progresses, the snowpack tends to become homogeneous with respect to temperature, liquid-water content, grain size, and density.

During the melt season, on clear nights, a relatively shallow surface layer cools considerably below 0°C due to the outgoing long-wave radiation. Below this nocturnal snow crust, the snowpack remains at 0°C, and liquid water continues to drain as long as the snowpack water content remains above its liquid-water-holding capacity. The thickness of the nocturnal crust is approximately 30 cm [10].

Heat Transfer Within the Snowpack. Heat transfer within the snowpack is among the processes responsible for its ripening or conditioning. During the spring melt season, the snowpack is normally at a temperature of 0°C, except for the nocturnal crust, and any heat reaching the surface is converted to melt. During the winter, however, the snowpack is often at temperatures below 0°C, and a certain amount of heat must be transferred from the surface and ground to the snowpack to meet the thermal deficiency before melting and runoff can occur. In hydrologic applications, the thermal deficiency of the snowpack can be taken as a *heat deficit* or initial heat loss, which must be satisfied before runoff can take place.

Water Transmission Through the Snowpack. The condition of the snowpack determines the amount of storage and the rate of downward movement of water. The temperature, size, shape, surface area, and spacing of the snow crystals, melt rates, and rainfall intensities control the retention and detention of water as it moves downward through the snowpack.

The time of travel of unprimed snow may be considerable, particularly when the snow is striated with ice planes that are flat or upwardly concave. If a water course is established within the snowpack, the travel time may be relatively short, being largely a function of the snowpack depth.

Energy Required for Ripening the Snowpack. The energy required for ripening the snowpack is [10]:

$$E_c = \frac{T_s}{1.6} + f_p \qquad (12\text{-}18)$$

in which E_c = equivalent energy required to ripen the snowpack, in percent of melt energy, T_s = average snow temperature below 0°C (positive value), and f_p = liquid-water-holding capacity in percent. For instance, for $T_s = -8°C$, and $f_p = 2\%$, the energy required to ripen the snowpack is 7 percent of the energy required to melt it. This energy is normally supplied during the transition between accumulation and melt periods, so that its effect on flows during the active melt season is relatively small. During the winter, however, the energy required for ripening the snowpack may be an appreciable fraction of the available energy.

12.5 SNOWMELT HYDROGRAPH SYNTHESIS

The synthesis of snowmelt or rain-on-snow hydrographs differs in several respects from that of rainfall hydrographs. For one thing, only a fraction of the basin may be covered by snow. Furthermore, for rain-on-snow floods, the losses (i.e., the hydrologic abstractions) of snow-covered areas may be quite different from those of snow-free areas.

Elevation Effects

Elevation is an important factor in snowmelt hydrograph determinations. Both rain and snow amounts tend to increase with elevation, whereas snowfall occurs more frequently at higher elevations. Moreover, melt rates have a tendency to decrease with an increase in elevation.

Two approaches are used in taking into account elevation effects in snowmelt computations: (1) the elevation-band method, and (2) the rational method. In the elevation-band method, the limiting (i.e., maximum and minimum) basin elevations are identified. Several elevation bands are chosen within these limits, and the subareas comprised within each elevation band are measured. Elevations limiting each band may be determined either by equal-elevation increments or by equal-area bands. A sufficient number of bands should be chosen to assure a smooth variation in snowmelt, rainfall, and loss amounts. Each elevation band is assumed to be either snow-covered or snow-free, melting or not melting, raining or not raining, and losing water through hydrologic abstractions at a constant rate. Snowmelt, rainfall, and loss amounts are estimated for each elevation band.

The rational method treats the drainage basin as a unit, making corrections for snow-free subareas and subareas with snowfall during rain-on-snow storms. In this approach, it is assumed that the snow-cover depletion progresses regularly upwards, from lower to higher elevations. The contribution to snowmelt originates in the area between the *snow line*, the average elevation of the lower limit of the snow-covered area, and the *melt line*, the average elevation of the upper limit of snowmelt. In the case of rain on snow, the contributing area for rainfall runoff has an upper limit at the elevation where the rain becomes snow. Typically there are three distinct bands in the rational method: (1) a lower band where rain falls on snow-free ground, (2) a middle band where rain falls on snow-covered ground, and (3) an upper band where snow falls on snow-covered ground.

Effect of Melt Period

The melt period has an effect on the synthesis of snowmelt hydrographs. Since snowmelt is diurnal in character, the daily snowmelt quantity is usually generated in less than 12 h. For large basins, particularly those for which no regular diurnal pattern of streamflow increase is discernible, daily melt amounts may be considered as daily increments. For smaller basins, especially those with faster response characteristics, the day may be divided into two 12-h or three 8-h increments, with all the melt attributed to one of the increments.

Rain-on-Snow Hydrographs

In rain-on-snow situations, the effect of the snowpack is twofold: (1) to add an increment of meltwater to rainfall and (2) to store and detain, in varying degrees, the rainwater and generated melt. The latter effect substantially increases the difficulty of calculating rain-on-snow flood hydrographs.

Depending on the condition of the snowpack, rainwater may be stored by the pack or pass through without depletion. A dry, subfreezing snowpack may be able to store considerable amounts of rainwater. Moreover, a deep snowpack may contribute an additional storage effect because of the increased travel time required for water to pass through it. On the other hand, a thoroughly conditioned snowpack may oppose very little resistance to the flow of water and may actually abet runoff by adding an increment of melt of its own. More importantly, however, a conditioned snowpack may have helped maintain soil moisture at high levels, which may result in the complete conversion of snowmelt into runoff. Between these two extremes lies an array of intermediate cases. Therefore, a knowledge of the initial condition of the snowpack is essential for the proper synthesis of rain-on-snow hydrographs.

Spring Snowmelt Runoff Hydrographs

Spring snowmelt runoff hydrographs tend to vary with orographic patterns. For instance, in the mountainous areas of the western United States, the annual spring snowmelt flood is caused by the sustained melting of deep snowpacks over a long period of time. On the other hand, in the northern Great Plains of the United States, snowmelt floods occur often in early spring, triggered by the comparatively fast melting of shallow snowpack covering extensive areas.

Three approaches are used to synthesize spring snowmelt runoff hydrographs. These are (1) the elevation-band method, (2) the rational method, and (3) the one-step method.

Elevation-band Method. In this method, the basin is subdivided into several (n) elevation bands, and snowmelt, rainfall, and losses are computed separately for each band. Melt and rain are assumed to be spatially uniform throughout each band, and the entire band is assumed to be either snow-covered or snow-free and melting or not melting. Snowmelt is computed by using temperature or other appropriate indexes. Excess basin water available for runoff (in centimeters per day, inches per day, or other suitable rate units) is obtained by weighing rainfall, snowmelt, and losses in proportion to the individual band subareas:

$$M_e = \frac{\sum\limits_{i=1}^{n} [(P_i + M_i - L_i)A_i]}{\sum\limits_{i=1}^{n} A_i} \tag{12-19}$$

in which M_e = excess basin water available for runoff, P_i = rainfall, M_i = melt, L_i = losses, and A_i = band subarea. The summation is for each of n bands.

Rational Method. In this method the basin is treated as a unit, and excess basin water available for runoff (in centimeters per day, inches per day, or other suitable rate units) is calculated as follows:

1. The melt line is established as the average elevation above which the air temperature is below freezing.
2. The snow line is established as the average elevation of the lower limit of the snow-covered area.
3. The contributing area, i.e., the fraction of the basin contributing to snowmelt, is the area between melt and snow lines.
4. Mean basin melt rate is obtained by multiplying the calculated snowmelt rate by the ratio of contributing area to basin area.
5. Excess basin water available for runoff is equal to mean basin melt rate plus mean basin precipitation minus mean basin loss rate.

One-step Method. This method makes use of a chart showing mean basin melt rate as a function of mean daily air temperature and snowline elevation. An example of this chart for the Boise River Basin above Twin Springs, Idaho, is shown in Fig. 12-8. The chart is site-specific, with the size of contributing area implicit in it. Excess basin water available for runoff is equal to mean basin melt rate plus mean basin precipitation minus mean basin loss rate.

Example 12-1.

Given the following rainfall, snowmelt, and losses data for a 950-km^2 catchment, calculate the excess basin water available for runoff by the elevation-band method.

Elevation Band (m)	Rainfall (cm/d)	Snowmelt (cm/d)	Losses (cm/d)	Subarea (km^2)
1000–1500	1.5	0.0	0.5	250
1500–2000	1.8	0.7	0.5	210
2000–2500	2.1	0.6	0.4	180
2500–3000	2.5	0.3	0.4	170
3000–3500	2.6	0.0	0.2	140

The application of Eq. 12-19 leads to: $M_e = [(1.0 \times 250) + (2.0 \times 210) + (2.3 \times 180) + (2.4 \times 170) + (2.4 \times 140)]/(250 + 210 + 180 + 170 + 140) = 1.924$ cm/d.

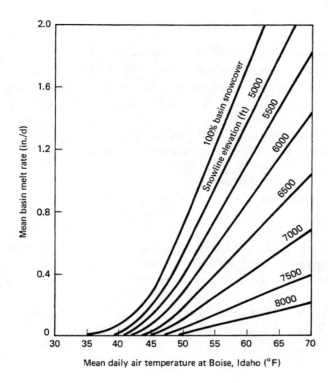

Figure 12-8 Snowmelt chart for Boise River Basin above Twin Springs, Idaho [10].

Example 12-2.

Given the following data, calculate the excess basin water available for runoff by the rational method: basin area 1000 km²; snow line at elevation 3000 m at 14°C mean daily temperature, with subarea below elevation 3000 m equal to 630 km²; melt line at elevation 4000 m at 6°C mean daily temperature, with subarea above elevation 4000 m equal to 150 km²; mean basin precipitation 0.36 cm/d, mean basin losses 0.23 cm/d, and degree-day factor 0.15 cm/(°C-d). Assume a linear decrease in mean daily temperature with elevation.

The contributing area is 1000 − 630 − 150 = 220 km². The mean daily temperature within the contributing area is 10°C. The snowmelt is 0.15 × 10 = 1.5 cm/d. Mean basin melt is 1.5 × (220/1000) = 0.33 cm/d. Total basin water is 0.36 + 0.33 = 0.69 cm/d. Excess basin water available for runoff is 0.69 − 0.23 = 0.46 cm/d.

Time Distribution of Runoff

Either unit hydrograph techniques (Chapter 5) or catchment routing methods (Chapter 10) are used in the generation of runoff hydrographs from snow-covered areas. Conventional unit hydrographs are used for rain-on-snow events. Special long-tail unit hydrographs may be necessary to synthesize runoff from excess spring snowmelt water [10].

QUESTIONS

1. What atmospheric and environmental conditions control snow accumulation?
2. What is albedo? What is a typical range of albedo for snow-covered surfaces?
3. What are the principal sources of heat energy involved in melting of the snowpack?
4. What is the thermal quality of a snowpack? What is a typical value of thermal quality for a ripe snowpack?
5. What is the solar constant? What is the value of the solar constant?
6. What is insolation? What is the value of daily insolation amount received at the outer limit of the earth's atmosphere for a 30°N latitude at winter solstice?
7. What is convective melt? What is condensation melt?
8. What is rain melt? What is ground melt?
9. What are the controlling factors in shortwave radiation melt?
10. What is an index in hydrologic practice?
11. What is the controlling factor for snowmelt runoff from an open site? What are the most important components in snowmelt runoff from forested areas?
12. What is a degree-day? What is the degree-day factor?
13. What is ripening of the snowpack?
14. How many types of water exist in the snowpack? Explain.
15. Describe the elevation-band method for determining spring snowmelt runoff hydrographs.

PROBLEMS

12-1. Calculate the daily snowmelt rate for a snowpack of 95% thermal quality subject to a snowmelt heat equivalent of 570 ly/d.

12-2. Calculate the shortwave radiation snowmelt for a snow-covered site with insolation under clear sky 590 ly/d, cloud cover coefficient 0.8, forest transmission coefficient 0.9, snow surface albedo 0.7, and thermal quality of the snowpack 97%.

12-3. The following mean daily temperatures (in degrees Celsius) were measured in a snow-covered forest site during a certain week: Monday, 10°; Tuesday, 11°; Wednesday, 12°; Thursday, 13°; Friday, 12°; Saturday, 14°, and Sunday, 11°. Calculate the melt for a degree-day factor of $0.25 \text{ cm}/(\text{°C-d})$. Assume a temperature base of 0°C.

12-4. The following mean monthly temperatures have been recorded in a snow-covered forest site during the spring: April, 44°F; May, 52°F; and June, 60°F. The snowpack water equivalent at the end of March is 36 in. Assuming a degree-day factor of $0.04 \text{ in.}/(\text{°F-d})$, determine the potential snowmelt during the 3-month period. Compare this potential snowmelt value with the water equivalent at the end of the snowfall season (assume end of March) to determine whether there is any melt in May or June. Assume a temperature base of 32°F.

12-5. Compute basin snowmelt using a generalized snowmelt equation for the following data: forested area; wind speed at 50-ft above the snow surface, $v = 30$ mph; air temperature at a 10-ft height, 65°F; dew point temperature at 10 ft height, 50°F; snow surface temperature, 32°F; and basin factor $k = 0.4$.

12-6. Compute basin snowmelt using a generalized snowmelt equation applicable for rainfall periods for the following data: heavily forested area; temperature of saturated air at a 10-ft height, $T_a = 60$°F; and rainfall rate $P_r = 1.5$ in./d.

12-7. Given the following area-elevation data and daily rainfall, snowmelt, and losses for a certain basin, use the elevation-band method to compute the excess basin water available for runoff.

Elevation (m)	Subarea (km²)	Rainfall (cm)	Snowmelt (cm)	Losses (cm)
2000–2500	150	4	0	3
2500–3000	100	5	4	2
3000–3500	50	6	2	1

12-8. Given the following area-elevation data for a certain basin, use the rational method to compute the excess basin water available for runoff. Assume snow line at elevation 2200 m; melt line at elevation 2600 m; mean basin rainfall, 7 cm/d; snowmelt, 3 cm/d; and mean basin losses, 2 cm/d.

Elevation (m)	2000	2200	2400	2600	2800	3000
Cumulative area (km²)	0	450	630	730	800	840

12-9. Given the following area-elevation data for a certain basin, use the rational method to compute the excess basin water available for runoff. Assume snow line at elevation 5000 ft; melt line at elevation 6000 ft; mean basin rainfall, 2 in./d; mean basin losses, 0.5 in./d; degree-day factor 0.075 in./(°F-d); temperature at the index station at elevation 5500 ft, 42°F; and a temperature gradient of −10°F per 500-ft increase in elevation.

Elevation (ft)	4500	5000	5500	6000	6500
Cumulative area (mi²)	0	500	900	1200	1400

12-10. Using Fig. 12-8, determine the mean basin melt rate for the Boise River Basin above Twin Springs, Idaho, for the following conditions: (a) mean air temperature of 55°F and snow line at elevation 5500 ft; and (b) mean air temperature of 65°F and snow line at elevation 7000 ft.

REFERENCES

1. Bruce, J. P., and R. H. Clark (1966). *Introduction to Meteorology.* London: Pergamon Press.

2. Clyde, G. D. (1931). "Snow Melting Characteristics," *Bulletin No.* 231, Utah Agricultural Experiment Station, Aug.

3. Church, J. E. (1941). "The Melting of Snow," *Proceedings,* Central Snow Conference, Vol. 1, Dec., pp. 21–32.

4. Gray, D. M., and D. H. Male. (1981). *Handbook of Snow.* Toronto: Pergamon Press.

5. Haurwitz, B. (1948). "Insolation in Relation to Cloud Type," *Journal of Meteorology,* Vol. 5, No. 3, June, pp. 110–113.

6. Horton, R. E. (1915). "The Melting of Snow," *Monthly Weather Review,* Vol. 43, No. 12, Dec., pp. 599–605.

7. Meiman, J. R. (1970). "Snow Accumulation Related to Elevation, Aspect and Forest Canopy," *Proceedings, Workshop and Seminar in Snow Hydrology,* Ottawa: Queen Printers of Canada, pp. 35–47.

8. Rhea, J. O., and L. O. Grant. (1974). "Topographic Influences on Snowfall Patterns in Mountainous Terrain," in *Advanced Concepts and Techniques in the Study of Snow and Ice Resources,* National Academy of Sciences, Washington, DC.

9. Spreen, W. C. (1947). "A Determination of the Effect of Topography Upon Precipitation," *Transactions, American Geophysical Union,* Vol. 28, No. 2, April, pp. 285–290.

10. U.S. Army Corps of Engineers, North Pacific Division. (1956). "Snow Hydrology, Summary Report of the Snow Investigations," Portland, Oregon, June.

11. U.S. Army Engineer, North Pacific Division. (1986). "Program Description and User Manual for SSARR Model, Streamflow Synthesis and Reservoir Regulation," Portland, Oregon, Draft, April.

12. Wilson, W. T. (1941). "An Outline of the Thermodynamics of Snowmelt," *Transactions, American Geophysical Union,* Part I, pp. 182–195.

SUGGESTED READINGS

Gray, D. M., and D. H. Male. (1981). *Handbook of Snow,* Toronto: Pergamon Press.

National Academy of Sciences. (1974). *Advanced Concepts and Techniques in the Study of Snow and Ice Resources,* Washington, DC.

U.S. Army Corps of Engineers, North Pacific Division. (1956). "Snow Hydrology, Summary Report of the Snow Investigations," Portland, Oregon, June.

CATCHMENT MODELING

Certain applications of engineering hydrology may require complex analyses involving temporal and/or spatial variations of precipitation, hydrologic abstractions, and runoff. Typically, such analyses involve a large number of calculations and are therefore suited for use with a digital computer. The use of computers in all aspects of engineering hydrology has led to increased emphasis on catchment modeling. Catchment modeling comprises the integration of key hydrologic processes into a modeling entity, i.e., a catchment model, for purposes of either analysis, design, long-term runoff-volume forecasting, or real-time flood forecasting.

A catchment (watershed or river basin) model is a set of mathematical abstractions describing relevant phases of the hydrologic cycle, with the objective of simulating the conversion of precipitation into runoff. In principle, the techniques of catchment modeling are applicable to catchments of any size, whether small (a few hectares), midsize (tens of square kilometers) or large (many thousands of square kilometers). In practice, however, catchment modeling applications are generally confined to the analysis of catchments for which the description of temporal and/or spatial variations of precipitation is warranted. Usually this is the case for midsize and large catchments.

A typical catchment modeling application consists of the following: (1) selection of model type, (2) model formulation and construction, (3) model testing, and (4) model application. Comprehensive catchment models include all relevant phases of the hydrologic cycle and, as such, are composed of one or more techniques for each phase. Commonly used methods and techniques for hydrologic modeling are described in Chapters 4 to 12. In practice, the hydrologic engineer would either (1) select an available model, with knowledge of its structure, operation, capabilities, and limitations, or (2) develop a model or modify an existing one, based on perceived needs, data availability, and budgetary constraints.

Most routine applications are of the first type, in which case it is necessary to become thoroughly familiar with the model's characteristics and features. Well-documented models have user manuals which describe the interaction between user and model. In addition, some models may have reference manuals providing additional information on the model's internal structure.

For research and development projects, the construction of a new model may be warranted. In this case the hydrologic engineer has a wider choice of methods and techniques, but the associated model development costs are comparatively high.

Hydrologic modeling techniques applicable to small and midsize catchments are discussed in Chapters 4 and 5. Routing methods for reservoirs, stream channels, and catchments are described in Chapters 8, 9, and 10. This chapter ties these concepts into a general framework for catchment modeling using a computer. Section 13.1 describes classification of catchment models. Section 13.2 describes model components and model construction. Section 13.3 discusses model calibration, verification, and sensitivity analysis. Section 13.4 contains a description of several computer catchment models in current use.

13.1 CLASSIFICATION OF CATCHMENT MODELS

Material versus Formal Models

Catchment models can be grouped into two general categories: (1) material and (2) formal. A material model is a physical representation of the prototype, simpler in structure but with properties resembling those of the prototype. Examples of material catchment models are rainfall simulators and experimental watersheds [28].

A formal model is a mathematical abstraction of an idealized situation that preserves the important structural properties of the prototype. Since formal models are invariably mathematical in nature, it is customary to refer to them as mathematical models. Mathematical models intended to be used with the aid of a computer are referred to as computer models. Many computer catchment models have been developed over the last three decades. A representative sample of United States catchment modeling practice is given in Section 13.4.

Material catchment models are expensive and of limited applicability. Conversely, formal models are readily available, highly flexible, and comparatively inexpensive to use. It is therefore not surprising that formal (i.e., mathematical) models are the preferred tool in the solution of catchment modeling problems.

Types of Mathematical Catchment Models

A mathematical catchment model consists of several components, each describing a certain phase or phases of the hydrologic cycle. A mathematical model can be of three types: (1) theoretical, (2) conceptual, or (3) empirical. Theoretical and empirical models are exactly opposite in meaning, with conceptual models lying somewhere in between them. In addition, a mathematical model can be either deterministic or probabilistic, linear or nonlinear, time-invariant or time-variant, lumped or distributed, continuous or discrete, analytical or numerical, and event-driven or continuous-process.

In catchment modeling practice, four general types of mathematical models are commonly recognized: (1) deterministic, (2) probabilistic, (3) conceptual, and (4) parametric. Deterministic models are formulated by following laws of physical and/or chemical processes as described by differential equations. A deterministic model is formulated in terms of a set of variables and parameters and equations relating them. A deterministic model implies a cause-effect relationship between chosen parameter values and results obtained from the solution of the equations. Ideally, a deterministic model should be able to provide the best detail in the simulation of physical or chemical processes. In practice, however, the application of deterministic models is often hindered by the model's (or modeler's) inability to resolve the temporal and spatial variability of natural phenomena into sufficiently small increments.

Probabilistic models are exactly opposite in meaning to deterministic models. A probabilistic model is formulated by following laws of chance or probability. Probabilistic models are of two types: (1) statistical and (2) stochastic. Statistical models deal with observed samples, whereas stochastic models deal with the random structure observed in certain hydrologic time series—for instance, daily streamflows in midsize catchments. The development of statistical models invariably requires the use of data; stochastic models place emphasis on the stochastic characteristics of hydrologic processes [29].

Conceptual models are simplified representations of the physical processes, usually relying on mathematical descriptions (either in algebraic form or by ordinary differential equations), which simulate complex processes in the mean by relying on a few key conceptual parameters. The extensive use of conceptual models in engineering hydrology reflects the inherent complexity of the phenomena and the practical inability to account for deterministic components in all instances. Therefore, conceptual models are useful and practical surrogates for deterministic models.

Parametric models (i.e., empirical, or *black-box*) are the simplest of all modeling approaches. As their name implies, the emphasis of parametric models is placed upon the key empirical parameter(s) on which the solution is based. Usually, a parametric model consists of an algebraic equation (or equations) containing one or more parameters to be determined by data analysis or other empirical means. The applicability of parametric models is restricted to the range of data used in the determination of the parameter values. Parametric models are useful where conceptual, deterministic, or probabilistic models are judged to be either impractical or too expensive.

Examples of types of mathematical catchment models and model components can be found in a variety of hydrologic applications. For instance, the kinematic wave-routing technique (Chapters 4, 9, and 10) is deterministic, being founded on basic principles of mass and momentum conservation. Once the parameters of the kinematic rating curve have been determined, analytical solutions of kinematic waves lead to predictable solutions. Numerical solutions, however, are subject to diffusion and dispersion caused by the finite nature of the grid. Therefore, careful evaluations are needed to ensure that all the relevant processes are being properly quantified.

The Gumbel method for flood frequency analysis (Chapter 6) is a typical example of the use of probabilistic methods in hydrology. The Gumbel method is statistical, since the parameters of the frequency distribution are evaluated from measured data. Stochastic methods (e.g., Monte Carlo simulation) have been used primarily in the synthetic generation of hydrologic time series such as daily streamflows from midsize catchments, which often show substantial random components [3, 29].

The cascade of linear reservoirs (Chapter 10) is a typical example of a conceptual model. In this case, the physical processes of runoff concentration and runoff diffusion are being simulated in the mean by the diffusion inherent in the mathematical solution of a linear reservoir. Two or more reservoirs in series produce enough diffusion so that translation (runoff concentration) and storage (runoff diffusion) are effectively simulated. As with any conceptual model, measured rainfall-runoff data is needed in order to determine appropriate values of the model parameters.

Regional analysis (Chapter 7) is a typical example of the parametric approach to hydrologic catchment modeling. In this case, statistical regression techniques are used to develop predictive equations having regional applicability. The parameters of the regression equation have regional meaning; therefore, extrapolation beyond the region of definition is usually unwarranted.

Linear versus Nonlinear Models

The choice between linear and nonlinear models is one that has far-reaching practical implications. In nature, physical processes are generally nonlinear; in modeling, however, linear models are often substituted for nonlinear processes in the interest of mathematical expediency. The simplicity of the linear models is a definite advantage, although it is usually achieved at the cost of a certain loss of detail. The nonlinear models are more complex but are generally better able to provide increased detail in the simulation of physical processes.

A linear model is formulated in terms of linear equations and processes; conversely, a nonlinear model is described by nonlinear equations and processes. A typical example of a linear model is the classical unit hydrograph. Examples of nonlinear equations used to model hydrologic processes are many—for instance, the nonlinear regression techniques used in regional analysis.

The cascade of linear reservoirs is another example of the use of linear models in hydrologic modeling practice. Since the model is conceptual, its parameters must be determined by calibration (using rainfall-runoff data) before attempting to use it in a predictive mode. Since the parameters are constant, extrapolation to flow regions other than that used in the calibration is usually unwarranted. In practice, it is often necessary to calibrate the conceptual model for each one of several flow levels (i.e., low, average, and high) and to use the calibrated parameter sets accordingly.

Deterministic models can have a complexity of their own. Partial differential equations can be either linear or quasi-linear, depending on whether the coefficients of the various terms are assumed to be constant or variable. The linear models have constant parameters; conversely, the quasi-linear models have variable parameters. Where simplicity is at stake, the use of constant parameters may be justified—for instance, the constant-parameter Muskingum-Cunge routing method (Section 9.4). For increased detail in the simulation of a wide range of flows, a variable-parameter model is the logical choice, i.e., the variable-parameter Muskingum-Cunge routing method.

Time-invariant versus Time-variant Models

In time-invariant models, the model parameter(s) remain constant in time. Conversely, in time-variant models, the model parameter(s) vary in time. A typical example is that of the conceptual model of a linear reservoir (Chapter 8). A time-invariant model is $S = KO$, whereas a time-variant model is $S = K(t)O$, in which the reservoir storage constant K is a function of time. In practice, most applications have been restricted to time-invariant models.

Lumped versus Distributed Models

The term *lumped-parameter* model—for short, lumped model—is used to refer to a model in which the parameters do not vary spatially within the catchment. Therefore, catchment response is evaluated only at the outlet, without explicit accounting of the response of individual subcatchments. A typical example of a lumped parameter model is the unit hydrograph.

The term *distributed-parameter* model—for short, distributed model—is used to refer to a model in which the parameters are allowed to vary spatially within the catchment. This enables the calculation not only of the overall catchment's response but also of the response of individual subcatchments. The increased detail with which simulations can be made with a distributed model renders it more computationally intensive than a lumped model. This permits the modeling of special features such as spatially varying rainfall and spatially varying hydrologic abstractions. However, for the results of distributed modeling to remain meaningful, the quality and quantity of available data must be commensurate with the increased level of detail.

The concepts of lumped and distributed modeling, while opposite in meaning, are not necessarily exclusive. Lumped catchment models can be used as components of larger distributed catchment models. In a typical application, a lumped model (i.e., the unit hydrograph) is used for hydrograph generation from individual subcatchments. Subsequently, these hydrographs are combined and routed through a network of stream channels and reservoirs. Since the parameters vary from subcatchment to subcatchment and hydrographs can be evaluated at any desired location within the stream channel network, the network model retains essentially a distributed structure.

Continuous versus Discrete Models

Continuous and discrete models are opposite in meaning. Mathematically, a continuous function is one that possesses a derivative at any point in the domain of the computation. Conversely, a discrete function lacks this property. Examples of the use of continuous and discrete functions are common in engineering hydrology. For instance, a streamflow hydrograph is continuous, but a rainfall hyetograph is discrete. In modeling, the term *continuous* is used to refer to models for which solutions can be obtained at any point. In discrete models, however, solutions can be obtained only at certain predetermined points.

A typical example of the difference between continuous and discrete models is afforded by reservoir theory. The differential equation of storage (Eq. 8-4) is an ordinary differential equation, and, therefore, a continuous solution can be obtained by analytical means. The same equation, however, can be discretized on the xt plane (Fig. 8-2) and solved by a method such as storage indication (Section 8.3) using numerical procedures. In general, functions described analytically lend themselves readily to continuous modeling. Conversely, functions of arbitrary shape are better handled with discrete models.

Analytical versus Numerical Models

The difference between analytical and numerical models closely parallels that of continuous and discrete models. Continuous functions and models can usually be solved by analytical means; discrete functions and models lend themselves readily to solution using numerical procedures. An analytical solution uses the tools of classical mathematics—for instance, perturbation theory, Laplace transforms, and so on. A numerical solution uses either finite differences, finite elements, the method of characteristics, or any other method based on the attendant discretization of the solution domain.

In general, analytical solutions can be obtained only for highly simplified problems, particularly those for which initial and boundary conditions can be expressed in analytical form. Numerical models are better suited for real-world applications, for which initial and boundary conditions are likely to be arbitrarily specified. Numerical models, schemes, and algorithms are often used as integral parts of computer models which simulate all relevant phases of the hydrologic cycle. Given the widespread use of computers, it seems certain that numerical models will continue to play an important role in catchment modeling practice.

Event-driven versus Continuous-process Models

Catchment models can be either (1) event-driven (or event), or (2) continuous-process. As their name implies, event models are short-term, designed to simulate individual rainfall-runoff events. Their emphasis is on infiltration and surface runoff; their objective is the evaluation of direct runoff. Event models are applicable to the calculation of flood flows, particularly in the cases where direct runoff is a major contributor to total runoff. Typically, event models have no provision for moisture recovery between storm events and, therefore, are not suited for the simulation of dry-weather (i.e., daily) flows.

Unlike event models, continuous-process models take explicit account of all runoff components, including direct runoff (surface flow) and indirect runoff (interflow and groundwater flow). Continuous-process models focus on evapotranspiration and other long-term hydrologic abstractions responsible for the rate of moisture recovery during periods of no precipitation. The objective of continuous-process models is the accounting of the catchment's overall moisture balance on a long-term basis. Continuous-process models are suited for simulation of daily, monthly, or seasonal streamflows, usually for long-term runoff-volume forecasting and estimates of water yield.

13.2 MODEL COMPONENTS AND MODEL CONSTRUCTION

The basic catchment model components are: (1) precipitation, (2) hydrologic abstractions, and (3) runoff. Usually, precipitation is the modeling input, hydrologic abstractions are determined by the catchment's properties, and runoff is the modeling output.

Precipitation

Precipitation, either in the form of rainfall or snowfall, is the process driving the catchment model. Surface runoff is a direct consequence of excess rainfall and/or snowmelt. Rainfall can be described in terms of the following: (1) intensity, (2) duration, (3) depth, (4) frequency, (5) temporal distribution, (6) spatial distribution, and (7) areal correction.

Rainfall Intensity. Rainfall intensity varies widely in time and space. In practice, it is necessary to resort to temporal and spatial averages in order to provide useful rainfall descriptions. For small catchments, average rainfall intensity during a period equal to the concentration time is usually the primary rainfall parameter. For midsize catchments, emphasis shifts from rainfall intensity to storm depth, storm duration, and a suitable temporal rainfall distribution. For large catchments (i.e., river basins), the rainfall spatial distribution may become the controlling factor.

High-intensity storms are usually of short duration and cover relatively small areas. Conversely, low-intensity storms are typically of longer duration and cover larger areas. Depending on the catchment's size, antecedent moisture condition, and storm areal coverage, both low- and high-intensity storms can produce runoff events of comparable magnitude. Therefore, a rainfall description relying exclusively on constant rainfall intensity is limited to small catchments.

Rainfall Duration. The duration of a rainfall event or storm varies widely, ranging from a few minutes to several days. The runoff concentration property (Section 2.4) indicates that all catchments, regardless of size, eventually reach an equilibrium runoff condition when subjected to constant effective rainfall. In practice, this implies that small catchments are likely to reach equilibrium runoff conditions much more readily than midsize and large catchments. This is the reason why small catchments are usually analyzed assuming concentrated catchment flow, i.e., an equilibrium runoff condition. The rational method is a typical example of the assumption of runoff concentration (Section 4.1).

For midsize catchments, runoff response is a function of cumulative storm depth and applicable temporal distribution. In this case, catchment response is usually of the subconcentrated type, with storm duration shorter than time of concentration. The design storm duration, to be determined by trial and error, is that which produces the highest peak rate of runoff for the given rainfall depth and temporal distribution.

Since runoff peaks are directly related to rainfall intensity and rainfall intensity decreases with an increase in storm duration, it follows that a longer storm duration does not necessarily lead to a higher runoff peak.

For large catchments, the storm's spatial distribution becomes important, although the storm duration continues to play a role. This is because long duration storms can often be regarded as consisting of two parts (i.e., a *storm couplet)*. The first part usually produces little runoff, the bulk of it going to increase the basin's overall moisture content. The second part, occurring in the wake of the first on a wet antecedent moisture condition, is almost entirely converted into runoff, resulting in abnormally high peak flows. This storm-couplet mechanism (see Section 14.5) is usually responsible for major floods experienced in large basins.

Rainfall Depth. For small catchments, a rainfall depth is implied by the assumption of a constant rainfall intensity lasting a specified duration. Rainfall depth per se becomes important in midsize catchment analysis, where it is used together with a chosen dimensionless temporal storm distribution to develop a design storm hyetograph. Isopluvial maps showing storm depth-duration-frequency data throughout the United States have been developed by the National Weather Service. References are shown in Table 13-1.

For large projects, especially those where structural failure due to hydrologic reasons (i.e., embankment dam overtopping) can result in loss of life, the concept of probable maximum precipitation, or PMP, is used in lieu of depth-duration-frequency. In practice, the PMP is used as an input to a catchment model to obtain the probable maximum flood, or PMF. Guidelines for the evaluation of PMP are given in the references shown in Table 13-2.

TABLE 13-1 NWS DEPTH-DURATION-FREQUENCY REFERENCES

U.S. Weather Bureau, "Generalized Estimates of Probable Maximum Precipitation and Rainfall Frequency Data for Puerto Rico and Virgin Islands for Areas to 400 Square Miles, Durations to 24 Hours, and Return Periods from 1 to 100 Years," *Technical Paper No.* 42, 1962.

U.S. Weather Bureau, "Rainfall Frequency Atlas of Hawaiian Islands for Areas to 200 Square Miles, Durations to 24 Hours, and Return Periods from 1 to 100 Years," *Technical Paper No.* 43, 1962.

U.S. Weather Bureau, "Rainfall Frequency Atlas of the United States for Durations from 30 Minutes to 24 Hours and Return Periods from 1 to 100 Years," *Technical Paper No.* 40, 1963, applicable to States East of the 105th Meridian.

U.S. Weather Bureau, "Probable Maximum Precipitation and Rainfall Frequency Data for Alaska for Areas to 400 Square Miles, Durations to 24 Hours and Return Periods from 1 to 100 Years," *Technical Paper No.* 47, 1963.

U.S. Weather Bureau, "Two-to-Ten Day Precipitation for Return Periods of 2 to 100 Years in the Contiguous United States," *Technical Paper No.* 49, 1964, applicable to the contiguous United States.

NOAA National Weather Service, "Atlas 2: Precipitation Atlas of the Western United States," 1973, applicable to the 11 Western States.

NOAA National Weather Service, "Five to 60-Minute Precipitation Frequency for the Eastern and Central United States," *Technical Memorandum NWS HYDRO-35,* 1977.

TABLE 13-2 NWS PROBABLE MAXIMUM PRECIPITATION REFERENCES

NOAA *Hydrometeorological Report No.* 36: "Interim Report, Probable Maximum Precipitation in California," 1961, reprinted with revisions of October 1969.

NOAA *Hydrometeorological Report No.* 39: "Probable Maximum Precipitation in the Hawaiian Islands," 1963.

NOAA *Hydrometeorological Report No.* 43: "Probable Maximum Precipitation, Northwest States," 1966.

NOAA *Hydrometeorological Report No.* 49: "Probable Maximum Precipitation Estimates, Colorado River and Great Basin Drainages," 1977.

NOAA *Hydrometeorological Report No.* 51: "Probable Maximum Precipitation Estimates, United States East of the 105th Meridian," 1978.

NOAA *Hydrometeorological Report No.* 52: "Application of Probable Maximum Precipitation Estimates—United States East of the 105th Meridian," 1982.

NOAA *Hydrometeorological Report No.* 54: "Probable Maximum Precipitation and Snowmelt Criteria for Southeast Alaska," 1983.

NOAA *Hydrometeorological Report No.* 55: "Probable Maximum Precipitation Estimates—United States Between the Continental Divide and the 103rd Meridian," 1984.

U.S. Weather Bureau, "Generalized Estimates of Probable Maximum Precipitation and Rainfall Frequency Data for Puerto Rico and Virgin Islands for Areas to 400 Square Miles, Durations to 24 Hours, and Return Periods from 1 to 100 Years," *Technical Paper No. 42,* 1962.

U.S. Weather Bureau, "Probable Maximum Precipitation and Rainfall Frequency Data for Alaska for Areas to 400 Square Miles, Durations to 24 Hours and Return Periods from 1 to 100 Years," *Technical Paper No. 47,* 1963.

Rainfall Frequency. In general, the larger the storm depth, the more infrequent its occurrence. Closely related to frequency is the concept of return period, defined as the average time elapsed between occurrences of two storm events of the same frequency. Return periods normally used in design practice vary from 5 to 10 y for small storm drainage facilities, to 50 to 100 y for more important structures. The choice of rainfall frequency is usually based on local practice and individual experience. For sizable projects, specially those where failure can result in loss of life, PMP is used in lieu of frequency as the basis for flood determinations.

Temporal Distribution. The temporal distribution of a storm has a leading role in the hydrologic response of midsize catchments. For a given storm depth and duration, the choice of dimensionless temporal rainfall distribution allows the development of a design storm hyetograph.

For a certain rainfall depth and duration, a uniform temporal distribution (Fig. 13-1(a)) will produce a slow response, with a relatively low peak and long time base. Conversely, a highly nonuniform temporal distribution (Fig. 13-1(b)) will produce a fast response, with a relatively high peak and short time base. In practice, a judicious choice of temporal distribution (or alternatively, a design storm) is necessary for the accurate calculation of peak flows using catchment modeling techniques.

A design temporal distribution can be either locally or regionally derived. Dimensionless temporal rainfall distributions are expressed as percent of rainfall duration in the abscissas and percent of rainfall depth in the ordinates. Alternatively, the duration can be fixed at a set value and only the ordinates expressed as percent of rainfall depth.

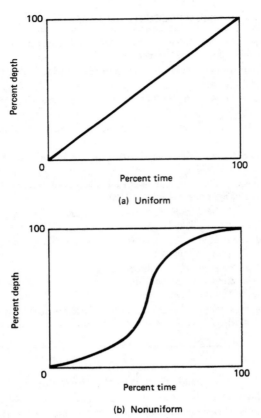

Figure 13-1 Types of temporal rainfall distributions.

The Soil Conservation Service has developed several generalized temporal distributions applicable to the United States, among them the 24-h Types I, IA, II and III (see Fig. 5-14 and Table A-9, Appendix A) and the 48-h Types I and II. The 24-h Type I temporal distribution is representative of the Pacific maritime climate, with wet winters and dry summers (California coast, approximately south of San Francisco). The Type IA distribution is representative of the low-intensity precipitation normally associated with frontal storms to the west of the Cascade Mountains (Washington and Oregon) and Sierra Nevada (California). The Type III distribution is representative of Gulf and Atlantic Coastal areas where tropical storms are responsible for large 24-h precipitation amounts. The Type II distribution is the most intense of all four SCS rainfall distributions and is representative of dominant thunderstorms that occur east of the Cascade Mountains (Washington and Oregon) and Sierra Nevada (California), excluding the areas where the Type III distribution is applicable (Fig. 5-15). The 48-h temporal distributions are applicable to situations that warrant the use of storm durations longer than 24 h.

The SCS 24-h Types I, IA, II, and III temporal distributions were designed primarily for use in midsize catchment analysis. However, in TR-55 their applicability has been extended to small urban catchments (see Section 5.3) . These 24-h temporal distributions are intended to encompass the rainfall intensities associated with storms of shorter duration, ranging from 30 min to 12 h. Therefore, they are typically high-intensity storms, suited for catchments similar in size to those used by the Soil Conservation Service, i.e., those with areas less than 250 km². Considerable judgment is required when using these SCS 24-h storms for catchment areas in excess of this limit.

Another widely used distribution is the SCS dimensionless temporal distribution for emergency spillway and freeboard-pool design (see Section 14.3). The Soil Conservation Service does not limit the storm depths and/or durations that could be used in conjunction with this temporal distribution. This distribution is shown graphically in Fig. 13-2 and in tabular form in Table A-10, Appendix A.

Regional alternatives to generalized temporal distributions can be derived based on a methodology developed by the National Weather Service [24, 25]. The method is based on the relation between *storms* and *rainfalls*. A storm is defined as the largest precipitation event at a given station for N consecutive hours (N is either 1, 2, 3, 6, 12, or 24) during a calendar year. The duration of this annual maximum storm is termed an independent-duration (ID) event. For each annual maximum storm, five other precipitation values are abstracted. These are the largest concurrent precipitation totals for M consecutive hours (M is either 1, 2, 3, 6, 12, or 24, but M ≠ N). These are events that include or are contained within the annual maximum storm. They are called dependent-duration (DD) events because they are dependent on the occurrence of the ID storm. The DD events are referred to as rainfalls. The sequential and quantitative relation between storms (ID) and rainfalls (DD) enables the development of temporal storm distributions based on regional rainfall characteristics.

Spatial Distribution.   For large catchments, the modeling hinges upon the assessment of the storm's spatial distribution. Storms that cover large areas tend to have elliptical shapes, with an *eye* of higher intensity located in the middle of the ellipse, surrounded by decreasing rainfall intensities and depths (Fig. 2-4(b)). Furthermore, the eye of the storm tends to move in a direction parallel to prevailing winds.

In certain cases it may be necessary to consider both local and general storms. For large catchments, local storms are typically high-intensity storms (i.e., thunder-

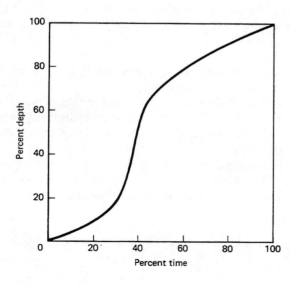

Figure 13-2   SCS dimensionless temporal rainfall distribution for emergency spillway and freeboard-pool design [21].

storms) covering only a fraction of the catchment area. Conversely, general storms cover the entire basin area, albeit with low-intensity, long-duration rainfall. For large catchments, both local and general storms can produce runoff and flooding of comparable magnitude.

Areal Correction. There is a limited amount of precipitable moisture in the atmosphere; therefore, the greater the storm's areal coverage, the smaller the storm's average depth. To account for this natural trend, storm depths shown in National Weather Service isopluvial maps are expressed as point depths, i.e., storm depths applicable to areas less than 25 km^2 (10 mi^2).

For catchments in excess of 25 km^2, an areal correction is needed in order to reduce the map's point depth to a smaller value that takes into account the areal coverage. Generalized depth-area relations have been developed by the National Weather Service (see Figs. 2-9(a) and 2-9(b)). These charts are strictly applicable to basins less than 1000 km^2 (400 mi^2). Other depth-area relations used by federal agencies are described in Section 14.1.

The NWS depth-area relations are applicable to the entire United States and are therefore conservative in their estimation of areal correction. Regional alternatives to the generalized NWS depth-area relations have been developed in certain geographic locations [26]. These charts are generally less conservative than the NWS charts, being more likely to depict regional rainfall patterns.

Experience and careful judgment are needed when extrapolating depth-area correction curves to catchment areas in excess of those included in the charts. In this regard, additional information can be obtained from depth-area-duration (DAD) charts included in reports dealing with PMP determinations [14, 15].

Hydrologic Abstractions

Hydrologic abstractions are the physical processes acting to reduce total precipitation into effective precipitation. Eventually, effective precipitation goes on to constitute surface runoff. There are many processes by which precipitation is abstracted by the catchment. Among them, those of interest to engineering hydrology are the following: (1) interception, (2) infiltration, (3) surface storage, (4) evaporation, and (5) evapotranspiration (Section 2.2).

The modeling objectives determine to a large extent which hydrologic abstractions are important in a certain application. For event models, the emphasis is on infiltration. For instance, the SCS runoff curve number method, which is widely used in event models, takes explicit account of infiltration. All other hydrologic abstractions are lumped into an initial abstraction parameter, defined as a fraction of the potential maximum retention (see Section 5.1).

Continuous-process models differ from event models in that they are designed to simulate daily flows, with or without the presence of precipitation. Accordingly, their emphasis is on evapotranspiration, their aim being to provide a detailed accounting of basin moisture at all times.

Interception. Interception is important in the modeling of high-frequency, low-intensity storms. Generally, storms having a high frequency of occurrence are substantially abstracted by interception. Conversely, for low-frequency, high-intensity storms, interception usually amounts to a very small fraction of the total rainfall. Detailed modeling of interception is usually warranted for continuous-process models. For event models, interception amounts are likely to be small compared to infiltration amounts.

Infiltration. Infiltration is regarded as the primary abstractive mechanism in event models. Rates and amounts of infiltration determine to a large extent the amounts of surface runoff. Either deterministic, conceptual, or parametric model components are used in modeling the infiltration process. For example, a deterministic model is the Green and Ampt formula (Section 2.2), which is based on the physics of the infiltration process. A typical example of a conceptual model is the SCS runoff curve method, which is based on an assumption of proportionality between retention and runoff (Section 5.1). Parametric models are also widely used, see, for instance, the general HEC loss-rate function (Section 13.4).

Infiltration rates vary widely in time and space, which makes the distributed modeling of infiltration a very complex task. For large catchments, and particularly for large floods, practical models of infiltration have been based on the concept of infiltration index, which lumps infiltration rates in time and space (Section 2.2). These indexes can be either estimated or obtained from analysis of rainfall-runoff data.

Surface Storage. The amount of surface storage is a function of catchment relief. In small upland catchments, surface storage is usually negligible. However, in urban and lowland catchments (swamps, marshes, and the like), surface storage may often be an important source of rainfall abstraction. The deterministic description of surface storage is very complex; therefore, surface storage is usually simulated with conceptual or empirical models (see Sections 2.2 and 5.3).

Evaporation. In catchment modeling, an evaporation component is used to quantify the loss of water from lakes and reservoirs. The process is particularly important for continuous modeling in arid and semiarid regions. In this case, reservoir and lake evaporation may represent a substantial contribution to the catchment's water balance.

Evaporation from water bodies can be estimated in several ways. The various approaches are classified as (1) water budget, (2) energy budget, (3) mass-transfer techniques and (4) pan-evaporation methods (Section 2.2). Given the largely empirical nature of evaporation estimations, the use of two or more methods is generally warranted to provide a basis for comparison.

Evapotranspiration. Evapotranspiration is regarded as the primary abstractive mechanism of continuous-process models. Evapotranspiration is taken into account in the form of either potential evapotranspiration (PET) or actual evapotranspiration. PET is modeled in much the same way as evaporation. Methods to calculate PET are classified as (1) temperature models, (2) radiation models, (3) combination (energy-budget and mass-transfer) models, and (4) pan-evaporation models (Section 2.2). Empirical formulas or other estimates of evapotranspiration are used in continuous-process modeling.

Runoff

Two distinct modes of runoff are recognized for modeling purposes: (1) catchment runoff and (2) stream channel runoff. Catchment runoff has three-dimensional features, but eventually this type of runoff concentrates at the catchment outlet. After it leaves the catchment, runoff enters the channel network, where it becomes stream channel flow. Unlike catchment runoff, the marked longitudinal orientation of stream channel flow generally justifies the assumption of one-dimensionality.

In practice, catchment runoff is modeled using either a lumped or a distributed approach. The lumped approach is based on the convolution of a unit hydrograph with an effective storm hyetograph (Section 5.2). The distributed approach is based on overland flow routing using kinematic or diffusion wave techniques (Sections 10.4 and 10.5). Both lumped and distributed approaches have advantages and disadvantages. The unit hydrograph is readily grasped and relatively easy to implement, although it does not take explicit account of the physical details inside the watershed. Kinematic wave routing is theoretically more appealing than the unit hydrograph, and, unlike the latter, can provide detailed information on surface runoff throughout the catchment. However, kinematic wave routing is generally more complex, difficult to implement, and requires substantial amounts of physical data for its successful operation. Moreover, the level of geometric abstraction required by operational kinematic wave models (e.g., the open-book representation) may in certain cases compromise the theoretical rigor of the technique.

An alternate approach to runoff modeling is the variable-source-area concept (Section 11.5) [19]. This approach is particularly applicable to hillslope hydrology, i.e., the study of runoff from upland and/or forested watersheds. This modeling concept is based on the assumption that the preferred path of rainfall is by infiltration through undisturbed forest soil, downslope migration, and maintenance of saturation or near-saturation at the lower slopes. As rainfall continues, the zone of saturated subsurface flow expands, with the saturated soil layers contributing substantial amounts of subsurface flow to runoff. The degree to which saturation and subsequent lateral expansion occurs is a function of antecedent moisture conditions, rainfall intensity, and duration. Variable-source-area modeling differs from overland flow modeling in that subsurface flow at the lower slope is considered to be the primary path to runoff (see Section 11.5).

In catchment modeling, the output from catchment runoff is the input to stream channel flow. The calculation of stream channel flow is accomplished by routing through the stream channel network. A distinction is made between *upland* subcatchments, which contribute upstream inflow to the stream network, and *reach* subcatchments, which contribute local inflow to the several reaches constituting the stream network. When using lumped models, runoff from reach subcatchments is concentrated at the downstream point. Conversely, with distributed models, runoff from reach subcatchments can be distributed laterally along the reach.

Routing through the stream network is accomplished by hydrologic or hydraulic routing techniques. Hydrologic techniques solve for discharge values directly; if desired, stages can be determined indirectly through the use of an appropriate rating curve. Hydraulic river-routing techniques (Section 9.5) generally solve for discharges and stages simultaneously, albeit at a substantial increase in complexity as compared to hydrologic techniques.

Distributed catchment models require a topological description of the stream network. A logical system of catchment and reach numbering is needed for proper hydrograph combination at network confluences.

Model Construction and Application

The construction of a catchment model begins with the selection of model components. Once these are chosen, they are assembled as parts of the overall model, following a logical sequence that resembles that of the natural processes. Rainfall and snowfall are considered first, followed by hydrologic abstractions, subcatchment hydrograph generation, reservoir and stream channel routing, and hydrograph combination at stream network confluences.

The issue of model resolution must be addressed at the outset of model construction and application. *Resolution* refers to the ability of the model to depict accurately certain scales of problems. Resolution is related to catchment scale and modeling objective. Modeling runoff from small catchments requires fine resolution, with typical time steps on the order of minutes and correspondingly small subcatchments and short channel reaches. On the other hand, modeling runoff from midsize catchments requires an average resolution, with typical time steps on the order of hours and correspondingly larger subcatchments and longer channel reaches. Moreover, modeling runoff from large catchments (i.e., river basins) may require a coarse resolution, with time steps on the order of one or more days and subcatchment size and channel reach lengths to match.

The modeling objective can have an influence on the choice of model resolution. Event models are short-term by definition and, therefore, are subject to relatively fast changes in model variables. A fine resolution, usually with time steps ranging from several minutes to a few hours, depending on catchment size, is usually required by event models. Continuous-process models are designed to account for long-term processes, with correspondingly lesser fluctuation in model variables. Therefore, a coarse model resolution is possible in continuous-process models.

13.3 MODEL CALIBRATION AND VERIFICATION

The essential ingredients of each model component are variables and parameters. Variables are the physical quantities themselves, i.e., discharge, stage, flow area, flow depth, mean velocity, and so on. Parameters are the quantities that control the behavior of the variables. Each model component may have one or more variables and parameters.

Parameters can have either a deterministic, conceptual, or empirical nature. Deterministic model parameters are based on laws of physical processes, usually in connection with distributed modeling applications, for which calculations are performed in the spatial and temporal domains. Conceptual parameters are part of conceptual models, i.e., those that simulate physical processes in a simple yet practical way. Empirical parameters are either calculated directly, based on measured data, or indirectly, based on related experience (as in the case of regionalization of model parameters).

In practice, it is likely that a certain catchment model will have more than one component type and, therefore, more than one parameter type. Generally speaking, a catchment model is referred to as either deterministic, conceptual, or empirical, depending on whether the majority of its components and parameters have a deterministic, conceptual, or empirical basis. The use and interpretation of catchment models, in particular the processes of calibration, verification, and sensitivity analysis, are largely a function of model type.

Calibration and Verification

Model calibration is the process by which the values of model parameters are identified for use in a particular application. It consists of the use of rainfall-runoff data and a procedure to identify the model parameters that provide the best agreement between simulated and recorded flows. Parameter identification can be accomplished either manually, by trial and error, or automatically, by using mathematical optimization techniques.

Calibration implies the existence of streamflow data; for ungaged catchments, calibration is simply not possible. The overall importance of calibration varies with the type of model. For instance, a deterministic model is generally regarded as highly predictive; therefore, it should require little or no calibration. In practice, however, deterministic models are usually not entirely deterministic, and therefore, a certain amount of calibration is often necessary.

In conceptual modeling, calibration is extremely important, since the parameters bear no direct relation to the physical processes. Therefore, calibration is required in order to determine appropriate values of these parameters. Practical estimates of conceptual model parameters, based on local experience, are sometimes used in lieu of calibration. However, such practice is risky and can lead to gross errors. Calibration also plays a major role in the determination of parameters of empirical models.

The calibration needs of time-invariant and time-variant processes and models are quite different. To evaluate the predictive accuracy of a time-invariant model, it is customary to divide the calibration process into two distinct stages: (1) calibration and (2) verification. For this purpose, two independent sets of rainfall-runoff data are assembled. The first set is used in the calibration per se, whereas the second set is used in model verification, i.e., a measure of the accuracy of the calibration. Once the model has been calibrated and the parameters verified, it is ready to be used in the predictive stage of the modeling.

With time-variant processes and models, the calibration is quite involved. Since the parameters vary in time (and with the model variables), a calibration and verification in the linear sense is only possible within a narrow variable range. A practical alternative is to select several variable ranges, e.g., low flow, average flow, and high flow, and to perform a calibration and verification for each flow level. In this way, a set of model parameters for each of several variable ranges can be identified. A typical example of multilevel (i.e., multistage) calibration is that of stream channel routing.

The routing parameters for inbank flow are likely to be quite different from those of overbank flow. Therefore, several calibrations are needed, encompassing a wide range of flow levels.

For certain processes and models, particularly those of deterministic nature, the model parameters can be explicitly related to model variables. In this case, the need for parameter calibration (in the linear sense) is circumvented. Variable-parameter models are, therefore, highly predictive.

Lumped and distributed models pose altogether different calibration problems. Lumped models have a relatively small number of parameters as compared to distributed models. For lumped conceptual models, calibration in the linear sense is possible. In this case, parameter estimations can often be obtained with automatic calibration techniques.

Unlike lumped models, distributed models have a large number of parameters, with most of them bearing some relation to the physical processes. The large parameter set, coupled with its enhanced physical basis, renders automatic calibration impractical and sometimes misleading. Accordingly, it is often advisable to limit the model parameters within physically realistic ranges and to perform trial-and-error calibrations.

Sensitivity Analysis

Uncertainties in catchment-modeling practice have led to increased reliance on *sensitivity analysis*, the process by which a model is tested to establish a measure of the relative change in model results caused by a corresponding change in model parameters. This type of analysis is a necessary complement to the modeling exercise, especially since it provides information on the level of certainty (or uncertainty) to be placed on the results of the modeling.

The issue of model sensitivity to parameter variations is particularly important in the case of deterministic models having some conceptual components. Because of the conceptual components, calibrations are strictly valid only within narrow variable ranges; therefore, errors in parameter estimation need to be ascertained in a qualitative way.

Sensitivity is usually analyzed by isolating the effect of a certain parameter. If a model is highly sensitive to a given parameter, small changes in the value of this parameter may cause correspondingly large changes in the model output. It is, therefore, necessary to concentrate the modeling effort into obtaining good estimates of this parameter. On the other hand, insensitive parameters can be relegated to a secondary role.

In catchment modeling, the choice of parameters for sensitivity analysis is largely a function of problem scale. For instance, in small catchments, the model's output is highly sensitive to the abstraction parameter(s), e.g., the runoff coefficient in the rational method. Therefore, it is imperative that the runoff coefficient be estimated in the best possible way. For low-frequency events, higher values of runoff coefficient are generally justified (Section 4.1).

In midsize catchment modeling, the model's sensitivity usually hinges on the temporal rainfall distribution, infiltration parameters, and unit hydrograph shape. The selection of rainfall distribution is crucial from the design standpoint. Catchment models are usually very sensitive to infiltration parameters, which need to be evaluated carefully, with particular attention to the physical processes. For instance, a high-intensity, short-duration storm may result in a high flow peak, due primarily to the high rainfall intensity. However, a low-intensity, long-duration storm may also result in a high flow peak, this time due to the long rainfall duration, which causes the hydrologic abstractions to be reduced to a minimum.

In large-catchment modeling, the model's sensitivity focuses on the spatial distribution of the storm, although the temporal distribution and infiltration parameters continue to play a significant role. In any case, a careful evaluation of model sensitivity is needed for increased confidence in the modeling results.

Sensitivity analyses provide an effective means of coping with the inherent complexities of catchment modeling, including the associated parameter uncertainties. In this sense, distributed models, while being widely regarded as deterministic, can often show a distinct probabilistic flavor [2].

13.4 CATCHMENT MODELS

Many catchment (watershed or basin) computer models have been developed in the last three decades. Six of these models, three event and three continuous-process, are described in this section. These models are considered to be representative of current U.S. modeling practice. Many other models could not be included here due to space limitations. A survey of catchment models is given in [17].

The event models are (1) HEC-1, developed and supported by the Hydrologic Engineering Center (HEC), U.S. Army Corps of Engineers, (2) TR-20, supported by the USDA Soil Conservation Service, and (3) SWMM (Stormwater Management Model), developed under the auspices of the U.S. Environmental Protection Agency. HEC-1 and TR-20 are used for the generation of flood hydrographs in analysis and design of flood control schemes. SWMM is suited for analysis and design applications involving simultaneous determinations of water quantity and quality.

The continuous-process models are (1) SSARR, developed and supported by the U.S. Army Corps of Engineers North Pacific Division, (2) Stanford Watershed Model (SWM), developed at Stanford University, and (3) Sacramento model, developed jointly by the U.S. National Weather Service and the California Department of Water Resources. These models have been used for hydrologic design, long-term runoff-volume forecasting, and real-time flood forecasting.

HEC-1

HEC-1, subtitled *Flood Hydrograph Package,* is designed to be used for the simulation of flood events in watersheds and river basins [10]. The river basin is represented as an interconnected system of hydrologic and hydraulic components. Each component models an aspect of the precipitation-runoff process within a portion of the basin referred to as subbasin. A component may be a surface runoff entity, a stream channel, or a reservoir. Its description requires the knowledge of a set of parameters and mathematical relationships describing the physical processes. The result of the modeling is the computation of streamflow hydrographs at desired locations within the river basin.

A river basin is represented as an interconnected group of subbasins. Within each subbasin, the hydrologic processes are represented by average parameter values. For hydrologically nonhomogeneous subbasins, further subdivision may be necessary to ensure that average parameter values are representative of each subbasin entity.

HEC-1 is an event model; therefore, it has no provision for soil moisture recovery during periods of no precipitation, with simulations being limited to a single-storm event. The model calculates discharges only, although stages can be indirectly determined through ratings supplied by the user. Alternatively, the results of HEC-1 can be used as input to HEC-2 , which calculates stages based on discharge by using steady gradually varied flow principles. In HEC-1, stream channel routing is accomplished by hydrologic methods. Therefore, the model does not account for the dynamic effects that are present in rivers of mild slope. Reservoir routing is based on the modified Puls technique (Section 8.3), which may not be applicable in cases where reservoirs are operated with controlled outflow (Section 8.4).

Stream Network Model Development. Using topographic maps and other geographic information, a river basin is configured into an interconnected system of stream network components (Fig. 13-3). A basin schematic diagram (Fig. 13-4), is developed by the following steps:

1. The boundaries of the catchment or basin under study are delineated with the aid of topographic maps. For urban catchments, municipal drainage maps may also be necessary.

2. The basin is subdivided into a number of subbasins in order to configure the stream network. In performing the subdivision, the following are taken into account: (1) the study purpose and (2) the spatial variability of precipitation and runoff response characteristics. The purpose of the study serves to pinpoint the areas of interest and, therefore, the location of subbasin boundaries. The spatial variability aids in the selection of the number of subbasins. Each subbasin is intended to represent an area of the basin which, on the average, has the same hydraulic and hydrologic properties. Usually, the assumption of uniform precipitation and infiltration over a subbasin becomes less accurate as the subbasin size increases.

3. Each subbasin is represented by a set of model components. The following components are available: (a) subbasin runoff, (b) river routing, (c) reservoir, (d) diversion, and (e) pumping.

4. The subbasins and their components are linked together to represent the connectivity or topology of the river basin. HEC-1 has several methods for combining or linking together outflow from the various components.

Model Components. The subbasin runoff component, such as subbasins 10, 20, 30, and so on (Figs. 13-3 and 13-4), is used to represent the flow over the land surface and in stream channels. The input to this component is a precipitation hyetograph. Effective rainfall is computed by abstracting infiltration and surface storage using an infiltration function. Within each subbasin, rainfall and infiltration rates are assumed to be uniformly distributed in space.

The outflow hydrograph at the subbasin outlet is generated using either unit

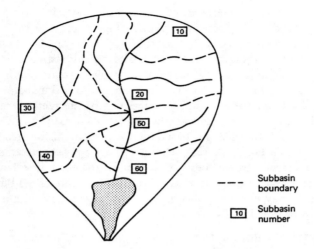

- - - Subbasin
boundary

[10] Subbasin
number

Figure 13-3 HEC-1 model: Example of river basin subdivision [10].

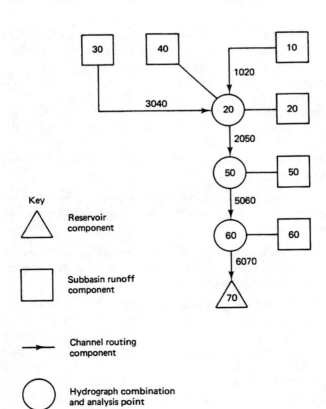

Key

△ Reservoir
component

▢ Subbasin runoff
component

→ Channel routing
component

○ Hydrograph combination
and analysis point

Figure 13-4 HEC-1 model: Example of basin schematic diagram [10].

hydrograph convolution (Section 5.2) or kinematic wave routing (Section 10.4). The hydrograph obtained by convolution is lumped at the subbasin outlet. However, surface runoff calculated by kinematic wave routing is distributed along the stream channel as lateral inflow. Baseflow is computed by an empirical method and added to the surface runoff hydrograph to obtain the flow at the subbasin outlet.

A river-routing component (element 1020, Fig. 13-4) is used to represent flood-wave movement in a stream channel. Input to this component is an upstream hydrograph resulting from individual or combined contributions of subbasin runoff, river routing, or diversions. This upstream hydrograph is routed through the stream channel by using one of several available methods of hydrologic routing.

With reference to Fig. 13-4, a typical HEC-1 computational sequence is as follows. Subbasin 10 runoff is calculated and routed to control point 20 via routing reach 1020. When using the unit hydrograph technique, subbasin 20 runoff is calculated and combined with the reach 1020 outflow hydrograph at control point 20. Alternatively, if subbasin 20 runoff is concentrated near the upstream end of reach 1020, runoff from subbasins 10 and 20 can be combined prior to routing through reach 1020. When using kinematic wave routing, subbasin 20 runoff is modeled as a uniformly distributed lateral inflow to reach 1020. Subbasin 10 runoff is routed in combination with this lateral inflow via reach 1020 to control point 20.

A suitable combination of subbasin runoff and river routing components can be used to represent the complexities of rainfall-runoff and stream network routing. The connection of the stream network is implied by the order in which the data components are arranged. Simulation must begin at the uppermost subbasin in a branch of the stream network, moving downstream until a confluence is reached. Before proceeding below the confluence, all upstream flows must be computed and routed to that point. The flows are combined at the confluence and the resulting flows are routed downstream.

The reservoir component can be used to represent the storage-outflow characteristics of a reservoir, lake, detention pond, highway culvert, and so on. The reservoir component operates by receiving upstream inflows and routing them through a reservoir using hydrologic routing methods, for which outflow is solely a function of storage (Chapter 8).

The diversion component is used to represent channel diversions, stream bifurcations, or any transfer of flow from one point of a river basin to another point in or out of the basin. The diversion component receives upstream inflow and divides the flow according to a user-prescribed rating.

The pumping component can be used to simulate the action of pumps used to lift water out of low-lying ponding areas, for instance, behind levees. Pump operation data describe the number of pumps, their capacities, and *on* and *off* elevations.

Rainfall-Runoff Simulation. HEC-1 model components simulate the rainfall-runoff process as it occurs in a river basin. Mathematical relations are intended to represent individual meteorological, hydrologic, and hydraulic processes encompassing the rainfall-runoff phenomena. The processes considered in HEC-1 are (1) precipitation, (2) interception/infiltration, (3) transformation of effective precipitation into subbasin runoff, (4) addition of baseflow, and (5) flood hydrograph routing, either in stream channels or reservoirs.

Precipitation. A precipitation hyetograph is used as input to HEC-1. The specified hyetograph is assumed to be a subbasin average, i.e., uniformly distributed over the subbasin. The following information can be used in deriving precipitation hyetographs: (1) historical storms, (2) synthetic storms based on either (a) depth-duration-frequency, (b) standard project storm (SPS), or (c) PMP, and (3) snowfall and snowmelt.

Interception/Infiltration. Interception, surface storage and infiltration are referred to in HEC-1 as precipitation losses. Two factors are important in the HEC-1 loss computation. First, precipitation losses do not contribute to surface runoff; therefore, they are considered to be lost from the system. Second, the equations for precipitation losses are not intended to account for soil moisture recovery during periods of no precipitation.

When using the unit hydrograph technique, the computed precipitation loss is considered to be a subbasin average. On the other hand, when using kinematic wave routing with two overland flow planes, a different precipitation loss can be specified for each flow plane. Within each flow plane, the computed precipitation loss is assumed to be uniformly distributed.

Precipitation losses can be calculated by one of the following methods: (1) initial loss and constant loss rate, (2) general HEC loss-rate function, (3) SCS runoff curve number, and (4) Holtan loss rate. Precipitation losses are subtracted from total rainfall/snowmelt, and the resulting precipitation excess is converted into surface runoff and a subbasin outflow hydrograph. A percent imperviousness factor can be used with any of the loss-rate methods to ensure 100 percent runoff from impervious surfaces in the subbasin.

In the initial loss and constant loss rate, an initial loss in millimeters (or inches) and a constant loss rate in millimeters per hour (or inches per hour) are specified. All precipitation is lost until the initial loss is satisfied; subsequently, precipitation is abstracted at the constant loss rate.

The general HEC loss-rate function is an empirical method that relates loss rate to precipitation intensity and accumulated losses. Accumulated losses are intended to represent soil-moisture storage. The loss rate is expressed as follows:

$$A = \left(\frac{S}{R^{0.1C}} + D \right) P^E \tag{13-1}$$

in which A = loss rate in millimeters per hour (inches per hour) during a time interval; P = precipitation intensity, in millimeters per hour (inches per hour) during a time interval; C = cumulative loss in millimeters (inches); and S, D, E, and R are parameters of the general HEC loss-rate function.

The parameter S, the starting value of the loss rate on an exponential decay function, is a function of soil type, land use, and vegetative cover. The parameter D, representing the amount of initial loss, is usually storm-dependent and a function of antecedent moisture condition. The parameter E, the exponent of precipitation intensity, varies in the range 0.0 to 1.0. For $E = 0$, precipitation intensity has no effect on loss rate; for values of $E \neq 0$, precipitation intensity has an effect on loss rate. The parameter R is the ratio of loss rate at a given C value to the loss rate at a value equal to $C + 10$ mm (in.). This parameter is a function of the ability of the basin surface to abstract precipitation.

The four-parameter general HEC loss-rate function can be reduced to the two-parameter initial loss and constant loss-rate method by setting $E = 0$ and $R = 1$. Likewise, it can be reduced to an exponential decay function with no initial loss by setting $D = 0$ and $E = 0$.

The use of the general HEC loss-rate function is illustrated by the following example.

Example 13-1.

Assume the following rainfall hyetograph:

Time (h)	0	2	4	6
Rainfall intensity (in./h)		1.0	1.0	0.5

Assume $S = 0.8$ in./h; $D = 0.0$ in.; $E = 0.7$; and $R = 2.5$. Calculate the basin losses using the general HEC loss-rate function.

The computations are shown in Table 13-3. Column 1 shows time in hours; Col. 2 shows the cumulative loss C in inches; Col. 3 shows $S/R^{0.1C}$; Col. 4 shows rainfall intensity P in inches per hour; Col. 5 shows P^E; Col. 6 shows loss rate A calculated by Eq. 13-1; Col. 7 shows the loss for each time interval in inches; and Col. 8 shows the effective rainfall or rainfall excess in inches, calculated by subtracting loss in inches (Col. 7) from total rainfall in inches.

The third method of hydrologic abstraction included in HEC-1 is the SCS runoff curve number method. This method expresses the amount of hydrologic abstraction in terms of CN, a runoff curve number varying in the range 1 to 100. The runoff curve number is a function of hydrologic soil group, land use and treatment, hydrologic condition of the catchment surface, and antecedent moisture condition (see Section 5.1).

The fourth method of hydrologic abstraction included in HEC-1 is the Holtan loss-rate method [8]. The infiltration equation is

$$F = 0.5(f_1 + f_2)\Delta t \tag{13-2}$$

in which $F =$ loss during the time interval Δt in millimeters (inches); $f_1 =$ infiltration

TABLE 13-3 APPLICATION OF GENERAL HEC LOSS-RATE FUNCTION: EXAMPLE 13-1

(1)	(2)	(3)	(4)	(5)	(6)	(7)	(8)
Time (h)	C (in.)	$S/R^{0.1C}$	P (in./h)	P^E	A (in/.h)	Loss (in.)	Excess (in.)
0	0						
		0.8	1.0	1.0	0.8	1.60	0.40
2	1.60						
		0.69	1.0	1.0	0.69	1.38	0.62
4	2.98						
		0.61	0.5	0.62	0.38	0.76	0.24
6	3.74						

rate at the beginning of time interval Δt in millimeters per hour (inches per hour); and f_2 = infiltration rate at the end of time interval Δt in millimeters per hour (inches per hour). The infiltration rates are calculated by the following formula:

$$f = GAS^b + f_c \qquad (13\text{-}3)$$

in which f = infiltration rate in millimeters per hour (inches per hour); G = a growth index representing the relative maturity of ground cover, varying from near 0.0 when the crops are planted to 1.0 when the crops are full-grown; A = infiltration capacity in millimeters per hour (inches per hour); S = equivalent depth of pore space in soil layer available for storage of infiltrated water in millimeters (inches); b = exponent with typical value equal to 1.4; and f_c = soil percolation rate, a function of hydrologic soil group (Section 5.1). For group A soils, f_c is in the range 0.45 to 0.30 in./h; for group B soils, 0.30 to 0.15 in./h; for group C soils, 0.15 to 0.05 in./h; and for group D soils, less than 0.05 in./h. The available storage S is depleted by the amount of infiltrated water and recovered at the percolation rate (f_c).

Unit Hydrograph.   Either user-supplied or synthetic unit hydrographs can be used with HEC-1. Three types of synthetic unit hydrographs are available: (1) Snyder unit hydrograph, (2) SCS dimensionless unit hydrograph, and (3) Clark unit hydrograph. The Snyder and SCS unit hydrographs are described in Section 5.2. The Clark unit hydrograph is described in Section 10.2.

The HEC-1 Clark unit hydrograph requires three parameters: (1) subbasin concentration time, (2) an estimated storage coefficient, and (3) a time-area curve, to be used in lieu of a time-area histogram. The time-area curve defines the cumulative area contributing runoff to the catchment outlet as a function of dimensionless time, i.e., the ratio of cumulative time to subbasin concentration time. HEC-1 incorporates a default parabolic time-area curve, to be used if a time-area curve is not supplied. The default time-area curve is

$$A = 1.414\,T^{1.5} \qquad \text{(for } 0 \leq T \leq 0.5) \qquad (13\text{-}4a)$$
$$A = 1 - 1.414(1 - T)^{1.5} \quad \text{(for } 0.5 \leq T \leq 1.0) \qquad (13\text{-}4b)$$

in which A = dimensionless area, the ratio of contributing area to subbasin area; and T = dimensionless time, the ratio of cumulative time to subbasin concentration time.

Since the Snyder method does not provide the complete shape of the unit hydrograph, HEC-1 uses the Clark method as a means of supplementing the Snyder method.

Kinematic Wave.   The kinematic wave technique transforms rainfall excess into subbasin runoff. Three conceptual elements are used in HEC-1 in connection with kinematic wave modeling: (1) overland flow planes, (2) collector channels, and (3) main channel.

As implemented in HEC-1, the kinematic wave scheme is similar to that described in Section 10.4. To maintain numerical stability, the HEC-1 kinematic wave formulation switches between a forward-in-time, backward-in-space scheme for Courant numbers $C \leq 1$ and a backward-in-time, forward-in-space scheme for $C \geq 1$.

Baseflow.   HEC-1 includes the effect of baseflow on the streamflow hydrograph as a function of the following three parameters: (1) initial flow in the stream, (2) threshold flow, which marks the beginning of the exponential recession on the receding limb of the computed outflow hydrograph, and (3) exponential decay rate, a characteristic of each individual basin or subbasin.

The initial flow in the stream is a function of antecedent moisture condition. In the absence of precipitation, the initial flow is affected by the long-term contribution of groundwater releases. Recession of initial flow and threshold flow follows the exponential decay rate.

Flood Routing.   Flood routing is used to simulate flood wave movement through reservoirs and river reaches. The methods included in HEC-1 are mostly hydrologic (storage) routing methods. An exception is the kinematic wave-routing technique, which, although based on a single-valued rating, can be considered a hydraulic routing technique.

The routing methods included in HEC-1 are the following: (1) Muskingum, (2) kinematic wave, 3) modified Puls, (4) working R & D, and (5) level-pool reservoir routing. The Muskingum method is described in Section 9.1; the kinematic wave method in Section 9.2; the modified Puls, or storage indication method, in Section 8.3. The working R & D method is a combination of modified Puls and Muskingum methods. The level-pool routing method is a discrete solution to the differential equation of storage. For all these methods, routing proceeds from upstream to downstream. There are no provisions for backwater effects or discontinuities (i.e., bores) in the water surface [10].

Parameter Calibration.   Rough estimates of HEC-1 model parameters can usually be obtained from individual experience or by other empirical means. Calibration using measured data, however, is the preferred way of estimating model parameters. With rainfall-runoff data from gaged catchments, the mathematical optimization algorithm included in HEC-1 can be used to estimate some model parameters. Using regional analysis, parameters obtained in this way can be transferred to ungaged catchments of similar hydrologic characteristics [10].

TR-20

Technical Release No. 20: "Computer Program for Project Formulation-Hydrology," or TR-20, was originally developed by the Hydrology Branch of the USDA Soil Conservation Service in cooperation with the Hydrology Laboratory, Agricultural Research Service. Since its original release in 1965, several modifications and additions have been made by SCS and others. The current version is dated 1983 [21].

The objective of TR-20 is to assist the engineer in the hydrologic evaluation of flood events for use in the analysis of water resource projects. TR-20 is an event model, with no provision for soil moisture recovery during periods of no precipitation. It computes direct runoff resulting from any synthetic or natural rainstorm, develops flood hydrographs from surface runoff, and routes the flow through reservoirs and stream channels. It combines the routed hydrograph with those from tributaries and computes the peak discharges, their times of occurrence, and the water surface elevations at any desired cross section or structure.

The program provides for the analysis of as many as nine different rainstorm distributions over a watershed, under various combinations of land treatment, flood-water retarding structures, diversions, and channel work.

TR-20 uses the procedures described in the SCS *National Engineering Handbook,* Section 4, Hydrology (NEH-4), except for the new reach-routing procedure (Attenuation-Kinematic, or Att-Kin method), which has superseded the convex method (Section 9.2).

Methodology. The catchment under study is divided into as many subcatchments as required to define hydrologic and alternative structural effects. Hydrologic effects are influenced by tributary confluences, catchment shape, valley slope changes, homogeneity of runoff curve number, and existing or proposed water-impoundment structures.

Each subcatchment is assumed to be hydrologically homogeneous. In addition, the chosen temporal rainfall distribution must be representative of the catchment or subcatchment under study. Either synthetic or natural rainfall distributions may be used.

A runoff hydrograph is developed for each subcatchment. The SCS runoff curve number, rainfall volume, and rainfall distribution are input variables needed by the program to calculate the runoff hydrograph. Subcatchment runoff curve numbers are based on hydrologic soil group, land use and treatment practices, and hydrologic surface condition. The runoff volume is computed using the SCS runoff equation, Eq. 5-6 (Section 5.1). At the user's request, the runoff curve number can be adjusted to account for antecedent moisture condition. The model provides for three levels of antecedent moisture: dry (AMC I), average (AMC II), and wet (AMC III).

Hydrograph Development. A unit hydrograph is developed for each subcatchment, following NEH-4 procedures described in Section 5.2. The unit hydrograph is convoluted with the effective rainfall to generate the composite hydrograph at each subcatchment outlet. The peak flow value is computed by a special routine that fits a second-degree polynomial through the three largest consecutive flood hydrograph ordinates.

The user-specified time interval should be such that it provides an adequate hydrograph definition. In practice, the user-specified time interval should be about one-fifth to one-tenth of the time of concentration of the smallest subcatchment. Time intervals exceeding these limits may result in a substantial loss of accuracy. Generally, however, the chosen time interval is not smaller than 0.1 h.

In selecting the time interval, the user must estimate the time base of the outflow hydrograph, which is related to catchment size and storm duration. For instance, if an outflow hydrograph with a 10-d (240 h) time base is expected, a time increment of 0.8 h—together with the maximum-possible 300 ordinates—would provide the required 240 h of simulation time.

Reservoir Routing. TR-20 routes a flood hydrograph through a reservoir using the storage indication method (Section 8.3). Either the starting elevation for routing or the pool elevation when runoff begins must be specified by the user. The outflow hydrograph can be printed in multiples of the user-specified time interval (subject to the 300-points maximum).

Channel Routing. TR-20 routes a flood hydrograph through a stream channel reach using the modified Att-Kin method, as described in the TR-20 manual [21]. The time interval and (computational) reach length must be specified by the user.

The modified Att-Kin method is a new SCS reach-routing procedure which has replaced the convex method, no longer supported as part of TR-20. The Att-Kin method consists of a two-step process, with the flood hydrograph routed through a reservoir as a first step (to simulate attenuation, i.e., storage), followed by routing with a kinematic wave method as a second step (to simulate pure, i.e., kinematic, translation).

Program Features. TR-20 has the following general programming capabilities:

1. Route flood hydrographs through up to 99 structures and 200 stream channel reaches.
2. Compute up to 300 flood hydrograph ordinates at a selected point in a catchment, and print the discharge and water surface elevation for each ordinate.
3. Consider up to nine different rainfall distributions, including 10 storms per rainfall distribution.
4. Combine hydrographs for a large number of tributaries and channel reaches.
5. Divide hydrographs into two separate hydrographs.
6. Route a discharge hydrograph through a channel network.
7. Store up to seven computed hydrographs at any time.

Program Structure. The general structure of TR-20 is shown in Fig. 13-5 [12]. It includes two types of operations (in addition to input and output): (1) control operations and (2) hydrograph computations. The control operations make it possible to obtain, in a single run, outputs for several combinations of storm rainfall and watershed conditions, variations in reservoir number, including size and location, channel characteristics, and land-use practices.

The hydrograph computations are contained in three subroutines: (1) RUNOFF, which calculates subcatchment flood hydrographs based on rainfall and hydrologic abstraction data; (2) RESVOR, which routes flood hydrographs through reservoirs or water impoundments; and (3) REACH, which routes flood hydrographs through stream channel reaches. Ancillary hydrograph manipulation operations consist of the following: (1) ADDHYD, the combination of two reach-routed hydrographs into one and subsequent saving in a third storage location; (2) SAVMOV, the saving and moving of a hydrograph from one storage location to another; and (3) DIVERT, the separation of one hydrograph into two hydrographs.

Applications. TR-20 is designed to be used in watersheds where peak flows are the result of thunderstorms or other high-intensity, short-duration storms. Snow-

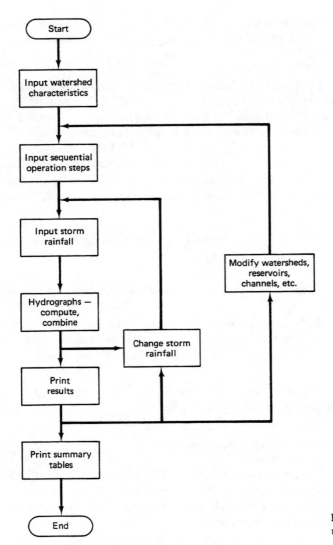

Figure 13-5 General structure of TR-20 model [12].

melt runoff is usually considered either as baseflow or interflow. The assumption of spatially uniform rainfall depth limits the applicability of TR-20 to small and midsize catchments. Typical applications have been with catchment sizes of 5 to 1000 km^2, with subcatchment areas varying between 0.25 and 25 km^2 [12].

SWMM

The Stormwater Management Model (SWMM) was originally developed in 1970 by a consortium led by Metcalf and Eddy, Inc., the University of Florida, and Water Resources Engineers, Inc., working under the auspices of the U.S. Environmental Protection Agency. SWMM is a computer model capable of representing urban stormwater runoff and combined sewer overflow. It portrays correctional devices in the form of user-selected options for storage and/or treatment, including associated cost estimates. Effectiveness is portrayed by computed treatment efficiencies and modeled changes in receiving water quality [27].

SWMM simulates storm events on the basis of rainfall (hyetograph) inputs and system characterization (catchment, conveyance, storage/treatment, and receiving water) and predicts outcomes in the form of water quantity and quality values. The simulation technique, i.e., the geometric and physical representation of the prototype system, enhances the ease of model interpretation while permitting the identification of remedial devices and other local phenomena (e.g., flooding). The model is oriented toward the description of spatial and temporal effects, with output in terms of inlet hydrographs and *pollutographs*.

Model Description. The model has the following structure:

1. The input sources:

   RUNOFF generates surface runoff based on arbitrary rainfall hyetographs, antecedent moisture conditions, land use, and topography. FILTH generates dry-weather sanitary flow based on land use, population density, and other factors. INFIL generates infiltration into the sewer system based on groundwater levels and sewer condition.

2. The central core:

   TRANS carries and combines the inputs through the sewer system using a modified kinematic wave approximation, based on the water continuity and Manning's equations (Section 9.2), assuming complete mixing at various inlet points.

3. The correctional devices:

   TSTRDT, TSTCST, STORAG, TREAT, and TRCOST modify hydrographs and pollutographs at selected points in the sewer system, accounting for retention time, treatment efficiency, and other parameters, including cost estimates.

4. The effect:

   RECEIV routes hydrographs and pollutographs through the receiving waters, which may consist of a stream, river, lake, estuary, or bay.

SWMM is structured into a main control block (executive), a service block (combine), and four computational blocks: (1) runoff, (2) transport, (3) storage, and (4) receiving.

Runoff Block. The runoff block simulates the quantity and quality of catchment runoff and the routing of flows and contaminants to the major sewer lines. The catchment is represented as a series of idealized subcatchments and gutters. Runoff accepts an arbitrary rainfall hyetograph and carries out a step-by-step accounting of rainfall infiltration losses in pervious areas, surface detention, overland flow, gutter flow, and contaminants washed into the inlet manholes. The result is the calculation of several inlet hydrographs and pollutographs.

The relationship among the subroutines that make up the runoff block are shown in Fig. 13-6. Subroutine RUNOFF is called by the executive block to gain access to the runoff block. Subroutine RUNOFF immediately calls subroutines RECAP

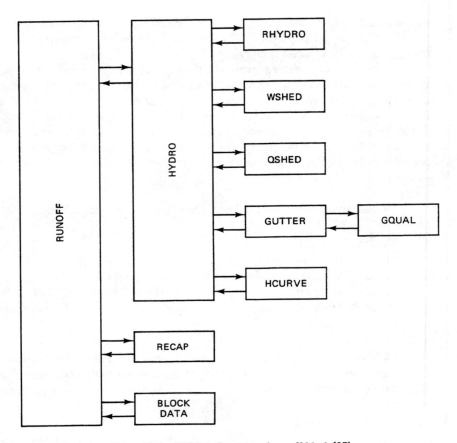

Figure 13-6 SWMM: Structure of runoff block [27].

to input and output table headings and HYDRO to compute hydrograph coordinates and water quality contributions. HYDRO is assisted by four subroutines: RHYDRO, WSHED, QSEHD, and GUTTER. RHYDRO reads in the rainfall hyetograph and information concerning the inlet catchment and sets up an ordering array to sequence the computational order for gutters/pipes according to upstream and downstream relationships.

HYDRO then computes the hydrograph coordinate for each time step. Subroutine WSHED is called to calculate the outflow rate from the idealized subcatchments. When simulating water quality, QSHED is called to compute the quality contributions from catchments, erosion, dust and dirt, and other sources. GUTTER is called to compute the instantaneous water depth and flow rate for the gutters/pipes and to route the flow downstream. Water flowing into the inlet point, whether from gutter/pipes or directly from the subcatchments, is added up to obtain a hydrograph coordinate.

The core of the RUNOFF model is the routing of hydrographs through the system. This is accomplished by a combination of overland flow and pipe routing. Three element types are available:

1. Subcatchment elements (overland flow)
2. Gutter elements (channel flow)
3. Pipe elements (special case of channel flow)

Flow from subcatchment elements is always into either gutter/pipe elements or inlet manholes. The subcatchment elements receive rainfall, account for infiltration losses using the Horton equation (Section 2.2), and permit surface storage such as ponding or retention on grass or shrubs. Gutter/pipe elements can be used to convey flow from the subcatchments to the entry to the main sewer system.

Geometric Representation of Catchment.
The catchment is conceptually represented by a network of hydraulic elements, i.e., subcatchments, gutters, and pipes. Hydraulic properties of each element are characterized by several parameters, such as size, slope, and roughness coefficients. Discretization begins with the identification of drainage boundaries, location of major sewer inlets, and selection of gutters/pipes to be included in the system. Either a coarse or fine discretization (i.e., spatial and temporal resolution) can be used, depending on the modeling objectives. In a small system, the downstream point of the runoff model is the outfall to a neighboring creek. In a large system, it could be an inflow to the transport model.

The transport model is used in the cases where:

1. Backwater effects are significant.
2. Hydraulic elements other than pipes and gutters, such as pumps, are involved.
3. Solids deposition or suspension is substantial.

Subcatchments represent idealized runoff areas with uniform slope. Parameters such as roughness values, detention depths, and infiltration values are taken as constant for the subcatchment and usually represent average values. However, within a subcatchment, pervious and impervious areas can have different characteristics.

While the discretization can be made as fine as desired, there is a practical limit imposed by computational requirements. A minimum of five subcatchments per catchment is recommended to allow flow routing between hydrographs.

Parameter Estimation.
Parameters necessary to characterize the hydraulic properties of a subcatchment include surface area, width of overland flow, ground slope, roughness coefficients, detention depths, infiltration data (maximum, minimum, and decay rates), and percent imperviousness. For a given amount of rainfall, these parameters determine the transient water depth over the catchment and the outflow rate. In practice, since real subcatchments are not regular areas experiencing spatially uniform overland flow, average values must be selected for computational purposes.

Values of Manning n applicable for overland flow conditions are estimated based on the condition of various surfaces, such as asphalt, clay, turf, and so on (Section 4.2). Unless otherwise specified, default values of detention depth for impervious areas and pervious areas are used. Limiting infiltration rates can be estimated either from standard tables or based on experience. For each time step, infiltration rates are calculated based on the Horton equation, with either user-specified or default parameter values. The infiltration rate is subtracted from the rainfall rate, with any excess going into detention storage. When detention storage is exceeded, the runoff depth is allowed to accumulate. The outflow rate is computed based on the runoff depth, using the Manning equation for a wide channel (overland flow plane).

For each subcatchment, the width of overland flow must be supplied by the model user. The total width of overland flow is twice the length of the drainage gutter, since two-plane catchments contribute flow along the side of the drainage gutter. Overland flow is assumed to be perpendicular to gutter flow (Section 10.4).

Time Discretization. The time step is usually 3 to 5 min, but it may range from 1 to 30 min, depending on the storm intensity and duration and the required accuracy. A sufficient number of time steps should be allowed to extend the simulation past the storm termination and thus account separately for the storm runoff.

The percentage of impervious area (i.e., that having zero detention) should be supplied to the model; otherwise, a default value of 25 percent is used. This ensures an immediate runoff response and a steep rising limb on the catchment's outflow hydrograph.

For the larger catchments, model predictions are sensitive to spatial variations of rainfall. For instance, summer thunderstorms may be very localized, with nearby gages having different readings. For increased modeling accuracy, it is essential that spatial and temporal variations of rainfall be properly assessed.

SSARR

The Streamflow Synthesis and Reservoir Regulation (SSARR) model has been in the process of development and application since 1956. It was developed to meet the needs of the U.S. Army Corps of Engineers North Pacific Division in the area of mathematical hydrologic simulation for the planning, design, and operation of water-control works. The SSARR model was first applied to operational flow forecasting and river-management activities in the Columbia River System. Later, it was used by the Cooperative Columbia River Forecasting Unit, consisting of the U.S. Army Corps of Engineers, National Weather Service, and Bonneville Power Administration. Numerous river systems in the United States and other countries have been modeled with SSARR. The current version of the model, SSARR-8, is dated April 1986 [20].

SSARR is a computer modeling system of the hydrology of a river basin. Streamflow at headwater points in the basin can be synthesized by evaluating rainfall, snow accumulation, and snowmelt. Streamflows throughout the basin can be synthesized by simulating the effects of channel routing, diversions, and reservoir regulation and storage. SSARR comprises (1) a watershed model and (2) a river system and reservoir-regulation model.

The watershed model accounts for the following processes: (1) interception, (2) evapotranspiration, (3) soil moisture, (4) baseflow infiltration, (5) routing of runoff into the stream network, and (6) snowmelt runoff. The river system and reservoir-regulation model routes (1) streamflows from upstream to downstream points through channel and lake storage and (2) flow through reservoirs under uncontrolled outflow conditions. Flows may be routed as a function of multivariable relationships involving backwater effects from tides and reservoirs. Diversions and overbank flows can also be simulated.

The simulation proceeds through time by computing the model state at successive intervals. The time interval may be as short as 0.1 h or as long as 24 h, depending on the modeling objectives and other factors such as drainage area, hydrologic response time, and the availability of hydrometeorological data with which to drive and calibrate the model.

Applicability. SSARR can be used for either runoff and streamflow forecasting or for long-term studies of the hydrologic response of a river system. Typical examples of SSARR model applications include the following:

1. Analysis of multipurpose reservoir operations for real-time reservoir regulation and management of water-control systems
2. Simulation of synthetically derived design storms
3. Analysis of streamflow diversion for irrigation purposes
4. Daily streamflow forecasting at selected points throughout a river system
5. Seasonal streamflow forecasting.

Calibration. The successful use of SSARR is dependent upon the identification of the various parameters and relationships for a specific river basin. Some of these relationships are of a general nature; therefore, they can be applied to several subbasins within a larger basin. Others can be derived for a specific watershed. Some are relationships which could be observed or derived; others must be considered as model parameters that are related to the physical processes by the use of indexes (see Section 12.3).

Calibration is performed manually, by trial and error, the aim being to obtain the best fit between simulated and recorded streamflows. Once this has been accomplished, the validity of model parameters and relationships are tested with an independent data set (model verification). The efficiency of the calibration and verification process is dependent upon the judgment and modeling skills of the user, who must evaluate the interaction between model parameters. A systematic approach to SSARR calibration is given in [7].

Basic Routing Method. SSARR uses the cascade of linear reservoirs (Section 10.3) as its basic algorithm for watershed and stream channel routing. A watershed or channel is represented as a series of *lakes* or reservoirs, which conceptually simulate the natural delay of runoff (translation and storage) when moving from upstream to downstream points. The model user specifies the routing characteristics of the lake (the ratio of time interval Δt to storage constant K) and the number of lake increments (number of linear reservoirs in series).

Data Requirements. Input data needed for model operation include the following:

1. Constant characteristics: physical features such as drainage area, reservoir stor-

age capacity, watershed characteristics affecting runoff, system configuration, mathematical functions, and so on.

2. Initial conditions: current conditions of all watershed indexes, initial flows, initial reservoir or lake elevations, and outflows.

3. Time series data: physical data expressed as time series—for example, precipitation, air temperature, streamflow, reservoir regulation data, and other hydrometeorological variables.

4. Job control data: total computation period, time interval, input/output instructions.

Watershed Models. The current (1986) SSARR version features two watershed models: (1) a depletion-curve model and (2) an integrated-snowband model.

The depletion-curve watershed model is essentially the same as that featured in the earlier (1975) version (SSARR-4). The model simulates snowmelt by using a depletion curve relating the percent of seasonal runoff to the percent of snow-covered area. However, the depletion-curve model lacks the snow-conditioning algorithms of the integrated-snowband model included in the current version (SSARR-8).

The integrated-snowband watershed model has the capability for continuous simulation of snow accumulation, snowmelt, and runoff from rainfall (it has essentially the same rainfall-runoff features of the depletion-curve model). The model is particularly useful for short-term runoff forecasting in coastal areas experiencing rapid changes in snow line. The snow-conditioning features of the model substantially enhance the forecast product. Also, the model's ability to compute snowpack from climatological data makes it suitable for runoff-volume forecasting.

Depletion-curve Watershed Model. The depletion-curve watershed model has been used in many basins around the world for simulating both runoff from rainfall only and runoff from combined rainfall and snowmelt. A schematic representation of the depletion-curve watershed model is shown in Fig. 13-7.

Precipitation Input. For a given watershed or basin, precipitation input is weighted according to the following formula:

$$\text{WP} = \sum_{i=1}^{n} P_i W_i \tag{13-5}$$

in which WP is the weighted precipitation input for a specified time period in inches (millimeters), based on n stations; P_i represents the period's precipitation values for each station i; and W_i is the weight applicable to each station ($\Sigma W = 1$). Station weights are usually calculated on the basis of percent area applicable to each station using mean annual precipitation, Thiessen polygons, or any other suitable means. Daily precipitation can be distributed over eight 3-h periods following temporal distribution functions specified by the user.

Soil Moisture-Runoff Relationships. Rain falling on a watershed will either (1) run off, (2) be retained in the soil system, (3) be intercepted and evaporated from trees and other vegetation, (4) evaporate from pond, lake, and stream surfaces,

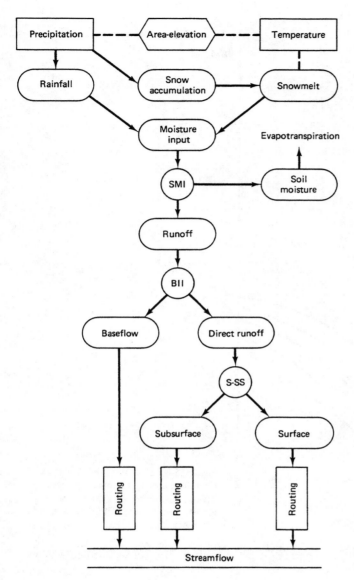

Figure 13-7 SSARR: Schematic of depletion-curve watershed model [20].

(5) return to the atmosphere by transpiration from trees and other vegetation and subsequent evaporation, or (6) percolate to the groundwater system and thereby be lost to the surface-water system. Rainfall input is divided into (1) runoff, (2) soil-moisture increases, (3) evapotranspiration losses, and (4) input to groundwater storage.

The fraction of rainfall input available for runoff is based on empirically derived relationships of soil-moisture index (SMI) versus runoff percent (ROP). As an option, rainfall intensity may be included as a third variable in the soil-moisture index-runoff percent relationship (SMI-ROP) (Fig. 13-8). For each basin, the SMI-ROP relationship may be specified either in tabular or polynomial-equation form.

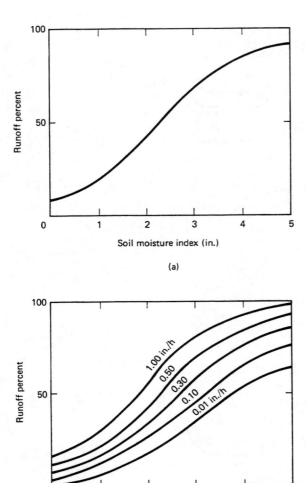

Figure 13-8 SSARR: (a) example of soil-moisture index versus runoff percent; (b) example of soil-moisture index versus runoff percent with rainfall intensity as a curve parameter [20].

The runoff generated in a period (RGP), in inches (millimeters), is calculated as follows:

$$RGP = \frac{(ROP)(WP)}{100} \qquad (13\text{-}6)$$

in which ROP is obtained from the appropriate SMI-ROP relationship as a function of soil-moisture index and, optionally, also of rainfall intensity (Fig. 13-8).

The SMI—in inches (millimeters)—is an indicator of relative soil wetness and, consequently, of watershed runoff potential. When the soil moisture is depleted to approximately the permanent wilting point, the SMI is a relatively small value that yields little or no runoff. When precipitation recharges soil moisture, the SMI increases until it reaches a maximum value considered to represent its field capacity, or water-holding capacity. As the maximum SMI is approached, ROP approaches 100.

The SMI is depleted only by the evapotranspiration index (ETI). The ETI can be specified either in tabular form, as month versus mean daily PET or as daily weighted pan-evaporation data at one or more stations. The tabular form is used either when pan-evaporation estimates are not available or when evapotranspiration amounts are not hydrologically significant. When monthly ETI values are used, the SMI is updated at the end of every period (PH hours) by the following formula:

$$SMI_2 = SMI_1 + (WP - RGP) - \left(\frac{PH}{24}\right)(KE)(ETI) \qquad (13\text{-}7)$$

in which SMI_1 is the soil-moisture index at the beginning of the period in inches (millimeters); SMI_2 is the soil-moisture index at the end of the period in inches (millimeters); WP is the weighted precipitation during the period in inches (millimeters); RGP is the runoff generated during the period in inches (millimeters); PH is the length of the time period in hours; ETI is the evaporation index in inches per day (millimeters per day); and KE is a factor for reducing ETI on rainy days, specified in tabular form (KE versus rainfall intensity). A typical KE versus rainfall intensity curve is shown in Fig. 13-9.

For a time period with zero precipitation, (WP − RGP) = 0, KE = 1, and SMI is reduced by a constant factor (PH/24)(ETI). The SMI increases when the rainfall not contributing to runoff (WP − RGP) exceeds the evapotranspiration amount [(PH/24)(KE)(ETI)].

Baseflow. Baseflow is modeled using the concept of baseflow infiltration index (BII). The percent of runoff contributing to baseflow (BFP) is related to the baseflow infiltration index (BII). The BFP versus BII relationship may be specified either in tabular or equation form. A relationship linking BII to baseflow input limit (BFL) may also be specified (Fig. 13-10).

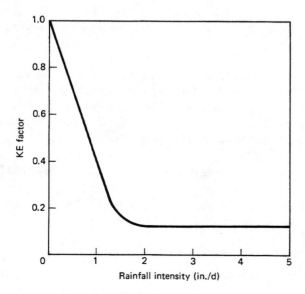

Figure 13-9 SSARR: Example of KE factor versus rainfall intensity [20].

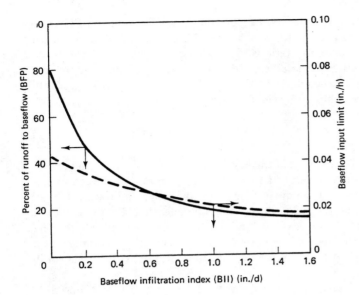

Figure 13-10 SSARR: Example of baseflow infiltration index function [20].

The BII is updated at the end of every period (PH) by the following formula:

$$BII_2 = BII_1 + (24RG - BII_1)[PH/(TSBII + 0.5PH)] \qquad (13\text{-}8)$$

in which BII_2 is the baseflow infiltration index at the end of period in inches per day (millimeters per day); BII_1 is the baseflow infiltration index at the beginning of period in inches per day (millimeters per day); RG is the generated runoff in inches per hour (millimeters per hour) ($RG = RGP/PH$); PH is the period in hours; and TSBII is the time of storage for the calculation of change in BII. Typical values of TSBII range from 30 to 60 hours.

The generated runoff is separated into two components: (1) baseflow (RB) and (2) surface and subsurface flow (RGS). The baseflow component (RB) is computed as the product of generated runoff (RG) and baseflow fraction (BFP/100). The surface and subsurface runoff component (RGS) is the difference between generated runoff (RG) and baseflow (RB).

Surface-Subsurface Flow Separation. The separation of the surface and subsurface flow component (RGS) into surface flow (RS) and subsurface flow (RSS) is accomplished by specifiying an RGS versus RS relationship in tabular form. An example of this relationship is shown in Fig. 13-11. Any such relationship may be specified, but the flow separation is usually based on the following criteria:

1. The minimum surface flow (RS) is 10 percent of the surface and subsurface flow (RGS)
2. The subsurface flow (RSS) eventually reaches a maximum value (KSS) and remains constant for values of RGS in excess of 2(KSS).

Formulas that satisfy these two conditions are the following:

$$RS = [0.1 + 0.2 \, (RGS/KSS)]RGS \qquad (13\text{-}9a)$$

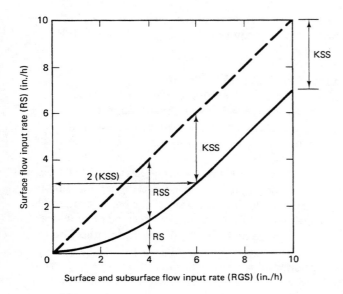

Figure 13-11 SSARR: Example of surface-subsurface flow separation [20].

Surface and subsurface flow input rate (RGS) (in./h)

applicable to values of RGS ≤ 2(KSS), and

$$RS = RGS - KSS \qquad (13\text{-}9b)$$

which is applicable to values of RGS > 2(KSS).

Routing of Surface, Subsurface, and Baseflow. Each runoff component, i.e., surface flow (RS), subsurface flow (RSS), and baseflow (RB), is taken as an input rate in inches (millimeters) per time period PH (in hours). Each input rate in inches per hour (millimeters per hour) is multiplied by the drainage area in square miles (square kilometers), and with the appropriate unit conversions, an input rate in cubic feet per second (cubic meters per second) is calculated.

Each input rate is routed through a specified number of lake increments (i.e., linear reservoirs). The conversion of the input rate into an outflow hydrograph is a function of the number of lake increments N, the *time of storage K* per increment in hours (storage constant), and the time period in hours (i.e., the time interval Δt). See Section 10.3 for a description of the SSARR routing procedure.

Generally, two to five increments (N) are used in typical river basins. The time of storage (K) varies from a few hours to several hundred hours, depending on whether surface, subsurface, or baseflow components are being simulated. The values of N and K determine the peak, shape, and time delay of the simulated outflow hydrograph. Usually, a set of N and K can be found for given peak flow and lag time. A typical example is that of a 208-mi^2 watershed in the Willamette River Basin, Oregon. In this case, the routing specifications were (a) surface flow, $N = 4$ and $K = 2.5$ h, (b) subsurface flow, $N = 3$ and $K = 11$ h, (c) baseflow, $N = 2$ and $K = 200$ h [20].

Snowmelt Computations. The computation of snowmelt with the depletion-curve watershed model is accomplished by using either (1) a temperature index or (2) a generalized basin snowmelt equation for a partly forested area (Eq. 12-16(c))

(Section 12.3). The temperature-index method is commonly used for daily forecast operations, whereas the generalized snowmelt equation is more appropriate for design applications.

Integrated-snowband Watershed Model. The integrated-snowband watershed model of the current SSARR version [20] has all the rainfall-runoff features of the depletion-curve model. In addition, it has the following three major features: (1) an interception algorithm to assist in the analysis of annual water balance, (2) an enhanced evapotranspiration simulation, and (3) a fourth routing component to simulate long-term return flow from groundwater.

Moisture accounting in the basin is accomplished by use of elevation bands. For each band, factors such as precipitation, snowpack water equivalent, and soil moisture are considered separately. The model allows subdivision of the basin into 1 to 20 bands, with each band considered to be either 100 percent snow-covered or snow-free. Runoff is calculated for each band and then combined to obtain basin runoff before channel routing (Section 12.5).

The loss of water by interception and subsequent evaporation from vegetated surfaces is usually quite small. However, for heavily vegetated areas, this loss may amount to a sizable fraction of annual precipitation. To account for interception in a simple yet meaningful way, the integrated-snowband model requires the user to specify a maximum interception quantity based on the type and density of vegetative cover. This interception quantity must be satisfied (i.e., filled) before water can become available for infiltration or runoff. The volume in interception storage is depleted by evapotranspiration. The interception value may be a constant or vary from month to month.

The snow-conditioning algorithm of the integrated-snowband model accounts for the *cold content* and liquid-water deficit of the snowpack, which must be satisfied before liquid water becomes available for runoff. The snowpack on a snowband can melt at both the snow-air interface and the snow-ground interface. Melt at the snow-air interface is subject to the snow-conditioning algorithm. Melt at the snow-ground interface is based on a monthly melt rate, entering runoff simulation directly, without being affected by the snow-conditioning algorithm.

River System and Reservoir-Regulation Model. The objective of the river system and reservoir model is to predict river flow at key points of interest for any given sequence of climatic events and any applicable reservoir regulation scheme.

Channel-routing characteristics must be developed so that river flow can be simulated. Often the routing characteristics of river channels can be more precisely determined than those of watersheds. Where appropriate, streamflow data allows the estimate of reach travel time for several streamflow levels.

For applications in flood control planning and design, the river system model can be used by itself, without the watershed model. In this case, historic streamflow data at the upstream boundary is used to drive the river system model, with streamflows being routed through the river system (channel network).

Channel Routing. Channel routing is accomplished by using the same procedure as that of watershed routing. The channel reach is conceptualized as a series of storage increments. For each storage increment and time period, outflow is computed by Eq. 10-11. The average outflow for each time period is used as the inflow to the next storage increment. The procedure is repeated for each storage increment and time period. The outflow from the last storage increment is taken as the outflow from the channel reach.

Time of storage for channel routing (the K value) is generally a function of discharge Q. Accordingly, the K value may be either specified either in tabular form (K versus Q), in the case of overbank-flow routing, or it may vary inversely or directly with discharge. For inbank flows, time of storage varies inversely with discharge; for overbank flows, it may vary directly or inversely, depending on the stage (Fig. 9-2). A convenient relationship has the following form:

$$K = \frac{\text{KTS}}{Q^n} \tag{13-10}$$

in which K is the time of storage per increment in hours; KTS is a constant determined by trial and error or estimated from physical measurements of flow and corresponding routing times; Q is the discharge; and n is an exponent usually varying between -1 and 1.

Lake Routing. The routing of flow through natural lakes is based on free-flow conditions, i.e., elevation-outflow relationships are fixed, and outflow is determined by hydraulic head. Routing is accomplished by an iterative solution of the equation of storage (Eq. 8-13).

Reservoir Routing and Regulation. Routing through artificial reservoirs is accomplished by procedures similar to those used for natural lakes, except that several controls can be exerted by the model user. Various types of reservoir operation can be specified. These include (1) free-flow, (2) outflow, (3) pool elevation or changes in pool elevation; and (4) storage quantities or changes in storage quantities. If no method is specified, outflow is made equal to inflow. These specifications are met unless violations of reservoir characteristics occur. For example: (1) the reservoir cannot be emptied below lower bounds, (2) discharge cannot exceed that obtained from the elevation-discharge relation, (3) when the reservoir elevation exceeds upper bounds, routing will be accomplished under free-flow conditions, or (4) the reservoir will pass inflow if no elevation-discharge relation is given.

Backwater Mode. The lake- and channel-routing methods described in the preceding sections assume that a relationship exists between the elevation in the lake or channel and the outflow from the water body. There are cases, however, where elevation and discharge are affected by backwater from a downstream time-variant source. Examples of such occurrences are river estuaries subject to tidal fluctuations, river reaches upstream from a junction with a major tributary, and the upstream reaches of reservoirs or lakes. In such cases, routing can be accomplished by the backwater mode of SSARR. This method utilizes a three-variable relationship between upstream elevation (E_1), downstream elevation (E_2) or discharge (Q_2), and upstream discharge (Q_1):

$$Q_1 = f(E_1, E_2) \qquad (13\text{-}11)$$

or

$$Q_1 = f(E_1, Q_2) \qquad (13\text{-}12)$$

The value of Q_1 is dependent on E_1 and E_2 (or Q_2). A sketch of this function is shown in Fig. 13-12. The elevation for the downstream section is that which was computed for the previous time period. Therefore, short time periods are required for increased accuracy. Each station must have a specified elevation-storage relationship. Outflows are then computed as a function of upstream elevation and downstream control.

System Configuration. The system configuration describes the physical layout and relationships of all components of the system. The user specifies the upstream to downstream order of all subbasins, lakes, reservoirs, channel reaches, and confluence points for a particular basin. The subbasin (or watershed) routing is performed first, followed by channel routing and hydrograph combination.

A hypothetical river system and the corresponding basin configuration are illustrated in Fig. 13-13. However, there may be more than one acceptable subdivision of a river basin. Basins are usually subdivided at the discretion of the user, as a function of subbasin characteristics and data availability.

Applications. The application of SSARR usually requires several years of data (rainfall, snowfall, streamflow) for model calibration and verification. Data sequences encompassing both low and high flows are usually necessary in order to assess the variability of model parameters with flow level.

SSARR has been applied to several large river basins throughout the world. Examples include the Salt River, Arizona (11,200 km²), Alto Paraguay River (370,000

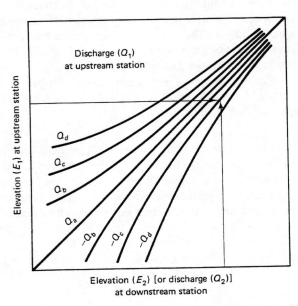

Elevation (E_2) [or discharge (Q_2)]
at downstream station

Figure 13-12 SSARR: Sketch of backwater relationship [20].

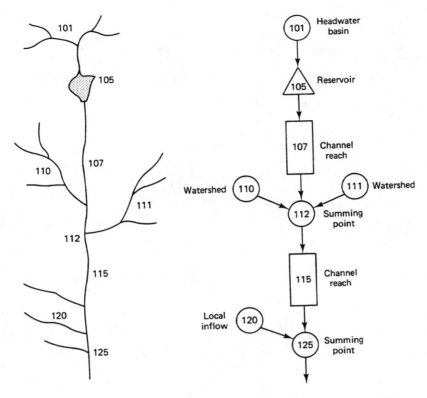

Figure 13-13 SSARR: Example of river system and basin configuration [20].

km²) and other river basins in Brazil, the Mekong River in Southeast Asia, and numerous rivers in Canada [12]. Most applications have been in flood forecasting and operational hydrology. The model is particularly suited to large basins where snowmelt is a major contributor to runoff, although it has also been applied to tropical and subtropical regions of the world.

Stanford Watershed Model (SWM)

The Stanford Watershed Model (SWM) IV was developed at Stanford University in the early sixties. It is a continuous-process, conceptual, lumped-parameter model for synthesizing hourly or daily streamflows at a watershed outlet. The model has served as the basis for several watershed models developed in the seventies [6, 13, 18, 23], including the widely used *Hydrologic Simulation Program-Fortran* (HSPF) [11].

The major components of SWM IV are shown in Fig. 13-14. The time series input data consist of precipitation and potential evapotranspiration. If snowmelt is significant, additional meteorological data are needed. Calculations begin from known or assumed initial conditions and are continued until the time series input data are exhausted.

Precipitation is stored in three soil-moisture storages and in the snowpack. The three soil-moisture storages are (1) upper-zone storage, (2) lower-zone storage, and (3)

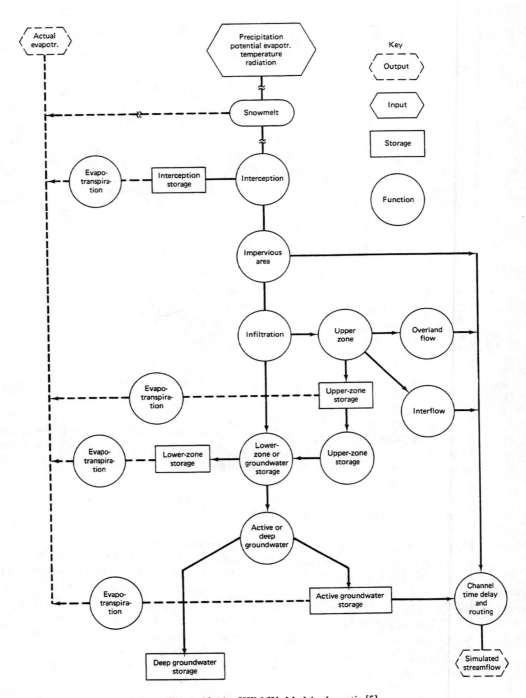

Figure 13-14 SWM IV: Model schematic [5].

groundwater storage. The upper- and lower-zone storages account for overland flow, infiltration, interflow, and inflow to groundwater storage. The upper zone simulates hydrologic abstractions and runoff resulting from minor (frequent) storms, including the first few hours of major (infrequent) storms. Conversely, the lower zone accounts for the hydrologic abstractions and runoff during major storms. Groundwater storage supplies baseflow to stream channels. Evaporation and transpiration may occur from any of these three storages. The runoff from overland flow, interflow, and baseflow enters the channel system and is routed downstream to the watershed outlet, where it is expressed as a continuous outflow hydrograph.

Land Surface Model.   The SWM land surface submodel consists of the following: (1) interception, (2) infiltration, (3) overland flow, (4) interflow, (5) groundwater, and (6) evapotranspiration. Calculations are made for a finite time interval, or time period. Within a time interval, the moisture supply is the volume of rainfall and/or snowmelt plus the surface detention carryover that is available for infiltration.

Interception.   For a given time interval, interception is a function of (1) type and extent of vegetative cover and (2) the current volume in interception storage. All moisture supply enters interception storage until a preassigned volume (EPXM) is filled. Evaporation from interception storage is assumed to be continuous and equal to the potential evapotranspiration rate.

Precipitation on the surface of lakes, reservoirs, and streams, as well as that in impervious areas adjacent to or directly connected to the channel system, is assumed to contribute entirely to surface runoff, without undergoing any hydrologic abstraction. The impervious fraction, an input to the model, is obtained by taking the sum of the areas with no abstraction and dividing the sum by the watershed area. Runoff from impervious areas that are not directly connected to the channel system is generally a function of the watershed soil-moisture conditions. Therefore, runoff from unconnected impervious areas is included in the infiltration component.

Infiltration.   Infiltration is divided into (1) direct infiltration into lower-zone and groundwater storages, and (2) indirect (or delayed) infiltration into upper-zone storage. Available moisture after interception is first subject to direct infiltration (long-term lower-zone and groundwater storage). Any moisture left after direct infiltration is subject to indirect infiltration (short-term upper-zone storage), designed to simulate depression storage, storage in soil fissures, and disturbed soils. The variable capacities of upper-zone, lower-zone, and groundwater storages are represented in terms of dimensionless storage ratios.

Direct Infiltration (Lower-zone and Groundwater Storages).   Spatial variations in infiltration capacity are simulated by the cumulative linear frequency distribution (infiltration capacity curve) shown in Fig. 13-15. For a given time interval (in hours), this curve shows the cumulative watershed area (in percent) having an infiltration capacity (inches) equal to or less than the indicated value.

As shown in Fig. 13-15(a), infiltration is assumed to go into (1) lower-zone and groundwater storage (shaded with solid lines) and (2) interflow (shaded with broken lines). Therefore, the fraction of infiltration going into interflow is assumed to be pro-

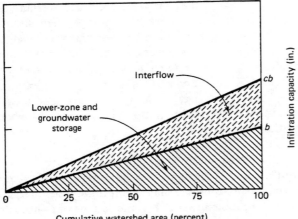

(a) Separation between lower-zone and
groundwater storage and interflow

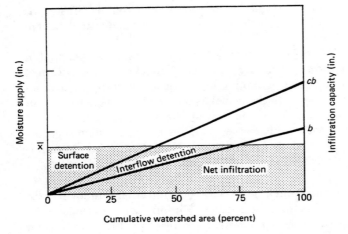

(b) Contributions to surface detention, interflow
detention, and net infiltration

Figure 13-15 SWM IV: Infiltration capacity curve [5].

portional to the local infiltration capacity. The median direct infiltration capacity (to lower-zone and groundwater) is $b/2$; the median total infiltration capacity (including interflow) is $(cb)/2$. Both b and c are functions of lower-zone dimensionless storage ratios. Given a constant moisture supply, Fig. 13-15(b) shows the contribution to (1) surface detention and (2) interflow detention. Modeling infiltration in this way assures a smooth variation in model response, as shown in Fig. 13-16.

The current value of b determines the amount of direct infiltration. i.e., moisture going into lower-zone and groundwater storages. The current value of c alters hydrograph shape and timing by controlling the contributions to surface and interflow detention and their ratio.

The current soil moisture in lower-zone storage is LZS; the nominal soil moisture

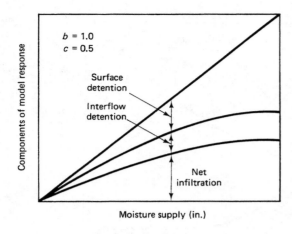

Figure 13-16 SWM IV: Model response to infiltration capacity function [5].

Moisture supply (in.)

in lower zone storage is LZSN (a median value of lower-zone storage is taken as the value of LZSN). The lower-zone dimensionless storage ratio is defined as LZSR = LZS/LSZN. For LZSR ≤ 1, the value of b is calculated as follows:

$$b = \frac{CB}{2^{4(LZSR)}} \qquad (13\text{-}13a)$$

Conversely, for LZSR > 1:

$$b = \frac{CB}{2^{4+2[(LZSR)-1]}} \qquad (13\text{-}13b)$$

In Eq. 13-13, CB is an input parameter that determines the overall level of direct infiltration. The minimum value of b is (CB)/64.

The value of c is calculated as follows:

$$c = CC(2^{LZSR}) \qquad (13\text{-}14)$$

in which CC is an input parameter that determines the relative contributions of overland (surface) flow and interflow. For the case of CB = CC = 1, Fig. 13-17 shows the following functions: (1) b versus LZSR and (2) c versus LZSR.

Indirect Infiltration (Upper-zone Storage). Moisture that does not contribute to lower-zone and groundwater storages will either contribute to overland flow or interflow or enter upper-zone storage. Depression storage and storage in highly permeable soils are modeled by the upper zone. Upper-zone storage capacity is relatively low and independent of rainfall intensity.

The current soil moisture in upper-zone storage is UZS; the nominal soil moisture in upper-zone storage is an input parameter referred to as UZSN. The upper-zone dimensionless storage ratio is defined as UZSR = UZS/UZSN. For UZSR ≤ 2, the percentage of surface detention contributed to upper-zone storage P_r is calculated as follows:

$$P_r = 100\left[1 - 0.5\,(UZSR)\,\frac{1}{(1 + UZI)^{UZI}}\right] \qquad (13\text{-}15a)$$

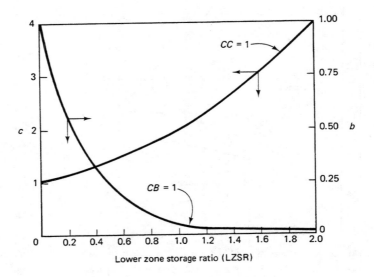

Figure 13-17 SWM IV: b and c versus lower-zone storage ratio LZSR [5].

with

$$UZI = 2[1 - 0.5(UZSR)] + 1 \qquad (13\text{-}15b)$$

Conversely, for UZSR > 2:

$$P_r = 100 \, \frac{1}{(1 + UZI)^{UZI}} \qquad (13\text{-}16a)$$

with

$$UZI = 2[(UZSR) - 2] + 1 \qquad (13\text{-}16b)$$

These relationships are shown in Fig. 13-18. Moisture is depleted from the upper zone by (1) evapotranspiration and (2) percolation to lower-zone and groundwater storages. Percolation (indirect or delayed infiltration) occurs from upper-zone storage to lower-zone and groundwater storages when the upper-zone storage ratio (UZSR) exceeds the lower-zone storage ratio (LZSR). The percolation amount PERC (inches per hour) is calculated as follows:

$$PERC = 0.003(CB)(USZN)[(UZSR) - (LZSR)]^3 \qquad (13\text{-}17)$$

Overland Flow.   Discharge from an overland flow plane is related to outflow depth, with outflow depth related to volume of surface detention. The volume of surface detention at equilibrium can be shown to be [5]:

$$D_e = \frac{0.000818 \, i^{3/5} n^{3/5} L^{8/5}}{S^{3/10}} \qquad (13\text{-}18)$$

in which D_e = volume of surface detention at equilibrium per unit of overland-flow

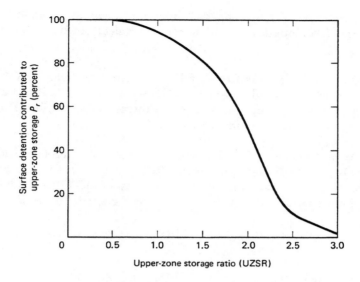

Figure 13-18 SWM IV: Contribution of surface detention to upper-zone storage [5].

plane width, in cubic feet per foot; i = supply rate (inflow to plane) in inches per hour; L = length of plane in feet; S = slope of plane in feet per foot.

The relationship between outflow depth and volume of surface detention is based on the following empirical formula:

$$y = \left(\frac{D}{L}\right)\left[1 + 0.6\left(\frac{D}{D_e}\right)^3\right] \tag{13-19}$$

in which y = outflow depth and D = volume of surface detention.

The discharge per unit of overland-flow plane width is calculated by Manning's equation:

$$q = \left(\frac{1.486}{n}\right)y^{5/3}S^{1/2} \tag{13-20}$$

in which y is calculated by Eq. 13-19.

The operation of the overland flow component is based on a discretized continuity equation:

$$D_2 = D_1 + \Delta D - q_{\text{ave}}\,\Delta t \tag{13-21}$$

in which D_2 = surface detention at the end of a time interval Δt; D_1 = surface detention at the beginning of a time interval Δt; ΔD = increase in surface detention in the time interval Δt (calculated as a byproduct of direct infiltration, as shown in Fig. 13-15 (b)); and q_{ave} = overland flow unit-width discharge, calculated with Eq. 13-20, using an average value of D in Eq. 13-19.

Interflow. Within a time interval Δt, the increase in interflow detention storage is calculated as a byproduct of direct infiltration, as shown in Fig. 13-15(b). The depletion from interflow detention storage is calculated every 15-min interval as:

$$\text{INTF} = [1 - (\text{IRC})^{1/96}](\text{SRGX}) \tag{13-22}$$

in which INTF = outflow from interflow detention storage, SRGX = interflow detention storage, and IRC = an interflow recession constant, defined as the ratio of interflow discharge at any time to interflow discharge 24 h earlier. The interflow recession constant (IRC) is an input parameter.

Groundwater.   The inflow to groundwater consists of both direct (lower-zone) and indirect (upper-zone) infiltration. The percentage of either direct or indirect infiltration that enters the groundwater storage is a function of the lower-zone storage ratio LZSR. For LZSR $\leq$ 1, the percentage to groundwater P_g is:

$$P_g = 100\left[(\text{LZSR})\frac{1}{(1 + \text{LZI})^{\text{LZI}}}\right] \tag{13-23a}$$

with

$$\text{LZI} = 1.5[1 - (\text{LZSR})] + 1 \tag{13-23b}$$

Conversely, for LZSR > 1:

$$P_g = 100\left[1 - \frac{1}{(1 + \text{LZI})^{\text{LZI}}}\right] \tag{13-24a}$$

with

$$\text{LZI} = 1.5[(\text{LZSR}) - 1] + 1 \tag{13-24b}$$

Groundwater storage is divided into (1) active groundwater storage, which contributes baseflow to streamflow, and (2) inactive, or deep groundwater storage, which does not contribute to streamflow.

The outflow from active groundwater storage is assumed to be a function of cross-sectional area and energy gradient. A representative cross-sectional area is assumed to be proportional to the groundwater level. The energy gradient is estimated as the sum of a base value plus a fluctuating value, which is a function of the groundwater accretion. The groundwater outflow at any time is given by the following equation:

$$\text{GWF} = \text{LKK4}[1 + (\text{KV})(\text{GWS})]\text{SGW} \tag{13-24}$$

in which GWF = groundwater flow, GWS = groundwater slope, SGW = groundwater storage, LKK4 = a groundwater recession parameter, and KV = a variable-rate groundwater recession parameter.

The variable GWS is an antecedent index, updated daily based on inflow to groundwater storage as follows:

$$\text{GWS}_{i+1} = 0.97[\text{GWS}_i + \Delta(\text{SGW})] \tag{13-25}$$

in which GWS_{i+1} = value of GWS at the beginning of day $i + 1$; GWS_i = value of GWS at the beginning of day i, and $\Delta(\text{SGW})$ = increase in groundwater storage during day i.

The groundwater recession constant LKK4 is defined as

$$LKK4 = 1 - (KK24)^{1/96} \qquad (13\text{-}26)$$

in which KK24 = minimum observed daily groundwater recession constant (the ratio of groundwater discharge at any time to the groundwater discharge 24 h earlier). When the parameter KV = 0 and inflow to groundwater storage is zero, Eq. 13-24 reduces to a logarithmic depletion curve, i.e., the semilogarithmic plot of discharge versus time is a straight line (see Section 11.5).

When KV ≠ 0, the groundwater recession rate is variable, and the semilogarithmic plot of discharge versus time is not linear. For example, assume a typical dry-season recession rate of 0.99 and a recharge-period recession rate of 0.98. Then, the value of KK24 can be set at 0.99, and the value of KV is adjusted so that the quantity [1 + (KV)(GWS)] will reduce the effective recession rate to 0.98 during recharge periods.

Percolation to inactive groundwater storage is modeled by allowing a fraction of the inflow to groundwater to percolate deep into inactive storage. This fraction is specified by the input parameter K24L.

Evapotranspiration.   Evapotranspiration from interception storage and upper-zone storage is assumed to occur at the potential rate. Evapotranspiration from lower-zone storage is controlled by the *evapotranspiration opportunity*. Minor amounts of evapotranspiration from groundwater storage and evaporation from stream surfaces are also simulated. Daily lake evaporation or PET are used as data. Hourly values are obtained from the daily values.

For a given time interval, potential evapotransporation is first satisfied from interception storage and second from upper-zone storage. Any remaining deficit of potential evapotranspiration is satisfied from lower-zone storage using the concept of evapotranspiration opportunity. An evapotranspiration opportunity curve is a cumulative linear frequency distribution similar to that shown in Fig. 13-15. For a given time interval (h), this curve shows the cumulative watershed area (in percent) having an evapotranspiration amount equal to or less than the indicated value, as shown in Fig. 13-19.

The quantity of water lost to evapotranspiration from the lower zone can be calculated from Fig. 13-19:

$$E = E_p - \frac{E_p^2}{2r} \qquad (13\text{-}27)$$

in which E = actual evapotranspiration from the lower zone; E_p is the excess PET (i.e. that not satisfied from interception and upper-zone storages), and r is the maximum value of evapotranspiration opportunity, estimated as follows:

$$r = K3(LZSR) \qquad (13\text{-}28)$$

in which K3 = an input parameter. Given Fig. 13-19, when the excess potential evapotranspiration E_p exceeds the evapotranspiration opportunity r, the actual evapotranspiration is equal to $(r/2)$ inches.

Evapotranspiration from groundwater storage is governed by K24EL, an input parameter equal to the fraction of watershed area from which evapotranspiration is

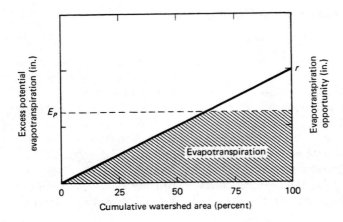

Figure 13-19 SWM IV: Evapotranspiration opportunity curve [5].

assumed to occur at the potential rate. Evaporation at the potential rate from stream surfaces is governed by ETL, an input parameter equal to the ratio of stream area to total watershed area.

Channel System Simulation. The operation of the land surface model produces continuous overland flow, interflow, and groundwater flow, which enter the stream channel system. Routing in the channel system is based on a concept similar to that of the Clark unit hydrograph (Section 10.2). As in the Clark procedure, the channel system simulation consists of two steps: (1) channel translation and (2) channel storage.

Channel Translation. In the Clark procedure, the time-area method is used with a unit effective-rainfall depth to produce the translated-only unit hydrograph. In SWM IV, the time-area histogram is replaced by the *time-discharge* histogram. The time-discharge histogram is obtained by estimating channel flows at successive points in the stream channel system and calculating their translation time to the watershed outlet. Following a routing procedure similar to that of the time-area method (Section 10.1), a translated-only outflow hydrograph is derived from the time-discharge histogram, as shown in the sketch of Fig. 13-20.

Channel Storage. In the Clark procedure, the translated-only hydrograph is routed through a linear reservoir. The same procedure is followed in SWM IV, with the following routing equation:

$$O_2 = I_{ave} - C(I_{ave} - O_1) \tag{13-29}$$

in which O_2 = outflow at time level 2, I_{ave} = average inflow between time levels 1 and 2, O_1 = outflow at time level 1, and C = routing constant, equal to:

$$C = \frac{2 - (\Delta t/K)}{2 + (\Delta t/K)} \tag{13-30}$$

in which K = storage constant. It can be shown that Eq. 13-29, with the routing coefficient of Eq. 13-30, is essentially the same as that of linear reservoir routing

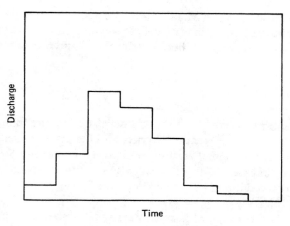

(a) Time-discharge histogram

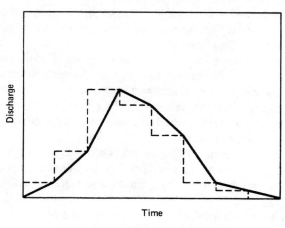

(b) Translated-only hydrograph

Figure 13-20 SWM IV: Channel translation [5].

model, Eq. 8-15 with the routing coefficients defined by Eqs. 8-16 to 8-18, or the SSARR routing equation, Eq. 10-11.

Snowmelt. The snowmelt component of SWM IV uses daily maximum and minimum temperatures, shortwave radiation, snow evaporation, and precipitation data. Two storage volumes are considered: (1) a liquid-water storage, and (2) a *negative heat* storage. Calculations are made at hourly intervals as incoming snowfall is added to the snowpack and incoming rainfall is added to liquid-water storage.

Temperature and radiation data are used to calculate the net heat transfer to the snowpack. Negative heat-transfer amounts (from snowpack to atmosphere) increase negative heat storage. Positive heat-transfer amounts (from atmosphere to snowpack) decrease negative heat storage. Snowmelt begins when the value of negative heat storage is depleted to zero. Melt amounts enter liquid-water storage until a maximum storage value is reached. Additional melt or rainfall is discharged from the snowpack and is converted into excess basin water available for runoff.

Applications.   Applications of SWM IV typically require 3 to 6 years of rainfall-runoff data for calibration of the various model parameters. Input parameters are sucessively adjusted until a close agreement between simulated and recorded flows is obtained.

Sacramento Model

The Sacramento model is a continuous simulation model developed by the Joint Federal-State River Forecast Center, National Weather Service, and the State of California Department of Water Resources. Its original version, entitled *A Generalized Streamflow Simulation System,* dates back to 1973 [4]. A slightly modified version of the Sacramento model was incorporated into the National Weather Service River Forecast System (NWSRFS) in 1976 [16].

The Sacramento model is a conceptual lumped-parameter system that can be used to simulate the headwater portion of the hydrologic cycle. The system consists of a set of percolation, soil-moisture storage, drainage, and evapotranspiration characteristics intended to represent the relevant hydrologic processes in a logical and consistent manner. The calculation of runoff from rainfall hinges upon a soil-moisture accounting procedure, hence the name *Sacramento soil-moisture accounting model.*

The soil mantle is divided into upper- and lower-zone storages. The upper-zone storage accounts for moisture retained in interception and the upper soil layer; the lower-zone accounts for the bulk of the soil moisture and longer groundwater storage. Each zone, upper and lower, has two substorages, one holding tension water and another holding free water. *Tension water* is that which is closely bound to the soil particles. Conversely, *free water* is not bound to the soil particles, being free to move either downward or laterally. For any zone, the maximum amounts of tension water and free water that it can hold are specified as model parameters. The amount of water at any time in each storage is a model variable. A sketch of zones and storages is shown in Fig. 13-21.

The catchment or basin is divided into two types of areas: (1) a permeable area, which produces runoff only when the rainfall rate exceeds a certain amount, and (2) an impermeable area (i.e., streams, lake surfaces, marshes, and other impervious areas directly connected to the channel system), which produces runoff for any amount of rainfall.

Permeable Areas.   In the permeable area, moisture initially enters into the upper-zone tension-water storage (UZTWS), which must be totally filled before any moisture becomes available to enter other storages. Water going into UZTWS provides sufficient moisture to the upper soil mantle so that percolation to deeper zones and horizontal drainage can begin. When the UZTWS has been filled, excess moisture is temporarily accumulated in the upper-zone free-water storage (UZFWS).

The water stored in UZFWS can move vertically (percolating to deeper zones) and horizontally (moving laterally and generating interflow). The demands on UZFWS are a function of available moisture. The rate of vertical drainage is a function of the available moisture in UZFWS and lower-zone storages. The preferred path for movement of moisture stored in UZFWS is by percolation.

Interflow can occur only when the moisture supply rate exceeds the rate at which

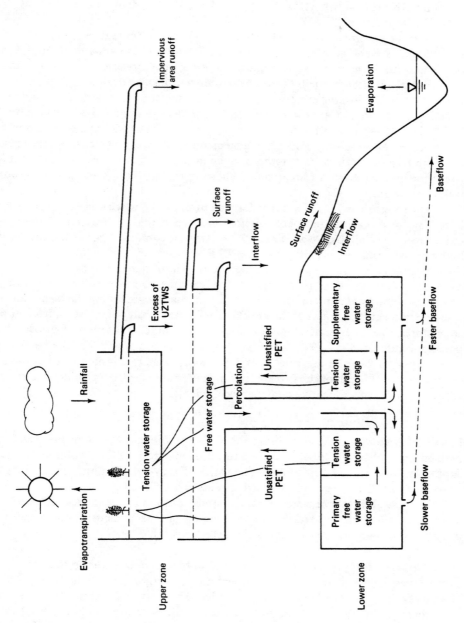

Figure 13-21 Conceptual diagram of Sacramento model [4].

moisture is percolating from water stored in UZFWS. Interflow is proportional to the available UZFWS after percolation:

$$INTER = (UZK)(UZFWC) \qquad (13\text{-}31)$$

in which INTER = interflow, UZK = the UZFWS depletion coefficient, and UZFWC = available UZFWS.

Surface runoff can occur only when the moisture supply rate exceeds the sum of percolation rate and maximum interflow rate. Therefore, the amount of surface runoff varies with the moisture supply rate and the degree of dryness of the various zones.

In the lower zone, the moisture held after wetting and drainage is referred to as lower-zone tension-water storage (LZTWS). The moisture available for drainage as baseflow is referred to as lower-zone free-water storage (LZFWS). The moisture in LZFWS is held in two different compartments: (1) primary storage, which simulates slower-draining baseflow, and (2) supplementary storage, which simulates faster-draining baseflow. These two storages fill simultaneously from percolated water but drain independently at different rates, allowing the simulation of variable ground-water recession.

The moisture in LZTWS is that claimed by dry soil particles when moisture from a wetting front reaches that depth. Before LZTWS is filled, a fraction of the percolated water goes into LZFWS. When LZTWS is full, continued percolation is divided between the primary and supplemental LZFWS volumes. Water available to enter these storages is distributed between them in response to their relative moisture deficiencies. The process of percolation in the Sacramento model follows the observed pattern of movement of moisture through the soil mantle, including the formation and transmission characteristics of the wetting front. The transfer of moisture from upper-zone storage to lower-zone storage is based upon the computation of a lower-zone percolation demand (LZPD). When the lower-zone storage is filled, the percolation rate is equal to the drainage rate. This limiting drainage rate is computed as the sum of the products of each of the two lower-zone free-water storages (primary and supplementary) and their respective depletion coefficients:

$$PBASE = (LZFM_s)(LZSK) + (LZFM_p)(LZPK) \qquad (13\text{-}32)$$

in which PBASE = limiting drainage rate from combined action of primary and supplementary LZFWS, $LZFM_s$ = maximum supplementary LZFWS volume, LZSK = supplementary-storage depletion coefficient, $LZFM_p$ = maximum primary LZFWS volume, and LZPK = primary-storage depletion coefficient.

The use of three free-water storage volumes, one in the upper zone and two in the lower zone, allows the generation of a wide variety of recession curves and is generally consistent with observed streamflow characteristics [4].

Variable Impervious Area. A portion of the water entering the basin is assumed to be deposited on impervious areas either directly connected or adjacent to the channel system and thus becomes channel flow. This portion is defined by two parameters representing its minimum and maximum values. The actual area used in the computation varies between these two limits as a function of the amount of water in storage.

Evaporation and Evapotranspiration.   PET may be computed from meteorological variables or evaporation pan measurements. The evapotranspiration demand curve is a product of the computed PET and a seasonal adjustment coefficient that reflects vegetation type and growth rate.

For areas covered by surface water, evaporation is computed at the potential rate. For areas of the soil mantle not covered with surface water, evapotranspiration is a function of the potential rate and the volume and distribution of tension-water storage. As the soil mantle dries from evapotranspiration, moisture is withdrawn from the upper zone at the potential rate multiplied by the proportional loading of UZTWS.

In the lower zone, evapotranspiration takes place at the unsatisfied potential rate (fraction of potential evapotranspiration unsatisfied by evapotranspiration from the upper zone) multiplied by the ratio of contents to capacity of tension-water storage. If evapotranspiration occurs at a rate such that the ratio of contents to capacity of free-water storage exceeds the ratio of contents to capacity of tension-water storage, moisture is transferred from free-water to tension-water storage, and the relative loadings are balanced in order to maintain a consistent moisture profile. Depending upon basin conditions, a fraction of lower-zone free water may be considered to be below the root zone, and therefore unavailable for such transfers.

Streamflow.   Streamflow is the result of processing precipitation through an algorithm representing the uppermost soil mantle and the lower zones. This algorithm produces runoff in five forms:

1. Direct runoff from variable impervious area
2. Surface runoff due to precipitation occurring at a faster rate than percolation and interflow can take place, when both UZTWS and UZFWS are full
3. Interflow resulting from the lateral drainage of UZFWS
4. Supplementary baseflow originating in lateral drainage from supplementary LZFWS
5. Primary baseflow originating in lateral drainage from primary LZFWS.

The first two types account largely for surface runoff, whereas the remaining three types account for subsurface runoff (interflow and baseflow).

Model Components.   The runoff characteristics are a function of the precipitation rate and soil moisture conditions. The model represents the basin as a set of storage volumes of specified capacities, which hold water temporarily and gradually recede as their contents are depleted by evapotranspiration, percolation, or lateral drainage. The basic components of the model are shown in schematic form in Fig. 13-22 [16].

Computational Technique.   The movement of moisture through the soil mantle is a continuous process. The flow rate at various points is a function of the supply rate and contents of the various storages. For a given time interval, all moisture-accounting computations are based on the assumption that the movement of

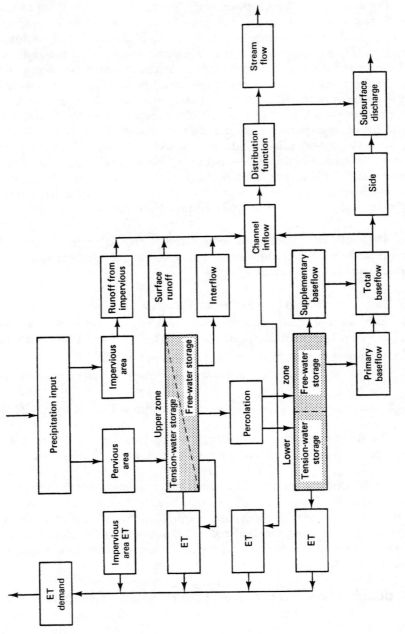

Figure 13-22 Sacramento model schematic [16].

moisture is defined by the conditions at the beginning of the time interval. The time interval is internally set so that changes in water storage are kept within 5 mm.

Model Parameters.   The soil-moisture-accounting component of the Sacramento model, exclusive of the evapotranspiration demand curve, involves 17 parameters. The evapotranspiration demand curve can be defined by a series of 12 ordinates or by a formula having 5 parameters. The temporal distribution function, which converts runoff volumes to a discharge hydrograph, is based on a unit hydrograph and, in some applications, a channel-routing function.

NWSRFS Modifications.   The original Sacramento model [4] applied the unit hydrograph to flow releases from the first three storages. The two lower-zone flow components were added to the channel flow in the time period in which they were released from the lower zone. In the National Weather Service River-Forecast System (NWSRFS) version, the unit hydrograph is applied to the sum of all five flow components [16].

The NWSRFS version requires an input/output time interval of 6 h. Though a smaller computational time interval may be used internally for increased accuracy, computations are accumulated over a 6-h period and used with a 6-h unit hydrograph.

Model Calibration.   The National Weather Service uses a combination of manual and automatic optimization techniques for model calibration. Manual calibration consists of subjective adjustments to model parameters on the basis of specific characteristics of previous model output. Automatic calibration techniques are those in which values of model parameters are adjusted based on changes in the value of a chosen error function.

A good set of parameters can usually be obtained by manual methods. However, the procedure is time-consuming and requires considerable judgment. Automatic methods, on the other hand, are faster and simple to use, although they have some disadvantages. Among them are (1) complete dependence on the error function, (2) failure to attain an optimal solution due to nonconvergence of the optimization algorithm, and (3) failure to recognize the effect of a simultaneous perturbation of a group of parameters. At worst, automatic optimization can degenerate into curve fitting, producing a set of parameters that may fit the calibration data reasonably well but are physically unrealistic [4].

In principle, both manual and automatic calibration procedures are applicable. In practice, it is better first to use a manual calibration and then follow with an automatic calibration after an initial fit has been obtained by manual means. The length of data base required for model calibration is a function of the meteorological characteristics of the catchment and the amount of hydrologic activity during the record period. Typically, 8 to 10 y of data are required for adequate model calibration [16].

QUESTIONS

1. Describe the differences between deterministic, probabilistic, conceptual, and parametric models.

2. Describe the difference between linear and nonlinear models.

3. Describe the difference between lumped and distributed models.

4. Describe the difference between continuous and discrete models.

5. Describe the differences between event and continuous-process models.

6. Comment on the importance of storm duration as a parameter in midsize catchment analysis.

7. When is PMP used in lieu of depth-duration-frequency data?

8. Discuss the role of the storm's temporal distribution in connection with midsize catchment modeling.

9. Discuss the role of the storm's spatial distribution in connection with large-catchment modeling.

10. What types of hydrologic abstraction are usually considered in catchment modeling for design and operational purposes?

11. Contrast the unit hydrograph and kinematic wave routing approaches to runoff modeling.

12. How does the variable-source-area concept differ from the overland flow concept?

13. Discuss the issue of model resolution in connection with catchment modeling.

14. Contrast model calibration and verification.

15. Why is the calibration of time-variant processes and models different from that of time-invariant models?

16. Discuss the role of sensitivity analysis in modeling catchment flows.

17. Why is HEC-1 a hydrologic rather than a hydraulic model? Why is HEC-1 an event model?

18. What methods of hydrologic abstractions are included as options in HEC-1?

19. What does the parameter E represent in the general HEC loss-rate function?

20. What options for synthetic unit hydrograph are available in HEC-1?

21. What methods of flood routing are considered in HEC-1?

22. What method of hydrologic abstraction is featured in TR-20? What reservoir routing method is featured in TR-20? What is the channel routing method included in the current version of TR-20?

23. Contrast HEC-1 and TR-20 in terms of their features and capabilities.

24. For what applications are the HEC-1 and TR-20 models best suited?

25. For what applications is the EPA Storm Water Management Model best suited?

26. What is the basic routing equation of the SSARR model?

27. Describe briefly the two watershed models included in the most recent (1986) version of SSARR.

28. How are hydrologic abstractions accounted for in the SSARR model?

29. What process augments the soil moisture index in the SSARR model? What process depletes it?

30. How is surface-subsurface flow separation accomplished in the SSARR model?

31. How are nonlinearities accounted for in channel routing with the SSARR model?

32. Describe the streamflow-backwater mode of SSARR.

33. For what applications is SSARR best suited?

34. Describe the soil-moisture storages considered in SWM IV.

35. What is an infiltration capacity curve as used in SWM IV?

36. Describe briefly the overland flow component of SWM IV.

37. What is an evapotranspiration opportunity curve as used in SWM IV?

38. For what applications is SWM IV best suited?

39. Why is the Sacramento model referred to as a *soil-moisture accounting model*?

40. Describe the various storage zones in the Sacramento model.

41. In the Sacramento model, what is the preferred path of movement for moisture stored in upper-zone free-water storage?

42. What storage zone is the source of interflow in the Sacramento model?

43. What is the difference between primary and supplementary lower-zone free-water storage?

44. What is the major difference between the original Sacramento model and the version implemented in the National Weather Service River Forecast System?

45. Contrast the use of manual versus automatic parameter-optimization techniques.

46. For what applications is the Sacramento model best suited?

PROBLEMS

13-1. Given the following rainfall hyetograph for a certain storm, calculate the hydrologic abstraction by the general HEC loss-rate function.

Time (h)	0	3	6	9	12
Rainfall intensity (in./h)		0.8	1.0	0.5	0.3

Assume $S = 0.5$ in./h, $D = 0.2$ in., $E = 0.6$ and $R = 2$. Report the cumulative loss (in inches) for the 12-h period.

13-2. Given the following rainfall hyetograph for a certain storm, calculate the hydrologic abstraction by the general HEC loss-rate function.

Time (h)	0	2	4	6
Rainfall intensity (in./h)		1.0	1.5	0.5

Assume $S = 0.8$ in./h, $D = 0$ in., $E = 0$ and $R = 1$. Report the cumulative loss (in inches) for the 6-h period.

13-3. Given the following rainfall hyetograph for a certain storm, calculate the hydrologic abstraction by the general HEC loss-rate function.

Time (h)	0	1	2	3	4
Rainfall intensity (in./h)		0.7	0.9	1.2	0.6

Assume $S = 0.6$ in./h, $D = 0$ in., $E = 0$ and $R = 2.5$. Report the cumulative loss (in inches) for the 4-h period.

13-4. Derive Eq. 13-27 for the actual evapotranspiration in terms of excess potential evapotranspiration E_p and maximum evapotranspiration opportunity r.

13-5. Show that the routing equation of the Stanford Watershed Model IV is the same as that of the SSARR model, Eq. 10-11.

REFERENCES

1. Anderson, M. G., and T. P. Burt. (1985). "Modeling Strategies," Chapter 1 in *Hydrological Forecasting*. M. G. Anderson and T. P. Burt, eds. New York: John Wiley.

2. Beven, K. (1985). "Distributed Models," Chapter 13 in *Hydrological Forecasting*. M G. Anderson and T. P. Burt, eds. New York: John Wiley.

3. Bras, R., and I. Rodriguez-Iturbe. (1985). *Random Functions and Hydrology*. Reading, Mass.: Addison-Wesley.

4. Burnash, J. C., R. L. Ferral, and R. A. McGuire. (1973). "A Generalized Streamflow Simulation System, Conceptual Modeling for Digital Computers," Joint Federal-State River Forecast Center, U.S. Department of Commerce, NOAA National Weather Service, and State of California Dept. of Water Resources, March.

5. Crawford, N. H., and R. K. Linsley. (1966). "Digital Simulation in Hydrology: Stanford Watershed Model IV," *Technical Report No.* 39, Department of Civil Engineering, Stanford University, Stanford, California, July.

6. Crawford, N. H. (1971). "Studies in the Application of Digital Simulation to Urban Hydrology," Hydrocomp International, Inc., Palo Alto, California, September.

7. Cundy, T. W., and K. N. Brooks. (1981). "Calibrating and Verifying the SSARR Model," *Water Resources Bulletin,* American Water Resources Association, October.

8. Holtan, H. N., G. J. Stitner, W. H. Henson, and N. C. Lopez. (1975). "USDAHL-74, Revised Model of Watershed Hydrology," *Technical Bulletin No.* 1518, Agricultural Research Service, U.S. Department of Agriculture, Washington, D.C.

9. Hydrologic Engineering Center, U.S. Army Corps of Engineers. (1981). "HEC-2, Water Surface Profiles, Users Manual," Davis, California.

10. Hydrologic Engineering Center, U.S. Army Corps of Engineers. (1985). "HEC-1, Flood Hydrograph Package, Users Manual," Davis, California, issued September, 1981, revised January.

11. Johanson, R. C., J. C. Imhoff, and H. H. Davis. (1980). "User Manual for the Hydrologic Simulation Program-Fortran (HSPF)," *Environmental Protection Agency Report* FPA-600/9-80.

12. Larson, C. L., C. A. Onstad, H. H. Richardson, and K. N. Brooks. (1982). "Some Particular Watershed Models," Chapter 10 in *Hydrologic Modeling of Small Watersheds,* C. T. Hann, H. P. Johnson, and D. L. Brakensiek, eds., ASAE Monograph Number 5, American Society of Agricultural Engineers, St. Joseph, Michigan.

13. Liou, E. Y. (1970). "OPSET: Program for Computerized Selection of Watershed Parameters Values for the Stanford Watershed Model," *Research Report No.* 34, Water Resources Institute, University of Kentucky, Lexington.

14. NOAA *Hydrometeorological Report No.* 49. (1977). "Probable Maximum Precipitation Estimates-Colorado River and Great Basin Drainages," September.

15. NOAA *Hydrometeorological Report No.* 55. (1984). "Probable Maximum Precipitation Estimates-United States Between the Continental Divide and the 103rd Meridian," March.

16. Peck, Eugene L. (1976). "Catchment Modeling and Initial Parameter Estimation for the National Weather Service River Forecast System," *NOAA Technical Memorandum* NWS HYDRO-31, U.S. Department of Commerce, Silver Spring, Maryland, June.

17. Renard, K. G., W. J. Rawls, and M. M. Fogel. (1982). "Currently Available Models," Chapter 13 in *Hydrologic Modeling of Small Watersheds,* C. T. Hann, H. P. Johnson, and D. L. Brakensiek, eds., ASAE Monograph Number 5, American Society of Agricultural Engineers, St. Joseph, Michigan.

18. Ricca, V. T. (1972). "The Ohio State University Version of the Stanford Streamflow Simulation Model, Part I- Technical Aspects," Ohio State University, Columbus.

19. Troendle, C. A. (1985). "Variable Source Area Models," Chapter 13 in *Hydrological Forecasting,* M. G. Anderson and T. P. Burt, eds., New York: John Wiley.

20. U.S. Army Corps of Engineers, North Pacific Division. (1986). "Program Description and User Manual for SSARR Model, Streamflow Synthesis and Reservoir Regulation," Portland, Oregon, issued September 1972, revised June 1975, revised April 1986 (Draft).

21. USDA Soil Conservation Service. (1983). "Computer Program for Project Formulation: Hydrology," Technical Release No. 20 (TR-20), Engineering Division, Washington, D.C., Draft, May.

22. USDA Soil Conservation Service. (1986). "Urban Hydrology for Small Watersheds," Technical Release No. 55 (TR-55), Engineering Division, Washington, D.C., June.

23. U.S. Department of Commerce, NOAA National Weather Service. (1972). "National Weather Service River Forecast System Procedures," *Technical Memorandum NWS HYDRO-14,* Silver Spring, Maryland, December.

24. U.S. Department of Commerce, NOAA National Weather Service. (1979). "Interduration Precipitation Relations for Storms-Southeast States," *NOAA Technical Report NWS 21,* Office of Hydrology, Silver Spring, Maryland, March.

25. U.S. Department of Commerce, NOAA National Weather Service. (1981). "Interduration Precipitation Relations for Storms-Western United States," *NOAA Technical Report NWS 27,* Office of Hydrology, Silver Spring, Maryland, September.

26. U.S. Department of Commerce, NOAA National Weather Service. (1984). "Depth-Area Ratios in the Semiarid Southwest United States," *Technical Memorandum NWS HYDRO-40,* Silver Spring, Maryland, August.

27. U.S. Environmental Protection Agency. (1975). "Storm Water Management Model, User's Manual, Version II," *Environmental Protection Technology Series,* EPA-670/2-75-017, March.

28. Woolhiser, D. A. (1982). "Hydrologic Modeling of Small Watersheds," Chapter 1 in *Hydrologic Modeling of Small Watersheds,* C. T. Hann, H. P. Johnson, and D. L. Brakensiek, eds., ASAE Monograph Number 5, American Society of Agricultural Engineers, St. Joseph, Michigan.

29. Yevjevich, V. (1972). *Stochastic Processes in Hydrology.* Fort Collins, Colorado: Water Resources Publications.

SUGGESTED READINGS

Anderson, M. G., and T. P. Burt. (1985). *Hydrological Forecasting.* New York: John Wiley.

Hann, C. T., H. P. Johnson, and D. L. Brakensiek. (1982). *Hydrologic Modeling of Small Watersheds.* ASAE Monograph Number 5, American Society of Agricultural Engineers, St. Joseph, Michigan.

Hydrologic Engineering Center, U.S. Army Corps of Engineers, (1985). "HEC-1, Flood Hydrograph Package, Users Manual," Davis, California, September, 1981, revised January.

U.S. Army Corps of Engineers, North Pacific Division. (1986). "Program Description and User Manual for SSARR Model, Streamflow Synthesis and Reservoir Regulation," Portland, Oregon, issued September 1972, revised June 1975, revised April 1986 (Draft).

USDA Soil Conservation Service. (1983). "Computer Program for Project Formulation: Hydrology," Technical Release No. 20 (TR-20), Engineering Division, Washington, D.C., Draft, May.

HYDROLOGIC DESIGN CRITERIA

The techniques of hydrologic analysis described in Chapters 4-13 are useful in hydrologic design and hydrologic forecasting. In hydrologic design, the objective is to predict the behavior of hydrologic variables under a hypothetical extreme condition such as the 100-y flood or the probable maximum flood. In hydrologic forecasting the aim is to predict the behavior of hydrologic variables within a shorter time frame, either daily, monthly, seasonally, or annually. These two types of hydrologic applications are quite different, paralleling the differences between event-driven and continuous-process catchment models (Section 13.1).

Hydrologic design precedes hydraulic design; i.e., the output of hydrologic design is the input to hydraulic design. Hydrologic design determines streamflows, discharges, and headwater levels, from which hydraulic design derives flow depths, velocities, and pressures acting on hydraulic structures and systems. Actual sizing of structures and appurtenances is obtained by balancing efficiency, practicality, and economy.

In practice, hydrologic design translates into hydrologic design criteria, i.e., a set of rules and procedures used by federal, state, or local agencies having cognizance with water resources projects. Of necessity, these criteria are likely to vary widely, reflecting the charter and jurisdiction of each agency, and the size and scope of individual projects.

The following federal agencies are actively engaged in developing and using hydrologic design criteria for engineering applications:

1. NOAA National Weather Service

2. U.S. Army Corps of Engineers
3. USDA Soil Conservation Service
4. U.S. Bureau of Reclamation
5. Tennessee Valley Authority
6. U.S. Geological Survey

This chapter describes selected design criteria and procedures used by these agencies. This information is intended to complement the subjects presented in Chapters 4 to 13.

14.1 NOAA NATIONAL WEATHER SERVICE

Probable Maximum Precipitation

The National Oceanic and Atmospheric Administration's (NOAA) National Weather Service (NWS) is the federal agency responsible for the development of design criteria to estimate probable maximum precipitation. The concepts and related methodologies are described in the NOAA Hydrometeorological Report (HMR) series. Reports of current applicability begin with HMR 36: *Probable Maximum Precipitation in California*, revised in 1969 [15–30].

In the United States, the use of extreme values of precipitation for hydrologic design dates back to the 1930s, as documented in the early issues of the HMR series (Nos. 1 to 35). The probable maximum precipitation (PMP) is the theoretically greatest depth of precipitation for a given duration that is physically possible over a given size storm area at a particular geographic location at a certain time of the year [26]. The probable maximum flood (PMF) is the flood associated with the PMP.

PMP values obtained from hydrometeorological reports are generalized estimates, that is, estimates that can be obtained by areal averaging of PMP values depicted on isoline maps (i.e., maps showing contours of equal PMP). The basic approach to generalized PMP estimates is described in the literature [37, 50, 52, 53, 54]. Procedures described in this section are applicable to (1) nonorographic regions, (2) orographic regions, and (3) regions without meteorological data.

PMP Estimates for Nonorographic Regions. For nonorographic regions, PMP estimates involve three operations on observed areal storm precipitation: (1) moisture maximization, (2) transposition, and (3) envelopment. Moisture maximization consists of increasing the storm precipitation to a value that is consistent with the maximum moisture in the atmosphere for the given geographic location and month of the year. Transposition refers to the relocation of storm precipitation within a meteorologically homogeneous region, with the aim of increasing the available data for evaluating maximum rainfall potential. Envelopment refers to the smooth interpolation between precipitation maxima for different durations and/or areas, intended to compensate for the random occurrence of large rainfall events.

PMP Estimates for Orographic Regions. Topography plays an important role in the production of rainfall, either by intensifying rainfall or sheltering regions from it. Usually, the forced lifting of air on the windward slope of a mountain range produces an increase in rainfall with elevation. The magnitude of this effect varies with wind velocity and direction, amount of moisture and temperature of the air masses, and height, extent, and slope of the mountain range. Precipitation due to horizontal convergence (i.e., convergence precipitation) is important in both orographic and nonorographic regions. Therefore, PMP estimates for orographic regions consist of two components: (1) orographic precipitation, which is dependent on topographic influences, and (2) convergence precipitation, which is presumed to be independent of topographic influences.

Statistical Estimates of PMP. An alternate way to estimate PMP is to use the statistical frequency formula

$$x_T = \bar{x}_n + K_T s_n \tag{14-1}$$

in which x_T is the rainfall associated with a return period T, $\bar{x}_n$ and s_n are the mean and standard deviation of a series of n annual maxima, and K_T is the frequency factor associated with T (Chapter 6).

In terms of maximum rainfall, Eq. 14-1 can be expressed as follows:

$$x_M = \bar{x}_n + K_M s_n \tag{14-2}$$

in which x_M is the maximum rainfall and K_M is the number of standard deviations that must be added to the mean in order to obtain x_M.

An empirical estimate of K_M can be obtained by enveloping a value of K_M based on a large number of computed values. The value $K_M = 15$ was considered initially by Hershfield [8] to be an appropriate enveloping value. This analysis was based on data from 2600 stations, approximately 90% of which were located in the United States. However, Hershfield's later studies [9] have shown that K_M actually varies with mean annual maximum rainfall and storm duration, as shown in Fig. 14-1. This procedure gives point estimates of PMP; areal estimates can be obtained through the use of an appropriate depth-area relation.

The step-by-step procedure to estimate the D-hr PMP based on statistical data is the following:

1. Compile the D-hour annual-maxima rainfall series.
2. Calculate the mean and standard deviation of the D-hour annual-maxima rainfall series.
3. Using Fig. 14-1, determine the value of K_M as a function of mean annual maximum rainfall and D-hour duration.
4. Using Eq. 14-2, calculate the D-hour point PMP.
5. For basins in excess of 25 km^2, determine the areal PMP by reducing the point PMP using an appropriate depth-area relation.

Generalized PMP Estimates. Generalized PMP estimates are obtained from isohyetal maps contained in the HMR series, which encompass the entire United States [15-30].

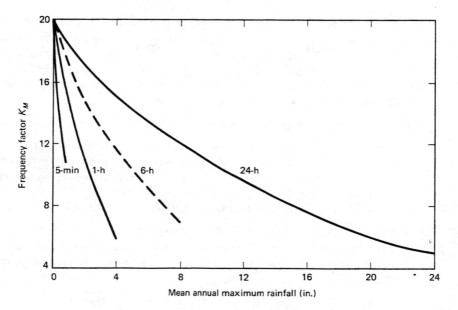

Figure 14-1 Frequency factor as a function of mean annual maximum rainfall and storm duration [9].

The procedure used for obtaining generalized PMP estimates in nonorographic regions involves storm transposition and maximization, requiring the analysis of a large number of major storms. Generalized estimates for large areas require both depth-area and depth-duration smoothing so that consistency can be maintained within and between the various charts. This is accomplished by determining PMP values for selected grid points, for several durations and areas, and smoothly enveloping the point values to determine PMP isolines.

For orographic regions, difficulties with storm transposition require the use of a number of indicators of orographic effects. For example, a chart of 2-y, 24-h rainfall may be compiled. Theoretical PMP values are then determined and related to the 2-y, 24-h indicator chart. Additional adjustments such as distance from the moisture source, direction of primary moisture inflow, barrier to moisture inflow, slope, and elevation, can be considered. Generalized PMP estimates for orographic regions tend to be less reliable indicators of the actual PMP than those of nonorographic regions. This is specifically the case for small basins whose topographic features may be very different from those of the large basins for which the generalized estimates are normally prepared.

In the United States, generalized PMP estimates for drainage basins located east of the 105th meridian are obtained using HMR 51 and HMR 52 [25, 26]. For the Tennessee valley region in particular, PMP estimates are developed using HMR 56 [30]. PMP estimates for drainage basins located west of the Continental Divide are developed using HMR 36 (California) [15], HMR 43 (Northwest States) [19], and HMR 49 (Colorado River and Great Basin drainages) [24]. PMP estimates for drainage basins located between the Continental Divide and the 103rd meridian are developed using HMR 55 [29]. Either HMR 51/HMR 52 or HMR 55 can be used for the overlapping area between the 103rd and 105th meridians (see Fig. 14-20).

PMP Estimates for Regions East of the 105th Meridian. Procedures for estimating PMP in U.S. locations east of the 105th meridian are described in HMR 51 and HMR 52 [25, 26]. HMR 51 contains a total of 30 maps, corresponding to combinations of five durations (6, 12, 24, 48, and 72 h) and six basin areas (10, 200, 1000, 5000, 10,000 and 20,000 mi²). The 24-h PMP maps are shown in Fig. 14-2. Generalized PMP estimates can be obtained by the following steps [51]:

1. Determine the geographic location and size of the drainage basin under study.

2. Using the maps, prepare a table of PMP depths applicable to the given geographic location. Use the maps of all five durations and of at least four areas surrounding the drainage basin size. For instance, for a basin size of 11,300 mi², tabulate $5 \times 4 = 20$ PMP depths (obtained from the 6-, 12-, 24-, 48-, and 72-hr maps, and from the 1000-, 5000-, 10,000-, and 20,000-mi² area maps).

3. For each duration, plot PMP depths versus basin area on semilogarithmic paper, with PMP depth in the abscissas (arithmetic scale) and basin area in the ordinates (logarithmic scale), to obtain a PMP depth-area-duration relation for the given geographic location.

4. Using the PMP depth-area-duration relation developed in step 3, determine, for the given basin area, the PMP depth for each duration.

5. Plot PMP depth versus duration on arithmetic scales, and draw a smooth curve connecting these points to determine the PMP depth-duration relation for the given basin area and geographic location. If necessary, interpolate to find the PMP depth for any intermediate duration.

6. The 6-h incremental PMP (i.e., the PMP expressed in 6-h increments) can be determined from the PMP depth-duration relation developed in step 5. This is accomplished by evaluating the PMP for successive durations in multiples of 6 h, and subtracting two consecutive PMP depths to determine an incremental PMP value. For instance, the difference between the 24- and 18-h PMP is the 6-h incremental PMP associated with the 18- to 24-h interval.

The application of HMR 51 to specific drainage basins, including the temporal and spatial distribution of the probable maximum storm is described in HMR 52 [26].

PMP Estimates for Regions West of the Continental Divide. The procedures for estimating PMP in U.S. locations west of the Continental Divide are described in HMR 36 (California) [15], HMR 43 (Northwest States) [19], and HMR 49 (Colorado River and Great Basin drainages) [24] (see Fig. 14-20). In these regionalized studies, the local storm (thunderstorm) is not enveloped with general storm depth-duration data, as in the case of the regions east of the 105th meridian.

To compute general storm PMP, the drainage basin characteristics such as size, width, elevation, and location must be known. Generally, the PMP estimate consists of two parts: (1) orographic component and (2) convergence component. The orographic component is determined by obtaining an orographic PMP index from a regional map and adjusting it to account for basin size, width, seasonal, and/or temporal variations. The convergence component is determined by obtaining a convergence PMP index from a regional map and adjusting it to account for basin size and tempo-

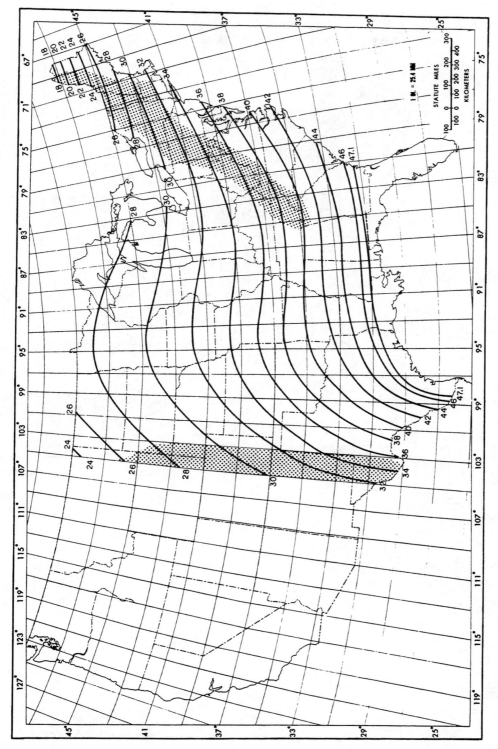

Figure 14-2(a) All-season PMP (in.) for 24-h 10-mi² rainfall [25].

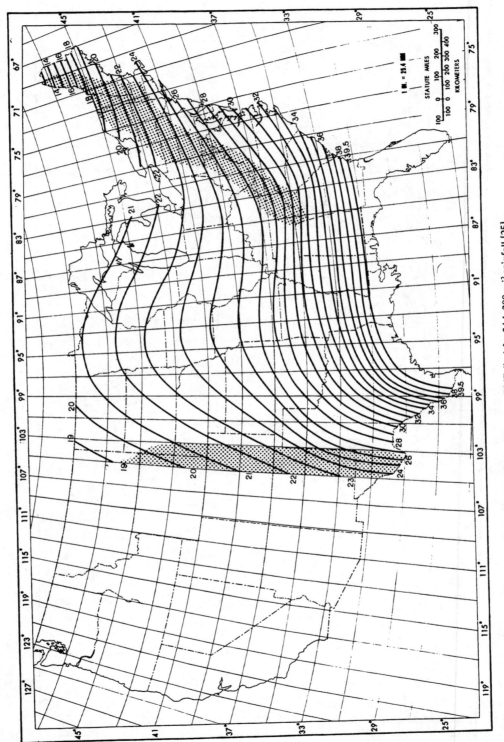

Figure 14-2(b) All-season PMP (in.) for 24-h 200-mi² rainfall [25].

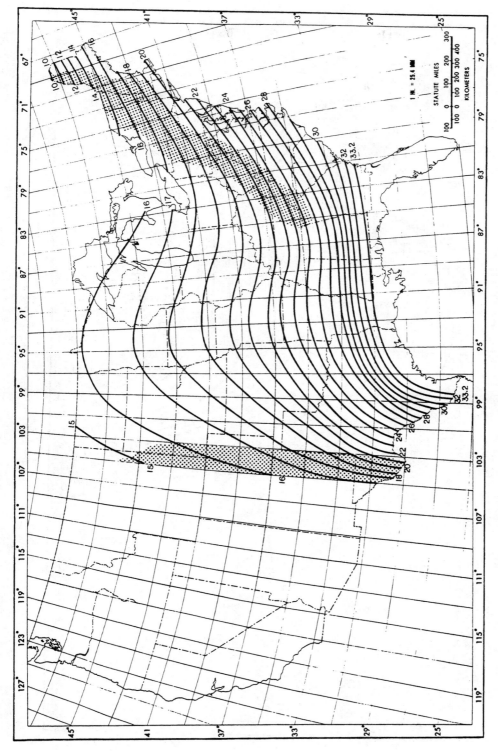

Figure 14-2(c) All-season PMP (in.) for 24-h 1000-mi² rainfall [25].

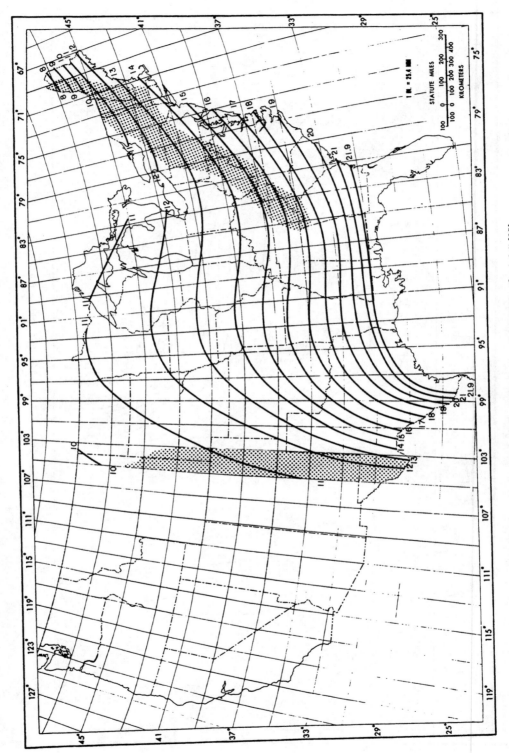

Figure 14-2(d) All-season PMP (in.) for 24-h 5000-mi² rainfall [25].

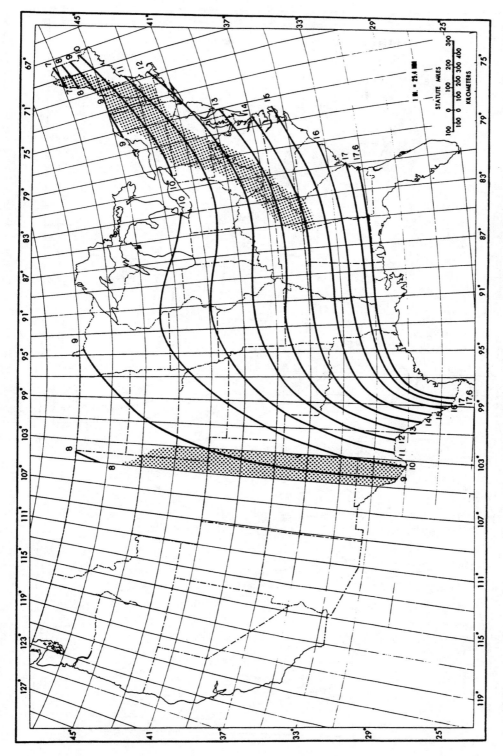

Figure 14-2(e) All-season PMP (in.) for 24-h 10,000-mi² rainfall [25].

461

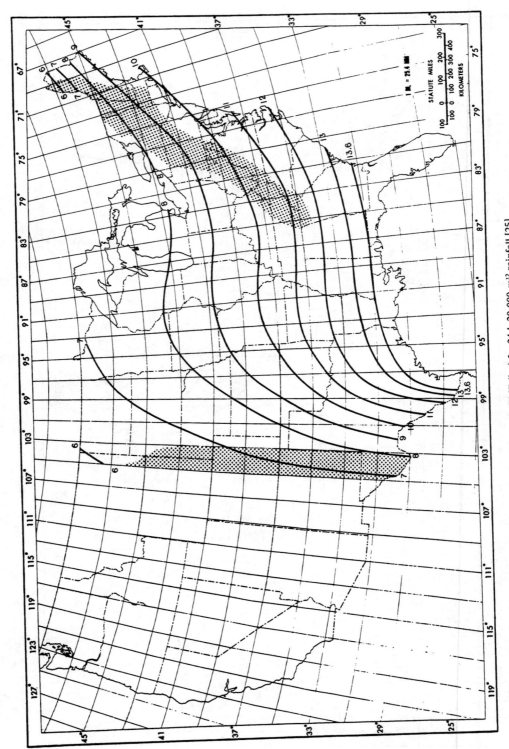

Figure 14-2(f)   All-season PMP (in.) for 24-h 20,000-mi² rainfall [25].

ral variations. The total PMP represents a combination of orographic and convergence PMP.

Estimates of local storm PMP for the Colorado River and Great Basin drainages and for California can be determined using HMR 49. Estimates of local storm PMP for the northwestern states are described in HMR 43. To derive a local storm for areas less than 500 mi^2 and durations less than 6 h, the average 1-h 1-mi^2 PMP is chosen from regionalized charts in the appropriate HMR. These values are then adjusted for basin elevation and size and are distributed over time. Elliptically shaped isohyetal patterns are used to account for basin shape and storm center location in the estimation of local storm PMP.

PMP Estimates for California. Procedures for estimating PMP in California are described in HMR 36 [15]. The total PMP consists of two parts: (1) orographic component and (2) restricted convergence component.

The orographic component is based on the following: (1) orographic PMP index map (Fig. 14-3), (2) orographic PMP computation areas (Fig. 14-4), (3) basin-width variation (Fig. 14-5), (4) seasonal variations (Table 14-1), and (5) durational variations (Table 14-2).

The restricted convergence component PMP is based on the following: (1) convergence PMP index map (Fig. 14-6) and (2) variation of convergence PMP with basin size and duration (Fig. 14-7). The PMP is the greatest of either (a) the sum of orographic plus restricted convergence PMP or (b) the unrestricted convergence PMP, calculated as four-thirds of the restricted convergence PMP. A detailed step-by-step procedure to develop a PMP estimate for California follows.

OROGRAPHIC PMP

1. Use Fig. 14-3 to determine a basin-average orographic PMP index (a grid average is adequate).

2. Determine the representative basin width, measured perpendicular to the parallel sides of one of the orographic computation areas shown in Fig. 14-4 that is closest to the basin under study.

3. Enter Fig. 14-5 with the representative basin width to determine the basin-width adjustment factor, in percent.

4. The adjusted basin-average orographic PMP index is obtained by multiplying the basin-width adjustment factor by the basin-average orographic PMP index and dividing by 100. The January 6-h orographic PMP is equal to the adjusted basin-average orographic PMP index.

5. Use Table 14-1 to determine the monthly 6-h orographic PMP for the months of October to April.

6. Use Table 14-2 to determine the 6-h orographic PMP incremental values (in percent of first 6-h period) for durations ranging from 12 to 72 h.

7. For small basins, the 1-h and 3-h orographic PMP values are 20 and 54 percent, respectively, of the 6-h orographic PMP.

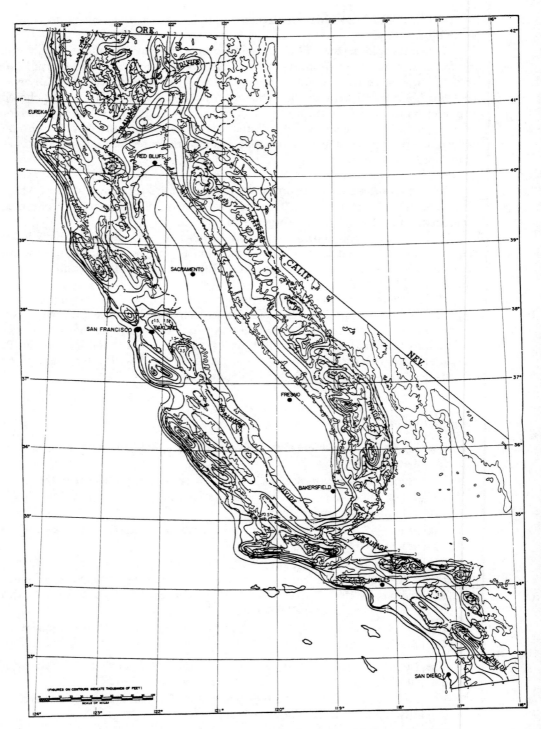

Figure 14-3 California: Orographic PMP Index, 6-h January (in.) [15].

Hydrologic Design Criteria   Chap. 14

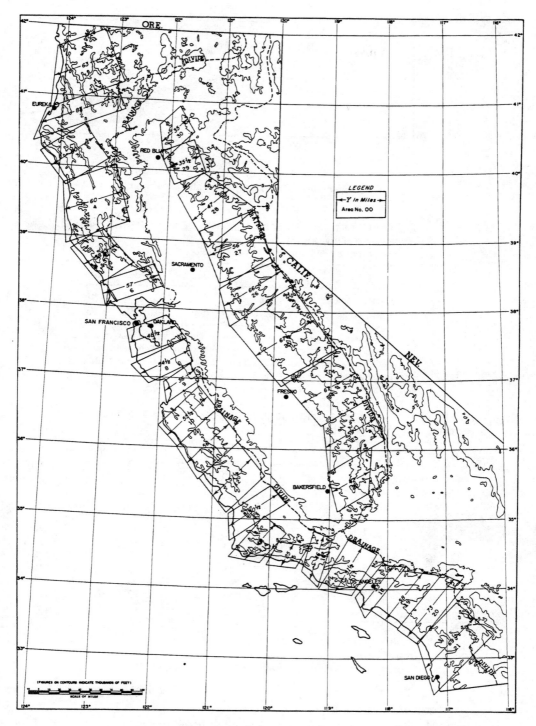

Figure 14-4 California: Orographic PMP Computation Areas [15].

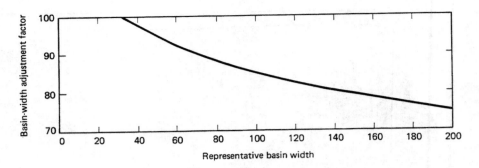

Figure 14-5 California: Basin-width adjustment [15].

TABLE 14-1 SEASONAL VARIATION OF OROGRAPHIC PMP
IN CALIFORNIA[1] [15]
(in percent of adjusted basin-average orographic PMP index)

	Oct	Nov	Dec	Jan	Feb	Mar	Apr
Coastal range	92	94	98	100	100	95	87
Sierra range	97	98	99	100	100	96	90

[1]The Sierra range values are for basins located to the east of a line
through the middle of the Central Valley between Redding and Bakersfield. Coastal range percentages apply to the remaining areas of interest
in California.

TABLE 14-2 DURATIONAL VARIATION OF OROGRAPHIC PMP IN CALIFORNIA [15]
(in percent of first 6-h period)

6-h period	1	2	3	4	5	6	7	8	9	10	11	12
Latitude North (degrees)												
42	100	89	80	71	63	56	50	44	39	34	30	26
39	100	87	78	69	61	54	48	42	37	33	29	25
38	100	86	76	66	58	51	45	39	34	30	26	23
37	100	85	74	63	54	47	41	36	31	27	23	20
36	100	83	70	59	50	43	37	32	28	24	21	18
35	100	80	66	55	46	39	33	28	24	21	18	15
34	100	77	61	50	42	35	30	25	21	18	15	13
33	100	74	58	47	39	32	27	22	18	15	13	12
32	100	72	56	45	36	29	24	20	16	13	11	10

CONVERGENCE PMP
(Restricted convergence PMP to be combined with orographic PMP)

1. Use Fig. 14-6 to determine a basin-average value of convergence PMP index (a grid average is adequate).
2. Use Fig. 14-7 to determine a table of 6-h PMP increments, in percent of convergence PMP index, for the months of October to April. After the third or fourth

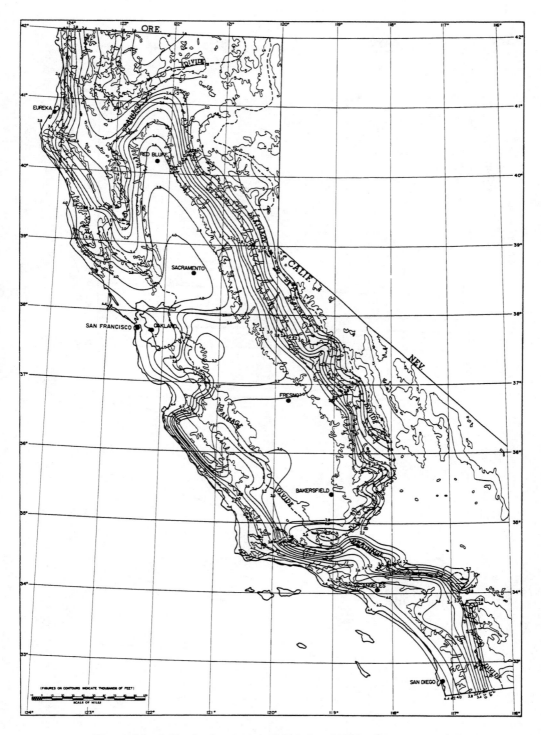

Figure 14-6 California: Convergence PMP Index, 6-h 200-mi² January (in.) [15].

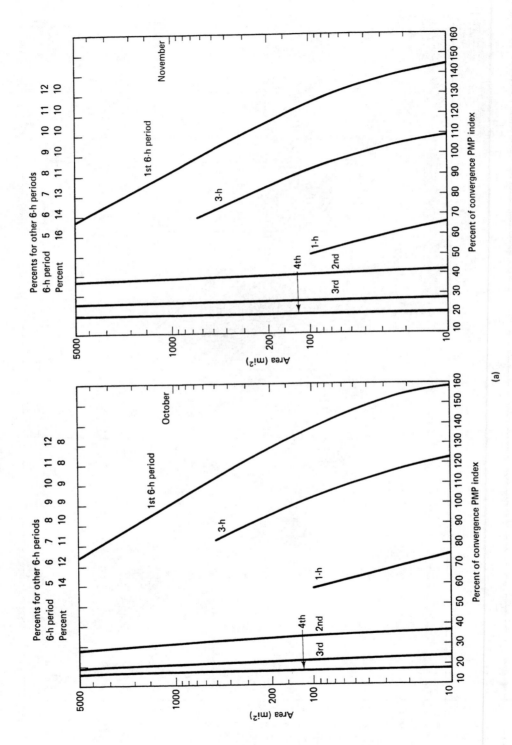

Figure 14-7(a) California: Variation of Convergence PMP with Basin Size and Duration, October and November [15].

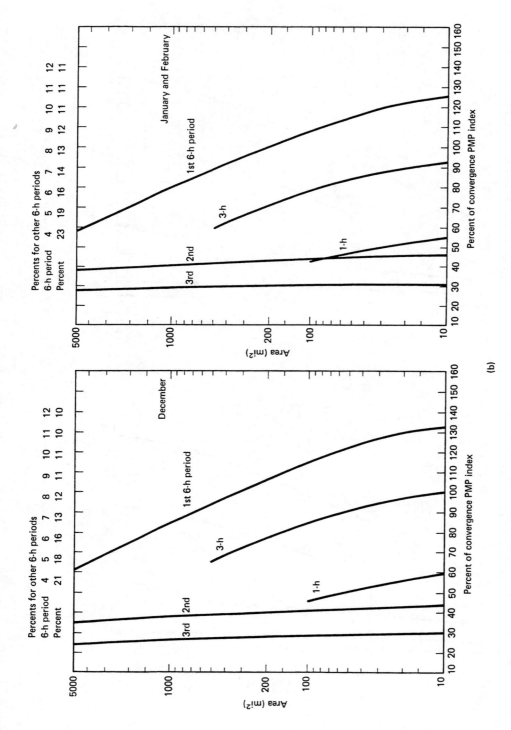

Figure 14-7(b) California: Variation of Convergence PMP with Basin Size and Duration, December, January and February [15].

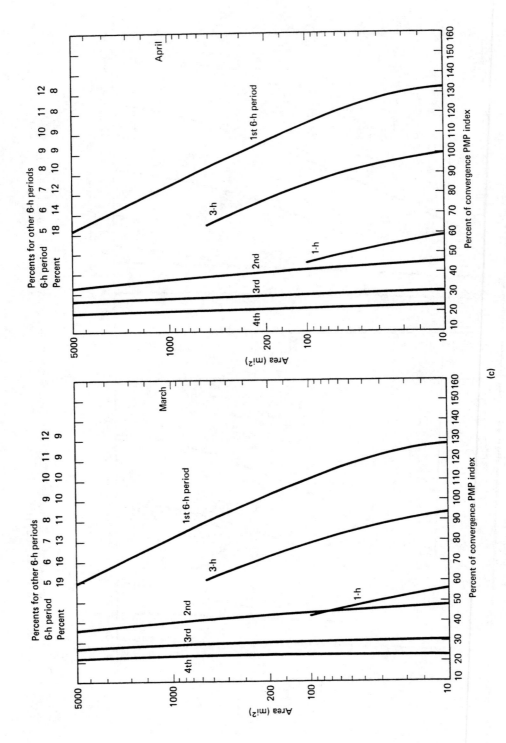

Figure 14-7(c) California: Variation of Convergence PMP with Basin Size and Duration, March and April [15].

(c)

increment, the 6-h increments are independent of basin size, as indicated in Fig. 14-7. The 1-h and 3-h convergence PMP percentages are also included in Fig. 14-7.

3. Use the percentages obtained in the previous step to determine the 1-h, 3-h, or the 6-h incremental restricted convergence PMP, on a monthly basis.

For durations in excess of 6-h, the 6-h increments are summed to obtain the PMP values for desired durations. For each duration, the all-season PMP is the highest monthly total. See [15] for additional temperature and wind criteria for determining snowmelt contribution to PMP.

Example 14-1.

Compute the PMP for a 1540-mi^2 basin located in the large Sierra Nevada slope basin, California, with coordinates 38° N, 119° 45′ W (adapted from [15]).

OROGRAPHIC PMP

1. From Fig. 14-3, the basin-average orographic PMP index is 4.86 in.
2. From Fig. 14-4, the representative basin width is measured perpendicular to the parallel sides of Area No. 26. Assume 30 mi for the basin of this example.
3. From Fig. 14-5, the basin-width adjustment factor is 100 percent.
4. The adjusted basin-average orographic PMP index is 4.86 × 100/100 = 4.86 in. Therefore, the 6-h January orographic PMP is 4.86 in.
5. From Table 14-1, the monthly 6-h orographic PMP values (Sierra Range), in inches, are: October, 4.71; November, 4.76; December, 4.81; January, 4.86; February, 4.86; March, 4.67; and April, 4.37.
6. From Table 14-2, for 38° N latitude, the 6-h incremental orographic PMP values are obtained, as shown in Table 14-3.
7. It is not necessary to compute 1-h and 3-h PMP values for this large-area basin.

CONVERGENCE PMP
(Restricted convergence PMP to be combined with orographic PMP)

1. From Fig. 14-6, the basin-average value of convergence PMP index is 2.26 in.
2. From Fig. 14-7, a table of 6-h PMP increments, in percent of convergence PMP index, for the months of October to April, is developed (see Table 14-4). It is not necessary to compute 1-h and 3-h PMP values for this large-area basin.

TABLE 14-3 6-H INCREMENTAL OROGRAPHIC PMP: EXAMPLE 14-1 (in.)

Increment	1	2	3	4	5	6	7	8	9	10	11	12
Month												
Oct	4.71	4.05	3.58	3.11	2.73	2.40	2.12	1.84	1.60	1.41	1.22	1.08
Nov	4.76	4.09	3.62	3.14	2.76	2.43	2.14	1.86	1.62	1.43	1.24	1.09
Dec	4.81	4.14	3.66	3.17	2.79	2.45	2.16	1.88	1.64	1.55	1.25	1.11
Jan	4.86	4.18	3.69	3.21	2.82	2.48	2.19	1.90	1.65	1.46	1.26	1.12
Feb	4.86	4.18	3.69	3.21	2.82	2.48	2.19	1.90	1.65	1.46	1.26	1.12
Mar	4.67	4.02	3.55	3.08	2.71	2.38	2.10	1.82	1.59	1.40	1.21	1.07
Apr	4.37	3.76	3.32	2.88	2.53	2.23	1.97	1.70	1.49	1.31	1.14	1.01

3. Table 14-5 shows the 6-h incremental restricted convergence PMP, in inches, calculated by multiplying the values in Table 14-4 by the convergence PMP index obtained in step 1 (2.26 in.).

Because of the pronounced orographic PMP component for this basin, it is not necessary to calculate the unrestricted convergence PMP. The total PMP for this basin is obtained by adding the values of Tables 14-3 and 14-5. Table 14-6 shows the 6-h incremental convergence and orographic PMP values. Table 14-7 shows the cumulative values of PMP for durations every 6-h interval. For each duration, the all-season PMP is the highest monthly total shown in Table 14-7.

Other PMP Estimates. HMR 51 and HMR 52 (United States east of the 105th Meridian) and HMR 36 (California) are typical of the methodologies used to obtain generalized PMP estimates. However, other HMR publications may differ from these, if not in principle, at least in detail. For a basin located in a given geographic region, an estimate of PMP should be based on the methodology outlined in the applicable HMR reference [15-30].

Time Distribution of PMP. In calculating the 6-h PMP increments, the first 6-h increment is interpreted as the first in magnitude rather that the first in sequence. The 6-h PMP increments are used to develop a critical storm pattern, i.e., a probable maximum storm. In turn, this probable maximum storm is used to calculate the PMF.

Critical storm patterns are obtained by ordering 6-h PMP increments based on the following sequencing rule: For each duration, the second-largest 6-h increment

TABLE 14-4 6-H CONVERGENCE PMP INCREMENTS: EXAMPLE 14-1
(in percent of convergence PMP index)

Increment	1	2	3	4	5	6	7	8	9	10	11	12
Month
Oct	94	30	20	16	14	12	11	10	9	9	8	8
Nov	85	38	26	20	16	14	13	11	10	10	10	10
Dec	79	37	26	21	18	16	13	12	11	11	10	10
Jan	74	40	28	23	19	16	14	13	12	11	11	11
Feb	74	40	28	23	19	16	14	13	12	11	11	11
Mar	76	39	28	22	19	16	13	11	10	10	9	9
Apr	78	36	27	21	18	14	12	10	9	9	8	8

TABLE 14-5 6-H INCREMENTAL RESTRICTED CONVERGENCE PMP: EXAMPLE 14-1 (in.)

Increment	1	2	3	4	5	6	7	8	9	10	11	12
Month
Oct	2.12	0.68	0.45	0.36	0.32	0.27	0.25	0.23	0.20	0.20	0.18	0.18
Nov	1.92	0.86	0.59	0.44	0.36	0.32	0.29	0.25	0.23	0.23	0.23	0.23
Dec	1.79	0.84	0.59	0.48	0.41	0.36	0.29	0.27	0.25	0.25	0.23	0.23
Jan	1.67	0.91	0.63	0.52	0.43	0.36	0.32	0.29	0.27	0.25	0.25	0.25
Feb	1.67	0.91	0.63	0.52	0.43	0.36	0.32	0.29	0.27	0.25	0.25	0.25
Mar	1.71	0.88	0.63	0.50	0.43	0.36	0.29	0.25	0.23	0.23	0.20	0.20
Apr	1.77	0.81	0.61	0.48	0.41	0.32	0.27	0.23	0.20	0.20	0.18	0.18

TABLE 14-6 6-H INCREMENTAL RESTRICTED CONVERGENCE PLUS OROGRAPHIC PMP: EXAMPLE 14-1 (in.)

Increment	1	2	3	4	5	6	7	8	9	10	11	12
Month												
Oct	6.83	4.73	4.03	3.47	3.05	2.67	2.37	2.07	1.80	1.61	1.40	1.26
Nov	6.68	4.95	4.21	3.58	3.12	2.75	2.43	2.11	1.85	1.66	1.47	1.32
Dec	6.60	4.98	4.25	3.65	3.20	2.81	2.45	2.15	1.89	1.80	1.48	1.34
Jan	6.53	5.09	4.32	3.73	3.25	2.84	2.51	2.19	1.92	1.71	1.51	1.37
Feb	6.53	5.09	4.32	3.73	3.25	2.84	2.51	2.19	1.92	1.71	1.51	1.37
Mar	6.38	4.90	4.18	3.58	3.14	2.74	2.39	2.07	1.82	1.63	1.41	1.27
Apr	6.14	4.57	3.93	3.36	2.94	2.55	2.24	1.93	1.69	1.51	1.32	1.19

TABLE 14-7 CUMULATIVE PMP FOR INDICATED DURATIONS: EXAMPLE 14-1 (PMP in inches and duration in hours)

Duration	6	12	18	24	30	36	42	48	54	60	66	72
Month												
Oct	6.8	11.6	15.6	19.1	22.1	24.8	27.1	29.2	31.0	32.6	34.0	35.3
Nov	6.7	11.6	15.8	19.4	22.5	25.3	27.7	29.8	31.7	33.3	34.8	36.1
Dec	6.6	11.6	15.8	19.5	22.7	25.5	27.9	30.1	32.0	33.8	35.3	36.6
Jan	6.5	11.6	15.9	19.7	22.9	25.8	28.3	30.5	32.4	34.1	35.6	37.0
Feb	6.5	11.6	15.9	19.7	22.9	25.8	28.3	30.5	32.4	34.1	35.6	37.0
Mar	6.4	11.3	15.5	19.0	22.2	24.9	27.3	29.4	31.2	32.8	34.2	35.5
Apr	6.1	10.7	14.6	18.0	20.9	23.5	25.7	27.7	29.3	30.9	32.2	33.4

must be adjacent to the largest in order to provide the most critical 12-h combination; the third-largest increment should be positioned immediately before or after this 12-h sequence in order to provide the most critical 18-h combination; and so on. Sample 24-h storm patterns following this critical sequencing rule are shown in Fig. 14-8(a) and (b).

The examination of 72-h hyetographs has shown that storms of this duration typically consist of two or more peaks or bursts. Time sequence patterns which show more than one peak are obtained by the following procedures:

1. Group the four largest 6-h increments of the 72-h PMP in a first 24-h sequence, the middle four increments in a second 24-h sequence, and the smallest four increments in a third 24-h sequence.

2. Within each of these three 24-h sequences, arrange the four increments in accordance with the critical sequencing rule, i.e., the second largest next to the largest, the third largest adjacent to these, and the fourth at either end.

3. Arrange the three 24-h sequences in accordance with the critical sequencing rule, that is, the second-largest 24-h period next to the largest, and the third at either end. Sample 72-h storm patterns obeying this critical sequencing rule are shown in Fig. 14-8(c) to (e). In practice, it may be necessary to experiment with several possible sequences in order to determine the most critical sequence for a particular basin.

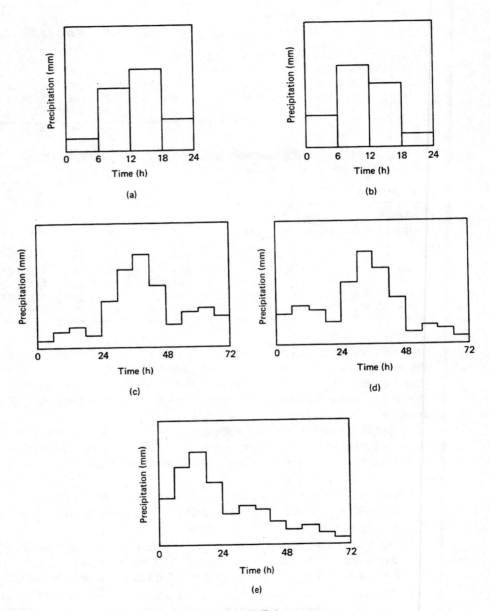

Figure 14-8 Sample PMP time sequences.

Hydrologic Design Criteria   Chap. 14

NWS Depth-Area Reduction Criteria

Several rainfall atlases containing isopluvial maps, applicable for storm durations up to 10 d, frequencies up to 100 y, and drainage areas up to 400 mi², have been published by the National Weather Service (see Table 13-1). These maps were developed by analyzing extensive point-rainfall measurements and, therefore, rainfall depths shown represent point-rainfall values.

Each atlas contains a chart of depth-area ratios designed to account for spatial averaging of rainfall depth with increasing basin area. These depth-area ratios allow the calculation of an areally averaged rainfall depth, given the basin area, storm duration, and point depth obtained from the appropriate isopluvial map. These ratios are referred to as *geographically fixed* depth-area ratios in order to distinguish them from *storm-centered* depth-area ratios, which are based on the morphological characteristics of individual storms [39].

Traditionally, the development of geographically fixed depth-area ratios has been based on empirical comparisons of point and areal rainfall at the same geographic location. Areal rainfall is approximated by averaging simultaneous gage measurements. This requirement of simultaneity limits the amount of usable data to that of recording gages. Therefore, depth-area ratios are developed based on data from a few existing networks of densely placed recording gages. Due to the sparseness of data, durational, frequency, and geographic variations of depth-area ratios cannot be readily ascertained. Therefore, the available data have been condensed into a single graph (based on a 2-y return period) applicable for depth-area reductions across the United States and for return periods up to 100 y. This graph, reproduced from NOAA Atlas 2 [31], is shown in Fig. 2-9(a).

A general methodology to compute regional depth-area ratios has been developed by the National Weather Service. In this method, the depth-area ratios are computed using the mean and standard deviations of the annual series of rainfall averages over circular areas. The method is summarized by the following formula:

$$DA(T, D, A) = \frac{\overline{x}'(D, A, n) + K(T, n)s'(D, A, n)C_v(D)}{1 + K(T, n)C_v(D)} \qquad (14\text{-}3)$$

in which DA = depth-area ratio; x' = relative mean; K = Gumbel frequency factor normalized for a 20-y record length; s' = relative standard deviation; C_v = variance coefficient; and A, D, n and T are area, duration, record length, and return period, respectively. The statistics used in Eq. 14-3 are obtained using procedures described in [32].

A depth-area reduction chart for Chicago based on Eq. 14-3 is shown in Fig. 14-9. The close resemblance between Figs. 2-9(a) and 14-9 is attributed to the fact that the Chicago data had a prominent role in the development of Fig. 2-9(a). Other geographic regions are likely to have different depth-area ratio patterns.

Figure 14-9(a) and (b) shows depth-area ratios decreasing with increasing area and return period, and decreasing with duration. The effect of return period on depth-area ratio is not accounted for by Fig. 2-9(a). In addition, Eq. 14-3 can be used to develop depth-area ratios for areas larger than those depicted in Fig. 2-9 (a). For example, a depth-area reduction chart for Chicago, applicable to areas up to 5000 mi², is shown in Fig. 14-10.

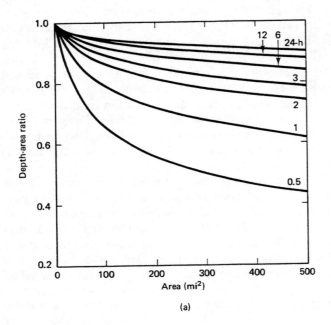

Figure 14-9(a) 2-y depth-area ratios for Chicago, Illinois [32].

(a)

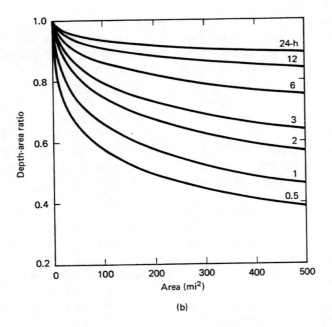

Figure 14-9(b) 100-y depth-area ratios for Chicago, Illinois [32].

(b)

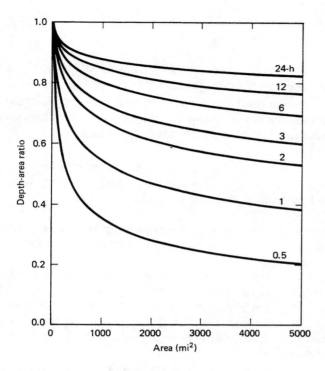

Figure 14-10 2-y depth-area ratios for Chicago, Illinois, for areas up to 5000 mi² [32].

14.2 U.S. ARMY CORPS OF ENGINEERS

Standard Project Flood Determinations

Hydrologic design criteria used by the U.S. Army Corps of Engineers are outlined in *Civil Engineer Bulletin No.* 52-8: "Standard Project Flood Determinations," revised March 1965 [45].

A *standard project storm* (SPS) for a particular drainage area and season of the year is the most severe flood-producing rainfall depth-area-duration relationship and related isohyetal pattern that is reasonably characteristic for the region. The *standard project flood* (SPF) is the flood hydrograph derived from the SPS. For areas where snowmelt may contribute a substantial volume to runoff, appropriate allowances are made to increase the value of the SPS. When floods are primarily caused by snowmelt, the calculation of SPF is based on estimates of the most critical combination of snow, temperature, and hydrologic abstractions.

Types of Flood Estimates.   Flood magnitudes are governed by a combination of several factors, among them, the quantity, intensity, temporal, and spatial distribution of precipitation, the infiltration capacity of the soil mantle, and the natural and artificial storage effects. When relatively long periods of streamflow records are available, statistical analyses can be used to develop flood estimates associated with frequencies bearing a reasonable relationship to the record length. However, for large projects, it is necessary to supplement statistical methods with hypothetical design flood estimates based on rainfall-runoff analysis.

Sec. 14.2   U.S. Army Corps of Engineers   **477**

Three types of flood estimates are used in flood-control planning and design investigations:

1. Statistical analysis of streamflow records, including individual-station flood-frequency estimates and regional flood-frequency analysis.
2. SPF estimates, which represent flood discharges that are expected to be caused by the most severe combination of meteorologic and hydrologic conditions that are considered to be reasonably characteristic of the region, excluding extremely rare combinations.
3. PMF estimates, which represent flood discharges that are expected to be caused by the most severe combination of critical meteorologic and hydrologic conditions that are reasonably possible in the region.

Statistical flood determinations are useful in project investigations, primarily as a basis for estimating the mean annual benefits that may be expected to accrue from the control or reduction of floods of relatively common occurrence.

An SPF serves the following purposes: (1) it represents a standard against which the selected degree of protection may be judged and compared with the protection provided at similar projects in different localities and (2) it represents the flood discharge that should be selected as the design flood for the project, where some small risk can be tolerated, but where an unusually high degree of protection is justified because of the risk to life and property.

A PMF is applicable to projects calling for a virtual elimination of the risk of failure. PMF applications are typically in the sizing of spillways for large dams or dams located immediately upstream of heavily populated areas.

Design Flood. The term *design flood* refers to the flood hydrograph or peak discharge value that is finally adopted as the basis for engineering design, after giving due consideration to flood characteristics, flood frequency, and flood damage potential, including economic and other related factors.

The selected design flood may be either greater or smaller than the SPF. However, in Corps of Engineers' practice, the SPF is intended to be a practical expression of the degree of protection sought in the design of flood control works. Since SPF estimates are based on generalized studies of meteorologic and hydrologic conditions, they provide a basis for comparing the degree of protection afforded by projects in different localities.

Generalized SPS Estimates for Small Basins. The procedures for obtaining generalized SPS estimates described in this section are applicable to areas east of the 105th meridian, and primarily to small basins (i.e., those less than 1000 mi^2, considered as small basins in Corps of Engineers' practice). They are based on data from storms that have occurred primarily in the spring, summer, and fall seasons, during which convective activity is prominent, and are not generally applicable to snow seasons without special adjustments.

Figure 14-11 shows generalized SPS estimates corresponding to a 24-h duration and a 200-mi^2 area, obtained by reducing PMP isohyets by 50 percent and reshaping them in certain regions to conform with supplementary studies of rainfall characteris-

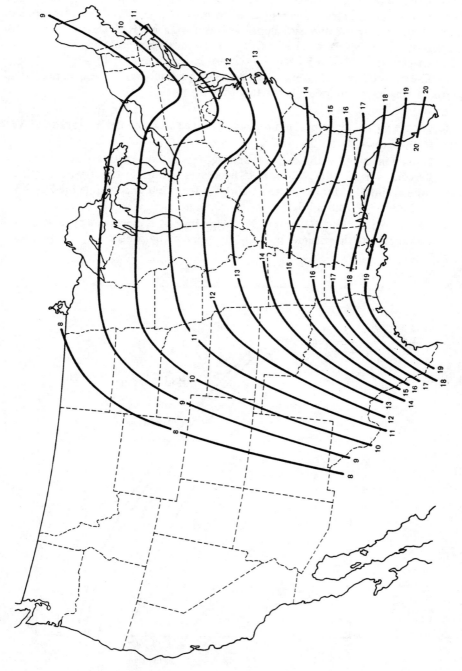

Figure 14-11 SPS index rainfall (in.) [45].

tics . These SPS estimates are approximately 40 to 60 percent of the associated PMP estimates.

The isohyetal map shown in Fig. 14-11 is termed *SPS index rainfall* in order to allow conversion to storms covering areas from 10 to 20,000 mi² and durations other than 24 h. The applicable SPS depth-area-duration chart is shown in Fig. 14-12. Criteria for the subdivision of 24-h SPS rainfall into 6-h increments is shown in Fig. 14-13. A standard 96-h SPS isohyetal pattern is shown in Fig. 14-14.

The procedure to develop an SPS estimate for small basins (less than 1000 mi² in Corps of Engineers' practice) is the following:

1. Locate the drainage basin in the map of Fig. 14-11 and determine the SPS index rainfall (inches).
2. Enter Fig. 14-12 with the basin area to obtain the SPS index-rainfall ratios (in percent) for 24-, 48-, 72- and 96-h periods. Multiply these ratios by the SPS index rainfall obtained in step 1 and divide by 100 to calculate the 24-, 48-, 72- and 96-h SPS rainfall depths.
3. The 24-h SPS rainfall depth is the first of four 24-h increments in a 96-h period. Calculate the three remaining 24-h increments by subtracting the 24-, 48-, 72-

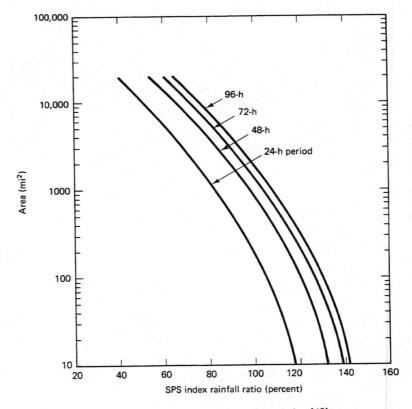

Figure 14-12 SPS depth-area-duration relation [45].

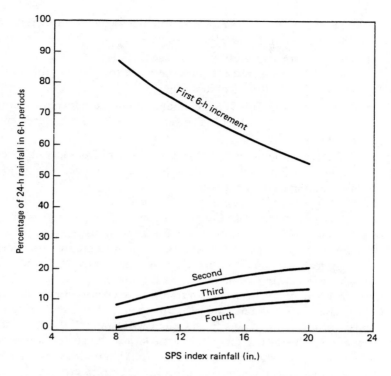

Figure 14-13 Time distribution of 24-h SPS rainfall [45].

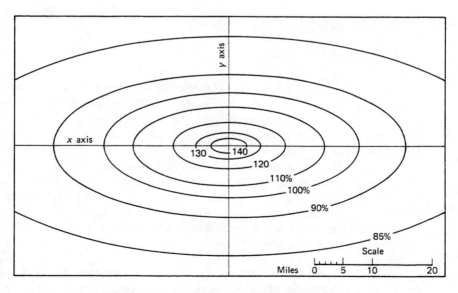

Figure 14-14 Standard 96-h SPS isohyetal pattern [45].

and 96-h SPS values. For instance, the second 24-h increment is equal to the 48-h depth minus the 24-h depth, and so on.

4. Based on an appraisal of hydrologic conditions within the basin, arrange the four 24-h SPS increments in a sequence that is favorable to the production of critical runoff at project locations.

5. Subdivide each 24-h SPS increment into four 6-h increments in accordance with the criteria of Fig. 14-13. The same sequence of 6-h increments is assumed for each day of the SPS.

6. Subtract estimates of hydrologic abstractions from the 6-h incremental SPS values obtained in step 5 to calculate the effective storm hyetograph to be used in the computation of the SPF.

SPS Estimates for Large Basins. The basic principles involved in the preparation of SPS and SPF estimates for large drainage basins (i.e., those in excess of 1000 mi^2, considered as large basins in Corps of Engineers' practice) are the same as those applicable to small basins. However, generalization of criteria becomes more difficult as the basin size increases. SPF estimates for small basins are usually governed by the maximum 6-h or 12-h rainfall associated with severe thunderstorms. For large basins, SPF estimates are generally the result of a succession of distinct rainfall events. Although the intensity and quantity of rainfall are important factors in the production of floods in a large basin, the location of successive increments of rainfall and the synchronization of intense bursts of rainfall with the progression of runoff are of equal or greater importance. Accordingly, an SPS estimate for a large basin must be based on a review of relevant meteorological data and an assessment of the hydrologic response characteristics of the basin, including the study of major floods and related historical accounts.

14.3 USDA SOIL CONSERVATION SERVICE

Spillway Design Criteria

USDA Soil Conservation Service design criteria for spillway discharges and floodwater storage volumes are described in *Technical Release No. 60: "Earth Dams and Reservoirs" (TR-60)*, revised October 1985 [47]. Additional procedures for hydrologic design of spillways are included in SCS *National Engineering Handbook*, Section No. 4: Hydrology (NEH-4), 1985 [48].

A *spillway* is an open channel, conduit, or drop structure used to convey water from a reservoir. It may have gates, either manually or automatically controlled, to regulate the flow of water through the spillway. SCS classifies spillways as: (1) principal spillways and (2) emergency spillways. The *principal spillway* is the lowest ungated spillway designed to convey water from the reservoir at predetermined release rates. The *emergency spillway* is the spillway designed to convey excess water through, over, or around a dam.

An earth spillway is an unlined spillway. A vegetated spillway is an open channel flow spillway lined with vegetative materials. A *ramp spillway* is a vegetated spillway constructed over an earth dam in such a way that the spillway is part of the embankment.

The following three elevations are used in spillway design: (1) the emergency-spillway crest elevation, (2) the maximum design-pool elevation, and (3) the minimum dam-crest elevation (after proper allowance for settling of the embankment). The emergency-spillway crest elevation is below the maximum design-pool elevation. In turn, the maximum design-pool elevation is below the minimum dam-crest elevation (Fig. 14-15).

Storage volume is the volume of the reservoir measured up to the emergency-spillway crest elevation. *Retarding pool storage* is the fraction of storage volume which is allocated to the temporary impoundment of flood waters. The emergency-spillway crest elevation is the upper limit of retarding pool storage. *Surcharge storage* is the portion of the reservoir located between emergency-spillway crest elevation and maximum design-pool elevation. *Freeboard* is the difference between minimum dam-crest elevation and maximum design-pool elevation (Fig. 14-15).

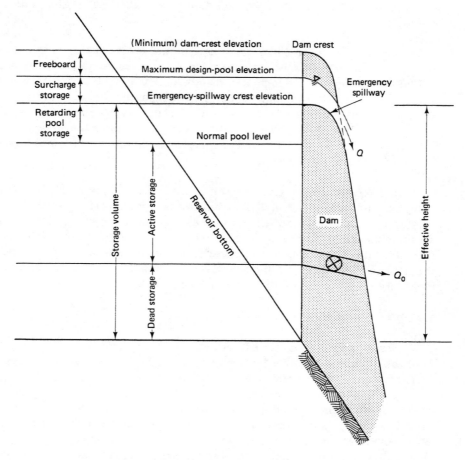

Figure 14-15 Elevation and storage features in SCS reservoirs.

The effective height of a dam is the difference between the emergency-spillway crest elevation and the lowest point in the original cross section on the dam centerline. When there is no open-channel emergency spillway, the dam crest is substituted for the emergency spillway crest in the definition of effective height (Fig. 14-15).

The *principal spillway hydrograph* (PSH) is used to determine the capacity of the principal spillway, the emergency-spillway crest elevation, and the volume of retarding pool storage. The *emergency spillway hydrograph* (ESH) is used to establish the capacity of the emergency spillway, the maximum design-pool elevation, and the volume of surcharge storage. The *freeboard hydrograph* (FBH) is used to set the minimum dam-crest elevation and to evaluate the structural integrity of the spillway system.

Quick-return flow is the gradually receding discharge directly associated with a specific storm, which occurs after surface runoff has reached its maximum. Quick-return flow consists of baseflow, interflow, and delayed surface runoff. *Design life* of a dam is the period of time during which it is designed to perform its assigned functions satisfactorily.

Dam Classification. The Soil Conservation Service classifies dams into the following three classes:

Class (a): Dams located in rural or agricultural areas, where failure may damage farm buildings, agricultural land, or township and country roads.

Class (b): Dams located in predominantly rural or agricultural areas, where failure may damage isolated homes, main highways, or minor railroads or cause interruption of use or service of relatively important public utilities.

Class (c): Dams located where failure may cause loss of life, serious damage to homes, industrial and commercial buildings, important public utilities, main highways, or railroads.

In determining dam classification, consideration is given to the assessment of potential damage in the event of dam failure due to either (1) hydrologic reasons (dam overtopping caused by insufficient emergency spillway capacity), (2) structural reasons (embankment piping), or (3) geologic reasons (landslide-generated waves). The stability of the dam and spillway structure, the physical characteristics of the site and the downstream valley, and the proximity to downstream industrial and residential areas all have a bearing on the analysis leading to dam classification.

Design Criteria for Principal Spillways

Precipitation and Runoff Volume. Precipitation data are obtained from applicable National Weather Service references. Frequency-based references are given in Table 13-1. PMP references are given in Table 13-2 and [15-30].

The return period to be selected for design is a function of dam class, purpose, relative size (product of storage volume and effective height), location with respect to other existing or planned upstream dams, and type of emergency spillway, either earth or vegetated. Minimum precipitation depths used to calculate principal spillway hydrographs are shown in Table 14-8. Minimum depth-area ratios for 1- and 10-d durations and areas from 10 to 100 mi^2 are shown in Table 14-9.

TABLE 14-8 MINIMUM DESIGN CRITERIA FOR PRINCIPAL SPILLWAYS [47]

Class of Dam	Purpose of Dam	$V_s H_e^1$	Existing or Planned Upstream	Precipitation Data for Maximum Frequency[2] of Use of Emergency Spillway Type:	
				Earth	Vegetated
(a)	Single[3] irrigation only	Less than 30,000	None	0.5DL[4]	0.5DL
		Greater than 30,000	None	0.75DL	0.75DL
	Single or multiple[5]	Less than 30,000	None	P_{50}	P_{25}^6
		Greater than 30,000	None	$0.5(P_{50} + P_{100})$	$0.5(P_{25} + P_{50})$
		All	Any[7]	P_{100}	P_{50}
(b)	Single or multiple	All	None or any	P_{100}	P_{50}
(c)	Single or multiple	All	None or any	P_{100}	P_{100}

[1]Product of reservoir storage volume V_s (acre-feet) times effective height of dam H_e (feet).

[2]Precipitation depths for indicated return periods (years). In some areas direct runoff amounts are determined using Figs. 14-16 and 14-17 or procedures described in Chapter 21 of NEH-4 [48].

[3]Applies to irrigation dams on ephemeral streams in areas where mean annual rainfall is less than 25 in.

[4]DL = design life (years).

[5]Class (a) dams involving industrial or municipal water are to use minimum criteria equivalent to that of class (b).

[6]In the case of a ramp spillway, the minimum criteria should be increased from P_{25} to P_{100}.

[7]Applies when the failure of the upstream dam may endanger the lower dam.

The runoff volume can be estimated by one or more procedures. These include:

1. The SCS runoff curve number method (Section 5.1). AMC II or greater should be chosen, unless otherwise justified by a special study. For a 10-d storm, a reduction in runoff curve number is indicated, as shown in Table 14-10.

2. The use of Fig. 14-16. Figures 14-16(a) and (b) show 100-y, 10-d runoff (inches), including appropriate conversion factors for 25- and 50-y return periods. Figures 14-16(c) and (d) show Q_1/Q_{10} ratios, allowing the conversion of 10-d runoff to 1-d runoff.

The quick-return flow, to be added to the direct runoff hydrograph, is obtained from either Fig. 14-17 or from Table 14-11 as a function of climatic index

$$C_i = \frac{100P_a}{T_a^2} \tag{14-4}$$

TABLE 14-9 PRINCIPAL SPILLWAY DESIGN
CRITERIA: MINIMUM DEPTH-AREA RATIOS [47]

Area (mi²)	Depth-Area Ratios for Given Duration	
	1-d	10-d
10	1.000	1.000
15	0.977	0.991
20	0.969	0.987
25	0.965	0.983
30	0.961	0.981
35	0.957	0.979
40	0.954	0.977
45	0.951	0.976
50	0.948	0.974
60	0.944	0.972
70	0.940	0.970
80	0.937	0.969
90	0.935	0.967
100	0.932	0.966

TABLE 14-10 PRINCIPAL SPILLWAY DESIGN CRITERIA: 10-d RUNOFF
CURVE NUMBER ADJUSTMENT[1] [47]

1-d	10-d	1-d	10-d	1-d	10-d
100	100	80	65	60	41
99	98	79	64	59	40
98	96	78	62	58	39
97	94	77	61	57	38
96	92	76	60	56	37
95	90	75	58	55	36
94	88	74	57	54	35
93	86	73	56	53	34
92	84	72	54	52	33
91	82	71	53	51	33
90	81	70	52	50	32
89	79	69	51	49	31
88	77	68	50	48	30
87	76	67	49	47	29
86	74	66	47	46	28
85	72	65	46	45	28
84	71	64	45	44	27
83	69	63	44	43	26
82	68	62	43	42	25
81	66	61	42	41	24

[1]This table is used only if the 100-y, 10-d point rainfall is 6 in. or more. If the 100-y,
10-d point rainfall is less than 6 in., the 10-d *CN* is the same as the 1-d CN.

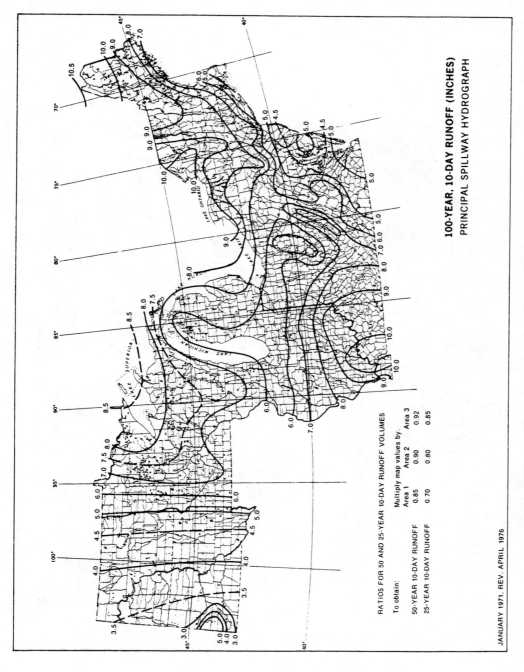

Figure 14-16(a) SCS principal spillway hydrograph: 100-y 10-d runoff (in.) [47].

100-YEAR, 10-DAY RUNOFF (INCHES)
PRINCIPAL SPILLWAY HYDROGRAPH

RATIOS FOR 50 AND 25-YEAR 10-DAY RUNOFF VOLUMES

To obtain:

| | Multiply map values by: | | |
	Area 1	Area 2	Area 3
50-YEAR 10-DAY RUNOFF	0.85	0.90	0.92
25-YEAR 10-DAY RUNOFF	0.70	0.80	0.85

JANUARY 1971, REV. APRIL 1976

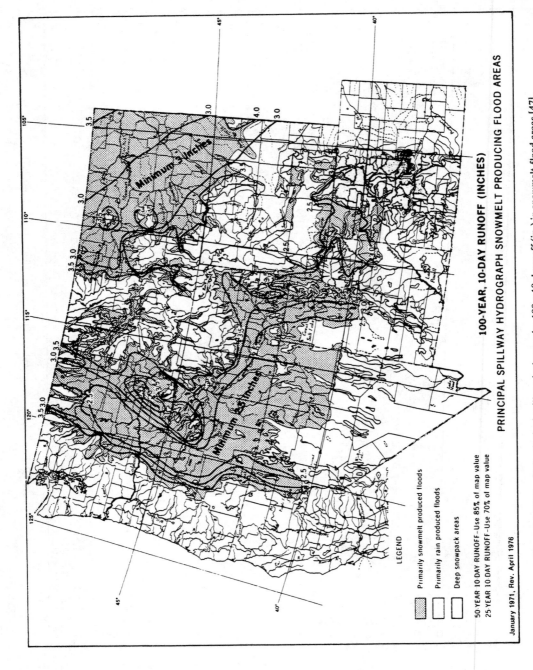

Figure 14-16(b) SCS principal spillway hydrograph: 100-y 10-d runoff (in.) in snowmelt-flood areas [47].

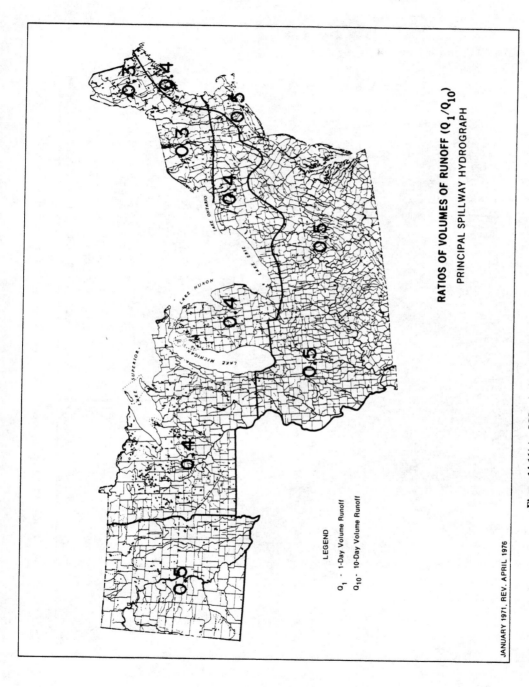

Figure 14-16(c) SCS principal spillway hydrograph: Q_1/Q_{10} ratios [47].

RATIOS OF VOLUMES OF RUNOFF (Q_1/Q_{10})
PRINCIPAL SPILLWAY HYDROGRAPH

LEGEND

Q_1 - 1-Day Volume Runoff

Q_{10} - 10-Day Volume Runoff

JANUARY 1971, REV. APRIL 1976

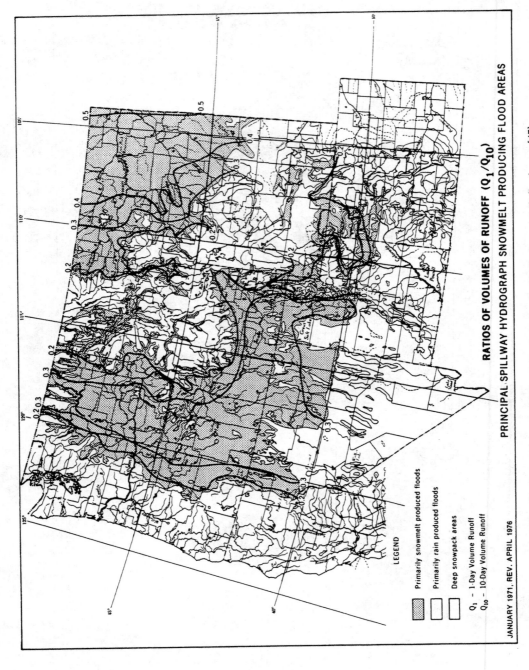

RATIOS OF VOLUMES OF RUNOFF (Q_1/Q_{10})

PRINCIPAL SPILLWAY HYDROGRAPH SNOWMELT PRODUCING FLOOD AREAS

LEGEND

Primarily snowmelt produced floods

Primarily rain produced floods

Deep snowpack areas

Q_1 – 1-Day Volume Runoff
Q_{10} – 10-Day Volume Runoff

JANUARY 1971, REV. APRIL 1976

Figure 14-16(d) SCS principal spillway hydrograph: Q_1/Q_{10} in snowmelt-flood areas [47].

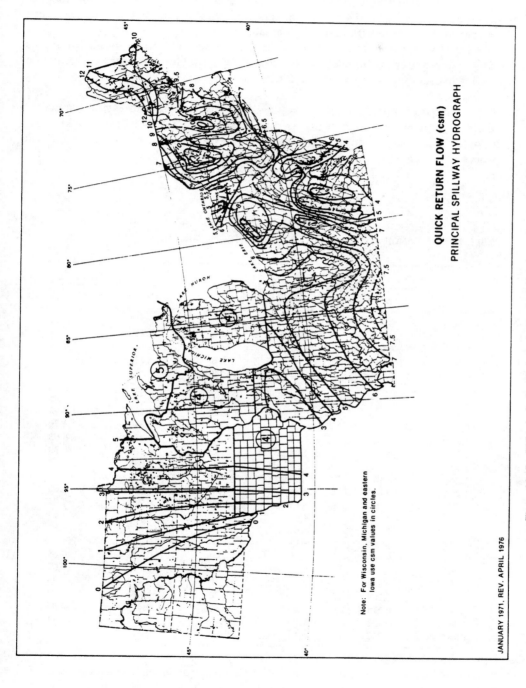

Figure 14-17 SCS principal spillway hydrograph: Quick-return flow (csm) [47].

in which C_i = climatic index, P_a = mean annual precipitation in inches, and T_a = mean annual temperature in degrees Fahrenheit. The values in Fig. 14-17 and Table 14-11 represent quick-return flow per unit area, in csm units (1 csm = 1 cubic foot per second per square mile). To obtain the quick-return flow, multiply the quick-return flow per unit area times the watershed area.

Given several estimates of runoff volume, the selected design runoff volume is that which requires the higher emergency-spillway crest elevation when the principal spillway hydrograph is routed through the structure.

Principal Spillway Hydrograph (PSH). NEH-4 procedures and the hydrologic computer program TR-20 are used to develop the PSH. For basins lacking hydrologic homegeneity, a subdivision into hydrologically homogeneous subbasins and associated stream channel routing may be necessary.

TABLE 14-11 PRINCIPAL SPILLWAY DESIGN CRITERIA: MINIMUM QUICK-RETURN FLOW [47]

C_i^a	QRF[b]		C_i	QRF	
	(in./d)	$(ft^3/s)/mi^2$		(in./d)	$(ft^3/s)/mi^2$
1.00	0.000	0.00	1.50	0.233	6.28
1.02	0.011	0.30	1.52	0.239	6.42
1.04	0.022	0.60	1.54	0.244	6.56
1.06	0.033	0.90	1.56	0.249	6.70
1.08	0.045	1.20	1.58	0.254	6.83
1.10	0.056	1.50	1.60	0.259	6.95
1.12	0.067	1.80	2.65	0.270	7.26
1.14	0.078	2.10	1.70	0.280	7.53
1.16	0.089	2.40	1.75	0.290	7.79
1.18	0.100	2.70	1.80	0.299	8.05
1.20	0.114	3.00	1.85	0.309	8.30
1.22	0.122	3.29	1.90	0.318	8.54
1.24	0.133	3.58	1.95	0.326	8.77
1.26	0.144	3.86	2.00	0.335	9.00
1.28	0.153	4.14	2.05	0.343	9.22
1.30	0.163	4.37	2.10	0.351	9.44
1.32	0.171	4.61	2.20	0.367	9.86
1.34	0.180	4.83	2.30	0.382	10.26
1.36	0.188	5.05	2.40	0.396	10.65
1.38	0.195	5.25	2.50	0.410	11.02
1.40	0.202	5.44	2.60	0.423	11.38
1.42	0.209	5.63	2.70	0.436	11.73
1.44	0.216	5.80	2.80	0.449	12.07
1.46	0.222	5.97	2.90	0.461	12.41
1.48	0.228	6.13	3.00[c]	0.473	12.73

[a]C_i = climatic index, Eq. 14-4.

[b]QRF = quick-return flow, in inches per day or cubic feet per second per square mile (csm).

[c]For C_i greater than 3, use QRF $[(ft^3/s)/mi^2] = 9 (C_i - 1)^{1/2}$.

Design Criteria for Emergency Spillways

Precipitation and Runoff Volume.   As in the case of principal spillways, precipitation data are obtained from the applicable National Weather Service reference. Figure 14-18 can be used to establish the 6-h, 100-y point precipitation (for drainage areas less than or equal to 10 mi^2). Figure 14-19 can be used to determine the 6-h point PMP depth. Figure 14-20 shows the regional applicability of HMR publications.

The return period to be selected for design is a function of dam class, purpose, relative size (product of storage volume and effective height), location with respect to other existing or planned upstream dams, and type of hydrograph, either (a) emergency spillway hydrograph or (b) freeboard hydrograph. Minimum precipitation depths used to calculate emergency spillway hydrographs are shown in Table 14-12. These precipitation depths range between a minimum value corresponding to the 100-y frequency and a maximum value corresponding to the PMP.

Depth-area reductions and temporal storm distributions are performed in accordance with the appropriate NWS references. In areas where NWS references do not contain an applicable procedure, depth-area ratios can be obtained from Fig. 14-21(a).

The minimum storm duration to be used in the design of emergency spillways is 6 h. Where the time of concentration exceeds 6 h, the storm duration is made equal to the time of concentration. Precipitation values are obtained from the appropriate NWS references. In areas where NWS references do not contain an applicable procedure, 100-y and PMP rainfall depths for durations up to 48-h (based on the 6-h depths) can be obtained from Fig. 14-21(b).

For locations where NWS references provide estimates of local (thunderstorm) and general storm PMP values, the design storm duration and temporal distribution are those resulting in the maximum design-pool elevation when the emergency spillway hydrograph is routed through the structure. For 6-h durations, unless a specific temporal distribution is recommended in a NWS reference, the chosen temporal distribution should closely approximate that shown in Fig. 14-21(c).

The runoff curve number method is used to determine runoff volumes, based on either AMC II or greater. The curve numbers are applicable for all storm durations.

Emergency Spillway Hydrograph (ESH).   As in the case of principal spillways, NEH-4 procedures and the hydrologic computer program TR-20 (Section 13.4) are used to develop the ESH. For basins lacking hydrologic homogeneity, a subdivision into hydrologically homogeneous subbasins and associated stream channel routing may be warranted.

Special Procedures for Large Areas

For dams with contributing drainage areas in excess of 50 mi^2, considered as large in SCS practice, it may be desirable to subdivide the drainage area into hydrologically homogeneous subbasins. Generally, a subbasin area should not exceed 20 mi^2 (50 km^2). The TR-20 computer model or a similar catchment model may be used for stream channel routing and hydrograph combination.

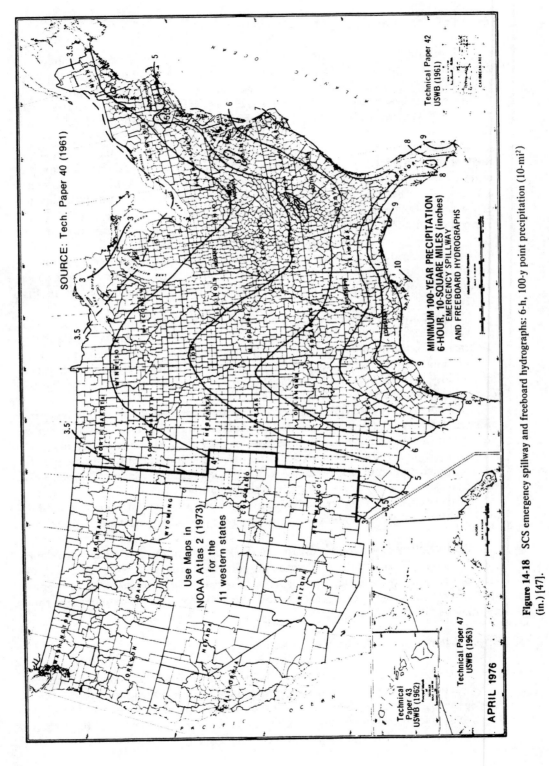

Figure 14-18 SCS emergency spillway and freeboard hydrographs: 6-h, 100-y point precipitation (in.) [47].

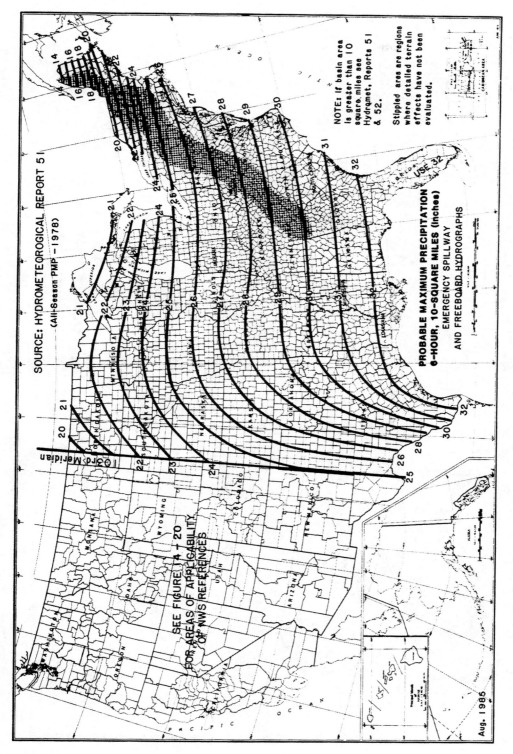

Figure 14-19 SCS emergency spillway and freeboard hydrographs: 6-h point PMP (10-mi²) (in.) [47].

495

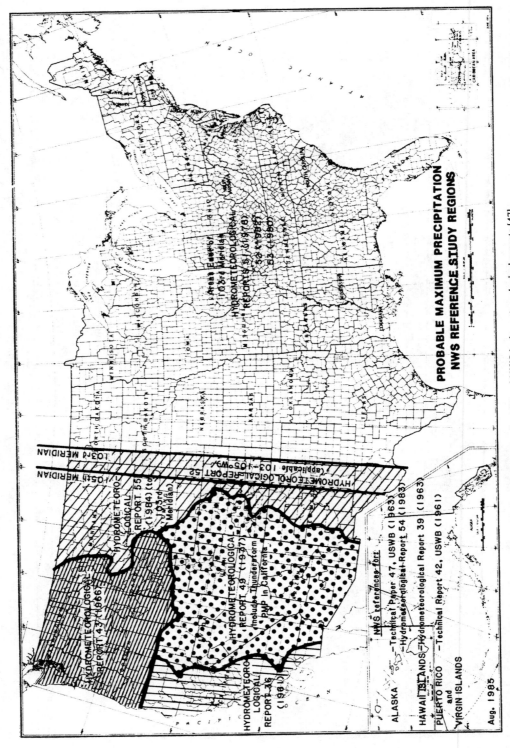

Figure 14-20 Regional applicability of NWS hydrometeorological reports [47].

TABLE 14-12 MINIMUM DESIGN CRITERIA FOR EMERGENCY SPILLWAYS [47]

Class of Dam	$V_s H_e^1$	Existing or Planned Upstream Dams	Precipitation Data[2] for Emergency Spillway Hydrograph	Freeboard Hydrograph
(a)[3]	Less than 30,000	None	P_{100}	$P_{100} + 0.12(\text{PMP} - P_{100})$
	Greater than 30,000	None	$P_{100} + 0.06(\text{PMP} - P_{100})$	$P_{100} + 0.26(\text{PMP} - P_{100})$
	All	Any[4]	$P_{100} + 0.12(\text{PMP} - P_{100})$	$P_{100} + 0.40(\text{PMP} - P_{100})$
(b)	All	None or any	$P_{100} + 0.12(\text{PMP} - P_{100})$	$P_{100} + 0.40 (\text{PMP} - P_{100})$
(c)	All	None or any	$P_{100} + 0.26(\text{PMP} - P_{100})$	PMP

[1]Product of reservoir storage volume V_s (acre-feet) times effective height of dam H_e (feet).
[2]Precipitation depths for either 100-y return period (P_{100}) or PMP.
[3]Class (a) dams involving industrial or municipal water are to use minimum criteria equivalent to that of class (b).
[4]Applies when the failure of the upstream dam may endanger the lower dam.

Spatial and temporal precipitation patterns may vary widely in a large basin. Topographic and meteorological factors such as basin shape, orientation, mean elevation, and storm movement may influence the overall storm distribution. Special studies may be warranted for basins greater than 100 mi² (250 km²) or experiencing significant snowmelt runoff.

14.4 U.S. BUREAU OF RECLAMATION

USBR Synthetic Unit Hydrographs

U.S. Bureau of Reclamation (USBR) procedures for the calculation of synthetic unit hydrographs are described in *Design of Small Dams*, third edition, 1987 [46]. In accordance with USBR's jurisdiction, the methodologies described in *Design of Small Dams* are primarily intended for use in the 11 western United States (Montana, Wyoming, Colorado, New Mexico, Idaho, Utah, Nevada, Arizona, Washington, Oregon, and California).

Unit Hydrograph Lag Time. The definitions of lag time (i.e., the catchment or basin lag) is similar to that of Eq. 5-16:

$$L_g = C \left[\frac{LL_c}{S^{1/2}} \right]^N \tag{14-5}$$

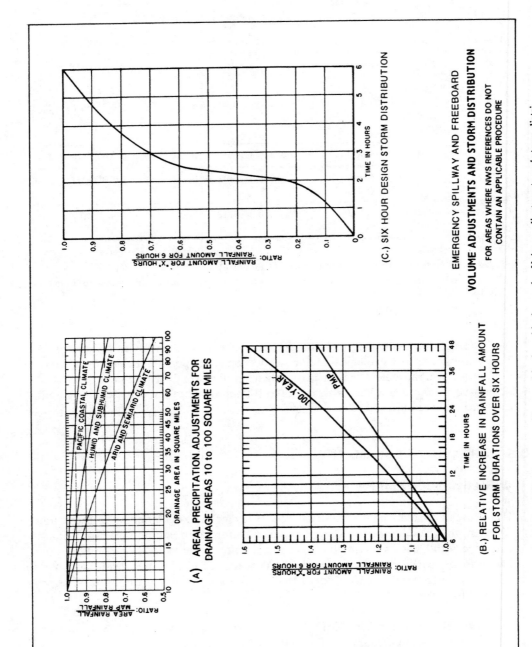

(C.) SIX HOUR DESIGN STORM DISTRIBUTION

RATIO: RAINFALL AMOUNT FOR 'X' HOURS / RAINFALL AMOUNT FOR 6 HOURS

(A) AREAL PRECIPITATION ADJUSTMENTS FOR DRAINAGE AREAS 10 to 100 SQUARE MILES

RATIO: AREA RAINFALL / MAP RAINFALL

PACIFIC COASTAL CLIMATE

HUMID AND SUBHUMID CLIMATE

ARID AND SEMIARID CLIMATE

DRAINAGE AREA IN SQUARE MILES

(B.) RELATIVE INCREASE IN RAINFALL AMOUNT FOR STORM DURATIONS OVER SIX HOURS

RATIO: RAINFALL AMOUNT FOR 'X' HOURS / RAINFALL AMOUNT FOR 6 HOURS

PMP

100-YEAR

TIME IN HOURS

EMERGENCY SPILLWAY AND FREEBOARD
VOLUME ADJUSTMENTS AND STORM DISTRIBUTION

FOR AREAS WHERE NWS REFERENCES DO NOT
CONTAIN AN APPLICABLE PROCEDURE

Figure 14-21 SCS emergency spillway and freeboard hydrographs: Volume adjustments and storm distribution [47].

in which L_g = unit hydrograph lag time in hours; L = distance measured along the longest watercourse, from basin outlet to divide in miles; L_c = distance measured along the longest watercourse, from basin outlet to a point closest to the basin centroid in miles; S = overall slope of the longest watercourse in feet per mile; C = coefficient; and N = exponent. The quantity $(LL_c)/S^{1/2}$ is referred to as the *basin factor*.

Experimental evidence has shown that the exponent N is approximately equal to 0.33. Furthermore, the coefficient C is equal to $26K_n$, with K_n being the average Manning n which is representative of the hydraulic and frictional characteristics of the drainage network.

Two procedures for unit hydrograph computations are described in *Design of Small Dams*. They are (1) the dimensionless unit hydrograph and (2) the dimensionless S-graph. The lag time is defined differently for each of these two procedures. For the dimensionless unit hydrograph, lag time is defined as the time elapsed from the centroid of effective rainfall to the centroid of runoff (i.e., $T3$ in Fig. 5-6). For the dimensionless S-graph, lag time is defined as the time elapsed from the start of effective rainfall to the time when runoff reaches 50 percent of the *ultimate discharge*. Ultimate discharge is the maximum discharge under a constant effective rainfall intensity achieved at the basin outlet when the entire basin is contributing to runoff (see runoff concentration, Section 2.4).

Unit Hydrograph Duration. The unit hydrograph duration is approximately equal to $\frac{2}{11}$ of the lag time (Snyder's method, see Eq. 5-24). The adopted value of unit hydrograph duration is rounded to the closest of the following: 5, 10, 15, 30 min, or 1, 2, or 6 h. For durations greater than 6 h, it may be necessary to subdivide the basin into subbasins and to develop a unit hydrograph for each subbasin. Hydrographs for each subbasin are then routed through the channel network, combined at confluences, and expressed at the basin outlet.

USBR Synthetic Unit Hydrograph Parameters

In USBR practice, hydrograph development is accomplished by using a regional value of parameter K_n. The analysis of 162 flood hydrographs, originating primarily in surface runoff as opposed to interflow or snowmelt, led to the range of K_n and corresponding C values for the six regions shown in Table 14-13. Within each region, a higher K_n indicates slower basin response; conversely, a lower K_n indicates faster basin response.

Region I applies to areas of the Great Plains west of the Mississippi River and east of the Rocky Mountains. The lower limit of K_n (0.030) reflects a well-defined drainage network with channels extending close to the basin boundary. The upper limit (0.069) generally reflects basins with substantial overland flow.

Region II applies to the Rocky Mountain region, including the Front, Sangre de Cristo, San Juan, Wasatch, Big Horn, Absoroka, Wind River, and Bitteroot ranges of New Mexico, Colorado, Wyoming, Utah, Idaho, Oregon, and Montana. The analysis does not include data from basins located at the higher elevations of these mountain ranges. Examination of the available data led to the conclusion that they represent two types of storms: (1) general (low-intensity) storms and (2) local storms (i.e., high-intensity thunderstorms). For general storms, low values of K_n (0.160 or less) are used

TABLE 14-13   USBR SYNTHETIC UNIT HYDROGRAPH PARAMETERS [46]

Region	Description	K_n	C
I	Great Plains (west of the Mississippi River and east of the Rocky Mountains)	0.030–0.069	0.78–1.80
II	Rocky Mountains (in New Mexico, Colorado, Wyoming, Utah, Idaho, Oregon, and Montana)		
	(a) General-storm hydrographs	0.130–0.260	3.40–6.80
	(b) Local-storm hydrographs	0.050–0.073	1.30–1.90
III	Southwestern desert, Great Basin and Colorado plateau (of southern California, Nevada, Utah, Arizona, and western Colorado and New Mexico)		
	(a) Rural basins	0.042–0.070	1.10–1.80
	(b) Partially urbanized basins	0.031	0.81
IV	Sierra Nevada (California)	0.064–0.150	1.65–3.90
V	Coast and Cascade ranges (California, Oregon, and Washington)	0.080–0.150	2.10–3.90
VI	Urban basins	0.013–0.034	0.34–0.88

for development of PMF hydrographs. Higher values (up to 0.260) are appropriate for development of frequency-based hydrographs. For local storms, the selection of K_n (in the range 0.050 to 0.073) is based on the type and extent of vegetation in the overland flow areas, the frictional characteristics of the channels, and the extent of drainage network development by erosional agents.

Region III applies to the southwestern desert, Great Basin, and Colorado plateau regions of southern California, Nevada, Utah, Arizona, and western Colorado and New Mexico. Basins in this arid region are generally characterized by sparse vegetation, fairly well defined drainage networks, and terrain varying from rolling to very rugged in the more mountainous areas. Lower values of K_n (in the range 0.042 to 0.070) are typical of desert terrain, whereas higher values are indicative of the slower response characteristics associated with coniferous forests located in higher elevations. A value of K_n equal to 0.031 corresponds to partially urbanized basins in desert regions, reflecting the faster response typical of urban areas.

Region IV applies to the Sierra Nevada region of California. Basins in this region normally have well-developed drainage networks and substantial coniferous forest growth at elevations above 2000 ft over mean sea level. River and stream channels are well incised into the bedrock. Values of K_n vary in the range 0.064 to 0.150, reflecting the variability in basin response characteristics. Given the paucity of data in this region, care should be exercised when selecting a value of K_n close to the lower end of the indicated range.

Region V applies to the Coast and Cascade ranges of California, Oregon and Washington. Values of K_n vary in the range 0.080 to 0.150, the lower values being typical of low-lying basins with considerably sparse vegetation, while the higher values are indicative of heavy coniferous forest growth extending into the overbank flood plain.

Region VI applies to urban conditions at several locations throughout the United States. The range in K_n values, from 0.013 to 0.034, reflects urban density and type of development and the extent to which floodwater-retarding structures may have affected the natural basin response characteristics. The lower K_n values are typical of high-density developments having extensive stormwater collection facilities. The higher K_n values represent low-density or partial developments with minor stormwater collection systems.

USBR Dimensionless Unit Hydrographs

The dimensionless unit hydrograph and the dimensionless S-graph lead to results which are roughly comparable. Therefore, the choice of method is a matter of individual preference.

Dimensionless Unit Hydrograph. The dimensionless unit hydrograph is a unit hydrograph with dimensionless time in the abscissas and dimensionless discharge in the ordinates. Dimensionless time is defined as follows:

$$t_* = \frac{t}{L_g + \dfrac{D}{2}} \tag{14-6}$$

in which t_* = dimensionless time; t = actual time in hours; L_g = lag time in hours; and D = unit hydrograph duration in hours. Dimensionless discharge is defined as follows:

$$Q_* = \frac{Q\left(L_g + \dfrac{D}{2}\right)}{U} \tag{14-7}$$

in which Q_* = dimensionless discharge; Q = discharge in cubic feet per second; L_g = lag time in hours; D = unit hydrograph duration in hours; and U = unit runoff volume in (cubic feet per second)-days. For 1 in. of runoff, the unit runoff volume per unit basin area is 26.89 [(ft^3/s)-d/(mi^2)]. The dimensionless discharge defined by Eq. 14-7 includes a factor of 24 h/d.

Dimensionless S-graph. The dimensionless S-graph is a cumulative unit hydrograph with time (in percent of lag time) in the abscissas and discharge (in percent of ultimate discharge) in the ordinates. Lag time is defined as the time from the start of effective rainfall to the time when runoff reaches 50 percent of the ultimate discharge. The latter is the product of basin area times effective rainfall intensity. For a unit hydrograph, the effective rainfall intensity is equal to a unit effective rainfall depth (1) divided by the unit hydrograph duration (D). Therefore,

$$Q_u = 645.3[A \times (1/D)] = 645.3\left(\frac{A}{D}\right) \tag{14-8}$$

in which Q_u = ultimate discharge in cubic feet per second; A = basin area in square miles; D = unit hydrograph duration in hours; and 645.3 is the conversion factor for the given units (square miles-inch/hour to cubic feet per second).

Unit Hydrograph Development. Tables 14-14 and 14-15 show the USBR dimensionless unit hydrographs and S-graphs for the six regions indicated in Table 14-13. The procedure to develop a unit hydrograph by the USBR method is illustrated by the following example.

Example 14-2.

A 300-mi^2 drainage basin located in Arizona has a time lag L_g = 9 h. Calculate a 2-h unit hydrograph by the USBR method, using: (a) a dimensionless unit hydrograph and (b) a dimensionless S-graph.

TABLE 14-14   USBR DIMENSIONLESS UNIT HYDROGRAPHS[1] [46]

(1)	(2)	(3)	(4)	(5)	(6)	(7)
Dimensionless time t_*	Dimensionless discharge Q_* for indicated region					
	I	II-G[2]	II-L[3]	III	IV–V	VI
0.10	0.20	0.90	0.21	0.32	1.30	1.56
0.20	1.66	3.00	0.51	0.74	2.60	3.57
0.30	4.83	6.00	1.62	1.81	4.23	5.80
0.40	9.18	9.00	6.38	3.68	7.17	8.39
0.50	14.03	18.11	10.94	8.41	12.17	11.52
0.60	18.07	24.01	15.70	16.50	15.83	15.18
0.70	21.40	21.21	20.76	23.97	20.59	19.27
0.80	24.02	16.91	25.83	28.91	21.54	20.00
0.90	20.59	14.21	26.53	26.38	18.13	19.27
1.00	16.65	12.71	22.68	21.55	15.26	16.12
1.10	13.52	11.21	18.84	16.08	12.53	13.08
1.20	11.40	10.01	14.99	12.61	10.29	11.31
1.30	9.59	8.80	11.04	9.99	8.73	9.63
1.40	8.26	7.70	8.41	8.20	7.65	8.27
1.50	6.96	6.80	6.69	6.78	6.69	7.22
1.60	5.95	6.00	5.47	5.83	5.99	6.27
1.70	5.05	5.35	4.55	5.15	5.36	5.55
1.80	4.39	4.80	3.89	4.57	4.85	4.92
1.90	3.78	4.30	3.34	4.10	4.43	4.39
2.00	3.30	3.90	2.93	3.68	4.06	3.93
2.50	2.08	2.52	1.69	2.10	2.79	2.47
3.00	1.42	1.70	1.09	1.21	1.98	1.71
3.50	0.98	1.13	0.71	0.69	1.42	1.21
4.00	0.67	0.77	0.46	0.40	1.03	0.84
4.50	0.46	0.52	0.29	0.23	0.75	0.58
5.00	0.32	0.35	0.19	0.13	0.46	0.41
5.50	0.23	0.24	0.13	0.00	0.14	0.29
6.00	0.16	0.17	0.08	0.00	0.00	0.21

[1]This table is condensed from [46]. See original reference for complete listing of hydrograph ordinates.

[2]Region II, general storm.

[3]Region II, local storm.

TABLE 14-15 USBR DIMENSIONLESS S-GRAPHS[1] [46]

(1)	(2)	(3)	(4)	(5)	(6)	(7)
Time (% of L_g)	Discharge (% of Q_u) for indicated region					
	I	II-G[2]	II-L[3]	III	IV–V	VI
10	0.06	0.23	0.07	0.10	0.43	0.48
20	0.52	1.20	0.24	0.34	1.44	1.82
30	2.01	3.46	0.70	0.91	3.13	4.11
40	5.02	6.72	2.57	2.08	6.04	7.49
50	9.70	12.74	6.31	4.57	10.94	12.21
60	16.20	21.47	11.88	9.79	17.64	18.51
70	24.17	30.58	19.41	18.03	26.42	26.47
80	33.23	38.32	28.93	28.90	35.72	34.95
90	42.39	44.35	39.99	40.15	43.50	43.09
100	50.00	50.00	50.00	50.00	50.00	50.00
110	56.25	54.87	58.48	57.86	55.32	55.64
120	61.43	59.21	65.42	63.83	59.66	60.42
130	65.81	63.08	70.83	68.53	63.35	64.50
140	69.53	66.50	74.86	72.34	66.56	68.51
150	72.73	69.51	78.02	75.47	69.38	71.06
160	75.46	72.17	80.55	78.10	71.89	73.71
170	77.80	74.53	82.65	80.40	74.13	76.04
180	79.80	76.64	84.44	82.43	76.15	78.10
190	81.54	78.54	85.95	84.25	78.00	79.94
200	83.07	80.26	87.27	85.88	79.70	81.58
250	88.61	86.89	91.96	91.88	86.40	87.73
300	92.30	91.32	94.86	95.39	91.08	91.85
350	94.87	94.33	96.77	97.45	94.41	94.69
400	96.66	96.38	98.01	98.66	96.81	96.66
450	97.91	97.78	98.82	99.37	98.51	98.03
500	98.78	98.74	99.36	99.81	99.59	98.97
550	99.48	99.46	99.71	100.00	100.00	99.62
600	100.00	100.00	100.00	100.00	100.00	100.00

[1]This table is condensed from [46]. See original reference for complete listing of hydrograph ordinates.
[2]Region II, general storm.
[3]Region II, local storm.

(a) Arizona is in Region III (Table 14-13). Therefore, Col. 5 of Table 14-14 contains the appropriate dimensionless unit hydrograph for this example. In this example, $[L_g + (D/2)] = 10$ h. The unit runoff volume is equal to $U = 26.89$ (ft³/s)-d/(mi²) $\times$ 300 mi² = 8067 (ft³/s)-d. Therefore, $U/[L_g + (D/2)] = 806.7$. The calculated unit hydrograph is shown in Table 14-16. Column 1, dimensionless time, is the same as Table 14-14, Col. 1. Column 2, dimensionless discharge, is the same as Table 14-14, Col. 5. Column 3, time, is obtained by multiplying the values of Col. 1 by $[L_g + (D/2)] = 10$. Column 4, discharge, is obtained by multiplying the values of Col. 2 by $U/[L_g + (D/2)] = 806.7$. The calculated unit hydrograph is shown in Col. 4.

(b) Arizona is in Region III (Table 14-13). Therefore, Col. 5 of Table 14-15 contains the appropriate dimensionless S-graph for this example. Using Eq. 14-8, the ultimate discharge is $Q_u = 645.3$ (300/2) = 96,795 ft³/s. The calculated unit hydrograph is shown in Table 14-17. Column 1 shows time at intervals equal to the unit hydrograph duration.

TABLE 14-16   UNIT HYDROGRAPH CALCULATED USING USBR DIMENSIONLESS UNIT HYDROGRAPH: EXAMPLE 14-2(a)

(1)	(2)	(3)	(4)
Dimensionless time t_*	Dimensionless discharge Q_*	Time t (h)	Discharge Q (ft³/s)
0.0	0.00	0	0
0.2	0.74	2	597
0.4	3.68	4	2,969
0.6	16.50	6	13,311
0.8	28.91	8	23,322
1.0	21.55	10	17,384
1.2	12.61	12	10,172
1.5	6.78	15	5,469
2.0	3.68	20	2,969
3.0	1.21	30	976
4.0	0.40	40	323
5.0	0.13	50	105

TABLE 14-17   UNIT HYDROGRAPH CALCULATED USING USBR DIMENSIONLESS S-GRAPH: EXAMPLE 14-2(b)

(1)	(2)	(3)	(4)	(5)
Time (h)	Time (% of L_g)	Discharge (% of Q_u)	S-graph ordinates (ft³/s)	Discharge Q (ft³/s)
0	0	0.00	0	0
2	22	0.43	416	416
4	44	2.88	2,788	2,372
6	67	15.34	14,849	12,061
8	89	39.05	37,800	22,951
10	111	58.49	56,618	18,818
12	133	69.73	67,499	10,881
14	156	77.09	74,623	7,144
16	178	82.03	79,405	4,782
18	200	85.88	83,132	3,727
20	222	88.92	86,075	2,943
22	244	91.32	88,393	2,318
24	267	93.29	90,300	1,907
26	289	94.77	91,733	1,433
28	311	95.94	92,865	1,132
30	333	96.86	93,756	891

Column 2 is obtained by expressing the values of Col. 1 in percent of time lag. For each value of time (in percent of time lag) shown in Col. 2, a discharge (in percent of ultimate discharge) is shown in Col. 3, obtained from Table 14-15, Col. 5, by linear interpolation. The S-graph ordinates shown in Col. 4 are obtained by multiplying the values of Col. 3 by the ultimate discharge Q_u and dividing by 100. Column 5 is obtained by subtracting two consecutive S-graph ordinates shown in Col. 4. Notice that the S-graph ordinates in Col. 4 *must* be defined at intervals equal to D, i.e., at 2-h intervals. The calculated unit hydrograph is shown in Col. 5. The unit hydrographs calculated by both methods—dimensionless unit hydrograph and dimensionless S-graph—are shown to be roughly comparable.

Flood Hydrograph Development

Hydrologic Abstractions. In USBR practice, infiltration losses are considered to be the most important source of hydrologic abstraction for extreme flood events. For PMF estimates, hydrologic analyses are based on the assumption that the minimum infiltration rate (i.e., the final rate f_c in the Horton equation, Eq. 2-13) prevails for the duration of the probable maximum storm. This assumption is consistent with conditions that have been shown to exist during extreme flood events. Historical records have shown that extreme floods are typically preceded by one or more antecedent storms. These antecedent storms have the effect of saturating the soil and producing minimum infiltration rates. Therefore, it can be assumed that antecedent storms satisfy all other precipitation-abstracting mechanisms.

Baseflow and Interflow. Baseflow is generally modeled as a recession curve, to represent a gradually decreasing flow rate (Section 11.5). This flow continues to decrease until the water surface in the stream is in equilibrium with the surface of the adjacent groundwater reservoir. Unlike baseflow, the interflow component of a hydrograph consists of water that reaches the watercourses after flowing below the surface for a relatively short distance.

The quantification of baseflow and interflow components of a hydrograph is usually based on the results of hydrograph reconstructions. Separation of an observed flood hydrograph into its surface, interflow, and baseflow components requires a substantial amount of judgment. If data are available, a complete recession curve representing the baseflow component for a given basin can be determined. For ungaged basins, results of flood hydrograph reconstructions on hydrologically similar basins are used. This is accomplished by expressing the observed baseflow component in terms of discharge per unit area (cubic feet per second per square mile). This *specific baseflow discharge* is then used to determine the appropriate baseflow value for the basin under study. In this case, it is appropriate to assume a constant baseflow for the entire duration of the flood.

After subtracting the baseflow component, the remaining portion of the observed flood hydrograph is composed of interflow and surface flow. Separation of interflow and surface flow components is accomplished on an empirical basis, with attention paid to estimating a physically realistic value of the interflow component. For ungaged basins, interflow calculations follow the same procedures as for baseflow, that is, the use of a *specific interflow discharge*, defined as the interflow rate per unit basin area.

Design Flood Hydrographs. In USBR practice, the PMF hydrograph is the standard design-flood hydrograph. In developing the PMF hydrograph, the following procedures are adhered to:

1. The PMF is based on the probable maximum storm. The temporal distribution of the storm should be such that the maximum peak discharge and the maximum concentration of discharge around the peak discharge are achieved.

2. Infiltration rates should be the lowest rates consistent with soil types and prevailing geologic conditions. These minimum rates are assumed to prevail for the entire duration of the probable maximum storm.

3. The unit hydrograph used to compute the PMF should represent extreme discharge conditions. Where flood hydrographs are available, care should be taken to ensure that the synthetic unit hydrograph parameter K_n reflects the conditions likely to prevail during the probable maximum storm. If necessary, the applicable value of K_n may be decreased in order better to simulate the basin response to a probable maximum storm. For ungaged basins, considerable judgment and experience are required to estimate an appropriate value of K_n.

4. The baseflow component should reflect the maximum flow consistent with the magnitude and timing of the antecedent flood event.

5. The interflow component should reflect the conditions likely to prevail during the probable maximum storm. However, the interflow component is not likely to vary substantially [46].

The hydrograph of surface flow is calculated by convoluting the unit hydrograph with the effective storm pattern (Section 5.2). The baseflow and interflow components are then added to the surface-runoff hydrograph to obtain the PMF hydrograph.

Snowmelt Contribution to PMF. USBR uses a snow-compaction method to compute snowmelt. The method requires the following data: (1) air temperature, (2) wind velocity, (3) forest cover density, (4) snowpack depth, and (5) snow density. The drainage basin is divided into several elevation bands selected at 500- or 1000-ft intervals, depending on the basin size and elevation range (Chapter 12). Several sets of rainfall, wind velocities, and air temperatures are tested to identify the most critical snowmelt condition. For each band, the snowmelt contribution is added to the PMP, and the result is averaged over the entire basin. Application of the method requires a substantial amount of judgment and experience [46].

As an alternative to the snow-compaction method, USBR uses a combination of the 100-y snowmelt flood with the PMF. The selected period of runoff is usually 15 d. The resulting snowmelt flood hydrograph is generally expressed in terms of mean daily flows for the 15-d period, neglecting diurnal fluctuations. The snowmelt flood hydrograph is then combined with the rain-only flood hydrograph, with the rain-only flood assumed to occur during the period of greatest snowmelt flood. The resulting rain-on-snow flood hydrograph is the PMF.

14.5 TENNESSEE VALLEY AUTHORITY

The Tennessee Valley Authority (TVA) is a U.S. government corporation with regional resource-development responsibilities in the Tennessee River basin, including agricultural and industrial development, navigation, and flood control.

The Tennessee River basin encompasses an area of 40,190 mi^2 extending from its mouth at the Ohio River near Paducah, Kentucky, upstream to its headwaters in the Blue Ridge and Great Smoky Mountains of western Virginia, western North Carolina, and northern Georgia. Mean annual rainfall in the basin is 52 in., varying from 40 in. in sheltered portions of the Appalachian Mountains to 90 in. along the Blue Ridge Mountains of northern Georgia and western North Carolina.

PMP and TVA Precipitation

TVA uses the concepts of PMP and TVA precipitation in the design and safety evaluation of its dams. PMP is defined in Section 14.1. TVA precipitation is the level of precipitation resulting from transposition and adjustment (without maximization) of outstanding storms that have occurred in the Tennessee Valley [30]. Unlike PMP, a few of the most important extreme events are excluded from TVA precipitation.

The hydrometeorological analysis to estimate PMP and TVA precipitation as needed by TVA for the design and evaluation of its projects has been developed by the National Weather Service. These include generalized studies for drainage areas less than 3000 mi^2 (HMR 56 [30]) and studies for specific larger basins (HMR 41 [18] and HMR 47 [22]).

PMP and TVA Precipitation for Drainage Areas Less Than 100 mi^2.
The rarest known storms, with moisture maximization and transposition, are used as guides in defining PMP. TVA precipitation is based on observed storms, without moisture maximization, and excludes the most extreme storm events.

Studies have shown that the local storm (thunderstorm) is the most appropriate PMP-type storm for drainage areas less than 100 mi^2 within the Tennessee River basin. The months of July and August are taken as the months of PMP and TVA precipitation for these small basins.

PMP and TVA precipitation maps for the Tennessee River basin, applicable to 6-h durations and 1-mi^2 areas, are shown in Figs. 14-22 and 14-23, respectively. Table 14-18 shows ratios for converting the 6-h, 1-mi^2 TVA precipitation shown in Fig. 14-23 to other durations. Figure 14-24 shows a depth-area reduction chart for PMP and TVA precipitation, applicable to 1- to 24-h durations and 1- to 100-mi^2 basins [30].

Orographic effects are particularly important in the rough topography of the eastern mountainous region of the Tennessee River basin. To take into account these effects, the following concepts are adopted:

1. Sheltered areas, defined as valleys having upwind barriers (from southeast through southwest) at elevations higher than 2000 ft above mean sea level.

2. First upslope, defined as the mountain slope facing the lowlands (from east through southwest), with no intervening mountains between the slope and the Gulf of Mexico or the Atlantic Ocean. In general, total summer precipitation on first upslope areas is approximately twice that of sheltered areas.

3. Secondary upslope, which is high and steep enough to increase precipitation but is partially shielded upwind by a lower mountain range, with an elevation difference between upper and lower ranges of at least 1500 ft. Total summer precipita-

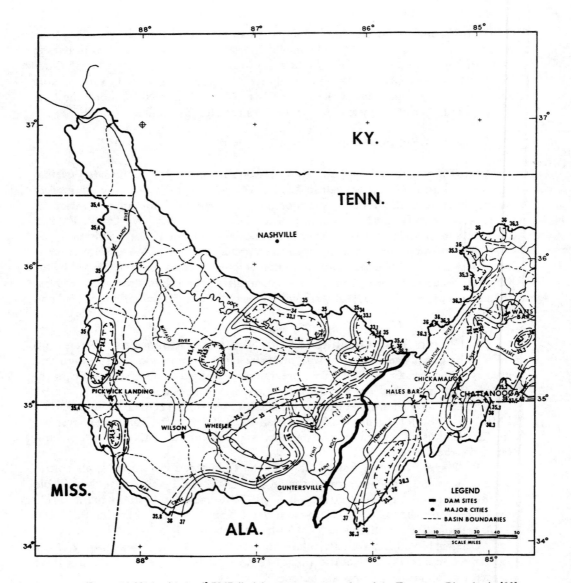

Figure 14-22(a) 6-h 1-mi^2 PMP (in.) for the western portion of the Tennessee River basin [30].

tion on secondary slopes is from 30 to 50 percent greater than the precipitation on sheltered areas.

4. Depression, defined as the elevation difference between the crest of a barrier and a low point within a sheltered area.

The following criteria are recommended to adjust PMP and TVA precipitation in basins less than 100 mi^2 to account for orographic effects [30]:

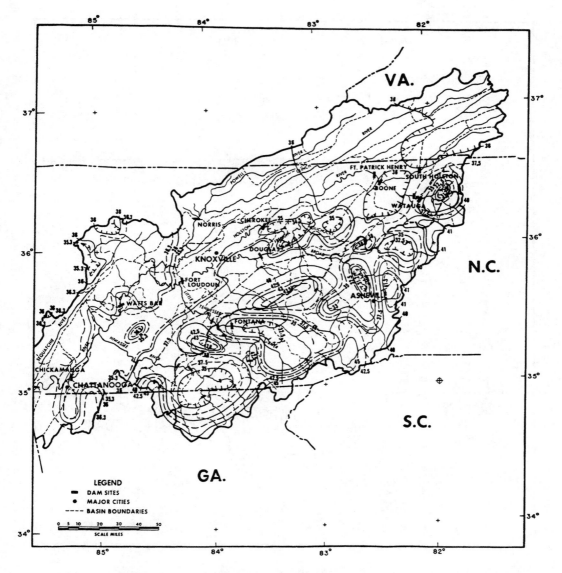

Figure 14-22(b) 6-h 1-mi² PMP (in.) for the eastern portion of the Tennessee River basin [30].

1. **First upslopes:** 10 percent increase per 1000 ft of elevation above mean sea level, limited to a maximum of 25 percent.

2. **Secondary upslopes:** 5 percent increase per 1000 ft of elevation above mean sea level.

3. **Sheltered areas:** 5 percent decrease per 1000 ft of depression.

The following criteria are recommended to develop temporal distributions of PMP and TVA precipitation in basins less than 100 mi² [30]:

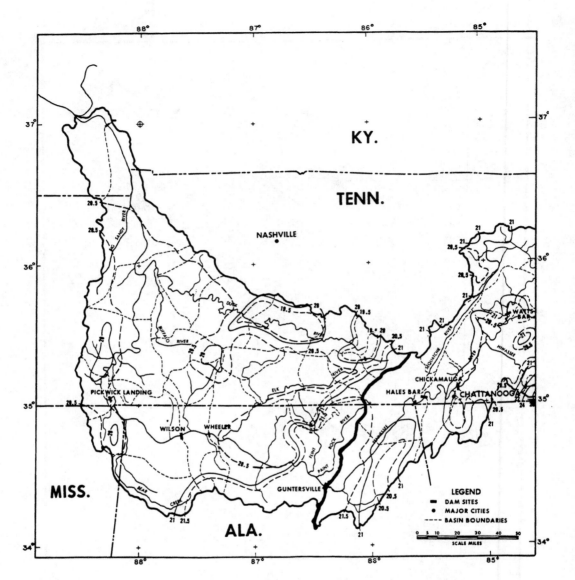

Figure 14-23(a) 6-h 1-mi² TVA precipitation (in.) for the western portion of the Tennessee River basin [30].

1. **Six-hour rainfall increments in 24-h storm:** The four 6-h increments are arranged in such a way that the second-largest increment is next to the highest, the third-largest increment adjacent to these, and the remaining increment at either end. Several arrangements are possible, with the critical storm being the one associated with the critical hydrograph.

2. **One-hour increments in 6-h maximum rainfall:** Any arrangement of 1-h increments is acceptable, provided that the two largest increments are adjacent, the three largest increments are adjacent, and so on.

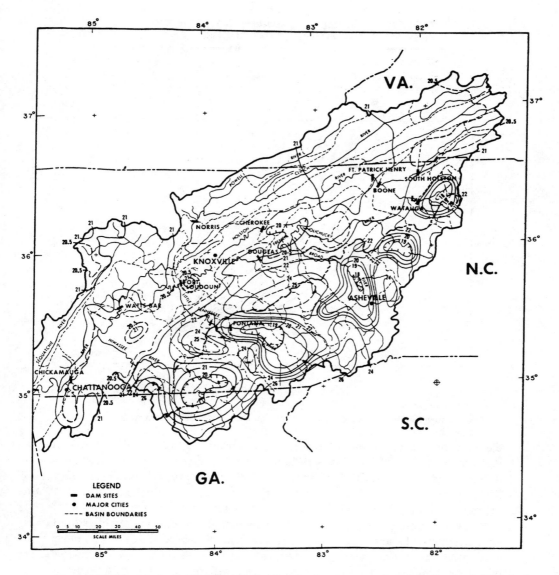

Figure 14-23(b) 6-h 1-mi² TVA precipitation (in.) for the eastern portion of the Tennessee River basin [30].

PMP and TVA Precipitation for Drainage Areas Between 100 and 3000 mi².   In drainage areas from 100 to 3000 mi², the primary rain-producing storms in the Tennessee Valley are derived from a combination of tropical storms and thunderstorms embedded in general storms.

The calculation of the nonorographic component of PMP and TVA precipitation is based on index values obtained from depth-area-duration charts for the Knoxville airport shown in Fig. 14-25(a) and (b). These charts are used together with the percentile map shown in Fig. 14-26 to determine PMP and TVA precipitation applica-

TABLE 14-18 RATIOS FOR ADJUSTING 6-H 1-MI² TVA PRECIPITATION TO OTHER DURATIONS [30]

Duration	1	2	3	6	12	24
Ratio	0.51	0.68	0.80	1.00	1.13	1.24

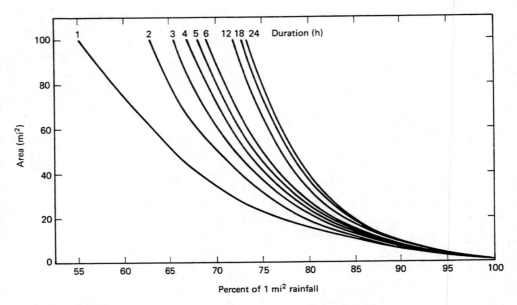

Figure 14-24 Depth-area relation for small-basin PMP and TVA precipitation [30].

ble for 6- to 72-h durations and drainage areas from 100 to 3000 mi². For example, a location in Fig. 14-26 determines the applicable percentile. Given the basin area and duration, the PMP and TVA precipitation depths are obtained from Fig. 14-25(a) and (b), respectively. These depths are multiplied by the applicable percentile to obtain the nonorographic PMP and TVA precipitation for the given location.

This nonorographic component does not include the effect of terrain stimulation on convective cells and/or thunderstorms in general storms. To take this effect into account, each quadrangle map is classified as either (a) smooth, (b) intermediate, or (c) rough. A smooth class corresponds to terrain with few elevation differences of 50 ft in 0.25 mi. An intermediate class terrain is that with frequent elevation differences ranging from 50 to 250 ft in 0.25 mi. A rough class terrain is that with extensive areas having elevation differences in excess of 150 ft in 0.25 mi.

For the Tennessee River valley, adjustment factors (percent of increase) for intermediate and rough terrain classes are shown in Fig. 14-27. Figure 14-28 shows the percent factor (as a function of basin area) to be applied to the sum of adjustment factors obtained from Fig. 14-27 to calculate the percent of increase in PMP and TVA precipitation. More elaborate adjustments for orographic influences in the mountainous eastern portion of the basin are described in HMR 56 [30]. The larger Tennessee River basins, with drainage areas exceeding 3000 mi², are covered by procedures described in HMR 41 [18] and HMR 47 [22].

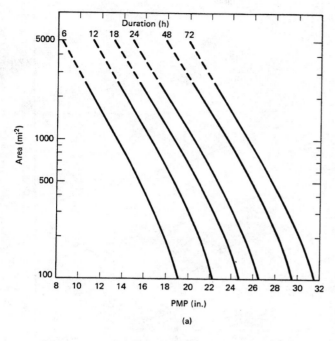

Figure 14-25(a) Depth-area-duration relation for PMP at Knoxville airport [30].

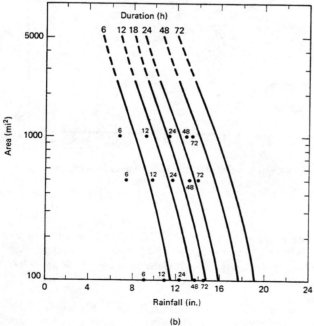

Figure 14-25(b) Depth-area-duration relation for TVA precipitation at Knoxville airport [30].

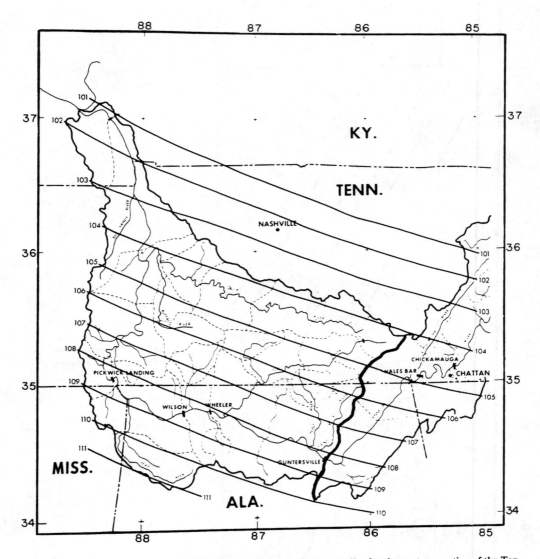

Figure 14-26(a)   24-h 1000-mi² PMP and TVA precipitation percentiles for the western portion of the Tennessee River basin [30].

Example 14-3.

Assume a 200-mi² basin located in the eastern half of the Tennessee River basin (non-mountainous eastern region), with 20 percent rough and 50 percent intermediate class terrain and a 95 percentile (obtained from Fig. 14-26). Determine 24-h PMP and TVA precipitation estimates.

From Fig. 14-25(a), the index PMP for 200-mi² and 24-h duration is 25 in. Likewise, from Fig. 14-25(b), the index TVA precipitation is 15 in. Therefore, the nonorographic values of PMP and TVA precipitation are 25 × 0.95 = 23.8 in. and 15 × 0.95 = 14.3 in., respectively. From Fig. 14-27, the adjustment for 20 percent rough terrain is 3.25 per-

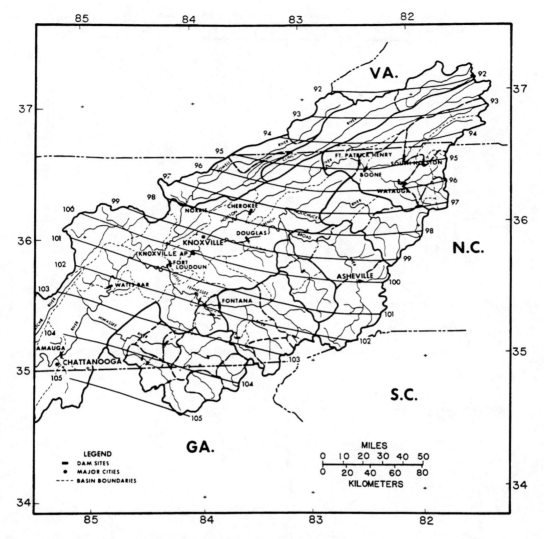

Figure 14-26(b) 24-h 1000-mi^2 PMP and TVA precipitation percentiles for the eastern portion of the Tennessee River basin [30].

cent, and the adjustment for 50 percent intermediate terrain is 10 percent. The 100-mi^2 roughness adjustment factor is $10 + 3.25 = 13.25$ percent. For 200 mi^2, the percent of areal adjustment (from Fig. 14-28) is 65 percent. Then, the increase in PMP and TVA precipitation is $(65/100) \times 13.25 = 8.6$ percent. Therefore, the 24-h PMP and TVA precipitation estimates are $23.8 \times 1.086 = 25.8$ in. and $14.3 \times 1.086 = 15.5$ in., respectively.

Areal Distribution of PMP and TVA Precipitation. Figure 14-29 shows the standard isohyetal pattern recommended for spatial distribution of nonorographic PMP for locations east of the 105th meridian, including the Tennessee River basin

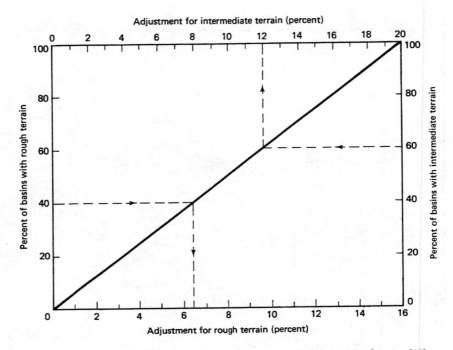

Figure 14-27 Terrain-roughness adjustments factors, applicable to 100-mi² basins [30].

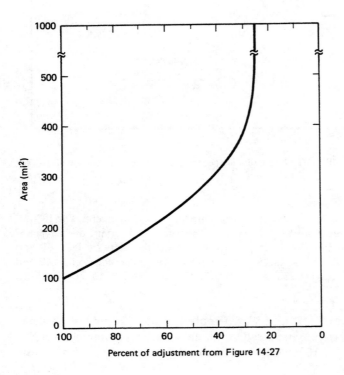

Figure 14-28 Variation of terrain-roughness adjustment factors with basin size [30].

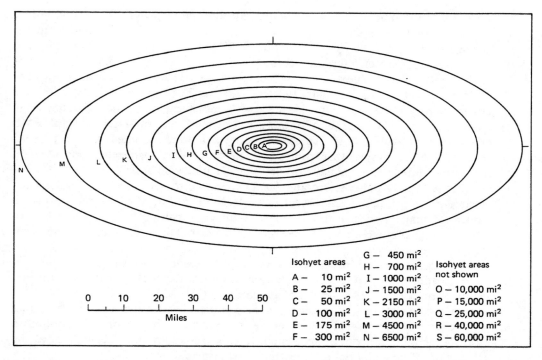

Isohyet areas

A — 10 mi²
B — 25 mi²
C — 50 mi²
D — 100 mi²
E — 175 mi²
F — 300 mi²

G — 450 mi²
H — 700 mi²
I — 1000 mi²
J — 1500 mi²
K — 2150 mi²
L — 3000 mi²
M — 4500 mi²
N — 6500 mi²

Isohyet areas
not shown

O — 10,000 mi²
P — 15,000 mi²
Q — 25,000 mi²
R — 40,000 mi²
S — 60,000 mi²

Figure 14-29 Standard isohyetal pattern for nonorographic PMP for locations east of the 105th meridian [26].

[26]. Methods for accounting for the orientation of an isohyetal pattern are described in HMR 56 [30].

TVA Procedures for PMF Determinations

The PMF provides an assessment of maximum flood potential considering both climatic and basin characteristics. The PMF is used to define maximum reservoir levels for purposes of dam design or safety evaluation. TVA's efforts have been directed towards finding an objective procedure to select a meteorologically and hydrologically reasonable sequence of events to calculate the PMF.

The major factors in a PMF determination are (1) the principal storm, (2) the antecedent storm, (3) temporal and spatial distribution of rainfall, (4) hydrologic abstractions (loss rates), and (5) unit hydrograph characteristics. TVA uses a two-step process to select the appropriate combination of events that lead to a PMF determination. The first step is to estimate the PMP and its associated exceedence probability. The second step is to choose a risk objective and to select the features of PMF computation and reservoir design to achieve the desired risk objective [36].

The principal storm in a PMF determination is the one associated with the PMP. When the basin shape is markedly different from a storm's isohyetal pattern, a storm centering that produces maximum basin rainfall is sought. For large basins, several alternate storms are explored to determine the critical spatial and temporal arrangement.

The adopted storm duration is the one that produces a critical combination of peak discharge and volume. In the Tennessee valley, 24-h storms are normally controlling for watersheds less than 100 mi^2, whereas 72-h storms are controlling for larger basins. Critical storms for locations with basins less than 7000 mi^2 occur in the months of June to October, and for larger basins they occur in March or April. PMP exceedence probability in the Tennessee Valley has been estimated at between 10^{-6} and 10^{-8} [36]. However, other studies have shown that PMP determinations based on NWS procedures are not equally probable on a nationwide basis [33].

TVA has evaluated storm and flood experience to determine meteorologically feasible storm conditions antecedent to the PMP. These studies have shown that 75 percent of major floods in the Tennessee Valley are caused by a sequence of two storms, the first antecedent and the second principal. The median antecedent storm was found to be about 26 percent of the principal storm, with an average of about 3 d between antecedent and principal storms [34]. Antecedent storm depths varying between 15 and 50 percent of PMP have been proposed, depending on storm duration, season of occurrence, and basin size and location [35].

A PMF determination based upon the postulated occurrence of a storm couplet is matched with the chosen risk objective. A conditional distribution is required to estimate the probability of a storm couplet. This probability can be calculated as the product of the probabilities of antecedent and principal storms [35].

For dam design or evaluation, given the design inflow hydrograph, computed reservoir levels are a function of initial pool level, operating assumptions, and windwave effects. For pool levels, the use of median levels on the starting date (as determined from reservoir operation studies or actual operating experience) is adequate. Seasonal variations in initial pool level and PMF are explored to determine the conditions which produce maximum reservoir levels.

Outlet capacity or spillway operation during a flood can significantly affect PMF reservoir pool levels. Assumptions about outlet capacity and spillway operations should be in accordance with the conditions expected to prevail during the PMF. Evaluation of the risk of failure requires an assessment of the probability of inoperability of a dam's outlets and spillways.

The potential effect of windwaves is considered in the hydrologic design of a major dam. An appropriate wind velocity is a function of site conditions and anticipated reservoir headwater levels. Wind velocities used in design vary from the median daily maximum at time of flood cresting to a maximum of 100 mi/h. When defining the time of flood cresting, two factors are considered: (1) the seasonal variations of PMF and (2) the number of hours that reservoir headwater levels are expected to be at or near the crest. In the Tennessee Valley, reservoirs are at crest level for periods of time not exceeding 1 d. Due to the prevailing topography, fetch lengths are relatively short; therefore, wind durations needed to create maximum wave heights usually do not exceed 60 min. TVA practice to calculate windwaves is based on the median of maximum winds for the day of cresting, from the critical direction, and with duration sufficient to generate maximum wave heights [35].

14.6 U.S. GEOLOGICAL SURVEY

USGS State Equations for Flood-peak Estimation at Ungaged Sites

The design of highway bridges, culverts, dams, levees, and other hydraulic structures requires a knowledge of the magnitude and frequency of floods. The U.S. Geological Survey (USGS) has collected annual peak flow data at over 22,000 stream-gaging stations throughout the U.S. For ungaged sites, USGS uses regional analysis of flood characteristics as the primary means of developing flood peak estimates.

Historical Perspective. In the 1960s, USGS used the index-flood method to estimate design floods at ungaged sites (Chapter 7). Using this method, a nationwide study of flood characteristics was completed and published as USGS Water-Supply Papers Nos. 1671 to 1689.

More recently, USGS has used multiple-regression techniques to determine flood frequency at ungaged sites. These techniques are based on statistical regression to relate flood magnitude and frequency to watershed and climatic characteristics. The predictive equations have the following general form [1, 2]:

$$Q = b_0 X_1^{b_1} X_2^{b_2} \cdots X_k^{b_k} \tag{14-9}$$

in which b_i $(i = 1, 2, \ldots, k)$ are regression constants determined by least squares, X_i are watershed and climatic characteristics, k is generally less than 7, and Q is a flood peak associated with a given frequency (or return period). This approach is currently used by most USGS district offices to estimate flood magnitudes and associated frequencies at ungaged sites [43].

The procedure consists of the following steps:

1. For a given frequency, compute estimates of design floods at several gaging stations using the guidelines of *Bulletin* 17B of the U.S. Interagency Advisory Committee on Water Data [49].
2. Determine watershed, climatic, or channel geometry characteristics applicable to the gaging stations under consideration.
3. Relate the design floods to the watershed, climatic, or channel geometry characteristics using multiple-regression techniques (Chapter 7).

The regression equations determined in this way are strictly applicable only to the region for which they were derived. Used within its established limits, a regression equation can provide a quick estimate of flood peaks for ungaged catchments.

State Equations for Flood-peak Estimation. Since 1973, USGS district offices have conducted regional regression studies, the results of which are contained in several USGS reports listed in Appendix C. These reports supersede USGS Water-Supply Papers Nos. 1671–1689, which dealt with the index-flood method. However, three states (Alaska, Idaho, and Rhode Island) still use the index-flood method in their estimation of peak discharges at ungaged sites.

USGS reports relating flood-peak estimates to watershed and climatic characteristics, on a state-by-state basis, are listed in Table C-1, Appendix C. In many arid and semiarid areas, flood discharges cannot be accurately estimated based on the traditional watershed characteristics such as drainage area and channel slope. In this case, it is necessary to use channel geometry characteristics such as channel width [7, 38, 51]. Nine states, mostly in the western United States, have developed estimation equations based on channel width. USGS reports relating flood peak estimates to channel geometry characteristics are listed in Table C-2, Appendix C.

For urban areas, multiple regression techniques have used watershed and climatic characteristics, with at least one characteristic indicative of urbanization effect. Percentage of impervious cover is the most frequently used urbanization parameter, although channel conveyance [12] and basin-development factor [40] have also been used. The basin-development factor is an index of the drainage features of the watershed, such as extent of channel improvements, curb-and-gutter streets, and storm sewers. In addition to the study by Sauer et al. [40], of nationwide applicability, 18 states have developed regression equations for estimating flood peaks in urban areas. The reports dealing with these studies are listed in Table C-3, Appendix C.

For certain projects, peak discharges may be needed at ungaged sites located in the same stream where there are one or more gaging stations. In these cases, a more accurate estimate is obtained by using information from both the gaging station and the regional regression equation. USGS uses several methods of combining station data with regional data. Invariably, these methods are based on weighing the station and regional estimates to determine a weighted estimate at the ungaged site [6, 10, 41, 42].

Predictor Variables Used in USGS State Equations. Procedures described in reports listed in Appendix C constitute a nationwide technique for estimating flood peaks at ungaged sites. The general features of these USGS state equations are described here.

The number of predictor variables (i.e., variables for which values are known) in the state equations are generally less than seven, with two or three being the most frequent number. When only one predictor variable is used, it is always the drainage area. A list of most frequently used predictor variables is shown in Table 14-19, including the number of states using each predictor variable in their equation.

Drainage area is used in all 50 states and Puerto Rico and is generally recognized as the most significant variable in explaining the variability of flood-peak estimates. Channel slope and precipitation index (such as mean annual precipitation, or the 2-y, 24-h precipitation) are frequently used predictor variables. Basin storage is expressed as percentage of drainage area covered by lakes, ponds, and swamps. Likewise, the hydrologic effect of forests is related to areal percentage of forest cover. Mean basin elevation is commonly used in the western United States, where a definite pattern of variation of flood characteristics with elevation can be documented. Channel length is occasionally used in conjunction with drainage area to describe channel shape.

The predictor variables shown in Table 14-19 can be readily determined from topographic maps or from climatic reports obtained from the National Weather Service, state, or local agencies. Most state equations are based on data from at least 200 gaging stations, representing an average of approximately one station per 300 mi^2.

TABLE 14-19 PREDICTOR VARIABLES IN USGS STATE EQUATIONS [43]

Predictor Variable	Number of States[1]
Drainage area	51
Channel slope	24
Mean annual precipitation	22
Basin storage (percentage)	18
Mean basin elevation	12
Precipitation intensity	11
Forest cover (percentage)	9
Channel length	5
Minimum January temperature	5
Soil characteristics	4

[1]Including Puerto Rico.

Gaging density is an important factor in assessing the accuracy of the regression-based state equations.

Thomas [43] has shown that the USGS state equations give reasonably good predictions for the midwestern and eastern United States, with Maryland, Minnesota, and Florida being exceptions. In the western United States, the quality of the prediction decreases. This is due to the greater variability in annual flood series (temporal sampling errors) and the sparsity of the gaging network (spatial sampling errors).

Examples of USGS State Equations. The following are examples of USGS state equations for flood estimation at ungaged sites. Common variables in these equations are Q_{100} = 100-y flood discharge in cubic feet per second; A = drainage area in square miles; S = channel slope in feet per mile, computed between two points located at 10 and 85 percent of the main channel, measured in the upstream direction; P = mean annual precipitation in inches; I = 2-y frequency 24-h duration precipitation in inches; L = main channel length in miles; T = storage factor, the fraction of drainage area covered by lakes, ponds, and marshes, in percent; and F = forest cover, the fraction of drainage area covered by forest vegetation, in percent.

INDIANA

The state of Indiana is divided into seven regions, as shown in Fig. 14-30. Equations are developed for 2-, 10-, 25-, 50- and 100-y frequencies. The 100-y frequency equations are as follows [5].

Region 1:

$$Q_{100} = 13.8A^{0.695}(T + 1)^{-0.243}(P - 30)^{1.132} \tag{14-10a}$$

Region 2:

$$Q_{100} = 127A^{0.608}(T + 1)^{-0.418}C^{0.902}(P - 30)^{0.708} \tag{14-10b}$$

Region 3:

$$Q_{100} = 181A^{0.779}S^{0.466}(I - 2.5)^{0.831} \tag{14-10c}$$

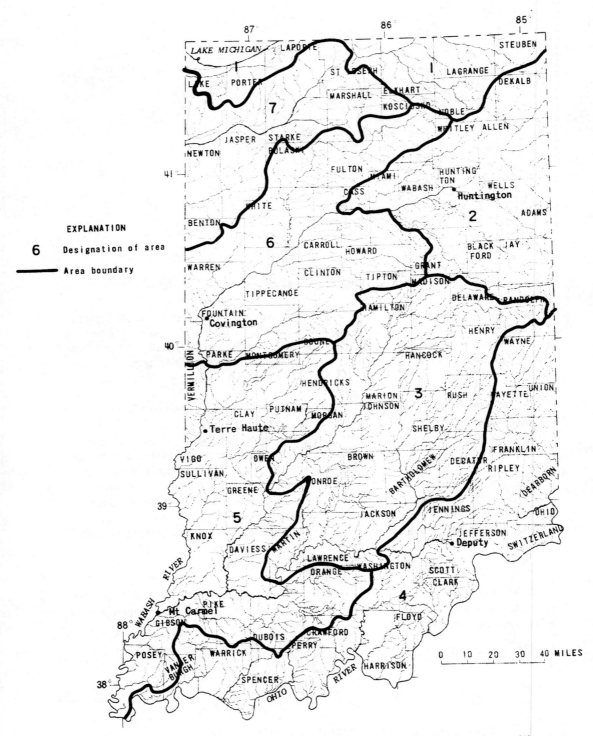

Figure 14-30 USGS state equations: Subdivision of Indiana into seven regions [5].

Hydrologic Design Criteria   Chap. 14

Region 4:

$$Q_{100} = 32A^{0.565}S^{0.705}L^{0.730}(I - 2.5)^{0.464} \qquad \text{(14-10d)}$$

Region 5:

$$Q_{100} = 91.2A^{0.811}S^{0.529} \qquad \text{(14-10e)}$$

Region 6:

$$Q_{100} = 4734A^{0.570}C^{0.834}(I - 2.5)^{2.068} \qquad \text{(14-10f)}$$

Region 7:

$$Q_{100} = 70.1A^{0.285}S^{0.488}L^{0.785}C^{0.854} \qquad \text{(14-10g)}$$

in which C = coefficient relating storm runoff to soil permeability. Values of C are obtained from soil maps, as follows: $C = 0.3$ for hydrologic soil group A, 0.5 for soil group B, 0.7 for soil group C, 0.8 for soil group D, and 1.0 for soil group E [5]. Equations 14-10(a)–(f) are used to estimate the magnitude and frequency of floods on unregulated rural streams in Indiana.

LOUISIANA

Equations are developed for 2-, 5-, 10-, 25-, 50- and 100-y frequencies. The 100-y frequency equation is [11]:

$$Q_{100} = 3.85A^{0.79}S^{0.84}(P - 35)^{1.13} \qquad \text{(14-11)}$$

To increase the accuracy of the estimate, S is limited to a value no greater than 16 ft/mi and P is limited to a value no greater than 60 in. If the actual S and P values exceed these limits, use $S = 16$ and $P = 60$. Equation 14-11 is applicable to basins less than 3000 mi^2, with no significant reservoir regulation.

MARYLAND

The state of Maryland is divided into three regions, as shown in Fig. 14-31: (1) northern region, (2) southern region, and 3) eastern region. Equations are developed for 2-, 5-, 10-, 25-, 50- and 100-y frequencies. The 100-y frequency equations are as follows [3].

Northern Region:

$$Q_{100} = 66.6A^{0.708}(F + 10)^{-0.336}I^{3.212} \qquad \text{(14-12a)}$$

Southern Region:

$$Q_{100} = 548A^{0.662} \qquad \text{(14-12b)}$$

Eastern Region:

$$Q_{100} = 4800A^{1.06}S^{1.035}(T + 10)^{-1.519}(F + 10)^{-0.963}(S_a + 10)^{-0.41}(S_d + 10)^{0.695} \qquad \text{(14-12c)}$$

in which S_a = percent of type A soils and S_d = percent of type D soils.

Equations 14-12(a)–(c) were developed based on data obtained from natural

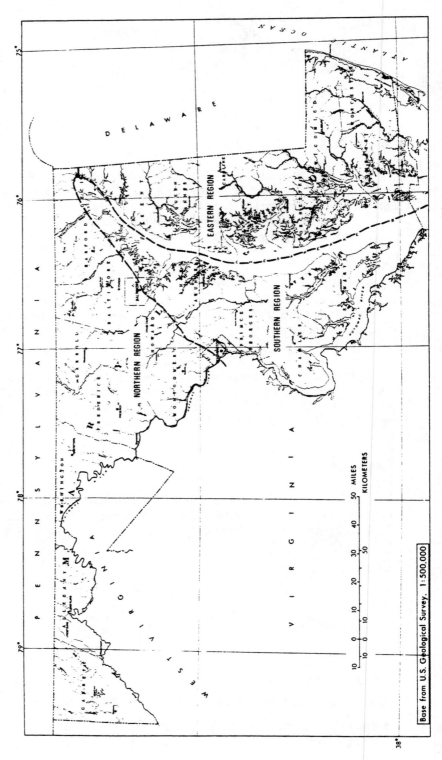

Figure 14-31 USGS state equations: Subdivision of Maryland into three regions [3].

streams. Therefore, they are not applicable to streams where flood peaks are significantly affected by urban development or regulation. Furthermore, the equations are not applicable to limestone regions or locations with tidal marshes.

OKLAHOMA

Equations are developed for 2-, 5-, 10-, 25-, 50-, 100- and 500-y frequencies. The 100-y frequency equation is [44]:

$$Q_{100} = 196A^{0.56}P^{0.68} \tag{14-13}$$

Equation 14-13 is restricted to drainage basins less than 2500 mi², with mean annual precipitation ranging from 14 to 59 in. It should be used with caution for watersheds less than 1 mi² and those significantly affected by urbanization or streamflow regulation.

WISCONSIN

The state of Wisconsin is divided into 5 regions, as shown in Fig. 14-32. Equations are developed for 2-, 5-, 10-, 25-, 50- and 100-y frequencies. The 100-y frequency equations as follows [4].

Region 1:

$$Q_{100} = 0.084A^{0.597}P^{1.80} \tag{14-14a}$$

Region 2:

$$Q_{100} = 34.8A^{0.939}S^{0.523}(T + 1)^{-0.204}K^{-0.703} \tag{14-14b}$$

Figure 14-32 USGS state equations: Subdivision of Wisconsin into five regions [4].

Region 3:
$$Q_{100} = 175.8 A^{0.840} S^{0.277} (F + 1)^{-0.146} I^{1.7} K^{-0.632} \qquad (14\text{-}14c)$$

Region 4:
$$Q_{100} = 7.56 A^{0.835} S^{0.285} (T + 1)^{-0.243} N^{0.567} K^{-0.313} \qquad (14\text{-}14d)$$

Region 5:
$$Q_{100} = 7.83 A^{1.08} S^{0.876} (T + 1)^{-0.423} K^{-0.378} \qquad (14\text{-}14e)$$

in which K = soil permeability in inches per hour, based on the least permeable soil horizon, and N = mean annual snowfall in inches.

Example 14-4.

Estimate the 100-y flood on Radcliffe Creek at State Highway 20, at Chestertown, Maryland (adapted from [3]).

The site is located in Maryland's eastern region (Fig. 14-31). Therefore, Eq. 14-12c is applicable. The drainage area is 4.3 mi². The channel slope is measured at 22 ft/mi. The storage factor (percentage of drainage area covered by lakes, ponds, and swamps) is measured at 1.7%. The forest cover is measured at 8%. The percent of type A soils is 0% and the percent of type D soil is 13%. Using Eq. 14-12(c), the 100-y peak discharge is

$$Q_{100} = 4800(4.3)^{1.06}(22)^{1.035}(11.7)^{-1.519}(18)^{-0.963}(10)^{-0.41}(23)^{0.695} = 2790 \ \text{ft}^3/\text{s}.$$

QUESTIONS

1. Define probable maximum precipitation.
2. What steps are involved in the development of PMP estimates for nonorographic regions?
3. What is the major difference between PMP estimates for U.S. regions east of the 105th meridian and elsewhere?
4. What is the difference between geographically fixed and storm-centered depth-area ratios? Explain.
5. What is a standard project storm? What is a standard project flood?
6. What is the design flood in U.S. Corps of Engineers' practice?
7. What is a principal spillway in SCS practice? An emergency spillway?
8. What is reservoir storage volume? What is retarding-pool storage?
9. What is surcharge storage? What is freeboard?
10. In SCS practice, what is the difference between emergency spillway hydrograph and freeboard hydrograph?
11. What is the basin factor in USBR unit hydrograph practice?
12. How is the synthetic unit hydrograph parameter K_n related to overall basin response?
13. From the computational standpoint, what is the difference between dimensionless unit hydrographs and S-graphs?
14. In USBR practice, what is a specific baseflow discharge? What is it used for?
15. What is TVA precipitation?
16. In TVA practice, what major factors are considered in PMF determinations?
17. What are the three variables most frequently used in the USGS state equations for flood-peak estimation at ungaged sites?

PROBLEMS

14-1. Calculate a 24-h PMP value for a 50-km² watershed in a humid climate, with the following 24-h annual-maxima rainfall data: mean, 300 mm; standard deviation, 120 mm. Use Fig. 14-21(a) to accomplish the areal correction.

14-2. Calculate a 1-h point PMP value for a geographic location with the following 1-h annual-maxima rainfall data: mean, 35 mm; standard deviation, 12 mm.

14-3. Determine a 24-h all-season PMP depth-area relation for Topeka, Kansas. Use this relation to determine the PMP for a 150-mi² basin in the region.

14-4. Determine a 24-h all-season PMP depth-area relation for Greensboro, North Carolina. Use this relation to determine the PMP for a 3,500-mi² basin in the region.

14-5. Compute the 24-, 48-, and 72-h PMP values for the month of December, for a 500-mi² basin located in San Diego County, California, at 33° N and 117° W.

14-6. Develop a critical storm pattern for the following 6-h storm PMP increments (in inches): (a) 6.2, 3.1, 2.1, 1.5 and (b) 6.1, 4.6, 3.9, 3.4, 2.9, 2.6, 2.2, 1.9, 1.7, 1.5, 1.3, 1.2.

14-7. Using Figs. 14-18 and 14-9(b), determine a 6-h, 100-y precipitation estimate for a 250-mi² basin near Chicago, Illinois.

14-8. Given an 800-mi² basin near Cedar Rapids, Iowa, at 42° N, 91°30' W, develop: (a) a 96-h SPS estimate, in increments of 24 h, and (b) a 24-h SPS estimate, in increments of 6 h.

14-9. Following SCS criteria for design of principal spillways, determine the 1-d and 10-d runoff (in inches) near Salt Lake City, Utah, for the following frequencies: (a) 100-y, (b) 50-y, and (c) 25-y.

14-10. Use Fig. 14-17 to determine the quick-return flow for a 48-mi² watershed in eastern Iowa.

14-11. Calculate the quick-return flow for a 105-mi² basin with 36 in. of mean annual precipitation and 48°F mean annual temperature.

14-12. Determine a freeboard hydrograph precipitation value for a class (a) storage reservoir with 6000 ac-ft of storage and 12 ft effective height, and $P_{100} = 6$ in., PMP = 20 in. Assume no existing or planned upstream dams.

14-13. A 525-mi² basin located in the Great Plains has the following properties: hydraulic length $L = 34.5$ mi, length to centroid $L_c = 14$ mi, and overall slope $S = 1$ ft/mi. Assume the basin has a well-defined drainage network. Calculate a 2-h unit hydrograph by the USBR method, using (a) the dimensionless unit hydrograph and (b) the dimensionless S-graph.

14-14. Calculate a 1-h (local storm) unit hydrograph by the USBR method for a 95-mi² watershed located in the Rocky Mountains, with lag time 4 h. Use (a) the dimensionless unit hydrograph, and (b) the dimensionless S-graph.

14-15. Calculate a 100-y peak discharge for a 21.8-mi² watershed near Terra Haute, Indiana, with channel slope 12.6 ft/mi.

14-16. Calculate a 100-y peak discharge for a 418-mi² basin located in Montgomery County, Maryland, with forest cover 43% and 2-y, 24-h precipitation 3.2 in.

14-17. Calculate a 100-y peak discharge for a 80-mi² catchment near Madison, Wisconsin, with channel slope 10 ft/mi, storage factor 4%, and soil permeability (based on the least permeable horizon) 1.08 in./h.

REFERENCES

1. Benson, M. A. (1962). "Factors Influencing the Occurrence of Floods in a Humid Region of Diverse Terrain," *U.S. Geological Survey Water Supply Paper* 1580-B.

2. Benson, M. A. (1964). "Factors Affecting the Occurrence of Floods in the Southwest," *U.S. Geological Survey Water Supply Paper* 1580-D.

3. Carpenter, D. H. (1980). "Technique for Estimating Magnitude and Frequency of Floods in Maryland," *U.S. Geological Survey Water Resources Investigations Open-File Report* 80-1016, Towson, Maryland.

4. Conger, D. H. (1981). "Techniques for Estimating Magnitude and Frequency of Floods for Wisconsin Streams," *U.S. Geological Survey Water Resources Investigations Open File Report* 80-1214, Madison, Wisconsin, March.

5. Glatfelter, D. R. (1984). "Techniques for Estimating Magnitude and Frequency of Floods of Streams in Indiana," *U.S. Geological Survey Water Resources Investigations Report* 84-4134, Indianapolis, Indiana.

6. Hannum, C. H. (1976). "Techniques for Estimating Magnitude and Frequency of Floods in Kentucky," *U.S. Geological Survey Water Resources Investigations* 76-62.

7. Hedman, E. R., and W. R. Osterkamp. (1982). "Streamflow Characteristics Related to Channel Geometry of Streams in Western United States," *U.S. Geological Survey Water Supply Paper* 2193.

8. Hershfield, D. M. (1961). "Estimating the Probable Maximum Precipitation," *Journal of the Hydraulics Division*, ASCE, Vol. 87, No. HY5, September, pp. 99–116.

9. Hershfield, D. M. (1965). "Method for Estimating Probable Maximum Precipitation," *Journal of the American Waterworks Association*, Vol. 57, No. 8, August, pp. 965–972.

10. Jordan, P. R. (1984). "Magnitude and Frequency of High Flows of Unregulated Streams in Kansas," *U.S. Geological Survey Open File Report* 84-453.

11. Lee, F. L. (1985). "Floods in Louisiana, Magnitude and Frequency, Fourth Edition," *U.S. Geological Survey Water Resources Technical Report No.* 36, Baton Rouge, Louisiana.

12. Liscum, F., and B. C. Massey. (1980). "Technique for Estimating the Magnitude and Frequency of Floods in the Houston, Texas, Metropolitan Area," *U.S. Geological Survey Water Resources Investigations* 80-17.

13. NOAA *Hydrometeorological Report No. 23.* (1947). "Generalized Estimates of Maximum Possible Precipitation Over the Unites States East of the 105th Meridian, for Areas of 10, 200 and 500 Square Miles."

14. NOAA *Hydrometeorological Report No. 33.* (1956). "Seasonal Variation of the Probable Maximum Precipitation East of the 105th Meridian for Areas from 10 to 1,000 Square Miles, and Durations of 6, 12, 24, and 48 Hours."

15. NOAA *Hydrometeorological Report No. 36.* (1969). "Interim Report, Probable Maximum Precipitation in California," issued 1961, revised 1969.

16. NOAA *Hydrometeorological Report No. 39.* (1963). "Probable Maximum Precipitation in the Hawaiian Islands."

17. NOAA *Hydrometeorological Report No. 40.* (1965). "Probable Maximum Precipitation, Susquehanna River Drainage above Harrisburg, Pennsylvania."

18. NOAA *Hydrometeorological Report No. 41.* (1965). "Probable Maximum and TVA Precipitation over the Tenneessee River Basin above Chattanooga."

19. NOAA *Hydrometeorological Report No. 43.* (1966). "Probable Maximum Precipitation, Northwest States."

20. NOAA *Hydrometeorological Report No. 44.* (1969). "Probable Maximum Precipitation over South Platte River, Colorado, and Minnesota River, Minnesota."

21. NOAA *Hydrometeorological Report No. 45.* (1969). "Probable Maximum and TVA Precipitation for Tennessee River Basins up to 3,000 Square Miles in Area and Durations to 72 Hours."

22. NOAA *Hydrometeorological Report No. 47.* (1973). "Meteorological Criteria for Extreme Floods for Four Basins in the Tennessee and Cumberland River Watersheds."

23. NOAA *Hydrometeorological Report No. 48.* (1973). "Probable Maximum Precipitation and Snowmelt Criteria for Red River of the North, Above Pembina, and Souris River Above Minot, North Dakota."

24. NOAA *Hydrometeorological Report No. 49.* (1977). "Probable Maximum Precipitation Estimates, Colorado River and Great Basin Drainages."

25. NOAA *Hydrometeorological Report No. 51.* (1978). "Probable Maximum Precipitation, United States East of 105th Meridian."

26. NOAA *Hydrometeorological Report No. 52.* (1982). "Application of Probable Maximum Precipitation Estimates, United States East of the 105th Meridian."

27. NOAA *Hydrometeorological Report No. 53.* (1980). "Seasonal Variation of 10-Square Mile Probable Maximum Precipitation Estimates, United States East of the 105th Meridian."

28. NOAA *Hydrometeorological Report No. 54.* (1954). "Probable Maximum Precipitation and Snowmelt Criteria for Southeast Alaska."

29. NOAA *Hydrometeorological Report No. 55.* (1984). "Probable Maximum Precipitation Estimates-United States Between the Continental Divide and the 103rd Meridian."

30. NOAA *Hydrometeorological Report No. 56.* (1986). "Probable Maximum and TVA Precipitation Estimates with Areal Distributions for Tennessee River Drainages Less Than 3,000 Square Miles in Area."

31. NOAA National Weather Service. (1973). "Precipitation-Frequency Atlas of the Western United States," *NOAA Atlas 2,* in 11 volumes, Silver Spring, Maryland.

32. NOAA National Weather Service. (1980). "A Methodology for Point-to-Area Rainfall Frequency Ratios," *NOAA Technical Report NWS* 24, February.

33. NOAA National Weather Service. (1980). "Comparison of Generalized Estimates of Probable Maximum Precipitation with Greatest Observed Rainfalls," *NOAA Technical Report NWS* 25, March.

34. Newton, D. W. and R. G. Lee. (1969). "Storms Antecedent to Major Floods," 50th Annual Meeting of the American Geophysical Union, Washington, D.C., April.

35. Newton, D. W. (1983). "Realistic Assessment of Maximum Flood Potential," *Journal of Hydraulic Engineering,* ASCE, Vol. 109, No. 6, June, pp. 905–918.

36. Newton, D. W. (1986). "TVA Practice in Flood Frequency and Risk Analysis," *Proceedings, International Symposium on Flood Frequency and Risk Analysis,* Louisiana State University, Baton Rouge, May 14–17.

37. Paulhus, J. L. H., and C. S. Gilman. (1953). "Evaluation of Probable Maximum Precipitation," *Transactions,* American Geophysical Union, Vol. 34, No. 5, pp. 701–708.

38. Riggs, H. C. (1978). "Streamflow Characteristics from Channel Size," *Journal of the Hydraulics Division,* ASCE, Vol. 104, No. HY1, January, pp. 87–96.

39. Rodriguez-Iturbe, I., and J. M. Mejia. (1974). "On the Transformation of Point Rainfall to Areal Rainfall," *Water Resources Research,* Vol. 10, pp. 729–735.

40. Sauer, V. B., W. O. Thomas, V. A. Stricker, and K. V. Wilson. (1983). "Flood Characteristics of Urban Watersheds in the United States," *U.S. Geological Survey Water Supply Paper* 2207.

41. Sauer, V. B. (1973). "Flood Characteristics of Oklahoma Streams," *U.S. Geological Survey Water Resources Investigations* 52–73.

42. Simmons, R. H., and D. H. Carpenter. (1978). "Technique for Estimating the Magnitude and Frequency of Floods in Delaware," *U.S. Geological Survey Water Resources Investigations Open File Report* 78–93.

43. Thomas, W. O. (1987). "Summary of U.S. Geological Survey Regional Regression Equations," 66th Annual Meeting of the Transportation Research Board, Washington, D.C., January.

44. Tortorelli, R. L., and D. L. Bergman. (1985). "Techniques for Estimating Flood Peak

Discharges for Unregulated Streams and Streams Regulated by Small Floodwater Retarding Structures in Oklahoma," *U.S. Geological Survey Water Resources Investigations Report* 84-4358, Oklahoma City, Oklahoma.

45. U.S. Army Corps of Engineers. (1965). "Standard Project Flood Determinations," *Civil Engineer Bulletin No. 52-8, Engineering Manual EM* 1110-2-1411, issued March 1952, revised 1965.

46. U.S. Bureau of Reclamation. (1987). *Design of Small Dams,* 3d. ed., Denver, Colorado.

47. USDA Soil Conservation Service. (1985). "Earth Dams and Reservoirs," *Technical Release No.* 60 (TR-60), 210-VI, revised October.

48. USDA Soil Conservation Service. (1985). *National Engineering Handbook,* Section 4: Hydrology," Washington, D.C.

49. U.S. Interagency Advisory Committee on Water Data, Hydrology Subcommittee. (1983). "Guidelines for Determining Flood Flow Frequency," *Bulletin No.* 17B, issued 1981, revised 1983, Reston, Virginia.

50. U.S. Weather Bureau. (1960). "Generalized Estimates of Probable Maximum Precipitation for the United States West of the 105th Meridian, for Areas to 400 Square Miles and Durations to 24 Hours," *Technical Paper No.* 38.

51. Wahl, K. L. (1977). "Accuracy of Channel Measurements and the Implications in Estimating Streamflow Characteristics," *U.S. Geological Survey Journal of Research,* Vol. 5, No. 6, pp. 811–814.

52. Wiesner, C. J. (1970). *Hydrometeorology.* London: Chapmand and Hall Ltd.

53. World Meteorological Organization. (1960). "Manual for Estimation of the Probable Maximum Precipitation," WMO No. 332, *Operational Hydrology Report No.* 1, Geneva, Switzerland.

54. World Meteorological Organization. (1969). "Estimation of Maximum Floods," WMO No. 233, TP 126, *Technical Note No.* 98, Geneva, Switzerland.

SUGGESTED READINGS

Hershfield, D. M. (1961). "Estimating the Probable Maximum Precipitation," *Journal of the Hydraulics Division,* ASCE, Vol. 87, No. HY5, September, pp. 99–116.

NOAA *Hydrometeorological Report No. 36.* (1969). "Interim Report, Probable Maximum Precipitation in California," issued 1961, revised 1969.

NOAA *Hydrometeorological Report No. 43.* (1966). "Probable Maximum Precipitation, Northwest States."

NOAA *Hydrometeorological Report No. 49.* (1977). "Probable Maximum Precipitation Estimates, Colorado River and Great Basin Drainages."

NOAA *Hydrometeorological Report No. 51.* (1978). "Probable Maximum Precipitation, United States East of 105th Meridian."

NOAA *Hydrometeorological Report No. 52.* (1982). "Application of Probable Maximum Precipitation Estimates, United States East of the 105th Meridian."

NOAA *Hydrometeorological Report No. 53.* (1980). "Seasonal Variation of 10-Square Mile Probable Maximum Precipitation Estimates, United States East of the 105th Meridian."

NOAA *Hydrometeorological Report No. 55.* (1984). "Probable Maximum Precipitation Estimates-United States Between the Continental Divide and the 103rd Meridian."

U.S. Army Corps of Engineers. (1965). "Standard Project Flood Determinations," *Civil Engineer Bulletin No. 52-8, Engineering Manual EM* 1110-2-1411, issued March 1952, revised 1965.

U.S. Bureau of Reclamation. (1987). *Design of Small Dams,* 3d. ed., Denver, Colorado.

USDA Soil Conservation Service. (1985). "Earth Dams and Reservoirs," *Technical Release No.* 60 (TR-60), 210-VI, revised October.

SEDIMENT IN THE HYDROLOGIC CYCLE

Rainfall and surface runoff are the agents responsible for the detachment and movement of soil particles on the land surface. These soil particles are referred to as *sediments*. The study of sediment detachment and movement is an important subject in engineering hydrology. Indeed, the subject of sediment transcends engineering hydrology to encompass the related fields of fluvial geomorphology, sediment transport, and sedimentation and river engineering [2, 4, 25, 37].

The study of sediments in the hydrologic cycle can be divided into the following three processes: (1) production, (2) transport, and (3) deposition. These can be linked to the various liquid-transport phases of the hydrologic cycle. At the catchment level, sediment production by soil particle detachment is primarily the result of raindrop impact. Once detachment has taken place, surface runoff acts to transport sediment downslope, first as overland flow (sheet and rill flow), and eventually as stream- and river flow. Deposition of sediment occurs at any point downstream where the kinetic energy of the flow is insufficient to support sediment entrainment in the flowing water.

Sediment production refers to the processes by which sediment is produced, the identification of sediment sources and amounts, and the determination of sediment yields. The source of sediment can usually be traced back to the upland catchments, although these are by no means the only source. In certain cases, streambank erosion in the lower valleys may constitute an important source of sediment.

Sediment from upland catchments is delivered to streams and rivers, wherein sediment transport takes place. Sediment transport refers to the mechanisms by which sediment is moved downstream by flowing water, either in suspension or by rolling and sliding along the river bottom.

The transport of sediment continues in the downstream direction until the flow is no longer able to carry the sediment, at which time sediment deposition occurs. Typically, the first opportunity for sediment deposition is at the entrance to reservoirs and water impoundments, where the flow is decelerated by the action of structures. Deposition is also likely to occur naturally, for instance, downstream of sudden decreases in energy slope or in situations where the capacity of the flow to carry sediment is substantially diminished. In the absence of these natural or human-made features, sediment transport by the flow may continue unabated until it reaches the ocean, at which time the flow loses its kinetic energy and sediment deposition goes on to contribute to delta growth.

This chapter is divided into five sections. Section 15.1 describes sediment properties. Section 15.2 describes sediment production, sediment sources, and sediment yield. Section 15.3 discusses sediment transport, sediment transport formulas, and sediment rating curves, including a brief introduction to sediment routing. Section 15.4 describes sediment deposition in reservoirs. Section 15.5 describes sediment measurement techniques.

15.1 SEDIMENT PROPERTIES

Sediment Formation

Sediments are the products of disintegration and decomposition of rocks. Disintegration includes all processes by which rocks are broken into smaller pieces without substantial chemical change. The disintegration of rocks is caused either by large temperature changes or by alternate cycles of freezing and thawing. Decomposition refers to the breaking down of mineral components of rocks by chemical reaction. Decomposition includes the processes of (1) carbonation, (2) hydration, (3) oxidation, and (4) solution.

Carbon dioxide (CO_2), present in the atmosphere and organic sources, readily unites with water to form carbonic acid (H_2CO_3). Carbonic acid reacts with feldspars to produce clay minerals, silica, calcite, and other relatively soluble carbonates containing potassium, sodium, iron, and magnesium. The addition of water to many of the minerals present in igneous rocks results in the formation of clay minerals such as aluminum silicates. Many secondary minerals are formed from igneous rocks by oxidation, which is accelerated by the presence of moisture in the air. Solution is another important mechanism in the alteration of igneous rock. Oxygen combines with other elements to form sulfates, carbonates, and nitrates, most of which are relatively soluble. The amount (by weight) of dissolved solids carried by streams in the contiguous United States has been estimated at more than 50 percent of the amount of suspended sediment [30].

Particle Characteristics

The characteristics of mineral grains help describe the properties of sediments. Among them are (1) size, (2) shape, (3) specific weight and specific gravity, and (4) fall velocity.

Size. Particle size is a readily measured sediment characteristic. A widely accepted classification of sediments according to size is shown in Table 15-1. Five groups of sizes are included in this table: (1) boulders and cobbles, (2) gravel, (3) sand, (4) silt, and (5) clay. Boulders and cobbles can be measured individually. Gravel-size particles can be measured individually or by sieving. Sand-size particles are readily measured by sieving. A No. 200 screen is used to separate sand particles from finer particles such as silt and clay. Silt and clay particles are separated by measuring the differences in their rate of fall in still water.

Shape. Particle shape is numerically defined in terms of its sphericity and roundness. True sphericity is the ratio of the surface area of a sphere having the same volume as the particle to the surface area of the particle. The practical difficulty of measuring true sphericity has led to an alternate definition of sphericity as the ratio of the diameter of a sphere having the same volume as the particle (i.e., the nominal diameter) to the diameter of a sphere circumscribing the particle. Accordingly, a sphere has a sphericity of 1, whereas all other shapes have a sphericity of less than 1.

TABLE 15-1 CLASSIFICATION OF SEDIMENTS ACCORDING TO SIZE [27]

Class	Size (mm)
Boulders and cobbles	
Very large boulders	4096–2048
Large boulders	2048–1024
Medium boulders	1024–512
Small boulders	512–256
Large cobbles	256–128
Small cobbles	128–64
Gravel	
Very coarse	64–32
Coarse	32–16
Medium	16–8
Fine	8–4
Very fine	4–2
Sand	
Very coarse	2.0–1.0
Coarse	1.0–0.5
Medium	0.50–0.25
Fine	0.250–0.125
Very fine	0.125–0.062
Silt	
Coarse	0.062–0.031
Medium	0.031–0.016
Fine	0.016–0.008
Very fine	0.008–0.004
Clay	
Coarse	0.0040–0.0020
Medium	0.0020–0.0010
Fine	0.0010–0.0005
Very fine	0.0005–0.00025

Roundness is defined as the ratio of the average radius of curvature of the particle edges to the radius of the largest inscribed circle. It refers to the sharpness of the edges of sediment particles and is commonly used as an indicator of particle wear.

In sediment studies, the *shape factor* is often used as an indicator of particle shape:

$$SF = \frac{c}{(ab)^{1/2}} \tag{15-1}$$

in which SF = shape factor and a, b, and c are three orthogonal particle length dimensions. According to Corey [12], a is the longest, b is the intermediate, and c is the shortest length dimension. However, according to McNown and Malaika [32], c is measured in the direction of motion, and a and b are perpendicular to c.

Specific Weight and Specific Gravity. The specific weight of a sediment particle is its weight per unit volume. The specific gravity of a sediment particle is the ratio of its weight to the weight of an equal volume of water. Most sediment particles consist of either quartz or feldspar, which are about 2.65 times heavier than water. Therefore, the specific gravity of sediments is generally considered to be about 2.65. Exceptions are heavy minerals (for instance, magnetite, with specific gravity of 5.18), but these occur rather infrequently.

Fall Velocity. The fall velocity of a sediment particle is its terminal rate of settling in still water. Fall velocity is a function of size, shape, and specific weight of the particle, and the specific weight and viscosity of the surrounding water. For spherical particles, the fall velocity (derived from a balance of submerged weight and drag) can be expressed as follows:

$$w = \left[\frac{4}{3} \frac{gd_s}{C_D} \frac{\gamma_s - \gamma}{\gamma} \right]^{1/2} \tag{15-2}$$

in which w = fall velocity, g = gravitational acceleration, d_s = particle diameter, C_D = drag coefficient (dimensionless), γ_s = specific weight of sediment particles, and γ = specific weight of water.

The drag coefficient is a function of the particle Reynolds number R, defined as:

$$R = \frac{wd_s}{\nu} \tag{15-3}$$

in which ν = kinematic viscosity of the fluid. For particle Reynolds numbers less than 0.1, the drag coefficient is equal to $C_D = 24/R$. Substituting this value of C_D into Eq. 15-2 leads to Stokes' law:

$$w = \left[\frac{gd^2}{18\nu} \right] \left(\frac{\gamma_s - \gamma}{\gamma} \right) \tag{15-4}$$

For particle Reynolds numbers greater than 0.1, the drag coefficient is still a function of Reynolds number, but the relationship cannot be expressed in analytical form. The relationship of C_D versus R for a wide range of particle Reynolds numbers is shown in Fig. 15-1 [35].

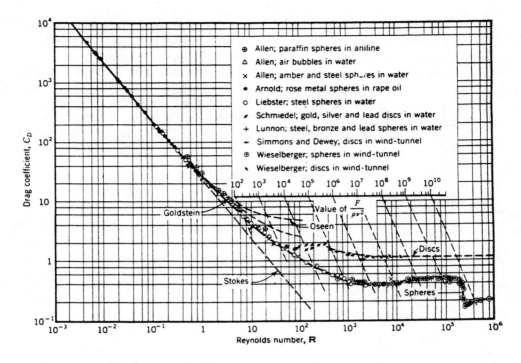

Figure 15-1 Drag coefficient of spheres versus Reynolds number [35].

Since fall velocities vary with fluid temperature and viscosity, two particles of the same size, shape, and specific gravity, falling on two fluids of different viscosity or in the same fluid at different temperatures, will have different fall velocities. To provide a measure of comparison, the concept of *standard fall velocity* was developed [23]. The standard fall velocity of a particle is the average rate of fall that it would attain if falling alone in quiescent water of infinite extent at the temperature of 24°C. Likewise, the standard fall diameter of a particle is the diameter of an equivalent sphere having the same standard fall velocity and specific gravity.

Example 15-1.

Calculate the fall velocity of a spherical quartz particle of diameter $d_s = 0.1$ mm and drag coefficient $C_D = 4$.

Using Eq. 15-2, with $\gamma_s = 2.65$ g/cm^3, and $\gamma = 1.0$ g/cm^3, $g = 9.81$ m/s^2, $d_s = 0.0001$ m: $w = 0.023$ m/s.

Size Distribution of Sediment Deposits

An important property of sediment deposits is the size distribution of its individual particles. Particle size distribution is a key to predicting the behavior of a sediment deposit and estimating its specific weight. A sediment sample containing a wide range of particle sizes is well graded, or poorly sorted. Conversely, a sediment sample consisting of particles in a narrow range of particle sizes is poorly graded, or well sorted.

The size distribution of sediments can be measured in several ways. The coarsest fraction can be separated by direct measurement for boulders and cobbles and by sieving for sands and gravels. For most applications involving sediments in the sand size, the visual accumulation (VA) tube is a fast, economical, and accurate method of determining the size distribution of sediment samples. In the VA tube method, the particles start falling from a common source and become stratified according to their relative settling velocities. At a given instant, the particles coming to rest at the bottom of the tube are of a certain *sedimentation size,* finer than particles that have already settled and coarser than those still remaining in suspension. See [19] for a description of laboratory methods for sediment analysis.

Specific Weight of Sediment Deposits

The specific weight of a sediment deposit is the dry weight of sedimentary material per unit volume. Due to the voids between sediment particles, the specific weight of a sediment deposit is always less than the specific weight of individual particles. A knowledge of the specific weight of a sediment deposit allows the conversion of sediment weights to sediment volumes and vice versa. In particular, the specific weight of a sediment deposit is useful in studies of reservoir storage depletion by deposition of fluvial sediments.

Factors influencing the specific weight of a sediment deposit are (1) its mechanical composition, (2) the environment in which the deposits are formed, and (3) time. Coarse materials, e.g., boulders, gravel, and coarse sand, are deposited with specific weights very nearly equal to their ultimate value and change very little with time. However, fine materials such as silts and clays may have initial specific weights that are only a fraction of their ultimate value.

Lane and Koelzer [28] have developed an empirical relationship to account for the variation of the specific weight of sediment deposits in reservoirs with time. Their relationship is

$$W = W_1 + B \log T \qquad (15-5)$$

in which W = specific weight of the deposit after T years; W_1 = initial specific weight of the deposit, measured after 1 y of consolidation; and B = a constant. Table 15-2 shows values of W_1 and B as a function of sediment size and mode of reservoir operation. For mixed deposits, a weighted average of specific weight is appropriate.

Drying or aeration of a sediment deposit helps to accelerate consolidation through removal of the water from the pore spaces. Table 15-3 shows the effect of aeration on the specific weight of sediment deposits for several types of soil mixtures [18].

Example 15-2.

Calculate the specific weight of a sediment deposit in a reservoir after an elapsed time of 50 y, with the sediment always submerged or nearly submerged. Assume the following size distribution: sand, 30%; silt, 45%, clay, 25%.

Using Table 15-2, the specific weights for the various sizes are: sand, 93 lb/ft³; silt, 75 lb/ft³; clay, 57 lb/ft³. Therefore, the weighted average is: $W = (93 \times 0.30) + (75 \times 0.45) + (57 \times 0.25) = 75.9$ lb/ft³.

TABLE 15-2 CONSTANTS FOR ESTIMATING SPECIFIC WEIGHT OF RESERVOIR SEDIMENT DEPOSITS, EQ. 15-5 (lb/ft³) [28]

Mode of Reservoir Operation	Sand		Silt		Clay	
	W_1	B	W_1	B	W_1	B
Sediment always submerged or nearly submerged	93	0	65	5.7	30	16.0
Normally a moderate reservoir drawdown	93	0	74	2.7	46	10.7
Normally considerable reservoir drawdown	93	0	79	1.0	60	6.0
Reservoir normally empty	93	0	82	0.0	78	0.0

Note: 1 lb/ft³ = 157.1 N/m³.

TABLE 15-3 RANGE IN SPECIFIC WEIGHT OF SEDIMENT DEPOSITS (lbs/ft³) [18]

Soil Description	Permanently Submerged	Aerated
Clay	40–60	60–80
Silt	55–75	75–85
Clay-silt mixture	40–65	65–85
Sand-silt mixture	75–95	95–110
Clay-silt-sand mixture	50–80	80–100
Sand	85–100	85–100
Gravel	85–125	85–125
Poorly-sorted sand and gravel	95–130	95–130

Note: 1 lb/ft³ = 157.1 N/m³.

15.2 SEDIMENT PRODUCTION

The presence of sediment in streams and rivers has its origin in soil erosion. Erosion encompasses a series of complex and interrelated natural processes that have the effect of loosening and moving away soil and rock materials under the action of water, wind, and other geologic factors. In the long term, the effect of erosion is the denudation of the land surface, i.e., the removal of soil and rock particles from exposed surfaces, their transport to lower elevations, and eventual deposition.

The rate of landscape denudation can be quantified from a geological perspective. For instance, the number of centimeters of denudation per 1000 y can be used as a measure of the erosive activity of a region. Geologic measures of landscape denudation appear insignificant when compared to the typical timespan of human activity, say 25 to 100 y. However, the quantities of sediment moved may be important when considering the impact that sediment loads have on the operation and design life of reservoirs and hydraulic structures.

At the outset of the study of sediment production, a distinction should be made between the amount of sediment eroded at the source(s) and the amount of sediment delivered to a downstream point. *Gross sediment production* refers to the amount of sediment eroded and removed from the source(s). *Sediment yield* refers to the actual delivery of eroded soil particles to a given downstream point. Since eroded particles may be deposited before they reach the downstream point of interest, sediment yield quantities are generally less than gross sediment production quantities. The ratio of sediment yield to gross sediment production is the *sediment-delivery ratio (SDR)*.

Gross sediment production is commonly measured in terms of weight of sediment per unit drainage area per unit time—for instance, metric tons per hectare per year, or tons per acre per year. Sediment yield is expressed in terms of weight per unit time past a certain point—for instance, metric tons per day at the catchment outlet.

Normal and Accelerated Erosion

According to the timespan involved, erosion can be classified as (1) normal, or geologic, and (2) accelerated, or human-induced. Normal erosion has been occurring at variable rates since the first solid materials formed on the surface of the earth. Normal erosion is extremely slow in most places and is largely a function of climate, parent rocks, precipitation, topography, and vegetative cover. Accelerated erosion occurs at a much faster rate than normal, usually through reduction of vegetative cover. Deforestation, cultivation, forest fires, and systematic destruction of natural vegetation result in accelerated erosion.

Sediment Sources

According to its source, erosion can be classified as (1) sheet erosion, (2) rill erosion, (3) gully erosion, and (4) channel erosion. *Sheet erosion* is the wearing away of a thin layer on the land surface, primarily by overland flow. *Rill erosion* is the removal of soil by small concentrations of flowing water (rills). *Gully erosion* is the removal of soil from incipient channels that are large enough so that they cannot be removed by normal cultivation. *Channel erosion* refers to erosion occurring in stream channels in the form of streambank erosion or streambed degradation. For practical purposes, a distinction is made between upland and channel erosion. *Upland erosion* is mostly made up of sheet and rill erosion, whereas channel erosion encompasses all other sediment sources, specifically excluding sheet and rill erosion.

Upland Erosion and the Universal Soil Loss Equation

In the United States, the prediction of upland erosion amounts (i.e., sheet and rill erosion) is commonly made by the universal soil loss equation (USLE), developed by the USDA Agricultural Research Service in cooperation with USDA Soil Conservation Service and certain state experiment stations.

The universal soil loss equation is [43]:

$$A = RKLSCP \tag{15-6}$$

in which A = (annual) soil loss due to sheet and rill erosion in tons per acre per year; R = rainfall factor; K = soil erodibility factor; L = slope-length factor; S = slope-gradient factor; C = crop-management factor; and P = erosion-control-practice factor.

Rainfall Factor.   When factors other than rainfall are held constant, soil losses from cultivated fields are shown to be directly proportional to the product of the storm's total kinetic energy E and its maximum 30-minute intensity I. The product EI reflects the combined potential of raindrop impact and runoff turbulence to transport dislodged soil particles.

The sum of EI products for a given year is an index of the erosivity of all rainfall for that year. The rainfall factor R is the average value of the series of annual sums of EI products. Values of R applicable to the contiguous United States are shown in Fig. 15-2.

Soil Erodibility Factor.   The soil erodibility factor K is a measure of the resistance of a soil surface to erosion. It is defined as the amount of soil loss (in tons per acre per year) per unit of rainfall factor R from a *unit plot*. A unit plot is 72.6 ft long, with a uniform lengthwise gradient of 9 percent, in continuous fallow, tilled up and down the slope.

Values of K for 23 major soils on which erosion plot studies were conducted since 1930 are listed in Table 15-4. Soil erodibility factors for other soils have been estimated by comparing their characteristics with those of the 23 soils listed in Table 15-4. A method for determining the soil erodibility factor based on soil characteristics has been proposed by Wischmeier et al. [44].

Slope-length and Slope-gradient Factors.   The rate of soil erosion by flowing water is a function of slope length (L) and gradient (S). For practical purposes, these two topographic characteristics are combined into a single topographic factor (LS). The topographic factor is defined as the ratio of soil loss from a slope of given length and gradient to the soil loss from the unit plot (of 72.6 ft length and 9 percent gradient). Figure 15-3 shows values of LS as a function of slope length and gradient.

Crop-management Factor.   The crop-management factor C is defined as the ratio of soil loss from a certain combination of vegetative cover and management practice to the soil loss resulting from tilled, continuous fallow. Values of C range from as little as 0.0001 for undisturbed forest land to a maximum of 1.0 for disturbed areas with no vegetation. Values of C for cropland are estimated on a local basis. Table 15-5 shows values of C for permanent pasture, grazed forest land, range, and idle land. Table 15-6 shows values of C for undisturbed forest land.

Erosion-control-practice Factor.   The erosion-control-practice factor P is defined as the ratio of soil loss under a certain erosion-control practice to the soil loss resulting from straight-row farming. Practices for which P have been established are contouring and contour strip-cropping. In contour strip-cropping, strips of sod or meadow are alternated with strips of row crops or small grains. Values of P used for contour strip-cropping are also used for contour-irrigated furrows. Table 15-7 shows values of P for contour-farmed terrace fields.

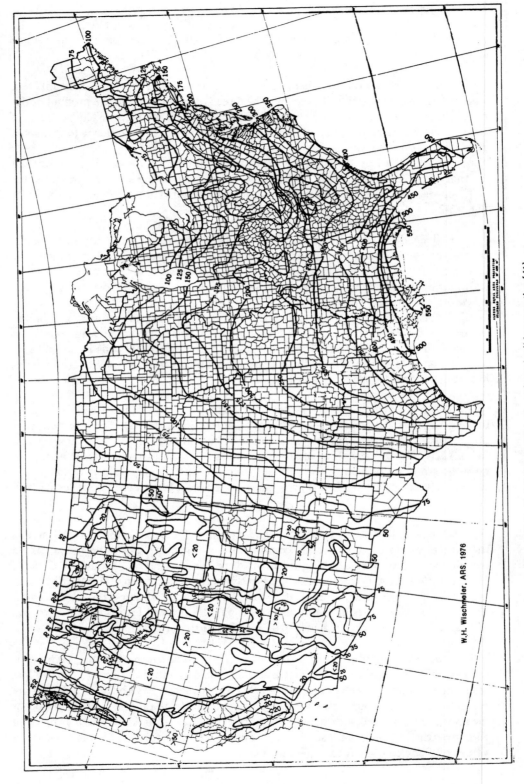

Figure 15-2 Rainfall factor R in the universal soil loss equation [41].

W.H. Wischmeler, ARS, 1976

TABLE 15-4 VALUES OF SOIL ERODIBILITY FACTOR K [43]

Soil Type	Source of Data	K
Dunkirk silt loam	Geneva, NY	0.69[1]
Keen silt loam	Zanesville, OH	0.48
Shelby loam	Bethany, MO	0.41
Lodi loam	Blacksbrug, VA	0.39
Fayette silt loam	LaCrosse, WI	0.38[1]
Cecil snady clay loam	Watkinsville, GA	0.36
Marshall silt loam	Clarinda, IO	0.33
Ida silt loam	Castana, IO	0.33
Mansic clay loam	Hays, KA	0.32
Hagerstown silty clay loam	State College, PA	0.31[1]
Austin clay	Temple, TX	0.29
Mexico silt loam	McCredie, MO	0.28
Honeoye silt loam	Marcellus, NY	0.28[1]
Cecil sandy loam	Clemson, SC	0.28[1]
Ontario loam	Geneva, NY	0.27[1]
Cecil clay loam	Watkinsville, GA	0.26
Boswell fine sandy loam	Tyler, TX	0.25
Cecil sand loam	Watkinsville, GA	0.23
Zaneis fine sandy loam	Guthrie, OK	0.22
Tifton loamy sand	Tifton, GA	0.10
Freehold loamy sand	Marlboro, NJ	0.08
Bath flaggy silt loam with surface stones greater than 2 in. removed	Arnot, NY	0.05[1]
Albia gravelly loam	Beemerville, NJ	0.03

[1]Evaluated from continuous fallow. All others were evaluated from row-crop data.

Use of the Universal Soil Loss Equation.

The USLE computes upland erosion from small watersheds on an average annual basis. It includes the detachment and transport components, but it does not account for the deposition component. Therefore, the USLE cannot be used to compute sediment yield. For example, in a 1000-mi² drainage basin, only 5 percent of the soil loss computed by the USLE may appear as sediment yield at the basin outlet. The remaining 95 percent is redistributed on uplands or flood plains and does not constitute a net soil loss from the drainage basin.

Example 15-3.

Assume a 600-ac watershed above a proposed floodwater-retarding structure in Fountain County, Indiana. Compute the average annual soil loss by the universal soil loss equation for the following conditions: (1) cropland, 280 ac, contour strip-cropped, soil is Fayette silt loam, slopes are 8% and 200 ft long; (2) pasture, 170 ac, 50% canopy cover, 80% ground cover with grass, soil is Fayette silt loam, slopes are 8% and 200 ft long; and (3) forest, 150 ac, soil is Marshall silt loam, 30% tree canopy cover, slopes are 12% and 100 ft long.

1. From Fig. 15-2, $R = 185$. From Table 15-4, $K = 0.38$. From Fig. 15-3, $LS = 1.4$. The value of C for cropland is obtained from local sources; assume $C = 0.12$ for this

TABLE 15-5 VALUES OF CROP-MANAGEMENT FACTOR C FOR PERMANENT PASTURE, GRAZED FOREST LAND, RANGE AND IDLE LAND [41]

Vegetative Canopy			Cover That Contacts the Soil Surface					
			Percent Ground Cover					
Type and Height[2]	% Cover[3]	Type[4]	0	20	40	60	80	100
No appreciable		G	0.45	0.20	0.10	0.042	0.013	0.003
canopy		W	0.45	0.24	0.15	0.091	0.043	0.011
Tall grass, weeds	25	G	0.36	0.17	0.09	0.038	0.013	0.003
or short brush		W	0.36	0.20	0.13	0.083	0.041	0.011
with average								
drop fall	50	G	0.26	0.13	0.07	0.035	0.012	0.003
of 20 in. or less		W	0.26	0.16	0.11	0.076	0.039	0.011
	75	G	0.17	0.10	0.06	0.032	0.011	0.003
		W	0.17	0.12	0.09	0.068	0.038	0.011
Appreciable brush	25	G	0.40	0.18	0.09	0.040	0.013	0.003
or bushes, with		W	0.40	0.22	0.14	0.087	0.042	0.011
average drop fall								
height of 6.5 ft	50	G	0.34	0.16	0.08	0.038	0.012	0.003
		W	0.34	0.19	0.13	0.082	0.041	0.011
	75	G	0.28	0.14	0.08	0.036	0.012	0.003
		W	0.28	0.17	0.12	0.078	0.040	0.011
Trees, but no	25	G	0.42	0.19	0.10	0.041	0.013	0.003
appreciable low		W	0.42	0.23	0.14	0.089	0.042	0.011
brush. Average								
drop fall height	50	G	0.39	0.18	0.09	0.040	0.013	0.003
of 13 ft		W	0.39	0.21	0.14	0.087	0.042	0.011
	75	G	0.36	0.17	0.09	0.039	0.012	0.003
		W	0.36	0.20	0.13	0.084	0.041	0.011

[1]The listed C values require that the vegetation and mulch be randomly distributed over the entire area. For grazed forest land, multiply these values by 0.7.

[2]Canopy height is measured as the average fall height of water drops falling from canopy to ground. Canopy effect is inversely proportional to drop fall height and is negligible if fall height exceeds 33 ft.

[3]Portion of total area surface that would be hidden from view by canopy in a vertical projection.

[4]G: cover at surface is grass, grasslike plants, decaying compacted duff, or litter. W: cover at surface is mostly broadleaf herbaceous plants (weeds) or undecayed residues or both.

example. From Table 15-7, $P = 0.25$. Using Eq. 15-6: $A = 185 \times 0.38 \times 1.4 \times 0.12 \times 0.25 = 2.95$ tons/ac/y.

2. $R = 185$; $K = 0.38$; $LS = 1.4$. From Table 15-5, $C = 0.012$. No value of P has been established for pasture; therefore, $P = 1$. Using Eq. 15-6: $A = 185 \times 0.38 \times 1.4 \times 0.012 \times 1.0 = 1.18$ tons/ac/y.

3. $R = 185$. From Table 15-4, $K = 0.33$. From Fig. 15-3, $LS = 1.8$. From Table 15-6, $C = 0.006$. No value of P has been established for forest. Using Eq. 15-6: $A = 185 \times 0.33 \times 1.8 \times 0.006 \times 1.0 = 0.66$ tons/ac/y.

The total sheet and rill erosion from the 600-ac watershed is $(280 \times 2.95) + (170 \times 1.18) + (150 \times 0.66) = 1126$ tons/y.

Percentage of Area Covered by Canopy of Trees and Undergrowth	Percentage of Area Covered by Litter[2]	C Value[3]
100–75	100–90	0.0001–0.001
70–45	85–75	0.002–0.004
40–20	70–40	0.003–0.009

[1]Where litter cover is less than 40% or canopy cover is less than 20%, use Table 15-5. Also, use Table 15-5 when woodlands are being grazed, harvested, or burned.

[2]Percentage of area covered by litter is dominant. Interpolate on basis of litter, not canopy.

[3]The ranges in listed C values are caused by the ranges in the specified forest litter and canopy cover, and by variations in effective canopy height.

TABLE 15-7   VALUES OF EROSION-CONTROL-PRACTICE FACTOR P
FOR CONTOURED-FARMED TERRACED FIELDS[1] [41]

Land Slope (percent)	For Farm Planning		For Computing Sediment Yield[2]	
	Contour Factor[3]	Strip-crop Factor	Graded Channels, Sod Outlets	Steep Backslope, Underground Outlets
1–2	0.60	0.30	0.12	0.05
3–8	0.50	0.25	0.10	0.05
9–12	0.60	0.30	0.12	0.05
13–16	0.70	0.35	0.14	0.05
17–20	0.80	0.40	0.16	0.06
21–25	0.90	0.45	0.18	0.06

[1]Slope length is the horizontal terrace interval. The listed values are for contour farming. No additional contour factor is used in the computation.

[2]These values include entrapment efficiency and are used for control of offsite sediment within limits and for estimating the field's contribution to watershed sediment yield.

[3]Use these values for control of interterrace erosion within specified soil-loss tolerances.

Channel Erosion

Channel erosion includes gully erosion, streambank erosion, streambed degradation, floodplain scour, and other sources of sediment, excluding upland erosion. Gullies are incipient channels in process of development. Gully growth is usually accelerated by severe climatic events, improper land use, or changes in stream base levels. Most of the significant gully activity, in terms of the quantities of sediment produced and delivered to downstream locations, is found in regions of moderate to steep topography having thick soil mantles. The total sediment outflow from gullies is usually less than sheet and rill erosion [31].

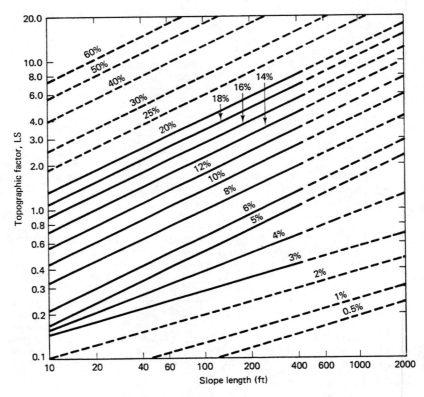

*The dashed lines represent estimates for slope dimensions beyond the
range of lengths and steepnesses for which data are available.

Figure 15-3 Topographic factor LS in the universal soil loss equation [41].

Streambank erosion and streambed degradation can be significant in certain
cases. Changes in channel alignment and/or removal of natural vegetation from
stream banks may cause increased bank erosion. Streambed degradation, typically
downstream of reservoirs, can also constitute an additional source of sediment.

Methods for determining soil loss due to the various types of channel erosion
include the following: (1) comparing aerial photographs taken at different times to
assess the growth rate of channels, (2) performing river cross-sectional surveys to determine changes in cross-sectional area, (3) assembling historical data to determine
the average age and growth rate of channels, and (4) performing field studies to evaluate the annual growth rate of channels.

Field surveys can often provide sufficient data to estimate streambank erosion as
follows [41]:

$$S = HLR \qquad (15\text{-}7)$$

in which S = annual volume of streambank erosion; H = average height of bank;
L = length of eroded bank, each side of channel if both sides are eroding; and R =
annual rate of bank recession (net rate if one side is eroding while the other is
depositing).

Streambed degradation can be estimated as follows [41]:

$$S = WLD \tag{15-8}$$

in which S = annual volume of streambed degradation; W = average bottom width of degrading channel reach; L = length of degrading channel reach; and D = annual rate of streambed degradation.

Accelerated Erosion Due to Strip-mining and Construction Activities

Strip-mining and construction activities greatly accelerate erosion rates. For instance, Collier et al. [10] found that a watershed with 10.4 percent of its area strip-mined eroded 76 times more sediment than a similar undisturbed watershed. Wolman and Shick [45] found that sediment concentrations in streams draining construction areas ranged from 3000 to 150,000 ppm, compared to concentrations of 2000 ppm in comparable natural settings. These studies indicate that human-induced land disturbances have a substantial impact on sediment production. With a careful choice of factors, the USLE can be used to compute soil loss from disturbed lands.

Sediment Yield

In engineering applications, the quantity of sediment eroded at the sources is not as important as the quantity of sediment delivered to a downstream point, i.e., the sediment yield.

The sediment yield is calculated by multiplying the gross sediment production, which includes all types of erosion (sheet, rill, gully, and channel erosion) by a sediment-delivery ratio that varies in the range 0 to 1 (it can also be expressed as a percentage). Therefore, a calculation of sediment yield hinges upon an estimate of gross sediment production (from the various sources) and an appropriate sediment-delivery ratio.

Sediment-delivery Ratio

The sediment-delivery ratio (SDR) is largely a function of (1) sediment source, (2) proximity of sediment source to the fluvial transport system, (3) density and condition of the fluvial transport system, (4) sediment size and texture, and (5) catchment characteristics.

The sediment source has an influence on the delivery ratio. Not all sediments originating in sheet and rill erosion are likely to enter the fluvial transport system. However, sediments produced by channel erosion are generally closer to the transport system and are more likely to be delivered to downstream points. The proximity of the sediment source to the transport system is also an important variable in the estimation of SDR. The amounts of sediment delivered to downstream points will depend to a large extent on the ability of the fluvial transport system to entrain and hold on to the sediment particles. Silt and clay particles can be transported much more readily than sand particles; therefore, the delivery of silts and clays is more likely to occur than that of sands. Catchment characteristics also affect sediment-delivery ratios. High relief often indicates both a high erosion rate and a high SDR. High channel density is usually an indication of an efficient transport system and, consequently, of a high SDR.

Estimation of Sediment-delivery Ratios. The SDR is the ratio of sediment yield to gross sediment production. Sediment yield can be evaluated by one of several methods. At reservoir locations, estimates of sediment yield can be obtained by reservoir sedimentation surveys. Alternatively, sediment yield can be evaluated by direct measurement of sediment load at the point of interest. Estimates of gross sediment production from upland sources can be obtained using either the USLE formula or a regionally derived formula for sheet and rill erosion. When warranted, this estimate can be augmented by field estimates of gully and channel erosion.

In the absence of actual measurements, statistical analysis can be used to develop regional regression equations to predict SDRs. The simplest SDR prediction equation is that based solely on drainage area, as shown in Fig. 15-4. This figure shows that SDR varies approximately in inverse proportion to the $\frac{1}{5}$ power of the drainage area. Other sources, however, have quoted values of this power as low as $\frac{1}{8}$ [2]. The fact remains that the greater the drainage area, the smaller the catchment relief and the greater the chances for sediment deposition within the catchment. Consequently, the smaller the catchment's SDR. Rough estimates of SDR can be obtained from Fig. 15-4, but caution is recommended for more refined studies.

An example of the use of statistical analysis for the estimation of sediment-delivery ratios is given by Roehl [34]. Using data from the southeast Piedmont region of the United States, he developed the following predictive equation:

$$\text{SDR} = 31{,}623(10A)^{-0.23}\left(\frac{L}{R}\right)^{-0.51} B^{-2.79} \tag{15-9}$$

in which SDR = sediment-delivery ratio in percentage; A = drainage area in square miles; L/R = dimensionless ratio of catchment length-to-relief (length measured parallel to main drainageway, relief measured as the elevation difference between drainage divide and outlet); and B = weighted mean bifurcation ratio, defined as the the ratio of the number of streams in a given order to the number of streams in the next higher order. Values of SDR measured by Roehl in the Piedmont area were in the range 3.7 to 59.4 percent.

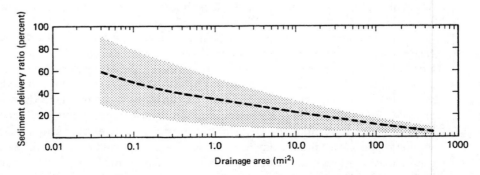

Figure 15-4 Relationship between sediment-delivery ratio and drainage area [41].

Empirical Formulas for Sediment Yield

As with the SDRs, statistical analysis can be used to develop regional equations for the prediction of sediment yield. A study by Dendy and Bolton [13] showed that sediment yield can be related to catchment area and mean annual runoff.

Sediment Yield versus Drainage Area.   Dendy and Bolton studied sedimentation data from about 1500 reservoirs, ponds, and sediment detention basins. In developing their formulas, they used data from about 800 of these reservoirs with drainage areas greater than or equal to 1 mi^2. The smaller watersheds—those of drainage area less than 1 mi^2—were excluded because of their large variability of sediment yield, reflecting the diverse effects of soils, local terrain, vegetation, land use, and agricultural practices.

For drainage areas between 1 and 30,000 mi^2, Dendy and Bolton found that the annual sediment yield per unit area was inversely related to the 0.16 power of the drainage area:

$$\frac{S}{S_R} = \left(\frac{A}{A_R}\right)^{-0.16} \tag{15-10}$$

in which S = sediment yield in tons per square mile per year; S_R = reference sediment yield corresponding to a 1-mi^2 drainage area, equal to 1645 tons per year; A = drainage area in square miles; and A_R = reference drainage area (1 mi^2).

Sediment Yield versus Mean Annual Runoff.   Dendy and Bolton studied sedimentation data from 505 reservoirs having mean annual runoff data. Annual sediment yield per unit area was shown to increase sharply as mean annual runoff Q increased from 0 to 2 in. Thereafter, for mean annual runoff from 2 to 50 in. annual sediment yield per unit area decreased exponentially. This led to the following equations.

For $Q < 2$ in.:

$$\frac{S}{S_R} = 1.07 \, (Q/Q_R)^{0.46} \tag{15-11a}$$

For $Q \geq 2$ in.:

$$\frac{S}{S_R} = 1.19 e^{-0.11(Q/Q_R)} \tag{15-11b}$$

in which Q_R = reference mean annual runoff, Q_R = 2 in.

Dendy and Bolton combined Eqs. 15-10 and 15-11 into a set of equations to express sediment yield in terms of drainage area and mean annual runoff.

For $Q < 2$ in.:

$$\frac{S}{S_R} = 1.07\left(\frac{Q}{Q_R}\right)^{0.46}\left[1.43 - 0.26 \log\left(\frac{A}{A_R}\right)\right] \tag{15-12a}$$

For $Q \geq 2$ in.:

$$\frac{S}{S_R} = 1.19 e^{-0.11(Q/Q_R)}\left[1.43 - 0.26 \log\left(\frac{A}{A_R}\right)\right] \tag{15-12b}$$

For $S_R = 1645$ tons/mi^2/y, $Q_R = 2$ in., and $A_R = 1$ mi^2, Eq. 15-12 reduces to the following:

For $Q < 2$ in.:

$$S = 1280 Q^{0.46}(1.43 - 0.26 \log A)$$

(15-13a)

For $Q \geq 2$ in.:

$$S = 1965 e^{-0.055Q}(1.43 - 0.26 \log A)$$

(15-13b)

Equations 5-12 and 5-13 are based on average values of grouped data; therefore, they should be used with caution. In certain cases, local factors such as soils, geology, topography, land use, and vegetation may have a greater influence on sediment yield than either mean annual runoff or drainage area. Nevertheless, these equations provide a first approximation to the regional assessment of sediment yield for watershed planning purposes.

Example 15-4.

Calculate the sediment yield by the Dendy and Bolton formula for a 150-mi^2 watershed with 3.5 in. of mean annual runoff.

The application of Eq. 15-13b leads to:

$$S = 1965 \times e^{(-0.055 \times 3.5)}[1.43 - 0.26 \log (150)] = 1400 \text{ ton/mi}^2/\text{y}$$

Therefore, the sediment yield is 210,000 ton/y.

Other widely used sediment yield prediction formulas are the modified universal soil loss equation (MUSLE), developed by Williams [42], and the Flaxman formula [17]. Unlike the USLE, which is based on annual values, the MUSLE is intended for use with individual storms. The Flaxman formula was developed using data from the western United States and is therefore particularly applicable to that region.

15.3 SEDIMENT TRANSPORT

Sediment transport refers to the entrainment and movement of sediment by flowing water. An understanding of the principles of sediment transport is essential for the interpretation and solution of many hydraulic, hydrologic, and water resources engineering problems.

The study of sediment transport can be divided into (1) sediment transport mechanics, (2) sediment transport prediction, and (3) sediment routing. Sediment transport mechanics refers to the fundamental processes by which sediment is entrained and transported by flowing water. Sediment transport prediction refers to the methods and techniques to predict the equilibrium or steady rate of sediment transport in streams and rivers. The prediction of sediment transport is accomplished by means of a sediment transport formula. *Sediment routing* refers to the nonequilibrium or unsteady sediment transport processes, the net result of which is either the aggradation or degradation of stream and river beds.

The description of sediment transport is based on principles of fluid mechanics, river mechanics, and fluvial geomorphology. The energy and turbulence of the flow gives streams and rivers the capacity to entrain and transport sediment. The sediment being transported can originate in either (a) upland sources or (b) channel sources.

A significant feature of sediment transport is the entrainment and transport by the flow of the material constituting the channel bed. Thus, sediment transport serves not only as the means for the movement of sediment from upstream to downstream but also as the mechanism by which streams and rivers determine their own cross-sectional shapes and boundary roughness. While the transport of sediment is a fluid mechanics subject, the interaction between flowing stream and its boundary is a river mechanics subject.

Sediment Transport Mechanics

Sediment load or sediment discharge is the total amount of sediment transported by a stream or river past a given point, expressed in terms of weight per unit time. Based on the predominant mode of transport, sediment load can be classified into (a) bed load and (b) suspended load. *Bed load* is the fraction of sediment load that moves by saltation and rolling along the channel bed, primarily by action of bottom shear stresses caused by vertical velocity gradients. *Suspended load* is the fraction of sediment load that moves in suspension by the action of turbulence. Particles transported as bed load are coarser than particles transported as suspended load. However, the distinction between bed load and suspended load is not all-exclusive; some particles may move as bed load at one point, as suspended load at another, and vice versa.

Based on whether the particle sizes are represented in the channel bed, sediment load can be classified into (a) bed-material load and (b) fine-material load. *Bed-material load* is the fraction of sediment load whose particle sizes are significantly represented in the channel bed. Conversely, *fine-material load*—commonly referred to as wash load—is the fraction of sediment load whose particle sizes are not significantly represented in the channel bed. Stated in other terms, bed-material load is the coarser fraction of sediment load that may have originated in the channel bed and that may be subject to deposition under certain flow conditions. Wash load is the finer fraction of sediment load that has not originated in the channel bed and that is not likely to deposit. Wash load is then, *washed* through the reach, largely unaffected by the hydraulics of the flow.

The relationship between the two classifications of sediment load is shown in Fig. 15-5 [11]. This figure shows that the concepts of bed load and wash load are mutually exclusive. The middle overlap is the *suspended bed-material load,* i.e., the fraction of sediment load that moves in suspended mode and is composed of particle sizes that are represented in the channel bed.

Initiation of Motion.   Water flowing over a streambed has a marked vertical velocity gradient near the streambed. This velocity gradient exerts a shear stress on the particles lying on the streambed, i.e., a bottom shear stress. For wide channels, the bottom shear stress can be approximated by the following formula [5]:

$$\tau_o = \gamma d S_o \tag{15-14}$$

in which τ_o = bottom shear stress; γ = specific weight of water; d = flow depth; and S_o = equilibrium or energy slope of the channel flow.

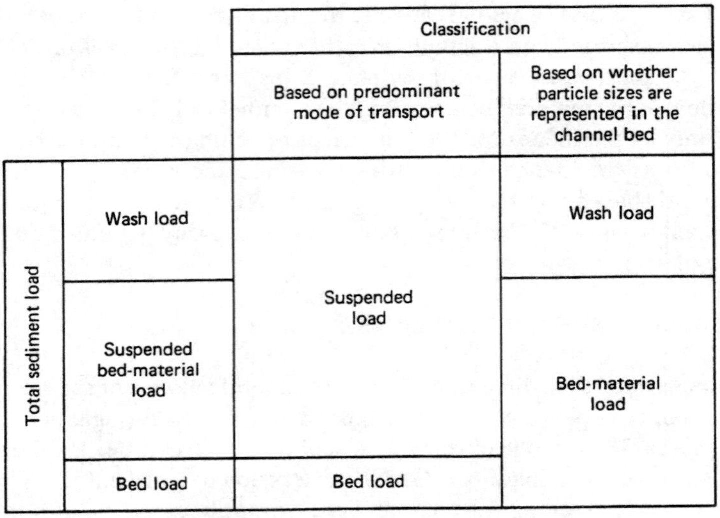

Figure 15-5 Relationship between the two classifications of sediment load [11].

There is a threshold value of bottom shear stress above which the particles actually begin to move. This threshold value is referred to as *critical bottom shear stress,* or critical tractive stress. Determinations of critical tractive stress for given flow and sediment conditions are largely empirical in nature. The Shields curve, shown in Fig. 15-6, represents the earliest attempt to combine theoretical and empirical approaches to estimate critical tractive stress [2]. The Shields curve depicts the threshold of motion, i.e., the condition separating motion (above the curve) from no motion (below the curve) [36].

The abscissa in the Shields diagram is the boundary Reynolds number, defined as:

$$R_* = \frac{U_* d_s}{\nu} \qquad (15\text{-}15)$$

in which R_* = boundary Reynolds number; U_* = shear velocity; d_s = mean particle diameter; and ν = kinematic viscosity of water. The shear velocity is defined as

$$U_* = \left(\frac{\tau_o}{\rho}\right)^{1/2} \qquad (15\text{-}16)$$

in which ρ = density of water. The ordinate in the Shields diagram is the dimensionless tractive stress, defined as:

$$\tau_* = \frac{\tau_o}{(\gamma_s - \gamma)d_s} \qquad (15\text{-}17)$$

in which τ_* = dimensionless tractive stress.

The Shields diagram indicates that, within a midrange of boundary Reynolds numbers (approximately 2–200), the dimensionless critical tractive stress can be taken as a constant for practical purposes. Therefore, within this range, the critical tractive stress is proportional to the sediment particle size. For instance, assuming a value of

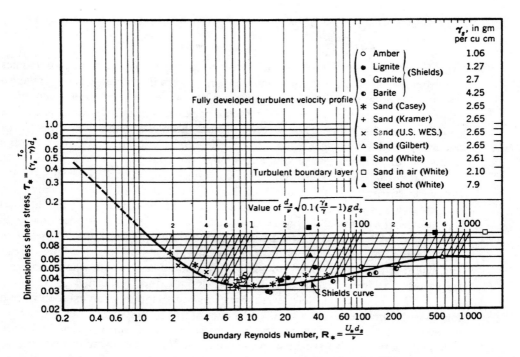

Figure 15-6 Shields diagram for initiation of motion [2].

dimensionless critical tractive stress $\tau_* = 0.04$ (from the Shields diagram, Fig. 15-6), Eq. 15-17 leads to:

$$\tau_c = 0.04(\gamma_s - \gamma)d_s \qquad (15\text{-}18)$$

in which τ_c = critical tractive stress. For quartz particles ($\gamma_s = 2.65 \times 62.4$ lb/ft³), Eq. 15-18 reduces to:

$$\tau_c = 0.34 d_s \qquad (15\text{-}19)$$

in which critical tractive stress is given in pounds per square foot and particle diameter in inches. Extensive experimental studies by Lane [29] have shown that the coefficient in Eq. 15-19 is around 0.5. Lane, however, used the d_{75} particle size (i.e., the diameter for which 75% by weight is finer) instead of the mean diameter (d_{50}) used by Shields.

Example 15-5.

Based on the Shields criterion for initiation of motion, determine whether a 3-mm diameter quartz particle is at rest or moving under the action of a 7-ft flow depth with channel slope $S_0 = 0.0001$. Assume water temperature 70°F.

From Eq. 15-14, the bottom shear stress is: $\tau_0 = 62.4 \times 7.0 \times 0.0001 = 0.0437$ lb/ft². From Eq. 15-17, the dimensionless tractive stress is $\tau_* = 0.0437/[(2.65 - 1.0) \times 62.4 \times 3.0/(25.4 \times 12)] = 0.0431$. From Eq. 15-16, the shear velocity is $U_* = (0.0437/1.94)^{1/2} = 0.150$ ft/s. From Table A-2, the kinematic viscosity is 1.058×10^{-5} ft²/s. From Eq. 15-15, the boundary Reynolds number is $R_* = 0.150 \times [3.0/(25.4 \times 12)]/(1.058 \times 10^{-5}) = 140$. For $R_* = 140$, the dimensionless critical tractive stress is obtained from the

Shields curve (Fig. 15-6): $\tau_{*_c} = 0.048$. Since $\tau_* = 0.0431$ is less than $\tau_{*_c} = 0.048$, it is concluded that the particle is at rest.

Forms of Bed Roughness. Streams and rivers create their own geometry. In particular, alluvial rivers determine to a large extent their cross-sectional shape and boundary friction as a function of the prevailing water and sediment discharge. An inherent property of river flows is their tendency to minimize changes in stage caused by changes in discharge. This is accomplished through a continuous adjustment in boundary friction in such a way that high values of friction prevail during low flows, while low values of friction prevail during high flows [26].

Bed forms are three-dimensional configurations of bed material, which are formed in streambeds by the action of flowing water. Adjustments in boundary friction are made possible by the existence of these bed forms, which develop during low flows (lower flow regime) only to be obliterated during high flows (upper flow regime) [37]. Boundary friction consists of two parts: (1) grain roughness and (2) form roughness. Grain roughness is a function of particle size; form roughness is a function of size and extent of bed forms. Grain roughness is essentially a constant when compared to the variation in form roughness that can be attributed to bed forms.

Studies have shown that several forms of bed roughness can exist on river bottoms, depending on the energy and bed-material transport capacity of the flow. In the absence of sediment movement, the bed configuration is that of plane bed with no sediment motion. With sediment movement, the following forms of bed roughness have been identified: (1) ripples, (2) dunes and superposed ripples, (3) dunes, (4) washed-out dunes or transition, (5) plane bed with sediment motion, (6) antidunes, and (7) chutes and pools. Sketches of these bed configurations are shown in Fig. 15-7 [37].

The occurrence of different forms of bed roughness can be shown to be related to the median fall diameter of the particles forming the bed and to the *stream power* of the flow. Stream power is defined as the product of bottom shear stress and mean velocity. Such a relationship is shown in Fig. 15-8. For low values of stream power, i.e., below the critical tractive stress, there is no bed load transport, and the streambed remains essentially flat. This is the condition of plane bed with no sediment motion. An increase in stream power leads first to ripples, then to dunes with superposed ripples, and, subsequently, to dunes. Ripples, however, are a rare occurrence for sediments coarser than 0.6 mm. Dunes are longer and bigger than ripples and occur at flow velocities and sediment loads that are generally greater than those of ripples. The plane bed with sediment motion represents the condition at which the flow's stream power is large enough to obliterate the dunes, essentially eliminating the form roughness. The plane bed, then, represents the condition of minimum boundary friction. For high values of stream power, antidunes (upper regime) form in conjunction with surface waves, with the tendency for upstream movement, usually under supercritical flow conditions. For even higher values of stream power, e.g., in very steep streams, the bed configuration resembles a sequence of chutes and pools [37].

The assessment of bed-form type has practical implications for engineering hydrology. Bed forms determine boundary roughness; in turn, boundary roughness determines river stages. For instance, ripples are associated with values of Manning n in the range 0.018 to 0.030, with form roughness usually a fraction of grain roughness.

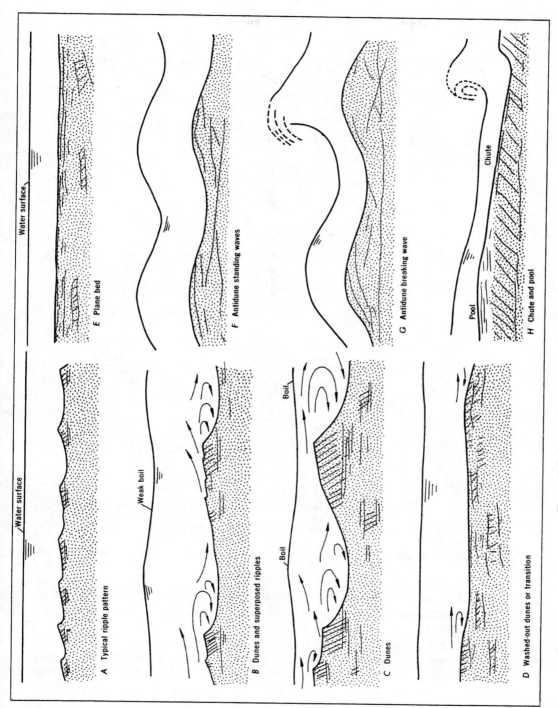

Figure 15-7 Forms of bed roughness in alluvial channels [37].

Water surface

A Typical ripple pattern

Weak boil

B Dunes and superposed ripples

Boil

Boil

C Dunes

D Washed-out dunes or transition

Water surface

E Plane bed

F Antidune standing waves

G Antidune breaking wave

Chute

Pool

H Chute and pool

553

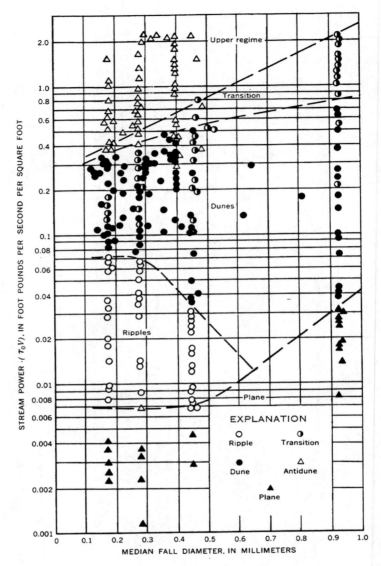

Figure 15-8 Form of bed roughness versus stream power and median fall diameter of bed material [37].

Dunes, however, are associated with n values in the range 0.020 to 0.040, with form roughness of the same order as grain roughness. Moreover, plane bed with sediment motion is associated with relatively low n values, in the range 0.012 to 0.015, and in this case, form roughness is minimal.

The proper assessment of boundary friction, including its variation as the flow changes from lower regime (ripples, ripples on dunes, and dunes) to upper regime (plane bed with sediment movement, antidunes, and chutes and pools) is an important subject in engineering hydrology and fluvial hydraulics.

Concentration of Suspended Sediment. For a given volume of water-sediment mixture, the suspended-sediment concentration is the ratio of the weight of dry sediment to the weight of the water-sediment mixture, expressed in parts per million. To convert the concentration in ppm to milligrams per liter (mg/L), the applicable factor ranges from 1.0 for concentrations between 0 and 15,900 ppm, to 1.5 for concentrations between 529,000 and 542,000 ppm, as shown in Table 15-8.

The suspended-sediment concentration varies with the flow depth, usually being higher near the stream bed and lower near the water surface. The coarsest sediment fractions, typically those in the sand size, exhibit the greatest variation in concentration with flow depth. The finer fractions, i.e., silt and clay particles, show a tendency for a nearly uniform distribution of suspended-sediment concentration with flow depth.

TABLE 15-8   FACTOR TO CONVERT CONCENTRATION IN PARTS PER MILLION (ppm) TO MILLIGRAMS PER LITER (mg/L) [19]

(1) $ppm = \dfrac{\text{Weight of Dry Sediment}}{\text{Weight of Water-and-Sediment Mixture}} \times 10^6$	(2) Factor
0–15,900	1.00
16,000–46,900	1.02
47,000–76,900	1.04
77,000–105,000	1.06
106,000–132,000	1.08
133,000–159,000	1.10
160,000–184,000	1.12
185,000–209,000	1.14
210,000–233,000	1.16
234,000–256,000	1.18
257,000–279,000	1.20
280,000–300,000	1.22
301,000–321,000	1.24
322,000–341,000	1.26
342,000–361,000	1.28
362,000–380,000	1.30
381,000–398,000	1.32
399,000–416,000	1.34
417,000–434,000	1.36
435,000–451,000	1.38
452,000–467,000	1.40
468,000–483,000	1.42
484,000–498,000	1.44
499,000–513,000	1.46
514,000–528,000	1.48
529,000–542,000	1.50

Note: To obtain concentration in mg/L, multiply concentration in ppm by applicable factor in Col. 2. This table is based on density of water 1 g/mL, plus or minus 0.005, in the temperature range 0°C-29°C, specific gravity of sediment 2.65, and dissolved solids concentration in the range 0 to 10,000 ppm.

Figure 15-9 shows the variation of suspended-sediment concentration along the flow depth. In this figure, y is the fraction of flow depth measured from the channel bottom, a is the reference distance measured from the channel bottom, and d is the flow depth. The abscissas show the dimensionless ratio C/C_a, in which C_a is the sediment concentration at the reference distance and C is the sediment concentration at a distance $y - a$. The value of a is small compared to d. The plot of Fig. 15-9 is specifically for the case of $a/d = 0.05$. The ordinates show the dimensionless ratio $(y - a)/(d - a)$. The curve parameter z is the Rouse number, defined as

$$z = \frac{w}{\beta \kappa U_*} \tag{15-20}$$

in which z = Rouse number (dimensionless); w = fall velocity of sediment particles; β = a coefficient relating mass and momentum transfer ($\beta \cong 1$ for fine sediments); κ = von Karman's constant ($\kappa = 0.4$ for clear fluids); and U_* = shear velocity, Eq. 15-16. From Fig. 15-9, it is seen that for high Rouse numbers the variation of suspended-sediment concentration along the flow depth is quite marked. Conversely, for low Rouse numbers there is a tendency for greater uniformity of suspended-sediment concentration along the flow depth.

Sediment Transport Prediction

Sediment load, sediment discharge, and sediment transport rate are synonymous in practice. However, bed load, suspended bed-material load, and wash load are mutually exclusive. Sediment transport prediction refers to the estimation of sediment transport rates under equilibrium (i.e., steady uniform) flow conditions.

There are numerous formulas for the prediction of sediment transport [2]. Most formulas compute only bed-material load, consisting of bed load and suspended bed-material load. A few compute total sediment load, which consists of bed load, suspended bed-material load, and wash load. Yet some may compute bed load and suspended bed-material load separately. Invariably, sediment transport formulas have some empirical components and, therefore, are most applicable within the range of laboratory and/or field data used in their development.

Duboys Formula. The Duboys formula is widely recognized as one of the earliest attempts to develop a sediment transport predictor. The Duboys formula is [14]:

$$q_s = \Psi_D \tau_o (\tau_o - \tau_c) \tag{15-21}$$

in which q_s = bed-material transport rate per unit channel width, in pounds per second per foot; Ψ_D = a parameter that is a function of particle size in cubic feet per pound per second; τ_o = bottom shear stress in pounds per square foot; and τ_c = critical tractive stress in pounds per square foot. Values of Ψ_D and critical tractive stress for use in the Duboys equation are shown in Fig. 15-10 [3].

Example 15-6.

Given a channel of mean flow depth $d = 12$ ft, mean width b = 320 ft; equilibrium channel slope $S_0 = 0.0001$, and median particle size $d_{50} = 0.6$ mm, calculate the bed-material transport rate by the Duboys formula.

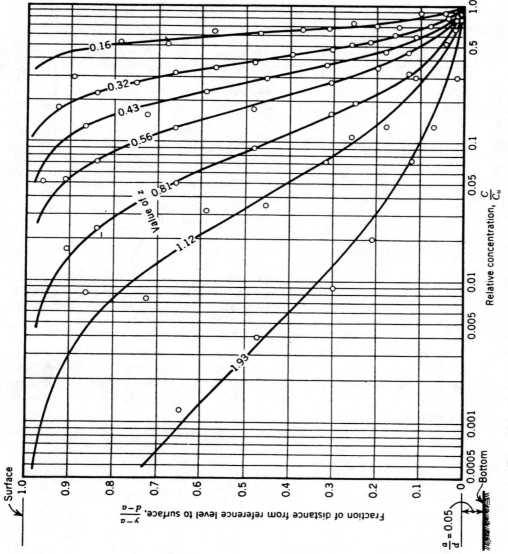

Figure 15-9 Variation of suspended-sediment concentration along the flow depth [2].

557

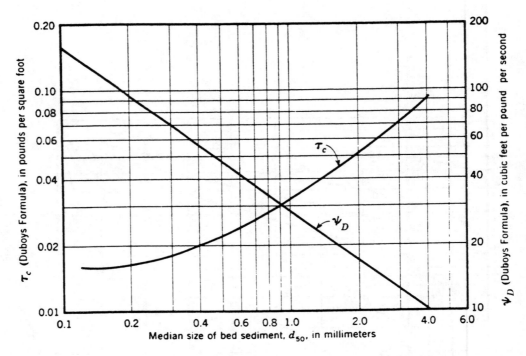

Figure 15-10 Values of Ψ_D and τ_c for use in the Duboys equation [2].

From Eq. 15-14, $\tau_o = 62.4 \times 12.0 \times 0.0001 = 0.07488$ lb/ft². From Fig. 15-10, $\Psi_D = 42$ ft³/lb/s; $\tau_c = 0.025$ lb/ft². From Eq. 15-21, $q_s = 42 \times 0.07488 \times (0.07488 - 0.025) = 0.157$ lb/s/ft. Therefore, $Q_s = q_s b = 0.157 \times 320 = 50.2$ lb/s.

Meyer-Peter Formula. The development of the Meyer-Peter formula was based on flume data, with uniform bed material size in the range 3 to 28 mm. Such coarse sediments do not produce appreciable bed forms; therefore, the formula is applicable to coarse sediment transport where form roughness is negligible. The Meyer-Peter formula is [2, 33]:

$$q_s = (39.25 q^{2/3} S_o - 9.95 d_{50})^{3/2} \tag{15-22}$$

in which q_s = bed-material transport rate per unit channel width in pounds per second per foot; q = water discharge per unit channel width in cubic feet per second per foot; S_0 = equilibrium channel slope; and d_{50} = median particle size in feet.

Example 15-7.

Given a channel of mean flow depth $d = 2$ ft; mean width $b = 25$ ft; mean velocity $v = 6$ fps; channel slope $S_0 = 0.008$; median particle size $d_{50} = 22$ mm, calculate the bed-material transport rate by the Meyer-Peter formula.

Discharge per unit width is $q = vd = 6 \times 2 = 12$ ft³/s/ft. From Eq. 15-22, $q_s = \{(39.25 \times 12^{2/3} \times 0.008) - [9.95 \times 22/(25.4 \times 12)]\}^{3/2} = 0.894$ lb/s/ft. Therefore, $Q_s = q_s b = 22.35$ lb/s.

Einstein Bed-load Function.

In 1950, Einstein published a procedure for the computation of bed material transport rate by size fractions [15]. The method was developed based on theoretical considerations of turbulent flow, supported by laboratory and field data. Einstein is credited with the introduction of several novel concepts in sediment transport theory, including the separation of boundary friction into grain and form roughness and the use of statistical properties of turbulence to explain the mechanics of sediment transport.

Einstein's bed-load function first computes the bed-load transport rate. Then it uses the bed-load transport rate to aid in the integration of the product of the suspended sediment concentration profile and the flow velocity profile, to determine the suspended bed-material transport rate, per individual size fraction. Several step-by-step procedures have been reported in the literature [2].

Modified Einstein Procedure.

The modified Einstein procedure was developed by Colby and Hembree [6] in order to include actual measurements of suspended load into the framework of the original Einstein method. Typically, measurements of suspended load do not include (a) the bed load and (b) the fraction of suspended load moving too close to the streambed to be effectively sampled. The modified Einstein procedure calculates the total sediment load by size fractions based on measurements of suspended load and relevant geometric and hydraulic characteristics of the stream or river. Details of the method are reported in the literature [2, 6].

Colby's 1957 Method.

Colby's 1957 method is based on some of the same measurements that led to the development of the modified Einstein procedure. However, unlike the latter, it does not account for sediment transport rate by size fractions. Instead, it provides the total bed-material discharge, i.e. the sum of measured and unmeasured bed-material discharges.

The following data are needed in an application of the Colby 1957 method: (1) mean flow depth d, (2) mean channel width b, (3) mean velocity v, and (4) measured concentration of suspended bed-material discharge C_m. The procedure is as follows [7]:

1. Use Fig. 15-11 to obtain the uncorrected unmeasured sediment discharge q_u' (in tons per day per foot of width) as a function of mean velocity.
2. Use Fig. 15-12 to obtain the relative concentration of suspended sands C_r (in parts per million) as a function of mean velocity and flow depth.
3. Calculate the availability ratio by dividing the measured concentration of suspended bed-material discharge C_m (ppm) by the relative concentration of suspended sands C_r (ppm).
4. Use the mean line of Fig. 15-13 and the availability ratio to obtain the correction factor C to be multiplied by the uncorrected unmeasured sediment discharge q_u' to obtain the unmeasured sediment discharge q_u (in tons per day per foot).
5. The total bed-material discharge q_s is the sum of measured and unmeasured sediment discharges:

$$q_s = 0.0027 \, C_m q + q_u \tag{15-23}$$

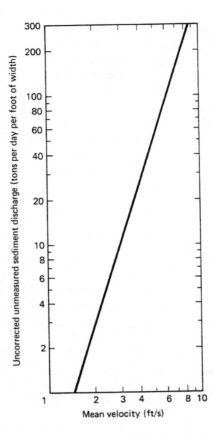

Figure 15-11 Uncorrected unmeasured sediment discharge versus mean velocity in Colby 1957 method [7].

in which q_s = bed-material discharge in tons per day per foot; q = water discharge in cubic feet per second per foot; C_m = measured concentration of suspended bed-material discharge in milligrams per liter; and 0.0027 is the conversion factor for the indicated units. Table 15-8 shows a factor to convert concentration in parts per million to milligrams per liter.

Example 15-8.

Given mean flow depth $d = 10$ ft, mean channel width $b = 300$ ft, mean velocity $v = 3$ fps, measured concentration of suspended bed material discharge $C_m = 100$ ppm, calculate the total bed material discharge by the Colby 1957 method.

From Fig. 15-11, the uncorrected unmeasured sediment discharge is $q_u' = 10$ ton/d/ft. From Fig. 15-12, the relative concentration of suspended sands is $C_r = 380$ ppm. The availability ratio is $100/380 = 0.26$. From Fig. 15-13, the correction factor is $C = 0.6$. Therefore, $q_u = 6$ ton/d/ft. The water discharge per unit width is $q = vd = 3 \times 10 = 30$ ft³/s/ft. From Eq. 15-23, the sediment discharge per unit width is $q_s = (0.0027 \times 100 \times 30) + 6 = 14.1$ ton/d/ft. Therefore, the bed-material discharge by the Colby 1957 method is $Q_s = q_s b = 14.1 \times 300 = 4230$ ton/d.

Colby's 1964 Method. In 1964, Colby published a method to calculate discharge of sands (i.e., bed-material discharge) in sand-bed streams and rivers. The

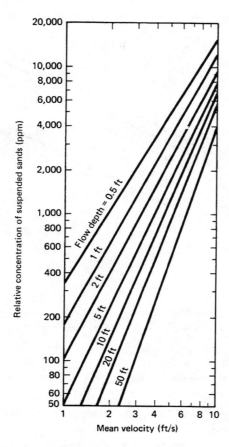

Figure 15-12 Relative concentration of suspended sands versus mean velocity and flow depth in Colby 1957 method [7].

development of the method was guided by the Einstein bed-load function and supported by large amounts of laboratory and field data. The method has been shown to provide a reasonably good prediction of sediment transport rates, particularly for sand-size particles.

The following data are needed in an application of the Colby 1964 method: (1) mean flow depth d, (2) mean channel width b, (3) mean velocity v, (4) water temperature, (5) concentration of fine-material load (i.e., wash load), and (6) median bed-material size. The procedure is as follows [8]:

1. Use Fig. 15-14 to determine the uncorrected discharge of sands q_u (in tons per day per foot of width) as a function of mean velocity, flow depth, and sediment size.

2. For water temperature of 60°F, negligible wash load concentration (less than 1000 ppm), and sediment size in the range 0.2 to 0.3 mm, no further calculations are required, and q_u is the discharge of sands q_s.

3. For conditions other than the preceding, use Fig. 15-15 to obtain the correction factor k_1 as a function of flow depth and water temperature, k_2 as a function of

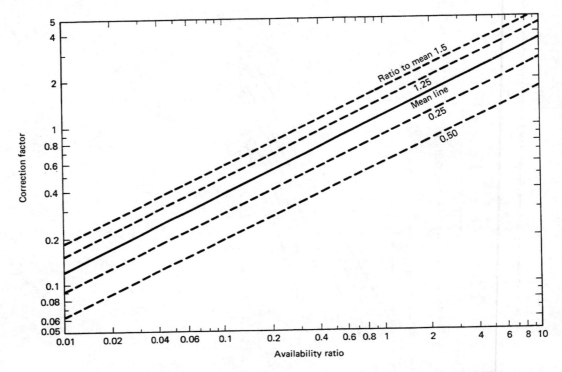

Figure 15-13 Correction factor versus availability ratio in Colby 1957 method [7].

flow depth and concentration of fine-material load, and k_3 as a function of median size of bed material.

4. The discharge of sands is given by the following formula:

$$q_s = [1 + (k_1 k_2 - 1)k_3]q_u \tag{15-24}$$

in which q_s discharge of sands in tons per day per foot.

Example 15-9.

Given mean flow depth $d = 1$ ft, mean channel width $b = 30$ ft, mean velocity $v = 2$ fps, water temperature 50°F, wash-load concentration $C_w = 10,000$ ppm, and median bed-material size $d_{50} = 0.1$ mm. Calculate the discharge of sands by the Colby 1964 method.

From Fig. 15-14, $q_u = 9.3$ ton/d/ft. From Fig. 15-15, $k_1 = 1.15$, $k_2 = 1.20$, $k_3 = 0.6$. From Eq. 15-24, $q_s = [1 + (1.15 \times 1.20 - 1) \times 0.6] \times 9.3 = 11.4$ ton/d/ft. Therefore, the discharge of sands is $Q_s = 11.4 \times 30 = 342$ ton/d.

Other Methods for the Calculation of Sediment Discharge. Many other methods have been proposed for the calculation of sediment discharge. Notable among them are the methods of Ackers and White [1], Engelund and Hansen [16], Toffaleti [39], and Yang [46]. The various procedures vary in complexity and range of applicability. For details on these and other sediment transport formulas, see [2, 4, 25, 38].

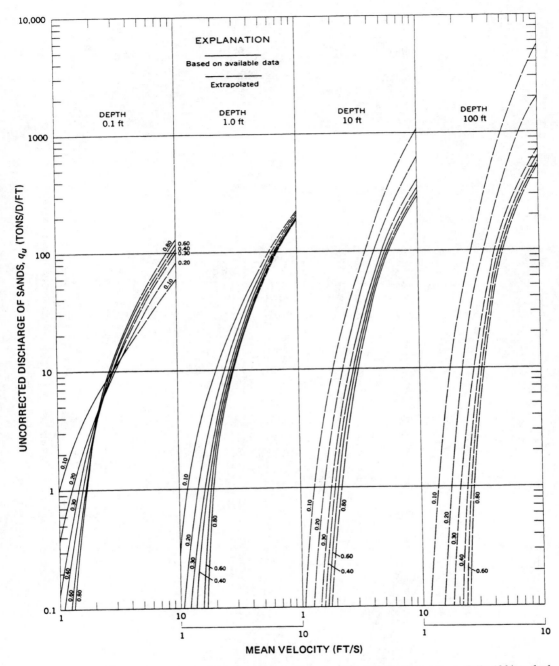

Figure 15-14   Discharge of sands versus mean velocity, flow depth and sediment size in Colby 1964 method [9].

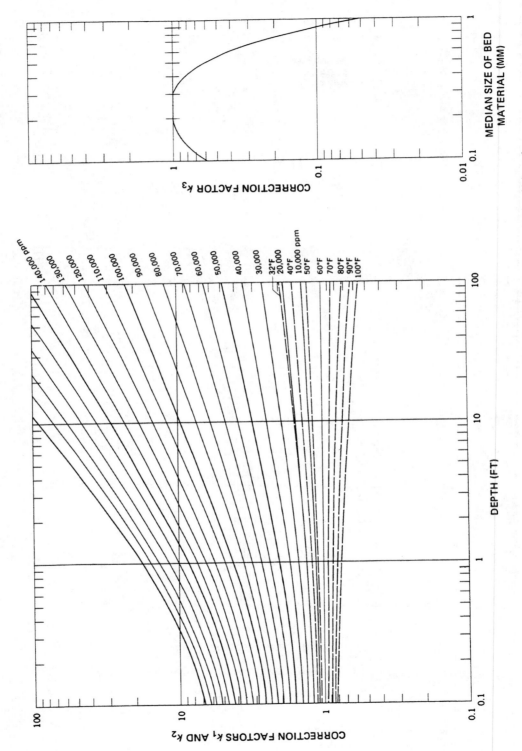

Figure 15-15 Correction factors for water temperature (k_1), fine-material-load concentration (k_2), and median size of bed material (k_3), in Colby 1964 method [9].

Sediment Rating Curves

A useful curve in sediment analysis is the sediment rating curve, defined as the relationship between water discharge and sediment discharge at a given gaging site. For a given water discharge, the sediment rating curve allows the estimation of sediment discharge, assuming steady equilibrium flow conditions.

The sediment rating curve is an *xy* plot showing water discharge in the abscissas and sediment discharge in the ordinates. This plot is obtained either by the simultaneous measurement of water and sediment discharge or, alternatively, by the use of sediment transport formulas. For low-water discharges, the sediment rating curve usually plots as a straight line on logarithmic paper, showing an increase of sediment concentration with water discharge. However, for high water discharges, the sediment rating curve has a tendency to curve slightly downward, approaching a line of equal sediment concentration (i.e., a line having a 45° slope in the *xy* plane) [2].

Like the single-valued stage-discharge rating, the single-valued sediment rating curve is strictly valid only for steady equilibrium flow conditions. For strongly unsteady flows, the existence of loops in both water and sediment rating curves has been demonstrated [2]. These loops are complex in nature and are likely to vary from flood to flood. In practice, loops in water and sediment rating are commonly disregarded.

Sediment Routing

The calculation of sediment yield is lumped, i.e., it does not provide a measure of the spatial or temporal variability of sediment production within the catchment. Sediment transport formulas are invariably based on the assumption of steady equilibrium flow. *Sediment routing*, on the other hand, refers to the distributed and unsteady calculation of sediment production, transport and deposition in catchments, streams, rivers, reservoirs, and estuaries.

Of necessity, sediment routing involves a large number of calculations and therefore is ideally suited for use with a computer. Sediment routing should be used—in addition to sediment yield and sediment transport evaluations—in cases where the description of spatial and temporal variations of sediment production, transport, and deposition is warranted. Sediment routing methods are particularly useful in the detailed analysis of sediment transport and deposition in rivers and reservoirs. For example, the computer model HEC-6, "Scour and Deposition in Rivers and Reservoirs," is a sediment routing model developed by the U.S. Army Corps of Engineers [21]. Several other sediment routing models have been developed in the last two decades; see, for instance, [4] and [25].

15.4 SEDIMENT DEPOSITION IN RESERVOIRS

The concepts of sediment yield and sediment transport are essential to the study of sediment deposition in reservoirs. Sediment is first produced at upland and channel sources and then transported downstream by the action of flowing water. If the flowing water is temporarily detained, as in the case of an instream reservoir, its ability to continue to entrain sediment is substantially impaired, and deposition takes place.

Sediment deposition occurs in the vicinity of reservoirs, typically as shown in Fig. 15-16 [20]. First, deposition of the coarser-size fractions takes place near the entrance to the reservoir. As water continues to flow into the reservoir and over the dam, the delta continues to grow in the direction of the dam until it eventually fills the entire reservoir volume. The process is quite slow but relentless. Typically, reservoirs may take 50 to 100 y to fill, and in some instances, up to 500 y or more.

The rate of sediment deposition in reservoirs is a matter of considerable economic and practical interest. Since reservoirs are key features of hydroelectric and water-resource development projects, the question of the design life of a reservoir is appropriate, given that most reservoirs will eventually fill with sediment. In an extreme example, the filling can occur in a single storm event, as in the case of a small sediment-retention basin located in a semiarid or arid region. On the other hand, the reservoir can take hundreds of years to fill, as in the case of a large reservoir located in a predominantly humid or subhumid environment.

Reservoir Trap Efficiency

The difference between incoming and outgoing sediment is the sediment deposited in the reservoir. The incoming sediment can be quantified by the sediment yield, i.e., the total sediment load entering the reservoir. The outgoing sediment can be quantified by the *trap efficiency*. Trap efficiency refers to the ability of the reservoir to entrap sediment being transported by the flowing water. It is defined as the ratio of trapped sediment to incoming sediment, in percentage, and is a function of (1) the ratio of reservoir volume to mean annual runoff volume and (2) the sediment characteristics.

The following procedure is used to determine trap efficiency [41]:

1. Determine the reservoir capacity C in cubic hectometers or acre-feet.

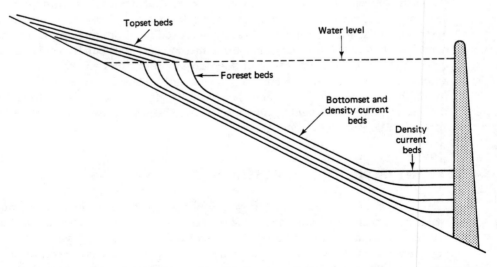

Figure 15-16 Longitudinal sedimentation patterns in a reservoir operating at constant water level [20].

2. Determine the mean annual (runoff volume) inflow I to the reservoir, in cubic hectometers or acre-feet.
3. Use Fig. 15-17 to determine the percentage trap efficiency as a function of the ratio C/I for any of three sediment characteristics. Estimate the texture of the incoming sediment by a study of sediment sources and/or sediment transport by size fractions. The upper curve of Fig. 15-17 is applicable to coarse sands or flocculated sediments; the middle curve, to sediments consisting of a wide range of particle sizes; and the lower curve, to fine silts and clays.

Reservoir Design Life

The design life of a reservoir is the period required for the reservoir to fulfill its intended purpose. For instance, structures designed by the Soil Conservation Service for watershed protection and flood prevention programs have a design life of 50 to 100 y. Due to reservoir sedimentation, provisions are made to guarantee the full-design reservoir water-storage capacity for the planned design life. This may entail either (1) cleaning out reservoir sediment deposits at predetermined intervals during the life of the structure or, as is more often the case, (2) providing a reservoir storage capacity large enough to store all the accumulated sediment deposits without encroachment on the designed water-storage volume. Typically, calculations of sediment-filling rates and sediment accumulation are part of the design of reservoir-storage projects.

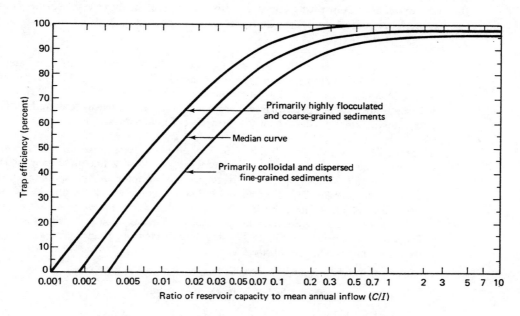

Figure 15-17 Trap efficiency of reservoirs versus capacity-inflow (C/I) ratio [41].

Sec. 15.4   Sediment Deposition in Reservoirs   **567**

Distribution of Sediment Deposits

The distribution of sediment deposits may be such as to materially affect the operation and maintenance of the dam and reservoir. The amount and types of sediment deposits vary with the nature of the sediment itself, the shape of the reservoir, the topography of the reservoir floor, the nature of the approach channel, detention time, and purpose of the reservoir. The coarser sediment sizes are the first to deposit in the vicinity of the reservoir entrance. Finer sediment sizes are able to travel longer distances inside the reservoir and deposit at locations close to the dam. However, very fine sediments are usually uniformly distributed in the reservoir bed.

Sediment-retention, or Debris, Basins

Sediment-retention basins, or debris basins, are small reservoirs located in upland areas with the specific purpose of trapping sediment and debris before they are able to reach the main fluvial network system. *Debris* is a general term used to describe the assortment of cobbles, boulders, branches, and other vegetative material that may clog channels and hydraulic structures, causing them to reach a critical design condition prematurely and often resulting in structural failure.

Debris basins are placed upstream of channels or reservoirs with the specific purpose of temporary detainment of debris. Debris basins are usually small and designed to be cleaned out from time to time. Some basins are sized to fill up during one or two major storms. Others may have a 50- or 100-y design life. Project costs and site conditions determine the size of debris basins.

Sediment-yield determinations for debris basin design should include both short-term and long-term analyses. The long-term sediment yield is determined from the appropriate sediment rating curve. For infrequent storms, however, sediment concentrations may exceed long-term averages by a factor of 2 or 3 [40].

Example 15-10.

A planned reservoir has a total capacity of 10 hm^3 and a contributing catchment area of 250 km^2. Mean annual runoff at the site is 400 mm, annual sediment yield is 1000 metric tons/km^2, and the specific weight of sediment deposits is estimated at 12,000 N/m^3. A sediment source study has confirmed that the sediments are primarily fine-grained. Calculate the time that it will take for the reservoir to fill up with sediments.

The calculations are shown in Table 15-9. Because of decreased reservoir capacity as it fills with sediment, an interval of storage equal to $\Delta V = 2$ hm^3 is chosen for this example. Column 2 shows the loss of reservoir capacity, and Col. 3 shows the accumulated sediment deposits. The mean annual inflow to the reservoir is 400 mm $\times$ 250 km^2 = 100 hm^3. Column 4 shows the capacity-inflow ratios at the end of each interval, and Col. 5 shows the average capacity-inflow ratios per interval. Column 6 shows the trap efficiencies T_i obtained from Fig. 15-17 using the curve for fine-grained sediments (lower curve). The annual sediment inflow I_s is:

$$I_s = \frac{1000 \text{ ton/km}^2/\text{y} \times 1000 \text{ kg/ton} \times 250 \text{ km}^2 \times 9.81 \text{ N/kg}}{12,000 \text{ N/m}^3 \times 10^6 \text{ m}^3/\text{hm}^3} \tag{15-25}$$

$$I_s = 0.204 \text{ hm}^3/\text{y}$$

TABLE 15-9 SEDIMENT ACCUMULATION IN RESERVOIRS: EXAMPLE 15-10

(1)	(2)	(3)	(4)	(5)	(6)	(7)
Interval i	Reservoir capacity C (hm³)	Accumulated volume (hm³)	C/I ratio	Average C/I in interval	Trap efficiency T_i (%)	Number of years to fill (y)
0	10	0	0.10			
1	8	2	0.08	0.09	77	13
2	6	4	0.06	0.07	72	14
3	4	6	0.04	0.05	66	15
4	2	8	0.02	0.03	55	18
5	0	10	0.00	0.01	30	33
Total						93

The number of years to fill each ΔV interval is $\Delta V / [I_s(T_i/100)]$, shown in Col. 7. The sum of Col. 7 is the total number of years required to fill up the reservoir: 93 y.

15.5 SEDIMENT MEASUREMENT TECHNIQUES

The measurement of fluvial sediments is often necessary to complement sediment yield and sediment transport studies. The accuracy of the measurement, however, is dependent not only on the equipment and techniques but also on the application of basic principles of sediment transport.

As sediment enters a stream or river, it separates itself into bed-material load and wash load. In turn, the bed-material load is transported as either bed load or suspended load. The suspended bed-material load plus the wash load constitutes the total suspended-sediment load of the stream or river.

The term *sampled suspended-sediment discharge* is used to describe the fraction of suspended-sediment load that can be sampled with available equipment. Generally, it excludes the *unsampled suspended-sediment discharge,* i.e., the fraction of suspended-sediment load that is carried too close to the stream bed to be effectively sampled. The suspended-sediment discharge is the sum of sampled and unsampled suspended-sediment discharges.

Sediment-sampling Equipment

Sediment-sampling equipment can be classified as the following:

1. Suspended-sediment samplers, which measure suspended-sediment concentration
2. Bed-load samplers, which measure bed load
3. Bed-material samplers, which sample the sediment in the top layer of the stream bed

Suspended-sediment Samplers. Suspended-sediment samplers can be classified as (1) depth-integrating, (2) point-integrating, (3) single-stage, or (4) pumping samplers. Depth-integrating samplers accumulate a water-sediment sample in a pint-size milk bottle as they are lowered to the stream bed and raised back to the surface at a uniform rate of transit. They are designed so that the velocity in the intake nozzle is nearly equal to the local stream velocity. Samples may be collected by wading in a stream, by hand from a suitable support, or mechanically with a cable-and-reel setup. The U.S. DH-48 sampler (4.5 lb) (Fig. 15-18) with wading-rod suspension is used in shallow streams when the product of flow depth (in feet) and mean velocity (in feet per second) does not exceed 10 [22]. The U.S. DH-59 sampler (24 lb) with hand-line suspension is used in streams with low velocities but with depths that do not permit samples to be collected by wading. The U.S. D-49 sampler (62 lb) with cable-and-reel suspension is designed for use in streams beyond the range of hand-operated equipment. Depth-integrating samplers were developed to improve sampling accuracy and to reduce the cost of collecting suspended-sediment data.

Point-integrating samplers accumulate a water-sediment sample that is representative of the mean concentration at any selected point in a stream during a short interval of time. The intake and exhaust characteristics of point-integrating samplers are identical to those of depth-integrating samplers. A rotary valve that opens and closes the sampler is operated by a solenoid energized by batteries at the surface. The current flows to the solenoid by a current-meter cable, which suspends the sampler. Point-integrating samplers can be used to collect depth-integrating samples by leaving the valve open as the sampler is moved through the stream vertical. This permits depth-integration in streams that are too deep to be appropriately sampled with a depth-integrating sampler. The U.S. P-46 and P-61 (100 lb), P-63 (200 lb), and P-50 (300 lb) point-integrating samplers are in current use.

The single-stage sampler was developed to obtain suspended sediment data in flashy streams, particularly those in remote areas [24]. It is used to sample sediment at a specific depth and on the rising stage only. The sampler operates on the siphon principle, and therefore the velocity in the intake is not equal to the stream velocity. With careful operation, the single-stage sampler can be used to obtain supplemental data on suspended-sediment concentration at selected points.

The pumping sampler does not require an operator and is designed to obtain a continuous record of sediment concentration by sampling at a fixed point at specific time intervals. The velocity in the intake is not equal to the stream velocity, and the intake does not meet the requirements of an ideal sampler, since it does not point into the flow. However, the pumping sampler can be calibrated by rating its measurements against those obtained with standard depth-integrating or point-integrating samplers.

Bed-load Samplers. Bed-load samplers are of three types: (1) basket type, (2) pan type, and (3) pressure-difference type. The basket and pan types cause an increase in resistance to flow and a reduction in stream velocity at the sampling location. The reduction in stream velocity interferes with the rate of bed-load transport, compromising the accuracy of the measurement. The pressure-difference bed-load sampler is designed to eliminate the reduction in velocity, resulting in increased sampling accuracy.

The efficiency of a bed-load sampler, i.e., the ratio of sampled bed load to that

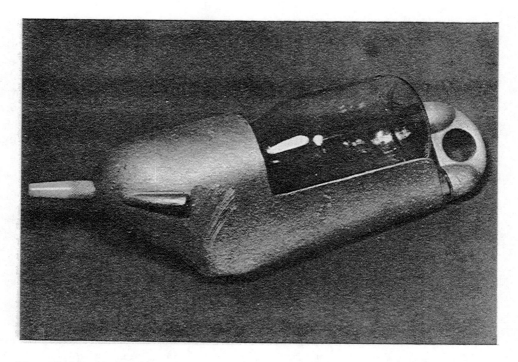

Figure 15-18 US DH-48 depth-integrating suspended-sediment sampler (photo courtesy of U.S. Geological Survey).

actually transported, varies with sample type, method of support, particle size, and bed configuration. Calibration of bed-load samplers has indicated a mean efficiency of about 45 percent for the basket and pan types and 70 percent for the pressure-difference type.

Bed-material Samplers. Bed-material samplers are of three types: (1) drag bucket, (2) grab bucket, and (3) vertical-pipe, or core sampler. The drag-bucket sampler consists of a weighted section of cylinder with an open mouth and cutting edge. As the sampler is dragged upstream along the bed, it collects a sample from the top layer of bed material. The grab-bucket sampler is similar to the drag-bucket, consisting of a cylinder section attached to a rod, and used primarily in shallow streams. The vertical-pipe, or core sampler, consists of a piece of metal or plastic pipe that can be forced into the stream by hand. Generally, the drag-bucket and grab-bucket samplers do not obtain representative samples of bed material because of the loss of fine material. The core sampler is satisfactory for use in shallow streams.

The U.S. BMH-53 sampler consists of a 9-in.-long, 2-in.-diameter brass or stainless steel pipe with a cutting edge and suction piston attached to a control rod. The piston is retracted as the cutting cylinder is forced into the stream bed. The partial vacuum that develops in the sampling chamber as the piston is withdrawn assists in holding the sample in the cylinder. The sampler can be used only in streams shallow enough to be waded.

The U.S. BMH-60 bed material sampler with both handline and cable suspension is designed to scoop up a sample of bed sediment about 3 in. wide and 2 in. deep. At the close of the sampling operation, the cutting edge rests against a rubber stop, which prevents any sediment from being lost. The aluminum sampler weighs 30 lb and the brass sampler, 40 lb. It is used to collect bed-material-sediment samples in streams with low velocities but with depths beyond the range of the BMH-53 sampler.

The U.S. BM-54 bed-material sampler (100 lb) with cable suspension is similar in design to the BHM-60 sampler. It is used in deep streams where a heavier sampler is necessary.

Suspended-sediment Discharge Measurements

Suspended-sediment samplers are used to determine sediment concentration at a point in a stream (i.e., a stream vertical), except for a small unmeasured zone located just above the stream bed. With wading equipment, measurements can generally be made down to within 0.3 ft of the stream bed. For cable-supported equipment, the unmeasured zone varies between 0.5 and 1.0 ft, depending on the size of sampler used.

Suspended-sediment discharge measurements include: (1) suspended-sediment concentration, (2) particle size, (3) specific gravity, (4) temperature of water-sediment mixture, (5) water discharge, and (6) distribution of flow in the stream cross section.

The streamflow depth and velocity and the facilities at the sampling site (bridge, cableway, and so on) have an influence on the choice of sampler. Stream depth determines whether hand samplers, such as the DH-48 or DH-59, or a cable-suspended sampler, such as the D-49, should be used. Flow depths over 15 ft require the use of point-integrating samplers to avoid overfilling of the sampling bottles. The larger the product of flow depth times mean velocity, the heavier the sample required for proper measurement.

The number of sampling verticals depends on the desired accuracy and the variation of sediment concentration across the stream. For streams with a stable cross section and essentially uniform sediment concentration across the width, sampling at a single vertical is usually adequate.

Depth-integrating samplers produce a suspended-sediment concentration, which can be measured in parts per million and converted to milligrams per liter. The suspended-sediment discharge is given by the following formula:

$$Q_s = 0.0027 C_s Q \qquad (15\text{-}26)$$

in which Q_s = suspended sediment discharge in tons per day; C_s = suspended-sediment concentration in milligrams per liter; Q = water discharge in cubic feet per second, and 0.0027 is the conversion factor for the indicated units. Table 15-8 shows a factor to convert concentration in parts per million to milligrams per liter.

There are two techniques to measure suspended-sediment discharge: (1) EDI, or equal-discharge increments, and (2) ETR, or equal-transit rate. In the EDI method, sampling is done at the centroids of equal-discharge increments. In the ETR method, sampling is done at the centroids of equal-length increments. The EDI method requires a knowledge of the lateral distribution of streamflow prior to the selection of sampling verticals. The ETR method is applicable to shallow streams where the cross-sectional distribution of streamflow is not stable. Generally, the EDI method requires fewer sampling verticals than the ETR method. The ETR method, however, does not require a prior discharge measurement. The suspended-sediment concentration in the EDI method is the average obtained from several depth-integrating samples. In the ETR method, the suspended sediment concentration is that of a composite sample encompassing several depth-integrating samples.

The error in suspended-sediment discharge provided by the measurement varies with the depth of the unsampled zone and the size distribution of suspended load. The error tends to be smallest in the cases where the vertical concentration gradient in the unsampled zone is small. The concentration gradient near the bed is small for silt and clay particles and large for coarser sand particles. Corrections in sampled suspended-sediment discharge to account for the unsampled portion are usually obtained through appropriate sediment transport predictors such as the Colby 1957 method or the modified Einstein procedure [7, 9].

QUESTIONS

1. Give two alternate definitions of particle sphericity.
2. What is the difference between specific weight and specific gravity?
3. What is standard fall velocity? What is standard fall diameter?
4. What is the difference between sediment production and sediment yield?
5. Describe the differences between normal and accelerated erosion.
6. Name four sources of sediment.
7. What is the rainfall factor R in the Universal Soil Loss Equation?
8. What is sediment-delivery ratio?
9. Why is sediment-delivery ratio inversely related to drainage-basin area?
10. Why are two formulas needed in the Dendy and Bolton approach to the computation of sediment yield?
11. Describe the classifications of sediment load based on (1) predominant mode of transport and (2) whether the particle sizes are represented on the channel bed.
12. What are possible forms of bed roughness in alluvial channels?
13. What is range of applicability of the Meyer-Peter formula for bed-load transport?
14. What is the basic difference between the Colby 1957 and Colby 1964 procedures for the computation of discharge of sands?
15. What is a sediment rating curve?
16. What is sediment routing?
17. What is the trap efficiency of a reservoir?
18. What is a debris basin?
19. Describe two techniques to measure suspended-sediment discharge. How do they differ in the evaluation of suspended-sediment concentration?

PROBLEMS

15-1. Calculate the fall velocity of a sediment particle using Stokes' law. Assume a diameter 0.1 mm, kinematic viscosity 1 centistoke, specific gravity 2.65.

15-2. Calculate the specific weight of a sediment deposit in a reservoir, after an elapsed time of 100 y, under moderate drawdown conditions. Assume the following mix of particle sizes: sand 55%, silt 30%, clay 15%.

15-3. Compute the average annual soil loss by the universal soil loss equation for a 300-ac watershed near Lexington, Kentucky, with the following conditions: (1) cropland, 250 ac, contoured, soil is Keen silt loam, slopes are 7% and 150 ft long, $C = 0.15$; (2) pasture, 50 acres, 75% canopy cover, 60% ground cover with grass, soil is Ida silt loam, slopes are 10% and 200 ft long.

15-4. Compute the average annual soil loss by the universal soil loss equation for a 1-mi^2 forested watershed near Bangor, Maine. The soil is Fayette silt loam, the slopes are 3% and 300 ft long, and the site is 80% covered by forest litter.

15-5. Using the Dendy and Bolton formula, calculate the sediment yield for a 25.9-km^2 watershed with 5 cm of mean annual runoff.

15-6. Determine whether a particle of 2-mm diameter is at rest under a 3-m flow depth and 0.0002 channel slope. Assume a specific gravity 2.65 and kinematic viscosity 1 centistoke.

15-7. Determine the form of bed roughness that is likely to prevail under the following flow conditions: mean velocity 3 ft/s, flow depth 8 ft, channel slope 0.0002, and mean particle diameter 0.3 mm.

15-8. Given the following flow characteristics: flow depth 9 ft, mean velocity 3 ft/s, channel slope 0.00015, mean particle diameter 0.4 mm, mean channel width 250 ft. Calculate the bed material transport rate by the Duboys formula.

15-9. Given the following flow characteristics: flow depth 3 ft, mean velocity 5 ft/s, energy slope 0.009, mean particle diameter 1.0 in., mean channel width 30 ft. Calculate the bed-material transport rate (in tons per day) by the Meyer-Peter formula.

15-10. Given the following flow characteristics: flow depth 5 ft, mean velocity 4 ft/s, mean channel width 180 ft, measured concentration of suspended bed-material discharge 200 ppm. Calculate the total bed-material discharge (in tons per day) by the Colby 1957 method.

15-11. Given the following flow characteristics: flow depth 5 ft, mean velocity 3 ft/s, median bed material size 0.3 mm, mean channel width 225 ft, water temperature 70°F, wash load concentration 300 ppm. Calculate the discharge of sands by the Colby 1964 method.

15-12. A reservoir is to be built with a total storage capacity of 50 hm^3. The contributing drainage basin is 800 km^2, and the mean annual runoff at the site is 200 mm. Assume well-graded sediment deposits with average specific weight 1400 kg/m^3. (a) How long will it take for the reservoir to lose 20% of its storage volume? (b) How long will it take for the reservoir to fill up with sediment? Estimate sediment yield by the Dendy and Bolton formula.

15-13. A reservoir is to be built with a total storage capacity of 120 hm^3. The contributing drainage basin is 425 km^2, and the mean annual runoff at the site is 45 mm. Assume coarse sediment deposits with average specific weight 13 kN/m^3. (a) How long will it take for the reservoir to lose 80% of its storage volume? (b) How long will it take for the reservoir to fill up with sediment? Estimate sediment yield by the Dendy and Bolton formula.

15-14. Derive the conversion factor 0.0027 in Eq. 15-26.

15-15. Calculate the suspended-sediment discharge (in tons per day) for the following cases: (1) suspended sediment concentration 100 ppm, water discharge 1200 ft^3/s, and (2) suspended sediment concentration 80,000 ppm, and water discharge 5000 ft^3/s.

15-16. Derive the unit conversion factor C in the following formula: $Q_s = CC_sQ$, in which Q_s is given in kilonewtons per day, C_s in milligrams per liter, and Q in cubic meters per second.

15-17. Calculate the suspended-sediment discharge (in kilonewtons per day) for a suspended-sediment concentration of 150 ppm and a flow of 68 m³/s.

15-18. Calculate the suspended-sediment discharge (in kilonewtons per day) for a suspended-sediment concentration of 22,000 ppm and a flow of 155 m³/s.

REFERENCES

1. Ackers, P., and W. R. White. (1973). "Sediment Transport: A New Approach and Analysis," *Journal of the Hydraulics Division*, ASCE, Vol. 99, No. HY11, Nov., pp. 2041–2060.

2. American Society of Civil Engineers. (1975). *Sedimentation Engineering*, Manual No. 54.

3. Brown, C. B. (1950). "Sediment Transportation," Chapter XII, *Engineering Hydraulics*, H. Rouse, ed.. New York: John Wiley.

4. Chang, H. H. (1988). *Fluvial Processes in River Engineering*, New York: Wiley Interscience.

5. Chow, V. T. (1959). *Open Channel Hydraulics*, New York: McGraw-Hill.

6. Colby, B. R., and C. H. Hembree. (1955). "Computations of Total Sediment Discharge, Niobrara River Near Cody, Nebraska," *U.S. Geological Survey Water-Supply Paper* 1357, Washington, D.C.

7. Colby, B. R. (1957). "Relations of Unmeasured Sediment Discharge to Mean Velocity," *Transactions*, American Geophysical Union, Vol. 38, No. 5, Washington, D.C., Oct., pp. 707–717.

8. Colby, B. R. (1964). "Practical Computations of Bed Material Discharge," *Journal of the Hydraulics Division*, ASCE, Vol. 90, No. HY2, March, pp. 217–246.

9. Colby, B. R. (1964). "Discharge of Sands and Mean Velocity Relations in Sand-Bed Streams," *U.S. Geological Survey Professional Paper* 462-A, Washington, D.C.

10. Collier, C. R., R. J. Pickering, and J. J. Musser. (1970). "Influences of Strip Mining on the Hydrologic Environment of Parts of Beaver Creek Basin, Kentucky, 1955–1966, *U.S. Geological Survey Professional Paper* 427-C, Washington, D.C.

11. Copper, R. H., and A. W. Peterson. (1970). Discussion of "Coordination in Mobile Bed Hydraulics," *Journal of the Hydraulics Division*, ASCE, Vol. 96, No. HY9, Sept., pp. 1880–1886.

12. Corey, A. T. (1949). "Influence of Shape on the Fall Velocity of Sand Grains," M.S. Thesis, Colorado State University, Fort Collins, Colorado.

13. Dendy, F. E., and G. C. Bolton. (1976). "Sediment Yield-Runoff-Drainage Area Relationships in the United States," *Journal of Soil and Water Conservation*, Vol. 31, No. 6, November-December, pp. 264–266.

14. DuBoys, P. (1879). "Le Rohne et les Rivieres a Lit Affouillable," *Annales de Ponts et Chausées*, Series 5, Vol. 18, pp. 141–195.

15. Einstein, H. A. (1950). "The Bed-Load Function for Sediment Transportation in Open Channel Flows," USDA Soil Conservation Service, *Technical Bulletin No.* 1026, Washington, D.C., September.

16. Engelund, F., and E. Hansen. (1967). "A Monograph on Sediment Transport in Alluvial Streams," *Teknisk Vorlag*, Copenhagen, Denmark.

17. Flaxman, E. M. (1972). "Predicting Sediment Yield in the Western United States," *Journal of the Hydraulics Division*, ASCE, Vol. 98, HY12, pp. 2073–2085.

18. Geiger, A. F. (1965). "Developing Sediment Storage Requirements for Upstream Retarding Reservoirs," Proceedings of Federal Interagency Sedimentation Conference, Jackson,

Miss., USDA Agricultural Research Service, *Miscellaneous Publication No.* 970, pp. 881–885.

19. Guy, H. P. (1969). "Laboratory Theory and Methods for Sediment Analysis," U.S. Geological Survey, *Techniques for Water Resources Investigations,* Book 5, Chapter C1.

20. Guy, H. P. (1970). "Fluvial Sediment Concepts," U.S. Geological Survey, *Techniques for Water Resources Investigations,* Book 3, Chapter C1, 1970.

21. Hydrologic Engineering Center, U.S. Army Corps of Engineers. (1976). "HEC-6, Scour and Deposition in Rivers and Reservoirs," Users' Manual.

22. Interagency Committee on Water Resources. (1952). "The Design of Improved Types of Sediment Samplers," *Report No.* 6, Subcommittee on Sedimentation, Federal Interagency River Basin Committee, Hydraulic Laboratory of the Iowa Institute of Hydraulic Research, Iowa City, Iowa.

23. Interagency Committee on Water Resources. (1957). "Some Fundamentals of Particle Size Analysis, A Study of Methods Used in Measurement and Analysis of Sediment Loads in Streams," *Report No.* 12, Subcommittee on Sedimentation, Saint Anthony Falls Hydraulic Laboratory, Minneapolis, Minnesota.

24. Interagency Committee on Water Resources. (1961). "The Single-Stage Sampler for Suspended Sediment," *Report No.* 13, Subcommittee on Sedimentation, Saint Anthony Falls Hydraulic Laboratory, Minneapolis, Minnesota.

25. Jansen, P. Ph., L. van Bendegom, J. van den Berg, M. de Vries, and A. Zanen. (1979). *Principles of River Engineering.* London: Pitman.

26. Kennedy, J. F. (1983). "Reflections on Rivers, Research, and Rouse," *Journal of Hydraulic Engineering,* ASCE, Vol. 109, No. 10, October, pp. 1254–1271.

27. Lane, E. W. (1947). "Report of the Subcommittee on Sediment Terminology," *Transactions,* American Geophysical Union, Vol. 28, No. 6, December, pp. 936–938.

28. Lane, E. W., and V. A. Koelzer. (1953). "Density of Sediments Deposited in Reservoirs," A Study of Methods Used in Measurement and Analysis of Sediment Loads in Streams, *Report No.* 9, St. Paul U.S. Army Engineer District, St. Paul, Minnesota.

29. Lane, E. W. (1955). "Design of Stable Channels," *Transactions,* ASCE, Vol. 120, paper No. 2776, pp. 1234–1279.

30. Leifeste, D. K. (1974). "Dissolved-Solids Discharge to the Oceans from the Conterminous United States," U.S. Geological Survey *Circular No.* 685.

31. Leopold, L. B., W. W. Emmett, and R. M. Myrick. (1966). "Channel and Hillslope Processes in a Semiarid Area in New Mexico," U.S. Geological Survey, *Professional Paper No.* 352-G, Washington, D.C.

32. McNown, J. S., and J. Malaika. (1950). "Effect of Particle Shape on Settling Velocity at Low Reynolds Number," *Transactions,* American Geophysical Union, Vol. 31, No. 1, Feb., pp. 74–82.

33. Meyer-Peter, E., and R. Muller. (1948). "Formulas for Bed Load Transport," Report on Second Meeting of International Association of Hydraulic Research, Stockholm, Sweden, pp. 39–64.

34. Roehl, J. W. (1962). "Sediment Source Areas, Delivery Ratios, and Influencing Morphological Factors," *Publication* 59, International Association of Scientific Hydrology, Commission on Land Erosion, pp. 202–213.

35. Rouse, H. (1937). "Nomograph for the Settling Velocity of Spheres," Division of Geology and Geography, Exhibit D of the Report of the Commission on Sedimentation, 1936–1937, National Research Council, Washington, D.C., October, pp. 57–64.

36. Rouse, H. (1939). "An Analysis of Sediment Transportation in the Light of Fluid Turbulence," USDA Soil Conservation Service, *Report No.* SCS-TP-25, Washington, D.C.

37. Simons, D. B., and E. V. Richardson. (1966). "Resistance to Flow in Alluvial Channels," U.S. Geological Survey, *Professional Paper No.* 422J, Washington, D.C.

38. Simons, D. B., and F. Senturk. (1976). *Sediment Transport Technology.* Fort Collins, Colorado: Water Resources Publications.

39. Toffaleti, F. B. (1969). "Definitive Computation of Sand Discharge in Rivers," *Journal of the Hydraulics Division,* ASCE, Vol. 95, No. HY1, Jan., pp. 225–248.

40. U.S. Army Corps of Engineers. (1987). "Sedimentation Investigations in Rivers and Reservoirs," *Engineer Manual* EM 1110-2-4000, Office of the Chief of Engineers, Washington, D.C., Draft.

41. USDA Soil Conservation Service. (1983). *National Engineering Handbook,* Section 3, Sedimentation, 2d. ed.

42. Williams, J. R. (1975). "Sediment Yield Prediction with the Universal Soil Loss Equation Using Runoff-Energy Factor," in *Present and Prospective Technology for Predicting Sediment Sources and Sediment Yields,"* USDA Agricultural Research Service, *Publication ARS-S-40.*

43. Wischmeier, W. H., and D. D. Smith. (1965). "Predicting Rainfall-Erosion Losses from Cropland East of the Rocky Mountains," USDA Agricultural Research Service, *Agriculture Handbook No. 282,* May.

44. Wischmeier, W. H., C. B. Johnson, and B. V. Cross. (1971). "A Soil Erodibility Nomograph for Farmland and Construction Sites," *Journal of Soil and Water Conservation,* Vol. 26, No. 5, Sept.-Oct.

45. Wolman, M. G., and A. P. Schick. (1967). "Effects of Construction on Fluvial Sediment, Urban and Suburban Areas of Maryland," *Water Resources Research,* Vol. 3, No. 2, pp. 451–464.

46. Yang, C. T. (1972). "Unit Stream Power and Sediment Transport," *Journal of the Hydraulics Division,* ASCE, Vol. 98, No. HY10, Oct., pp. 1805–1826.

SUGGESTED READINGS

American Society of Civil Engineers. (1975). *Sedimentation Engineering,* Manual No. 54.

Colby, B. R. (1964). "Discharge of Sands and Mean Velocity Relations in Sand-Bed Streams," *U.S. Geological Survey Professional Paper 462-A,* Washington, D.C.

Guy, H. P. (1970). "Fluvial Sediment Concepts," U.S. Geological Survey, *Techniques for Water Resources Investigations,* Book 3, Chapter C1, 1970.

Simons, D. B., and F. Senturk. (1976). *Sediment Transport Technology,* Fort Collins, Colorado: Water Resources Publications.

U.S. Army Corps of Engineers. (1987). "Sedimentation Investigations in Rivers and Reservoirs," *Engineer Manual* EM 1110-2-4000, Office of the Chief of Engineers, Washington, D.C., Draft.

USDA Soil Conservation Service. (1983). *National Engineering Handbook,* Section 3, Sedimentation, 2d. ed.

TABLES

TABLE A-1 PROPERTIES OF WATER IN SI UNITS

Temperature (°C)	Specific Gravity	Density (g/cm³)	Heat of Vaporization (cal/g)	Viscosity Absolute (cp)	Viscosity Kinematic (cs)	Vapor Pressure (mm Hg)	Vapor Pressure (mb)	Vapor Pressure (g/cm²)
0	0.99987	0.99984	597.3	1.790	1.790	4.58	6.11	6.23
5	0.99999	0.99996	594.5	1.520	1.520	6.54	8.72	8.89
10	0.99973	0.99970	591.7	1.310	1.310	9.20	12.27	12.51
15	0.99913	0.99910	588.9	1.140	1.140	12.78	17.04	17.38
20	0.99824	0.99821	586.0	1.000	1.000	17.53	23.37	23.83
25	0.99708	0.99705	583.2	0.890	0.893	23.76	31.67	32.30
30	0.99568	0.99565	580.4	0.798	0.801	31.83	42.43	43.27
35	0.99407	0.99404	577.6	0.719	0.723	42.18	56.24	57.34
40	0.99225	0.99222	574.7	0.653	0.658	55.34	73.78	75.23
50	0.98807	0.98804	569.0	0.547	0.554	92.56	123.40	125.83
60	0.98323	0.98320	563.2	0.466	0.474	149.46	199.26	203.19
70	0.97780	0.97777	557.4	0.404	0.413	233.79	311.69	317.84
80	0.97182	0.97179	551.4	0.355	0.365	355.28	473.67	483.01
90	0.96534	0.96531	545.3	0.315	0.326	525.89	701.13	714.95
100	0.95839	0.95836	539.1	0.282	0.294	760.00	1013.25	1033.23

Source: Linsley, R. K. *et al.* (1982). *Hydrology for Engineers.* 3d. ed. New York: McGraw-Hill.

TABLE A-2 PROPERTIES OF WATER IN U.S. CUSTOMARY UNITS

Temperature (°F)	Specific Gravity	Density (lb/ft³)	Heat of Vaporization (Btu/lb)	Viscosity[1] Absolute (lb-s/ft²)	Viscosity[1] Kinematic (ft²/s)	Vapor Pressure (in Hg)	Vapor Pressure (mb)	Vapor Pressure (lb/in.²)
32	0.99986	62.418	1075.5	3.746	1.931	0.180	6.11	0.089
40	0.99998	62.426	1071.0	3.229	1.664	0.248	8.39	0.122
50	0.99971	62.409	1065.3	2.735	1.410	0.362	12.27	0.178
60	0.99902	62.366	1059.7	2.359	1.217	0.522	17.66	0.256
70	0.99798	62.301	1054.0	2.050	1.058	0.739	25.03	0.363
80	0.99662	62.216	1048.4	1.799	0.930	1.032	34.96	0.507
90	0.99497	62.113	1042.7	1.595	0.826	1.422	48.15	0.698
100	0.99306	61.994	1037.1	1.424	0.739	1.933	65.47	0.950
120	0.98856	61.713	1025.6	1.168	0.609	3.448	116.75	1.693
140	0.98321	61.379	1014.0	0.981	0.514	5.884	199.26	2.890
160	0.97714	61.000	1002.2	0.838	0.442	9.656	326.98	4.742
180	0.97041	60.580	990.2	0.726	0.386	15.295	517.95	7.512
200	0.96306	60.121	977.9	0.637	0.341	23.468	794.72	11.526
212	0.95837	59.828	970.3	0.593	0.319	29.921	1013.25	14.696

[1]To obtain values of viscosity, multiply values shown in table by 10^{-5}.

Source: Linsley, R. K. *et al.* (1982). *Hydrology for Engineers.* 3d. ed. New York: McGraw-Hill.

TABLE A-3 FACTOR p IN BLANEY-CRIDDLE METHOD[1]

Latitude North / South	Jan Jul	Feb Aug	Mar Sep	Apr Oct	May Nov	Jun Dec	Jul Jan	Aug Feb	Sep Mar	Oct Apr	Nov May	Dec Jun
60°	0.15	0.20	0.26	0.32	0.38	0.41	0.40	0.34	0.28	0.22	0.17	0.13
50	0.19	0.23	0.27	0.31	0.34	0.36	0.35	0.32	0.28	0.24	0.20	0.18
40	0.22	0.24	0.27	0.30	0.32	0.34	0.33	0.31	0.28	0.25	0.22	0.21
30	0.24	0.25	0.27	0.29	0.31	0.32	0.31	0.30	0.28	0.26	0.24	0.23
20	0.25	0.26	0.27	0.28	0.29	0.30	0.30	0.29	0.28	0.26	0.25	0.25
10	0.26	0.27	0.27	0.28	0.28	0.29	0.29	0.28	0.28	0.27	0.26	0.26
0	0.27	0.27	0.27	0.27	0.27	0.27	0.27	0.27	0.27	0.27	0.27	0.27

[1]$p = \dfrac{\text{mean daily daytime hours for a given month}}{\text{total daytime hours in the year}} \times 100$

TABLE A-4 CONSTANT *K* IN THORNTHWAITE METHOD[1]

Latitude	Jan	Feb	Mar	Apr	May	Jun	Jul	Aug	Sep	Oct	Nov	Dec
60°N	0.54	0.67	0.97	1.19	1.33	1.56	1.55	1.33	1.07	0.84	0.58	0.48
50°N	0.71	0.84	0.98	1.14	1.28	1.36	1.33	1.21	1.06	0.90	0.76	0.68
40°N	0.80	0.89	0.99	1.10	1.20	1.25	1.23	1.15	1.04	0.93	0.83	0.78
30°N	0.87	0.93	1.00	1.07	1.14	1.17	1.16	1.11	1.03	0.96	0.89	0.85
20°N	0.92	0.96	1.00	1.05	1.09	1.11	1.10	1.07	1.02	0.98	0.93	0.91
10°N	0.97	0.98	1.00	1.03	1.05	1.06	1.05	1.04	1.02	0.99	0.97	0.96
0°	1.00	1.00	1.00	1.00	1.00	1.00	1.00	1.00	1.00	1.00	1.00	1.00
10°S	1.05	1.04	1.02	0.99	0.97	0.96	0.97	0.98	1.00	1.03	1.05	1.06
20°S	1.10	1.07	1.02	0.98	0.93	0.91	0.92	0.96	1.00	1.05	1.09	1.11
30°S	1.16	1.11	1.03	0.96	0.89	0.85	0.87	0.93	1.00	1.07	1.14	1.17
40°S	1.23	1.15	1.04	0.93	0.83	0.78	0.80	0.89	0.99	1.10	1.20	1.25
50°S	1.33	1.19	1.05	0.89	0.75	0.68	0.70	0.82	0.97	1.13	1.27	1.36

[1]*K* is a constant to correct PET for latitudes other than 0° (Eq. 2-47).

TABLE A-5 VALUES OF F(z) (AREAS UNDER HALF OF THE NORMAL PROBABILITY DENSITY FUNCTION, Eq. 6-11) VERSUS FREQUENCY FACTOR z. (DOUBLE-ENTRY TABLE)

z	.00	.01	.02	.03	.04	.05	.06	.07	.08	.09
0.0	0.0000	0.0040	0.0080	0.0120	0.0159	0.0199	0.0239	0.0279	0.0319	0.0359
0.1	0.0398	0.0438	0.0478	0.0517	0.0557	0.0596	0.0636	0.0675	0.0714	0.0753
0.2	0.0793	0.0832	0.0871	0.0910	0.0948	0.0987	0.1026	0.1064	0.1103	0.1141
0.3	0.1179	0.1217	0.1255	0.1293	0.1331	0.1368	0.1406	0.1443	0.1480	0.1517
0.4	0.1554	0.1591	0.1628	0.1664	0.1700	0.1736	0.1772	0.1808	0.1844	0.1879
0.5	0.1915	0.1950	0.1985	0.2019	0.2054	0.2088	0.2123	0.2157	0.2190	0.2224
0.6	0.2257	0.2291	0.2324	0.2357	0.2389	0.2422	0.2454	0.2486	0.2518	0.2549
0.7	0.2580	0.2611	0.2642	0.2673	0.2704	0.2734	0.2764	0.2794	0.2823	0.2852
0.8	0.2881	0.2910	0.2939	0.2967	0.2995	0.3023	0.3051	0.3078	0.3106	0.3133
0.9	0.3159	0.3186	0.3212	0.3238	0.3264	0.3289	0.3315	0.3340	0.3365	0.3389
1.0	0.3413	0.3438	0.3461	0.3485	0.3508	0.3531	0.3554	0.3577	0.3599	0.3621
1.1	0.3643	0.3665	0.3686	0.3708	0.3729	0.3749	0.3770	0.3790	0.3810	0.3830
1.2	0.3849	0.3869	0.3888	0.3907	0.3925	0.3944	0.3962	0.3980	0.3997	0.4015
1.3	0.4032	0.4049	0.4066	0.4082	0.4099	0.4115	0.4131	0.4147	0.4162	0.4177
1.4	0.4192	0.4207	0.4222	0.4236	0.4251	0.4265	0.4279	0.4292	0.4306	0.4319
1.5	0.4332	0.4345	0.4357	0.4370	0.4382	0.4394	0.4406	0.4418	0.4430	0.4441
1.6	0.4452	0.4463	0.4474	0.4485	0.4495	0.4505	0.4515	0.4525	0.4535	0.4545
1.7	0.4554	0.4564	0.4573	0.4582	0.4591	0.4599	0.4608	0.4616	0.4625	0.4633
1.8	0.4641	0.4649	0.4656	0.4664	0.4671	0.4678	0.4686	0.4693	0.4699	0.4706
1.9	0.4713	0.4719	0.4726	0.4732	0.4738	0.4744	0.4750	0.4756	0.4762	0.4767
2.0	0.4772	0.4778	0.4783	0.4788	0.4793	0.4798	0.4803	0.4808	0.4812	0.4817
2.1	0.4821	0.4826	0.4830	0.4834	0.4838	0.4842	0.4846	0.4850	0.4854	0.4857
2.2	0.4861	0.4865	0.4868	0.4871	0.4875	0.4878	0.4881	0.4884	0.4887	0.4890
2.3	0.4893	0.4896	0.4898	0.4901	0.4904	0.4906	0.4909	0.4911	0.4913	0.4916
2.4	0.4918	0.4920	0.4922	0.4925	0.4927	0.4929	0.4931	0.4932	0.4934	0.4936
2.5	0.4938	0.4940	0.4941	0.4943	0.4945	0.4946	0.4948	0.4949	0.4951	0.4952
2.6	0.4953	0.4955	0.4956	0.4957	0.4959	0.4960	0.4961	0.4962	0.4963	0.4964
2.7	0.4965	0.4966	0.4967	0.4968	0.4969	0.4970	0.4971	0.4972	0.4973	0.4974
2.8	0.4974	0.4975	0.4976	0.4977	0.4977	0.4978	0.4979	0.4980	0.4980	0.4981
2.9	0.4981	0.4982	0.4983	0.4983	0.4984	0.4984	0.4985	0.4985	0.4986	0.4986
3.0	0.4987	0.4987	0.4987	0.4988	0.4988	0.4989	0.4989	0.4989	0.4990	0.4990

Source: Weatherburn, C. E. (1957). *Mathematical Statistics.* Cambridge University Press.

TABLE A-6   FREQUENCY FACTORS *K* FOR PEARSON TYPE III DISTRIBUTIONS VERSUS SKEW COEFFICIENT C_s AND RETURN PERIOD *T* (OR PROBABILITY OF EXCEEDENCE *P*)

Skew Coefficient C_s	Return Period *T* (y)									
	1.05	1.11	1.25	2	5	10	25	50	100	200
	Probability of Exceedence *P* (percent)									
	95	90	80	50	20	10	4	2	1	0.5
3.0	−0.665	−0.660	−0.636	−0.396	0.420	1.180	2.278	3.152	4.051	4.970
2.8	−0.711	−0.702	−0.666	−0.384	0.460	1.210	2.275	3.114	3.973	4.847
2.6	−0.762	−0.747	−0.696	−0.368	0.499	1.238	2.267	3.071	3.889	4.718
2.4	−0.819	−0.795	−0.725	−0.351	0.537	1.262	2.256	3.023	3.800	4.584
2.2	−0.882	−0.844	−0.752	−0.330	0.574	1.284	2.240	2.970	3.075	4.444
2.0	−0.949	−0.895	−0.777	−0.307	0.609	1.302	2.219	2.912	3.605	4.398
1.8	−1.020	−0.945	−0.799	−0.282	0.643	1.318	2.193	2.848	3.499	4.147
1.6	−1.093	−0.994	−0.817	−0.254	0.675	1.329	2.163	2.780	3.388	3.990
1.4	−1.168	−1.041	−0.832	−0.225	0.705	1.337	2.128	2.706	3.271	3.828
1.2	−1.243	−1.086	−0.844	−0.195	0.732	1.340	2.087	2.626	3.149	3.661
1.0	−1.317	−1.128	−0.852	−0.164	0.758	1.340	2.043	2.542	3.022	3.489
0.8	−1.388	−1.166	−0.856	−0.132	0.780	1.336	1.993	2.453	2.891	3.312
0.6	−1.458	−1.200	−0.857	−0.099	0.800	1.328	1.939	2.359	2.755	3.132
0.4	−1.524	−1.231	−0.855	−0.066	0.816	1.317	1.880	2.261	2.615	2.949
0.2	−1.586	−1.258	−0.850	−0.033	0.830	1.301	1.818	2.159	2.472	2.763
0.0	−1.645	−1.282	−0.842	0.000	0.842	1.282	1.751	2.054	2.326	2.576
−0.2	−1.700	−1.301	−0.830	0.033	0.850	1.258	1.680	1.945	2.178	2.388
−0.4	−1.750	−1.317	−0.816	0.066	0.855	1.231	1.606	1.834	2.029	2.201
−0.6	−1.797	−1.328	−0.800	0.099	0.857	1.200	1.528	1.720	1.880	2.016
−0.8	−1.839	−1.336	−0.780	0.132	0.856	1.166	1.448	1.606	1.733	1.837
−1.0	−1.877	−1.340	−0.758	0.164	0.852	1.128	1.366	1.492	1.588	1.664
−1.2	−1.910	−1.340	−0.732	0.195	0.844	1.086	1.282	1.379	1.449	1.501
−1.4	−1.938	−1.337	−0.705	0.225	0.832	1.041	1.198	1.270	1.318	1.351
−1.6	−1.962	−1.329	−0.675	0.254	0.817	0.994	1.116	1.166	1.197	1.216
−1.8	−1.981	−1.318	−0.643	0.282	0.799	0.945	1.035	1.069	1.087	1.097
−2.0	−1.996	−1.302	−0.609	0.307	0.777	0.895	0.959	0.980	0.990	0.995
−2.2	−2.006	−1.284	−0.574	0.330	0.752	0.844	0.888	0.900	0.905	0.907
−2.4	−2.011	−1.262	−0.537	0.351	0.725	0.795	0.823	0.830	0.832	0.833
−2.6	−2.013	−1.238	−0.499	0.368	0.696	0.747	0.764	0.768	0.769	0.769
−2.8	−2.010	−1.210	−0.460	0.384	0.666	0.702	0.712	0.714	0.714	0.714
−3.0	−2.003	−1.180	−0.420	0.393	0.636	0.660	0.666	0.666	0.667	0.667

Source: U.S. Interagency Advisory Committee on Water Data, Hydrology Subcommittee (1983). "Guidelines for Determining Flood Flow Frequency," *Bulletin No.* 17B, issued 1981, revised 1983.

TABLE A-7 OUTLIER FREQUENCY FACTORS K_n FOR LOG PEARSON III DISTRIBUTIONS

Record Length (y)	K_n	Record Length (y)	K_n	Record Length (y)	K_n	Record Length (y)	K_n
10	2.036	45	2.727	80	2.940	115	3.064
15	2.247	50	2.768	85	2.961	120	3.078
20	2.385	55	2.804	90	2.981	125	3.092
25	2.486	60	2.837	95	3.000	130	3.104
30	2.563	65	2.866	100	3.017	135	3.116
35	2.628	70	2.893	105	3.033	140	3.129
40	2.682	75	2.917	110	3.049	145	3.140

Source: U.S. Interagency Advisory Committee on Water Data, Hydrology Subcommittee (1983). "Guidelines for Determining Flood Flow Frequency," *Bulletin No.* 17B, issued 1981, revised 1983.

TABLE A-8 MEAN $\bar{y}_n$ AND STANDARD DEVIATION σ_n OF GUMBEL VARIATE *(y)* VERSUS RECORD LENGTH *(n)*

n	$\bar{y}_n$	σ_n	n	$\bar{y}_n$	σ_n	n	$\bar{y}_n$	σ_n
8	0.4843	0.9043	35	0.5403	1.1285	64	0.5533	1.1793
9	0.4902	0.9288	36	0.5410	1.1313	66	0.5538	1.1814
10	0.4952	0.9497	37	0.5418	1.1339	68	0.5543	1.1834
11	0.4996	0.9676	38	0.5424	1.1363	70	0.5548	1.1854
12	0.5035	0.9833	39	0.5430	1.1388	72	0.5552	1.1873
13	0.5070	0.9972	40	0.5436	1.1413	74	0.5557	1.1890
14	0.5100	1.0095	41	0.5442	1.1436	76	0.5561	1.1906
15	0.5128	1.0206	42	0.5448	1.1458	78	0.5565	1.1923
16	0.5157	1.0316	43	0.5453	1.1480	80	0.5569	1.1938
17	0.5181	1.0411	44	0.5458	1.1499	82	0.5572	1.1953
18	0.5202	1.0493	45	0.5463	1.1519	84	0.5576	1.1967
19	0.5220	1.0566	46	0.5468	1.1538	86	0.5580	1.1980
20	0.5236	1.0628	47	0.5473	1.1557	88	0.5583	1.1994
21	0.5252	1.0696	48	0.5477	1.1574	90	0.5586	1.2007
22	0.5268	1.0754	49	0.5481	1.1590	92	0.5589	1.2020
23	0.5283	1.0811	50	0.5485	1.1607	94	0.5592	1.2032
24	0.5296	1.0864	51	0.5489	1.1623	96	0.5595	1.2044
25	0.5309	1.0915	52	0.5493	1.1638	98	0.5598	1.2055
26	0.5320	1.0961	53	0.5497	1.1653	100	0.5600	1.2065
27	0.5332	1.1004	54	0.5501	1.1667	150	0.5646	1.2253
28	0.5343	1.1047	55	0.5504	1.1681	200	0.5672	1.2360
29	0.5353	1.1086	56	0.5508	1.1696	250	0.5688	1.2429
30	0.5362	1.1124	57	0.5511	1.1708	300	0.5699	1.2479
31	0.5371	1.1159	58	0.5515	1.1721	400	0.5714	1.2545
32	0.5380	1.1193	59	0.5518	1.1734	500	0.5724	1.2588
33	0.5388	1.1226	60	0.5521	1.1747	750	0.5738	1.2651
34	0.5396	1.1255	62	0.5527	1.1770	1000	0.5745	1.2685

Source: Gumbel, E. J. (1958). *Statistics of Extremes.* Irvington, New York: Columbia University Press.

TABLE A-9 SCS 24-HR RAINFALL TABLES (AT HALF-HOUR INCREMENTS)

Time (h)	Fraction of 24-h rainfall depth			
	Type I	Type IA	Type II	Type III
0.0	0.00000	0.00000	0.00000	0.00000
0.5	0.00871	0.01000	0.00513	0.00500
1.0	0.01745	0.02000	0.01050	0.01000
1.5	0.02621	0.03500	0.01613	0.01500
2.0	0.03500	0.05000	0.02200	0.02000
2.5	0.04416	0.06600	0.02813	0.02519
3.0	0.05405	0.08200	0.03450	0.03075
3.5	0.06466	0.09800	0.04113	0.03669
4.0	0.07600	0.11600	0.04800	0.04300
4.5	0.08784	0.13500	0.05525	0.04969
5.0	0.09995	0.15600	0.06300	0.05675
5.5	0.11234	0.18000	0.07125	0.06419
6.0	0.12500	0.20600	0.08000	0.07200
6.5	0.13915	0.23700	0.08925	0.08063
7.0	0.15600	0.26800	0.09900	0.09050
7.5	0.17460	0.31000	0.10925	0.10163
8.0	0.19400	0.42500	0.12000	0.11400
8.5	0.21900	0.48000	0.13225	0.12844
9.0	0.25400	0.52000	0.14700	0.14575
9.5	0.30300	0.55000	0.16300	0.16594
10.0	0.51500	0.57700	0.18100	0.18900
10.5	0.58300	0.60100	0.20400	0.21650
11.0	0.62300	0.62400	0.23500	0.25000
11.5	0.65550	0.64500	0.28300	0.29800
12.0	0.68400	0.66400	0.66300	0.50000
12.5	0.70925	0.68300	0.73500	0.70200
13.0	0.73200	0.70100	0.77200	0.75000
13.5	0.75225	0.71900	0.79900	0.78350
14.0	0.77000	0.73600	0.82000	0.81100
14.5	0.78625	0.75281	0.83763	0.83406
15.0	0.80200	0.76924	0.85350	0.85425
15.5	0.81725	0.78529	0.86763	0.87156
16.0	0.83200	0.80096	0.88000	0.88600
16.5	0.84625	0.81625	0.89119	0.89838
17.0	0.86000	0.83116	0.90175	0.90950
17.5	0.87325	0.84569	0.91169	0.91938
18.0	0.88600	0.85984	0.92100	0.92800
18.5	0.89825	0.87361	0.92969	0.93581
19.0	0.91000	0.88700	0.93775	0.94325
19.5	0.92125	0.90001	0.94519	0.95031
20.0	0.93200	0.91264	0.95200	0.95700
20.5	0.94225	0.92489	0.95844	0.96336
21.0	0.95200	0.93676	0.96475	0.96944
21.5	0.96125	0.94825	0.97094	0.97523
22.0	0.97000	0.95936	0.97700	0.98075
22.5	0.97825	0.97009	0.98294	0.98598
23.0	0.98600	0.98044	0.98875	0.99094
23.5	0.99325	0.99041	0.99444	0.99561
24.0	1.00000	1.00000	1.00000	1.00000

Source: Roger G. Cronshey, USDA Soil Conservation Service, Engineering Division, Washington, D.C., June 1988.

Appendix A

TABLE A-10 STANDARD SCS DIMENSIONLESS TEMPORAL RAINFALL DISTRIBUTION FOR EMERGENCY SPILLWAY AND FREEBOARD-POOL DESIGN

Fraction of storm duration	Fraction of storm depth	Fraction of storm duration	Fraction of storm depth
0.00	0.0000		
0.02	0.0080		
0.04	0.0162	0.52	0.7240
0.06	0.0246	0.54	0.7420
0.08	0.0333	0.56	0.7590
0.10	0.0425	0.58	0.7750
0.12	0.0524	0.60	0.7900
0.14	0.0630	0.62	0.8043
0.16	0.0743	0.64	0.8180
0.18	0.0863	0.66	0.8312
0.20	0.0990	0.68	0.8439
0.22	0.1124	0.70	0.8561
0.24	0.1265	0.72	0.8678
0.26	0.1420	0.74	0.8790
0.28	0.1595	0.76	0.8898
0.30	0.1800	0.78	0.9002
0.32	0.2050	0.80	0.9103
0.34	0.2550	0.82	0.9201
0.36	0.3450	0.84	0.9297
0.38	0.4370	0.86	0.9391
0.40	0.5300	0.88	0.9483
0.42	0.6030	0.90	0.9573
0.44	0.6330	0.92	0.9661
0.46	0.6600	0.94	0.9747
0.48	0.6840	0.96	0.9832
0.50	0.7050	0.98	0.9916
		1.00	1.0000

Source: USDA Soil Conservation Service. (1983). "Computer Program for Project Formulation: Hydrology," *Technical Release No.* 20 (TR-20), Engineering Division, Washington, D.C., May.

DERIVATION OF
THE NUMERICAL DIFFUSION COEFFICIENT OF
THE MUSKINGUM-CUNGE METHOD

APPENDIX B: DERIVATION OF THE NUMERICAL DIFFUSION
COEFFICIENT OF THE MUSKINGUM-CUNGE METHOD

Expanding the grid function $Q(j\Delta x, n\Delta t)$ (Fig. 9-8) in Taylor series about point $(j\Delta x, n\Delta t)$ leads to:

$$Q_j^{n+1} = Q_j^n + \left[\frac{\partial Q}{\partial t}\right]_j \Delta t + \frac{1}{2}\left[\frac{\partial^2 Q}{\partial t^2}\right]_j \Delta t^2 + o(\Delta t^3) \qquad \text{(B.1)}$$

$$Q_{j+1}^{n+1} = Q_{j+1}^n + \left[\frac{\partial Q}{\partial t}\right]_{j+1} \Delta t + \frac{1}{2}\left[\frac{\partial^2 Q}{\partial t^2}\right]_{j+1} \Delta t^2 + o(\Delta t^3) \qquad \text{(B.2)}$$

$$Q_{j+1}^n = Q_j^n + \left[\frac{\partial Q}{\partial x}\right]_n \Delta x + \frac{1}{2}\left[\frac{\partial^2 Q}{\partial x^2}\right]_n \Delta x^2 + o(\Delta x^3) \qquad \text{(B.3)}$$

$$Q_{j+1}^{n+1} = Q_j^{n+1} + \left[\frac{\partial Q}{\partial x}\right]_{n+1} \Delta x + \frac{1}{2}\left[\frac{\partial^2 Q}{\partial x^2}\right]_{n+1} \Delta x^2 + o(\Delta x^3) \qquad \text{(B.4)}$$

Substituting Eqs. B.1 to B.4 into Eq. 9-61 and neglecting third-order terms yields:

$$X\left\{\left[\frac{\partial Q}{\partial t}\right]_j \Delta t + \frac{1}{2}\left[\frac{\partial^2 Q}{\partial t^2}\right]_j \Delta t^2\right\}$$

$$+ (1 - X) \left\{ \left[\frac{\partial Q}{\partial t} \right]_{j+1} \Delta t + \frac{1}{2} \left[\frac{\partial^2 Q}{\partial t^2} \right]_{j+1} \Delta t^2 \right\}$$

$$+ \frac{C}{2} \left\{ \left[\frac{\partial Q}{\partial x} \right]_n \Delta x + \frac{1}{2} \left[\frac{\partial^2 Q}{\partial x^2} \right]_n \Delta x^2 \right\}$$

$$+ \frac{C}{2} \left\{ \left[\frac{\partial Q}{\partial x} \right]_{n+1} \Delta x + \frac{1}{2} \left[\frac{\partial^2 Q}{\partial x^2} \right]_{n+1} \Delta x^2 \right\} = 0 \qquad \text{(B.5)}$$

in which $C = c(\Delta t / \Delta x)$ is the Courant number.

Expressing the derivatives at grid point $[(j + 1)\Delta x, (n + 1)\Delta t]$ in terms of the derivatives at grid point $(j\Delta x, n\Delta t)$ by means of Taylor series:

$$\left[\frac{\partial Q}{\partial t} \right]_{j+1} = \left[\frac{\partial Q}{\partial t} \right]_j + \left[\frac{\partial^2 Q}{\partial x \partial t} \right]_{j,n} \Delta x + o(\Delta x^2) \qquad \text{(B.6)}$$

$$\left[\frac{\partial Q}{\partial x} \right]_{n+1} = \left[\frac{\partial Q}{\partial x} \right]_n + \left[\frac{\partial^2 Q}{\partial x \partial t} \right]_{j,n} \Delta t + o(\Delta t^2) \qquad \text{(B.7)}$$

$$\left[\frac{\partial^2 Q}{\partial t^2} \right]_{j+1} = \left[\frac{\partial^2 Q}{\partial t^2} \right]_j + \left[\frac{\partial^3 Q}{\partial t^3 \partial x} \right]_j \Delta x + o(\Delta x^2) \qquad \text{(B.8)}$$

$$\left[\frac{\partial^2 Q}{\partial x^2} \right]_{n+1} = \left[\frac{\partial^2 Q}{\partial x^2} \right]_n + \left[\frac{\partial^3 Q}{\partial x^2 \partial t} \right]_n \Delta t + o(\Delta t^2) \qquad \text{(B.9)}$$

Substituting Eqs. B.6 to B.9 into B.5 and neglecting third-order terms:

$$X \left\{ \left[\frac{\partial Q}{\partial t} \right]_j \Delta t + \frac{1}{2} \left[\frac{\partial^2 Q}{\partial t^2} \right]_j \Delta t^2 \right\}$$

$$+ (1 - X) \left\{ \left[\frac{\partial Q}{\partial t} \right]_j \Delta t + \left[\frac{\partial^2 Q}{\partial x \partial t} \right]_{j,n} \Delta x \Delta t + \frac{1}{2} \left[\frac{\partial^2 Q}{\partial t^2} \right]_j \Delta t^2 \right\}$$

$$+ \frac{C}{2} \left\{ \left[\frac{\partial Q}{\partial x} \right]_n \Delta x + \frac{1}{2} \left[\frac{\partial^2 Q}{\partial x^2} \right]_n \Delta x^2 \right\}$$

$$+ \frac{C}{2} \left\{ \left[\frac{\partial Q}{\partial x} \right]_n \Delta x + \left[\frac{\partial^2 Q}{\partial x \partial t} \right]_{j,n} \Delta x \Delta t + \frac{1}{2} \left[\frac{\partial^2 Q}{\partial x^2} \right]_n \Delta x^2 \right\} = 0 \qquad \text{(B.10)}$$

In Eq. B.10, dividing by Δt and simplifying,

$$\left[\frac{\partial Q}{\partial t} \right]_j + c \left[\frac{\partial Q}{\partial x} \right]_n$$

$$+ \frac{\Delta t}{2} \left[\frac{\partial^2 Q}{\partial t^2} \right]_j + \frac{c \, \Delta x}{2} \left[\frac{\partial^2 Q}{\partial x^2} \right]_n$$

$$+ \Delta x \left\{ (1 - X) + \frac{C}{2} \right\} \left[\frac{\partial^2 Q}{\partial x \, \partial t} \right]_{j,n} = 0 \tag{B.11}$$

The first two terms of Eq. B.11 constitute the kinematic wave equation, Eq. 9-18. The remaining terms are the error R of the first-order-accurate numerical scheme:

$$R = \frac{\Delta t}{2} \left[\frac{\partial^2 Q}{\partial t^2} \right]_j + \frac{c \, \Delta x}{2} \left[\frac{\partial^2 Q}{\partial x^2} \right]_n$$

$$+ \Delta x \left\{ (1 - X) + \frac{C}{2} \right\} \left[\frac{\partial^2 Q}{\partial x \, \partial t} \right]_{j,n} = 0 \tag{B.12}$$

From Eq. 9-18

$$\frac{\partial Q}{\partial t} = -c \frac{\partial Q}{\partial x} \tag{B.13}$$

Therefore

$$\frac{\partial^2 Q}{\partial x \, \partial t} = -c \frac{\partial^2 Q}{\partial x^2} \tag{B.14}$$

$$\frac{\partial^2 Q}{\partial t^2} = c^2 \frac{\partial^2 Q}{\partial x^2} \tag{B.15}$$

Substituting Eqs. B.14 and B.15 into B.12 and simplifying:

$$R = c \, \Delta x \left(X - \frac{1}{2} \right) \frac{\partial^2 Q}{\partial x^2} \tag{B.16}$$

Comparing Eq. B.16 with the right-hand side of the diffusion wave equation, repeated here:

$$\frac{\partial Q}{\partial t} + c \frac{\partial Q}{\partial x} = \nu_h \frac{\partial^2 Q}{\partial x^2} \tag{B.17}$$

it follows that the numerical diffusion coefficient of the Muskingum-Cunge method is:

$$\nu_n = c \, \Delta x \left(\frac{1}{2} - X \right) \tag{B.18}$$

U.S. GEOLOGICAL SURVEY
FLOOD HYDROLOGY REPORTS

APPENDIX C: U.S. GEOLOGICAL SURVEY
FLOOD HYDROLOGY REPORTS

TABLE C-1 USGS REPORTS FOR ESTIMATING RURAL FLOOD PEAKS USING WATERSHED AND CLIMATIC CHARACTERISTICS

Alabama	Olin, D. A. (1984). "Magnitude and frequency of floods in Alabama," U.S. Geological Survey Water-Resources Investigations 84-4191.
Alaska	Lamke, R. D. (1978). "Flood characteristics of Alaskan streams," U.S. Geological Survey Water-Resources Investigations 78-129.
Arizona	Eychaner, J. H. (1984). "Estimation of magnitude and frequency of floods in Pima County, Arizona, with comparisons of alternative methods," U.S. Geological Survey Water-Resources Investigations 84-4142.
	Roeske, R. H. (1978). "Methods for estimating the magnitude and frequency of floods in Arizona," U.S. Geological Survey Open-File Report 78-711.
Arkansas	Neely, B. L., Jr. (1986). "Magnitude and frequency of floods in Arkansas," U.S. Geological Survey Water-Resources Investigations 86-4335.
California	Waananen, A. O., and J. R. Crippen. (1977). "Magnitude and frequency of floods in California," U.S. Geological Survey Water-Resources Investigations 77-21.
Colorado	Kircher, J. E., A. F. Choquette, and B. D. Richter. (1985). "Estimation of natural streamflow characteristics in Western Colorado," U.S. Geological Survey Water-Resources Investigations 85-4086.
	Livingston, R. K. (1980). "Rainfall-runoff modeling and preliminary regional flood characteristics of small rural watersheds in the Arkansas River Basin in Colorado," U.S. Geological Survey Water-Resources Investigations 80-112.

TABLE C-1 Continued

	Livingston, R. K., and D. R. Minges. (1987). "Techniques for estimating regional flood characteristics of small rural watersheds in the plains regions of eastern Colorado," U.S. Geological Survey Water-Resources Investigations 87-4094.
Connecticut	Weiss, L. A. (1975). "Flood-flow formula for urbanized and non-urbanized areas of Connecticut," Watershed Management Symposium of ASCE Irrigation and Drainage Division, pp. 658–675, August 11–13.
Delaware	Simmons, R. H., and D. H. Carpenter. (1978). "Technique for estimating the magnitude and frequency of floods in Delaware," U.S. Geological Survey Water-Resources Investigations Open-File Report 78-93.
Florida	Bridges, W. C. (1982). "Technique for estimating magnitude and frequency of floods on natural-flow streams in Florida," U.S. Geological Survey Water-Resources Investigations 82-4012.
Georgia	Price, M. (1979). "Floods in Georgia, magnitude and frequency," U.S. Geological Survey Water-Resources Investigations 78-137.
Hawaii	Nakahara, R. H. (1980). "An analysis of the magnitude and frequency of floods on Oahu, Hawaii," U.S. Geological Survey Water-Resources Investigations 80-45.
Idaho	Kjelstrom, L. C., and R. L. Moffatt. (1981). "Method of estimating flood-frequency parameters for streams in Idaho," U.S. Geological Survey Open-File Report 81-909.
Illinois	Curtis, G. W. (1977). "Technique for estimating flood-peak discharges and frequencies on rural streams in Illinois," U.S. Geological Survey Water-Resources Investigations 87-4207.
Indiana	Glatfelter, D. R. (1984). "Techniques for estimating magnitude and frequency of floods in Indiana," U.S. Geological Survey Water-Resources Investigations 84-4134.
Iowa	Lara, O. (1987). "Methods for estimating the magnitude and frequency of floods at ungaged sites on unregulated rural streams in Iowa," U.S. Geological Survey Water-Resources Investigations 87-4132.
Kansas	Clement, R. W. (1987). "Floods in Kansas and techniques for estimating their magnitude and frequency," U.S. Geological Survey Water-Resources Investigations 87-4008.
Kentucky	Choquette, A. F. (1987). "Regionalization of peak discharges for streams in Kentucky," U.S. Geological Survey Water-Resources Investigations 87-4209.
Louisiana	Lee, F. N. (1985). "Floods in Louisiana, Magnitude and frequency," Fourth Edition, Department of Transportation and Development, Water Resources Technical Report No. 36.
Maine	Morrill, R. A. (1975). "A technique for estimating the magnitude and frequency of floods in Maine," U.S. Geological Survey Open-File Report.
Maryland	Carpenter, D. H. (1980). "Technique for estimating magnitude and frequency of floods in Maryland," U.S. Geological Survey Water-Resources Investigations Open-File Report 80-1016.
Massachusetts	Wandle, S. W. (1983). "Estimating peak discharges of small rural streams in Massachusetts," U.S. Geological Survey Water-Supply Paper 2214.
Michigan	Holtschlag, D. J., and H. M. Croskey. (1984). "Statistical models for estimating flow characteristics of Michigan streams," U.S. Geological Survey Water Resources Investigations 84-4207.
Minnesota	Jacques, J. E., and D. L. Lorenz. (1987). "Techniques for estimating the magnitude and frequency of floods in Minnesota," U.S. Geological Survey Water-Resources Investigations 87-4170.

TABLE C-1 Continued

Mississippi	Colson, B. E., and J. W. Hudson. (1976). "Flood frequency of Mississippi streams," Mississippi State Highway Department.
Missouri	Hauth, L. D (1974). "A technique for estimating the magnitude and frequency of Missouri floods," U.S. Geological Survey Open-File Report.
Montana	Omang, R. J., Parrett, C., and J. A. Hull. (1986). "Methods of estimating magnitude and frequency of floods in Montana based on data through 1983," U.S. Geological Survey Water-Resources Investigations 86-4027.
Nebraska	Beckman, E. W. (1976). "Magnitude and frequency of floods in Nebraska," U.S. Geological Survey Water-Resources Investigations 76-109.
Nevada	Moore, D. O. (1976). "Estimating peak discharges from small drainages in Nevada according to basin areas within elevation zones," Nevada State Highway Department Hydrologic Report No. 3.
New Hampshire	LeBlanc, D. R. (1978). "Progress report on hydrologic investigations of small drainage areas in New Hampshire—Preliminary relations for estimating peak discharges on rural, unregulated streams," U.S. Geological Survey Water-Resources Investigations 78-47.
New Jersey	Stankowski, S. J. (1974). "Magnitude and frequency of floods in New Jersey with effects of urbanization," New Jersey Department of Environmental Protection Special Report 38.
New Mexico	Waltmeyer, S. D. (1986). "Techniques for estimating flood-flow frequency for unregulated streams in New Mexico," U.S. Geological Survey Water-Resources Investigations 86-4104.
New York	Zembrzuski, T. J., and B. Dunn. (1979). "Techniques for estimating magnitude and frequency of floods on rural unregulated streams in New York, excluding Long Island," U.S. Geological Survey Water-Resources Investigations 79-83.
North Carolina	Gunter, H. C., R. R. Mason, and T. C. Stamey. (1987). "Magnitude and frequency of floods in rural and urban basins of North Carolina," U.S. Geological Survey Water-Resources Investigations 87-4096.
North Dakota	Crosby, O. A. (1975). "Magnitude and frequency of floods in small drainage basins in North Dakota," U.S. Geological Survey Water-Resources Investigations 19-75.
Ohio	Webber, E. E., and W. P. Bartlett., Jr. (1977). "Floods in Ohio, magnitude and frequency," State of Ohio, Department of Natural Resources, Division of Water, Bulletin 45.
Oklahoma	Tortorelli, R. L., and D. L. Bergman. (1984). "Techniques for estimating flood-peak discharge for unregulated streams and streams regulated by small floodwater retarding structures in Oklahoma," U.S. Geological Survey Water-Resources Investigations 84-4358.
Oregon	Harris, D. D., and L. E. Hubbard. (1982). "Magnitude and frequency of floods in eastern Oregon," U.S. Geological Survey Water-Resources Investigations 82-4078.
	Harris, JD. D., L. L. Hubbard, and L. E. Hubbard. (1979). "Magnitude and frequency of floods in western Oregon," U.S. Geological Survey Open-File Report 79-553.
Pennsylvania	Flippo, H. N., Jr. (1977). "Floods in Pennsylvania: A manual for estimation of their magnitude and frequency," Pennsylvania Department of Environmental Resources Bulletin No. 13.
Puerto Rico	Lopez, M. A., E. Colon-Dieppa, and E. D. Cobb. (1978). "Floods in Puerto Rico: Magnitude and frequency," U.S. Geological Survey Water-Resources Investigations 78-141.
Rhode Island	Johnson, C. G., and G. A. Laraway. (1976). "Flood magnitude and frequency of

TABLE C-1 Continued

	small Rhode Island streams—Preliminary estimating relations," U.S. Geological Survey Open-File Report.
South Carolina	Whetstone, B. H. (1982). "Floods in South Carolina—Techniques for estimating magnitude and frequency of floods with compilation of flood data," U.S. Geological Survey Water-Resources Investigations 82-1.
South Dakota	Becker, L. D. (1974). "A method for estimating the magnitude and frequency of floods in South Dakota," U.S. Geological Survey Water-Resources Investigations 35-74.
	Becker, L. D. (1980). "Techniques for estimating flood peaks, volumes, and hydrographs on small streams in South Dakota," U.S. Geological Survey Water-Resources Investigations 80-80.
Tennessee	Randolph, W. J., and C. R. Gamble. (1976). "A technique for estimating magnitude and frequency of floods in Tennessee," Tennessee Department of Transportation.
Texas	Schroeder, E. E., and B. C. Massey. (1977). "Techniques for estimating the magnitude and frequency of floods in Texas," U.S. Geological Survey Water-Resources Investigations Open-File Report 77-110.
Utah	Thomas, B. E., and K. L. Lindskov. (1983). "Methods for estimating peak discharges and flood boundaries of streams in Utah," U.S. Geological Survey Water-Resources Investigations 83-4129.
Vermont	Johnson, C. G., and G. D. Tasker. (1974). "Flood magnitude and frequency of Vermont streams," U.S. Geological Survey Open-File Report 74-130.
Virginia	Miller, E. M. (1978). "Technique for estimating the magnitude and frequency of floods in Virginia," U.S. Geological Survey Water-Resources Investigations Open-File Report 78-5.
Washington	Cummans, J. E., M. R. Collins, and E. G. Nassar. (1974). "Magnitude and frequency of floods in Washington," U.S. Geological Survey Open-File Report 74-336.
	Haushild, W. L. (1978). "Estimation of floods of various frequencies of the small ephemeral streams in eastern Washington," U.S. Geological Survey Water-Resources Investigations 79-81.
West Virginia	Runner, G. S. (1980). "Technique for estimating magnitude and frequency of floods in West Virginia," U.S. Geological Survey Open-File Report 80-1218.
Wisconsin	Conger, D. H. (1980). "Techniques for estimating magnitude and frequency of floods for Wisconsin streams," U.S. Geological Survey Water-Resources Investigations Open-File Report 80-1214.
Wyoming	Craig, G. S., Jr., and J. G. Rankl. (1977). "Analysis of runoff from small drainage basins in Wyoming," U.S. Geological Survey Water-Supply Paper 2056.
	Lowham, H. W. (1976). "Techniques for estimating flow characteristics of Wyoming streams," U.S. Geological Survey Water-Resources Investigations 76-112.

Source: Wilbert Thomas Jr., Water Resources Division, U.S. Geological Survey, Reston, Va.

TABLE C-2 USGS REPORTS FOR ESTIMATING RURAL FLOOD-PEAK DISCHARGES
USING CHANNEL-GEOMETRY CHARACTERISTICS

Colorado	Hedman, E. R., D. O. Moore, and R. K. Livinston. (1972). "Selected streamflow characteristics as related to channel geometry of perennial streams in Colorado," U.S. Geological Survey Open-File Report.
Idaho	Harenberg, W. A. (1980). "Using channel geometry to estimate flood flows at ungaged sites in Idaho," U.S. Geological Survey Water-Resources Investigations 80-32.
Kansas	Hedman, E. R., W. M. Kastner, and H. R. Hejl. (1974). "Selected streamflow characteristics as related to active-channel geometry of streams in Kansas," Kansas Water Resources Board Technical Report No. 10.
Montana	Omang, R. J. (1983). "Mean annual runoff and peak-flow estimates based on channel geometry of streams in southeastern Montana," U.S. Geological Survey Water-Resources Investigations Report 82-4092.
	Parrett, C. (1983). "Mean annual runoff and peak-flow estimates based on channel geometry of streams in northeastern and western Montana," U.S. Geological Survey Water-Resources Investigations Report 83-4046.
	Parrett, C., J. A. Hull, and R. J. Omang. (1987). "Revised techniques for estimating peak discharges from channel width in Montana," U.S. Geological Survey Water-Resources Investigations 87-4121.
Nevada	Moore, D. O. (1974). "Estimating flood discharges in Nevada using channel-geometry measurements," Nevada State Highway Department Hydrologic Report No. 1.
New Mexico	Scott, A. G., and J. L. Kunkler. (1976). "Flood discharges of streams in New Mexico as related to channel geometry," U.S. Geological Survey Open-File Report.
Ohio	Roth, D. K. (1985). "Estimation of flood peaks from channel characteristics in Ohio," U.S. Geological Survey Water-Resources Investigations 85-4175.
	Webber, E. E., and J. W. Roberts. (1981). "Floodflow characteristics related to channel geometry in Ohio," U. S. Geological Survey Open-File Report 81-1105.
Utah	Fields, F. K. (1974). "Estimating streamflow characteristics for streams in Utah using selected channel-geometry parameters," U.S. Geological Survey Water-Resources Investigations 34-74.
Wyoming	Lowham, H. W. (1976). "Techniques for estimating flow characteristics of Wyoming streams," U.S. Geological Survey Water-Resources Investigations 76-112.

Source: Wilbert Thomas Jr., Water Resources Division, U.S. Geological Survey, Reston, Va.

TABLE C-3   USGS REPORTS FOR ESTIMATING URBAN FLOOD-PEAK DISCHARGES

Alabama	Olin, D. A., and R. H. Bingham. (1982). "Synthesized flood frequency of urban streams in Alabama," U.S. Geological Survey Water-Resources Investigations 82-683.
California	Waananen, A. O., and J. R. Crippen. (1977). "Magnitude and Frequency of Floods in California, "U.S. Geological Survey Water-Resources Investigations 77-21.
Connecticut	Weiss, L. A. (1975). "Flood-flow formula for urbanized and non-urbanized areas of Connecticut," Watershed Management Symposium of ASCE Irrigation and Drainage Division, pp. 658–675, August 11–13.
Florida	Franklin, M. A. (1984). "Magnitude and frequency of floods from urban streams in Leon County, Florida," U.S. Geological Survey Water-Resources Investigations 84-4004.
	Lopez, M. A., and W. M. Woodham. (1982). "Magnitude and frequency of flooding on small urban watersheds in the Tampa Bay area, west-central Florida," U.S. Geological Survey Water-Resources Investigations 82-42.
Georgia	Inman, E. J. (1983). "Flood-frequency relations for urban streams in metropolitan Atlanta, Georgia," U.S. Geological Survey Water-Resources Investigations 83-4203.
Illinois	Allen, H. E., Jr., and R. M. Bejcek. (1979). "Effects of urbanization on the magnitude and frequency of floods in northeastern Illinois," U.S. Geological Survey Water-Resources Investigations 79-36.
Iowa	Lara, O. (1978). "Effects of urban development on the flood-flow characteristics of Walnut Creek basin, Des Moines metropolitan area, Iowa," U.S. Geological Survey Water-Resources Investigations 78-11.
Kansas	Peek, C. O., and P. R. Jordan. (1978). "Determination of peak discharge from rainfall relations for urbanized basins, Wichita, Kansas," U.S. Geological Survey Open-File Report 78-974.
Missouri	Becker, L. D. (1986). "Techniques for estimating flood-peak discharges for urban streams in Missouri," U.S. Geological Survey Water-Resources Investigations 86-4322.
New Jersey	Stankowski, S. J. (1974). "Magnitude and frequency of floods in New Jersey with effects of urbanization," New Jersey Department of Environmental Protection Special Report 38.
North Carolina	Gunter, H. C., R. R. Mason, and T. C. Stamey. (1987). "Magnitude and frequency of floods in rural and urban basins of North Carolina," U.S. Geological Survey Water-Resources Investigations 87-4096.
Ohio	Sherwood, J. M. (1986). "Estimating peak discharges, flood volumes, and hydrograph stages of small urban streams in Ohio," U.S. Geological Survey Water-Resources Investigations 86-4197.
Oklahoma	Sauer, V. B. (1974). "An approach to estimating flood frequency for urban areas in Oklahoma," U.S. Geological Survey Water-Resources Investigations 23-74.
Oregon	Laenen, A. (1980). "Storm runoff as related to urbanization in the Portland, Oregon-Vancouver, Washington, area," U.S. Geological Survey Water-Resources Investigations Open-File Report 80-689.
Pennsylvania	Bailey, J. F., W. O. Thomas, K. L. Wetzel, and T. J. Ross. (1987). "Estimation of flood frequency characteristics and the effects of urbanization for streams in the Philadelphia, Pennsylvania, area," U.S. Geological Survey Water-Resources Investigations 87-4194.
Tennessee	Neely, B. L., Jr. (1984). "Flood frequency and storm runoff of urban areas of Memphis and Shelby County, Tennessee," U.S. Geological Survey Water-Resources Investigations 84-4110.

TABLE C-3 Continued

	Robbins, C. H. (1984). "Synthesized flood frequency for small urban streams in Tennessee," U.S. Geological Survey Water-Resources Investigations 84-4182.
	Wibben, H. D. (1976). "Effects of urbanization on flood characteristics in Nashville-Davidson County, Tennessee," U.S. Geological Survey Water-Resources Investigations 76-121.
Texas	Land, L. F., E. E. Schroeder, and B. B. Hampton. (1982). "Techniques for estimating the magnitude and frequency of floods in the Dallas-Fort Worth Metropolitan Area, Texas," U.S. Geological Survey Water-Resources Investigations 82-18.
	Liscum, F., and B. C. Massey. (1980). "Technique for estimating the magnitude and frequency of floods in the Houston, Texas, metropolitan area," U.S. Geological Survey Water-Resources Investigations 80-17.
	Veenhuis, J. E., and D. G. Garrett. (1986). "The effects of urbanization on floods in the Austin metropolitan area, Texas," U.S. Geological Survey Water-Resources Investigations 86-4069.
Virginia	Anderson, D. G. (1970). "Effects of urban development on floods in Northern Virginia," U.S. Geological Survey Water-Supply Paper 2001-C.
Wisconsin	Conger, D. H. (1986). "Estimating magnitude and frequency of floods for ungaged urban streams in Wisconsin," U.S. Geological Survey Water-Resources Investigations 86-4005.

Source: Wilbert Thomas Jr., U.S. Geological Survey, Water Resources Division, Reston, Va.

COMPUTER PROGRAMS
DOCUMENTATION AND LISTING

APPENDIX D: COMPUTER PROGRAMS DOCUMENTATION AND LISTING

This appendix contains documentation and listing of ten EH series computer programs. These programs illustrate the application of hydrologic concepts and methods described in this book. These programs are intended for instructional use only; any other use is expressly at the user's own risk. They are written in VAX/VMS Release V4.7 FORTRAN 77 language. Minor changes may be required when adapting these programs for use with other FORTRAN compilers.

The following device logical units are preset in these programs:

LU5 = 5: To read from the screen
LU6 = 6: To write to the screen
LU7 = 7: To read from an input file
LU8 = 8: To write to an output file

Program EH500 (Chapter 5)
SCS TR-55 Graphical Method for Peak Discharge Determinations in Small and Midsize Urban Catchments

This program can be used with either SI or U.S. customary units. The program is written in an interactive mode, with all input and output directly to the screen.

Input. At running time, EH500 asks you if you are using SI units. If the answer is Y (or YE or YES), input will be queried in SI units. If the answer is not Y (or YE or YES), input will be queried in U.S. customary units. EH500 proceeds to ask you for the following input data: (1) catchment area (km^2 or mi^2); (2) rainfall depth (cm or in.); (3) rainfall frequency (y); (4) runoff curve number; (5) storm type; (6) time of concentration (h); and (7) percentage of pond and swamp areas.

Logic. EH500 follows the computational procedure of Example 5-9, with the exception of the runoff curve number, which is treated as input data. Figure 5-19 is digitized and stored in DATA statements, and a piecewise linear fit is used to obtain values of unit peak flow. Equation 5-47 is used to calculate the peak discharge.

Output. EH500 output is the peak discharge (m^3/s or ft^3/s) for the specified frequency.

Program EH600A (Chapter 6)
Flood Frequency Analysis by the Log Pearson III Method

This program can be used with either SI or U.S. customary units. The flood series can be read either from the screen or from an input file. Output can be sent to the screen or to an output file.

Size. Variable sizes are set in the PARAMETER statement as follows: NZ = 200, the length of the flood series; MZ = 10, the number of return periods (or exceedence probability) being calculated; LZ = 31, the number of points used to linearize Table A-6 of Appendix A. The dimension NZ can be changed if necessary.

Input. At running time EH600A asks you to enter the number of years in the flood series. Then it asks you if you want to input the flood series interactively. If the answer is Y (or YE or YES), EH600A will read the flood series interactively. If the answer is not Y (or YE or YES), EH600A will read the flood series from the appropriate input file (i.e., that linked to logical unit LU7).

Logic. EH600A calculates the mean, standard deviation, and skew coefficient of the logarithms of the flood series. Table A-6 is digitized and stored in DATA statements, and a piecewise linear fit is used to obtain values of the frequency factor of the Log Pearson III distribution. Values of flood discharge are calculated using Eq. 6-32. See Example 6-4 in the text for an illustration of the computational procedure.

Output. EH600A output is a table containing corresponding values of (1) return period (y); (2) exceedence probability (percent); (3) frequency factor, and (4) flood discharge (m^3/s or ft^3/s). Return periods vary from 1.05 to 200 y.

Program EH600B (Chapter 6)
Flood Frequency Analysis by the Gumbel Method

This program can be used with either SI or U.S. customary units. The flood series can be read either from the screen or from an input file. Output can be sent to the screen or to an output file.

Size. EH600A variable sizes are set in the PARAMETER statement as follows: NZ = 200, the length of the flood series; MZ = 10, the number of return periods (or exceedence probability) being calculated; LZ = 16, the number of points used to linearize Table A-8 of Appendix A. The dimension NZ can be changed if necessary.

Input. At running time EH600B asks you to enter the number of years in the flood series. Then it asks you if you want to input the flood series interactively. If the answer is Y (or YE or YES), EH600B will read the flood series interactively. If the answer is not Y (or YE or YES), EH600B will read the flood series from the appropriate input file (i.e., that linked to logical unit LU7).

Logic. EH600B calculates the mean and standard deviation of the flood series. Table A-8 is digitized and stored in DATA statements, and a piecewise linear fit is used to obtain values of the mean and standard deviation of the Gumbel variate. The frequency factors K are calculated using Eq. 6-40. The flood discharges are calculated using Eq. 6-29.

Output. EH600B output is a table containing corresponding values of (1) return period (y); (2) exceedence probability (percent); (3) Gumbel variate; (4) frequency factor; and (5) flood discharge (m^3/s or ft^3/s). Return periods vary from 1.05 to 200 y.

Program EH700A (Chapter 7)
Correlation Coefficient of the Joint Probability Distribution of Monthly
(or Seasonal) Runoff Volumes

This program can be used with either SI or U.S. customary units. The input data can be read either from the screen or from an input file. The output is sent to the screen.

Size. EH700A variable sizes are set in the PARAMETER statement, as follows: NZ = 20, maximum number of runoff volume groupings (either for stream X or stream Y). The dimension NZ can be changed if necessary.

Input. EH700A asks you to enter the number of runoff volume classes: NDX for stream X and NDY for stream Y. Then it asks you if you want to input the data interactively. If the answer is Y (or YE or YES), EH700A will read the data interactively. If the answer is not Y (or YE or YES), EH700A will read the data from the appropriate input file (i.e., that linked to logical unit LU7). The data is read in the following order: (1) VX(J) class array, (2) VY(K) class array, and (3) FXY(J,K) array, with *J* varying faster than *K*.

Logic. EH700A calculates the marginal distributions (Eqs. 7-5 and 7-6), the means, standard deviations, covariance (Eq. 7-13), and correlation coefficient (Eq. 7-15). See Example 7-1 in the text for an illustration of the computational procedure.

Output. EH700A output is the correlation coefficient of the joint probability distribution of monthly (or seasonal) runoff volumes of streams X and Y.

Program EH700B (Chapter 7)
Two-Predictor-Variable Nonlinear Regression

This program solves the two-predictor-variable nonlinear regression fit of the following form:

$$y = a x_1^{b_1} x_2^{b_2}$$

The program can be used with either SI or U.S. customary units. The input data can be read either from the screen or from an input file. The output can be sent to the screen or to an output file.

Size. Array size is set in the PARAMETER statement. The maximum array size is NZ = 200. The dimension NZ can be changed if necessary.

Input. At running time EH700B asks you to enter the number of data sets. Then it asks you if you want to input the data interactively. If the answer is Y (or YE or YES), EH700B will read the data sets interactively. If the answer is not Y (or YE or YES), EH700B will read the data sets from the appropriate input file (i.e., that linked to logical unit LU7). It proceeds to read the Y(J), X_1(J), and X_2(J) arrays, in this order.

Logic. The program is based on Eq. 7-45 in the text, with the regression constants calculated using Eqs. 7-40 to 7-42.

Output. EH700B output consists of the regression constants a, b_1, and b_2.

Program EH800 (Chapter 8)
Reservoir Routing By Storage Indication Method

This program solves the storage indication method of reservoir routing. The outflow structure is assumed to be a broad-crested overflow spillway of rectangular cross section.

EH800 can be used with either SI or U.S. customary units. In SI units, use meters, cubic meters per second, and cubic hectometers for distances, flows, and storage volumes, respectively. In U.S. customary units, use feet, cubic feet per second, and acre-feet.

Coefficients and other constants are read interactively. Elevation-storage data and inflow hydrograph ordinates can be read either from the screen or from an input file. The output can be sent to the screen or to an output file.

Size. Variable sizes are set in the PARAMETER statement, as follows: $NZ = 200$, the maximum number of inflow or outflow hydrograph ordinates; $MZ = 21$, the maximum number of data points in calculated storage indication-outflow relation; and $LZ = 20$, the maximum number of data points in input elevation-storage array. The dimension NZ can be changed if necessary.

Input. The program routes a flood wave through a natural reservoir assuming that the initial reservoir elevation is at or above spillway crest elevation. At running time EH800 asks you if you are using SI units. If the answer is Y (or YE or YES), input will be queried in SI units. If the answer is not Y (or YE or YES), input will be queried in U.S. customary units. EH800 proceeds to ask you for the following input data: (1) width of the (rectangular) emergency spillway (m or ft); (2) rating coefficient of the spillway, (3) exponent of the spillway rating, (4) spillway crest elevation (m or ft); (5) dam crest elevation (m or ft); (6) initial reservoir elevation (m or ft); (7) number of elevation-storage data pairs; (8) elevation-storage data pairs; (9) time interval (h); (10) number of inflow hydrograph ordinates; and (11) inflow hydrograph ordinates. Queries 8 and 11 can be read interactively (logical unit LU5) or from the appropriate input file (logical unit LU7).

EH800 calculates the baseflow discharge based on the input initial reservoir elevation and spillway rating data. Initial inflow and initial outflow are set equal to the calculated baseflow discharge. Hydrograph inflows following the last input inflow hydrograph ordinate are set equal to the calculated baseflow discharge.

Logic. EH800 follows the computational procedure illustrated in Example 8-3. Elevation-storage and storage indication-outflow relations are handled through the use of a piecewise linear fit (subroutine PWLIN).

Output. EH800 output is a table containing corresponding values of (1) time (h); (2) inflow (m^3/s of ft^3/s); 3) outflow (m^3/s or ft^3/s); 4) storage (hm^3 or ac-ft); and (5) elevation (m or ft). EH800 also calculates the water surface elevation corresponding to the peak discharge, i.e., the maximum water surface elevation (m or ft).

Program EH900 (Chapter 9)
Stream Channel Routing with the Constant-Parameter Muskingum-Cunge Method

This program can be used only with SI units. Minor changes are required to convert it to U.S. customary units. All input data is read interactively. Output can be sent to the screen or to an output file.

Size. Variable sizes are set in the PARAMETER statement: $NZ = 200$, the maximum number of inflow or outflow hydrograph ordinates. The dimension NZ can be changed if necessary.

Input. EH900 uses a triangular inflow hydrograph. At running time EH900 asks you for the following data: (1) peak discharge (m^3/s); (2) baseflow (m^3/s); (3) time-to-peak (of the triangular inflow hydrograph) (h); (4) time base (of the triangular inflow hydrograph) (h); (5) channel bed slope (m/m); (6) flow area corresponding to the peak discharge (m^2); (7) channel top width corresponding to the peak discharge (m); (8) rating exponent β (in Eq. 9-13), (9) reach length (km); and (10) time interval (h). For ratios of time-to-peak to time interval less than 5, EH800 prints a warning message on the screen (logical unit LU6).

Logic. EH900 follows the computational procedure illustrated in Example 9-9. Equation 9-62 is used as the Muskingum routing equation, with the routing coefficients calculated by Eqs. 9-74 to 9-76.

Output. EH900 output is a table containing corresponding values of (1) time (h); (2) inflow (m^3/s); and (3) outflow (m^3/s). The output includes the calculated Courant and cell Reynolds numbers to enable a check of resolution accuracy (Eq. 9-77).

Program EH1000A (Chapter 10)
Catchment Routing by the Method of Cascade of Linear Reservoirs

This program can be used only with SI units. Minor changes are required to convert it to U.S. customary units. All input data is read interactively. The output can be sent to the screen or to an output file.

Size. Variable sizes are set in the PARAMETER statement: NZ = 200, the maximum number of inflow or outflow hydrograph ordinates. The dimension NZ can be changed if necessary.

Input. At running time EH1000A asks you for the following data: (1) catchment area (km^2); (2) time interval (h); (3) number of rainfall increments (NI), (4) NI values of net rainfall increments (cm); (5) reservoir storage constant (RK) (h); and (6) number of (linear) reservoirs (NR).

Logic. EH1000A follows the computational procedure illustrated in Example 10-3. Equation 10-11 is used as a routing equation for each of NR linear reservoirs of equal storage constant RK.

Output. EH1000A output is a table containing corresponding values of (1) time (h); and 2) outflow (m^3/s).

Program EH1000B (Chapter 10)
Two-plane Linear Kinematic Catchment Routing Model

This program can be used only with SI units. Minor changes are required to convert it to U.S. customary units. Input data can be read from the screen or from an input file. Output can be sent to the screen or to an output file.

Size. Variable sizes are set in the PARAMETER statement: NT = 480, maximum number of time intervals; NX = 32, maximum number of spatial increments per plane in the x-direction (see Fig. 10-5); NY = 32, maximum number of spatial increments in the y-direction (channel) (see Fig. 10-5). The dimensiones NT, NX and NY can be changed if necessary.

Input. At running time EH1000B asks if you want to input the data interactively. If the answer is Y (or YE or YES), EH1000B will query interactively for the input data. If the answer is not Y (or YE or YES), EH1000B will read the input from the appropriate input file (i.e., that linked to logical unit LU7). The program requests the following data: (1) plane length (m); (2) number of plane increments (per plane), (3) channel length (m); (4) number of channel increments, (5) total simulation time (min); (6) number of time intervals, (7) net rainfall intensity (cm/h); (8) rainfall duration (min); (9) average wave celerity in the planes (m/s); and (10) average wave celerity in the channel (m/s).

Logic. EH1000B follows the computational procedure illustrated in Section 10.4. Depending on the Courant number, either Eqs. 10-16 or 10-17 are used as routing equations for plane(s) and channel.

Output. EH1000B output is a table containing corresponding values of (1) time step number, (2) time (s); (3) time (min); and (4) outflow (m^3/s). Also shown are the net rainfall volume (m^3) and the volume under the outflow hydrograph (m^3).

Program EH1000C (Chapter 10)
Two-plane Linear Diffusion Catchment Routing Model

This program can be used only with SI units. Minor changes are required to convert it to U.S. customary units. Input data can be read from the screen or from an input file. Output can be sent to the screen or to an output file.

Size. Variable sizes are set in the PARAMETER statement: NT = 480, maximum number of time intervals; NX = 32, maximum number of spatial increments per plane in the x-direction (see Fig. 10-5); NY = 32, maximum number of spatial increments in the y-direction (channel) (see Fig. 10-5). The dimensions NT, NX and NY can be changed is necessary.

Input. At running time EH1000C asks if you want to input the data interactively. If the answer is Y (or YE or YES), EH1000C will query interactively for the input data. If the answer is not Y (or YE or YES), EH1000C will read the input from the appropriate input file (i.e., that linked to logical unit LU7). The program requests the following data: (1) plane length (m); (2) number of plane increments (per plane), (3) channel length (m); (4) number of channel increments, (5) total simulation time (min); (6) number of time intervals, (7) net rainfall intensity (cm/h); (8) rainfall duration (min); (9) average wave celerity in the planes (m/s); and (10) average wave celerity in the channel (m/s); (11) slope of the planes (m/m); (12) slope of the channel (m/m);

(13) average unit-width flow over the planes (m^2/s); and (14) average unit-width flow over the channel (m^2/s).

Logic. EH1000C follows the computational procedure illustrated in Section 10.5.

Output. EH1000C output is a table containing corresponding values of (1) time step number, (2) time (s); (3) time (min); and (4) outflow (m^3/s). Also shown are the net rainfall volume (m^3) and the volume under the outflow hydrograph (m^3).

```
      PROGRAM EH500
C     THIS PROGRAM IS PART OF CHAPTER 5, "ENGINEERING HYDROLOGY,
C     PRINCIPLES AND PRACTICES," BY V. M. PONCE.
C     THIS PROGRAM IS INTENDED FOR INSTRUCTIONAL USE ONLY;
C     ANY OTHER USE IS EXPRESSLY AT THE USER'S OWN RISK.
C     THE PROGRAM IS WRITTEN IN VAX/VMS FORTRAN 77.
C     IT SOLVES THE SCS TR-55 GRAPHICAL METHOD FOR PEAK DISCHARGE
C     DETERMINATIONS IN SMALL AND MIDSIZE URBAN CATCHMENTS.
C     SEE SECTION 5.3 IN THE TEXT FOR A DESCRIPTION
C     OF THE COMPUTATIONAL PROCEDURE.
C     THIS PROGRAM CAN BE USED WITH EITHER SI OR U.S. CUSTOMARY UNITS.
C     THE PROGRAM IS WRITTEN IN AN INTERACTIVE MODE,
C     WITH ALL INPUT AND OUTPUT DIRECTLY TO THE SCREEN.
C     THE FOLLOWING DEVICE LOGICAL UNITS ARE PRESET IN THIS PROGRAM:
C     LU5= 5:  TO READ FROM THE SCREEN;
C     LU6= 6:  TO WRITE TO THE SCREEN.
      PARAMETER (NX=5,NY=9,NZ=8)
      PARAMETER (LU5=5,LU6=6)
      DIMENSION TC(NY),UQ(NY,NZ,0:3),XIP(NZ)
      DIMENSION UQL(NY),UQR(NY),PER(NX),FT(NX)
      DATA TC /0.1,0.2,0.4,0.7,1.,2.,4.,7.,10./
      DATA XIP /0.1,0.2,0.25,0.3,0.35,0.4,0.45,0.5/
      DATA PER /0.,0.2,1.,3.,5./
      DATA FT /1.0,0.97,0.87,0.75,0.72/
      DATA C,R /2*1./
      DATA ((UQ(J,K,0),J=1,9),K=1,8) /162.,154.,138.,120.,108.,84.,62.,
     147.,38.5,136.,121.,105.,92.,83.,68.,53.,43.,37.5,118.,101.,87.,
     276.,69.,58.,47.5,40.,36.5,90.,76.,65.,57.5,53.5,47.,42.,38.,36.5,
     380.,68.,59.,54.,50.,45.,41.,35.,36.,70.,62.,55.,51.,48.,43.5,39.5,
     436.5,35.5,60.,56.,51.,47.5,45.,42.,38.5,37.,35.2,53.,50.,47.,45.,
     543.,40.5,38.,36.5,35./
      DATA ((UQ(J,K,1),J=1,9),K=1,8) /510.,410.,315.,240.,201.,138.,90.,
     161.5,47.5,450.,360.,262.,205.,172.,120.,80.,52.,44.5,400.,310.,
     2235.,180.,152.,108.,73.5,53.,43.5,345.,260.,192.,150.,128.,92.5,
     366.5,50.,42.5,270.,200.,148.,117.,101.,77.,58.,47.5,42.,181.,136.,
     4103.5,85.,76.,61.,51.,45.,42.,85.,75.,67.,61.5,58.,52.5,47.5,43.5,
     542.,56.,54.,51.,49.,48.,46.,44.,42.,41./
      DATA ((UQ(J,K,2),J=1,9),K=1,8) /1010.,800.,600.,445.,360.,227.,
     1133.,82.5,60.,1000.,770.,560.,420.,340.,215.,128.,79.,58.,970.,
     2730.,530.,385.,315.,200.,120.,75.,56.,950.,700.,500.,365.,295.,
     3185.,112.,72.5,54.,900.,650.,450.,328.,265.,180.,104.,69.,53.,
     4810.,575.,395.,287.,235.,152.,97.,66.,52.,700.,480.,330.,240.,
     5197.,132.,87.5,62.5,51.,545.,375.,260.,193.,160.,112.5,79.,60.,
     650./
      DATA ((UQ(J,K,3),J=1,9),K=1,8) /655.,550.,445.,350.,295.,200.,
     1125.,82.,61.,640.,530.,425.,330.,280.,190.,120.,79.,60.,620.,
     2510.,395.,315.,265.,180.,115.,77.,58.5,590.,480.,378.,295.,250.,
     3170.,132.,74.,57.5,535.,438.,340.,265.,225.,156.,102.,71.,55.,
     4455.,378.,298.,238.,202.,143.,97.,68.,54.,350.,300.,248.,203.,
     5177.,130.,91.,66.,53.,280.,245.,204.,170.,150.,114.,83.,63.,52./
      CHARACTER*6 A
      WRITE(LU6,'(3(/,X,A),$)')'THANK YOU FOR RUNNING EH500.',
     1'THIS PROGRAM SOLVES THE SCS TR-55 GRAPHICAL METHOD.',
     2'ARE YOU USING SI UNITS (Y/N)?:  '
      READ(LU5,'(A)') A
      IF(A.EQ.'YES'.OR.A.EQ.'yes'.OR.A.EQ.'YE'.OR.A.EQ.'ye'.OR.A.EQ.'Y'.
     1OR.A.EQ.'y') THEN
      INU= 1
10    WRITE(LU6,'(X,A,$)')'ENTER CATCHMENT AREA (KM2):  '
      READ(LU5,*) AREA
      IF(AREA.LE.0.) THEN
      WRITE(LU6,*)'CATCHMENT AREA CANNOT BE ZERO OR NEGATIVE.  PLEASE TR
     1Y AGAIN.'
      GOTO 10
      ENDIF
20    WRITE(LU6,'(X,A,$)')'ENTER RAINFALL DEPTH (CM):  '
      READ(LU5,*) PDEPTH
      IF(PDEPTH.LE.0.) THEN
      WRITE(LU6,*)'RAINFALL DEPTH CANNOT BE ZERO OR NEGATIVE.  PLEASE TR
     1Y AGAIN.'
      GOTO 20
      ENDIF
      ELSE
30    WRITE(LU6,'(X,A,$)')'ENTER CATCHMENT AREA (MI2):  '
      READ(LU5,*) AREA
      IF(AREA.LE.0.) THEN
      WRITE(LU6,*)'CATCHMENT AREA CANNOT BE ZERO OR NEGATIVE.  PLEASE TRY
     1 AGAIN.'
```

```
            GOTO 30
            ENDIF
   40   WRITE(LU6,'(X,A,$)')'ENTER RAINFALL DEPTH (IN.):  '
            READ(LU5,*) PDEPTH
            IF(PDEPTH.LE.0.) THEN
            WRITE(LU6,*)'RAINFALL DEPTH CANNOT BE ZERO OR NEGATIVE.  PLEASE TRY
         1 AGAIN.'
            GOTO 40
            ENDIF
            ENDIF
   50   WRITE(LU6,'(X,A,$)')'ENTER RAINFALL FREQUENCY (Y):  '
            READ(LU5,*) FREQ
            IF(FREQ.LT.1.) THEN
            WRITE(LU6,*)'RAINFALL FREQUENCY OUT OF RANGE.  PLEASE TRY AGAIN.'
            GOTO 50
            ENDIF
   60   WRITE(LU6,'(X,A,$)')'ENTER RUNOFF CURVE NUMBER:  '
            READ(LU5,*) CN
            IF(CN.LT.1.OR.CN.GT.100.) THEN
            WRITE(LU6,*)'INVALID CURVE NUMBER.  PLEASE TRY AGAIN.'
            GOTO 60
            ENDIF
   70   WRITE(LU6,'(X,A,/,X,A,$)')'ENTER STORM TYPE.','(ENTER 0 FOR TYPE I
         1A, 1 FOR TYPE I, 2 FOR TYPE II, 3 FOR TYPE III):  '
            READ(LU5,*) IT
            IF(IT.LT.0.OR.IT.GT.3) THEN
            WRITE(LU6,*)'INVALID STORM TYPE CODE.  PLEASE TRY AGAIN.'
            GOTO 70
            ENDIF
   80   WRITE(LU6,'(X,A,$)')'ENTER TIME OF CONCENTRATION (H):  '
            READ(LU5,*) TCON
            IF(TCON.LT.0.1.OR.TCON.GT.10.) THEN
            WRITE(LU6,*)'TIME OF CONCENTRATION IS OUT OF PERMISSIBLE RANGE.  P
         1LEASE TRY AGAIN.'
            GOTO 80
            ENDIF
   90   WRITE(LU6,'(X,A,$)')'ENTER PERCENTAGE OF POND AND SWAMP AREAS:  '
            READ(LU5,*) PF
            IF(PF.LT.0..OR.PF.GT.5.) THEN
            WRITE(LU6,*)'PERCENTAGE OF POND AND SWAMP AREAS IS OUT OF PERMISSI
         1BLE RANGE.  PLEASE TRY AGAIN.'
            GOTO 90
            ENDIF
            IFREQ= FREQ
            IF(INU.EQ.1) THEN
            C= 0.0043
            R= 2.54
            ENDIF
C*****CALCULATION OF SURFACE-STORAGE CORRECTION FACTOR (F)
            CALL PWLIN(PER,FT,NX,PF,F)
C*****CALCULATION OF RUNOFF DEPTH (Q)
            TEMP= CN*(PDEPTH/R + 2.) - 200.
            IF(TEMP.GT.0.) THEN
            Q= R*TEMP**2/(CN*(CN*(PDEPTH/R - 8.) + 800.))
            ENDIF
C*****CALCULATION OF UNIT PEAK FLOW (UPQ)
            XIABS= 200.*R/CN - 2.*R
            XIABSP= XIABS/PDEPTH
            DO 100 L=1,NZ
            IF(XIABSP.LE.XIP(L)) THEN
            LOC= L
            GOTO 110
            ENDIF
  100   CONTINUE
  110   CONTINUE
            IF(LOC.EQ.0) LOC= 9
            IF(LOC.EQ.1) THEN
            DO 120 J=1,NY
  120   UQR(J)= UQ(J,1,IT)
            CALL PWLINLOG(TC,UQR,NY,TCON,UPQ)
            ELSEIF(LOC.EQ.9) THEN
            DO 130 J=1,NY
  130   UQL(J)= UQ(J,8,IT)
            CALL PWLINLOG(TC,UQL,NY,TCON,UPQ)
            ELSE
            DO 140 J=1,NY
            UQL(J)= UQ(J,LOC-1,IT)
  140   UQR(J)= UQ(J,LOC,IT)
```

```
        CALL PWLINLOG(TC,UQL,NY,TCON,UPQL)
        CALL PWLINLOG(TC,UQR,NY,TCON,UPQR)
        UPQ= 10.**(ALOG10(UPQL) - (ALOG10(UPQL)-ALOG10(UPQR))*
     1      (ALOG10(XIABSP)-ALOG10(XIP(LOC-1)))/
     2      (ALOG10(XIP(LOC))-ALOG10(XIP(LOC-1))))
        ENDIF
C*****CALCULATION OF PEAK DISCHARGE (QPEAK)
        QPEAK= C*UPQ*AREA*Q*F
C*****PRINTING RESULTS
        IF(INU.EQ.1) THEN
        WRITE(LU6,'(/,X,A,I4,A,F10.3,A)')
     1'THE',IFREQ,'-Y PEAK DISCHARGE IS:   ',QPEAK,
     2' CUBIC METERS PER SECOND.'
        ELSE
        WRITE(LU6,'(/,X,A,I4,A,F10.3,A)')
     1'THE',IFREQ,'-Y PEAK DISCHARGE IS:   ',QPEAK,
     2' CUBIC FEET PER SECOND.'
        ENDIF
        WRITE(LU6,'(/,X,A)')
     1'THANK YOU FOR RUNNING EH500.  PLEASE CALL AGAIN.'
        STOP
        END
        SUBROUTINE PWLIN(X,Y,N,XL,YL)
        DIMENSION X(N),Y(N)
        IF(XL.LT.X(1)) THEN
        YL= ((Y(2)-Y(1))*XL+(Y(1)*X(2)-Y(2)*X(1)))/(X(2)-X(1))
        ELSEIF(XL.GE.X(N)) THEN
        K=N-1
        YL= ((Y(N)-Y(K))*XL+(Y(K)*X(N)-Y(N)*X(K)))/(X(N)-X(K))
        ELSE
        DO 1 J=2,N
        IF(XL.GE.X(J-1).AND.XL.LT.X(J)) M=J
    1   CONTINUE
        L=M-1
        YL= ((Y(M)-Y(L))*XL+(Y(L)*X(M)-Y(M)*X(L)))/(X(M)-X(L))
        ENDIF
        RETURN
        END
        SUBROUTINE PWLINLOG(X,Y,N,XL,YL)
        DIMENSION X(N),Y(N)
        IF(XL.LT.X(1)) THEN
        YL= 10.**(((ALOG10(Y(2))-ALOG10(Y(1)))*
     1      ALOG10(XL)+(ALOG10(Y(1))*ALOG10(X(2))-ALOG10(Y(2))*
     2      ALOG10(X(1))))/(ALOG10(X(2))-ALOG10(X(1))))
        ELSEIF(XL.GE.X(N)) THEN
        K=N-1
        YL= 10.**(((ALOG10(Y(N))-ALOG10(Y(K)))*
     1      ALOG10(XL)+(ALOG10(Y(K))*ALOG10(X(N))-ALOG10(Y(N))*
     2      ALOG10(X(K))))/(ALOG10(X(N))-ALOG10(X(K))))
        ELSE
        DO 1 J=2,N
        IF(XL.GE.X(J-1).AND.XL.LT.X(J)) M=J
    1   CONTINUE
        L=M-1
        YL= 10.**(((ALOG10(Y(M))-ALOG10(Y(L)))*
     1      ALOG10(XL)+(ALOG10(Y(L))*ALOG10(X(M))-ALOG10(Y(M))*
     2      ALOG10(X(L))))/(ALOG10(X(M))-ALOG10(X(L))))
        ENDIF
        RETURN
        END




        PROGRAM EH600A
C       THIS PROGRAM IS PART OF CHAPTER 6, "ENGINEERING HYDROLOGY,
C       PRINCIPLES AND PRACTICES," BY V. M. PONCE.
C       THIS PROGRAM IS INTENDED FOR INSTRUCTIONAL USE ONLY;
C       ANY OTHER USE IS EXPRESSLY AT THE USER'S OWN RISK.
C       THE PROGRAM IS WRITTEN IN VAX/VMS FORTRAN 77.
C       IT PERFORMS FLOOD FREQUENCY ANALYSIS
C       BY THE LOG PEARSON III METHOD.
C       SEE EXAMPLE 6-4 IN THE TEXT
C       FOR AN ILLUSTRATION OF THE COMPUTATIONAL PROCEDURE.
```

```
C        THIS PROGRAM CAN BE USED WITH EITHER SI OR U.S. CUSTOMARY UNITS.
C        THE FLOOD SERIES CAN BE READ EITHER FROM THE SCREEN
C        OR FROM AN INPUT FILE.
C        OUTPUT CAN BE SENT TO THE SCREEN OR TO AN OUTPUT FILE.
C        IF READING FROM AN INPUT FILE,
C        PLEASE ASSIGN PARAMETER LU7 TO THE APPROPRIATE FILE NAME.
C        IF WRITING TO AN OUTPUT FILE,
C        PLEASE ASSIGN PARAMETER LU8 TO THE APPROPRIATE FILE NAME.
C        THE FOLLOWING DEVICE LOGICAL UNITS ARE PRESET IN THIS PROGRAM:
C        LU5= 5:  TO READ FROM THE SCREEN;
C        LU6= 6:  TO WRITE TO THE SCREEN;
C        LU7= 7:  TO READ FROM A FILE;
C        LU8= 8:  TO WRITE TO A FILE.
C        THE CURRENT MAXIMUM ARRAY SIZE IS NZ= 200 (200 YEARS OF RECORD).
         PARAMETER (NZ=200)
         PARAMETER (MZ=10,LZ=31)
         PARAMETER (LU5=5,LU6=6,LU7=7,LU8=8)
         DIMENSION QIN(NZ),T(MZ),P(MZ),FK(MZ),QF(MZ)
         DIMENSION RCS(LZ),RYK(LZ,MZ),CS(LZ),YK(LZ,MZ),X1T(LZ),X2T(NZ)
         CHARACTER*6 A
         DATA T /1.05,1.11,1.25,2.,5.,10.,25.,50.,100.,200./
         DATA P /95.,90.,80.,50.,20.,10.,4.,2.,1.,0.5/
         DATA RCS /3.0,2.8,2.6,2.4,2.2,2.0,1.8,1.6,1.4,1.2,1.0,0.8,0.6,
        10.4,0.2,0.0,-0.2,-0.4,-0.6,-0.8,-1.0,-1.2,-1.4,-1.6,-1.8,-2.0,
        2-2.2,-2.4,-2.6,-2.8,-3.0/
         DATA ((RYK(J,K),K=1,MZ),J=1,19)/
        1-0.665,-0.660,-0.636,-0.396,0.420,1.180,2.278,3.152,4.051,4.970,
        2-0.711,-0.702,-0.666,-0.384,0.460,1.210,2.275,3.114,3.973,4.847,
        3-0.762,-0.747,-0.696,-0.368,0.499,1.238,2.267,3.071,3.889,4.718,
        4-0.819,-0.795,-0.725,-0.351,0.537,1.262,2.256,3.023,3.800,4.584,
        5-0.882,-0.844,-0.752,-0.330,0.574,1.284,2.240,2.970,3.075,4.444,
        6-0.949,-0.895,-0.777,-0.307,0.609,1.302,2.219,2.912,3.605,4.398,
        7-1.020,-0.945,-0.799,-0.282,0.643,1.318,2.193,2.848,3.499,4.147,
        8-1.093,-0.994,-0.817,-0.254,0.675,1.329,2.163,2.780,3.388,3.990,
        9-1.168,-1.041,-0.832,-0.225,0.705,1.337,2.128,2.706,3.271,3.828,
        *-1.243,-1.086,-0.844,-0.195,0.732,1.340,2.087,2.626,3.149,3.661,
        1-1.317,-1.128,-0.852,-0.164,0.758,1.340,2.043,2.542,3.022,3.489,
        2-1.388,-1.166,-0.856,-0.132,0.780,1.336,1.993,2.453,2.891,3.312,
        3-1.458,-1.200,-0.857,-0.099,0.800,1.328,1.939,2.359,2.755,3.132,
        4-1.524,-1.231,-0.855,-0.066,0.816,1.317,1.880,2.261,2.615,2.949,
        5-1.586,-1.258,-0.850,-0.033,0.830,1.301,1.818,2.159,2.472,2.763,
        6-1.645,-1.282,-0.842, 0.000,0.842,1.282,1.751,2.054,2.326,2.576,
        7-1.700,-1.301,-0.830, 0.033,0.850,1.258,1.680,1.945,2.178,2.388,
        8-1.750,-1.317,-0.816, 0.066,0.855,1.231,1.606,1.834,2.029,2.201,
        9-1.797,-1.328,-0.800, 0.099,0.857,1.200,1.528,1.720,1.880,2.016/
         DATA ((RYK(J,K),K=1,MZ),J=20,LZ)/
        *-1.839,-1.336,-0.780, 0.132,0.856,1.166,1.448,1.606,1.733,1.837,
        1-1.877,-1.340,-0.758, 0.164,0.852,1.128,1.366,1.492,1.588,1.664,
        2-1.910,-1.340,-0.732, 0.195,0.844,1.086,1.282,1.379,1.449,1.501,
        3-1.938,-1.337,-0.705, 0.225,0.832,1.041,1.198,1.270,1.318,1.351,
        4-1.962,-1.329,-0.675, 0.254,0.817,0.994,1.116,1.166,1.197,1.216,
        5-1.981,-1.318,-0.643, 0.282,0.799,0.945,1.035,1.069,1.087,1.097,
        6-1.996,-1.302,-0.609, 0.307,0.777,0.895,0.959,0.980,0.990,0.995,
        7-2.006,-1.284,-0.574, 0.330,0.752,0.844,0.888,0.900,0.905,0.907,
        8-2.011,-1.262,-0.537, 0.351,0.725,0.795,0.823,0.830,0.832,0.833,
        9-2.013,-1.238,-0.499, 0.368,0.696,0.747,0.764,0.768,0.769,0.769,
        *-2.010,-1.210,-0.460, 0.384,0.666,0.702,0.712,0.714,0.714,0.714,
        1-2.003,-1.180,-0.420, 0.393,0.636,0.660,0.666,0.667,0.667,0.667/
         WRITE(LU6,'(4(/,X,A),$)') 'THANK YOU FOR RUNNING EH600A.',
        1'THIS PROGRAM PERFORMS FLOOD FREQUENCY ANALYSIS',
        2'BY THE LOG PEARSON III METHOD.',
        3'ARE YOU USING SI UNITS (Y/N)?:  '
         READ(LU5,'(A)') A
         IF(A.EQ.'YES'.OR.A.EQ.'yes'.OR.A.EQ.'YE'.OR.A.EQ.'ye'.OR.A.EQ.'Y'.
        1OR.A.EQ.'y') INU= 1
      10 WRITE(LU6,'(X,A,$)')'ENTER NUMBER OF YEARS IN FLOOD SERIES:  '
         READ(LU5,*) ND
         IF(ND.GT.NZ) THEN
         WRITE(LU6,'(X,A,I3,A)')'SORRY, LENGTH OF FLOOD SERIES CANNOT BE GR
        1EATER THAN ',NZ,'.  PLEASE TRY AGAIN.'
         GOTO 10
         ENDIF
         WRITE(LU6,'(X,A,$)')'DO YOU WANT TO INPUT THE FLOOD SERIES INTERAC
        1TIVELY (Y/N)?:  '
         READ(LU5,'(A)') A
         IF(A.EQ.'YES'.OR.A.EQ.'yes'.OR.A.EQ.'YE'.OR.A.EQ.'ye'.OR.A.EQ.'Y'.
        1OR.A.EQ.'y') THEN
         WRITE(LU6,'(X,A,I3,A)')'ENTER ',ND,' VALUES OF FLOOD SERIES:  '
```

```
      READ(LU5,*) (QIN(J),J=1,ND)
      ELSE
      WRITE(LU6,'(X,A,I3,A)')'READING ',ND,' VALUES FROM YOUR INPUT FILE
     1.'
      READ(LU7,*) (QIN(J),J=1,ND)
      ENDIF
      DO 20 J=1,LZ
      CS(J)= RCS(LZ+1-J)
      DO 20 K=1,MZ
   20 YK(J,K)= RYK(LZ+1-J,K)
C*****CALCULATION OF MEAN, STANDARD DEVIATION, AND SKEW COEFFICIENT
C*****OF THE LOGARITHMS OF THE FLOOD SERIES.
      DO 30 J=1,ND
   30 SUM1= SUM1 + ALOG10(QIN(J))
      AVE= SUM1/ND
      DO 40 J=1,ND
      SUM2= SUM2 + (ALOG10(QIN(J)) - AVE)**2
   40 SUM3= SUM3 + (ALOG10(QIN(J)) - AVE)**3
      STD= SQRT(SUM2/(ND-1))
      SK= SUM3*ND/((ND-1)*(ND-2))
      CSK= SK/STD**3
      IF(ABS(CSK).GT.3.) THEN
      WRITE(LU6,'(X,A)') 'ERROR (WARNING):  SKEW COEFFICIENT IS OUT OF THE
     1 RANGE OF TABLE A-6 (APPENDIX A).  RESULTS MAY NOT BE ACCURATE.'
      ENDIF
C*****CALCULATION OF FREQUENCY FACTORS (FK(I))
      DO 70 I=1,MZ
      DO 60 K=1,MZ
      DO 50 J=1,LZ
   50 X1T(J)= YK(J,K)
   60 CALL PWLIN(CS,X1T,LZ,CSK,X2T(K))
      CALL PWLIN(T,X2T,MZ,T(I),FK(I))
C*****CALCULATION OF FLOOD DISCHARGES (QF(I))
   70 QF(I) = 10.**(AVE + FK(I)*STD)
C*****PRINTING RESULTS
      WRITE(LU6,'(X,A,$)')'DO YOU WANT YOUR OUTPUT TO THE SCREEN (Y/N)?:
     1 '
      READ(LU5,'(A)') A
      IF(A.EQ.'YES'.OR.A.EQ.'yes'.OR.A.EQ.'YE'.OR.A.EQ.'ye'.OR.A.EQ.'Y'.
     1OR.A.EQ.'y') THEN
      LU= LU6
      ELSE
      LU= LU8
      ENDIF
      WRITE(LU,'(2(/,X,A))')
     1'    RETURN          EXCEEDENCE           FREQUENCY          FLOOD',
     2'    PERIOD          PROBABILITY           FACTOR          DISCHARGE'
      IF(INU.EQ.1) THEN
      WRITE(LU,'(X,A)')
     2'     (Y)            (PERCENT)                               (M3/S)'
      ELSE
      WRITE(LU,'(X,A)')
     2'     (Y)            (PERCENT)                               (FT3/S)'
      ENDIF
      DO 80 J= 1,MZ
   80 WRITE(LU,'(X,F9.2,F14.1,F16.3,F16.0)') T(J),P(J),FK(J),QF(J)
      WRITE(LU6,'(/,X,A)')'THANK YOU FOR RUNNING EH600A.  PLEASE CALL AG
     1AIN.'
      STOP
      END
      SUBROUTINE PWLIN(X,Y,N,XL,YL)
      DIMENSION X(N),Y(N)
      IF(XL.LT.X(1)) THEN
      YL= ((Y(2)-Y(1))*XL+(Y(1)*X(2)-Y(2)*X(1)))/(X(2)-X(1))
      ELSEIF(XL.GE.X(N)) THEN
      K=N-1
      YL= ((Y(N)-Y(K))*XL+(Y(K)*X(N)-Y(N)*X(K)))/(X(N)-X(K))
      ELSE
      DO 1 J=2,N
      IF(XL.GE.X(J-1).AND.XL.LT.X(J)) M=J
    1 CONTINUE
      L=M-1
      YL= ((Y(M)-Y(L))*XL+(Y(L)*X(M)-Y(M)*X(L)))/(X(M)-X(L))
      ENDIF
      RETURN
      END
```

```
      PROGRAM EH600B
C     THIS PROGRAM IS PART OF CHAPTER 6, "ENGINEERING HYDROLOGY,
C     PRINCIPLES AND PRACTICES," BY V. M. PONCE.
C     THIS PROGRAM IS INTENDED FOR INSTRUCTIONAL USE ONLY;
C     ANY OTHER USE IS EXPRESSLY AT THE USER'S OWN RISK.
C     THE PROGRAM IS WRITTEN IN VAX/VMS FORTRAN 77.
C     IT PERFORMS FLOOD FREQUENCY ANALYSIS
C     USING THE GUMBEL (EXTREME VALUE 1) METHOD.
C     SEE EXAMPLE 6-6 IN THE TEXT
C     FOR AN ILLUSTRATION OF THE COMPUTATIONAL PROCEDURE.
C     THIS PROGRAM CAN BE USED WITH EITHER SI OR U.S. CUSTOMARY UNITS.
C     THE FLOOD SERIES CAN BE READ EITHER FROM THE SCREEN
C     OR FROM AN INPUT FILE.
C     OUTPUT CAN BE SENT TO THE SCREEN OR TO AN OUTPUT FILE.
C     IF READING FROM AN INPUT FILE,
C     PLEASE ASSIGN PARAMETER LU7 TO THE APPROPRIATE FILE NAME.
C     IF WRITING TO AN OUTPUT FILE,
C     PLEASE ASSIGN PARAMETER LU8 TO THE APPROPRIATE FILE NAME.
C     THE FOLLOWING DEVICE LOGICAL UNITS ARE PRESET IN THIS PROGRAM:
C     LU5= 5:  TO READ FROM THE SCREEN;
C     LU6= 6:  TO WRITE TO THE SCREEN;
C     LU7= 7:  TO READ FROM A FILE;
C     LU8= 8:  TO WRITE TO A FILE.
C     THE CURRENT MAXIMUM ARRAY SIZE IS NZ= 200 (200 YEARS OF RECORD).
      PARAMETER (NZ=200)
      PARAMETER (MZ=10,LZ=16)
      PARAMETER (LU5=5,LU6=6,LU7=7,LU8=8)
      DIMENSION QIN(NZ),T(MZ),P(MZ),Y(MZ),FK(MZ),QF(MZ)
      DIMENSION XN(LZ),YN(LZ),SN(LZ)
      CHARACTER*6 A
      DATA T /1.05,1.11,1.25,2.,5.,10.,25.,50.,100.,200./
      DATA P /95.,90.,80.,50.,20.,10.,4.,2.,1.,0.5/
      DATA Y/-1.113,-0.838,-0.476,0.376,1.5,2.25,3.199,3.902,4.6,5.296/
      DATA XN /8.,10.,15.,20.,30.,40.,50.,60.,
     1         70.,80.,90.,100.,250.,500.,750.,1000./
      DATA YN /0.4843,0.4952,0.5128,0.5236,0.5362,0.5436,0.5485,0.5521,
     1         0.5548,0.5569,0.5586,0.5600,0.5688,0.5724,0.5738,0.5745/
      DATA SN /0.9043,0.9497,1.0206,1.0628,1.1124,1.1413,1.1607,1.1747,
     1         1.1854,1.1938,1.2007,1.2065,1.2429,1.2588,1.2651,1.2685/
      WRITE(LU6,'(4(/,X,A),$)') 'THANK YOU FOR RUNNING EH600B.',
     1'THIS PROGRAM PERFORMS FLOOD FREQUENCY ANALYSIS',
     2'USING THE GUMBEL (EXTREME VALUE 1) METHOD.',
     3'ARE YOU USING SI UNITS (Y/N)?:  '
      READ(LU5,'(A)') A
      IF(A.EQ.'YES'.OR.A.EQ.'yes'.OR.A.EQ.'YE'.OR.A.EQ.'ye'.OR.A.EQ.'Y'.
     1OR.A.EQ.'y') INU= 1
  10  WRITE(LU6,'(X,A,$)')'ENTER NUMBER OF YEARS IN FLOOD SERIES:  '
      READ(LU5,*) ND
      IF(ND.GT.NZ) THEN
      WRITE(LU6,'(X,A,I3,A)')'SORRY, LENGTH OF FLOOD SERIES CANNOT BE GR
     1EATER THAN ',NZ,'.  PLEASE TRY AGAIN.'
      GOTO 10
      ENDIF
      WRITE(LU6,'(X,A,$)')'DO YOU WANT TO INPUT THE FLOOD SERIES INTERAC
     1TIVELY (Y/N)?:  '
      READ(LU5,'(A)') A
      IF(A.EQ.'YES'.OR.A.EQ.'yes'.OR.A.EQ.'YE'.OR.A.EQ.'ye'.OR.A.EQ.'Y'.
     1OR.A.EQ.'y') THEN
      WRITE(LU6,'(X,A,I3,A)')'ENTER ',ND,' VALUES OF FLOOD SERIES:  '
      READ(LU5,*) (QIN(J),J=1,ND)
      ELSE
      WRITE(LU6,'(X,A,I3,A)')'READING ',ND,' VALUES FROM YOUR INPUT FILE
     1.'
      READ(LU7,*) (QIN(J),J=1,ND)
      ENDIF
C*****CALCULATION OF MEAN AND STANDARD DEVIATION OF THE FLOOD SERIES.
      DO 20 J=1,ND
  20  SUM1= SUM1 + QIN(J)
      AVE= SUM1/ND
      DO 30 J=1,ND
  30  SUM2= SUM2 + (QIN(J) - AVE)**2
      STD= SQRT(SUM2/(ND-1))
C*****CALCULATION OF MEAN AND STANDARD DEVIATION OF THE GUMBEL VARIATE.
      XL= ND
      CALL PWLIN(XN,YN,LZ,XL,YNL)
      CALL PWLIN(XN,SN,LZ,XL,SNL)
C*****CALCULATION OF FREQUENCY FACTORS (FK(J))
      DO 40 J= 1,MZ
```

```
         FK(J)= (Y(J)-YNL)/SNL
C****CALCULATION OF FLOOD DISCHARGES (QF(J))
   40 QF(J)= AVE + FK(J)*STD
C*****PRINTING RESULTS
         WRITE(LU6,'(X,A,$)')'DO YOU WANT YOUR OUTPUT TO THE SCREEN (Y/N)?:
   1 '
         READ(LU5,'(A)') A
         IF(A.EQ.'YES'.OR.A.EQ.'yes'.OR.A.EQ.'YE'.OR.A.EQ.'ye'.OR.A.EQ.'Y'.
   1OR.A.EQ.'y') THEN
         LU= LU6
         ELSE
         LU= LU8
         ENDIF
         WRITE(LU,'(2(/,X,A))')
   1'    RETURN      EXCEEDENCE      GUMBEL     FREQUENCY      FLOOD',
   2'    PERIOD     PROBABILITY     VARIATE     FACTOR      DISCHARGE'
         IF(INU.EQ.1) THEN
         WRITE(LU,'(X,A)')
   1'    (Y)          (PERCENT)                              (M3/S)'
         ELSE
         WRITE(LU,'(X,A)')
   1'    (Y)          (PERCENT)                              (FT3/S)'
         ENDIF
         DO 50 J= 1,MZ
         WRITE(LU,'(X,F9.2,F12.1,F13.3,F12.3,F12.0)')
   1T(J),P(J),Y(J),FK(J),QF(J)
   50 CONTINUE
         WRITE(LU6,'(/,X,A)')'THANK YOU FOR RUNNING EH600B.  PLEASE CALL AG
   1AIN.'
         STOP
         END
         SUBROUTINE PWLIN(X,Y,N,XL,YL)
         DIMENSION X(N),Y(N)
         IF(XL.LT.X(1)) THEN
         YL= ((Y(2)-Y(1))*XL+(Y(1)*X(2)-Y(2)*X(1)))/(X(2)-X(1))
         ELSEIF(XL.GE.X(N)) THEN
         K=N-1
         YL= ((Y(N)-Y(K))*XL+(Y(K)*X(N)-Y(N)*X(K)))/(X(N)-X(K))
         ELSE
         DO 1 J=2,N
         IF(XL.GE.X(J-1).AND.XL.LT.X(J)) M=J
   1 CONTINUE
         L=M-1
         YL= ((Y(M)-Y(L))*XL+(Y(L)*X(M)-Y(M)*X(L)))/(X(M)-X(L))
         ENDIF
         RETURN
         END




         PROGRAM EH700A
C      THIS PROGRAM IS PART OF CHAPTER 7, "ENGINEERING HYDROLOGY,
C      PRINCIPLES AND PRACTICES," BY V. M. PONCE.
C      THIS PROGRAM IS INTENDED FOR INSTRUCTIONAL USE ONLY;
C      ANY OTHER USE IS EXPRESSLY AT THE USER'S OWN RISK.
C      THE PROGRAM IS WRITTEN IN VAX/VMS FORTRAN 77.
C      IT CALCULATES THE CORRELATION COEFFICIENT
C      OF THE JOINT PROBABILITY DISTRIBUTION
C      OF MONTHLY (OR SEASONAL) RUNOFF VOLUMES OF STREAMS X AND Y.
C      SEE EXAMPLE 7-1 IN THE TEXT
C      FOR AN ILLUSTRATION OF THE COMPUTATIONAL PROCEDURE.
C      THE JOINT PROBABILITY DISTRIBUTION
C      CAN BE READ EITHER FROM THE SCREEN OR FROM AN INPUT FILE.
C      IF READING FROM AN INPUT FILE,
C      PLEASE ASSIGN PARAMETER LU7 TO THE APPROPRIATE FILE NAME.
C      THE FOLLOWING DEVICE LOGICAL UNITS ARE PRESET IN THIS PROGRAM:
C      LU5= 5:  TO READ FROM THE SCREEN;
C      LU6= 6:  TO WRITE TO THE SCREEN;
C      LU7= 7:  TO READ FROM A FILE;
C      THE CURRENT MAXIMUM ARRAY SIZE IS NZ= 20.
C      (20 CLASSES PER X OR Y ARRAY).
         PARAMETER (NZ=10)
         PARAMETER (LU5=5,LU6=6,LU7=7)
```

```
            DIMENSION VX(NZ),VY(NZ),FXY(NZ,NZ),FX(NZ),FY(NZ)
            CHARACTER*6 A
            DATA SUM,AVEX,AVEY,SUMX,SUMY,SUMXY /6*0./
            WRITE(LU6,'(3(/,X,A))') 'THANK YOU FOR RUNNING EH700A.',
           1'THIS PROGRAM CALCULATES THE CORRELATION COEFFICIENT',
           2'OF MONTHLY (OR SEASONAL) RUNOFF VOLUMES OF STREAMS X AND Y'
        10 WRITE(LU6,'(X,A,$)') 'ENTER NUMBER OF X CLASSES: '
            READ(LU5,*) NDX
            IF(NDX.GT.NZ) THEN
            WRITE(LU6,'(X,A,I2,A,$)')'SORRY, NUMBER OF CLASSES CANNOT BE GREAT
           1ER THAN ',NZ,'.  PLEASE TRY AGAIN.'
            GOTO 10
            ENDIF
        20 WRITE(LU6,'(X,A,$)') 'ENTER NUMBER OF Y CLASSES:  '
            READ(LU5,*) NDY
            IF(NDY.GT.NZ) THEN
            WRITE(LU6,'(X,A,I2,A,$)')'SORRY, NUMBER OF CLASSES CANNOT BE GREAT
           1ER THAN ',NZ,'.  PLEASE TRY AGAIN.'
            GOTO 20
            ENDIF
            WRITE(LU6,'(X,A,$)') 'DO YOU WANT TO INPUT THE DATA INTERACTIVELY
           1(Y/N)?: '
            READ(LU5,'(A)') A
            IF(A.EQ.'YES'.OR.A.EQ.'yes'.OR.A.EQ.'YE'.OR.A.EQ.'ye'.OR.A.EQ.'Y'.
           1OR.A.EQ.'y') THEN
            WRITE(LU6,'(X,A,I2,A,$)')'ENTER ',NDX,' VALUES OF VX(J) ARRAY:  '
            READ(LU5,*) (VX(J),J=1,NDX)
            WRITE(LU6,'(X,A,I2,A,$)')'ENTER ',NDY,' VALUES OF VY(K) ARRAY:  '
            READ(LU5,*) (VY(K),K=1,NDY)
            WRITE(LU6,'(X,A,I3,A,$)')'ENTER ',NDX*NDY,' VALUES OF FXY(J,K) AR
           1RAY (J VARIES FASTER THAN K):  '
            READ(LU5,*) ((FXY(J,K),J=1,NDX),K=1,NDY)
            ELSE
            WRITE(LU6,'(X,A,I2,A,$)')'READING ',NDX,' VALUES OF VX(J) ARRAY'
            READ(LU7,*) (VX(J),J=1,NDX)
            WRITE(LU6,'(X,A,I2,A,$)')'READING ',NDY,' VALUES OF VY(K) ARRAY'
            READ(LU7,*) (VY(K),K=1,NDY)
            WRITE(LU6,'(X,A,I2,A,$)')'READING ',NDX*NDY,' VALUES OF FXY(J,K) A
           1RRAY (J VARIES FASTER THAN K)'
            READ(LU7,*) ((FXY(J,K),J=1,NDX),K=1,NDY)
            ENDIF
            DO 30 J=1,NDX
            DO 30 K=1,NDY
        30 SUM= SUM + FXY(J,K)
            IF(SUM.LT.0.9999.OR.SUM.GT.1.0001)THEN
            WRITE(LU6,*)'SUM OF PROBABILITIES IS NOT EQUAL TO 1.  PLEASE CHECK
           1 YOUR INPUT DATA.'
            ENDIF
C*****CALCULATION OF MARGINAL DISTRIBUTIONS AND MEANS
            DO 50 J=1,NDX
            FX(J)=0.
            DO 40 K=1,NDY
        40 FX(J)= FX(J) + FXY(J,K)
        50 AVEX= AVEX + VX(J)*FX(J)
            DO 70 K=1,NDY
            FY(K)=0.
            DO 60 J=1,NDX
        60 FY(K)= FY(K) + FXY(J,K)
        70 AVEY= AVEY + VY(K)*FY(K)
C*****CALCULATION OF STANDARD DEVIATIONS
            DO 80 J=1,NDX
        80 SUMX= SUMX + (VX(J) - AVEX)**2*FX(J)
            DO 90 K=1,NDY
        90 SUMY= SUMY + (VY(K) - AVEY)**2*FY(K)
            STAX= SQRT(SUMX)
            STAY= SQRT(SUMY)
C*****CALCULATION OF COVARIANCE
            DO 100 J=1,NDX
            DO 100 K=1,NDY
       100 SUMXY= SUMXY + (VX(J)-AVEX)*(VY(K)-AVEY)*FXY(J,K)
C*****CALCULATION OF CORRELATION COEFFICIENT
            CORR= SUMXY/(STAX*STAY)
C*****PRINTING CORRELATION COEFFICIENT
            WRITE(LU6,'(/,X,A,F7.3)')'THE CORRELATION COEFFICIENT IS =  ',CORR
            WRITE(LU6,'(/,X,A)')'THANK YOU FOR RUNNING EH700A.  PLEASE CALL AG
           1AIN.'
            STOP
            END
```

```
      PROGRAM EH700B
C     THIS PROGRAM IS PART OF CHAPTER 7, "ENGINEERING HYDROLOGY,
C     PRINCIPLES AND PRACTICES," BY V. M. PONCE.
C     THIS PROGRAM IS INTENDED FOR INSTRUCTIONAL USE ONLY;
C     ANY OTHER USE IS EXPRESSLY AT THE USER'S OWN RISK.
C     THE PROGRAM IS WRITTEN IN VAX/VMS FORTRAN 77.
C     THIS PROGRAM SOLVES THE MULTIPLE NONLINEAR REGRESSION FIT
C     WITH TWO PREDICTOR VARIABLES.
C     PREDICTION EQUATION:  Y= A * X1**B1 * X2**B2
C     SEE SECTION 7.2 OF THE TEXT
C     FOR AN EXPLANATION OF THE COMPUTATIONAL PROCEDURE.
C     THE INPUT DATA CAN BE READ EITHER FROM THE SCREEN
C     OR FROM AN INPUT FILE.
C     THE OUTPUT CAN BE SENT TO THE SCREEN OR TO AN OUTPUT FILE.
C     IF READING FROM AN INPUT FILE,
C     PLEASE ASSIGN PARAMETER LU7 TO THE APPROPRIATE FILE NAME.
C     IF WRITING TO AN OUTPUT FILE,
C     PLEASE ASSIGN PARAMETER LU8 TO THE APPROPRIATE FILE NAME.
C     THE FOLLOWING DEVICE LOGICAL UNITS ARE PRESET IN THIS PROGRAM:
C     LU5= 5:  TO READ FROM THE SCREEN.
C     LU6= 6:  TO WRITE TO THE SCREEN.
C     LU7= 7:  TO READ FROM A FILE.
C     LU8= 8:  TO WRITE TO A FILE.
C     THE CURRENT MAXIMUM ARRAY SIZE IS NZ= 200.
C     (200 SETS OF VALUES OF Y, X1 AND X2).
      PARAMETER (NZ=200)
      PARAMETER (LU5=5,LU6=6,LU7=7,LU8=8)
      DIMENSION X1(NZ),X2(NZ),Y(NZ),X1L(NZ),X2L(NZ),YL(NZ)
      CHARACTER*6 A
      WRITE(LU6,'(2(/,X,A))') 'THANK YOU FOR RUNNING EH700B.',
     1'THIS PROGRAM SOLVES THE TWO-PREDICTOR-VARIABLE NONLINEAR REGRESSI
     2ON FIT.'
   10 WRITE(LU6,'(X,A,$)')'ENTER NUMBER OF DATA SETS:  '
      READ(LU5,*) ND
      IF(ND.GT.NZ) THEN
      WRITE(LU6,'(X,A,I3,A)')'SORRY, NUMBER OF SETS CANNOT BE GREATER TH
     1AN ',NZ,'.  PLEASE TRY AGAIN.'
      GOTO 10
      ENDIF
      WRITE(LU6,'(X,A,$)')'DO YOU WANT TO INPUT THE DATA INTERACTIVELY
     1(Y(J),X1(J),X2(J)) (Y/N)?:  '
      READ(LU5,'(A)') A
      IF(A.EQ.'YES'.OR.A.EQ.'yes'.OR.A.EQ.'YE'.OR.A.EQ.'ye'.OR.A.EQ.'Y'.
     1OR.A.EQ.'y') THEN
      READ(LU5,*) (Y(J),X1(J),X2(J),J=1,ND)
      ELSE
      WRITE(LU6,'(X,A,I3,A)')'READING ',ND,' DATA SETS FROM YOUR INPUT F
     1ILE.'
      READ(LU7,*) (Y(J),X1(J),X2(J),J=1,ND)
      ENDIF
C*****CALCULATION OF SUMS OF PRODUCTS
      DO 20 J=1,ND
      YL(J)= ALOG10(Y(J))
      X1L(J)= ALOG10(X1(J))
      X2L(J)= ALOG10(X2(J))
      SUMX1= SUMX1 + X1L(J)
      SUMX2= SUMX2 + X2L(J)
      SUMY= SUMY + YL(J)
      SUMYX1= SUMYX1 + YL(J)*X1L(J)
      SUMYX2= SUMYX2 + YL(J)*X2L(J)
      SUMX1S= SUMX1S + X1L(J)**2
      SUMX2S= SUMX2S + X2L(J)**2
      SUMX1X2= SUMX1X2 + X1L(J)*X2L(J)
   20 CONTINUE
C*****CALCULATION OF REGRESSION CONSTANTS
      B1= ((ND*SUMYX2-SUMY*SUMX2)*(ND*SUMX1X2-SUMX1*SUMX2)
     1    -(ND*SUMX2S-SUMX2**2)*(ND*SUMYX1-SUMY*SUMX1))/
     2    ((ND*SUMX1X2-SUMX1*SUMX2)**2
     3    -(ND*SUMX1S-SUMX1**2)*(ND*SUMX2S-SUMX2**2))
      B2= ((ND*SUMYX1-SUMY*SUMX1) - B1*(ND*SUMX1S-SUMX1**2))/
     1    (ND*SUMX1X2-SUMX1*SUMX2)
      ACOEF= 10.**((SUMY - B1*SUMX1 - B2*SUMX2)/ND)
C*****PRINTING RESULTS
      WRITE(LU6,'(X,A,$)')'DO YOU WANT YOUR OUTPUT TO THE SCREEN (Y/N)?:
     1  '
      READ(LU5,'(A)') A
      IF(A.EQ.'YES'.OR.A.EQ.'yes'.OR.A.EQ.'YE'.OR.A.EQ.'ye'.OR.A.EQ.'Y'.
     1OR.A.EQ.'y') THEN
```

```
         LU= LU6
         ELSE
         LU= LU8
         ENDIF
         WRITE(LU,'(/,X,A,//,X,A)')'TWO-PREDICTOR-VARIABLE NONLINEAR REG
        1RESSION FIT:','Y = A * X1**B1 * X2**B2'
         WRITE(LU,'(/,X,A,F15.3,2(/,X,A,F8.3))')'COEFFICIENT A= ',ACOEF,
        2'EXPONENT OF X1:  B1=  ',B1,'EXPONENT OF X2:  B2=  ',B2
         WRITE(LU6,'(/,X,A)')'THANK YOU FOR RUNNING EH700B.  PLEASE CALL AG
        1AIN.'
         STOP
         END
```

```
         PROGRAM EH800
C        THIS PROGRAM IS PART OF CHAPTER 8, "ENGINEERING HYDROLOGY,
C        PRINCIPLES AND PRACTICES," BY V. M. PONCE.
C        THIS PROGRAM IS INTENDED FOR INSTRUCTIONAL USE ONLY;
C        ANY OTHER USE IS EXPRESSLY AT THE USER'S OWN RISK.
C        THE PROGRAM IS WRITTEN IN VAX/VMS FORTRAN 77.
C        THIS PROGRAM SOLVES THE STORAGE INDICATION METHOD OF RESERVOIR
C        ROUTING, FOR A SPILLWAY OF RECTANGULAR CROSS SECTION.
C        THE PROGRAM CAN BE USED WITH EITHER SI OR U.S. CUSTOMARY UNITS.
C        SEE EXAMPLE 8-3 IN THE TEXT
C        FOR AN ILLUSTRATION OF THE COMPUTATIONAL PROCEDURE.
C        COEFFICIENTS AND OTHER CONSTANTS ARE READ INTERACTIVELY.
C        ELEVATION-STORAGE AND INFLOW HYDROGRAPH DATA
C        CAN BE READ EITHER FROM THE SCREEN OR FROM AN INPUT FILE.
C        OUTPUT CAN BE SENT TO THE SCREEN OR TO AN OUTPUT FILE.
C        IF READING FROM AN INPUT FILE,
C        PLEASE ASSIGN PARAMETER LU7 TO THE APPROPRIATE FILE NAME.
C        IF WRITING TO AN OUTPUT FILE,
C        PLEASE ASSIGN PARAMETER LU8 TO THE APPROPRIATE FILE NAME.
C        THE FOLLOWING DEVICE LOGICAL UNITS ARE PRESET IN THIS PROGRAM:
C        LU5- 5:  TO READ FROM THE SCREEN;
C        LU6- 6:  TO WRITE TO THE SCREEN;
C        LU7- 7:  TO READ FROM A FILE;
C        LU8- 8:  TO WRITE TO A FILE.
C        INFLOW/OUTFLOW MAXIMUM ARRAY SIZE:  NZ= 200
C        INPUT ELEVATION/STORAGE MAXIMUM ARRAY SIZE:  LZ=20
C        STORAGE INDICATION/OUTFLOW ARRAY SIZE:  MZ= 21
         PARAMETER (NZ=200,LZ=20)
         PARAMETER (MZ=51)
         PARAMETER (LU5=5,LU6=6,LU7=7,LU8=8)
         DIMENSION TIME(NZ),QIN(NZ),QOUT(NZ),STOROUT(NZ),ELEVOUT(NZ),STORIN
        1D(NZ),STORINC(NZ),ELEV(MZ),HEAD(MZ),OUTF(MZ),STOR(MZ),STIN(MZ),
        2ELEVIN(LZ),STORAGEIN(LZ)
         DATA HEAD(1),OUTF(1),TIME(1),QMAX /4*0./
         DATA UNIT /43560./
         CHARACTER*6 A
         WRITE(LU6,'(4(/,X,A),$)')'THANK YOU FOR RUNNING EH800.',
        1'THIS PROGRAM ROUTES A FLOOD WAVE THROUGH A NATURAL RESERVOIR',
        2'ASSUMING THE INITIAL POOL LEVEL AT OR ABOVE SPILLWAY CREST.',
        3'ARE YOU USING SI UNITS (Y/N)?:  '
         READ(LU5,'(A)') A
         IF(A.EQ.'YES'.OR.A.EQ.'yes'.OR.A.EQ.'YE'.OR.A.EQ.'ye'.or.A.EQ.'Y'.
        1OR.A.EQ.'y') THEN
         INU= 1
         UNIT= 1.E06
         WRITE(LU6,*)'IN SI UNITS USE M, M3/S, AND HM3.'
         WRITE(LU6,*)'ANSWERS WILL BE GIVEN IN M AND M3/S.'
         ELSE
         WRITE(LU6,*)'IN U.S. CUSTOMARY UNITS USE FT, FT3/S, AND AC-FT.'
         WRITE(LU6,*)'ANSWERS WILL BE GIVEN IN FT AND FT3/S.'
         ENDIF
     10  WRITE(LU6,'(/,X,A,$)')'ENTER WIDTH OF THE EMERGENCY SPILLWAY (M OR
        1 FT):  '
         READ(LU5,*) XLES
         IF(XLES.LE.0.) THEN
         WRITE(LU6,*)'WIDTH OF EMERGENCY SPILLWAY CANNOT BE ZERO OR NEGATIV
        1E.  PLEASE TRY AGAIN.'
         GOTO 10
```

```
      ENDIF
   11 WRITE(LU6,'(X,A,$)')'ENTER COEFFICIENT OF THE SPILLWAY RATING:   '
      READ(LU5,*) RATC
      IF(RATC.LE.0.) THEN
      WRITE(LU6,*)'COEFFICIENT OF THE SPILLWAY RATING CANNOT BE ZERO OR
     1NEGATIVE.  PLEASE TRY AGAIN.'
      GOTO 11
      ENDIF
   12 WRITE(LU6,'(X,A,$)')'ENTER EXPONENT OF THE SPILLWAY RATING:   '
      READ(LU5,*) EXPR
      IF(EXPR.LE.0.) THEN
      WRITE(LU6,*)'EXPONENT OF THE SPILLWAY RATING CANNOT BE ZERO OR NEG
     1ATIVE.  PLEASE TRY AGAIN.'
      GOTO 12
      ENDIF
   13 WRITE(LU6,'(X,A,$)')'ENTER SPILLWAY CREST ELEVATION (M OR FT):   '
      READ(LU5,*) SCEL
      IF(SCEL.LE.0.) THEN
      WRITE(LU6,*)'SPILLWAY CREST ELEVATION CANNOT BE ZERO OR NEGATIVE.
     1 PLEASE TRY AGAIN.'
      GOTO 13
      ENDIF
   14 WRITE(LU6,'(X,A,$)')'ENTER DAM CREST ELEVATION (M OR FT):   '
      READ(LU5,*) DCEL
      IF(DCEL.LE.0.) THEN
      WRITE(LU6,*)'DAM CREST ELEVATION CANNOT BE ZERO OR NEGATIVE.  PLEA
     1SE TRY AGAIN.'
      GOTO 14
      ELSEIF(DCEL.LE.SCEL) THEN
      WRITE(LU6,'(X,A,/,X,A)')'SORRY, DAM CREST ELEV. CANNOT BE LESS THA
     1N SPILLWAY CREST ELEV.','PLEASE TRY AGAIN.'
      GOTO 14
      ENDIF
   15 WRITE(LU6,'(X,A,$)')'ENTER INITIAL RESERVOIR LEVEL (M OR FT):   '
      READ(LU5,*) ELEVI
      IF(ELEVI.LT.SCEL.OR.ELEVI.GT.DCEL)THEN
      WRITE(LU6,'(X,A,/,X,A)')'SORRY, INITIAL RESERVOIR LEVEL IS RESTRIC
     1TED BETWEEN SPILLWAY CREST ','AND DAM CREST ELEVATIONS.  PLEASE TR
     2Y AGAIN.'
      GOTO 15
      ENDIF
   16 WRITE(LU6,'(X,A,$)')'ENTER NUMBER OF ELEVATION-STORAGE DATA PAIRS:
     1  '
      READ(LU5,*) NE
      IF(NE.LE.0) THEN
      WRITE(LU6,*)'NUMBER OF ELEVATION-STORAGE PAIRS CANNOT BE ZERO OR N
     1EGATIVE.  PLEASE TRY AGAIN.'
      GOTO 16
      ELSEIF(NE.GT.LZ) THEN
      WRITE(LU6,'(X,A,I3,A,/,X,A)')'SORRY, NUMBER OF ELEVATION-STORAGE D
     1ATA PAIRS CANNOT BE GREATER THAN ',LZ,'.','PLEASE TRY AGAIN.'
      GOTO 16
      ENDIF
      WRITE(LU6,'(X,A,$)')'ARE YOU READING ELEVATION-STORAGE PAIRS INTER
     1ACTIVELY (Y/N)?:   'A
      READ(LU5,'(A)') A
      IF(A.EQ.'YES'.OR.A.EQ.'yes'.OR.A.EQ.'YE'.OR.A.EQ.'ye'.OR.A.EQ.'Y'.
     1OR.A.EQ.'y')THEN
      WRITE(LU6,'(X,A,I3,A,$)')'ENTER',NE,' ELEVATION-STORAGE DATA PAIRS
     1 (M OR FT, AND HM3 OR AC-FT):   '
      READ(LU5,*) (ELEVIN(J),STORAGEIN(J),J=1,NE)
      ELSE
      WRITE(LU6,'(X,A,I3,A,/,X,A)')'READING ',NE,' ELEVATION-STORAGE DAT
     1A PAIRS (M OR FT, AND HM3 OR AC-FT)','FROM YOUR INPUT FILE.'
      READ(LU7,*) (ELEVIN(J),STORAGEIN(J),J=1,NE)
      ENDIF
   17 WRITE(LU6,'(X,A,$)')'ENTER TIME INTERVAL (H):   '
      READ(LU5,*) DT
      IF(DT.LE.0.) THEN
      WRITE(LU6,*)'TIME INTERVAL CANNOT BE ZERO OR NEGATIVE.  PLEASE TRY
     1 AGAIN.'
      GOTO 17
      ENDIF
   18 WRITE(LU6,'(X,A,$)')'ENTER NUMBER OF INFLOW VALUES:   '
      READ(LU5,*) NI
      IF(NI.LE.0) THEN
      WRITE(LU6,*)'NUMBER OF INFLOW VALUES CANNOT BE ZERO OR NEGATIVE.
     1PLEASE TRY AGAIN.'
```

```
        GOTO 18
        ELSEIF(NI.GT.NZ) THEN
        WRITE(LU6,'(X,A,I3,A,/,A)')'SORRY, NUMBER OF INFLOW VALUES CANNOT
       1BE GREATER THAN ',NZ,'.','PLEASE TRY AGAIN.'
        GOTO 18
        ENDIF
        WRITE(LU6,'(X,A,$)')'ARE YOU READING INFLOW VALUES INTERACTIVELY (
       1Y/N)?:  '
        READ(LU5,'(A)') A
        IF(A.EQ.'YES'.OR.A.EQ.'yes'.OR.A.EQ.'YE'.OR.A.EQ.'ye'.OR.A.EQ.'Y'.
       1OR.A.EQ.'y')THEN
        READ(LU5,*) (QIN(J),J=2,NI+1)
        ELSE
        WRITE(LU6,'(X,A,I3,A)')'READING',NI,' INFLOW VALUES (M3/S OR FT3/S
       1) FROM YOUR INPUT FILE.'
        READ(LU7,*) (QIN(J),J=2,NI+1)
        ENDIF
C*****CALCULATION OF STORAGE-INDICATION FUNCTION
        DELEV= (DCEL-SCEL)/(MZ-1)
        ELEV(1)= SCEL
        CALL PWLIN (ELEVIN,STORAGEIN,NE,ELEV(1),STOR(1))
        STIN(1)= (2*UNIT/86400.)*STOR(1)/(DT/24.) + OUTF(1)
        DO 20 L=2,MZ
        ELEV(L)= ELEV(L-1) + DELEV
        HEAD(L)= ELEV(L) - SCEL
        OUTF(L)= RATC*XLES*HEAD(L)**EXPR
        CALL PWLIN (ELEVIN,STORAGEIN,NE,ELEV(L),STOR(L))
        STIN(L)= (2*UNIT/86400.)*STOR(L)/(DT/24.) + OUTF(L)
     20 CONTINUE
        QBASE= RATC*XLES*(ELEVI-SCEL)**EXPR
        QIN(1)= QBASE
        NC= 3*NI + 1
        IF(NC.GT.NZ) NC=NZ
        DO 30 J= NI+2,NC
     30 QIN(J)= QBASE
        CALL PWLIN(ELEVIN,STORAGEIN,NE,ELEVI,STOROUT(1))
        CALL PWLIN(OUTF,STIN,MZ,QBASE,STORIND(1))
        QOUT(1)= QBASE
        ELEVOUT(1)= ELEVI
        STORINC(1)= STORIND(1) - 2.*QOUT(1)
C*****RESERVOIR ROUTING CALCULATIONS
        DO 40 J=2,NC
        TIME(J) = TIME(J-1) + DT
        STORIND(J)= QIN(J) + QIN(J-1) + STORINC(J-1)
        CALL PWLIN (STIN,OUTF,MZ,STORIND(J),QOUT(J))
        CALL PWLIN(OUTF,STOR,MZ,QOUT(J),STOROUT(J))
        CALL PWLIN(STORAGEIN,ELEVIN,NE,STOROUT(J),ELEVOUT(J))
        STORINC(J)= STORIND(J) - 2*QOUT(J)
     40 CONTINUE
        DO 50 J=1,NC
        IF(QOUT(J).GT.QMAX) QMAX= QOUT(J)
     50 CONTINUE
        ELEVMAX= SCEL + (QMAX/(RATC*XLES))**(1./EXPR)
C*****PRINTING RESULTS
        WRITE(LU6,'(X,A,$)')'DO YOU WANT YOUR OUTPUT TO THE SCREEN (Y/N)?:
       1 '
        READ(LU5,'(A)') A
        IF(A.EQ.'YES'.OR.A.EQ.'yes'.OR.A.EQ.'YE'.OR.A.EQ.'ye'.OR.A.EQ.'Y'.
       1OR.A.EQ.'y')THEN
        LU= LU6
        ELSE
        LU= LU8
        ENDIF
        WRITE(LU,'(/,X,A)')
       1'      TIME            INFLOW           OUTFLOW           STORAGE
       2 ELEVATION'
        IF(INU.EQ.1) THEN
        WRITE(LU,*)
       1'     (H)             (M3/S)            (M3/S)            (HM3)
       2   (M)'
        ELSE
        WRITE(LU,*)
       1'     (H)             (FT3/S)           (FT3/S)           (AC-FT)
       2   (FT)'
        ENDIF
        DO 60 J=1,NC
        WRITE(LU,'(X,F9.2,3F16.3,F18.3)')
       1TIME(J),QIN(J),QOUT(J),STOROUT(J),ELEVOUT(J)
```

```
 60 CONTINUE
    IF(INU.EQ.1) THEN
    WRITE(LU,'(/,X,A,F12.3,A)')
   1'MAXIMUM POOL ELEVATION= ',ELEVMAX,' M.'
    ELSE
    WRITE(LU,'(/,X,A,F12.3,A)')
   1'MAXIMUM POOL ELEVATION= ',ELEVMAX,' FT.'
    ENDIF
    WRITE(LU6,'(/,X,A)')'THANK YOU FOR RUNNING EH800.  PLEASE CALL AGA
   1IN.'
    STOP
    END
    SUBROUTINE PWLIN(X,Y,N,XL,YL)
    DIMENSION X(N),Y(N)
    IF(XL.LT.X(1)) THEN
    YL= ((Y(2)-Y(1))*XL+(Y(1)*X(2)-Y(2)*X(1)))/(X(2)-X(1))
    ELSEIF(XL.GE.X(N)) THEN
    K=N-1
    YL= ((Y(N)-Y(K))*XL+(Y(K)*X(N)-Y(N)*X(K)))/(X(N)-X(K))
    ELSE
    DO 1 J=2,N
    IF(XL.GE.X(J-1).AND.XL.LT.X(J)) M=J
  1 CONTINUE
    L=M-1
    YL= ((Y(M)-Y(L))*XL+(Y(L)*X(M)-Y(M)*X(L)))/(X(M)-X(L))
    ENDIF
    RETURN
    END



    PROGRAM EH900
C   THIS PROGRAM IS PART OF CHAPTER 9, "ENGINEERING HYDROLOGY,
C   PRINCIPLES AND PRACTICES," BY V. M. PONCE.
C   THIS PROGRAM IS INTENDED FOR INSTRUCTIONAL USE ONLY;
C   ANY OTHER USE IS EXPRESSLY AT THE USER'S OWN RISK.
C   THE PROGRAM IS WRITTEN IN VAX/VMS FORTRAN 77.
C   IT SOLVES THE STREAM CHANNEL ROUTING PROBLEM WITH THE
C   MUSKINGUM-CUNGE METHOD, USING A TRIANGULAR INFLOW HYDROGRAPH.
C   SEE EXAMPLE 9-9 IN THE TEXT
C   FOR AN ILLUSTRATION OF THE COMPUTATIONAL PROCEDURE.
C   THE PROGRAM IS WRITTEN IN SI UNITS.
C   MINOR MODIFICATIONS ARE NEEDED TO CONVERT IT TO U.S. CUSTOMARY UNITS.
C   ALL INPUT IS READ INTERACTIVELY.
C   OUTPUT CAN BE SENT TO THE SCREEN OR TO AN OUTPUT FILE.
C   IF WRITING TO AN OUTPUT FILE,
C   PLEASE ASSIGN PARAMETER LU8 TO THE APPROPRIATE FILE NAME.
C   THE FOLLOWING DEVICE LOGICAL UNITS ARE PRESET IN THIS PROGRAM:
C   LU5= 5:  TO READ FROM THE SCREEN;
C   LU6= 6:  TO WRITE TO THE SCREEN;
C   LU8= 8:  TO WRITE TO A FILE.
C   INFLOW/OUTFLOW ARRAY SIZE: NZ= 200
    PARAMETER (NZ=200)
    PARAMETER (LU5=5,LU6=6,LU8=8)
    DIMENSION TIME(0:NZ),QIN(0:NZ),QOUT(0:NZ)
    DOUBLE PRECISION QIN,QOUT
    CHARACTER*6 A
    WRITE(LU6,'(3(/,X,A))')'THANK YOU FOR RUNNING PROGRAM EH900.',
   1'THIS PROGRAM SOLVES THE STREAM CHANNEL ROUTING PROBLEM',
   2'BY THE MUSKINGUM-CUNGE METHOD, USING A TRIANGULAR INFLOW HYDROGRA
   3PH.'
 10 WRITE(LU6,'(X,A,$)')'ENTER THE PEAK DISCHARGE (M3/S):  '
    READ(LU5,*) QPEK
    IF(QPEK.LE.0.) THEN
    WRITE(LU6,*)'PEAK DISCHARGE CANNOT BE ZERO OR NEGATIVE.  PLEASE TR
   1Y AGAIN.'
    GOTO 10
    ENDIF
 11 WRITE(LU6,'(X,A,$)')'ENTER THE BASEFLOW (M3/S):  '
    READ(LU5,*) QBAS
    IF(QBAS.LT.0.) THEN
    WRITE(LU6,*)'BASE FLOW CANNOT BE NEGATIVE.  PLEASE TRY AGAIN.'
    GOTO 11
```

```fortran
      ENDIF
   12 WRITE(LU6,'(X,A,$)')'ENTER THE TIME-TO-PEAK (H):  '
      READ(LU5,*) TPEK
      IF(TPEK.LE.0.)THEN
      WRITE(LU6,*)'TIME-TO-PEAK CANNOT BE ZERO OR NEGATIVE.   PLEASE TRY
     1AGAIN.'
      GOTO 12
      ENDIF
   13 WRITE(LU6,'(X,A,$)')'ENTER THE TIME BASE (H):  '
      READ(LU5,*) TBAS
      IF(TBAS.LE.TPEK)THEN
      WRITE(LU6,*)'TIME BASE CANNOT BE LESS THAN OR EQUAL TO TIME-TO-PEA
     1K.   PLEASE TRY AGAIN.'
      GOTO 13
      ENDIF
   14 WRITE(LU6,'(X,A,$)')'ENTER THE CHANNEL BED SLOPE (M/M):   '
      READ(LU5,*) BSLO
      IF(BSLO.LE.0.)THEN
      WRITE(LU6,*)'BED SLOPE CANNOT BE ZERO OR NEGATIVE.   PLEASE TRY AGA
     1IN.'
      GOTO 14
      ENDIF
   15 WRITE(LU6,'(X,A,$)')'ENTER THE FLOW AREA CORRESPONDING TO THE PEAK
     1 DISCHARGE (M2):   '
      READ(LU5,*) FARE
      IF(FARE.LE.0.)THEN
      WRITE(LU6,*)'FLOW AREA CANNOT BE ZERO OR NEGATIVE.   PLEASE TRY AGA
     1IN.'
      GOTO 15
      ENDIF
   16 WRITE(LU6,'(X,A,$)')'ENTER THE CHANNEL TOP WIDTH CORRESPONDING TO
     1THE PEAK DISCHARGE (M):   '
      READ(LU5,*) TWQP
      IF(TWQP.LE.0.)THEN
      WRITE(LU6,*)'CHANNEL TOP WIDTH CANNOT BE ZERO OR NEGATIVE.   PLEASE
     1 TRY AGAIN.'
      GOTO 16
      ENDIF
   17 WRITE(LU6,'(X,A,$)')'ENTER THE RATING EXPONENT (BETA):   '
      READ(LU5,*) BETA
      IF(BETA.LE.0.)THEN
      WRITE(LU6,*)'EXPONENT BETA CANNOT BE ZERO OR NEGATIVE.   PLEASE TRY
     1 AGAIN.'
      GOTO 17
      ENDIF
   18 WRITE(LU6,'(X,A,$)')'ENTER THE REACH LENGTH (KM):   '
      READ(LU5,*) DXKM
      IF(DXKM.LE.0.)THEN
      WRITE(LU6,*)'REACH LENGTH CANNOT BE ZERO OR NEGATIVE.   PLEASE TRY
     1AGAIN.'
      GOTO 18
      ENDIF
   19 WRITE(LU6,'(X,A,$)')'ENTER THE TIME INTERVAL (H):   '
      READ(LU5,*) DTH
      IF(DTH.LE.0.)THEN
      WRITE(LU6,*)'TIME INTERVAL CANNOT BE ZERO OR NEGATIVE.   PLEASE TRY
     1 AGAIN.'
      GOTO 19
      ELSEIF((TPEK/DTH).LT.5.)THEN
      WRITE(LU6,'(X,A,/,X,A)')'CAUTION.   RATIO OF TIME-TO-PEAK TO TIME I
     1NTERVAL IS LESS THAN 5.','THE RESULTS MAY NOT BE ACCURATE.'
      ENDIF
      DTS= DTH*3600.
      DXM= DXKM*1000.
      DO 20 J=0,NZ
      QIN(J)= QBAS
   20 QOUT(J)= QBAS
      NU= TPEK/DTH + 0.01
      ND= (TBAS-TPEK)/DTH + 0.01
      NT= NU+ND
      DO 30 J=1,NU
   30 QIN(J)= QBAS + (QPEK-QBAS)*FLOAT(J)/NU
      DO 40 J=NT-1,NU+1,-1
   40 QIN(J)= QBAS + (QPEK-QBAS)*FLOAT(NT-J)/ND
C*****CALCULATION OF ROUTING PARAMETERS
      CPEK= BETA*QPEK/FARE
      COU= CPEK*DTS/DXM
      REY= QPEK/(TWQP*BSLO*CPEK*DXM)
```

```fortran
      C- 1./(1.+COU+REY)
      C0- C*(-1+COU+REY)
      C1- C*(1.+COU-REY)
      C2- C*(1.-COU+REY)
C*****STREAM CHANNEL ROUTING CALCULATIONS
      DO 50 J-1,NZ
   50 QOUT(J)- C0*QIN(J) + C1*QIN(J-1) + C2*QOUT(J-1)
      DO 60 J-1,NZ
      NP- NP+1
      IF((QOUT(J)-QBAS).LT.0.001.AND.J.GT.10)THEN
      GOTO 70
      ENDIF
   60 CONTINUE
   70 CONTINUE
C*****PRINTING RESULTS
      DO 80 J-1,NP
   80 TIME(J)- TIME(J-1) + DTH
      WRITE(LU6,'(X,A,$)')'DO YOU WANT YOUR OUTPUT TO THE SCREEN (Y/N)?:
     1 '
      READ(LU5,'(A)') A
      IF(A.EQ.'YES'.OR.A.EQ.'yes'.OR.A.EQ.'YE'.OR.A.EQ.'ye'.OR.A.EQ.'Y'.
     1OR.A.EQ.'y')THEN
      LU- LU6
      ELSE
      LU- LU8
      ENDIF
      WRITE(LU,*)
      WRITE(LU,*)'COURANT NUMBER C- ',COU
      WRITE(LU,*)'CELL REYNOLDS NUMBER D- ',REY
      WRITE(LU,'(X,A)')'PLEASE CHECK RESOLUTION ACCURACY CRITERIA:  '
      WRITE(LU,'(X,A,/)!)'(C + D) GREATER THAN OR EQUAL TO 1.'
      WRITE(LU,*)'      TIME           INFLOW           OUTFLOW '
      WRITE(LU,*)'       (H)           (M3/S)           (M3/S) '
      DO 90 J-0,NP
   90 WRITE(LU,'(X,F9.2,2F20.3)') TIME(J),QIN(J),QOUT(J)
      WRITE(LU6,'(/,X,A)')'THANK YOU FOR RUNNING EH900.  PLEASE CALL AG
     1AIN.'
      STOP
      END
```

```fortran
      PROGRAM EH1000A
C     THIS PROGRAM IS PART OF CHAPTER 10, "ENGINEERING HYDROLOGY,
C     PRINCIPLES AND PRACTICES," BY V. M. PONCE.
C     THIS PROGRAM IS INTENDED FOR INSTRUCTIONAL USE ONLY;
C     ANY OTHER USE IS EXPRESSLY AT THE USER'S OWN RISK.
C     THE PROGRAM IS WRITTEN IN VAX/VMS FORTRAN 77.
C     IT SOLVES THE CATCHMENT ROUTING PROBLEM
C     BY THE METHOD OF CASCADE OF LINEAR RESERVOIRS.
C     SEE EXAMPLE 10-3 IN THE TEXT
C     FOR AN ILLUSTRATION OF THE COMPUTATIONAL PROCEDURE.
C     THE PROGRAM IS WRITTEN IN SI UNITS.
C     MINOR MODIFICATIONS ARE NEEDED TO CONVERT IT TO U.S. CUSTOMARY UNITS.
C     ALL INPUT IS READ INTERACTIVELY.
C     OUTPUT CAN BE SENT TO THE SCREEN OR TO AN OUTPUT FILE.
C     IF WRITING TO AN OUTPUT FILE,
C     PLEASE ASSIGN PARAMETER LU8 TO THE APPROPRIATE FILE NAME.
C     THE FOLLOWING DEVICE LOGICAL UNITS ARE PRESET IN THIS PROGRAM:
C     LU5- 5:  TO READ FROM THE SCREEN;
C     LU6- 6:  TO WRITE TO THE SCREEN;
C     LU8- 8:  TO WRITE TO A FILE.
C     INFLOW/OUTFLOW ARRAY SIZE:  NZ- 200
      PARAMETER (NZ-200)
      PARAMETER (LU5-5,LU6-6,LU8-8)
      DIMENSION TIME(0:NZ),QOUT(0:NZ),RADE(NZ),AVIN(NZ)
      DATA TIME(0),NP,C /0.,0,2.777778/
      CHARACTER*6 A
      WRITE(LU6,'(3(/,X,A))')'THANK YOU FOR RUNNING EH1000A.',
     1'THIS PROGRAM SOLVES THE CATCHMENT ROUTING PROBLEM',
     2'BY THE METHOD OF CASCADE OF LINEAR RESERVOIRS.'
   10 WRITE(LU6,'(X,A,$)')'ENTER CATCHMENT AREA (KM2): '
      READ(LU5,*) AREA
```

```
      IF(AREA.LE.0.) THEN
      WRITE(LU6,*)'CATCHMENT AREA CANNOT BE ZERO OR NEGATIVE.  PLEASE TR
     1Y AGAIN.'
      GOTO 10
      ENDIF
   11 WRITE(LU6,'(X,A,$)')'ENTER TIME INTERVAL (H):  '
      READ(LU5,*) DTH
      IF(DTH.LE.0.) THEN
      WRITE(LU6,*)'TIME INTERVAL CANNOT BE ZERO OR NEGATIVE.  PLEASE TRY
     1AGAIN.'
      GOTO 11
      ENDIF
   12 WRITE(LU6,'(X,A,$)')'ENTER NUMBER OF RAINFALL INCREMENTS:  '
      READ(LU5,*) NI
      IF(NI.GT.NZ) THEN
      WRITE(LU6,*)'NUMBER OF RAINFALL INCREMENTS CANNOT BE GREATER THAN
     1',NZ,'PLEASE TRY AGAIN.'
      GOTO 12
      ENDIF
      WRITE(LU6,'(X,A,I3,A)')'ENTER ',NI,' NET RAINFALL INCREMENT(S) (CM
     1), IN FREE FORMAT:  '
      READ(LU5,*) (RADE(J),J=1,NI)
   13 WRITE(LU6,'(X,A,$)')'ENTER RESERVOIR STORAGE CONSTANT K (H):  '
      READ(LU5,*) RK
      IF(RK.LE.0.) THEN
      WRITE(LU6,*)'STORAGE CONSTANT K CANNOT BE LESS THAN OR EQUAL TO ZE
     1RO.  PLEASE TRY AGAIN.'
      GOTO 13
      ENDIF
   14 WRITE(LU6,'(X,A,$)')'ENTER NUMBER OF RESERVOIRS:  '
      READ(LU5,*) NR
      IF(NR.LE.0) THEN
      WRITE(LU6,*)'NUMBER OF RESERVOIRS CANNOT BE LESS THAN OR EQUAL TO
     1 ZERO.  PLEASE TRY AGAIN.'
      GOTO 14
      ENDIF
C*****CATCHMENT ROUTING CALCULATIONS
      COU= DTH/RK
      CF= 2*COU/(2+COU)
      DO 20 J=1,NI
   20 AVIN(J)= C*RADE(J)*AREA/DTH
      DO 50 N=1,NR
      DO 30 J=1,NZ
   30 QOUT(J)= CF*(AVIN(J)-QOUT(J-1)) + QOUT(J-1)
      IF(N.LT.NR) THEN
      DO 40 J=1,NZ
   40 AVIN(J)= 0.5*(QOUT(J-1) + QOUT(J))
      ENDIF
   50 CONTINUE
      DO 60 J=1,NZ
      NP= NP+1
      IF(QOUT(J).LT.0.001) THEN
      GOTO 70
      ENDIF
   60 CONTINUE
   70 CONTINUE
C*****PRINTING RESULTS
      DO J=1,NP
      TIME(J)= TIME(J-1) + DTH
      ENDDO
      WRITE(LU6,'(X,A,$)')'DO YOU WANT YOUR OUTPUT TO THE SCREEN (Y/N)?:
     1 '
      READ(LU5,'(A)') A
      IF(A.EQ.'YES'.OR.A.EQ.'yes'.OR.A.EQ.'YE'.OR.A.EQ.'ye'.OR.A.EQ.'Y'.
     1OR.A.EQ.'y') THEN
      LU= LU6
      ELSE
      LU= LU8
      ENDIF
      WRITE(LU,'(2(/,X,A))')'      TIME              OUTFLOW',
     1'       (H)             (M3/S)'
      DO 80 J=0,NP
   80 WRITE(LU,'(X,F9.2,F20.3)') TIME(J),QOUT(J)
      WRITE(LU6,'(/,X,A)')'THANK YOU FOR RUNNING EH1000A.  PLEASE CALL A
     1GAIN.'
      STOP
      END
```

```
      PROGRAM EH1000B
C     THIS PROGRAM IS PART OF CHAPTER 10, "ENGINEERING HYDROLOGY,
C     PRINCIPLES AND PRACTICES," BY V. M. PONCE.
C     THIS PROGRAM IS INTENDED FOR INSTRUCTIONAL USE ONLY;
C     ANY OTHER USE IS EXPRESSLY AT THE USER'S OWN RISK.
C     THE PROGRAM IS WRITTEN IN VAX/VMS FORTRAN 77.
C     IT SOLVES THE TWO-PLANE LINEAR KINEMATIC CATCHMENT ROUTING MODEL.
C     SEE SECTION 10.4 OF THE TEXT
C     FOR A DESCRIPTION OF THE COMPUTATIONAL PROCEDURE.
C     THE PROGRAM IS WRITTEN IN SI UNITS.
C     MINOR MODIFICATIONS ARE NEEDED TO CONVERT IT TO U.S. CUSTOMARY UNITS.
C     INPUT DATA CAN BE READ FROM THE SCREEN OR FROM AN INPUT FILE.
C     OUTPUT CAN BE SENT TO THE SCREEN OR TO AN OUTPUT FILE.
C     IF READING FROM AN INPUT FILE,
C     PLEASE ASSIGN PARAMETER LU7 TO THE APPROPRIATE FILE NAME.
C     IF WRITING TO AN OUTPUT FILE,
C     PLEASE ASSIGN PARAMETER LU8 TO THE APPROPRIATE FILE NAME.
C     THE FOLLOWING DEVICE LOGICAL UNITS ARE PRESET IN THIS PROGRAM:
C     LU5= 5:   TO READ FROM THE SCREEN;
C     LU6= 6:   TO WRITE TO THE SCREEN;
C     LU7= 7:   TO READ FROM A FILE;
C     LU8= 8:   TO WRITE TO A FILE.
C     ARRAY SIZES (SUBJECT TO CHANGE IF NECESSARY):
C     MAXIMUM NUMBER OF SPACE INTERVALS IN PLANE(S):   NX=32;
C     MAXIMUM NUMBER OF SPACE INTERVALS IN THE CHANNEL:  NY=32;
C     MAXIMUM NUMBER OF TIME INTERVALS:  NT=480.
      PARAMETER (NT=480,NX=32,NY=32)
      PARAMETER (LU5=5,LU6=6,LU7=7,LU8=8)
      DIMENSION QP(0:NX,0:NT),QC(0:NY,0:NT),QLP(NX,NT),QLC(NT)
      CHARACTER*6 A
      WRITE(LU6,'(4(/,X,A),$)')'THANK YOU FOR RUNNING EH1000B.',
     1'THIS PROGRAM SOLVES THE TWO-PLANE LINEAR KINEMATIC',
     2'CATCHMENT ROUTING MODEL.',
     3'DO YOU WANT TO INPUT THE DATA INTERACTIVELY (Y/N)?:  '
      READ(LU5,'(A)') A
      IF(A.EQ.'YES'.OR.A.EQ.'yes'.OR.A.EQ.'YE'.OR.A.EQ.'ye'.OR.A.EQ.'Y'.
     1OR.A.EQ.'y')THEN
10    WRITE(LU6,'(X,A,$)')'ENTER PLANE LENGTH (M):  '
      READ(LU5,*) XLE
      IF(XLE.LE.0.) THEN
      WRITE(LU6,*)'PLANE LENGTH CANNOT BE ZERO OR NEGATIVE.  PLEASE TRY
     1AGAIN.'
      GOTO 10
      ENDIF
11    WRITE(LU6,'(X,A,$)')'ENTER NUMBER OF PLANE INCREMENTS:  '
      READ(LU5,*) NDX
      IF(NDX.LE.0) THEN
      WRITE(LU6,*)'NUMBER OF PLANE INCREMENTS CANNOT BE ZERO OR NEGATIVE
     1.  PLEASE TRY AGAIN.'
      GOTO 11
      ELSEIF(NDX.GT.NX) THEN
      WRITE(LU6,*)'ARRAY SIZE EXCEEDED.  PLEASE TRY AGAIN.'
      GOTO 11
      ENDIF
12    WRITE(LU6,'(X,A,$)')'ENTER CHANNEL LENGTH (M):   '
      READ(LU5,*) YLE
      IF(YLE.LE.0.) THEN
      WRITE(LU6,*)'CHANNEL LENGTH CANNOT BE ZERO OR NEGATIVE.  PLEASE TR
     1Y AGAIN.'
      GOTO 12
      ENDIF
13    WRITE(LU6,'(X,A,$)')'ENTER NUMBER OF CHANNEL INCREMENTS:  '
      READ(LU5,*) NDY
      IF(NDY.LE.0) THEN
      WRITE(LU6,*)'NUMBER OF CHANNEL INCREMENTS CANNOT BE ZERO OR NEGATI
     1VE.  PLEASE TRY AGAIN.'
      GOTO 13
      ELSEIF(NDY.GT.NY) THEN
      WRITE(LU6,*)'ARRAY SIZE EXCEEDED.  PLEASE TRY AGAIN.'
      GOTO 13
      ENDIF
14    WRITE(LU6,'(X,A,$)')'ENTER TOTAL SIMULATION TIME (MIN):   '
      READ(LU5,*) TST
      IF(TST.LE.0.) THEN
      WRITE(LU6,*)'TOTAL SIMULATION TIME CANNOT BE ZERO OR NEGATIVE.  PL
     1EASE TRY AGAIN.'
      GOTO 14
      ENDIF
```

```
   15 WRITE(LU6,'(X,A,$)')'ENTER NUMBER OF TIME INTERVALS:  '
      READ(LU5,*) NDT
      IF(NDT.LE.0) THEN
      WRITE(LU6,*)'NUMBER OF TIME INTERVALS CANNOT BE ZERO OR NEGATIVE.
     1PLEASE TRY AGAIN.'
      GOTO 15
      ELSEIF(NDT.GT.NT) THEN
      WRITE(LU6,*)'ARRAY SIZE EXCEEDED.  PLEASE TRY AGAIN.'
      GOTO 15
      ENDIF
   16 WRITE(LU6,'(X,A,$)')'ENTER EFFECTIVE RAINFALL INTENSITY (CM/H):  '
      READ(LU5,*) RAINT
      IF(RAINT.LE.0.) THEN
      WRITE(LU6,*)'RAINFALL INTENSITY CANNOT BE ZERO OR NEGATIVE.  PLEAS
     1E TRY AGAIN.'
      GOTO 16
      ENDIF
   17 WRITE(LU6,'(X,A,$)')'ENTER EFFECTIVE RAINFALL DURATION (MIN)?:  '
      READ(LU5,*) RADUR
      IF(RADUR.LE.0.) THEN
      WRITE(LU6,*)'RAINFALL DURATION CANNOT BE ZERO OR NEGATIVE.  PLEASE
     1TRY AGAIN.'
      GOTO 17
      ENDIF
   18 WRITE(LU6,'(X,A,$)')'ENTER AVERAGE WAVE CELERITY IN THE PLANES (M/
     1S):  '
      READ(LU5,*) CELP
      IF(CELP.LE.0.) THEN
      WRITE(LU6,*)'AVERAGE WAVE CELERITY IN THE PLANES CANNOT BE ZERO OR
     1 NEGATIVE.  PLEASE TRY AGAIN.'
      GOTO 18
      ENDIF
   19 WRITE(LU6,'(X,A,$)')'ENTER AVERAGE WAVE CELERITY IN THE CHANNEL (M
     1/S)?:  '
      READ(LU5,*) CELC
      IF(CELC.LE.0.) THEN
      WRITE(LU6,*)'AVERAGE WAVE CELERITY IN THE CHANNEL CANNOT BE ZERO O
     1R NEGATIVE.  PLEASE TRY AGAIN.'
      GOTO 19
      ENDIF
      ELSE
      READ(LU7,*) XLE,NDX,YLE,NDY,TST,NDT,RAINT,RADUR,CELP,CELC
      ENDIF
      RADURS= RADUR*60.
      RFVOL= (RAINT/360000.)*RADURS*XLE*YLE*2
      TSTS= TST*60.
      DX= XLE/NDX
      DY= YLE/NDY
      DT= TSTS/NDT
      NRDT= RADURS/DT + 0.01
C*****CALCULATION OF ROUTING PARAMETERS
      COUP= CELP*DT/DX
      COUC= CELC*DT/DY
      COUP1= 1.-COUP
      COUP2= (COUP-1.)/COUP
      COUP3= 1./COUP
      COUC1= 1.-COUC
      COUC2= (COUC-1.)/COUC
      COUC3= 1./COUC
      RCMS= (RAINT/360000.)*DX*DY
C*****CATCHMENT ROUTING CALCULATIONS
      DO 20 J=1,NDX
      DO 20 N=1,NRDT
   20 QLP(J,N)= RCMS
      DO 30 N=1,NDT
      DO 30 J=1,NDX
      IF(COUP.LT.1.)THEN
      QP(J,N)= COUP*QP(J-1,N-1) + COUP1*QP(J,N-1) + COUP*QLP(J,N)
      ELSEIF(COUP.GT.1.)THEN
      QP(J,N)= COUP2*QP(J-1,N) + COUP3*QP(J-1,N-1) + QLP(J,N)
      ELSE
      QP(J,N)= QP(J-1,N-1) + QLP(J,N)
      ENDIF
   30 CONTINUE
      DO 40 N=1,NDT
   40 QLC(N)= QP(NDX,N-1) + QP(NDX,N)
      DO 50 N=1,NDT
      DO 50 K=1,NDY
```

```
      IF(COUC.LT.1.)THEN
      QC(K,N)= COUC*QC(K-1,N-1)+ COUC1*QC(K,N-1) + COUC*QLC(N)
      ELSEIF(COUC.GT.1.)THEN
      QC(K,N)= COUC2*QC(K-1,N) + COUC3*QC(K-1,N-1) + QLC(N)
      ELSE
      QC(K,N)= QC(K-1,N-1) + QLC(N)
      ENDIF
   50 CONTINUE
      DO 60 L=1,NDT-1,2
   60 SUMA= SUMA + QC(NDY,L)
      DO 70 L=2,NDT-2,2
   70 SUMB= SUMB + QC(NDY,L)
      VOL= (DT/3.)*(QC(NDY,0) + 4*SUMA + 2*SUMB + QC(NDY,NDT))
C*****PRINTING RESULTS
      TIMESEC= -DT
      WRITE(LU6,'(X,A,$)')'DO YOU WANT YOUR OUTPUT TO THE SCREEN (Y/N)?:
     1 '
      READ(LU5,'(A)') A
      IF(A.EQ.'YES'.OR.A.EQ.'yes'.OR.A.EQ.'YE'.OR.A.EQ.'ye'.OR.A.EQ.'Y'.
     1OR.A.EQ.'y')THEN
      LU= LU6
      ELSE
      LU= LU8
      ENDIF
      WRITE(LU,'(2(/,A))')
     1'  TIME INTERVAL        TIME            TIME           OUTFLOW',
     2'                       (S)            (MIN)           (M3/S)'
      DO 80 N=0,NDT
      TIMESEC= TIMESEC + DT
      TIMEMIN= TIMESEC/60.
      IF(N.GT.10.AND.QC(NDY,N).LT.0.0001) THEN
      GOTO 90
      ELSE
      WRITE(LU,'(X,I10,2F16.2,F17.4)') N,TIMESEC,TIMEMIN,QC(NDY,N)
      ENDIF
   80 CONTINUE
   90 CONTINUE
      WRITE(LU,'(2(/,X,A,F12.3,A))')'RAINFALL VOLUME:     ',RFVOL,' M3.',
     1'VOLUME UNDER OUTFLOW HYDROGRAPH:     ',VOL,' M3.'
      WRITE(LU6,'(/,X,A)')'THANK YOU FOR RUNNING EH1000B.  PLEASE CALL A
     1GAIN.'
      STOP
      END




      PROGRAM EH1000C
C     THIS PROGRAM IS PART OF CHAPTER 10, "ENGINEERING HYDROLOGY,
C     PRINCIPLES AND PRACTICES," BY V. M. PONCE.
C     THIS PROGRAM IS INTENDED FOR INSTRUCTIONAL USE ONLY;
C     ANY OTHER USE IS EXPRESSLY AT THE USER'S OWN RISK.
C     THE PROGRAM IS WRITTEN IN VAX/VMS FORTRAN 77.
C     IT SOLVES THE TWO-PLANE LINEAR DIFFUSION CATCHMENT ROUTING MODEL.
C     SEE SECTION 10.5 OF THE TEXT
C     FOR A DESCRIPTION OF THE COMPUTATIONAL PROCEDURE.
C     THE PROGRAM IS WRITTEN IN SI UNITS.
C     MINOR MODIFICATIONS ARE NEEDED TO CONVERT IT TO U.S. CUSTOMARY UNITS.
C     INPUT DATA CAN BE READ FROM THE SCREEN OR FROM AN INPUT FILE.
C     OUTPUT CAN BE SENT TO THE SCREEN OR TO AN OUTPUT FILE.
C     IF READING FROM AN INPUT FILE,
C     PLEASE ASSIGN PARAMETER LU7 TO THE APPROPRIATE FILE NAME.
C     IF WRITING TO AN OUTPUT FILE,
C     PLEASE ASSIGN PARAMETER LU8 TO THE APPROPRIATE FILE NAME.
C     THE FOLLOWING DEVICE LOGICAL UNITS ARE PRESET IN THIS PROGRAM:
C     LU5= 5:  READ FROM THE SCREEN;
C     LU6= 6:  WRITE TO THE SCREEN;
C     LU7= 7:  READ FROM A FILE;
C     LU8= 8:  WRITE TO A FILE.
C     ARRAY SIZES (SUBJECT TO CHANGE IF NECESSARY):
C     MAXIMUM NUMBER OF SPACE INTERVALS IN PLANE(S):  NX=32;
C     MAXIMUM NUMBER OF SPACE INTERVALS IN THE CHANNEL:  NY=32;
C     MAXIMUM NUMBER OF TIME INTERVALS:  NT=480.
      PARAMETER (NT=480,NX=32,NY=32)
      PARAMETER (LU5=5,LU6=6,LU7=7,LU8=8)
```

```
            DIMENSION QP(0:NX,0:NT),QC(0:NY,0:NT),QLP(NX,NT),QLC(NT)
            CHARACTER*6 A
            WRITE(LU6,'(4(/,X,A),$)')'THANK YOU FOR RUNNING EH1000C.',
           1'THIS PROGRAM SOLVES THE TWO-PLANE LINEAR DIFFUSION',
           2'CATCHMENT ROUTING MODEL.',
           3'DO YOU WANT TO INPUT THE DATA INTERACTIVELY (Y/N)?: '
            READ(LU5,'(A)') A
            IF(A.EQ.'YES'.OR.A.EQ.'yes'.OR.A.EQ.'YE'.OR.A.EQ.'ye'.OR.A.EQ.'Y'.
           1OR.A.EQ.'y')THEN
        10  WRITE(LU6,'(X,A,$)')'ENTER PLANE LENGTH (M): '
            READ(LU5,*) XLE
            IF(XLE.LE.0.) THEN
            WRITE(LU6,*)'PLANE LENGTH CANNOT BE ZERO OR NEGATIVE.  PLEASE TRY
           1AGAIN.'
            GOTO 10
            ENDIF
        11  WRITE(LU6,'(X,A,$)')'ENTER NUMBER OF PLANE INCREMENTS: '
            READ(LU5,*) NDX
            IF(NDX.LE.0) THEN
            WRITE(LU6,*)'NUMBER OF PLANE INCREMENTS CANNOT BE ZERO OR NEGATIVE
           1.  PLEASE TRY AGAIN.'
            GOTO 11
            ELSEIF(NDX.GT.NX) THEN
            WRITE(LU6,*)'ARRAY SIZE EXCEEDED.  PLEASE TRY AGAIN.'
            GOTO 11
            ENDIF
        12  WRITE(LU6,'(X,A,$)')'ENTER CHANNEL LENGTH (M):  '
            READ(LU5,*) YLE
            IF(YLE.LE.0.) THEN
            WRITE(LU6,*)'CHANNEL LENGTH CANNOT BE ZERO OR NEGATIVE.  PLEASE TR
           1Y AGAIN.'
            GOTO 12
            ENDIF
        13  WRITE(LU6,'(X,A,$)')'ENTER NUMBER OF CHANNEL INCREMENTS: '
            READ(LU5,*) NDY
            IF(NDY.LE.0) THEN
            WRITE(LU6,*)'NUMBER OF CHANNEL INCREMENTS CANNOT BE ZERO OR NEGATI
           1VE.  PLEASE TRY AGAIN.'
            GOTO 13
            ELSEIF(NDY.GT.NY) THEN
            WRITE(LU6,*)'ARRAY SIZE EXCEEDED.  PLEASE TRY AGAIN.'
            GOTO 13
            ENDIF
        14  WRITE(LU6,'(X,A,$)')'ENTER TOTAL SIMULATION TIME (MIN):  '
            READ(LU5,*) TST
            IF(TST.LE.0.) THEN
            WRITE(LU6,*)'TOTAL SIMULATION TIME CANNOT BE ZERO OR NEGATIVE.  PL
           1EASE TRY AGAIN.'
            GOTO 14
            ENDIF
        15  WRITE(LU6,'(X,A,$)')'ENTER NUMBER OF TIME INTERVALS: '
            READ(LU5,*) NDT
           .IF(NDT.LE.0) THEN
            WRITE(LU6,*)'NUMBER OF TIME INTERVALS CANNOT BE ZERO OR NEGATIVE.
           1PLEASE TRY AGAIN.'
            GOTO 15
            ELSEIF(NDT.GT.NT) THEN
            WRITE(LU6,*)'ARRAY SIZE EXCEEDED.  PLEASE TRY AGAIN.'
            GOTO 15
            ENDIF
        16  WRITE(LU6,'(X,A,$)')'ENTER EFFECTIVE RAINFALL INTENSITY (CM/H): '
            READ(LU5,*) RAINT
            IF(RAINT.LE.0.) THEN
            WRITE(LU6,*)'RAINFALL INTENSITY CANNOT BE ZERO OR NEGATIVE.  PLEAS
           1E TRY AGAIN.'
            GOTO 16
            ENDIF
        17  WRITE(LU6,'(X,A,$)')'ENTER EFFECTIVE RAINFALL DURATION (MIN):  '
            READ(LU5,*) RADUR
            IF(RADUR.LE.0.) THEN
            WRITE(LU6,*)'RAINFALL DURATION CANNOT BE ZERO OR NEGATIVE.  PLEASE
           1TRY AGAIN.'
            GOTO 17
            ENDIF
        18  WRITE(LU6,'(X,A,$)')'ENTER AVERAGE WAVE CELERITY IN THE PLANES (M/
           1S): '
            READ(LU5,*) CELP
            IF(CELP.LE.0.) THEN
            WRITE(LU6,*)'AVERAGE WAVE CELERITY IN THE PLANES CANNOT BE ZERO OR
```

```
                1 NEGATIVE.  PLEASE TRY AGAIN.'
                  GOTO 18
                  ENDIF
              19 WRITE(LU6,'(X,A,$)')'ENTER AVERAGE WAVE CELERITY IN THE CHANNEL (M
                1/S):  '
                  READ(LU5,*) CELC
                  IF(CELC.LE.0.) THEN
                  WRITE(LU6,*)'AVERAGE WAVE CELERITY IN THE CHANNEL CANNOT BE ZERO O
                1R NEGATIVE.  PLEASE TRY AGAIN.'
                  GOTO 19
                  ENDIF
              20 WRITE(LU6,'(X,A,$)')'ENTER SLOPE OF THE PLANES (M/M):  '
                  READ(LU5,*) SLOPEP
                  IF(SLOPEP.LE.0.) THEN
                  WRITE(LU6,*)'SLOPE OF THE PLANES CANNOT BE ZERO OR NEGATIVE.  PLEA
                1SE TRY AGAIN.'
                  GOTO 20
                  ENDIF
              21 WRITE(LU6,'(X,A,$)')'ENTER SLOPE OF THE CHANNEL (M/M):  '
                  READ(LU5,*) SLOPEC
                  IF(SLOPEC.LE.0.) THEN
                  WRITE(LU6,*)'SLOPE OF THE CHANNEL CANNOT BE ZERO OR NEGATIVE.  PLE
                1ASE TRY AGAIN.'
                  GOTO 21
                  ENDIF
              22 WRITE(LU6,'(X,A,$)')'ENTER AVERAGE UNIT-WIDTH FLOW OVER THE PLANES
                1 (M2/S):  '
                  READ(LU5,*) UQP
                  IF(UQP.LE.0.) THEN
                  WRITE(LU6,*)'UNIT-WIDTH FLOW OVER PLANES CANNOT BE ZERO OR NEGATIV
                1E.  PLEASE TRY AGAIN.'
                  GOTO 22
                  ENDIF
              23 WRITE(LU6,'(X,A,$)')'ENTER AVERAGE UNIT-WIDTH FLOW IN THE CHANNEL
                1(M2/S):  '
                  READ(LU5,*) UQC
                  IF(UQC.LE.0.) THEN
                  WRITE(LU6,*)'UNIT-WIDTH FLOW OVER CHANNEL CANNOT BE ZERO OR NEGATI
                1VE.  PLEASE TRY AGAIN.'
                  GOTO 23
                  ENDIF
                  ELSE
                  READ(LU7,*) XLE,NDX,YLE,NDY,TST,NDT,RAINT,RADUR,CELP,CELC,
                1SLOPEP,SLOPEC,UQP,UQC
                  ENDIF
                  TSTS= TST*60.
                  RADURS= RADUR*60.
                  RFVOL= (RAINT/360000.)*RADURS*XLE*YLE*2
C*****CALCULATION OF ROUTING PARAMETERS
                  DX= XLE/NDX
                  DY= YLE/NDY
                  DT= TSTS/NDT
                  NRDT= RADURS/DT + 0.01
                  COUP= CELP*DT/DX
                  COUC= CELC*DT/DY
                  REYP= UQP/(SLOPEP*CELP*DX)
                  REYC= UQC/(SLOPEC*CELC*DY)
                  CP= 1+COUP+REYP
                  CC= 1+COUC+REYC
                  C0P= (-1+COUP+REYP)/CP
                  C1P=  (1+COUP-REYP)/CP
                  C2P=  (1-COUP+REYP)/CP
                  C3P= 2*COUP/CP
                  C0C= (-1+COUC+REYC)/CC
                  C1C=  (1+COUC-REYC)/CC
                  C2C=  (1-COUC+REYC)/CC
                  C3C= 2*COUC/CC
                  RCMS= (RAINT/360000.)*DX*DY
C*****CATCHMENT-ROUTING CALCULATIONS
                  DO 30 J=1,NDX
                  DO 30 N=1,NRDT
              30 QLP(J,N)= RCMS
                  DO 40 N=1,NDT
                  DO 40 J=1,NDX
                  QP(J,N)= C0P*QP(J-1,N) + C1P*QP(J-1,N-1) + C2P*QP(J,N-1)
                1         + C3P*QLP(J,N)
              40 CONTINUE
                  DO 50 N=1,NDT
```

```
   50 QLC(N)= QP(NDX,N-1) + QP(NDX,N)
      DO 60 N=1,NDT
      DO 60 K=1,NDY
      QC(K,N)= C0C*QC(K-1,N) + C1C*QC(K-1,N-1) + C2C*QC(K,N-1)
     1         + C3C*QLC(N)
   60 CONTINUE
      DO 70 L=1,NDT-1,2
   70 SUMA= SUMA + QC(NDY,L)
      DO 75 L=2,NDT-2,2
   75 SUMB= SUMB + QC(NDY,L)
      VOL= (DT/3.)*(QC(NDY,0) + 4*SUMA + 2*SUMB + QC(NDY,NDT))
C*****PRINTING RESULTS
      TIMESEC= -DT
      WRITE(LU6,'(X,A,$)')'DO YOU WANT YOUR OUTPUT TO THE SCREEN (Y/N)?:
     1 '
      READ(LU5,'(A)') A
      IF(A.EQ.'YES'.OR.A.EQ.'yes'.OR.A.EQ.'YE'.OR.A.EQ.'ye'.OR.A.EQ.'Y'.
     1OR.A.EQ.'y')THEN
      LU= LU6
      ELSE
      LU= LU8
      ENDIF
      WRITE(LU,'(2(/,A))')
     1'   TIME INTERVAL       TIME            TIME            OUTFLOW',
     2'                        (S)             (MIN)           (M3/S)'
      DO 80 N=0,NDT
      TIMESEC= TIMESEC + DT
      TIMEMIN= TIMESEC/60.
      IF(N.GT.10.AND.QC(NDY,N).LT.0.0001) THEN
      GOTO 90
      ELSE
      WRITE(LU,'(X,I10,2F16.2,F17.4)') N,TIMESEC,TIMEMIN,QC(NDY,N)
      ENDIF
   80 CONTINUE
   90 CONTINUE
      WRITE(LU,'(2(/,X,A,F12.3,A))')'RAINFALL VOLUME:   ',RFVOL,' M3.',
     1'VOLUME UNDER OUTFLOW HYDROGRAPH:  ',VOL,' M3.'
      WRITE(LU6,'(/,X,A)')'THANK YOU FOR RUNNING EH1000C.  PLEASE CALL A
     1GAIN.'
      STOP
      END
```

ABBREVIATIONS

ac	acre(s)
ac-ft	acre-feet
atm/cm	atmosphere(s) per centimeter
Btu	British thermal unit(s)
Btu/in.^{2}/d	British thermal unit(s) per square inch per day
Btu/lb	British thermal unit(s) per pound
cal	calorie(s)
cal/cm^2	calorie(s) per square centimeter
cal/cm^2/d	calorie(s) per square centimeter per day
cal/g	calorie(s) per gram
cm	centimeter(s)
cm/d	centimeter(s) per day
cm/h	centimeter(s) per hour
cm/(°C-d)	centimeter(s) per degree Celsius-day
cp	centipoise(s)
cs	centistoke(s)
csm	cubic feet per second per square mile
d	day(s)
ft	foot (feet)
ft/mi	foot/mile (feet/mile)
ft^2/s	square feet per second
ft^3/lb/s	cubic feet per pound per second
ft^3/s	cubic feet per second
(ft^3/s)-d	(cubic feet per second)-days
(ft^3/s)-d/mi^2	(cubic feet per second)-days per square mile
ft^3/s/ft	cubic feet per second per foot
(ft^3/s)/mi^2	cubic feet per second per square mile
ft^3/(s-mi^2-in.)	cubic feet per second per square mile per inch
g/cm^3	gram(s) per cubic centimeter
h	hour(s)
ha	hectare(s)
hm^3	cubic hectometer(s)
in.	inch(es)
in./d	inch(es) per day
in./(°F-d)	inch(es) per degree Fahrenheit-day
in./h	inch(es) per hour
in. Hg	inch(es) of mercury
kg	kilogram(s)
kg/m^3	kilogram(s) per cubic meter
km	kilometer(s)
km^2	square kilometer(s)
km^3	cubic kilometer(s)
km^2-cm/h	square kilometer-(centimeter per hour)
km/d	kilometer(s) per day
kN/m^3	kilonewton(s) per cubic meter
kN/d	kilonewton(s) per day
L	liter(s)

L/s	liter(s) per second
lb/ft^2	pound(s) per square foot
lb/ft^3	pound(s) per cubic foot
lb/s	pound(s) per second
lb/s/ft	pound(s) per second per foot
ly	langley(s)
ly/d	langley(s) per day
ly/min	langley(s) per minute
m	meter(s)
mb	millibar(s)
mb/°C	millibar(s) per degree Celsius
m/d	meter(s) per day
min	minute(s)
mi	mile(s)
mi/h	mile(s) per hour
mi^2	square mile(s)
m/m	meter(s) per meter
mm	millimeter(s)
mm Hg	millimeter(s) of mercury
mm/d	millimeter(s) per day
mm/h	millimeter(s) per hour
mg/L	milligram(s) per liter
m/s	meter(s) per second
m/s^2	meter(s) per square second
m^2/N	square meter(s) per newton
m^2/s	square meter(s) per second
m^3/s	cubic meter(s) per second
m^3/(s-km^2-cm)	cubic meter(s) per second per square kilometer per centimeter
m^3/s/m	cubic meter(s) per second per meter
(m^3/s)-d	(cubic meter(s) per second)-day
(m^3/s)-h	(cubic meter(s) per second)-hour
N	newton(s)
N/m^2	newton(s) per square meter
N/m^3	newton(s) per cubic meter
Pa	pascal(s)
Pa^{-1}	reciprocal of pascal
ppm	parts per million (by weight)
s	second(s)
ton/ac/y	ton(s) per acre per year
ton/d	ton(s) per day
ton/d/ft	ton(s) per day per foot
ton/km^2/y	metric ton(s) per square kilometer per year
ton/mi^2/y	ton(s) per square mile per year
ton/y	ton(s) per year
W	watt(s)
wk	week(s)
μs	microsecond(s)

INDEX

Meyer, 44
Penman, **45–47**, 52
effect of climate, 39
lake, 40–41
measurements, 102
Colorado sunken pan, 53–54
NWS Class A pan, 102
pan coefficient, 47, 54
pans:
Colorado sunken, 53–54
NWS Class A, 53, 102
reservoir:
combination methods, 45–47
determinations using pans, 47
energy budget method, 42–43
mass-transfer approach, 43–44
water budget method, 39, 42
Sacramento model, 445
Evapotranspiration, **47**, 401–2
actual, 104
formulas:
Blaney-Criddle, 48–51, 580
Priestley and Taylor, 52
Thornthwaite, 51–52, 581
measurements, 102
evapotranspirometers, 103
lysimeters, 104
models:
combination, 52–53
pan-evaporation, 53
Penman, 52
radiation, 52
temperature, 48–52
opportunity, 439
potential, **48**, 103
reference crop, 48
Sacramento model, 445
Stanford watershed model, 439–40
Exposure sector, 362
Extreme value type I method. (*See* Gumbel method)

Fall:
slope-area method, 112
velocity, 534
Field capacity, 106, **337–38**

Finite difference schemes, 280
backward, 280
central, 280
forward, 280
Flaxman formula, 548
Flood:
forecasting, 97, 108
frequency analysis, 213
comparison between methods, 226–27
selection of data series, 213
hydrograph development:
composite hydrographs, 185–87
USBR practice, 505
hydrology:
approaches, 8
USGS reports listing, 589–95
plain, 77
probable maximum. (*See* Probable maximum flood)
routing, 270
HEC-1 model, 413
series, 213
annual exceedence, 213
annual maxima, 213
annual minima, 213, 229
extreme value, 213
partial duration, 213
peaks-over-a-threshold (POT), 213
standard project, 14
wave:
dam break, 300
Flow through porous media, 338–43
Flow-duration curve, 80, 229
Flow-mass curve, 81
Forms:
bed roughness, 552–53
antidunes, 552–53
chutes and pools, 552–53
dunes, 552–53
plane bed, 552–53
ripples, 552–53
precipitation, 13
Free water, 442
Freeboard, 483
hydrograph, 484
Frequency, 214
factor, 217, 582–83
analysis, 205
Front, 12

Frontal surface, 12

Gated spillway:
rating, 265–66
General HEC-1 loss-rate function. (*See* HEC-1 loss-rate function)
Geological Survey. (*See* U.S. Geological Survey)
Gradient. (*See* Slope)
Gravitational water, 380
Groundwater:
extent of resources, 333
flow, **64**, 332
equations, 343
steady-state saturated, 343
transient saturated, 343
reservoir, 335
permeability, 334
porosity, 334
replenishment, 334
Stanford watershed model, 438
table. (*See* Water table)
Gully, 62
erosion, 538, 543
Gumbel:
method, **223–26**, 229
modifications, 225–26
probability paper, 224
program EH600B, **598**, 609–10

Heat of vaporization, 42, 579–80
HEC loss-rate function, 410–11
HEC models:
HEC-1, 154, **406–13**
HEC-2, 407
HEC-6, 565
Hillslope hydrology, 332, 350
partial-area concept, 350
variable-source-area model, 351
HMR series. (*See* Hydrometeorological report series)
Holtan loss-rate method, 411–12
Horton infiltration equation, 34
SWMM model, 419
Hortonian flow, 332, 350

SPECIFIC WEIGHT

1 newton per cubic meter = 0.0063636 pounds per cubic foot
1 pound per cubic foot = 157.14 newtons per cubic meter

VELOCITY

1 meter per second = 3.28 feet per second
1 kilometer per hour = 0.6215 miles per hour

1 foot per second = 0.3048 meters per second
1 mile per hour = 1.609 kilometers per hour

VISCOSITY (ABSOLUTE)

1 poise = 0.1 newton-second per square meter
1 poise = 0.0020878 lb-second per square foot
1 centipoise = 0.01 poise

1 lb-second per square foot = 478.97 poise
1 lb-second per square foot = 47897 centipoise

VISCOSITY (KINEMATIC)

1 stoke = 1 square centimeter per second
1 centistoke = 0.01 stoke
1 centistoke = 0.000001 square meter per second
1 centistoke = 0.000010764 square feet per second

1 square foot per second = 929.03 stoke
1 square foot per second = 92,903 centistoke

VOLUME

1 liter = 0.2642 U.S. gallons
1 liter = 0.035315 cubic feet
1 liter = 0.001 cubic meter
1 cubic meter = 35.315 cubic feet
1 cubic meter = 1.308 cubic yards
1 cubic meter = 264.17 U.S. gallons

1 U.S. gallon = 3.7854 liters
1 cubic foot = 28.317 liters
1 U.S. gallon = 0.13368 cubic feet
1 cubic foot = 0.028317 cubic meters
1 cubic yard = 0.7646 cubic meters
1 million U.S. gallons = 3785.4 cubic meters

1 cubic hectometer = 1 million cubic meters
1 cubic hectometer = 810.71 acre-feet
10,000 cubic meters = 4.0873 (ft^3/s)-day

1 million cubic feet = 28,317 cubic meters
1 acre-foot = 1233.5 cubic meters
1 (ft^3/s)-day = 2446.6 cubic meters
1 (ft^3/s)-day = 1.983 acre-feet